PERIODIC TABLE

1 H 1.008																	2 He 4.00
3 Li 6.94	4 Be 9.01											5 B 10.8	6 C 12.01	7 N 14.01	8 O 16.00	9 F 19.0	10 Ne 20.2
11 Na 23.0	12 Mg 24.3											13 Al 27.0	14 Si 28.1	15 P 31.0	16 S 32.1	17 Cl 35.5	18 Ar 39.9
19 K 39.1	20 Ca 40.1	21 Sc 45.0	22 Ti 47.9	23 V 50.9	24 Cr 52.0	25 Mn 54.9	26 Fe 55.8	27 Co 58.9	28 Ni 58.7	29 Cu 63.5	30 Zn 65.4	31 Ga 69.7	32 Ge 72.6	33 As 74.9	34 Se 79.0	35 Br 79.9	36 Kr 83.8
37 Rb 85.5	38 Sr 87.6	39 Y 88.9	40 Zr 91.2	41 Nb 92.9	42 Mo 95.9	43 Tc (99)	44 Ru 101.1	45 Rh 102.9	46 Pd 106.4	47 Ag 107.9	48 Cd 112.4	49 In 114.8	50 Sn 118.7	51 Sb 121.8	52 Te 127.6	53 I 126.9	54 Xe 131.3
55 Cs 132.9	56 Ba 137.3	57-71 See below	72 Hf 178.5	73 Ta 180.9	74 W 183.9	75 Re 186.2	76 Os 190.2	77 Ir 192.2	78 Pt 195.1	79 Au 197.0	80 Hg 200.6	81 Tl 204.4	82 Pb 207.2	83 Bi 209.0	84 Po (209)	85 At (210)	86 Rn (222)
87 Fr (223)	88 Ra (226)	89- See below															

57 La 138.9	58 Ce 140.1	59 Pr 140.9	60 Nd 144.2	61 Pm (147)	62 Sm 150.4	63 Eu 152.0	64 Gd 157.3	65 Tb 158.9	66 Dy 162.5	67 Ho 164.9	68 Er 167.3	69 Tm 168.9	70 Yb 173.0	71 Lu 175.0
89 Ac (227)	90 Th 232.0	91 Pa (231)	92 U 238.0	93 Np (237)	94 Pu (242)	95 Am (243)	96 Cm (247)	97 Bk (245)	98 Cf (251)	99 Es (254)	100 Fm (253)	101 Md (256)	102 (254)	103 Lw (257)

Parenthetical values are mass numbers of the isotopes with longest half lives

P9-BYO-667

INTERNATIONAL ATOMIC WEIGHTS

NAME	SYMBOL	ATOMIC NUMBER	ATOMIC WEIGHT	NAME	SYMBOL	ATOMIC NUMBER	ATOMIC WEIGHT
Actinium	Ac	89	(227)	Mercury	Hg	80	200.6
Aluminum	Al	13	27.0	Molybdenum	Mo	42	95.9
Americium	Am	95	(243)	Neodymium	Nd	60	144.2
Antimony	Sb	51	121.8	Neon	Ne	10	20.2
Argon	Ar	18	39.9	Neptunium	Np	93	(237)
Arsenic	As	33	74.9	Nickel	Ni	28	58.7
Astatine	At	85	(210)	Niobium	Nb	41	92.9
Barium	Ba	56	137.3	Nitrogen	N	7	14.01
Berkelium	Bk	97	245	Osmium	Os	76	190.2
Beryllium	Be	4	9.01	Oxygen	O	8	16.00
Bismuth	Bi	83	209.0	Palladium	Pd	46	106.4
Boron	B	5	10.8	Phosphorus	P	15	31.0
Bromine	Br	35	79.9	Platinum	Pt	78	195.1
Cadmium	Cd	48	112.4	Plutonium	Pu	94	(242)
Calcium	Ca	20	40.1	Polonium	Po	84	210
Californium	Cf	98	(251)	Potassium	K	19	39.1
Carbon	C	6	12.01	Praseodymium	Pr	59	140.9
Cerium	Ce	58	140.1	Promethium	Pm	61	(147)
Cesium	Cs	55	132.9	Protactinium	Pa	91	(231)
Chlorine	Cl	17	35.5	Radium	Ra	88	(226)
Chromium	Cr	24	52.0	Radon	Rn	86	(222)
Cobalt	Co	27	58.9	Rhenium	Re	75	186.2
Copper	Cu	29	63.5	Rhodium	Rh	45	102.9
Curium	Cm	96	(247)	Rubidium	Rb	37	85.5
Dysprosium	Dy	66	162.5	Ruthenium	Ru	44	101.1
Einsteinium	Es	99	(254)	Samarium	Sm	62	150.4
Erbium	Er	68	167.3	Scandium	Sc	21	45.0
Europium	Eu	63	152.0	Selenium	Se	34	79.0
Fermium	Fm	100	(253)	Silicon	Si	14	28.1
Fluorine	F	9	19.0	Silver	Ag	47	107.9
Francium	Fr	87	(223)	Sodium	Na	11	23.0
Gadolinium	Gd	64	157.3	Strontium	Sr	38	87.6
Gallium	Ga	31	69.7	Sulfur	S	16	32.1
Germanium	Ge	32	72.6	Tantalum	Ta	73	180.9
Gold	Au	79	197.0	Technetium	Tc	43	(99)
Hafnium	Hf	72	178.5	Tellurium	Te	52	127.6
Helium	He	2	4.00	Terbium	Tb	65	158.9
Holmium	Ho	67	164.9	Thallium	Tl	81	204.4
Hydrogen	H	1	1.008	Thorium	Th	90	232.0
Indium	In	49	114.8	Thulium	Tm	69	168.9
Iodine	I	53	126.9	Tin	Sn	50	118.7
Iridium	Ir	77	192.2	Titanium	Ti	22	47.9
Iron	Fe	26	55.8	Tungsten	W	74	183.9
Krypton	Kr	36	83.8	Uranium	U	92	238.0
Lanthanum	La	57	138.9	Vanadium	V	23	50.9
Lead	Pb	82	207.2	Xenon	Xe	54	131.3
Lithium	Li	3	6.94	Ytterbium	Yb	70	173.0
Lutetium	Lu	71	175.0	Yttrium	Y	39	88.9
Magnesium	Mg	12	24.3	Zinc	Zn	30	65.4
Manganese	Mn	25	54.9	Zirconium	Zr	40	91.2
Mendelevium	Md	101	(256)				

Parenthetical names refer to radioactive elements; the mass number (not the atomic weight) of the isotope with largest half-life is usually given.

* *Latest values recommended by the International Union of Pure and Applied Chemistry, 1961.*

CHEMISTRY

AN EXPERIMENTAL SCIENCE

CHEMISTRY

Prepared by

CHEMICAL EDUCATION MATERIAL STUDY

Under a grant from

THE NATIONAL SCIENCE FOUNDATION

Editor: GEORGE C. PIMENTEL, University of California, Berkeley, California

Associate Editors
BRUCE H. MAHAN, University of California, Berkeley, California
A. L. McCLELLAN, California Research Corporation, Richmond, California
KEITH MacNAB, Sir Francis Drake High School, San Anselmo, California
MARGARET NICHOLSON, Acalanes High School, Lafayette, California

An Experimental Science

Contributors

ROBERT F. CAMPBELL
Miramonte High School, Orinda, California
JOSEPH E. DAVIS, JR.
Miramonte High School, Orinda, California
SAUL L. GEFFNER
Forest Hills High School, Forest Hills, New York
THEODORE A. GEISSMAN
University of California, Los Angeles, California
MELVIN GREENSTADT
Fairfax High School, Los Angeles, California
CARL GRUHN
South Pasadena High School, South Pasadena, California
EDWARD L. HAENISCH
Wabash College, Crawfordsville, Indiana
ROLFE H. HERBER
Rutgers University, New Brunswick, New Jersey
C. ROBERT HURLEY
Sacramento State College, Sacramento, California
LAWRENCE D. LYNCH, JR.
Beverly Hills High School, Beverly Hills, California
LLOYD E. MALM
University of Utah, Salt Lake City, Utah
CLYDE E. PARRISH
Cubberley Senior High School, Palo Alto, California
ROBERT W. PARRY
University of Michigan, Ann Arbor, Michigan
EUGENE ROBERTS
Polytechnic High School, San Francisco, California
MICHELL J. SIENKO
Cornell University, Ithaca, New York
ROBERT SILBER
American Chemical Society, Washington, D.C.
HARLEY L. SORENSEN
San Ramon Valley Union High School, Danville, California
LUKE E. STEINER
Oberlin College, Oberlin, Ohio
MODDIE D. TAYLOR
Howard University, Washington, D.C.
ROBERT L. TELLEFSEN
Napa High School, Napa, California

Director: J. ARTHUR CAMPBELL, Harvey Mudd College, Claremont, California

Chairman: GLENN T. SEABORG, University of California, Berkeley, California

W. H. FREEMAN AND COMPANY, *Cooperating Publishers*
SAN FRANCISCO AND LONDON

© *Copyright 1960, 1961, 1962, 1963 by The Regents of the University of California.*

The University of California reserves all rights to reproduce this book, in whole or in part, with the exception of the right to use short quotations for review of the book.

Printed in the United States of America.
Library of Congress Catalog Card Number: 63-18323.

Preface

Chemistry deals with all of the *substances* that make up our environment. It also deals with the *changes* that take place in these substances—changes that make the difference between a cold and lifeless planet and one that teems with life and growth. Chemistry helps us *understand* and *benefit* from nature's wondrous ways.

Chemistry is an important part of what is called science. Since every phase of our daily life is affected by the fruits of scientific activity, we all should know what scientific activity is, what it can do, and how it works. The study of chemistry will help you learn these things.

CHEMISTRY—An Experimental Science presents chemistry as it is today. It does so with emphasis upon the most enjoyable part of chemistry: experimentation. Unifying principles are developed, as is appropriate in a modern chemistry course, with the laboratory work providing the basis for this development. When we are familiar with these widely applicable principles we no longer have need for endless memorization of innumerable chemical facts. To see these principles grow out of observations you have made in the laboratory gives you a valid picture of how all scientific advances begin. It permits you to engage in scientific activity and thus, to some extent, to become a scientist yourself.

At the end of this course you won't know all of chemistry. We hope that you will know enough chemistry and enough about science to feel that the part you don't know is understandable, not mysterious. Perhaps you will appreciate the great power of scientific methods and appreciate their limitations. We hope that you will have become practiced in making unexpected observations, in weighing facts, and in framing valid conclusions. We hope that you will have formed the habit of questioning and of seeking understanding rather than being satisfied with blind acceptance of dogmatic assertions. We expect that you will share in the excitement of science and that you will feel the rich pleasure that comes with discovery. If most of these hopes are fulfilled, then you have had an optimum introduction to science through chemistry. Nothing could be a more important part of your education at a time when science is molding our age.

January 1963

GEORGE C. PIMENTEL
Editor for the Chemical Education Material Study

Foreword

This textbook was prepared over a three year period by a group of university and high school chemistry teachers under a grant from the National Science Foundation. The project, called CHEM Study, was organized and directed on broad policy lines by a Steering Committee of nationally known teachers and pre-eminent scientists from a variety of chemical fields. The Steering Committee, headed by Nobel Laureate Glenn T. Seaborg, attempted to staff the study with the country's most able university scientists and high school teachers. The university professors were drawn from all over the United States on the basis of demonstrated understanding of science and recognized leadership in teaching it. The names of the contributors to this text already appear on more than a dozen widely accepted college level textbooks. An equal number of outstanding high school teachers were named as contributors, each one individually selected on the basis of enthusiastic recommendations by his peers. These teachers participated in every phase of the preparation of this course. The effort of these highly qualified persons, totaling over fifteen man-years, is summed in the CHEM Study course. The National Science Foundation deserves commendation for making such activities possible; never before has such an array of talent been assembled to construct a high school chemistry course.

The textbook, *CHEMISTRY—An Experimental Science*, is designed for a high school introductory chemistry course and it is meshed closely with an accompanying *Laboratory Manual* and a set of pertinent films. A comprehensive *Teachers Guide* is available to aid teachers in gaining familiarity with the course. The first editions of the textbook and laboratory manual, written during the summer of 1960, were used during 1960–1961 in 23 high schools and one junior college by about 1300 students. During this first year, there was weekly staff contact with the pioneering teachers. On the basis of their experience, the materials were revised during the summer of 1961 and the *Teachers Guide* was written. This second edition was used in 123 high schools and 3 junior colleges scattered over the country and involving 13,000 students. Again the closely monitored field experience founded the third and final revision. The course, essentially in the form presented here, was used during 1962–1963 in 560 high schools in 46 states by about 45,000 randomly selected students. Its teachability is assured.

The title, *CHEMISTRY—An Experimental Science*, states the theme of this one year course. A clear and valid picture of the steps by which scientists proceed is carefully presented and repeatedly used. Observations and measurements lead to the development of unifying principles and then these principles are used to interrelate diverse phenomena. Heavy reliance is placed upon laboratory work so that chemical principles can be drawn directly from student experience. Not only does this give a correct and nonauthoritarian view of the origin of chemical principles, but it gives maximum opportunity for discovery, the most exciting part of scientific activity. This experimental theme is supported by a number of films to provide experimental evidence that is needed but not readily available in the classroom because of inherent danger, rarity, or expense.

The initial set of experiments and the first few textbook chapters lay down a foundation for the course. The elements of scientific activity are immediately displayed, including the role of uncertainty. The atomic theory, the nature of matter in its various phases, and the mole concept are developed. Then an extended section of the course is devoted to the extraction of important chemical principles from relevant laboratory experience. The principles considered include energy, rate and equilibrium characteristics of chemical reactions, chemical periodicity, and chemical bonding in gases, liquids, and solids. The course concludes with several chapters of descriptive chemistry in which the applicability and worth of the chemical principles developed earlier are seen again and again.

There are a number of differences from more traditional courses. The most obvious are, of course, the shift of emphasis from descriptive chemistry toward chemical principles to represent properly the change of chemistry over the last two decades. Naturally, this reconstruction of the entire course gives a unique opportunity to delete obsolete terminology and out-moded material. Less obvious but perhaps more important is the systematic development of the relationship between experiment and theory. Chemistry is gradually and logically unfolded, not presented as a collection of facts, dicta, and dogma. We hope to convey an awareness of the significance and capabilities of scientific activities that will help the future citizen assess calmly and wisely the growing impact of technological advances on his social environment. Finally, we have striven for closer continuity of subject matter and pedagogy between high school and modern freshman chemistry courses for those students who will continue their science training.

We do believe that the CHEM Study course achieves the goals we have set. Experience has shown that the course is interesting to and within the grasp of the average high school chemistry student and that it challenges and stimulates the gifted student. The course content provides a strong foundation for the college-bound student. Inevitably the question arises, "Is this course better than (or, as good as) the traditional one?" An answer is not readily found in comparative tests. A CHEM Study student might be handicapped in a test that has little emphasis upon principles, that is heavily laden with descriptive "recall questions," or that uses obsolete terminology. Conversely, a test designed specifically for the modern CHEM Study course content would surely prejudice against a student with a traditional preparation. The issue cannot be completely resolved "objectively" because value judgments are ultimately involved. Whether the CHEM Study goals are valid and the approach is reasonable must be decided with due consideration to the reported experience of teachers and to the credentials of those who developed the materials.

There are numberless ways in which CHEM Study is indebted to the University of California and to Harvey Mudd College for contributions of facilities, personnel, and encouragement. We acknowledge with thanks the stimulation and support we have received from the National Science Foundation. Finally, the Staff feels a heavy debt of gratitude to all of those who participated so energetically and enthusiastically in the preparation of the CHEM Study materials. We thank the Steering Committee for their valued and helpful guidance. We thank the contributors listed on the title page for their dedication of time, interest, and their ample talents to this effort. We acknowledge especially the key roles of Mr. Joseph Davis, Mr. Saul Geffner, Mr. Keith MacNab, Miss Margaret Nicholson, and Mr. Harley Sorensen. These individuals not only used the CHEM Study materials in the classroom but also served continuously as staff members. Their contributions and critiques have greatly increased the teachability of the CHEM Study course. We thank the many teachers who used the trial editions in their classrooms; their careful scrutiny of the text and laboratory manual and their many valuable suggestions provided a firm basis for revisions. Finally, we thank the many students who labored through the trial versions of CHEM Study; their every reaction—pain or pleasure, enthusiasm or ennui, spark or sputter—was noted and lent to the improvement of the course.

J. ARTHUR CAMPBELL,
Director, Chemical Education Material Study
Harvey Mudd College

GEORGE C. PIMENTEL,
Editor, *Textbook*
University of California

LLOYD E. MALM
Editor, *Laboratory Manual*
University of Utah

A. L. MC CLELLAN
Editor, *Teachers Guide*
California Research Corporation

DAVID RIDGWAY
Producer, *Films*

Berkeley, California
January, 1963

Acknowledgments

Quotations appearing on the following pages are used, with permission, from the indicated sources.

Page 1 *History of Science*, W. Dampier. New York: Cambridge University Press, 1949.

17 *Principia*, Isaac Newton. Mott's translation revised by F. Cajori. Berkeley: University of California Press, 1934, p. 673.

38 *New Systems of Chemical Philosophy*, John Dalton. Manchester, England, 1810.

49 *Readings in the Literature of Science*, W. C. Dampier and M. Dampier. New York: Harper and Row, 1959, p. 100.

65 *Solutions*, W. Ostwald. London: Longmans, Green and Co., 1891.

85 Letter by J. A. R. Newlands, *Chemical News*, Vol. 10, 1864, p. 94.

108 *Chemical Thermodynamics, A Course of Study*, Frederick T. Wall. San Francisco: W. H. Freeman and Company, 1958, p. 2.

124 The Drift Toward Equilibrium, H. Eyring, from *Science in Progress*, Fourth Series, edited by G. A. Baitsell, New Haven: Yale University Press, 1945, p. 169.

142 *Thermodynamics*, G. N. Lewis and M. Randall. New York: McGraw-Hill Book Co., Inc., 1923, p. 18.

163 *Solubility of Non-electrolytes*, J. H. Hildebrand. New York: Reinhold Publishing Corp., 1936, p. 13.

179 *Elements of Chemistry*, A. Lavoisier. New York, 1806, p. 14.

199 Predictions and Speculation in Chemistry, W. M. Latimer, *Chemical and Engineering News*, Vol. 31, 1953, p. 3366.

224 *Textbook of Quantitative Inorganic Analysis*, I. M. Kolthoff and E. B. Sandell. New York: Macmillan, 1936, p. 2.

233 *The Rise of Scientific Philosophy*, Hans Reichenbach. Berkeley: University of California Press, 1956, p. 168.

252 *Valence*, C. A. Coulson. New York: Oxford University Press, 1961, p. 3.

274 *Chemical Analysis by Infrared*, Bryce Crawford, Jr., New York: *Scientific American*, Oct. 1953.

300 *The Nature of the Chemical Bond*, L. Pauling. Ithaca: Cornell University Press, 1939, p. 422.

321 *Les Prix Nobel*, 1947, Nobel lecture by R. Robinson. Stockholm: Norstedt and Söner, 1947, p. 110.

421 From Quantum Chemistry to Quantum Biochemistry, Alberte Pullman and Bernard Pullman, in *Albert Szent-Gyoergyi and Modern Biochemistry*, edited by Rene Wurmser. Paris: Institute of Biology, Physics, Chemistry, 1962.

436 Genesis of Life, J. B. S. Haldane, in *The Earth and Its Atmosphere*, edited by D. R. Bates. New York: Basic Books, Inc., 1960.

The following photographs are used with permission from the indicated source.

Frontispiece The Candle—Illuminating Chemistry, by Bernard Abramson.

Page 5 Ice melting, by Ross H. McGregor.

5 Aluminum melting, courtesy Aluminum Corporation of America.

5 Solder melting, by Charles L. Finance.

48 G. N. Lewis, courtesy the Hagemeyer Collection, Bancroft Library, University of California.

94 Cutting potassium, by Charles L. Finance.

107 D. Mendeleev, courtesy the University of Leningrad.

141 H. Eyring, courtesy H. Eyring.

198 S. Arrhenius, courtesy The Bettmann Archive.

299 L. Pauling, courtesy The California Institute of Technology.

310 Network silicates, by Charles L. Finance.

312 Sodium chloride crystals, by Charles L. Finance.

320 P. Debye, courtesy Cornell University.

351 R. Robinson, courtesy Canadian Industries Limited.

386 A. Stock, courtesy The American Chemical Society.

420 G. T. Seaborg, courtesy California Research Corporation, Richmond Laboratory, Richmond, California.

435 R. B. Woodward, courtesy The American Chemical Society.

Color plate I Elements and compounds, by Charles L. Finance.

II Indicator colors, by Charles L. Finance.

III Spectrograph, by Charles L. Finance.

Contents

THE CANDLE—ILLUMINATING CHEMISTRY

CHAPTER 1

Chemistry: An Experimental Science

. . . those sciences are vain and full of errors which are not born from experiment, the mother of all certainty. . . .

LEONARDO DA VINCI, 1452–1519

Many words have been spoken and written in answer to the questions:

"What is the nature of scientific study?"
"What is the nature of chemistry?"

We shall try to find the answers in this course, not through words alone, but through experience. No one can completely convey through words the excitement and interest of scientific discovery. Hence we shall see the nature of science by engaging in scientific activity. We shall see the nature of chemistry by considering problems which interest chemists.

Our starting point will be based on examples of the activities of science, rather than on definitions. We will *perform* these activities, beginning on familiar ground. On such ground, where *you* know the answer, you will best see the steps by which science advances.

1-1 THE ACTIVITIES OF SCIENCE

Every form of life "feels" its surroundings in one way or another. In response to the feel of the surroundings, it behaves according to a pattern which tends to prolong its existence.

A tree is illuminated by the morning sunshine. In response, the leaves of the tree turn on their stems to present full surface to the light. This movement causes the leaves to intercept more light, and light is the source of energy which runs the amazing chemical factory operated by the tree. The tree grows.

A bear feels that summer is over—perhaps by

the length of the day or by the color of fall leaves, perhaps by some ursine almanac humans cannot read. In response, he seeks a secluded spot and takes a winter-long nap. During this hibernation, his blood pressure and body temperature drop, his digestion closes shop. The bear uses the minimum energy necessary to stay alive. It is not a coincidence that this occurs during the season when food is most difficult to find and the weather is quite unbearable.

Of all living things, man feels his surroundings and responds to them in the most complex way. He is more curious than the most inquisitive kitten. Through his intellect he uses his senses more effectively than an antelope avoiding a stalking lion. He has developed communication far beyond the warning quack of a sentry duck or the mating call of a lonely moose. Man's intellect, together with his communicative ability, permit him to respond to his environment in uniquely beneficial ways. He *accumulates* information about his surroundings, he *organizes* this information and *seeks regularities* in it, he *wonders why* the regularities exist, and he *transmits* his findings to the next generation. These are the basic activities of science:

to accumulate information through observation;
to organize this information and to seek regularities in it;
to wonder why the regularities exist;
to communicate the findings to others.

Fig. 1-1. **A scientist makes careful observations.**

So the activities of science begin with observation. Observation is most useful when the conditions which affect the observation are controlled carefully. A condition is controlled when it is fixed, known, and can be varied deliberately if desired. This control is best obtained in a special locale—a laboratory. When the observation is brought under careful control, it is dignified by a special name—*a controlled sequence of observations is called an* ***experiment.*** **All science is built upon the results of experiments.**

1-1.1 Observation and Description

Everyone thinks of himself as a good observer. Yet there is much more to it than meets the eye. It takes concentration, alertness to detail, ingenuity, and often just plain patience. It even takes practice! Consider an example from your own experience. Think how much can be written about an object as familiar as a burning candle! Of course, it takes careful observation—a careful experiment. This means the candle must be observed in a *laboratory*, that is, in *a place where conditions can be controlled.* But, how do we know which conditions need be controlled? Be ready for surprises here! Sometimes the important conditions are difficult to discover. Here are some conditions that *are* important in *some* experiments but are *not* important here.

The experiment is done on the second floor.
The experiment is done in the daytime.
The room lights are on.

Here are some conditions that might be important here.

The lab bench is near the door.
The windows are open.
You are standing close enough to the candle to breathe on it.

Why are these conditions important? Do they have something in common? Yes, there is the common factor that a candle does not operate well in a draft. The conditions are important because they influence the result of the experiment. Important conditions are often not as easily recognized as these. A good experimentalist pays much attention to the discovery of conditions that must be controlled. His success is often determined by his ability to control them.

Review your own description of a burning candle and compare your essay with the one in Appendix 1. How many of your observations are included there? How many listed in the appendix are not in your description? We see that the burning candle is a complicated and fascinating object when subjected to *careful observation* and *detailed description.*

1-1.2 The Search for Regularities

Observation inevitably leads to questions. One of the first questions that usually arises is "What regularities appear?" The discovery of regularities permits simplification of the observations. Instead of each observation standing alone, several observations can be classed together and, hence, can be used more effectively.

You must become aware of the pitfalls that exist in the search for regularities. The search is a meandering one, frequently taking wrong turns. It is inherent in the exploration of the unknown that not every step is an advance. Yet there is no other way to advance than by taking steps. How the search proceeds is best seen in a fable. The development of such a transparent example may help you see how a scientist searches for regularities.

Fable: A Lost Child Keeping Warm

Once upon a time a small child became lost. Because the weather was cold, he decided to gather materials for a fire. As he brought objects back to his campfire, he discovered that some of them burned and some of them didn't burn. To avoid collecting useless substances, the child began to keep track of those objects that burned and those that did not. (He *organized* his information.) After a few trips, his classification contained the information that is shown in Table 1-I.

Table 1-I. **FLAMMABILITY**

WILL BURN	WON'T BURN
Tree limbs	Rocks
Broom handles	Blackberries
Pencils	Marbles
Chair legs	Paperweights
Flagpoles	

This organization of the information was quite an aid in his quest for warmth. However, as tree limbs and broom handles became scarce, the child tried to find a regularity that would guide him to new burnable materials. Looking at the pile of objects that failed to burn and comparing it with the pile of objects that would burn, the child noticed that a regularity appeared. He proposed a possible "generalization."

Perhaps: "*Cylindrical objects burn.*"

This procedure is one of the elementary logical thought processes by which information is systematized. It is called **inductive reasoning,** and it means that *a general rule is framed on the basis of a collection of individual observations* (or "facts"). Of what use is the inductive process? It is an efficient way of remembering.

The next day the child went looking for burnable materials, but he forgot to bring along his list. However, he remembered his generalization. So, he returned to his hearthside hauling a tree limb, an old cane, and three baseball bats (successful predictions!). What's more, he reflected with pleasure that he hadn't bothered to carry back some other objects: an automobile radiator, a piece of chain, and a large door. Since these objects weren't cylindrical there was no reason to expect them to burn.

No doubt you are ready to complain that this generalization isn't really true! Quite the opposite! The generalization states a regularity discovered among *all* the observations available, and as long as observations are restricted to objects in the list, the generalization *is* applicable. **A generalization is reliable within the bounds defined by the experiments that led to the rule.**

As long as we restrict ourselves to the objects in Table 1-I (together with canes and baseball bats) it is surely true that all of the cylindrical objects burn!

Fig. 1-2. **"Cylindrical objects burn."**

Because of his successful predictions, the child became confident of his generalization. The next day he deliberately left the list at his campsite. This time, with the aid of his rule, he came back heavily laden with three pieces of pipe, two ginger ale bottles, and the axle from an old car, while spurning a huge cardboard box full of newspapers.

During the long cold night that followed he drew these conclusions:

(1) The cylindrical shape of a burnable object may not be intimately associated with its flammability after all.
(2) Even though the "cylindrical" rule is no longer useful, tree limbs, broom handles, pencils, and the other burnables in Table 1-I still burn.
(3) He'd better bring the list along tomorrow.

But, thinking over the longer list, he saw a *new* regularity that fitted Table 1-I and the newly acquired information as well:

Perhaps: "*Wooden objects burn.*"

What good is this rule in the light of the earlier disappointment? Well, it caused the child to go back and get that door he had passed up two days earlier, but it didn't lead him to go after the chain, the automobile radiator, or the cardboard box full of newspapers.

Don't think this is facetious—it is exactly what science is all about! We make some observations, organize them, and seek regularities to aid us in the effective use of our knowledge. The regularities are stated as generalizations that are called *theories.* A theory is retained as long as it is consistent with the known facts of nature or as long as it is an aid in systematizing our knowledge. We can be sure that some day a number of our present scientific views will seem as absurd as "Cylindrical objects burn." But on that day we will be proud of better views that have been substituted. If you are discouraged by the child's faltering progress—he hasn't yet decided that the box of newspapers will burn—be reassured. This child is a scientist and his faltering steps *will* lead him to the newspapers. They are the same steps that led us to our present understanding of relativity, to our discovery of polio vaccine, and to our propulsion of rockets to the moon.

A GENERALIZATION ABOUT THE MELTING OF SOLIDS

Through experiment, you yourself have discovered an important regularity in the behaviors of solid substances.

A solid melts to a liquid when the temperature is raised sufficiently. **The temperature at which a solid melts is characteristic.** *When the warm liquid is recooled, it solidifies at this same temperature.*

This generalization is of great value. It is based upon exactly the type of experiment you have performed. We have confidence in the rule because this type of experiment has been conducted successfully on hundreds of thousands of substances. *The melting behavior is one of the most commonly used methods of characterizing a substance.* It leads us to wonder if *every* solid can be converted to a liquid if the temperature is raised sufficiently. Further, it leads us to wonder

Fig. 1-3. **A solid melts to a liquid** *at a characteristic temperature.*

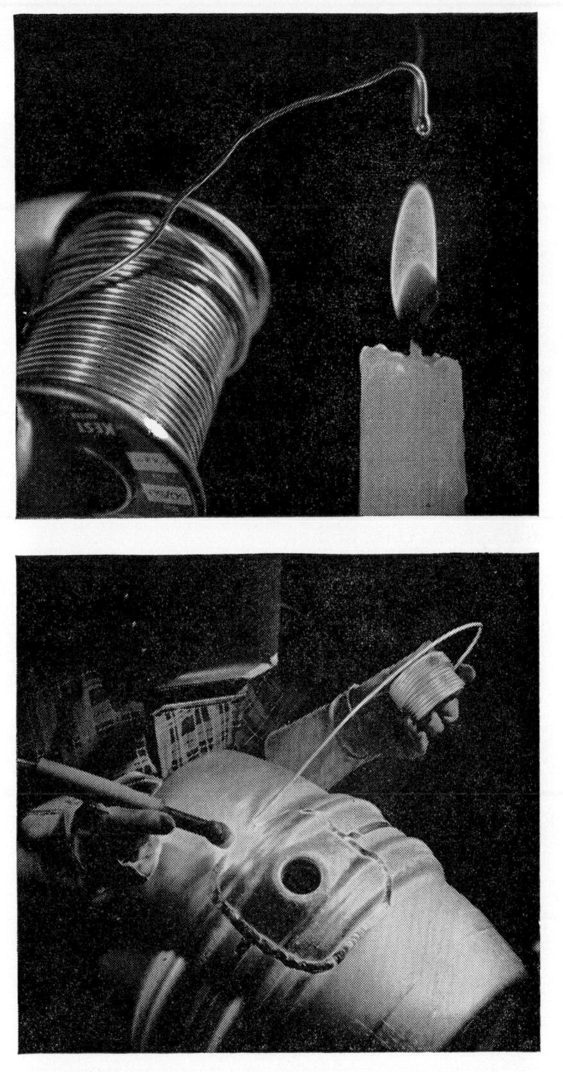

if *every* liquid can be converted to a solid if the temperature is lowered sufficiently.

SOME TERMINOLOGY

We have discovered that a solid can be converted to a liquid by warming it at or above its melting point. Then the solid can be restored merely by recooling. The solid and the liquid are similar in many respects and one is easily obtained from the other. Hence they are called different *phases* of the same substance. Ice is the solid phase of water and, at room temperature, water is in the liquid phase. The change that occurs when a solid melts or a liquid freezes is called a **phase change.**

1-1.3 Wondering Why

We have already experienced some of the activities of science. First came careful observation under controlled conditions, then organization of the information and the search for regularities of behavior. There is one more activity that, like dessert, fittingly comes last. This activity may be called "wondering why," and it arises from our irresistible urge to know more than merely "What happens?" We must also seek the answer to "*Why* does it happen?" This activity is probably the most creative and the most rewarding part of a science. What is the process? What does it mean to answer a question beginning "Why"?

EXPLANATIONS

Let us see what it means to search for an explanation. Consider a child blowing up a balloon. As he blows into the balloon again and again it expands and it becomes "harder." Evidently the gas is "pushing" on the inside of the balloon, stretching its elastic walls. Why does the gas push outward more and more on the walls of the balloon as it is inflated? Why does the gas continue to push outward without "tiring" or "running down"? These are "wondering why" questions.

There are two ways to proceed in trying to answer these questions. We have already examined one of these ways—to look more closely at the balloon, to record carefully what we see, and to seek regularities in what we observe. The second way is to look *away* from the balloon and to seek similar behavior in another situation that we understand better. Maybe this will enable us to frame an explanation of the gas pressure in terms of the better-understood situation. Sometimes a useful explanation turns up in a quite unexpected direction.

Consider the motion of a billiard ball. After it is struck by the cue, it moves until it strikes a cushion, from which it bounces, apparently with undiminished velocity. It rolls along in a new direction until it strikes another cushion, changing its direction again. It may continue to roll until it has hit the cushions six or seven times. The billiard ball seems almost tireless as it rebounds time and again from the "walls" of the billiard table. Could there be a connection between the "untiring" motion of a billiard ball and the "untiring" pressure of a gas in a balloon?

Billiard balls have long fascinated both idle and curious men. The latter group has found that the motion of a billiard ball can be described by assuming each collision with a cushion is perfectly elastic. When the ball strikes the cushion, pushing on it, the cushion pushes back, and the ball leaves again without any loss of velocity. Its movement can be reasonably foretold on the basis of this *elastic collision* assumption. Perhaps the behavior of a gas can be explained in these same terms. Suppose we picture a gas as a collection of particles bouncing around in a container, endlessly, making elastic collisions with the walls, just like miniature billiard balls. Every time one of these particles strikes a wall, it "pushes" on the wall and bounces away again. If there are many particles, there will be many such collisions per second, which accounts for the pressure of the gas. If gas is added to the balloon, there will be even more particles, hence more wall collisions per second, hence higher pressure. Thus the billiard ball model *does* offer a possible answer to our question.

With this example, we can now see the meaning of an explanation. It began with a "wondering why" question:

Question: Why does a balloon expand as it is inflated?

Possible Answer: Perhaps the gas put in the balloon consists of a collection of small particles that rebound from the wall of the balloon just as billiard balls rebound from the cushions of a billiard table. As the gas particles rebound from the balloon wall, they push on it. When more gas particles are added, the number of such wall collisions per second increases, hence the outward push on the balloon wall increases. The balloon expands.

Fig. 1-4. **A rebounding billiard ball** *suggests a possible explanation of gas pressure.*

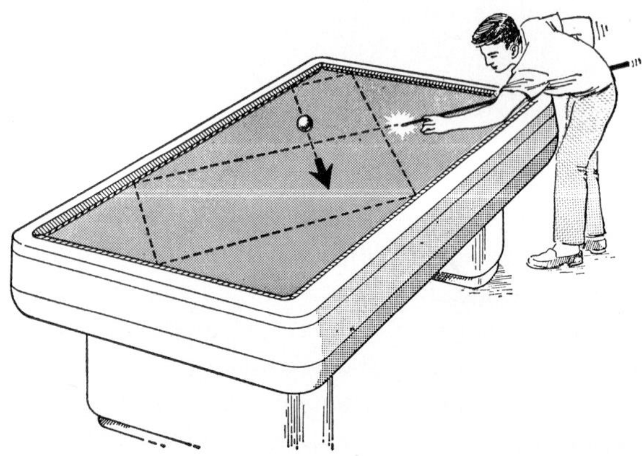

This is the characteristic pattern of an explanation. It begins with a "Why?" question that asks about a process that is *not* well understood. An answer is framed in terms of a process that *is* well understood. In our example, the origin of gas pressure in the balloon is the process we wish to clarify. It is difficult even to sense the presence of a gas. The air around us usually cannot be seen, tasted, nor smelled (take away smog); it cannot be heard or felt if there is no wind. So we attempt to explain the properties of a gas in terms of the behavior of billiard balls. These objects *are* readily seen and felt; their behavior has been thoroughly studied and is well understood.

The search for explanation is, then, the search for likenesses that connect the system under study with a model system which has been studied earlier. The explanation is considered to be "good" when:

(1) the model system is well understood (that is, when the regularities in the behavior of the model system have been thoroughly explored); and
(2) the connection is a strong one (that is, when there are close similarities between the studied system and the model system).

Our example constitutes a good explanation, because:

(1) how a billiard ball rebounds is well understood; we can calculate in mathematical detail just how much push the billiard ball exerts on the cushion at each bounce; and
(2) the connection to gas pressure is close; exactly the same mathematics describes the pressure behavior if the gas is pictured as a collection of many small particles, endlessly in motion, bouncing elastically against the walls of the container.

Therefore, the particle explanation of gas pressure is a good one.

Now, perhaps, you can see that answering the question "Why?" is merely a highly sophisticated form of seeking regularities. It is indeed a regularity of nature that gases and billiard balls have properties in common. The special creativity shown in the discovery of this regularity is that the likeness is not readily apparent. Fittingly, there is a special reward for the discovery of such hidden likeness. The discoverer can bring to bear on the studied system all of the experience and knowledge accumulated from the well understood system.

HEATING STEEL WOOL

These ideas are best learned by using them yourself. Let us take an example from your own laboratory work. You have observed the heating of a variety of solid materials: sulfur, wax, tin, lead, silver chloride, and copper. Each melts at a characteristic temperature. This fact led us to the generalization "A solid melts to a liquid at a characteristic temperature." Your confidence in the generalization was bolstered by the additional information (communicated to you, not experienced) that this applies to "hundreds of thousands of substances."

Your teacher demonstrated the effect of heating steel wool, and it proved to be a spectacular exception to the generalization about melting. An inexperienced observer might note this special behavior in his notebook, cross reference it under "Sparklers," and proceed to the next substance. A curious person, however, cannot resist wondering, "Why does steel wool behave in this special manner?"

In quite another direction of thought you investigated the burning of a candle. You discovered that air plays a role and that the products are different from the starting materials.

Here we have two areas that have been investigated thoroughly and that are well understood (and that you have examined to some extent yourself). Stated briefly, our starting point of knowledge is:

(1) Solids melt on heating.
(2) A candle burns, consuming oxygen from the air.

We are wondering why steel wool makes such a brilliant display on heating. Perhaps the explanation is to be found in 2. Though there are striking visual differences, perhaps we can *explain* the behavior of steel wool by likening it to

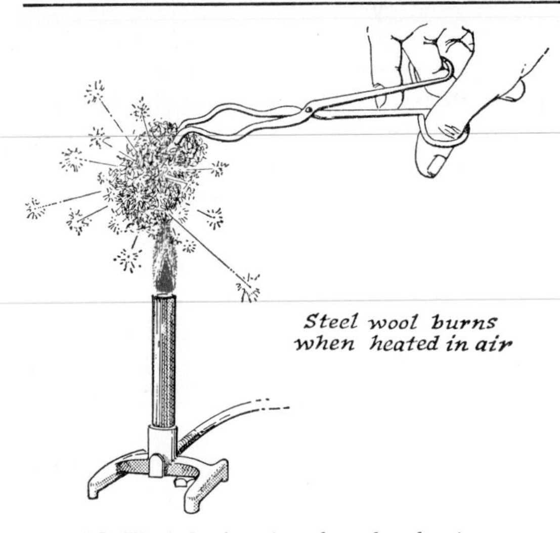

Fig. 1-5. **The behavior** *of steel wool on heating.*

a candle. Can we substitute the words "steel wool" for "candle" in 2?

Perhaps: (3) Steel wool burns, consuming oxygen from the air.

If 3 is a useful connection with the behavior of a burning candle, then we should enjoy the special reward for this discovery. The connection we propose implies that the knowledge accumulated on the candle can be brought to bear on the new system.

Since: A candle does not burn if it is denied oxygen,
then: steel wool should not sparkle if it is denied oxygen.

Here is a proposal subject to direct test. We can place steel wool under an atmosphere that excludes oxygen gas, and look for a change in the heating behavior. We go to the laboratory, heat steel wool under carbon dioxide and, lo, the steel wool melts!

The special behavior of steel wool on heating can now be said to be *explained:*

"Steel wool, like a candle, burns when heated in air."

"Steel wool, like other solids, melts to a liquid when heated under conditions that prevent burning."

This type of understanding makes possible the metallurgical processing of steel (and other metals). This type of reasoning makes possible an increase in our perception of the regularities of nature. *It begins with wondering why.*

1-2 UNCERTAINTY IN SCIENCE

Here are three statements concerning the melting behavior of *para*dichlorobenzene:

(1) The melting point is 53°C.
(2) The melting point is 53.2°C.
(3) The melting point is 53.203°C.

Apparently the third statement says more than the second, and the second statement says more than the first. Would it surprise you to learn that the second statement could be the most informative of the three? It is so. To understand why, we must consider uncertainty in measurement.

1-2.1 Uncertainty in a Measurement

A scientific statement conveys knowledge about the environment. The statement is careless if it says less than is known. It is misleading if it says more than is known. *The most accurate statement clearly conveys just what is known and no more.* Thus, a scientist will decide whether to list the melting point as 53°C, 53.2°C or 53.203°C by considering which value tells just what is known about the melting behavior and no more.

Consider your own laboratory measurement of the melting point of *para*dichlorobenzene (Experiment 3). Do your temperature measurements permit you to say that the melting point is 53°C, not 54°C? Probably, yes. It isn't too difficult to read the thermometer with this certainty. Can you read the thermometer so as to distinguish 53.0°C and 53.2°C? This is more difficult. It depends upon the thermometer and your skill in its use. It also depends upon whether the temperature of the solid during melting is uniform throughout. Still, a magnifying glass permits more certainty in the reading of the scale, and slower heating would increase the uniformity of the sample temperature. With this type of extra care, it is possible to tell that the melting point of *para*dichlorobenzene is 53.2°C, not 53.0°C. Consider, however, the chance of refining your thermometry sufficiently to distinguish between 53.200°C and 53.203°C. With the equipment available to you, it just isn't possible.

We conclude, then, that a careful measurement could establish the melting point to be 53.2°C. Then the second statement (m.p. = 53.2°C) would tell just what is known. The first statement (m.p. = 53°C) does not tell all that is known since only two figures are given, though three were measured. The third statement (m.p. = 53.203°C) tells far more than is known since the last two figures were not learned experimentally. We see that the measurement gives us three figures which are meaningful and significant. The number 53.2°C is said to have three significant figures.

Fig. 1-6. **Every measurement involves some uncertainty.** *The thermometer on the left can be read to ±0.2°C. The one on the right can be read to ±0.002°C.*

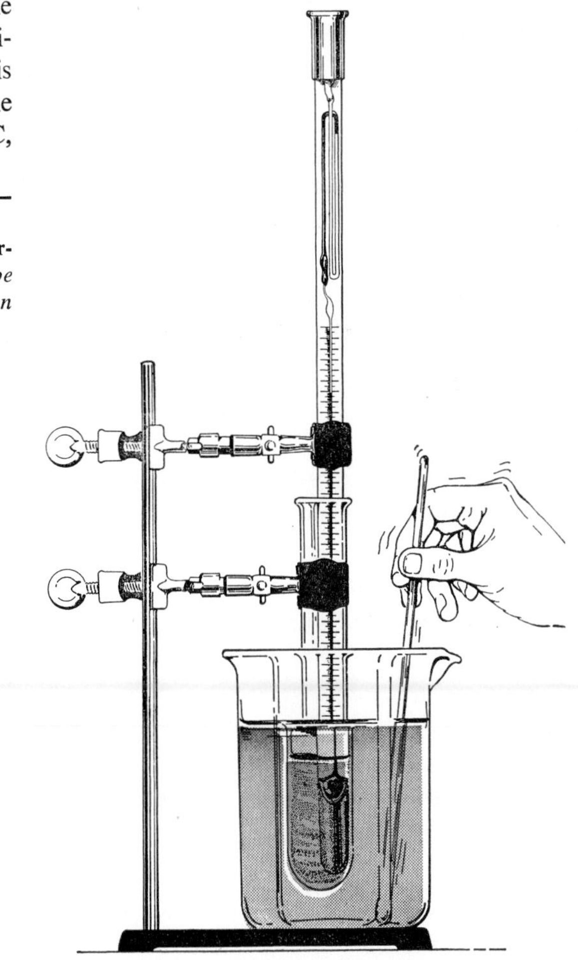

1-2.2 Uncertainty in a Derived Quantity: Addition and Subtraction

Results of scientific observations are often combined. For example, in Experiment 5 you will determine the change of water temperature during the combustion of a candle (or during the solidification of candle wax). The change of temperature, which we called Δt, is the result of *two* measurements, not just one—it is a *derived quantity:*

Temperature after heating = 38.5°C
Temperature before heating = 9.3°C
Difference (temperature change), Δt = 29.2°C

In accordance with good scientific practice, we would like to express the temperature change so as to include just what is known but no more. To do this, we must investigate how the uncertainties in the two temperature readings fix the uncertainty in the difference, Δt. Suppose the final temperature is measured by a second student who finds the temperature to be 38.3°C. Then a third student finds it to be 38.7°C. Apparently, different students making the same measurement may differ by a few tenths of a degree. Apparently, the measurement recorded by any one student could be in error by this amount, about 0.2°C. Perhaps the temperature recorded as 38.5°C could be as high as 38.7°C (0.2° higher) or as low as 38.3°C (0.2° lower)! This can be expressed briefly as follows:

Temperature after heating = 38.5 ± 0.2°C*

Presumably, the same uncertainty is present in the first temperature measurement, so our calculation becomes:

Temperature after heating = 38.5 ± 0.2°C
Temperature before heating = 9.3 ± 0.2°C
Difference (temperature change), Δt = 29.2 ± ??°C

To decide what uncertainty to ascribe to 29.2, consider the worst possible combination of the uncertainties. The first temperature could be as low as 9.1°C and the final temperature could be as high as 38.7°C. Then the difference would be 29.6°C. Thus the worst possible combination of errors would place an error in the difference equal to the sum of the uncertainties in the parts, 0.2 + 0.2 = 0.4. Hence our derived result can be written

Difference (temperature change) = 29.2 ± 0.4°C

We see that the uncertainty in a derived quantity is fixed by the uncertainties in the measurements that must be combined. For an addition or a subtraction, the maximum uncertainty is simply the sum of the uncertainties in the components: 0.2 + 0.2 = 0.4.

* The symbol ± is read "plus or minus."

EXERCISE 1-1

In Experiment 5, the weight of a sample of water is determined by subtracting the weight of the empty can from the weight of the can containing the water.

(wt. water) = (wt. can + water) − (wt. empty can)

Suppose the weight of the can is 61 ± 1 grams and the weight of the can plus water is 406 ± 1 grams. Calculate the weight of the water and the maximum uncertainty in the weight caused by the uncertainties in each of the two weighings.

1-2.3 Uncertainty in a Derived Quantity: Multiplication and Division

The temperature measurements made in Experiment 5 enable you to calculate the quantity of heat liberated when a known weight of candle is burned. Heat is measured in units called calories; **one calorie** *is the heat needed to raise the temperature of one gram of water one degree Centigrade.*† To raise the temperature of two grams of water one degree would require two calories—10 grams would require 10 calories. In general,

quantity of heat necessary to raise w grams of water one degree = w calories

But in the experiment, the temperature of the

† The heat necessary to raise the temperature of one gram of water one degree is constant within ±0.2% between 8 and 80°C. The calorie has been, in the past, defined to be the heat necessary to raise the temperature of one gram of water from 14.5 to 15.5°C.

water rises several degrees—we call the temperature change Δt. If it takes one calorie to warm one gram of water one degree Centigrade, it takes five calories to warm it five degrees. In general,

quantity of heat necessary to raise w grams of water Δt degrees

$$= w \times \Delta t \text{ calories} \quad (1)$$

Again we are faced with the dual problem: first, to calculate the quantity of heat, q; and, second, to decide the uncertainty in q.

The quantity of heat, q, is calculated with the aid of equation (*1*); it is simply the product of the weight of water times the temperature difference. Referring to our data, we find,

$$\text{wt. of water} = 345 \pm 2 \text{ grams}$$
$$\Delta t = 29.2 \pm 0.4°\text{C}$$

The quantity of heat is equal to the product,

```
     345
  × 29.2
     690
    3105
    690
10,074.0 calories
```

$$q = 10{,}074.0 \pm \text{ ? calories}$$

As in the last section, we can estimate the uncertainty in q by considering the worst possible combination of uncertainties. Suppose the weight is actually 343 grams and Δt is actually 28.8°. Then the product would be below 10,074.0 calories. But perhaps the weight is actually 347 grams and Δt is actually 29.6°. Then the product would be above 10,074.0 calories. These extremes determine the uncertainty in the product:

Minimum value	*Maximum value*
343	347
× 28.8	× 29.6
2744	2082
2744	3123
686	694
9878.4 calories	10,271.2 calories

We see that the product 10,074.0 could be in error by almost 200 calories. Now we can express our result together with its uncertainty:

$$q = 10{,}074.0 \pm \text{about } 196 \text{ calories}$$

In view of this large uncertainty, we can round off the answer sensibly:

$$q = 10{,}100 \pm 200 \text{ calories}$$

Once again the uncertainty in the product, a derived quantity, is fixed by the uncertainties in the measurements that must be combined.

EXERCISE 1-2

Calculate the uncertainty in the product $w \times \Delta t$ caused by the temperature measurement alone (assuming the uncertainty in the 345 gram weight of water has been made negligible by more careful weighings). Calculate the uncertainty caused by the ± 2 gram weighing uncertainty alone (assuming the uncertainty in the temperature change, 29.2°C, has been made negligible by use of a more sensitive thermometer). Compare these two contributions to the total uncertainty, about 200 calories.

The uncertainty in the product, ± 200 calories, is not simply the sum of the uncertainties in the factors, ± 0.4°C and ± 2 grams. Instead, the sum of the *percentage* uncertainties in the factors determines the uncertainty in a product or a quotient.* Fortunately, there is an easy method for estimating it roughly without calculating percentages. This method, based upon the number of figures written, is described in Section 1-2.5.

1-2.4 The Absence of Certainty in Science

Each measuring device has limitations that fix its accuracy. Hence every individual observation has some uncertainty associated with it. Since every regularity of nature is discovered through observations, every regularity (law, rule, theory) has uncertainty attached to it.

Every scientific statement involves some uncertainty.

A corollary:

No scientific statement is absolutely certain.

* The calculation based upon percentage uncertainty is presented in Appendix 4 of the Laboratory Manual.

1-2.5 How Uncertainty Is Indicated

We now have seen two methods for indicating uncertainty in a number. The most informative is to follow the number by the symbol ± and then the best estimate available of the uncertainty. The less informative but more widely used method is to indicate crudely the uncertainty by the number of figures shown. The last figure shown is generally the one in which there is some uncertainty. Thus, the number 53.2°C indicates there may be uncertainty in the figure 2, but there is none in either of the figures 5 or 3. *The digits that are certain and one more are called* **significant figures.** The correct number of significant figures should always be used and, wherever possible, the more definite indication ± should be added.

We need convenient rules for estimating the maximum uncertainty in derived quantities. This is rather easy for a sum or a difference. In either case, merely add up the uncertainties in the components. Fortunately there is an equally simple rule for estimating roughly the uncertainty in a product or quotient. *A product ($a \times b$) or quotient (a/b) has the same number of significant figures as the less precise component (a or b).**

EXERCISE 1-3

In Section 1-2.3 we multiplied 345 × 29.2 to obtain 10,074.0 calories.

(a) How many significant figures are there in the factor 345? In the factor 29.2?
(b) How many significant figures should be retained in the product, 10,074.0?
(c) Six figures are specified in the number 10,074.0—more than are warranted. "Round off" this number in accordance with your answer to (b). Compare your answer with the final result derived in Section 1-2.3, $q = 10{,}100 \pm 200$ calories.

1-3 COMMUNICATING SCIENTIFIC INFORMATION

One of the most important reasons for man's progress in understanding and controlling his environment is his ability to communicate knowledge to the next generation. It isn't necessary for each twentieth century scientist to invent the atomic description of matter. This was invented by John Dalton in the nineteenth century, and Dalton recorded his ideas in the scientific literature together with the observations that led him to the model. By study of this and subsequent literature a modern scientist can appraise the nature of the description, the facts it will explain, and the limitations. He is quickly able to approach the frontiers of knowledge—the frontiers defined by the limitations in our accepted models of the behavior of matter.

One can almost say that a scientific advance is important *only* if it is communicated to others. If Dalton had not told others of his ideas nor tried to convince them (through his supporting arguments), then someone else would have had to do it all again.

Communication of knowledge is, then, an important part of scientific activity. The first requirement is good use of language. If an idea is not well expressed, whether orally or in writing, it is not likely to be clearly understood. An argument loses its force if it is stated in an ambiguous way. An essential thought can be lost in a maze of excess words. *Choose and use your language carefully.*

The manner in which you present an idea depends to some extent upon the intended use, to some extent upon the type of information available. Usually, the more precise the statement of the regularity, the more valuable it is. In general, there will be more than one way to express a generalization, and judgment is needed in mak-

* The use of significant figures is discussed in Appendix 4 in the Laboratory Manual.

Fig. 1-7. **A regularity in the behavior** *of a fixed amount of gas. As pressure rises, volume decreases.*

ing a choice among them. This is best seen in terms of an example.

Figure 1-7 shows a tire pump with a pressure gauge attached to the hose so that the gauge prevents escape of the gas in the pump. Pushing on the handle of the pump causes the plunger to go down, and reduces the volume occupied by the gas. The gauge shows that the pressure is higher. Pushing still harder on the handle increases the pressure still more. Again the increase in pressure causes a reduction in volume. We see that *as the pressure rises, the volume decreases.* This qualitative statement describes a regularity in the behavior of a fixed amount of gas. Such a qualitative statement is the crudest form in which a regularity can be expressed.

A curious person, attempting to understand this regularity, might see a necessity for more careful measurements. He might build a new piece of apparatus, one that would be more suitable for measuring volumes and pressures over a wide range. After carrying out a series of measurements, he would conclude with a table of data, such as Table 1-II. A table of data is a second way in which the regularity can be expressed. In a third mode of expression, the measurements can be presented graphically in a plot of pressure against volume, as in Figure 1-8.

With these careful measurements, we might also seek a mathematical statement of the behavior. Sometimes inspection of the data suggests a relationship. Sometimes the appearance of a plot, such as Figure 1-8, reveals a mathematical expression. In the example we're studying, the curve resembles a hyperbola, a curve described by the simple equation, xy = a constant. This similarity leads us to multiply each pressure and volume pair, as shown in the third column of Table 1-II. We find that during a ten-fold increase in pressure, the product $P \times V$ is fairly constant. There are some variations in the product, as seen both in Table 1-II and, as well, in the scatter of the points around the smooth curve in Figure 1-8. The random nature of the deviations from constancy suggests that they measure the uncertainty due to experimental technique. We can use these deviations to provide an estimate of the uncertainty in the average, ± 0.6. (How this is done is shown in Exercise 1-4.)

Hence, with reasonable confidence we can state the regularity mathematically:

$$P \times V = 22.4 \pm 0.6$$

Thus we have found four ways to express the regularity between the pressure and volume of oxygen gas:

(a) *Qualitatively:* As the pressure rises, the volume decreases.

(b) *Quantitatively:* List the original data that show how pressure and volume are related, as in Table 1-II.

(c) *Graphically:* Plot the relationship between pressure and volume of 32.0 grams of oxygen gas at 0°C, as in Figure 1-8.

(d) *Mathematically:*

$P \times V = 22.4 \pm 0.6$

P = pressure (in atmospheres)

V = volume (in liters) of 32.0 grams of oxygen gas at 0°C

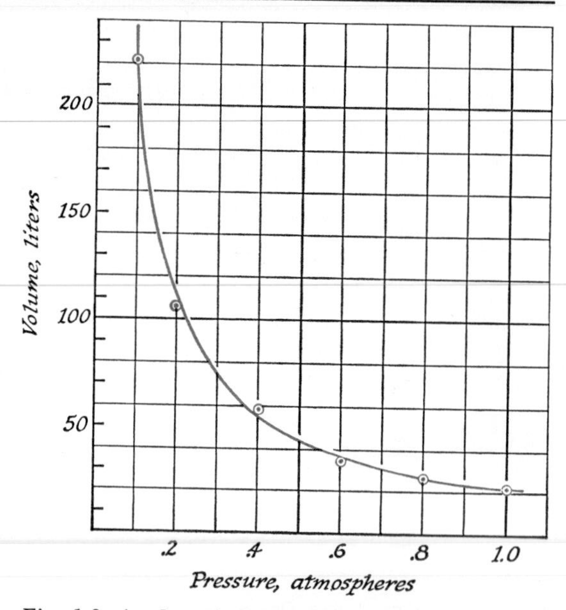

Fig. 1-8. **A plot of the pressure versus volume** *of 32.0 grams of oxygen gas; $t = 0°C$.*

Obviously the regularity expressed in the qualitative form (a) is far less informative than any one of the quantitative presentations, (b), (c), or (d). The relative merits of the expressions (b), (c), and (d) depend upon the use. Table 1-II tells in most detail exactly how much is known about the pressure-volume behavior of oxygen gas (from this experiment). In the graphical presentation of Figure 1-8 the trend of the data is shown by the smooth curve drawn to pass near as many points as possible. Uncertainties caused by experimental errors cause the data points to fall above and below this smooth curve. Hence the graphical presentation reveals reliability of the measurements. The smooth curve "irons out" these uncertainties and provides a convenient basis for predicting volumes at intermediate pressures (that is, for interpolating). However, from the standpoint of usefulness, the mathematical expression (d) is often the best. It is the most compact way of stating the regularity together with its uncertainty. Mathematics is one of the most important tools of chemistry.

No matter how expressed, *all* scientific "rules," "laws," and "theories" are statements of regularities of nature. Their usefulness depends upon the amount of experimental evidence that shows that the "rule," "law," or "theory" corresponds to experimental reality. Within the bounds that it is known to correspond to experimental reality, the relation can be used for prediction.

Table 1-II.

PRESSURE AND VOLUME OF 32.0 GRAMS OF OXYGEN GAS $t = 0°C$

PRESSURE (in units called atmospheres)	VOLUME (in units called liters)	$P \times V$
0.100	224	22.4
0.200	109	21.8
0.400	60.0	24.0
0.600	35.7	21.4
0.800	27.7	22.2
1.00	22.4	22.4
	Average	22.4 ± 0.6

EXERCISE 1-4

(a) Add the six values of $P \times V$ in Table 1-II and divide by 6 to obtain the average, $(P \times V)_{av}$ or, $(PV)_{av}$.

(b) Now add a fourth column to Table 1-II showing the deviation of each $P \times V$ product from $(PV)_{av}$. Head this column with the word "Deviation," and calculate each entry by subtracting $(PV)_{av}$ from the measured value. For example, the second entry will be -0.6 (since, $21.8 - 22.4 = -0.6$).

(c) When you have completed the column of deviations, add the column (disregarding algebraic signs) and divide by 6 to obtain an average deviation.

(d) Compare your calculations in (a) and (c) with the result given in Table 1-II,

$$\text{Average} = 22.4 \pm 0.6$$

1-4 REVIEW

This chapter began with the statement that we would find through experience what science is all about. Already you have had opportunities to do so in the laboratory. We see that **science is man's systematic investigation of his environment.** Chapter 1 has told how this investigation proceeds. The remainder of the book is concerned with those parts of this investigation that are carried out by chemists. Before going on to see what chemistry is, let us review your laboratory accomplishments so far with emphasis on the activities of science.

1-4.1 Accumulating Information Through Observation

Observation of a burning candle reveals an astonishing complexity. It also reveals the importance and value of careful study and attention to detail.

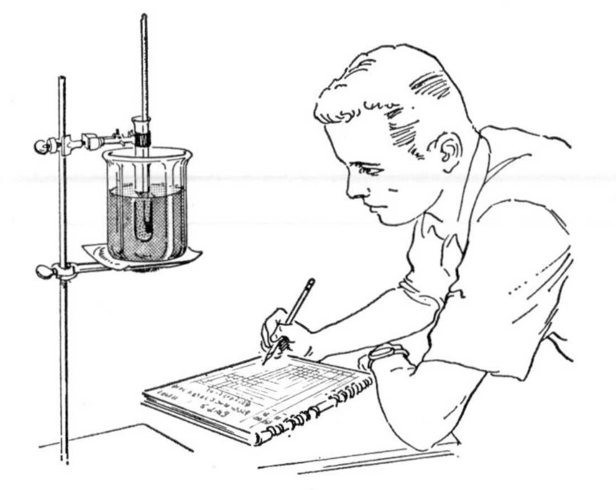

Fig. 1-9. **A good experimentalist is a good observer.** *Record in your notebook at the moment of observation. Prepare tables in advance.*

In your experimentation, be alert and ready for unexpected developments. Record in your notebook at the moment of observation a description of everything you see. The time of the observation frequently has importance. Completeness is, by far, the most important property of a good notebook. Next in importance, legibility, neatness, and organization make your notebook a more valuable record. Whenever possible, prepare tables in advance for the results of measurements you can anticipate. This guarantees that you won't forget to note important information, and it frees you from clerical work during an experiment.

Remember, *chemistry is built upon the results of experiments. An experiment is a controlled sequence of observations. A good experimentalist is a good observer.*

1-4.2 Organizing Information and Seeking Regularities in It

The mere cataloging of observations is not science. In fact, the advance of our knowledge of nature would long ago have ground to a halt if we merely made observations. The multitude of known facts can be dealt with only if it is packaged efficiently. This packaging we have called "organizing the information" and "seeking regularities."

There is no single recipe for seeking regularities. That is probably why the search is so interesting and why the scientist receives so much personal satisfaction from his work. Here is opportunity for originality, opportunity for testing one's wit and cleverness. You can experience the pleasure a scientist derives from clarifying a previously mystifying behavior by careful experiments of your own.

In our study of the candle the presence of liquid at the top of the candle caught our attention. It led us to wonder about the behavior of other familiar solids given a similar treatment. In this case, we went looking for regularity—we sought, through experiment, to discover how other solids behave on heating. Our first studies, when organized, led us to the generalization that solids melt at a characteristic temperature when heated. We made two gains thereby: we found an efficient expression of the results of a number of experiments, and we provided a basis for expectation on the effect of heating solids which we have not studied before. *The confidence this expectation deserves is fixed by the amount of evidence supporting the generalization.*

1-4.3 Wondering Why

The culmination of the investigation of our environment we have dubbed "wondering why." We seek *explanations.* Through an example, we have seen that an explanation is the discovery of likenesses connecting a process we do not understand with processes we do understand. This is the most rewarding activity of science. It leads to exploration. Learn to ask yourself questions beginning with "Why" when you observe—both in and out of the chemistry laboratory. It is a good habit to have, and it frequently makes life more interesting.

You have had opportunities to ask many "Why" questions already from your work in the laboratory. In fact, there are enough such questions to provide the basis for the rest of the course. Some of the questions that have been raised in your experiments are listed at the end of the chapter. Can you add to this list? How many of the questions can you answer now? We will find the answers to many of them in our subsequent study. Some may not yet have satisfactory answers. These are the most interesting questions because they point into the future—*your* future.

SOME QUESTIONS RAISED DURING THE STUDY OF A BURNING CANDLE

Why does a solid absorb heat when it melts?

Why is heat liberated in the burning of a candle?

Why is the heat effect so much larger in the chemical reaction than in the phase change you studied?

Why does the candle react with air to give carbon dioxide and water rather than the reverse, carbon dioxide and water reacting to give candle and air?

Why didn't the candle react with air (that is, burn) while the candle was stored in your desk drawer? Why did it wait until you wanted it to burn? What is the role of the match you used to light the candle?

Why does a candle burn slowly when you light the wick, in contrast to what happens when you light the "wick" of a firecracker?

What is the role of the wick of the candle?

How much water and carbon dioxide are produced from a burning candle?

Why does carbon dioxide cause limewater to become cloudy?

Why does the burning of sulfur produce a bad smell while the burning of steel wool produces sparks?

Why does a flame emit colored light?

Why is the base of the flame blue?

What is the dark zone in the candle flame?

Why does the candle flame smoke more in a breeze?

Why don't we run out of questions?

CHAPTER 2

A Scientific Model: The Atomic Theory

. . . hypotheses ought to be fitted merely to explain the properties of things and not attempt to predetermine them except in so far as they can be an aid to experiments.

ISAAC NEWTON, 1689

One of the activities of science is the search for regularity. Regularities that directly correlate experimental results are generally called rules or laws. A more abstract regularity, expressing a hidden likeness, is generally called a model, theory, or principle. Thus, the behavior of oxygen gas summarized in the equation $P \times V = a$ *constant* is called a law.* The explanation of this same regular gas behavior in terms of the motion of particles is called a theory. It is a greater abstraction to connect the PV product with the mathematical equations that describe rebounding billiard balls. Nevertheless, rules, laws, models, theories, and principles all have a common aim—they all systematize our experimental knowledge. They all state regularities among known facts.

The more abstract regularities come from the discovery of a hidden likeness. When the likeness involves a real physical system (such as rebounding billiard balls), the explanation is usually called a model. When the likeness involves an abstract idea (such as a mathematical equation), the explanation is usually called a theory. There is no real distinction to be made, however, and we shall use the words model and theory interchangeably.

When seeking an explanation, we sometimes find more than one explanation. When this happens the model (or theory) that is most useful is most used. A model that proves to be useful generally points to new directions of thought. The new directions guide us to new experiments and, thereby, new facts come to light. Often, the new facts will require growth of the model. Occasionally the new facts forcibly contradict the model and it must be abandoned in favor of another. Both the growth and abandonment of models or theories reflect increase in our understanding of the environment.

Let us see how a model grows.

* It is called Boyle's Law, after Robert Boyle, the scientist who first discovered this particular regularity.

2-1 IMPLICATIONS AND GROWTH OF A SCIENTIFIC MODEL

As an example, we can explore the implications of our explanation of the behavior of gases.

Question: Why does a balloon expand as it is inflated?

Possible Answer: Perhaps the gas put in the balloon consists of a collection of small particles that rebound from the wall of the balloon just as billiard balls rebound from the cushions of a billiard table. As the gas particles rebound from the balloon wall, they push on it.

This model is useful, first, because we can calculate in mathematical detail just how much push a billiard ball exerts on a cushion at each rebound, and, second, because exactly the same mathematics describes the pressure behavior of gas in a balloon. The success of the model leads to new directions of thought. For example, we might now wonder whether the pressure-volume behavior of oxygen, as shown in Table 1-II (p. 14), can be explained in terms of the particle model of a gas.

2-1.1 The Pressure-Volume Behavior of Oxygen Gas

The experimental data in Table 1-II show that decreasing the volume by one-half doubles the pressure (within the uncertainty of the measurements). How does the particle model correlate with this observation? We picture particles of oxygen bounding back and forth between the walls of the container. The pressure is determined by the push each collision gives to the wall and by the frequency of collisions. If the volume is halved without changing the number of particles, then there must be twice as many particles per liter. With twice as many particles per liter, the frequency of wall collisions will be doubled. Doubling the wall collisions will double the pressure. Hence, our model is consistent with observation: Halving the volume doubles the pressure.

Fig. 2-1. **In the particle model,** *wall collisions determine pressure. Halving the volume doubles the pressure.*

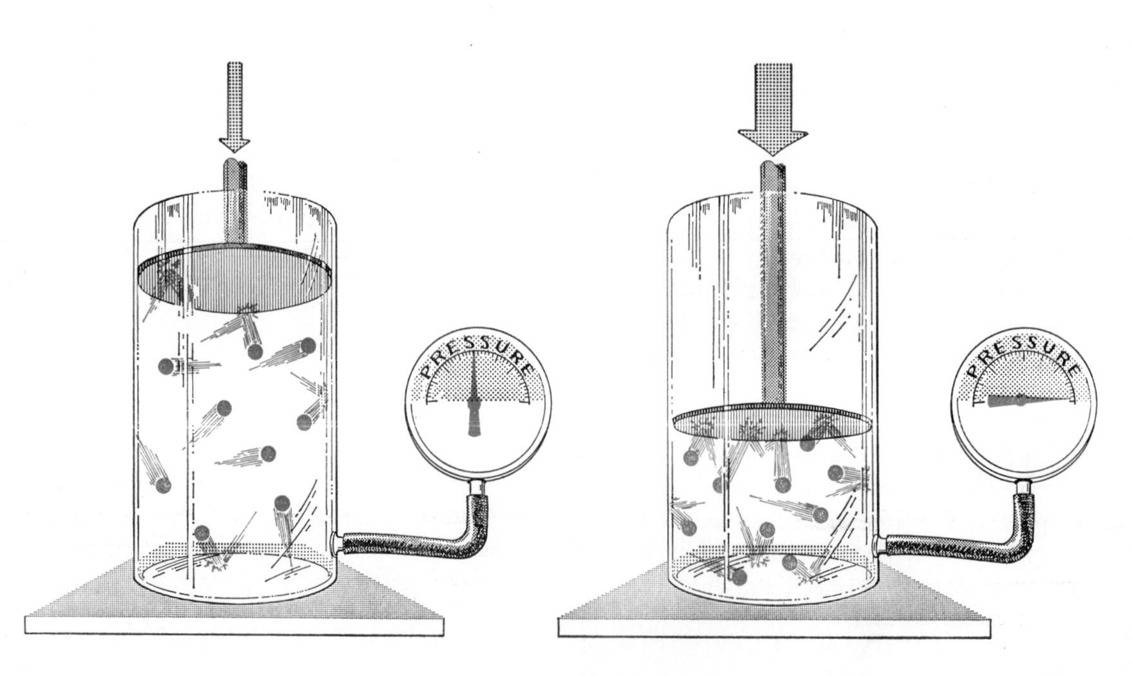

2-1.2 The Pressure-Volume Behavior of Other Gases

Having gained this understanding of the pressure-volume behavior of oxygen, it is natural to wonder whether the same model is applicable to other gases. Thus, the development of the theory leads us to perform new experiments. Such experiments provide a systematic growth of our knowledge of our environment. They are generally much more effective than random, "shot-in-the-dark" experiments.

Two other gases on the chemist's shelf are ammonia and hydrogen chloride. Is the particle model applicable to them as well as to oxygen? To find out, we must perform experiments that duplicate the conditions used in the study of oxygen. Table 2-I shows pressure-volume measurements for 32.0 grams of gaseous ammonia at 0°C. Table 2-II shows the same type of data for 32.0 grams of gaseous hydrogen chloride at this same temperature.

Table 2-I

PRESSURE AND VOLUME OF 32.0 GRAMS OF AMMONIA GAS t = 0°C

PRESSURE (atmospheres)	VOLUME (liters)	$P \times V$
0.100	421	42.1
0.500	84.2	42.1
1.00	42.1	42.1

We see that these two gases also show the behavior at a fixed temperature, $PV = a\ constant$. The particle model should be useful for these gases as well as for oxygen. On the other hand, the numerical value of the constant varies from one gas to another, if the same weight of gas is considered. Thus 32.0 grams of oxygen at 0°C and one atmosphere occupy 22.4 liters. The same weight of ammonia at this temperature and pressure occupies 42.1 liters. The same weight of hydrogen chloride occupies only 19.6 liters. The particle model of gases must be modified to explain these differences.

Table 2-II

PRESSURE AND VOLUME OF 32.0 GRAMS OF HYDROGEN CHLORIDE GAS t = 0°C

PRESSURE (atmospheres)	VOLUME (liters)	$P \times V$
0.100	196	19.6
0.500	39.2	19.6
1.00	19.6	19.6

To explain this behavior, chemists have found it convenient to consider a different weight of each gas; they select that amount that gives the same PV product as 32.0 grams of oxygen gas. Consider, first, ammonia gas. At 0°C and a pressure of one atmosphere, 32.0 grams of ammonia occupies 42.1 liters. We have taken too large a weight of ammonia. The weight of ammonia needed to occupy only 22.4 liters at this pressure is smaller by a factor 22.4/42.1:

$$\text{wt. ammonia} = 32.0 \text{ g} \times \frac{22.4}{42.1} = 17.0 \text{ g}$$

Pressure-volume data for this weight of ammonia are shown in Table 2-III.

Table 2-III

PRESSURE AND VOLUME OF 17.0 GRAMS OF AMMONIA GAS

PRESSURE (atmospheres)	VOLUME (liters)	$P \times V$
0.100	224	22.4
0.500	44.8	22.4
1.00	22.4	22.4

EXERCISE 2-1

If 32.00 grams of hydrogen chloride gas (at 0°C and one atmosphere) occupy 19.65 liters, then a larger weight of hydrogen chloride is needed to occupy the larger volume, 22.4 liters. Show that the weight needed is 36.5 grams.

Now the regularity between pressure and volume of these three gases can be expressed as follows:

For 32.0 grams of oxygen at 0°C,

$$P \times V = 22.4;$$

for 17.0 grams of ammonia at 0°C,

$$P \times V = 22.4;$$

for 36.5 grams of hydrogen chloride at 0°C,

$$P \times V = 22.4.$$

Each of the gases has a behavior consistent with the particle model of a gas (*PV = a constant*). However, the particles of the gas called oxygen must differ from the particles of gas called ammonia. These, in turn, must differ from the particles of the gas called hydrogen chloride. How do the particles differ? Why is it that 32.0 grams of oxygen give the same *PV* product as 17.0 grams of ammonia and 36.5 grams of hydrogen chloride (all at 0°C)? Do the particles have different weights? Again, we are led to new questions and new questions lead to new experiments.

2-1.3 Some Properties of Gases

What gases do we find in the chemical stockroom? How do they compare in other properties? Looking down the row of tanks we find the names ammonia, chlorine, hydrogen, hydrogen chloride, nitric oxide, nitrogen dioxide, and oxygen, among others. Two of these gases are colored—chlorine is yellow-green and nitrogen dioxide is reddish-brown—and the other five are colorless. The colorless gases can be further sorted according to their solubilities in water. Figure 2-2 shows what happens if a stoppered test tube full of each gas is opened with the mouth of the test tube under water. In the tubes containing ammonia and hydrogen chloride the water rises rapidly, filling the tubes. These two gases dissolve readily in water. In each of the other three test tubes the liquid level rises very little, showing that little gas dissolves.

Fig. 2-2. **Gases have different solubilities in water.**

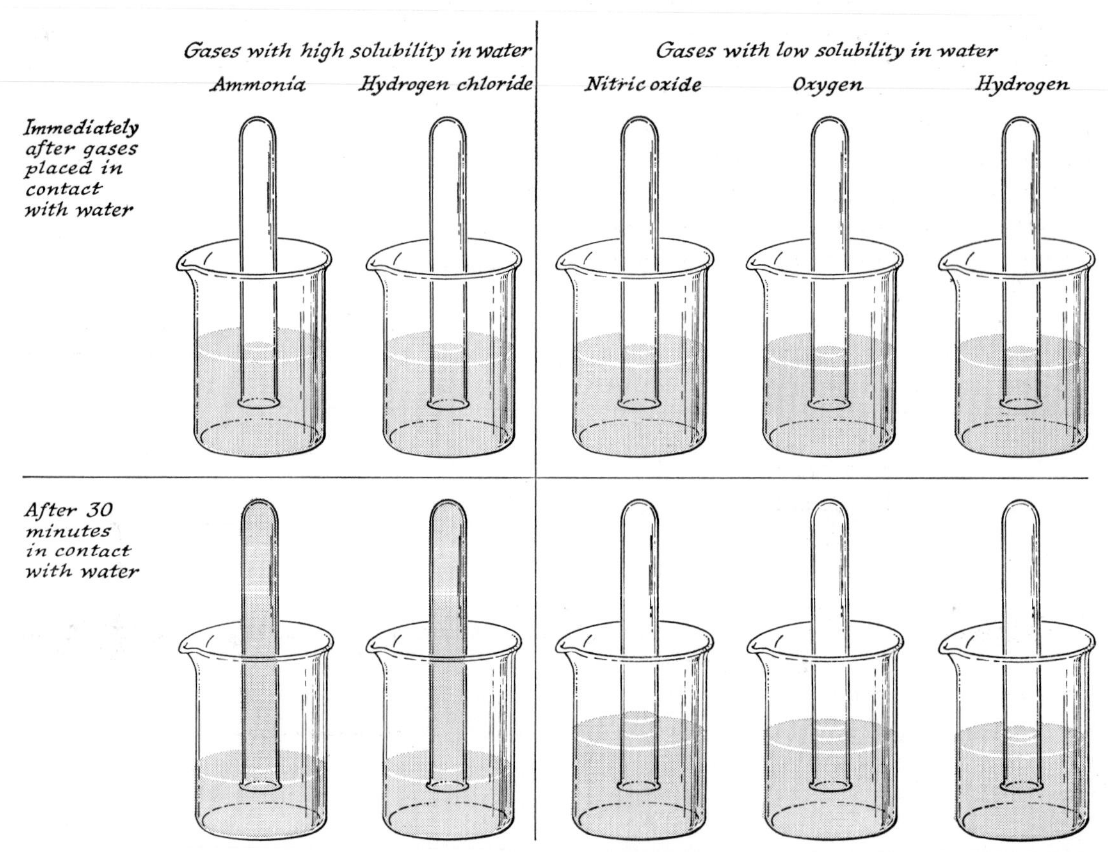

Though ammonia and hydrogen chloride both dissolve in water, these two gases are very different in other properties. For example, they behave differently when placed in contact with the dye, litmus. This dye, when moistened, turns red if it is placed in hydrogen chloride. However, if it is placed in ammonia, it turns blue.

We have not yet distinguished the gases nitric oxide, hydrogen, and oxygen. Nitric oxide has its own personality. Immediately upon exposure to air, the colorless nitric oxide becomes reddish-brown—exactly the color of nitrogen dioxide. Neither oxygen nor hydrogen behaves this way.

The gases oxygen and hydrogen are readily distinguished by their combustion properties. When a glowing splint is plunged into oxygen, the splint bursts into flame. When a brightly glowing splint is plunged into hydrogen, the glow is either extinguished or, if air has mixed with hydrogen, it produces a small explosion.

Thus we find that each of these gases has distinctive properties. If these gases are made up of particles, then the particles must be distinctive. The particles that are present in ammonia cannot be like the particles in hydrogen chloride, or like those in the other gases. The nature of the ammonia particles, then, is the key to the properties of ammonia. The particles that make up a gas determine its chemistry. They are so important to the chemist that they are given a special name. A gas is described as a collection of particles called **molecules**.

2-2 MOLECULES AND ATOMS

The particles, or molecules, of the gas nitric oxide cannot be exactly like those of nitrogen dioxide. There must be differences that account for the fact that one gas is colorless and the other reddish-brown. Yet, when nitric oxide and air are mixed, color appears, suggesting that nitrogen dioxide has been formed. Apparently molecules present in air somehow combine with the molecules of nitric oxide to form molecules of nitrogen dioxide. We would like to develop our picture of molecules so it will aid us in discussing these changes.

To explain how molecules can rearrange and change, we assume they must be built of smaller fragments. These smaller fragments, or building blocks, are called **atoms**. With this assumption, we can explain differences between two molecules in terms of the atoms present in each molecule. Nitric oxide is different from hydrogen chloride because it contains different atoms. Nitric oxide exposed to air forms nitrogen dioxide by some rearrangement of the available atoms. In general, the properties of a gas are fixed by the number and types of atoms it contains.

Fig. 2-3. **Models of molecules.** *The properties of a molecule are fixed by the number and types of atoms it contains.*

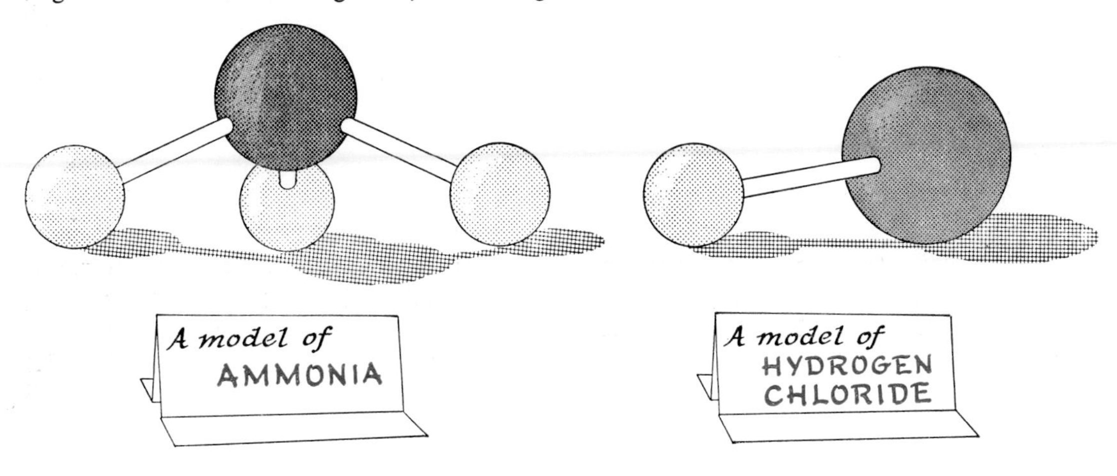

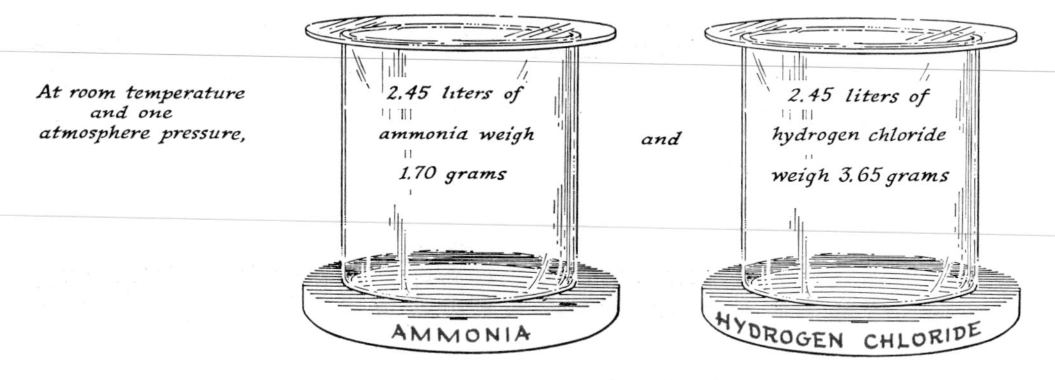

Fig. 2-4. **Why** *do equal volume of ammonia and hydrogen chloride have different weights?*

Chemists construct models of molecules to show how many atoms they contain. Figure 2-3 shows some examples. The model of ammonia represents three atoms of one kind attached to one atom of another kind. The model of hydrogen chloride contains only two atoms, and of different kind. We shall spend the entire year discussing why these and other models are constructed as they are. We shall see that a molecule of ammonia is pictured as shown in Figure 2-3 because this model helps us explain the properties of ammonia. Throughout the course, we shall investigate properties of substances found in nature or prepared in the laboratory and then we will seek explanations in terms of the numbers, types, and arrangements of the atoms present. These explanations are called the **atomic theory.** The atomic theory is regarded as the cornerstone of chemistry.

2-2.1 The Weights of Molecules

We have discovered that 32.0 grams of oxygen, 17.0 grams of ammonia, and 36.5 grams of hydrogen chloride each exhibits the regular behavior,

$$PV = 22.4 \qquad \text{at } 0°\text{C} \qquad (1)$$

For any of the three gases, if the pressure is given, the volume occupied can be calculated. At one atmosphere, each of the specified weights of gas occupies 22.4 liters. At two atmospheres, each gas is compressed into a smaller volume, 11.2 liters. We wonder, Why do 22.4 liters of ammonia weigh 17.0 grams when the same volume of hydrogen chloride weighs 36.5 grams?

There are two factors to consider. These are the same two factors we would be concerned with if we were to ask why a bag full of beans weighs 17.0 grams whereas the same bag full of marbles weighs 36.5 grams. The explanation would be found by comparing the number of beans in the bag and the weight per bean to the number and individual weights of marbles in the same bag. In our gas problem we make the same kind of comparison; the weight per molecule and number of ammonia molecules in 22.4 liters must be compared to the weight per molecule and number of hydrogen chloride molecules in 22.4 liters. There are two particularly simple possibilities:

(A) Perhaps:
   (1) Equal volumes of these two gases contain the same number of molecules, and
   (2) ammonia molecules weigh less, per molecule, than hydrogen chloride molecules by the factor 17.0/36.5.

(B) Perhaps:
   (1) Equal volumes of these two gases contain different numbers of molecules, ammonia containing fewer by the factor 17.0/36.5, and
   (2) ammonia molecules weigh exactly the same as hydrogen chloride molecules.

SEARCH FOR AN EXPLANATION:

Why do equal volumes of ammonia and hydrogen chloride have different weights?

At room temperature and one atmosphere pressure,

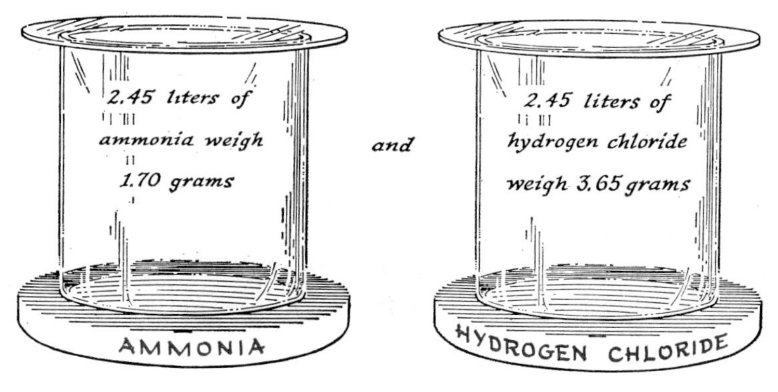

WHY?

MODEL A

Perhaps:

1. *2.45 liters of ammonia contain the same number of molecules as 2.45 liters of hydrogen chloride.*

and

2. *Ammonia molecules weigh less than hydrogen chloride molecules. (less by a factor, 1.70/3.65).*

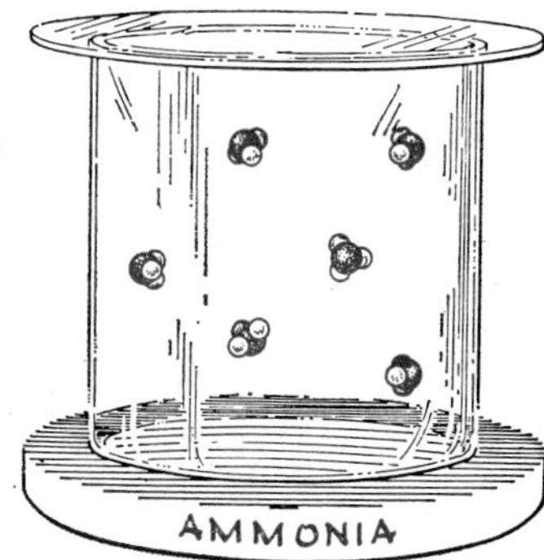

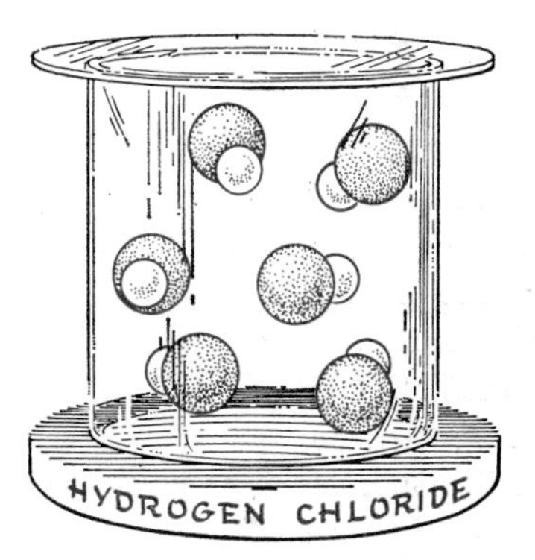

MODEL B

Perhaps:

1. *2.45 liters of ammonia contain fewer molecules than do 2.45 liters of hydrogen chloride. (less by a factor, 1.70/3.65).*

and

2. *Ammonia molecules weigh the same as hydrogen chloride molecules*

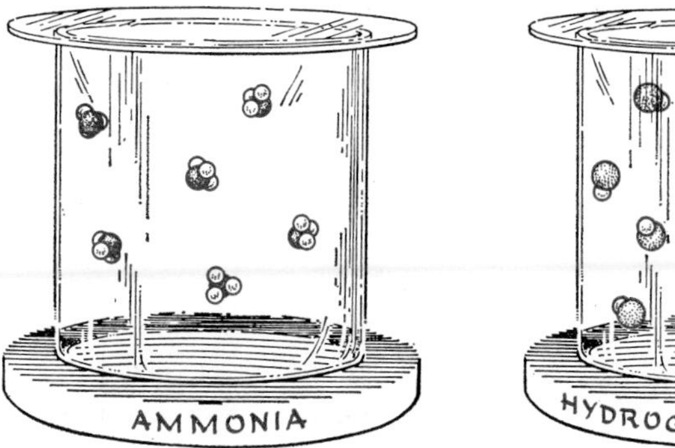

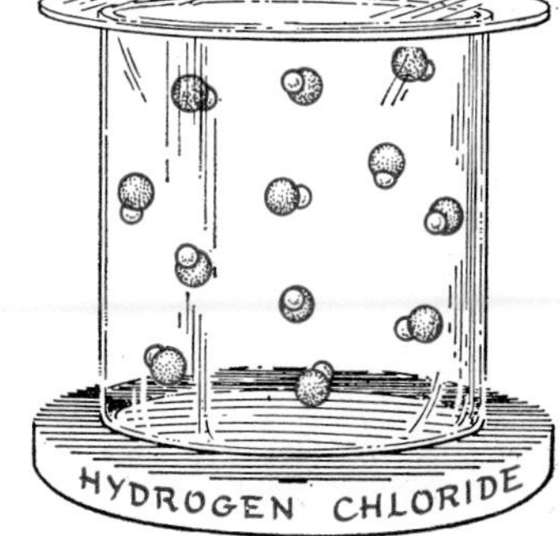

Fig. 2-5. **Two simple models** *to explain the weights of equal volumes of gases.*

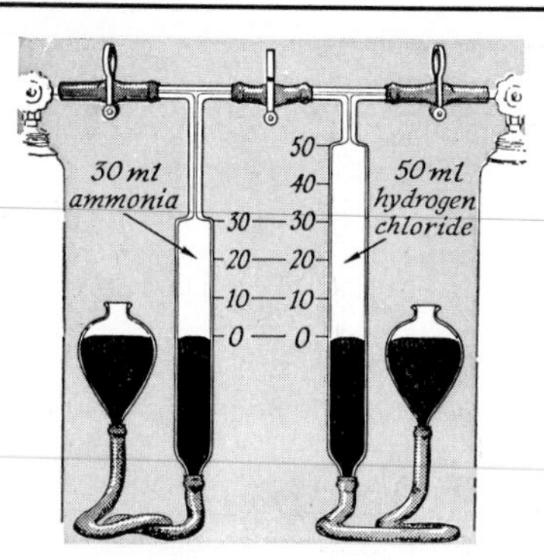

Fig. 2-6. **An apparatus suited to the measurement** *of volumes of gases.*

These two possibilities are attractive because they are simple—one factor alone is held responsible for the weight difference. We must be prepared, however, for disappointment. There is the third possibility that neither of these proposals, A or B, accounts for the properties of gases. After all, neither A nor B applies to the beans and marbles example. The bag probably wouldn't contain the same number of beans as marbles (as in B) but, in addition, beans and marbles don't weigh the same (as in A). We need more information to decide if either proposal A or B applies to gases. More information is obtained by observing how some gases behave when mixed.

2-2.2 Mixtures of Ammonia and Hydrogen Chloride

When hydrogen chloride and ammonia gases are mixed, a white powder is formed. When hydrogen chloride molecules and ammonia molecules are mixed, the atoms are rearranged and an entirely different substance, a solid, results. A quantitative study of this process is informative.

Figure 2-6 shows an apparatus suited to the measurement of volumes of gases. Thirty milliliters of ammonia have been admitted to the left tube from the ammonia storage tank. Next, 50 ml of hydrogen chloride were admitted to the right tube. The leveling bulbs were used to adjust the pressure of each gas to one atmosphere. The apparatus is ready. The hydrogen chloride sample can be transferred slowly into the tube containing the ammonia. Figure 2-7 shows the progress of the experiment.

In Figure 2-7A we see the situation after 20 ml of hydrogen chloride gas have been transferred. A cloud of the white solid fills the left tube where the gases mix and, after the leveling bulbs are adjusted, we find that just 10 ml of ammonia remain. Twenty milliliters of hydrogen chloride combined with just 20 ml of ammonia, forming the white solid. Ten milliliters more hydrogen chloride are just enough to consume the last of the ammonia, forming the solid of insignificant volume, as shown in Figure 2-7B. Continued addition of hydrogen chloride causes no further solid formation but merely leaves an excess of hydrogen chloride in the left tube (Figure 2-7C).

This is a significant and simple result. Thirty milliliters of hydrogen chloride combine with just 30 ml of ammonia, measured at the same temperature and pressure. Therefore, one liter of hydrogen chloride would combine with just one liter of ammonia. Though a given volume of ammonia weighs less than the same volume of

Fig. 2-7. **Mixing measured volumes** *of hydrogen chloride gas and ammonia gas.*

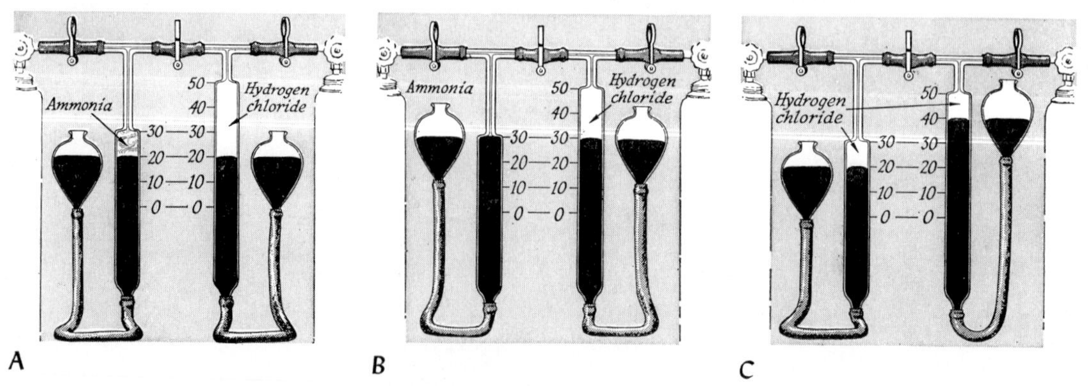

hydrogen chloride (less by the factor 17.0/36.5), these equal volumes combine. This simple situation suggests that we should seek a simple explanation. Our proposal A in Section 2-2.1 fits nicely. If we propose that equal volumes contain equal numbers of molecules, then 30 ml of ammonia contain the same number of molecules as do 30 ml of hydrogen chloride. Proposal A leads us to conclude that *one* molecule of ammonia combines with *one* molecule of hydrogen chloride to form the white solid. Through proposal A, the combining volumes tell us the numbers of molecules that combine. In contrast, there is no correspondingly simple way to explain the new data with proposal B.

Of course, a single example hardly furnishes compelling evidence that equal volumes of any pair of gases (at the same temperature and pressure) contain equal numbers of molecules. Neither can we be convinced on the basis of simplicity when we have but one example. However, many gases behave as simply as a mixture of hydrogen chloride and ammonia. For example,

two liters of the gas nitric oxide	combine with	*one* liter of the gas oxygen	to form	nitrogen dioxide
one liter of the gas oxygen	will burn coal to form	*one* liter of the gas carbon dioxide		
two liters of the gas hydrogen	combine with	*one* liter of the gas oxygen	to form	water

These simple, integer volume ratios confirm the usefulness of the interpretation that equal volumes contain equal numbers of molecules. This proposal was first made in 1811 by an Italian scientist, Amadeo Avogadro; hence it is called Avogadro's Hypothesis. It has been used successfully in explaining the properties of gases for a century and a half.

Avogadro's Hypothesis: *Equal volumes of gases, measured at the same temperature and pressure, contain equal numbers of molecules.*

2-2.3 The Relative Weights of Molecules

The importance of Avogadro's Hypothesis is that it furnishes a basis for weighing molecules. Two equal volumes of gas (at the same temperature and pressure) are weighed. If we assume these two volumes contain identical numbers of molecules, then we must also conclude that the gas that weighs more must have heavier molecules. Furthermore, the ratio of the weights of the molecules must be exactly the ratio of the weights of the two gas samples.

For example, in Table 1-II (p. 14) data were given that show that 32.0 grams of oxygen at 0°C and one atmosphere pressure occupy 22.4 liters. This same volume of ammonia (also at 0°C and one atmosphere pressure) weighs 17.0 grams. By Avogadro's Hypothesis, these two volumes contain equal numbers of molecules. Hence each ammonia molecule must weigh less than an oxygen molecule by the factor 17.0/32.0. By the same argument, each hydrogen chloride molecule must weigh more than an oxygen molecule by the factor 36.5/32.0. We say "*must* weigh" but, of course, this is a valid statement only if Avogadro's Hypothesis is applicable.

By many such weighings, scientists have learned the relative weights of many gases. The experiment is fairly simple. A carefully measured volume of oxygen is weighed at a fixed pressure and temperature. Then the same volume of another gas is weighed at this same pressure and temperature. The relative weights of the gases indicate the relative weights of the molecules, provided Avogadro's Hypothesis is applicable. Neither the pressure nor the temperature need be measured, provided they are held constant.

Table 2-IV

WEIGHTS OF EQUAL VOLUMES OF GASES UNDER FIXED TEMPERATURE AND PRESSURE (BASED ON THE VOLUME OCCUPIED BY 32.0 GRAMS OF OXYGEN)

GAS	WEIGHT (grams)
oxygen	(32.0)
ammonia	17.0
chlorine	71.0
hydrogen	2.02
hydrogen chloride	36.5
nitric oxide	30.0
nitrogen dioxide	46.0

Table 2-IV shows for some other gases the weights that have the same volume as 32.0 grams of oxygen (at the same pressure and temperature).

2-2.4 The Number of Atoms in a Molecule

Figure 2-3 shows a model of an ammonia molecule and a model of a hydrogen chloride molecule. These models show how chemists picture the molecule of ammonia: it contains four atoms. A hydrogen chloride molecule contains only two atoms. Chemists decide how to construct these molecules from the same type of information described in Section 2-2.2, by the volumes of gases that combine.

Consider the combination of nitric oxide and oxygen. Nitric oxide (a colorless gas) when mixed with oxygen gas (also colorless) becomes reddish-brown. The color is identical to that of another gas, nitrogen dioxide. All the properties of the nitric oxide-oxygen mixture are consistent with the conclusion that the gas nitrogen dioxide has been formed. How can this change be discussed in terms of our molecular model of a gas?

First, we explain the differences between nitric oxide, oxygen, and nitrogen dioxide by asserting that the molecules of nitric oxide, oxygen, and nitrogen dioxide are somehow different. They must be composed of smaller components that we call atoms. The numbers and kinds of atoms in a molecule of nitric oxide must be different from the numbers and kinds of atoms in a molecule of oxygen.

Now we find that nitrogen dioxide can be formed from a mixture of nitric oxide and oxygen. This means that the atoms in nitrogen dioxide must have come from those in nitric oxide together with those in oxygen.

Finally, we discover that exactly *two* volumes of nitric oxide combine with *one* volume of oxygen and that exactly *two* volumes of nitrogen dioxide are formed. According to Avogadro's Hypothesis, this indicates that

two molecules of nitric oxide combine with *one* molecule of oxygen to form *two* molecules of nitrogen dioxide

All of the atoms in the two molecules of nitrogen dioxide came from two molecules of nitric oxide and one molecule of oxygen. Of course, the two molecules of nitrogen dioxide have twice as many atoms as does a single molecule of nitrogen dioxide. Hence, no matter how many atoms one molecule of nitrogen dioxide might contain (for example, one, two, three, four, . . .), two molecules of nitrogen dioxide must contain an *even* number of atoms (for example, two, four, six, eight, . . .). The same statement is applicable to the two molecules of nitric oxide that were combined. No matter how many atoms one molecule of nitric oxide contains, two molecules must contain an *even* number of atoms.

Thus we see that after the even number of atoms in two molecules of nitric oxide have combined with the atoms in one molecule of oxygen, there is still an even number of atoms. This can be so only if a molecule of oxygen *also* contains an even number of atoms. We are led to the conclusion that *a molecule of oxygen contains an even number of atoms.*

This can be demonstrated clearly in algebraic language.

Suppose:
one molecule of nitric oxide contains X atoms,
one molecule of oxygen contains Y atoms, and
one molecule of nitrogen dioxide contains Z atoms,
where X, Y, and Z are integers.

Then:
two molecules of nitric oxide contain $2X$ atoms, and
two molecules of nitrogen dioxide contain $2Z$ atoms.

Also:
all of the atoms in two molecules of nitrogen dioxide ($2Z$ atoms) came from two molecules of nitric oxide ($2X$ atoms) plus one molecule of oxygen (Y atoms), or

$$2X + Y = 2Z \qquad (2)$$

So:
we can solve for Y by subtracting $2X$ from each side of this equation:

$$Y = 2Z - 2X \qquad (3)$$
$$Y = 2(Z - X) \qquad (4)$$

Thus no matter what the integer values of Z and X are, their difference $(Z - X)$ is an integer. Since doubling *any* integer produces an *even* number, $Y = 2(Z - X)$ must be an even number. We have proved that *a molecule of oxygen must contain an even number of atoms.*

The simplest acceptable structure we can picture for oxygen is that it contains two atoms. More experiments are needed before we can eliminate the possibility that oxygen contains four, six, or a higher (but even) number of atoms.

EXERCISE 2-2

In Section 2-2.2 (p. 24) it was noted that two volumes of hydrogen gas combine with one volume of oxygen gas and two volumes of gaseous water are produced. According to Avogadro's Hypothesis, this means that *two* molecules of hydrogen combine with *one* molecule of oxygen to form *two* molecules of water. If we define

X = the number of atoms in a molecule of hydrogen,
Y = the number of atoms in a molecule of oxygen,
Z = the number of atoms in a molecule of water,

then we have the algebraic relationship

$$2X + Y = 2Z$$

(a) Convince yourself that from these data alone we can conclude that Y must be *even* (that is, oxygen molecules must contain an even number of atoms).
(b) By solving for X in terms of Y and Z, convince yourself that from the above data alone, X could be odd or even.

2-2.5 Atoms in Liquids and Solids

When a candle is burned, a gas is produced—a gas containing carbon dioxide and water vapor. It is useful to describe such a gas as a collection of molecules, each molecule containing smaller units called atoms. Each carbon dioxide molecule contains one carbon atom and two oxygen atoms. Each water molecule contains one oxygen atom and two hydrogen atoms. Where did these atoms come from? Were they present in the candle before it burned?

Similar questions are raised if we consider the effect of cooling the gases produced from the candle. Cooling these gases results in condensation—drops of liquid water appear. If the water vapor contains molecules, made up of atoms, what happens to these molecules (and atoms) when the gas condenses? Are they still present in the liquid?

Scientists always seek the simplest explanation that fits the known facts. Since we find it convenient to describe water vapor as a collection of groups of atoms (called molecules), the simplest assumption we can make about the con-

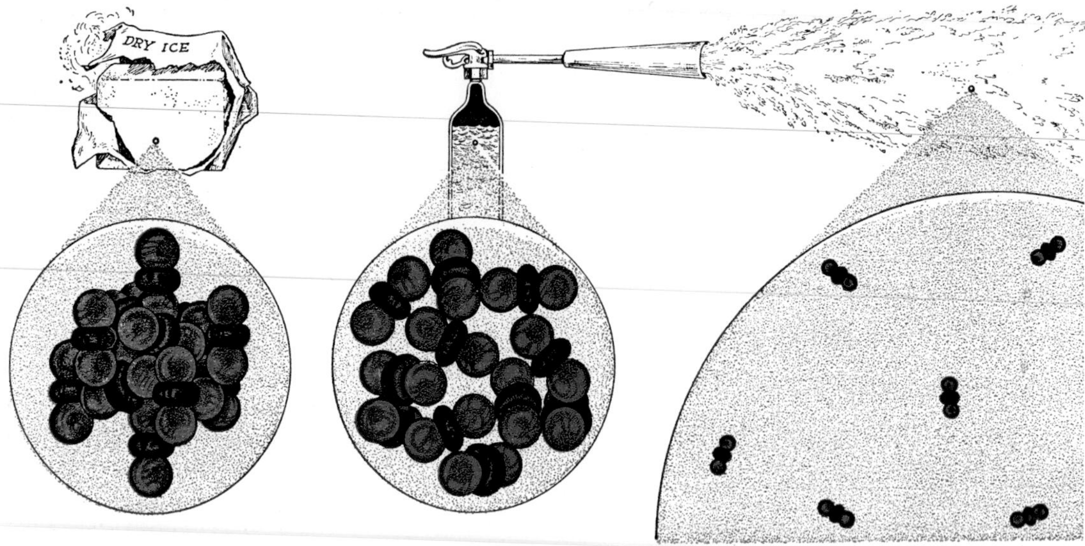

Fig. 2-8. **All matter consists of particles.** *In a gas, the particles are far apart; in a liquid or solid, they are close together.*

densation of this vapor is that the liquid still contains the atoms. Since a candle burns to produce gases—collections of molecules—the simplest assumption we can make is that the candle already contained the atoms that formed the gaseous molecules during combustion. These simplifying assumptions—that liquids and solids are made up of atoms—are acceptable and convenient as long as they prove to be consistent with all that is known about liquids and solids.

Thus, we are led to the view that all matter consists of particles. We can state this as a proposal.

Proposal: All matter, whether solid, liquid, or gas, consists of particles. In a gas these particles are far apart; in a liquid or a solid, they are packed close together.

This proposal is called the **atomic theory.** As with any theory, its value depends upon its ability to aid us in explaining facts of nature. There is no more valuable theory in science than the atomic theory. We shall use it throughout this course. Later, in Chapter 14, we shall review many of the types of experiments which cause chemists to regard the atomic theory as the cornerstone of their science.

2-3 SUBSTANCES: ELEMENTS AND COMPOUNDS

Molecules are clusters of atoms. Two types of molecules are possible. Some molecules are clusters of atoms in which all the atoms in a cluster are identical; some molecules contain two or more different kinds of atoms. These two kinds of molecules are given different names.

An *element* or *elementary substance* contains only *one* kind of atom.

A *compound* or *compound substance* contains *two or more* kinds of atoms.

Usually a good deal of experimentation is needed before a substance can be considered to be pure. Even then, much more work and study are needed before one can decide with confidence that a given pure substance is an element or a compound. Consider the substance water. Water is probably the most familiar substance in our environment and all of us recognize it easily. We are familiar with its appearance and feel, its density (weight per unit volume), the way in which it flows, the temperature at which it freezes and boils, and the way in which it dissolves sugar and salt. Because water is identified by constant and characteristic properties, it is a pure substance. In a later experiment you will see how we can change the pure substance water into two other substances, hydrogen gas and oxygen gas. The hydrogen and oxygen are produced in definite amounts. Since water can be decomposed into two other substances, it must contain at least two kinds of atoms. Hence water is a *compound.*

Notice the pattern here. First, we established the characteristic properties of water that cause us to identify it as a *pure substance*. Second, we found a change in which two other substances were formed in definite amounts from water alone. This second piece of information shows that water contains more than one kind of atom and that, hence, *water is a compound.*

Common sugar is another example of a substance. Most commercial samples of white sugar are rather pure; that is, they contain only very small amounts of substances other than sugar. One characteristic property of sugar is its sweetness. Another characteristic property is the way it dissolves in water. Still another is the way it behaves when heated. At a definite temperature sugar not only begins to melt to a liquid but it also begins to decompose. The liquid darkens and gaseous water bubbles off. Finally, a black solid (charcoal) remains in the container. We recognize the black solid as a form of carbon. Thus, pure sugar, identified by its characteristic properties, can be decomposed to form water and charcoal in definite amounts. *Sugar is a compound.*

Water and sugar are compounds. What about hydrogen and oxygen? Hydrogen, for example, is a gas at normal conditions. It can be liquefied at a characteristic temperature by cooling. By further cooling it can be solidified at a second characteristic temperature. It is a pure substance. No treatment, however, causes it to form two other substances. Hydrogen must contain only one kind of atom, hence *hydrogen is an element.* We call this kind of atom, the *hydrogen atom.* Oxygen, too, has characteristic properties but cannot be caused to form two other substances. *Oxygen, then, is an element*—it contains only one kind of atom, called the oxygen atom.

Now we can return to the decomposition of water. Water can be decomposed to give hydrogen and oxygen. Since hydrogen contains only hydrogen atoms and oxygen contains only oxygen atoms, water molecules must contain some hydrogen atoms and some oxygen atoms, but *no other kind of atom.*

This type of problem is one of the most important in chemistry—deciding what atoms are present in a given substance. How important this is can be seen by comparing the three substances water, oxygen, and hydrogen. Both water and oxygen contain oxygen atoms but these substances are very different in their properties. Both water and hydrogen contain hydrogen atoms but these substances are no more alike than are water and oxygen. The properties of water are fixed by the combination of the two kinds of atoms and these properties are distinctive.

EXERCISE 2-3

What differences between water, oxygen, and hydrogen can you point out from your own experience? For example, you might consider

(a) boiling and melting points,
(b) role in combustion,
(c) role in supporting life.

Sugar is another substance that contains both oxygen and hydrogen atoms but it contains carbon atoms as well. Sugar does not resemble water, oxygen, or hydrogen. The presence of the various types of atoms and their arrangement accounts for the distinctive properties which identify sugar. In any substance, the atoms present, their numbers, and their arrangement fix the properties of that substance.

2-3.1 The Elements

An **element** *is a pure substance that contains only one kind of atom.* There are about one hundred different elements known today—hence there are about one hundred kinds of atoms that are chemically different. Some of these elements occur pure in nature and hence have been known for thousands of years. Such elements as iron, silver, gold, mercury, and sulfur were known to the ancients and were given Latin names by the alchemists. For example, iron was called ferrum, silver was called argentum, and gold was called aurum.

During the nineteenth century the discovery of elements increased as chemists began to adopt quantitative methods. At the beginning of the nineteenth century, perhaps 26 elements were known. One hundred years later, at the beginning of the present century, over 81 elements were known. Over twice as many elements were discovered in that one century as were discovered in *all of time* before.

Figure 2-9 shows, as a function of time, our knowledge of the elements. In Figure 2-9A we see how the total number of elements known has increased since 1700. Figure 2-9B re-expresses this same information in terms of the number of elements discovered in each half-century since 1700. Both graphs show that the rate of discovery of new elements is declining. The plots suggest that there is a limited number of elements to be found in nature.

Fig. 2-9. **The discovery of the elements. A.** *The total number of elements known as a function of time.* **B.** *The number of elements discovered in each half-century since 1700.*

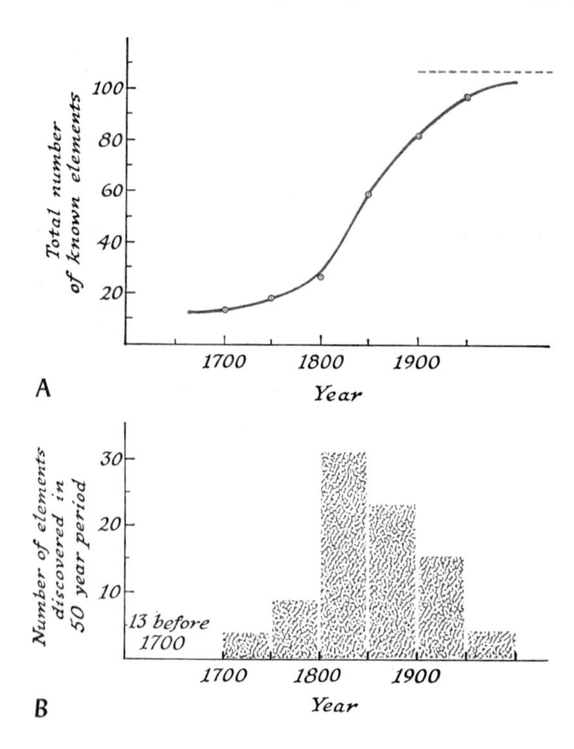

Each element has been named and, for convenience, has been given a nickname—a shorthand symbol of one or two letters. Thus the element carbon is symbolized by the letter C, the element neon by the letters Ne. The symbols are adopted by international agreement among chemists. Eleven of the elements have names derived from the capitalized first letter of the Latin name of the element and, if necessary, by a second letter (uncapitalized).* These eleven include seven common metals known to the ancients. (See Table 2-V.)

The elements discovered more recently have the same names in all languages, again by international agreement. Except for the eleven elements listed in Table 2-V, all of the elements have symbols that can be derived from their English names. For example, the symbols for hydrogen (H), helium (He), carbon (C), nitrogen (N), oxygen (O), calcium (Ca), and chlorine (Cl) are easily obtained from the names. Notice that He is used for helium to distinguish it from H for hydrogen. Again, since C is used for carbon, the symbols for calcium and chlorine each have a second letter added to the first. The table of elements (inside the back cover of the book) contains a complete list of the chemical symbols.

2-3.2 Chemical Formulas

Molecules are made up of atoms in definite numbers and definite arrangements. Models and symbols for the elements aid us in showing the

* These symbols were adopted as a result of the urging of an outstanding Swedish chemist, Jons Berzelius, during the first half of the nineteenth century. Until Berzelius lent his prestige as a chemist to this system, many of the elements had several symbols in accordance with their different names in different languages.

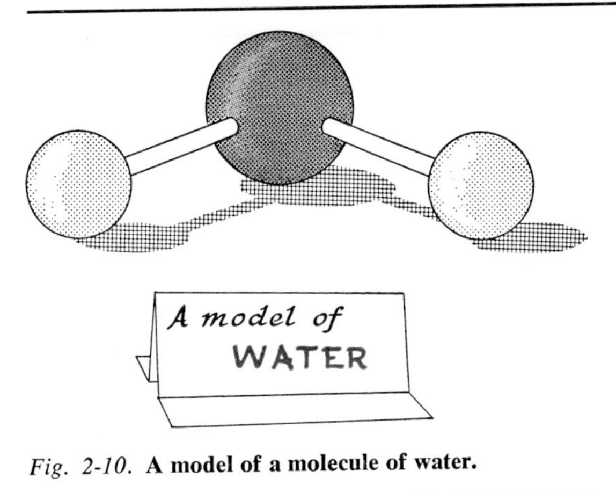

Fig. 2-10. **A model of a molecule of water.**

composition of a molecule. Figure 2-10 shows a model of a water molecule. Experiments have shown that the model should contain two atoms of hydrogen and one atom of oxygen. The advantage of such a model is that it shows also the spatial arrangement of the atoms. In a molecule of water, each of the two hydrogen atoms is connected to the oxygen atom in a triangular arrangement. How the shape is determined and how important it is in chemistry will be treated later in the course.

The number and kinds of atoms in a molecule can also be shown in a *molecular formula*. For example, the water molecule is symbolized "H_2O." In this molecular formula, "H" means "hydrogen atom," "O" means "oxygen atom," and the subscript "2" following "H" indicates there are two hydrogen atoms bound to the single oxygen atom. The molecular formula of ammonia, NH_3, indicates that one molecule of ammonia contains one atom of nitrogen (N) and three atoms of hydrogen (H). Experiments show that oxygen is **diatomic** (each molecule contains two atoms), hence its molecular formula is O_2. Hydrogen gas is diatomic; its formula is H_2.

Both the numbers and the arrangement of the atoms in the molecule are shown by a *structural formula*. The structural formulas, like the models we have seen, show which atoms are attached to each other. Thus, H_2O has the structural formula

H—O—H

not

H—H—O

or

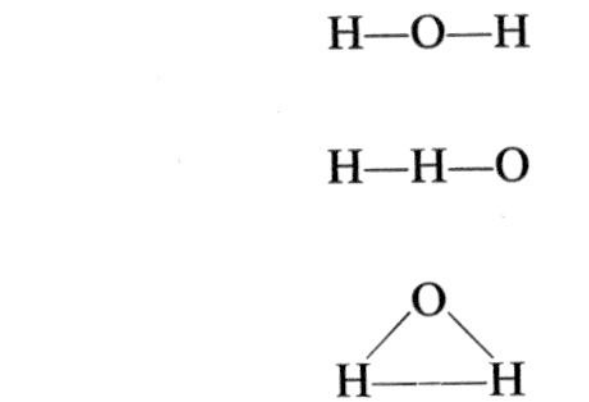

In the structural formula H—O—H, the dashes indicate the connections between the atoms. *The connections between atoms are called* ***chemical bonds.*** We see that each of the two hydrogen atoms is bound to the oxygen atom. Both of the alternate arrangements,

H—H—O

and

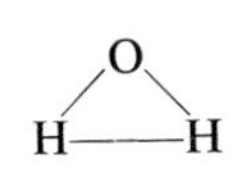

agree with the molecular formula H_2O but the properties of water show that the atoms are not so bonded.

No written formula is quite as effective as a molecular model to help us visualize molecular shape. Since chemists find that the shape of a molecule strongly influences its chemical behavior, pictures and models of molecules are important aids. A variety of types of models are

Table 2-V. **CHEMICAL SYMBOLS THAT ARE NOT DERIVABLE FROM THE COMMON ENGLISH NAME OF THE ELEMENT**

COMMON NAME	SYMBOL	SYMBOL SOURCE	COMMON NAME	SYMBOL	SYMBOL SOURCE
antimony	Sb	stibnum	potassium	K	kalium
copper	Cu	cuprum	silver	Ag	argentum
gold	Au	aurum	sodium	Na	natrium
iron	Fe	ferrum	tin	Sn	stannum
lead	Pb	plumbum	tungsten	W	wolfram
mercury	Hg	hydrargyrum			

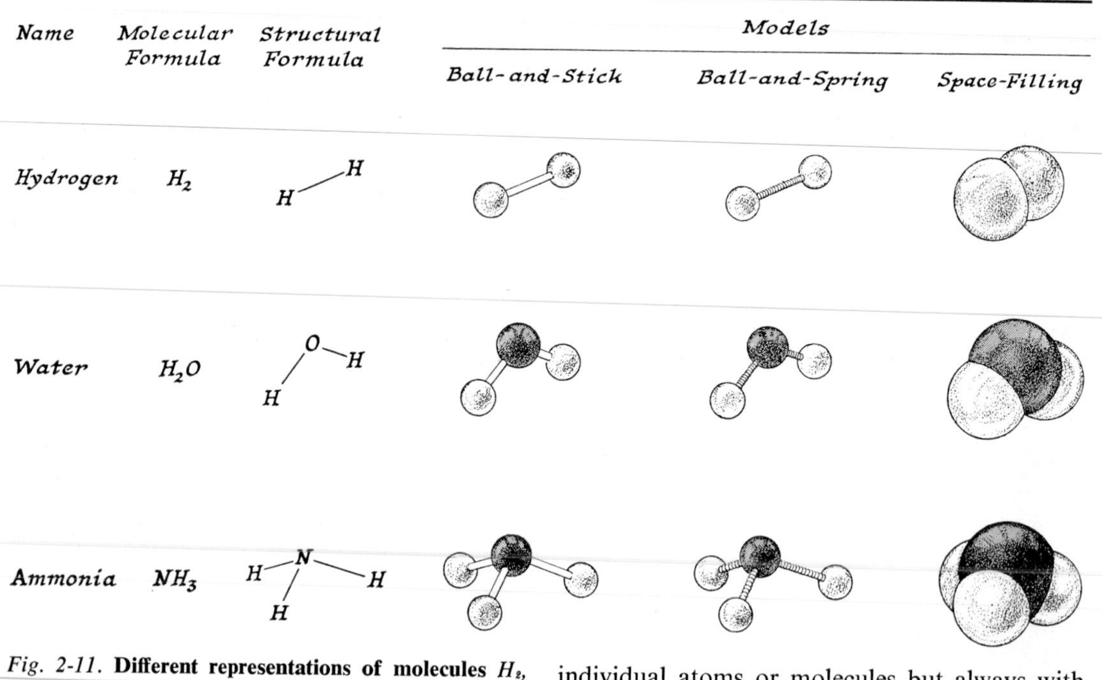

Fig. 2-11. **Different representations of molecules** *H_2, H_2O, and NH_3.*

commonly used, depending upon the emphasis needed. Figure 2-11 shows some representations of molecules of hydrogen, water, and ammonia. The ball-and-stick and ball-and-spring models display clearly the bonds and their orientations. The ball-and-spring models indicate molecular flexibility. The space-filling models provide a more realistic view of the spatial relationships and crowding among nonbonded atoms.

EXERCISE 2-4

Carbon dioxide has the formula CO_2. Remembering that the prefix "di" means two, and "tri" means three, write the molecular formula for each of the following substances: carbon disulfide, sulfur dioxide, sulfur trioxide. (If you don't know the symbol for an element, use the table inside the back cover of the book.)

2-3.3 The Mole

Any sample of matter we examine contains a very large number of atoms. We never work with individual atoms or molecules but always with collections of these particles. Chemists, therefore, have selected a unit larger than a single atom or molecule for comparing amounts of different materials. This unit, called the **mole,** contains a very large number of particles, 6.02×10^{23}. A mole of oxygen atoms, or a mole of hydrogen atoms, or a mole of copper atoms contains 6.02×10^{23} atoms of the specified kind. A mole of oxygen molecules, O_2, contains 2 moles of oxygen atoms ($2 \times 6.02 \times 10^{23}$ oxygen atoms) because each oxygen molecule contains two atoms. A mole of P_4 molecules contains $4 \times 6.02 \times 10^{23}$ phosphorus atoms, that is, four moles of phosphorus atoms.*

A baker counts biscuits in dozens—a convenient number. Money is counted in dollars—one hundred cents is, again, a convenient number. How did chemists choose to count in terms of moles—the number 6.02×10^{23} seems an odd choice. Why, for instance, didn't they settle on some simpler number, such as exactly one billion particles? There is a reason. Chemists prefer a definition in terms of a quantity that can be measured readily and with high accuracy. Weigh-

* If you have difficulty expressing numbers in terms of powers of ten, refer to Appendix 5 in the Laboratory Manual.

ing is easier than counting when the number of particles to be counted is so very large. Consequently, chemists based the definition of the mole upon a chosen *weight* rather than a chosen number of particles. During the nineteenth century, chemists decided that the number of molecules in a sample of oxygen weighing exactly 32 grams would be taken as a standard number. Thus defined, *a* ***mole*** *is the number of oxygen molecules in exactly thirty-two grams of oxygen.** The significance of a mole is most usefully connected with this number of particles, rather than the weight. The number, found later to be 6.02×10^{23}, *is called* ***Avogadro's number.*** (Avogadro was the first to propose how to obtain equal numbers of molecules of different substances.)

weighed on an ordinary balance. For practical purposes, the weight of a mole of atoms is a valuable number. This weight is called the atomic weight. *The* ***atomic weight*** *of an element is the weight in grams of Avogadro's number of atoms.*

Now consider compounds. Again, a useful number to the chemist is the weight of a mole of molecules. This weight is called the molecular weight (or the molar weight). *The* ***molecular weight*** *of a compound is the weight in grams of Avogadro's number of molecules.*

Consider the substance hydrogen chloride. This compound has the molecular formula HCl. A chemist working with hydrogen chloride, HCl, must often know the weight of a mole of molecules (the molecular weight). This weight is readily calculated from the atomic weights of the two kinds of atoms, H and Cl:

one molecule of HCl	contains	*one* atom of hydrogen	and	*one* atom of chlorine
one mole of HCl molecules	contains	*one* mole of hydrogen atoms	and	*one* mole of chlorine atoms
the weight of one mole of HCl molecules	is the same as	the weight of one mole of hydrogen atoms	+	the weight of one mole of chlorine atoms
molecular weight of HCl	=	atomic weight of hydrogen	+	atomic weight of chlorine
molecular weight of HCl	=	1.01 g	+	35.5 g
molecular weight of HCl	=		36.5 g	

2-3.4 Atomic and Molecular Weights

Chemists deal with amounts of substances that are readily measured. Although a chemist is aware that the mass of a single oxygen atom is 2.656×10^{-23} gram, he finds it much more useful to know that a mole of oxygen atoms weighs 16.00 grams. This is an amount that can be

* Recently the definition of a mole has been altered to put it in terms of measurements made with higher accuracy. The change implied, −.0045%, is unimportant from a chemist's point of view.

In a similar way, the weight of a mole of H_2O molecules is the weight of two moles of hydrogen atoms plus the weight of one mole of oxygen atoms. Hence

$$\begin{aligned}\text{molecular wt. of } H_2O &= 2 \times (\text{atomic wt. of H}) \\ &\quad + 1 \times (\text{atomic wt. of O}) \\ &= 2(1.01) + (16.00)\text{ g} \\ &= 18.02\text{ g}\end{aligned}$$

Atomic weights have everyday importance to a chemist. Therefore, the atomic weights must be readily available. They are listed both in the periodic table (inside the front cover) and in the table of atomic weights (inside the back cover).

EXERCISE 2-5

Show that the weight of a mole (the molecular weight) of CO_2 is 44.0 grams and that the weight of a mole of SO_2 is 64.1 grams.

EXERCISE 2-6

In Experiment 6 you calculated the ratio of the weight of carbon dioxide to the weight of the same volume of oxygen. Oxygen has been assigned a molecular weight of 32.0. From the molecular weight of oxygen and your measured ratio, calculate the molecular weight of carbon dioxide. Estimate the uncertainty in your result. Compare to the value obtained in Exercise 2-5.

EXERCISE 2-7

What is the molecular weight of each of the substances sulfur (formula, S_8), ammonia (formula, NH_3), and nitrogen (a diatomic molecule)?

EXERCISE 2-8

Calculate the weight of 6.02×10^{23} molecules of carbon monoxide, CO.

EXERCISE 2-9

How many moles of iron atoms are present in 1.73 grams of iron?

2-4 REVIEW: THE ATOMIC THEORY

"Gases are composed of particles" was proposed in Chapter 1 as a useful model to aid us in discussing certain properties of oxygen gas. We have continued to use this model in discussing other types of information. First it was tested for applicability to other gases, ammonia and hydrogen chloride. The assumption of molecular particles turned out to fit the properties of these two gases as well, but it was discovered that the weights of the particles of different gases must be different. Further studies of a variety of properties of gases (color, solubility in water, behavior toward litmus, etc.) caused us to propose that the molecules differ in their structures. We were thus led to the view that molecules contain smaller building blocks, called atoms.

Having proposed the existence of atoms, we began representing the structure of molecules through molecular models and molecular formulas. These models and formulas picture what is known about the numbers, types, and arrangements of atoms in the molecules represented. Our success in treating gases, using this atomic theory, led us to propose that the atoms are present, as well, in solids and liquids. Now we postulate that atoms are present in all matter.

Thus a model (a theory) grows. As it is tested in an ever widening range of experience, the model often grows more complex. This is offset by the advantage of developing interrelationships among diverse phenomena (that is, by discovering hidden likenesses). The atomic theory, as developed to correlate chemical behavior, is much more complicated than is needed to explain the simple gas behavior mentioned in Chapter 1. Nevertheless, connections developed between billiard balls rebounding from a cushion and the pressure in a balloon have provided us with a substantial start in understanding chemistry.

QUESTIONS AND PROBLEMS

1. Hydrogen, helium, and carbon dioxide are all gases at normal temperatures. What differences among the properties of these gases account for the following?

   (a) Hydrogen and helium are used in balloons whereas carbon dioxide is not.
   (b) Helium is less dangerous to use in balloons than is hydrogen.

2. Four differences between helium gas and nitrogen gas are listed below.

   (a) Dry air contains 80% nitrogen but only 0.0005% helium (by volume).
   (b) Helium is much less dense than nitrogen.
   (c) Helium has much lower solubility in water than nitrogen has.
   (d) Helium is much more expensive than nitrogen.

   Which difference could account for the fact that a diver is much less likely to suffer from the bends if he breathes a mixture of 80% helium and 20% oxygen than if he breathes air? (The bends is a painful, sometimes fatal, disease caused by the formation of gas bubbles in the veins and consequent interruption of blood flow. The bubbles form from gas dissolved in the blood at high pressure.)

3. The most important step in the process for the conversion of atmospheric nitrogen into important commercial compounds such as fertilizers and explosives involves the combination of one volume of nitrogen gas with three volumes of hydrogen gas to form two volumes of ammonia gas.

   From these data alone and Avogadro's Hypothesis, how many *molecules* of hydrogen combine with one *molecule* of nitrogen? How many *molecules* of ammonia are produced from one *molecule* of nitrogen?

4. Gaseous uranium hexafluoride is important in the preparation of uranium as a source of "atomic energy."

   A flask filled with this gas is weighed under certain laboratory conditions (temperature and pressure), and the weight of the gas is found to be 3.52 grams. The same flask is filled with oxygen gas and is weighed under the same laboratory conditions. The weight of the oxygen in the flask is found to be 0.32 gram.

   What is the ratio of the weight of one uranium hexafluoride molecule to the weight of an oxygen molecule? State any guiding principles needed in answering this question.

5. Two volumes of hydrogen fluoride gas combine with one volume of the gas dinitrogen difluoride to form two volumes of a gas G.

   (a) According to Avogadro's Hypothesis, how many molecules of G are produced from one molecule of dinitrogen difluoride?
   (b) If X = number of atoms in a molecule of hydrogen fluoride,
   Y = number of atoms in a molecule of dinitrogen difluoride,
   Z = number of atoms in a molecule of G,
   write the relation among X, Y, and Z appropriate to the combining volumes given.
   (c) For each of the following possible values of X and Y, calculate the required value of Z.

If X is	and	Y is	then	Z must be
1		2		
1		4		
2		2		
2		4		

   (d) No odd value of Y is suggested in question (c). Prove that Y *must* be an even integer.

6. There are a number of gas mixtures that combine in the general pattern

one volume of gas *A*	combines with	one volume of gas *B*	to form	two volumes of gas *C*

For example,

one volume of carbon dioxide	combines with	one volume of carbon disulfide	to form	two volumes of carbon oxysulfide
one volume of hydrogen	combines with	one volume of chlorine	to form	two volumes of hydrogen chloride

If X = number of atoms in a molecule of A,
Y = number of atoms in a molecule of B, and
Z = number of atoms in a molecule of C,

(a) Show that if X is even, Y must be even.
(b) Show that if X is odd, Y must be odd.

7. A pure white substance, on heating, forms a colorless gas and a purple solid. Is the substance an element or a compound?

8. What do the following symbols represent? K, Ca, Co, CO, Pb.

9. Write formulas for

silicon dioxide (common sand, or silica),
sulfur dichloride,
nitrogen trifluoride,
aluminum trifluoride,
dinitrogen difluoride.

10. (a) Write formulas for

hydrogen chloride,
hydrogen bromide,
hydrogen iodide,
boron trichloride,
carbon tetrachloride,
nitrogen trichloride,
oxygen dichloride.

(b) Locate in the table inside the front cover the symbol for each element involved in these compounds.

11. For each of the following substances give the name of each kind of atom present and the total number of atoms represented in the formula shown.

Name	Formula
(a) graphite (pencil lead)	C
(b) diamond	C
(c) sodium chloride (table salt)	$NaCl$
(d) sodium hydroxide	$NaOH$
(e) calcium hydroxide	$Ca(OH)_2$
(f) potassium nitrate	KNO_3
(g) magnesium nitrate	$Mg(NO_3)_2$
(h) sodium sulfate	Na_2SO_4
(i) calcium sulfate	$CaSO_4$

12. All of the following substances are called "acids." What element do they have in common?

(a) nitric acid	HNO_3
(b) hydrochloric acid (or, hydrogen chloride)	HCl
(c) hydrofluoric acid (or, hydrogen fluoride)	HF
(d) sulfuric acid	H_2SO_4
(e) phosphoric acid	H_3PO_4

13. Here are the names of some common chemicals and their formulas. What elements does each contain?

(a) hydrogen peroxide	H_2O_2
(b) jeweler's rouge	Fe_2O_3
(c) light bulb filament	W
(d) tetraethyl lead	$Pb(C_2H_5)_4$
(e) baking soda	$NaHCO_3$
(f) octane	C_8H_{18}
(g) household gas	CH_4

14. (a) What does the molecular formula CBr_4 mean?
(b) What information is added by the following structural formula?

```
     Br
     |
 Br—C—Br
     |
     Br
```

15. How many particles are there in a mole?

16. A stone about the size of a softball weighs roughly a kilogram. How many moles of such stones would be needed to account for the entire mass of the Earth, about 6×10^{27} grams?

17. If we had one mole of dollars to divide among all the people in the world, how much would each of the three billion inhabitants receive?

18. How many moles of atoms are in
(a) 9.0 grams of aluminum
(b) 0.83 gram of iron

Answer. (a) $\frac{1}{3}$ mole.

19. The most delicate balance can detect a change of about 10^{-8} gram. How many atoms of gold would be in a sample of that weight?

20. How many moles of oxygen atoms are in one mole of nitric acid molecules? Of sulfuric acid molecules?

21. Determine the weight, in grams, of one silver atom.

Answer. 1.79×10^{-22} gram.

22. Write the formulas for the following compounds and give the weight of one mole of each: carbon disulfide, sulfur hexafluoride, nitrogen trichloride, osmium tetroxide.

23. Consider the following data

Element	Atomic Weight
A	12.01
B	35.5

A and *B* combine to form a new substance, *X*. If four moles of *B* atoms combine with one mole of *A* to give one mole of *X*, then the weight of one mole of *X* is

(a) 47.5 grams
(b) 74.0 grams
(c) 83.0 grams
(d) 154.0 grams
(e) 166.0 grams

24. Calculate the molecular weight for each of the following: SiF_4, HF, Cl_2, Xe, NO_2.

25. If $1\frac{1}{2}$ moles of hydrogen gas (H_2) react in a given experiment, how many grams of H_2 does this represent?

26. How many moles are contained in 49 grams of pure H_2SO_4?

Answer. 0.50 mole.

27. A chemist weighs out 10.0 grams of water, 10.0 grams of ammonia, and 10.0 grams of hydrogen chloride (hydrochloric acid). How many moles of each substance does he have?

28. (a) The ratio of the weight of a liter of chlorine gas to the weight of a liter of oxygen gas, both measured at room temperature and pressure, is 2.22. Calculate the molecular weight of chlorine.
(b) How does this value compare with the atomic weight of chlorine found in the chart of atomic weights? What is the formula of a molecule of chlorine?

29. A flask of gaseous CCl_4 was weighed at a measured temperature and pressure. The flask was flushed and then filled with oxygen at the same temperature and pressure. The weight of the CCl_4 vapor will be about

(a) the same as that of oxygen,
(b) one-fifth as heavy as the oxygen,
(c) five times as heavy as the oxygen,
(d) twice as heavy as the oxygen,
(e) one-half as heavy as the oxygen.

30. Suppose chemists had chosen a billion billion (10^{18}) as the number of particles in one mole. What would the molecular weight of oxygen gas be?

31. One volume of hydrogen gas combines with one volume of chlorine gas to produce two volumes of hydrogen chloride gas (all measured at the same temperature and pressure). A variety of other types of evidence suggests that hydrogen is an element and that its molecules are diatomic.

(a) Which one of the following possible molecular formulas for the substance chlorine is *not* consistent with the volumes that combine? (Use only the data given here; do not presume the molecular formula of hydrogen chloride.)

(i) Cl_2
(ii) Cl_4
(iii) H_2Cl_2
(iv) H_3Cl_2
(v) H_4Cl_2

(b) For each formula in part (a) that *is* consistent with the combining volumes data (and the formula H_2 for hydrogen), calculate the molecular weight indicated by that formula.
(c) For each acceptable formula in part (a) predict the molecular formula for the substance hydrogen chloride.

CHAPTER 3

Chemical Reactions

We might as well attempt to introduce a new planet into the solar system, or to annihilate one already in existence, as to create or destroy a particle of hydrogen. All the changes we can produce consist in separating particles that are in a state of · · · combination, and joining those that were previously at a distance.

JOHN DALTON, 1810

In Experiment 5 you compared the magnitude of the heat released when melted wax solidifies to the heat released when the same amount of wax is burned. In each case an obvious change occurs (liquid changes to solid or solid burns to a gas) and it is accompanied by a measurable heat release. These are likenesses between solidification and combustion. More apparent, however, are striking differences. Through experiments on warming and cooling wax, you know that after wax melts to a liquid the solid wax can be recovered merely by recooling. After combustion of the wax, however, cooling the gases produced does *not* restore the wax. Instead, the products, carbon dioxide and liquid water (and perhaps some soot), bear no resemblance to wax. Equally dramatic is the contrast of the heat effects. The heat released during combustion is well over two hundred times greater than the heat released during solidification.

Because of these differences, chemists differentiate these two kinds of change. We have already named the solidification of wax—in Section 1-1.2 we called this type of change a phase change. A change like combustion, with its much larger heat effects, is called a *chemical change* or a *chemical reaction*.

3-1 PRINCIPLES OF CHEMICAL REACTIONS

The central activity of the chemist is to explore and exploit chemical changes. Sometimes his wish is to cause a change, sometimes to prevent it, in a system of interest. Always he wishes to understand and control the chemical changes that might occur.

We shall begin our study of chemical changes with a simple chemical reaction that forms a familiar substance—water.

3-1.1 Formation of Water from Hydrogen and Oxygen

In Section 2-2.2 it was mentioned that two volumes of hydrogen combine with one volume of oxygen. Water is produced. The reaction gives off heat—a large amount of heat, as in the combustion of a candle. The product, water, is not at all like the starting materials, hydrogen and oxygen. Hence, the change that occurs when hydrogen and oxygen combine should be classified as a chemical reaction.

In terms of the atomic theory, we begin with molecules of hydrogen and molecules of oxygen. After reaction, we find molecules of water. The bonds between atoms in the reacting substances are broken, and the atoms rearrange to form the new bonds in the product molecules. These changes are readily pictured with the aid of our molecular models. In Figure 3-1 two hydrogen molecules (four atoms) and one oxygen molecule (two atoms) are represented on the left. If these molecules are to react to form water, the bonds between the atoms in the oxygen molecule and in the hydrogen molecules must be broken. Then, the atoms can rearrange themselves to form two water molecules. Notice that the atoms are rearranged as a result of the reaction but that the total number of atoms does not change.

EXERCISE 3-1

Suppose ten hydrogen molecules and ten oxygen molecules are mixed. How many molecules of water could be formed? What would be left over?

EXERCISE 3-2

One million oxygen molecules react with sufficient hydrogen molecules to form water molecules. How many water molecules are formed? How many hydrogen molecules are consumed?

Extending Figure 3-1, you can reason that the production of 100 molecules of water requires 100 molecules of hydrogen and 50 molecules of oxygen. Also, to produce one mole of water (6.02×10^{23} molecules) you need one mole of hydrogen gas (6.02×10^{23} molecules) and one-half mole of oxygen (3.01×10^{23} molecules). These results are summarized in Table 3-I.

Table 3-I.

REACTING AMOUNTS OF HYDROGEN AND OXYGEN TO FORM WATER

	Amounts of Reacting Substances		*Amounts of Product*
	HYDROGEN	OXYGEN	WATER
(a) In numbers of molecules	2	1	2
	4	2	4
	100	50	100
	6.02×10^{23}	3.01×10^{23}	6.02×10^{23}
(b) In numbers of moles	1	$\frac{1}{2}$	1
	2	1	2
	10	5	10

The reaction between hydrogen gas and oxygen gas proceeds more quickly if we mix the gases and then ignite the mixture with a spark. A violent explosion results. Even so, the quantity of product, water, and the heat evolved are the same per mole of hydrogen reacting as in controlled burning.

Fig. 3-1. **Formation of water molecules** *from hydrogen molecules and oxygen molecules.*

reacts with

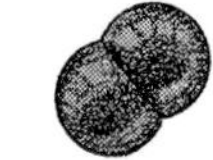

to form

If we react one mole of pure hydrogen and one-half mole of pure oxygen, one mole of water is produced. The quantity of heat produced when one mole of water is formed is 68,000 calories. If we mix only 0.025 mole of pure hydrogen, only ($\frac{1}{2}$) × (0.025) mole of oxygen is needed. The amount of water produced is 0.025 mole. If only 0.025 mole of water is produced, only (0.025) × (68,000) calories or about 1700 calories of heat are released.

The source of this heat energy must be the reactants (hydrogen and oxygen) themselves since no heat was supplied externally other than that needed to ignite the mixture. We may conclude that the water has less energy than did the reactants used to make it. Such a reaction in which energy is released is called an **exothermic reaction.** The quantity of energy produced when one mole of hydrogen is burned, 68,000 calories (68 kcal*), is called the **molar heat of combustion of hydrogen.**

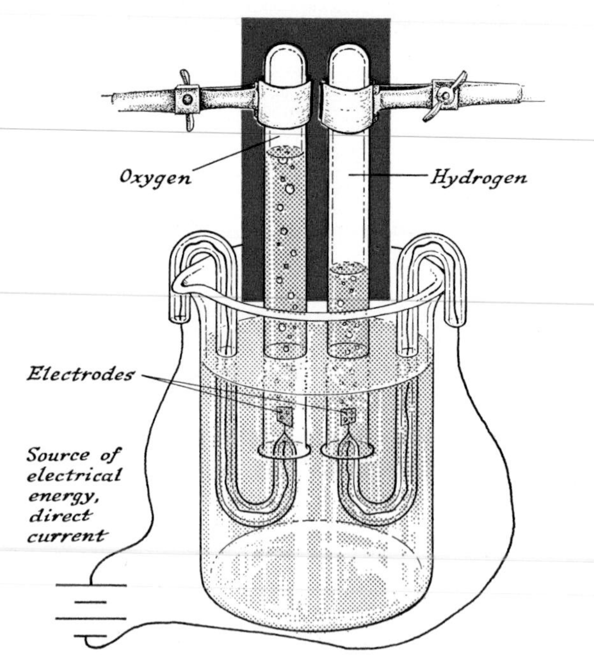

Fig. 3-2. **Electrolytic decomposition of water.**

EXERCISE 3-3

How much heat is released when two moles of hydrogen burn? One-half mole?

3-1.2 Decomposition of Water

We can decompose the water in a solution of water and sulfuric acid by passing an electric current through the solution in an electrolysis apparatus, as shown in Figure 3-2. In this apparatus, two conductors (called electrodes) are immersed in the liquid. When the electrodes are connected to a source of electrical energy hydrogen gas appears at one electrode and oxygen gas appears at the other. If we operate the apparatus until one mole of water has decomposed, one mole of hydrogen gas and one-half mole of oxygen gas are produced. We observe also that energy (electrical energy in this case) is needed to cause the water to decompose. If energy is absorbed in a reaction, the reaction is called **endothermic.**

Now we can compare the formation and decomposition of water. As shown graphically in Figure 3-3, the chemical change involved in the formation of water is exactly the reverse of that involved in the decomposition of water. These changes are governed by simple rules. On the left we find two atoms of oxygen, and on the right we find two atoms of oxygen. On the left we find four atoms of hydrogen, and on the right we find four atoms of hydrogen. We see that atoms are neither gained, nor are they lost. *In chemical reactions, atoms are conserved.*

The number of molecules of H_2 needed to react with one molecule of O_2 is the number needed to produce two molecules of H_2O. If two molecules of H_2O are formed, four atoms of hydrogen are needed. Two molecules of H_2 contain four atoms of hydrogen. Remember, in *chemical reactions, atoms are conserved.*

3-1.3 Conservation of Mass

Belief in the conservation of atoms is based upon a generalization that has stood the test of many

---

* 1 kilocalorie (1 kcal) = 1000 calories. The prefix "kilo-" always means 1000. Thus, 1 kilogram = 1000 grams; 1 kilometer = 1000 meters.

decades. *Matter can be neither created nor destroyed.* Since we often measure a quantity of matter in terms of its mass (by weighing, for example) we may say that *mass is conserved.* Thus, one mole of liquid water weighs 18.0 grams; in the decomposition of one mole of water, 2.0 grams of hydrogen and 16.0 grams of oxygen are produced:

1 mole of liquid water, H_2O	$\longrightarrow$	1 mole of hydrogen, H_2	+	$\frac{1}{2}$ mole of oxygen, $\frac{1}{2}O_2$	(*1*)
18.0 g	$\longrightarrow$	2.0 g	+	16.0 g	
18.0 g	$\longrightarrow$	18.0 g			

Since the mass of a mole of water is the sum of the masses of the atoms in the mole of water, the conservation of mass implies conservation of atoms.

Fig. 3-3. **The reactions of formation and decomposition of water** *shown with molecular models.*

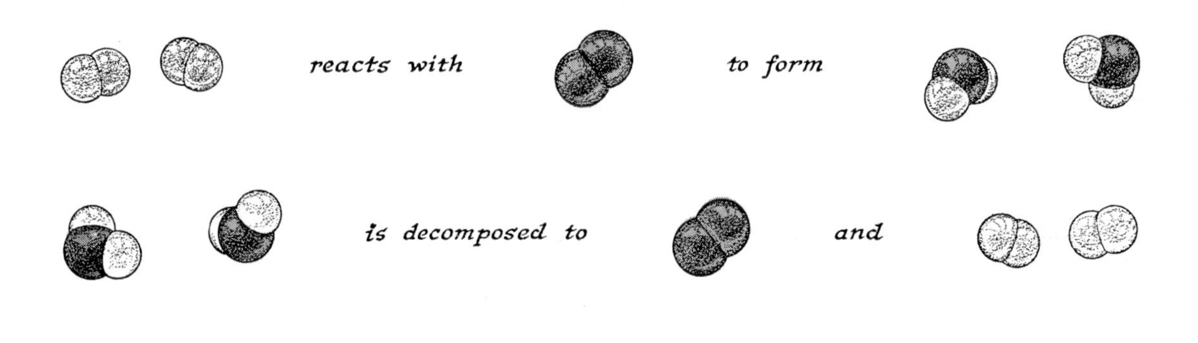

3-2 EQUATIONS FOR CHEMICAL REACTIONS

The graphic representations we have used for reactions help us visualize the rearrangement of atoms in the reactions. By a slight change we can show the results in a less detailed but simpler way. Chemical formulas can be used rather than drawings of the atoms and molecules. Thus, the formula of elementary hydrogen is H_2, the formula of elementary oxygen is O_2, and the formula of water is H_2O. By using the formulas to represent the molecules, we can replace the diagram of Figure 3-3 by the following expressions:*

$$2\,H_2 + 1\,O_2 \longrightarrow 2\,H_2O \qquad (2)$$
$$4\,H_2 + 2\,O_2 \longrightarrow 4\,H_2O \qquad (3)$$
$$2\,H_2O \longrightarrow 1\,O_2 + 2\,H_2 \qquad (4)$$

Such expressions are called *chemical equations.* Notice that we show two molecules of a substance by writing the coefficient 2 before the formula. The coefficient 2 before the formula H_2O means two molecules—it applies to every symbol in the formula. In two molecules of water, there are six atoms, four of hydrogen and two of oxygen. Notice also that we can change equation (*2*) to equation (*3*) merely by doubling all the coefficients. We can change equation (*2*) to equation (*4*) by reversing it. Equation (*2*) represents the formation of water; equation (*4*) represents the decomposition of water.

All equations are based upon the conservation of atoms. Every symbol, when multiplied by the subscript after it and the coefficient before the formula, must appear as often on the left side of the equation as on the right.

Natural gas contains mainly the substance methane, with the formula CH_4. The chemical equation for the burning of methane is

$$1\,CH_4 + 2\,O_2 \longrightarrow 1\,CO_2 + 2\,H_2O \qquad (5)$$

The number of atoms in the reactants equals the

* Many chemists use an "equals" sign in place of the arrow. Thus, equation (*2*) would be written:

$$2\,H_2 + 1\,O_2 = 2\,H_2O$$

number of atoms in the products. One carbon atom, four hydrogen atoms, and four oxygen atoms are represented on each side of the equation.

Because of the relation between a molecule and a mole, we can read equation (*5*) in either of two ways: (1) "one *molecule* of methane reacts with two *molecules* of oxygen to form one *molecule* of carbon dioxide and two *molecules* of water," or (2) "one *mole* of methane reacts with two *moles* of oxygen to form one *mole* of carbon dioxide and two *moles* of water."

EXERCISE 3-4

Write an equation containing the information expressed in your answer to Question 1 of Experiment 7.

3-2.1 Writing Equations for Reactions

How can you write the equations for a reaction without first drawing a picture? Remember that to do either you must:

(1) know what reactants are consumed and what products form;
(2) know the correct formula of each reactant and each product;
(3) satisfy the law of conservation of atoms.

Consider the reaction for the burning of magnesium to form magnesium oxide. Magnesium metal and magnesium oxide are solids. They have the formulas Mg and MgO, respectively. In preparation for writing the equation, we write the formulas for the reactants on the left and the formula for the product on the right:

$$Mg + O_2 \longrightarrow MgO \qquad (a)$$

Expression (*a*) does not yet conserve atoms. We must find numerical coefficients to place before each formula so that there are the same number of atoms of each element on the left side of the equation as there are on the right. The process of finding these coefficients is called *balancing* the equation. For simple reactions, it is an easy and logical process.

First, we may begin by choosing one mole of oxygen as the amount of this reactant consumed in (*a*):

$$Mg + 1O_2 \longrightarrow MgO \qquad (b)$$

But one mole of O_2, with its two moles of oxygen atoms, will form two moles of magnesium oxide:

$$Mg + 1O_2 \longrightarrow 2MgO \qquad (c)$$

Two moles of magnesium oxide require two moles of magnesium metal. Thus,

$$2Mg + 1O_2 \longrightarrow 2MgO \qquad (d) \qquad (6)$$

Equation (*6*) is a chemical equation—since atoms are conserved, it is said to be *balanced*.

We could have decided to begin by choosing one mole of magnesium metal as the amount of reactant consumed in (*a*):

$$1Mg + O_2 \longrightarrow MgO \qquad (b')$$

One mole of magnesium contains a mole of atoms, hence will form one mole of magnesium oxide:

$$1Mg + O_2 \longrightarrow 1MgO \qquad (c')$$

One mole of magnesium oxide contains one mole of oxygen atoms, the number contained in one-half mole of oxygen molecules. Thus,

$$1Mg + \tfrac{1}{2}O_2 \longrightarrow MgO \qquad (d') \qquad (7)$$

Equation (*7*) is also a chemical equation—again atoms are conserved. It is just as correct an expression for the burning of magnesium as is (*6*). To show this, we can multiply (*7*) by 2 to obtain equation (*6*). We can always multiply all the coefficients by a common factor or divide by a common factor and obtain equally valid equations.

In equation (*6*) the coefficient 1 may be dropped but it is never wrong to retain it.

3-2.2 Other Examples of Chemical Equations

Hydrogen gas, H_2, and chlorine gas, Cl_2, react to form hydrogen chloride gas, HCl.* The reactants are H_2 and Cl_2; the product is HCl:

$$1H_2 + Cl_2 \longrightarrow HCl$$

* Hydrogen chloride, HCl, dissolved in water, is commonly called hydrochloric acid.

If we base the reaction upon one mole of H_2, we see that conservation of hydrogen atoms requires that two moles of HCl be produced:

$$1H_2 + Cl_2 \longrightarrow 2HCl$$

Now the product, 2HCl, contains two moles of chlorine atoms. This is just the number of chlorine atoms in one mole of chlorine. The reaction is balanced. We may write

$$1H_2 + 1Cl_2 \longrightarrow 2HCl \quad (8)$$

or

$$H_2 + Cl_2 \longrightarrow 2HCl$$

or

$$H_2 + Cl_2 = 2HCl$$

EXERCISE 3-5

Ammonia gas, NH_3, can be burned with oxygen gas, O_2, to give nitrogen gas, N_2, and water, H_2O. See if you can follow the logic of the following steps in balancing this reaction.

$$NH_3 + O_2 \longrightarrow N_2 + H_2O$$
$$NH_3 + O_2 \longrightarrow 1N_2 + H_2O$$
$$2NH_3 + O_2 \longrightarrow 1N_2 + H_2O$$
$$2NH_3 + O_2 \longrightarrow 1N_2 + 3H_2O$$
$$2NH_3 + \tfrac{3}{2}O_2 \longrightarrow 1N_2 + 3H_2O \quad (9)$$

State briefly what was done in each step.

The molecular formula for the substance formaldehyde is H_2CO. Formaldehyde burns to form carbon dioxide and water. What equation represents this reaction?

Again we begin by writing the formulas for reactants and products:

$$1H_2CO + O_2 \longrightarrow CO_2 + H_2O$$

Suppose we burn one molecule of formaldehyde. The one carbon atom, the two hydrogen atoms, and one oxygen atom in the H_2CO molecule must appear in the products. Since, among the products, carbon atoms appear only in carbon dioxide, there must be one molecule of CO_2:

$$1H_2CO + O_2 \longrightarrow 1CO_2 + H_2O$$

Since hydrogen atoms appear in only one of the products, water, there must be one molecule of water to accommodate the two atoms of hydrogen. Now we have

$$1H_2CO + yO_2 \longrightarrow 1CO_2 + 1H_2O$$

Notice that we have not yet determined the coefficient of O_2; we have designated it as y to remind ourselves of this. Since the oxygen atoms must also be conserved and three are required for the products, three oxygen atoms must have been present in the reactants. One oxygen atom was present in the molecule of formaldehyde, so two more are required. It follows that y must be 1.

We now have the balanced equation

$$1H_2CO + 1O_2 \longrightarrow 1CO_2 + 1H_2O \quad (10)$$

EXERCISE 3-6

A paraffin candle burns in air to form water and carbon dioxide. Paraffin is made up of molecules of several sizes. We shall use the molecular formula $C_{25}H_{52}$ as representative of the molecules present. One mole of candle contains the Avogadro number of these molecules.

Formulas of Reactants		Formulas of Products
$C_{25}H_{52} + O_2$	$\longrightarrow$	$H_2O + CO_2$

Suppose one mole of paraffin (which weighs 353 grams) is burned. Using the method shown in the preceding example, we obtain

$$1C_{25}H_{52} + yO_2 \longrightarrow 26H_2O + 25CO_2$$

We still have not determined the coefficient for O_2. Since 76 O's are required for the products $[26 + (2 \times 25) = 76]$, they must have been present in the reactants. Show that it follows that y must be 38:

$$C_{25}H_{52} + 38O_2 \longrightarrow 26H_2O + 25CO_2 \quad (11)$$

Usually it is more useful to think of equations in terms of moles rather than molecules since a mole is a weighable amount. In equation (*6*) two moles of magnesium weigh 48.6 grams; one mole of oxygen weighs 32.0 grams; two moles of MgO weigh 80.6 grams. Mass is conserved: 48.6 grams + 32.0 grams = 80.6 grams.

In equation (*11*) the $38O_2$ is usually read as

38 moles, not 38 molecules; "38 moles of oxygen gas" has experimental meaning. They weigh 38×32 grams = 1216 grams.

3-2.3 Calculations Based upon Chemical Equations

Equations give us all the information we need for computing the weights of the substances consumed or produced in chemical reactions. Suppose we wish to know how many moles of water are produced when 68 grams of ammonia are burned. Equation (*9*) represents the reaction:

$$2NH_3 + \tfrac{3}{2}O_2 \longrightarrow 1N_2 + 3H_2O \qquad (9)$$

One mole of ammonia weighs 17 grams. Two moles of ammonia, weighing 34 grams, produces three moles of water. We wish to burn 68 grams of ammonia. How many moles is this?

$$\frac{68 \text{ grams}}{17 \text{ grams/mole}} = 4.0 \text{ moles of ammonia}$$

Hence we can write

$2NH_3$	$+ \tfrac{3}{2}O_2$	$\longrightarrow$	$1N_2 +$	$3H_2O$
two moles of ammonia		produce		three moles of water

so,

four moles of ammonia	produce	six moles of water

We see that 68 grams (four moles) of ammonia produce six moles of water.

Suppose we wish to know how many grams of water are produced from the burning of one-half mole of paraffin. Equation (*11*) shows the reaction:

$1C_{25}H_{52}$	$+ 38O_2$	$\longrightarrow$	$26H_2O + 25CO_2$ (*11*)
1 mole of $C_{25}H_{52}$		produces	26 moles of H_2O
$\tfrac{1}{2}$ mole of $C_{25}H_{52}$		produces	13 moles of H_2O

Since one mole of water weighs 18 grams, the weight of 13 moles of water is (13 moles) $\times$ (18g/mole) = 234 g.

EXERCISE 3-7

Show that 3.80 moles of oxygen are needed to burn 35.3 g of paraffin by reaction (*11*).

EXERCISE 3-8

How many moles of oxygen, O_2, are required to produce 242 grams of magnesium oxide by equation (*6*)?

$$2Mg + 1O_2 \longrightarrow 2MgO \qquad (6)$$

EXERCISE 3-9

Write the equation for the reaction which took place in Experiment 8, Part II. What was the residue you obtained on evaporation of the solution in beaker number 2?

EXERCISE 3-10

In Experiment 8 you determined the number of moles of silver chloride formed in the reaction of some sodium chloride with a known amount of silver nitrate. How many moles of sodium chloride reacted with the silver nitrate? Compare this with the number of moles of sodium chloride you added.

QUESTIONS AND PROBLEMS

1. One volume of hydrogen gas combines with one volume of chlorine gas to give two volumes of hydrogen chloride gas. On the basis of many reactions, we have learned that the molecular formulas are, for hydrogen, H_2, for chlorine, Cl_2, and for hydrogen chloride, HCl. The reaction, in symbols, is

$$H_2 + Cl_2 \longrightarrow 2HCl$$

   (a) According to this reaction, how many *molecules* of hydrogen chloride, HCl, can be formed from one *molecule* of hydrogen, H_2?
   (b) How many *moles* of hydrogen chloride, HCl, can be formed from one *mole* of hydrogen, H_2?
   (c) Four *molecules* of chlorine, Cl_2, will produce how many *molecules* of HCl?

(d) Eight *moles* of hydrogen chloride are formed from how many *moles* of Cl_2?

2. The reaction between nitric oxide, NO, and oxygen, O_2, is written

$$2NO + O_2 \longrightarrow 2NO_2$$

(a) Two *molecules* of nitric oxide give how many *molecules* of nitrogen dioxide, NO_2?
(b) Two *moles* of NO give how many *moles* of NO_2?
(c) How many moles of oxygen atoms are there in two moles of NO?
(d) How many moles of oxygen atoms are there in one mole of O_2?
(e) How many moles of oxygen atoms are there in two moles of NO_2?
(f) Use the answers to parts (c), (d), and (e) to verify that the reaction is written so as to conserve oxygen atoms.

3. (a) Write the equation for the reaction between nitrogen and hydrogen to give ammonia on the basis of your answer to Problem 3 of Chapter 2, and assuming the following molecular formulas: nitrogen, N_2; hydrogen, H_2; ammonia, NH_3.
(b) Verify that your equation conserves nitrogen atoms.
(c) Verify that your equation conserves hydrogen atoms.

4. When ammonia is decomposed into nitrogen and hydrogen, the reaction absorbs heat. Written in terms of moles, the equation is

$$2NH_3 + 22 \text{ kcal} \longrightarrow N_2 + 3H_2$$

(a) Two moles of ammonia produce how many moles of nitrogen?
(b) The production of one mole of nitrogen absorbs how much heat?
(c) The production of nine moles of hydrogen, H_2, absorbs how much heat?
(d) Calculate the weight of two moles of ammonia and compare it to the sum of the weights of one mole of nitrogen, N_2, plus three moles of hydrogen, H_2.

5. In the manufacture of nitric acid, HNO_3, nitrogen dioxide reacts with water to form HNO_3 and nitric oxide, NO:

$$3NO_2 + H_2O \longrightarrow 2HNO_3 + NO$$

(a) Verify that the equation conserves oxygen atoms.
(b) How many molecules of nitrogen dioxide are required to form 25 molecules of nitric oxide?
(c) How many moles of nitric oxide are formed from 0.60 mole of nitrogen dioxide?

6. If 3 grams of substance *A* combine with 4 grams of substance *B* to make 5 grams of substance *C* and some *D*, how many grams of *D* would you expect?

7. One step in the manufacture of sulfuric acid is to burn sulfur (formula, S_8) in air to form a colorless gas with a choking odor. The name of the gas is sulfur dioxide and it has the molecular formula SO_2. On the basis of this information:

(a) Write the balanced equation for this reaction.
(b) Interpret the equation in terms of molecules.
(c) Interpret the equation in terms of moles.
(d) Two moles of sulfur, S_8, would produce how many moles of sulfur dioxide, SO_2?

8. When iron rusts, it combines with oxygen of the air to form iron oxide, Fe_2O_3. Which of the following is FALSE?

(a) The equation is

$$3O_2 + 4Fe \longrightarrow 2Fe_2O_3$$

(b) There are five atoms represented by the formula, Fe_2O_3.
(c) Oxygen gas is triatomic.
(d) The mass of the reactants equals the mass of the products.
(e) Atoms are conserved.

9. Balance the equations for each of the following reactions. Begin on the basis of one mole of the substance underscored.

(a) $Li + \underline{Cl_2} \longrightarrow LiCl$
(b) $Na + \underline{Cl_2} \longrightarrow NaCl$
(c) $Na + \underline{F_2} \longrightarrow NaF$
(d) $Na + \underline{Br_2} \longrightarrow NaBr$
(e) $\underline{O_2} + Cl_2 \longrightarrow Cl_2O$
(f) $O_2 + \underline{Cl_2} \longrightarrow Cl_2O$

Show that your answers to parts (e) and (f) contain the same information.

10. Balance the equations for each of the following reactions involving oxygen. Begin on the basis of one mole of the substance underscored.

(a) With metallic nickel:

$$Ni + \underline{O_2} \longrightarrow NiO$$

(b) With metallic nickel:

$$\underline{Ni} + O_2 \longrightarrow NiO$$

(c) With metallic lithium:

$$Li + \underline{O_2} \longrightarrow Li_2O$$

(d) With the rocket fuel hydrazine, N_2H_4:

$$\underline{N_2H_4} + O_2 \longrightarrow N_2 + H_2O$$

(e) With acetylene, C_2H_2, in an acetylene torch flame:

$$\underline{C_2H_2} + O_2 \longrightarrow CO_2 + H_2O$$

Answer. $C_2H_2 + \frac{5}{2}O_2 \longrightarrow 2CO_2 + H_2O$

(f) With the important copper ore, chalcocite, Cu_2S (the process called "roasting" the ore):

$$\underline{Cu_2S} + O_2 \longrightarrow Cu_2O + SO_2$$

(g) With the important iron ore, iron pyrites, FeS_2 (again, "roasting" the ore):

$$FeS_2 + O_2 \longrightarrow \underline{Fe_2O_3} + SO_2$$

11. (a) Balance the equations for the decomposition (to elements) of ammonia, NH_3, nitrogen trifluoride, NF_3, and nitrogen trichloride, NCl_3. Base each equation upon the production of one mole of N_2.

$$NH_3 \longrightarrow 1N_2 + H_2$$
$$NF_3 \longrightarrow 1N_2 + F_2$$
$$NCl_3 \longrightarrow 1N_2 + Cl_2$$

(b) Rewrite the equations to include the information that the decomposition of ammonia is endothermic, absorbing 22.08 kcal/mole N_2, the decomposition of NF_3 is endothermic, absorbing 54.4 kcal/mole N_2, and the decomposition of NCl_3 is exothermic, releasing 109.4 kcal/mole N_2.

(c) One of the three compounds NH_3, NF_3, and NCl_3 is dangerously explosive. Which would you expect to be the explosive substance? Why?

12. Graphite, a form of carbon, C, burns in air to produce the colorless gas, carbon dioxide. On the basis of this information:

(a) Write the equation for the reaction.

(b) If one mole of graphite is burned, how many moles of carbon dioxide are produced? What is the weight in grams of this amount of carbon dioxide?

(c) If two moles of graphite were burned, how many moles of carbon dioxide would be produced? What is the weight in grams?

(d) If five moles of graphite were burned in a vessel containing 10 moles of oxygen gas, what is the maximum number of moles of carbon dioxide that could be produced?

13. If a piece of sodium metal is lowered into a bottle of chlorine gas, a reaction takes place. Table salt, NaCl, is formed.

(a) Write the equation for the reaction.

(b) How many moles of NaCl could be formed from one mole of Na?

(c) How many moles of NaCl could be formed from 2.30 grams of Na?

14. Methane, the main constituent of natural gas, has the formula CH_4. Its combustion products are carbon dioxide and water.

(a) Write the equation for the combustion of methane. Compare your answer with equation (5), p. 41.

(b) One mole of methane produces how many moles of water vapor?

(c) One-eighth mole of methane would produce how many moles of carbon dioxide?

(d) How many moles of water vapor would be produced by 4.0 grams of methane?

15. If potassium chlorate, $KClO_3$, is heated gently, the crystals will melt. Further heating will decompose it to give oxygen gas and potassium chloride, KCl.

(a) Write the equation for the decomposition.

(b) How many moles of $KClO_3$ are needed to give 1.5 moles of oxygen gas?

(c) How many moles of KCl would be given by $\frac{1}{3}$ mole of $KClO_3$?

(d) How many moles of oxygen gas would be produced by 122.6 grams of $KClO_3$?

16. One gallon of gasoline can be considered to be about 25 moles of octane, C_8H_{18}.

(a) How many moles of oxygen must be used to burn this gasoline, assuming the only products are carbon dioxide and water?

(b) How many moles of carbon dioxide are formed?

(c) How much does this carbon dioxide weigh? (Express your answer in kilograms.)

(d) What weight of carbon dioxide is released into the atmosphere when your automobile consumes 10 gallons of gasoline? Express this answer in pounds (1 kg = 2.2 pounds).

17. Iron (Fe) burns in air to form a black, solid oxide (Fe_3O_4).

(a) Write the equation for the reaction.
(b) How many moles of oxygen gas are needed to burn one mole of iron?
(c) How many grams of oxygen gas is that?
(d) Can a piece of iron weighing 5.6 grams burn completely to Fe_3O_4 in a vessel containing 0.05 mole of O_2?

18. Problem 5 relates to the manufacture of nitric acid.

(a) According to the equation given in that problem, how many grams of nitric acid are formed from one mole of nitrogen dioxide?
(b) How many *more* grams of nitric acid could be formed if the nitric oxide formed could be completely converted into nitric acid (assume one mole of nitric oxide gives one mole of nitric acid)?

19. Hydrazine, N_2H_4, can be burned with oxygen to provide energy for rocket propulsion. The energy released is 150 kcal per mole of hydrazine burned.

(a) How much energy is released if 10.0 kg of hydrazine fuel are burned?
(b) Compare the energy that would be released if the same weight of hydrogen, 10.0 kg, were burned as a fuel instead (see Section 3-1.1).

GILBERT NEWTON LEWIS, 1875–1946

Gilbert Newton Lewis, one of the greatest American chemists of the twentieth century, began his career teaching high school chemistry. Born near Boston and reared in Nebraska, young Lewis returned to the East to study and graduated from Harvard University. After a year teaching high school, he returned to Harvard and received the Ph.D. in 1899. There followed a year at universities in Germany, another as Superintendent of Weights and Measures in the Philippine Islands. Then, in seven years at the Massachusetts Institute of Technology he rose to the rank of Professor. Finally, in 1912 he accepted the position of Chairman of what was then a little known chemistry department far from the recognized scientific centers. He moved to the University of California and spent the remainder of his career at Berkeley, building one of the most powerful chemistry departments in the world.

Lewis devoted most of his career to the understanding of the structures of molecules and of thermodynamics, the energy relations in chemical changes. His thinking was far ahead of his time and his theories have had profound influence on chemistry. His understanding of chemical bonding has strongly influenced modern thinking on this subject. Lewis was one of the first to recognize that energy effects provide a basis for predicting what chemical reactions can occur. Thus he awakened chemists to the crucial importance of thermodynamics. His book on this subject, published in 1923, became a classic of the chemical literature. He published over 150 research publications on topics extending from the phases of sulfur to quantum mechanics.

G. N. Lewis enjoyed chemistry. Throughout his distinguished career he remained active in the laboratory and never tired of the thrill of discovery. He favored simple and direct experiments—many of his important discoveries were performed with a few test tubes and simple chemicals. His enthusiasm and burning interest in chemistry were contagious—many of his students became great scientists. Lewis virtually eliminated graduate courses, relying instead upon the open debate of research seminars. He encouraged his colleagues to think critically, to challenge his ideas, and to welcome challenge of their own.

G. N. Lewis died March 23, 1946, in the laboratory he loved, surrounded by the beakers and books that were the tools of his trade. He is remembered and respected by chemists the world over.

CHAPTER 4

The Gas Phase: Kinetic Theory

. . . it is my intention to make known some new properties in gases, the effects of which are regular, by showing that these substances combine amongst themselves in very simple proportions . . .

JOSEPH L. GAY-LUSSAC, 1808

We have already seen that the behavior of gases is important to a chemist. The pressure-volume behavior leads to the particle model of a gas. Differences among gases (in properties such as color, odor, and solubility) show that the particles of one gas differ from the particles of another gas. In chemical reactions, the simple combining volume relationships support Avogadro's Hypothesis and, hence, give us a way to measure molecular weights.

Thus we see that the properties of gases provide a substantial basis for developing the atomic theory. The gaseous state is, in many ways, the simplest state of matter for us to understand. The regularities we discover are susceptible to detailed mathematical interpretation. We shall examine these regularities in this chapter. We shall find that their interpretation, called the kinetic theory, provides an understanding of the meaning of temperature on the molecular level.

4-1 THE VOLUME OCCUPIED BY ONE MOLE OF GAS

To a chemist, one of the most important regularities displayed by a gas relates to the volume occupied by one mole of a gas. We shall begin investigating this subject by comparing the sizes of gaseous particles with the average spacing between them under normal conditions of temperature and pressure. The comparison can be based upon the volumes occupied by a mole of nitrogen first as a solid, then as a liquid, and finally as a gas.

4-1.1 The Volume Occupied by a Mole of Nitrogen, N_2

The molecular formula of nitrogen is N_2; the nitrogen molecule is diatomic. One mole of N_2

molecules contains, then, two moles of nitrogen atoms. The weight of one mole of nitrogen molecules is 28.0 grams.

At a sufficiently low temperature, below −210°C, nitrogen is a solid with a density of 1.03 grams per milliliter. The volume occupied by one mole of the solid, called the **molar volume,** is

$$\text{Molar volume, } solid\text{:} \quad \frac{28.0 \text{ g/mole}}{1.03 \text{ g/ml}} = 27.2 \text{ ml/mole}$$

Now if we warm the solid to −210°C, the solid melts to form liquid nitrogen. The density of this liquid is 0.81 grams per milliliter. Now the volume of a mole is

$$\text{Molar volume, } liquid\text{:} \quad \frac{28.0 \text{ g/mole}}{0.81 \text{ g/ml}} = 34.6 \text{ ml/mole}$$

If we raise the temperature still further, the liquid vaporizes to form nitrogen gas, *taking whatever density is necessary to fill the container.* The density now depends upon the volume of the container and the temperature. For the sake of comparison, suppose the gaseous nitrogen is placed in that volume that gives a pressure of one atmosphere when the container is placed in an ice bath at 0°C. Then the density is found to be only 0.00125 gram per milliliter. This means that the volume required for one mole of gas is

Molar volume, *gas* at 0°C, 1 atm:

$$\frac{28.0 \text{ g/mole}}{0.00125 \text{ g/ml}} = 22.4 \times 10^3 \text{ ml/mole}$$
$$= 22.4 \text{ liters/mole}$$

The volume of this gas is almost 1000 times as great as the volume of the same weight of solid. Experiments with other gases lead to similar results. If the size of a single molecule is assumed to be the same in the solid and gas, then the molecules must have separated from each other in the gas. The free space between gaseous molecules is on the order of 1000 times the volume a molecule occupies in the solid.

EXERCISE 4-1

How many molecules of nitrogen are present in one liter of the gas at 0°C and one atmosphere pressure?

EXERCISE 4-2

(a) Calculate the volume (in milliliters) occupied by one nitrogen molecule in the solid phase.
(b) Recognizing that one milliliter is 1.00 cubic centimeter, estimate the size (in centimeters) of a cube that has the volume calculated in part (a). Use one significant figure. Now express your answer in Ångstroms (1 Å = 10^{-8} cm).

4-1.2 A Comparison of Molar Volumes of Gases

The volume calculated above, 22.4 liters, we have seen before. In Table 1-II the pressure-volume product of 32.0 grams of oxygen, O_2, was found to be 22.4 at 0°C. (Notice that 32.0 grams of O_2 is the weight of one mole of oxygen.) So we can use this relation,

$$P \times V = 22.4 \frac{\text{liters} \times \text{atmospheres}}{\text{mole}} \text{ (at 0°C)}$$

to solve for the volume of a mole of O_2 at one atmosphere pressure:

$$1 \text{ atm} \times V = 22.4 \text{ liters} \times \text{atm/mole}$$
$$V = 22.4 \frac{\text{liters} \times \text{atm/mole}}{1 \text{ atm}}$$
$$= 22.4 \text{ liters/mole}$$

This is the same volume as that just calculated for a mole of nitrogen at 0°C and one atmosphere pressure (in Section 4-1.1). Furthermore, it is the same volume occupied by 17.0 grams of ammonia at 0°C and one atmosphere pressure (See Table 2-III, p. 19). Since the molecular formula of ammonia is NH_3, its molecular weight is $(14.0 + 3 \times 1.0) = 17.0$ grams. Thus one mole of ammonia, 17.0 grams, also occupies 22.4 liters at 0°C and one atmosphere pressure. Experiments on many other gases are in agreement and lead to the generalization:

A mole of gas occupies 22.4 liters at 0°C and one atmosphere pressure. (*1*)

What happens to a gas as the temperature is changed? An experiment provides the answer. Table 4-I shows some pressure-volume measure-

Fig. 4-1. **A mole of gas** *occupies 22.4 liters at 0°C, 1 atmosphere.* **A mole of gas** *occupies 24.5 liters at 25°C, 1 atmosphere.*

ments for ammonia gas at 25°C (approximately room temperature). Although the data shown contain some experimental uncertainty, we find again the regularity, *PV = a constant.*

Table 4-I

PRESSURE AND VOLUME OF 17.0 GRAMS OF AMMONIA GAS, NH_3 t = 25°C

PRESSURE (atmospheres)	VOLUME (liters)		$P \times V$
0.200	123		24.6
0.400	60.0		24.0
0.600	43.0		25.8
0.800	29.3		23.4
1.00	25.7		25.7
1.50	15.9		23.9
2.00	12.1		24.2
		Average	24.5 ± 0.7

This time, however, the pressure-volume product for a mole of ammonia is 24.5 ± 0.7. We can compare this with our earlier result:

For one mole of ammonia at 0°C,

$$P \times V = 22.4 \text{ liter-atm} \quad (2)$$

For one mole of ammonia at 25°C,

$$P \times V = 24.5 \text{ liter-atm} \quad (3)$$

From (*3*) we see that the molar volume of ammonia at 25°C and one atmosphere pressure is 24.5 liters, whereas it is 22.4 liters at 0°C. The molar volume of ammonia depends upon the temperature. This result is no surprise—a sample of gas expands when heated at constant pressure. So when we compare the molar volumes of different gases, they should be at the same temperature (and, by the same sort of argument, at the same pressure).

Consider the following experiment. The air is removed from a one-liter flask and it is weighed

Table 4-II. **THE VOLUME OF A MOLE OF GAS AT 25°C AND ONE ATMOSPHERE PRESSURE**

GAS	WT. OF FLASK EMPTY W_1 (g)	WT. OF FLASK + GAS W_2 (g)	WT. OF 1 LITER OF GAS $W_2 - W_1$ (g/liter)	MOLECULAR WEIGHT MW (g/mole)	VOLUME (liter/mole) $MW/(W_2 - W_1)$
oxygen, O_2	157.35	158.66	1.31	32.0	24.5
nitrogen, N_2	157.35	158.50	1.15	28.0	24.3
carbon monoxide, CO	157.35	158.50	1.15	28.0	24.4
carbon dioxide, CO_2	157.35	159.16	1.81	44.0	24.3

empty. Then the flask is weighed again filled with a gas at one atmosphere pressure and at 25°C. The difference in weight is the weight of one liter of the gas. From this, we can calculate the volume of a mole of that gas. Table 4-II shows the results. We find that all the gases have about the same molar volume at 25°C and one atmosphere. Whether the gas is O_2, N_2, CO, or CO_2, the same volume, 24.5 ± 0.2 liters, contains 6.02×10^{23} molecules (at 25°C, one atmosphere). Whether the gas is N_2 or CO, a volume of 22.4 ± 0.1 liter contains 6.02×10^{23} molecules at 0°C and one atmosphere pressure.

4-1.3 Avogadro's Hypothesis

When different gases are compared at the same temperature and pressure, they have the same volume per mole. This is true at 0°C and one atmosphere but, more important, it is true at other temperatures and pressures as well.

When the two gases ammonia, NH_3, and hydrogen chloride, HCl, react, one liter of ammonia reacts with one liter of hydrogen chloride if the two volumes are measured at the same temperature and pressure. This simple one-to-one volume ratio is observed at 0°C and one atmosphere pressure but, more important, this simple ratio is observed at other temperatures and pressures as well.

These results and many similar ones led Avogadro to propose his famous hypothesis, as discussed in Section 2-2.3. The hypothesis states that *equal volumes of gases contain equal numbers of molecules* (*at the same temperature and pressure*). Therefore, the molecular weight of a gas can be determined by comparing the weight of a known volume of the gas with the weight of the same volume of another gas of known molecular weight. It does not matter what t and P are as long as they are the same for the two gases.

"Avogadro's Hypothesis" is often called "Avogadro's Law" because it has such wide applicability. It is one of the important generalizations of chemistry. It is important, not because it is exact but because it applies to all gases, regardless of whether their molecules are large or small. The molecules of different gases actually have different sizes and slightly different attractions for one another. As a result, different gases do not have exactly the same number of molecules in a given volume. Such variations are small (usually less than 1%) and do not impair the usefulness of Avogadro's Hypothesis as a method for determining the molecular weight of a gas.

It is an interesting commentary on the progress of science that this important regularity, now often called a "Law," was not generally accepted for half a century after it was proposed. Though Avogadro published his idea in 1811, its validity was not widely recognized until the proposal was reintroduced at an international conference of chemists at Karlsruhe, Germany, in 1858. Today, we find it easy to "discover" or "confirm" Avogadro's Hypothesis because we can draw upon a wealth of accumulated quantitative weight and volume relations to develop a tightly knit pattern of self-consistency. In contrast, even atomic weights were in doubt early in the nineteenth century and quantitative methods were relatively crude.

4-2 THE KINETIC THEORY

Avogadro's Hypothesis provides a method for identifying the molecules present in a gas. Also, it explains why the volumes of gases that react with each other are in the same simple ratio as are the moles in the balanced equation. The importance of these results makes the explanation of the properties of gases important to a chemist.

We have already observed that there are many and close similarities between a gas and a collection of particles in endless motion. It is essential in the particle model that each particle possesses energy of motion, called **kinetic energy.** Hence, the mathematical expression that describes this model is called the **kinetic theory of gases.** According to this theory, the molecules of a gas are in rapid motion. They travel in straight lines until they meet other molecules of the gas or the atoms in the walls of the container. They are then deflected and scattered. The net result is a helter-skelter movement of molecules traveling in all directions and at different speeds.

At room temperature, the average speed of a nitrogen molecule is found to be about one-quarter mile per second. In one second, however, the nitrogen molecule has collided with many other molecules, so its motion follows a zig-zag path. Although the average distance between molecules is small, the molecule passes by many other molecules without hitting them, so the distance it travels between collisions is about fifteen times the average distance between the molecules (at room pressure and temperature).

4-2.1 Gas Pressure

Pressure is an important quantity in a discussion of gas behavior. The applicability of the kinetic theory to an understanding of gas pressure is, then, an important success (see Section 2-1.1). We shall investigate this success in more detail, but first we should investigate how pressure is measured.

MEASURING THE PRESSURE OF A GAS

A gas exerts pressure equally on all the walls of its container. The standard method of measuring this pressure is by measuring the height of a mercury column supported by the gas. An instrument for measuring the pressure of the air is illustrated in Figure 4-2A; it is called a barometer. We can make a barometer by filling a long tube (closed at one end) with mercury and inverting it in a dish of mercury. Mercury will flow from the tube into the dish until the column of mercury exerts a downward pressure which is exactly balanced by the pressure of the air. In the illustration, the pressure of the air is expressed as "755 millimeters of mercury" (written mm Hg or mm). This is the height of the mercury column. (Notice that only mercury vapor is present to exert pressure in the space at the top of the column. At room temperature, this vapor pressure is negligible, about 10^{-3} mm.)

The pressure of a gas sample can be measured in a device similar to a barometer, called a manometer. Figures 4-2B and 4-2C show two types. Figure 4-2B shows a closed-end manometer. Here the downward pressure exerted by the column of mercury is balanced by the pressure of the gas sample placed in the flask. The gas pressure is, in the example shown, 105 mm. As in the barometer, only mercury vapor is present in the right-hand tube.

The apparatus shown in Figure 4-2C differs in that the right-hand tube is open. In this type of manometer, atmospheric pressure is exerted on the right-hand mercury column. Hence the pressure in the flask plus the height of the mercury column equals atmospheric pressure. In the example shown, the pressure is $755 - 650 = 105$ mm, the same as pictured in the closed-end manometer, Figure 4-2B.

STP

Two conditions that are often important in chemical experiments are temperature and pressure. Consequently, chemists usually control and measure these conditions during experiments. In addition, it is useful to refer many experimental results to a standard and generally accepted set of temperature and pressure conditions. This facilitates comparison of results of different types and from different laboratories.

The temperature 0°C is readily obtained and maintained with an ice water bath. The temperature is one at which thermometers are calibrated; this aids measurement. A temperature that is easy to maintain and easy to measure makes a good standard temperature.

Air pressure varies somewhat from day to day and from place to place. Nevertheless, air pressure is always reasonably near 760 mm Hg, so atmospheric pressure furnishes a convenient, though approximate, reference pressure. However, it is not sufficiently constant for many purposes. So, by international agreement, *a*

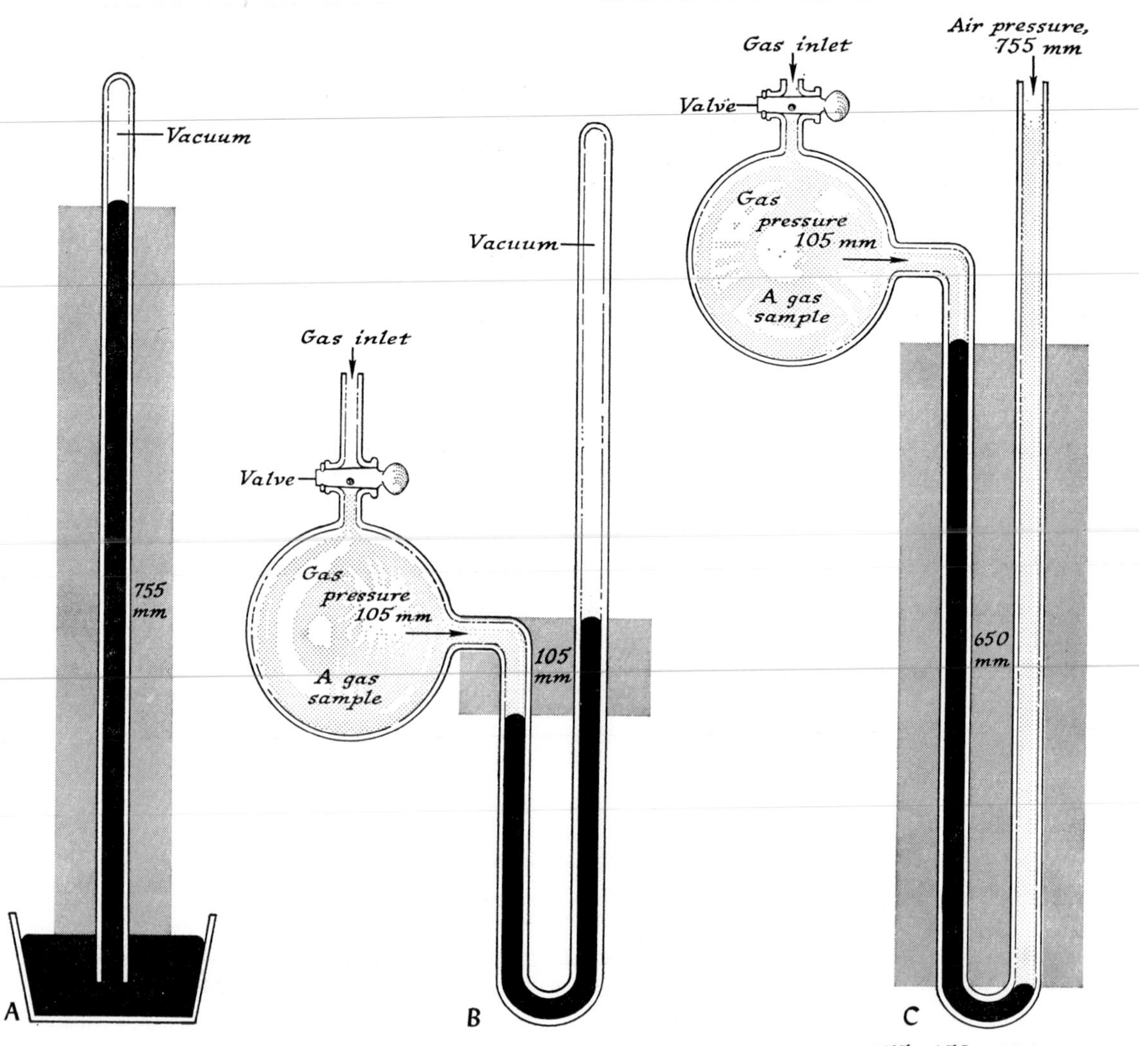

Fig. 4-2. **Pressure measurement. A.** *Barometer: pressure = 755 mm.* **B.** *Closed-end manometer: pressure = 105 mm.* **C.** *Open-end manometer: pressure = 755 − 650 = 105 mm.*

standard pressure for gases is represented by a height of 760 mm of mercury. This standard pressure is often expressed merely as **one atmosphere** (1 atm).

Thus chemists have accepted 0°C and one atmosphere as convenient standard conditions. *These conditions,* ***0°C and 760 mm pressure,*** *are called* ***standard temperature and pressure*** *and are abbreviated* ***STP.***

The standard pressure is defined in terms of a pressure reading on a standard barometer. A standard barometer takes account of the fact that the gravitational attraction of the earth on the mercury varies slightly from place to place, and the fact that mercury expands and becomes less dense when it is heated. Thus, the mercury column of a barometer is several millimeters longer at 20°C than at 0°C. In the standard barometer, the mercury is at 0°C. You can find in published tables how much to subtract from (or add to) your barometer reading to obtain the same pressure value a standard barometer would give. The correction is seldom more than 1 or 2 mm, and is often negligible compared to other possible errors. Unless your other experimental measurements are rather precise, you need not convert your readings in this way.

THE CAUSE OF GAS PRESSURE

In Chapter 1 we explained how gases exert pressure in terms of collisions of particles with the container walls: this model of gas pressure is part of the kinetic theory. Each time a gas molecule strikes a wall, or a mercury surface, it exerts a small push or force, just as a ball thrown at a wall exerts a force upon it. The force per unit area, called the **pressure,** depends directly upon the number of molecules that strike the unit area of the surface. Twice as many molecules in a given volume result in twice as many collisions per unit area, hence twice the original pressure. Thus we explain why the pressure goes up as we pump air into an automobile tire. If the volume and temperature of the tire remain unchanged, the pressure goes up in direct proportion to the number of moles of air pumped in.

EXERCISE 4-3

A container of fixed volume contains two moles of gas at room temperature. The pressure in the container is four atmospheres. Three more moles of gas are added to the container at the same temperature. Use the result just stated to show that the pressure is now 10 atmospheres.

4-2.2 Partial Pressure

Figure 4-3 shows three one-liter bulbs at 25°C. The first bulb contains 0.0050 mole of air. The manometer shows that the pressure is 93 mm Hg (93 mm). The second bulb contains 0.0011 mole of water vapor. The pressure in this bulb is 20 mm Hg. The third bulb contains 0.0050 mole of air and also 0.0011 mole of water vapor. The third manometer shows that the pressure in the last bulb is 113 mm Hg.

This experiment shows that the pressure exerted by the mixture of gases is just the sum of the pressure the air exerts when alone in the flask and the pressure the water vapor exerts when alone in the flask:

$$113 \text{ mm} = 93 \text{ mm} + 20 \text{ mm} \qquad (4)$$

The total pressure can be regarded as a sum of the parts furnished by the individual pressures exerted by each of the components of the gas mixtures. *The pressure exerted by each of the gases in a gas mixture is called the* ***partial pressure*** *of that gas.* The partial pressure is the pressure that the gas would exert if it were alone in the container. In the example of Figure 4-3, the total pressure in the third bulb is 113 mm. The partial pressure of water vapor in this bulb is 20 mm and the partial pressure of air is 93 mm.

EXERCISE 4-4

Assume that 0.0050 mole of air contains 0.0040 mole of nitrogen, N_2, and 0.0010 mole of oxygen, O_2. What is the partial pressure of oxygen in the first bulb in Figure 4-3? What is the partial pressure of oxygen in the third bulb? Use three significant figures.

Fig. 4-3. **Pressure of a mixture of gases.**

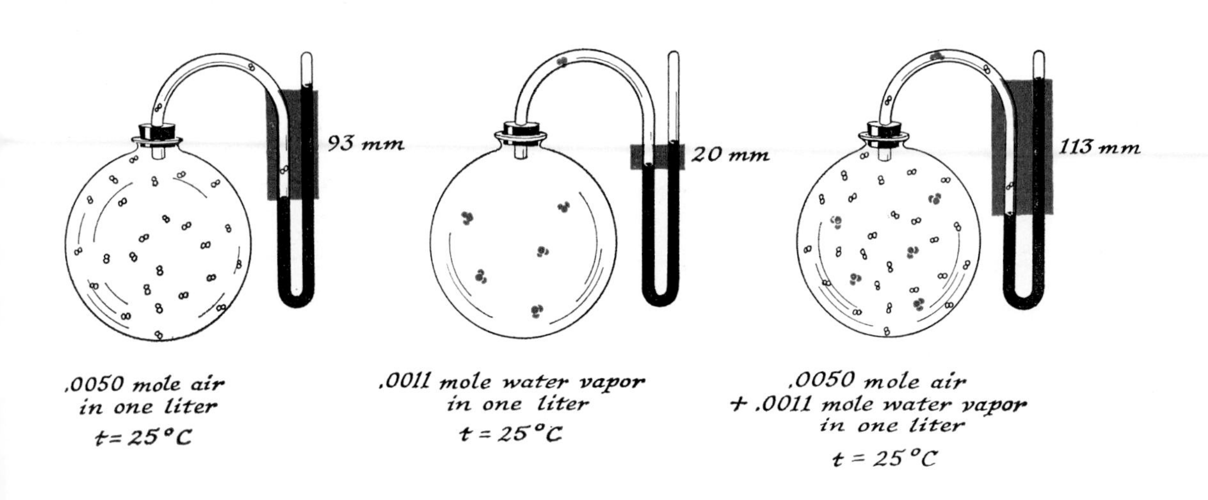

The pressure behavior shown in Figure 4-3 is readily explained in terms of the kinetic theory of gases. There is so much space between the molecules that each behaves independently, contributing its share to the total pressure through its occasional collisions with the container walls. The water molecules in the third bulb are seldom close to each other or to molecules provided by the air. Consequently, they contribute to the pressure exactly the same amount they do in the second bulb—the pressure they would exert if the air were not present. The 0.0011 mole of water vapor contributes 20 mm of pressure whether the air is there or not. The 0.0050 mole of air contributes 93 mm of pressure whether the water vapor is there or not. Together, the two partial pressures, 20 mm and 93 mm, determine the measured total pressure.

4-2.3 Temperature and Kinetic Energy

If the kinetic theory is applicable to gases, we should expect pressure to be affected by other factors than the number of moles per unit volume. For example, the mass of the molecules and their velocities should be important, as well. After all, a baseball exerts more "push" on a catcher's mitt than would a ping-pong ball thrown with the same velocity. Also, a baseball exerts more "push" on the mitt if a "fast ball" is thrown rather than a "slow ball." To see how the mass of the molecules and their velocities are dealt with in the kinetic theory, we must consider temperature.

To measure the temperature of a gas we immerse some kind of thermometer in it. If the thermometer is colder than the system, heat flows into the thermometer until the gas and the thermometer are at the same temperature. Then we read the thermometer to get a numerical value for the temperature. If the thermometer were hotter than the gas, heat would flow *from* the thermometer. When there is no net flow of heat, the thermometer is said to be in **thermal equilibrium** with the gas.

There are many kinds of thermometers. Any substance can be fashioned into a thermometer if it has a readily measured property that is sensitive to a change in temperature. The familiar mercury thermometer depends upon the expansion of the liquid as temperature is raised. Solids and gases also change volume with temperature change. Hence either can be (and both are) used as a basis for a thermometer. A gas held at constant volume also responds to a change in temperature, the pressure rising with rising temperature. This is the more common way in which a gas is used in a thermometer: the volume is fixed and the pressure varies with temperature.

So let us measure the temperature of a sample of gas A by placing it in thermal contact with a sample of gas B (our thermometer). There will be heat flow between the two gas samples if they are initially at different temperatures. Energy is transferred from the hotter gas to the cooler gas. When heat flow ceases, the gases have reached thermal equilibrium. Then the gases have the same temperature.

We can visualize what is going on with the aid of the kinetic theory of gases. Suppose sample A is initially at a high temperature relative to the thermometer gas B. We interpret this to mean that the molecules in gas A have more energy of motion than those of gas B—the molecules of gas A have higher kinetic energies (on the average). When the samples are brought into thermal contact, collisions permit the rapidly moving A molecules to transfer kinetic energy through the thermal connection to the slowly moving B molecules. This transfer of kinetic energy from gas A to gas B is the process that raises the temperature of gas B and lowers the temperature of gas A. When the thermal contact between molecules of A and B no longer results in a net transfer of kinetic energy from one gas to the other, then gases A and B are in thermal equilibrium: they have the same temperatures.

Thus, we picture heat flow between two gas samples as a transfer of kinetic energy. The process continues until the molecules of both gases have the *same* average kinetic energy. Then the gases are at the same temperature. This is a basic premise of the kinetic theory: *When gases are at the same temperature, the molecules of the gases have the same kinetic energy* (on the average).

4-2.4 Absolute Temperature

The quantitative effects of temperature on gases were first studied by Jacques Charles, a French scientist, in 1787. He found that all gases expand by the same fraction of their original volumes when they are heated over the same temperature range. (In these experiments the pressure remained the same.) A simple experiment shows the relations. Into a small-bore glass tube, one-half meter in length and closed at one end, we place a drop of mercury. This mercury falls and finally traps a sample of air in the bottom of the tube. (See Figure 4-4.) Since the tube has a uniform bore, we can use the length of the air sample as a measure of its volume. The mercury plug moves up or down and maintains a constant pressure.

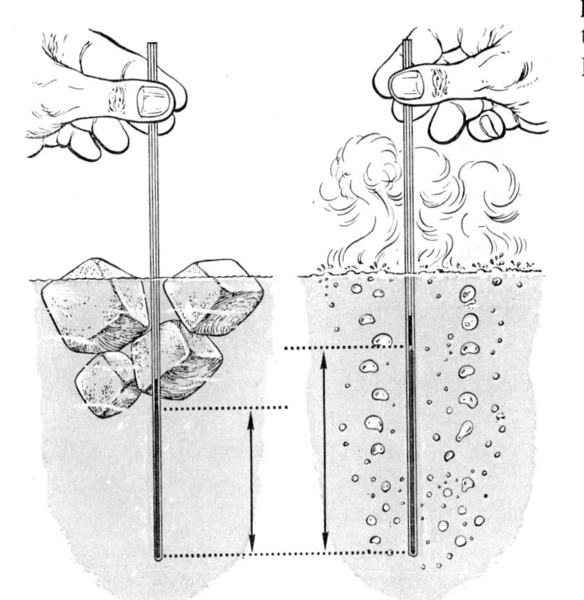

Fig. 4-4. **Apparatus for demonstrating the effect of temperature** *on the volume of a gas.*

We may place the tube in ice water (0°C) and measure the relative volume of the air sample. If the tube is immersed in water boiling at one atmosphere pressure (100°C), the relative volume has a higher value. From these data and from similar measurements at other temperatures, we collect data such as those in Table 4-III.

Table 4-III.

CHANGE OF VOLUME OF A GAS WITH CHANGE IN TEMPERATURE

TEMPERATURE (°C)	RELATIVE VOLUME (as measured by length of sample)
200	1.73
100	1.37
50	1.18
0	1.00

When we plot these results with relative volumes on the ordinate (vertical axis) and temperatures on the abscissa (horizontal axis), we obtain the graph shown in Figure 4-5. The straight line passes through the experimental points. When extrapolated upward, it shows that the volume at 273°C is double that at 0°C. Extrapolated downward, the line shows that the volume would become zero at −273°C. The volume change per degree centigrade is $\frac{1}{273}$ of the volume at 0°C. Actually, all gases liquefy before their temperature reaches −273°C.

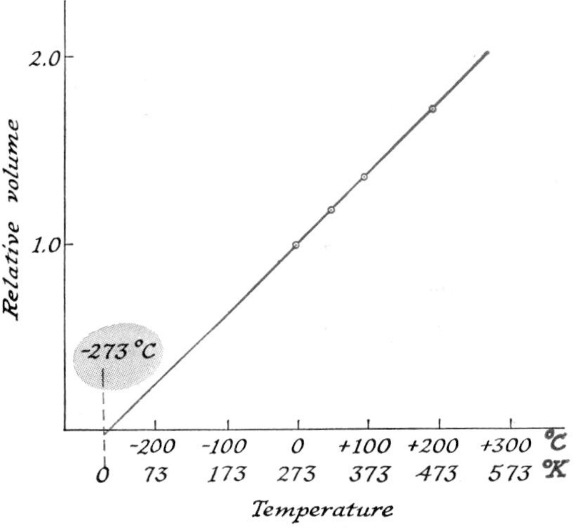

Fig. 4-5. **An absolute temperature scale** *from the change of volume of a gas with temperature.*

If gases are heated or cooled at constant volume, the pressure changes, also at the rate of $\frac{1}{273}$ of its value at 0°C. Then the pressure of a gas would become zero at −273°C. In terms of the kinetic theory, the motion of the molecules ceases at this temperature. The kinetic energy has become zero.

There are great advantages to an *absolute* temperature scale that has its zero point at −273°C. Whereas the "zero" of temperature in the Centigrade scale is based upon an arbitrary temperature, selected because it is easily measured, the zero point of the absolute scale has inherent significance in the kinetic theory. If we express temperatures on an absolute temperature scale, we find that *the volume of a fixed amount of gas* (at constant pressure) *varies directly with temperature*.* Also, *the pressure of a fixed amount of gas* (at constant volume) *varies directly with temperature*. And, according to the kinetic theory, the kinetic energy of the molecules varies directly with the absolute temperature. For these reasons, in dealing with gas relations, we shall usually express temperature on an absolute temperature scale.

This temperature scale, with the same size degrees as the Centigrade scale, is called the **Kelvin** scale and values on this scale are expressed in degrees Kelvin (°K). Both Kelvin and Centigrade temperatures are shown in Figure 4-5. Notice that all numerical values on the Kelvin scale are 273 degrees higher than the corresponding temperatures on the Centigrade scale.

EXERCISE 4-5

(a) Express the following temperatures in degrees Kelvin:

Boiling point of water:	100°C
Freezing point of mercury:	−38.9°C
Boiling point of liquid nitrogen:	−196°C

(b) Express the following temperatures in degrees Centigrade:

Melting point of lead:	600°K
A normal room temperature:	298°K
Boiling point of liquid helium:	4°K

* This direct relation between volume and temperature (at constant pressure) is called **Charles' Law**.

EXERCISE 4-6

In Experiment 9 a student obtained the result that 2.00×10^{-3} mole of magnesium produced a volume of hydrogen that would occupy 49.0 ml at 25°C and one atmosphere pressure.

(a) If one mole of magnesium produces one mole of hydrogen, use these data to calculate the volume of one mole of hydrogen at 25°C (298°K) and one atmosphere.
(b) Calculate the volume one mole of hydrogen would occupy at 0°C (273°K) and one atmosphere.

We have remarked that a temperature of zero on the absolute temperature scale would correspond to the absence of all motion. The kinetic energy would become zero. Very interesting phenomena occur at temperatures near 0°K (the superconductivity of many metals and the superfluidity of liquid helium are two examples). Hence, scientists are extremely interested in methods of reaching temperatures as close to absolute zero as possible. Two low temperature coolants commonly used are liquid hydrogen (which boils at 20°K) and liquid helium (which boils at 4°K). Helium, under reduced pressure, boils at even lower temperatures and provides a means of reaching temperatures near 1°K. More exotic techniques have been developed to produce still lower temperatures (as low as 0.001°K) but even thermometry becomes a severe problem at such temperatures.

4-2.5 Avogadro's Hypothesis and the Kinetic Theory

The kinetic theory is based upon the premise that if two gases are at the same temperature, the molecules of the gases have the same average kinetic energy. The ability of this kinetic theory to explain Avogadro's Hypothesis is one of its most important successes.

We may state Avogadro's Hypothesis in this form: If two gases at the same temperature have the same number of particles in a given volume, they must exert the same pressure. Yet, as remarked in Section 4-2.3, the mass of a molecule, as well as its velocity, should influence the pressure exerted. If the molecules of our two gas samples have different masses, they must have different speeds in order to have the same kinetic energies. The lighter molecules must travel faster, so they will strike the container walls more times per second. The effect of the more frequent collisions exactly counteracts the lower "push" per collision from these lower-mass molecules. The result is in perfect accord with Avogadro's Hypothesis: Two gases at the same concentration and at the same temperature exert the same pressure even though their molecules have different masses.

Avogadro's Hypothesis can be shown quite readily in an approximate way. The kinetic energy of a moving particle is expressed by the equation

$$KE = \tfrac{1}{2}mv^2 \tag{5}$$

where m is the mass of the particle and v is the velocity. Therefore, for gas A and gas B at the same temperature, we have

$$(KE)_A = (KE)_B \tag{6}$$

$$\tfrac{1}{2}m_Av_A^2 = \tfrac{1}{2}m_Bv_B^2$$

or

$$m_Av_A^2 = m_Bv_B^2 \tag{7}$$

Now suppose we place n molecules in a cubical box of dimension d. The pressure is fixed by the number of wall collisions per second on each square centimeter times the momentum transferred per collision:

$$\text{Pressure} = \left(\frac{\text{collisions}}{\text{second}}\right)\left(\frac{1}{\text{area}}\right)\left(\frac{\text{momentum}}{\text{collision}}\right) \tag{8}$$

Momentum depends upon mass and velocity. The particle approaches the wall with momentum mv and leaves with this same momentum in the opposite direction. The momentum transferred to the wall is, then

$$\text{Momentum} = 2mv \tag{9}$$

The collisions per second with the wall, on the other hand, depend upon the container dimension and the velocity (as the molecule bounces back and forth between the walls). We can assume that one-third of the molecules bounce back and forth in a given direction between two opposite walls. Then if there are n molecules in the container, there are $n/3$ hitting these two walls. One of these walls receives a collision each time one of the molecules travels the box dimension d and back, a distance of $2d$.

$$\frac{\text{collisions}}{\text{second}} = \left(\frac{\text{no. particles bouncing back and forth}}{\text{time for a particle to travel distance } 2d}\right)$$

$$\frac{\text{collisions}}{\text{second}} = \left(\frac{n/3}{2d/v}\right) = \left(\frac{n}{3}\right)\left(\frac{v}{2d}\right) = \frac{nv}{6d} \tag{10}$$

Combining (*8*), (*9*), and (*10*), we find

$$\text{Pressure} = \left(\frac{\text{collisions}}{\text{second}}\right)\left(\frac{1}{\text{area}}\right)\left(\frac{\text{momentum}}{\text{collision}}\right)$$

$$= \frac{1}{6}\frac{nv}{d} \times \frac{1}{d^2} \times 2mv$$

$$= \frac{1}{3}\left(\frac{n}{d^3}\right)(mv^2) \tag{11}$$

Applying equation (*11*) to each of the gases A and B,

$$P_A = \frac{1}{3}\left(\frac{n_A}{d^3}\right)(m_Av_A^2) \tag{12}$$

$$P_B = \frac{1}{3}\left(\frac{n_B}{d^3}\right)(m_Bv_B^2) \tag{13}$$

If the gases have the same pressure, $P_A = P_B$, we can equate (*12*) and (*13*) so that

$$\frac{1}{3}\left(\frac{n_A}{d^3}\right)(m_Av_A^2) = \frac{1}{3}\left(\frac{n_B}{d^3}\right)(m_Bv_B^2) \tag{14}$$

If the temperatures of the gases are the same, equation (*7*) is applicable and equation (*14*) becomes

$$\frac{n_A}{d^3} = \frac{n_B}{d^3} \tag{15}$$

Thus we see that at the same temperature and pressure, the two gases have the same number of molecules per unit volume. This is Avogadro's Hypothesis.

4-2.6 The Perfect Gas

We have examined experimental pressure-volume data for oxygen gas (Table 1-II), ammonia gas (Table 4-I), and hydrogen chloride gas (Table 2-II). In each case, within the experimental uncertainty of the data shown, the gases have the regular behavior, $PV = a\ constant$. We find from many such experiments that many gases follow this simple behavior. Of course such a generalization is subject to uncertainty, as is any other scientific statement. The generalization was derived from a set of measurements, each of which involves some uncertainty, and, hence, the constancy of the PV product is established only within corresponding bounds of uncertainty. What's more, there are limits to the pressure range over which the behavior has been tested.

To be specific, consider the data for 17.0 grams of ammonia gas at 0°C, as presented in Table 4-I (p. 51). These data show that $PV = 24.5$, but a complete statement should include both the uncertainty and the range over which the data are known to apply. In this case the uncertainty is ± 0.7 and the range is 0.2–2 atmospheres pressure. It would not be safe, from these data alone, to assume that the pressure-volume product is constant to four significant figures, $PV = 24.50$. Neither would it be safe to assume that the pressure-volume product is constant outside the range of pressure studied, 0.2–2.0 atmospheres. Remember, *a generalization is reliable within the bounds defined by the experiments that led to the rule.* If we need four significant figures, or wish to know the behavior at a higher pressure, more experiments are needed.

More accurate pressure-volume measurements extending to much higher pressures have been performed. Table 4-IV shows the results of such experiments.

Table 4-IV

ACCURATE PRESSURE-VOLUME MEASUREMENTS FOR 17.00 GRAMS OF AMMONIA GAS AT 25°C

PRESSURE (atmospheres)	VOLUME (liters)	$P \times V$	
0.1000	244.5	24.45	
0.2000	122.2	24.44	
0.4000	61.02	24.41	
0.8000	30.44	24.35	
2.000	12.17	24.34	
4.000	5.975	23.90	
8.000	2.925	23.40	
9.800	2.360	23.10	condensation beginning
9.800	0.020	0.20	no gas left; liquid only
20.00	0.020	0.40	liquid only present
50.00	0.020	1.0	liquid only present

The most startling fact revealed in Table 4-IV is the drastic deviation from $PV = 24.5$ that occurs when the pressure is raised above 9.800 atmospheres. Suddenly the relation $PV = a\ constant$ is no longer applicable. Here is dramatic evidence of the danger lurking in careless extrapolation beyond the range of experience.

Even below the condensation pressure the pressure-volume product was not perfectly constant. With measurements of sufficient accuracy and precision, we can see that the PV product of ammonia at 25°C is not really constant after all. It varies systematically from 24.45 at 0.1000 atmospheres to 23.10 at 9.800 atmospheres, just before condensation begins. Similar measurements on 28.0 grams of carbon monoxide at 0°C show that the PV product is 22.410 at 0.2500 atmospheres pressure, but if the pressure is raised to 4.000 atmospheres, the PV product becomes 22.308. This type of deviation is common. Careful measurements reveal the fact that *no* gas follows perfectly the generalization $PV = a\ constant$ at all pressures. On the other hand, *every* gas follows this rule approximately, and the fit becomes better and better as the pressure is lowered. So we find that every gas *approaches* the behavior $PV = a\ constant$ as pressure is lowered.

There is a reasonable explanation for this type of deviation. The kinetic theory, which "explains" the pressure-volume behavior, is based upon the assumption that the particles exert no force on each other. But real molecules *do* exert force on each other! The condensation of every gas on cooling shows that there are always attractive forces. These forces are not very important when the molecules are far apart (that is, at low pressures) but they become noticeable at higher pressures. With this explanation, we see that the kinetic theory is based on an "idealized" gas—one for which the molecules exert no force on each other whatsoever. Every gas approaches such ideal behavior if the pressure is low enough. Then the molecules are, on the average, so far apart that their attractive forces are negligible. A gas that behaves as though the molecules exert no force on each other is called an ***ideal* gas** or a ***perfect* gas.**

Table 4-V. **MOLAR VOLUMES OF SOME GASES**

GAS	FORMULA	MOLAR WEIGHT (grams)	MOLAR VOLUME AT 0°C AND 1 ATM (liters)
hydrogen	H_2	2.0160	22.430
helium	He	4.003	22.426
("perfect" gas)	—	—	(22.414)
nitrogen	N_2	28.016	22.402
carbon monoxide	CO	28.011	22.402
oxygen	O_2	32.000	22.393
methane	CH_4	16.043	22.360
carbon dioxide	CO_2	44.011	22.262
hydrogen chloride	HCl	36.465	22.248
ammonia	NH_3	17.032	22.094
chlorine	Cl_2	70.914	22.063
sulfur dioxide	SO_2	64.066	21.888

Avogadro's Hypothesis is consistent with the kinetic theory. Therefore a perfect gas follows Avogadro's Hypothesis. At one atmosphere pressure and 0°C, one mole (6.02×10^{23} molecules) of a perfect gas occupies 22.414 liters. How closely real gases approximate a perfect gas at one atmosphere pressure and 0°C is shown by measuring the *volume occupied by one mole of that gas, the* **molar volume.** Table 4-V shows the molar volumes of a number of gases. We see that real gases do approximate closely (to three significant figures) the perfect gas behavior at one atmosphere and 0°C. Every gas becomes a perfect gas as the pressure is reduced toward zero.

4-3 REVIEW

Regularities observed in the behavior of gases have contributed much to our understanding of the structure of matter. One of the most important regularities is Avogadro's Hypothesis: Equal volumes of gases contain equal numbers of particles (at the same pressure and temperature). This relationship is valuable in the determination of molecular formulas—these formulas must be known before we can understand chemical bonding.

We have explored the meaning of temperature. According to the kinetic theory, when two gases are at the same temperature, the molecules of the two gases have the same average kinetic energies. Changing the temperature of a sample of gas at constant pressure reveals that the volume is directly proportional to the temperature if the temperature is expressed in terms of a new, absolute scale. The melting point of ice (0°C) on this new scale (called the Kelvin scale) is 273°K. The boiling point of water at one atmosphere (100°C) is 373°K. The zero temperature on the Kelvin scale corresponds to the hypothetical loss of all molecular motion.

This progress gives us substantial basis for confidence in the usefulness of the atomic theory and it encourages us to develop the model further. We shall see that the concepts we have developed in our consideration of gases are also useful when we consider the behavior of condensed phases—liquids and solids.

QUESTIONS AND PROBLEMS

1. How many molecules are there in a molar volume of a gas at 100°C? At 0°C?

2. What is the molar volume of water under each of the following conditions?

   (a) Solid, 0°C;
   density of ice = 0.915 g/ml.
   (b) Liquid, 0°C;
   density of water (liquid, 0°C) = 1.000 g/ml.
   (c) Gas, 100°C;
   density of water vapor (100°C, 1 atm) = 5.88×10^{-4} g/ml.

3. What is the molecular weight of a gas if at 0°C and one atmosphere pressure, 1.00 liter of the gas weighs 2.00 grams?

   Answer. 44.8 g/mole

4. The gas sulfur dioxide combines with oxygen to form the gas sulfur trioxide:

$$2SO_2(gas) + O_2(gas) \longrightarrow 2SO_3(gas)$$

   What ratio would you expect for the following?

   (a) $\dfrac{\text{number of } SO_3 \text{ molecules produced}}{\text{number of } O_2 \text{ molecules consumed}}$.

   (b) $\dfrac{\text{volume of } SO_3 \text{ gas produced}}{\text{volume of } O_2 \text{ gas consumed}}$.

5. A glass bulb weighs 108.11 grams after all of the gas has been removed from it. When filled with oxygen gas at atmospheric pressure and room temperature, the bulb weighs 109.56 grams. When filled at atmospheric pressure and room temperature with a gas sample obtained from the mouth of a volcano, the bulb weighs 111.01 grams. Which of the following molecular formulas for the volcano gas could account for the data?

   CO_2
   OCS
   Si_2H_6
   SO_2
   NF_3
   SO_3
   S_8
   A gas mixture, half CO_2, half Kr

6. Compressed oxygen gas is sold at a pressure of 130 atm in steel cylinders of 40 liters volume.

   (a) How many moles of oxygen does such a filled cylinder contain?
   (b) How many kilograms of oxygen are in the cylinder?

   Answer. 6.7 kg.

7. A carbon dioxide fire extinguisher of 3 liters volume contains about 10 pounds (4.4 kg) of CO_2. What volume of gas could this extinguisher deliver at room conditions?

8. Hydrogen for weather balloons is often supplied by the reaction between solid calcium hydride, CaH_2, and water to form solid calcium hydroxide, $Ca(OH)_2$, and hydrogen gas, H_2.

   (a) Balance the equation for the reaction and decide how many moles of CaH_2 would be required to fill a weather balloon with 250 liters of hydrogen gas at normal conditions.
   (b) What weight of water would be consumed in forming the hydrogen?

   Answer. 0.18 kg.

9. Gas is slowly added to the empty chamber of a closed-end manometer (see Figure 4-2B). Draw a picture of the manometer mercury levels, showing in millimeters the difference in heights of the two mercury levels:

   (a) before any gas has been added to the empty gas chamber;
   (b) when the gas pressure in the chamber is 300 mm;
   (c) when the gas pressure in the chamber is 760 mm;
   (d) when the gas pressure in the chamber is 865 mm.

10. Repeat Problem 9 but with an open-end manometer (see Figure 4-2C). Atmospheric pressure is 760 mm.

11. The balloons that are used for weather study are quite large. When they are released at the surface of the earth they contain a relatively small volume of gas compared to the volume they acquire when aloft. Explain.

12. A 1.50 liter sample of dry air in a cylinder exerts a pressure of 3.00 atm at a temperature of 25°C. Without change in temperature, a piston is moved in the cylinder until the pressure in the cylinder is reduced to 1.00 atm. What is the volume of the gas in the cylinder now?

13. Suppose the total pressure in an automobile tire is 30 pounds/inch^2 and we want to increase the pressure to 40 pounds/inch^2. What change in the amount of air in the tire must take place? Assume that the temperature and volume of the tire remain constant.

14. The density of liquid carbon dioxide at room temperature is 0.80 grams/ml. How large a cartridge of liquid CO_2 must be provided to inflate a life jacket of 4.0 liters capacity at *STP?*

15. A student collects a volume of hydrogen over water. He determines that there is 2.00×10^{-3} mole of hydrogen and 6.0×10^{-5} mole of water vapor present. If the total pressure inside the collecting tube is 760 mm, what is the partial pressure of each gas?

    Answer. Partial pressure H_2 = 738 mm.
    Partial pressure H_2O = 22 mm.

16. A sample of nitrogen is collected over water at 18.5°C. The vapor pressure of water at 18.5°C is 16 mm. When the pressure on the sample has been equalized against atmospheric pressure, 756 mm, what is the partial pressure of nitrogen? What will be the partial pressure of nitrogen if the volume is reduced by a factor 740/760?

17. A candle is burned under a beaker until it extinguishes itself. A sample of the gaseous mixture in the beaker contains 6.08×10^{20} molecules of nitrogen, 0.76×10^{20} molecules of oxygen, and 0.50×10^{20} molecules of carbon dioxide. The total pressure is 764 mm. What is the partial pressure of each gas?

18. A cylinder contains nitrogen gas and a small amount of liquid water at a temperature of 25°C (the vapor pressure of water at 25°C is 23.8 mm). The total pressure is 600.0 mm Hg. A piston is pushed into the cylinder until the volume is halved. What is the final total pressure?

    Answer. 1176 mm.

19. Consider two closed glass containers of the same volume. One is filled with hydrogen gas, the other with carbon dioxide gas, both at room temperature and pressure.

    (a) How do the number of moles of the two gases compare?

(b) How do the number of molecules of the two gases compare?
(c) How do the number of grams of the two gases compare?
(d) If the temperature of the hydrogen container is now raised, how do the two gases now compare in:

(i) pressure,
(ii) volume,
(iii) number of moles,
(iv) average molecular kinetic energy.

20. The boiling points and freezing points in degrees Centigrade of certain liquids are listed below. Express these temperatures on the absolute temperature (degree Kelvin) scale.

Liquid helium, boiling point = −269
Liquid hydrogen, freezing point = −259
Answer. 14°K.

Liquid hydrogen, boiling point = −253
Answer. 20°K.

Liquid nitrogen, freezing point = −210
Liquid nitrogen, boiling point = −196

Liquid oxygen, freezing point = −219
Liquid oxygen, boiling point = −183

21. If exactly 100 ml of a gas at 10°C are heated to 20°C (pressure and number of molecules remaining constant), the resulting volume of the gas will be which of the following?

(a) 50 ml,
(b) 1000 ml,
(c) 100 ml,
(d) 375 ml,
(e) 103 ml.

22. Why is it desirable to express all temperatures in degrees Kelvin when working with problems dealing with gas relationships?

23. A gaseous reaction between methane, CH_4, and oxygen, O_2, is carried out in a sealed container. Under the conditions used, the products are hydrogen, H_2, and carbon dioxide, CO_2. Energy is released, so the temperature rises during the reaction.

(a) Will the final pressure be greater or lower than the original pressure?
(b) By what factor does the pressure change if one mole of methane and one mole of oxygen are mixed and reacted (with the temperature changing from 25°C to 200°C)?

Answer. 2.38.

24. Automobiles are propelled by burning gasoline, typical formula C_8H_{18}, inside a container (the cylinder) that can change volume and drive the wheels. Oxygen reacts with the gasoline to form carbon dioxide and water, releasing enough energy to heat the gas from about 300°K to about 1500°K.

Balance the equation for the reaction and decide whether the work done by the gas in the cylinder is mainly due to pressure rise caused by change in number of moles of gas or due to pressure rise resulting from heating.

25. Why does the pressure build up in a tire on a hot day? Answer in terms of the kinetic theory.

26. A vessel contains equal numbers of oxygen and of hydrogen molecules. The pressure is 760 mm Hg when the volume is 50 liters. Which of the following statements is FALSE?

(a) On the average, the hydrogen molecules are traveling faster than the oxygen molecules.
(b) On the average, more hydrogen molecules strike the walls per second than oxygen molecules.
(c) If the oxygen were removed from the system, the pressure would drop to 190 mm Hg.
(d) Equal numbers of moles of each gas are present.
(e) The average kinetic energies of oxygen and hydrogen are the same.

27. The vapor pressure of a molten metal can be measured with a device called a Knudsen cell. This is a container closed across the top by a thin foil pierced by a small, measured hole. The cell is heated in a vacuum, until the vapor above the melt streams from the small hole (it *effuses*). The weight of the material escaping per second tells the rate at which gaseous atoms leave.

Two identical Knudsen cells are heated at 1000°C, one containing lead and the other containing magnesium.

(a) Contrast the average kinetic energies of the lead and magnesium atoms within each cell.
(b) Contrast the average velocities of the lead and magnesium atoms leaving each cell.
(c) At this fixed temperature, the rate at which atoms leave is determined by two factors, the vapor pressure and the mass of the gaseous particles. Explain.

28. The following table indicates the boiling points and the molar volumes (0°C and 1 atm) of some common gases.

(a) What regularity is suggested in the relationship between the boiling points and molar volumes?
(b) Account for this regularity.

GAS	FORMULA	BOILING POINT (°C)	MOLAR VOLUME (liters)
helium	He	−269	22.426
nitrogen	N_2	−196	22.402
carbon monoxide	CO	−190	22.402
oxygen	O_2	−183	22.393
methane	CH_4	−161	22.360
hydrogen chloride	HCl	−84.0	22.248
ammonia	NH_3	−33.3	22.094
chlorine	Cl_2	−34.6	22.063
sulfur dioxide	SO_2	−10.0	21.888

CHAPTER 5

Liquids and Solids: Condensed Phases of Matter

Almost all the chemical processes which occur in nature, whether in animal or vegetable organisms, or in the non-living surface of the earth, . . . take place between substances in solution.

W. OSTWALD, 1890

Only a handful of substances are gases under normal conditions of temperature and pressure. Of the hundred or so elements, most are normally solids; two or three are liquids. As for compound substances, more than a million have been prepared by chemists, yet, more than 99% of these are liquids or solids, each with distinctive and characteristic properties. It is no surprise, then, that there is great variety among all of these substances. Rather, it is remarkable that they can be classified into a small number of types and that the wealth of information represented by the diversity of all of these compounds can be treated within a simple framework. We shall begin our study of this framework by considering the properties of pure substances in their liquid and solid phases.

EXERCISE 5-1

The dozen or so elements that are normally found as gases include nitrogen, oxygen, fluorine, helium, neon, argon, krypton, xenon, and chlorine. Where are these placed in the periodic table (see inside front cover)?

5-1 PURE SUBSTANCES

A gas, when cooled, condenses to a liquid. Further cooling causes solidification. The condensation of ammonia under pressure, the condensation of water vapor (steam) on cooling, and the freezing of liquid water to ice are familiar cases. These changes are called phase changes. We shall consider liquid-gas changes first and then, solid-liquid phase changes.

5-1.1 Liquid–Gas Phase Changes

When a pan of water is warmed, the input of heat causes the water temperature to rise. At a certain point, however, the water begins to boil. Then the temperature is constant as long as liquid water remains, and continued heating causes the formation of water vapor. Water changes from the liquid phase to the gas phase, absorbing energy though the temperature remains constant. The energy of the liquid is less than the energy of the same weight of gas.

Let us consider how much energy is needed for this particular phase change.

$$H_2O(\text{liquid}) \longrightarrow H_2O(\text{gas}) \qquad (1)$$

or, in abbreviation,

$$H_2O(l) \longrightarrow H_2O(g) \qquad (1)$$

Suppose we wish to evaporate one mole of water, as expressed in equation (*1*). One mole contains the Avogadro number of molecules (6.02×10^{23}) and has a weight of 18.0 grams. Using a calorimeter, as you did in Experiment 5, you could measure the quantity of heat required to evaporate one mole of water. It is 10 kilocalories per mole. This value is called the **molar heat of vaporization** of water. This is the energy required to separate 6.02×10^{23} molecules of water from one another, as pictured in Figure 5-1.

EXERCISE 5-2

When two moles of water are evaporated, how much heat is required? One-half mole of water?

When water vapor condenses to liquid water, the molecules release the energy it took to separate them. A mole of gaseous water, therefore, will release 10 kilocalories of heat when condensed to liquid water at the same temperature. The amount of heat released is numerically equal to the molar heat of vaporization.

Other liquid–gas phase changes are similar, though boiling points vary over a wide range. Table 5-I shows the boiling points and heats of vaporization of a variety of liquids. In each case, energy is absorbed as the particles that make up the liquid are separated into the molecules of the gas. We shall see in Chapter 17 that the extreme range of heats of vaporization shown in this table can be explained using rather simple principles. These principles provide a basis for qualitative predictions of boiling point, heat of vaporization, and other properties.

5-1.2 Liquid–Gas Equilibrium: Vapor Pressure

Our knowledge of gas behavior helps us interpret the evaporation of liquids. We have considered, thus far, vaporization of a liquid at its usual

Fig. 5-1. **Evaporation of liquid water.**

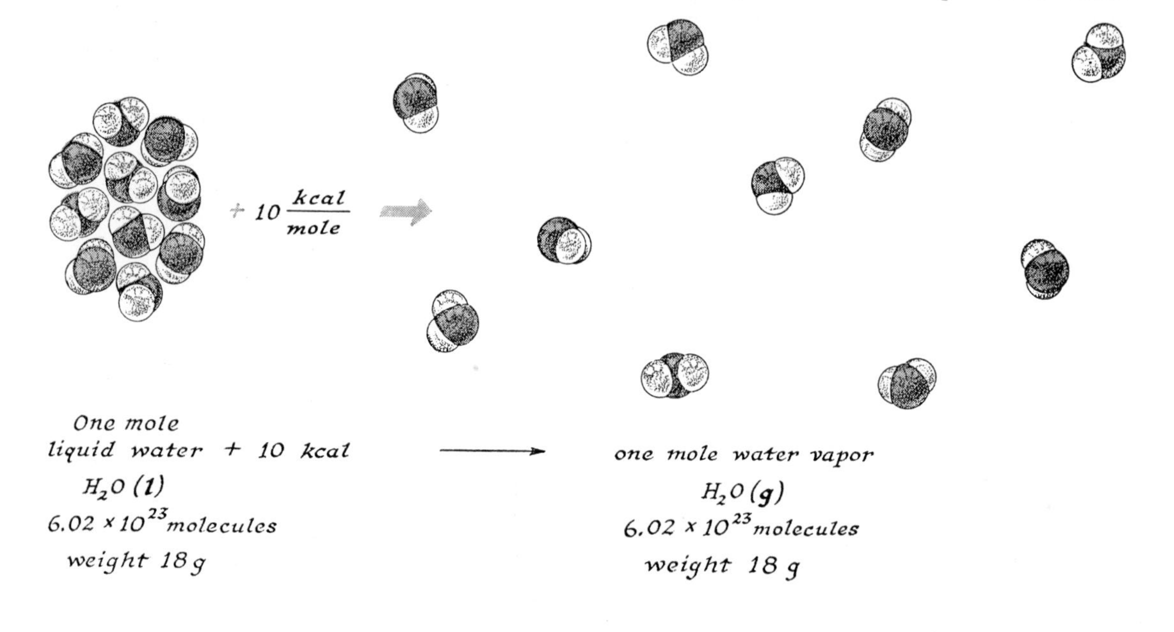

Table 5-I. **THE NORMAL BOILING POINTS AND MOLAR HEATS OF VAPORIZATION OF SOME PURE SUBSTANCES**

SUBSTANCE	PHASE CHANGE (liquid) ⟶ (gas)	BOILING POINT °K	BOILING POINT °C	MOLAR HEAT OF VAPORIZATION (kcal/mole)
neon	$Ne(l) \longrightarrow Ne(g)$	27.2	−245.8	0.405
chlorine	$Cl_2(l) \longrightarrow Cl_2(g)$	238.9	−34.1	4.88
water	$H_2O(l) \longrightarrow H_2O(g)$	373	100	9.7
sodium	$Na(l) \longrightarrow Na(g)$	1162	889	24.1
sodium chloride	$NaCl(l) \longrightarrow NaCl(g)$	1738	1465	40.8
copper	$Cu(l) \longrightarrow Cu(g)$	2855	2582	72.8

boiling point. But liquids vaporize at all temperatures. Let us consider this process, beginning with liquid water again.

If we place some liquid water in a flask at 20°C and seal the flask, some water molecules leave the liquid and enter the gas phase. The partial pressure of water vapor in the flask rises, but when it reaches 17.5 mm no more change can be observed. The amount of excess liquid remains constant thereafter, and the partial pressure of water vapor in the flask remains at 17.5 mm, as long as the temperature is maintained at 20°C. This partial pressure is called the **vapor pressure** of water at 20°C. At this vapor pressure, liquid and gaseous water can coexist indefinitely at 20°C. This vapor pressure, 17.5 mm, is the same whether air is present or not; it is a property of water. If the flask were originally evacuated, liquid would evaporate until the pressure rose from 0 mm to 17.5 mm. If the flask originally contained dry air at a pressure of 750 mm, liquid would evaporate until the pressure rose from 750 mm to 767.5 mm (the partial pressure of water vapor changing from 0 mm to 17.5 mm). *When a liquid is in contact with its vapor at the vapor pressure, the liquid and gas are said to be in equilibrium.* **At equilibrium, no measurable changes are taking place.**

EFFECT OF TEMPERATURE

The vapor pressure of water at 20°C is 17.5 mm. At 40°C, the vapor pressure is 55.3 mm; at 60°C, it is 149.4 mm. The vapor pressure of water increases with increasing temperature.

Ethyl alcohol is also a liquid at room temperature. Its vapor pressure at 20°C is 44 mm, higher than the vapor pressure of water at this same temperature. At 40°C, ethyl alcohol has a vapor pressure of 134 mm; at 60°C, the vapor pressure is 352 mm. Again we find that the vapor pressure increases rapidly with increasing temperature. This is always so. *The vapor pressure of every liquid increases as the temperature is raised.*

THE BOILING POINT

At any temperature, molecules can escape from the surface of a liquid (vaporizing or evaporating) to enter the gas phase as vapor. At the special temperature at which the vapor pressure just equals the atmospheric pressure, a new phenomenon occurs. There, bubbles of vapor can form *anywhere within the liquid.* At this temperature, the liquid boils.

We see that the boiling point is fixed by the surrounding pressure. For example, if the surrounding pressure is 760 mm, water boils at 100°C. This is the temperature at which the vapor pressure of water is just 760 mm. Ethyl alcohol, having a higher vapor pressure, achieves a vapor pressure of 760 mm at 78.5°C. Ethyl alcohol boils at 78.5°C with this surrounding pressure. Suppose, however, that the atmospheric pressure drops to 750 mm (as it might just before a storm). Then bubbles of vapor could form anywhere in liquid water at a temperature of 99.6°C since the vapor pressure of water is 750 mm at 99.6°C. Water boils at 99.6°C when the surrounding pressure is 750 mm.

The **normal boiling point** of a liquid is defined as the temperature at which the vapor pressure of that liquid is exactly one standard atmosphere, 760 mm Hg.

Fig. 5-2. **Vapor pressure increases** *as temperature is raised.*

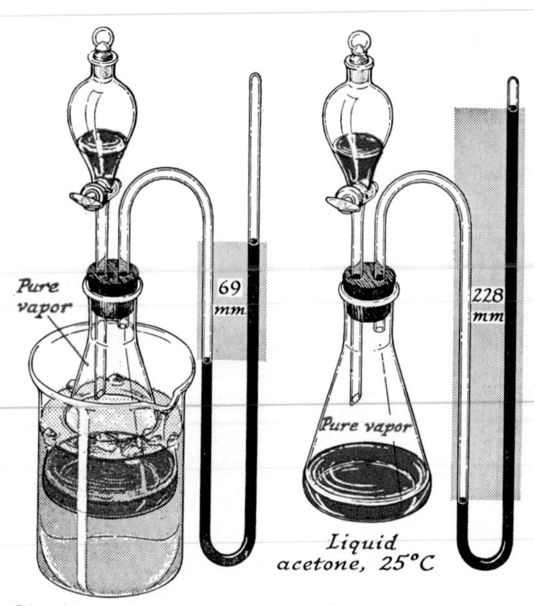

Liquid acetone in ice bath, 0°C

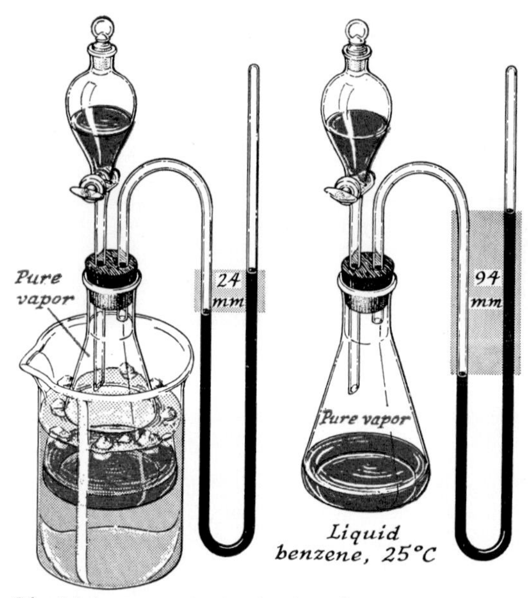

Liquid benzene in ice bath, 0°C

EXERCISE 5-3

What is the normal boiling point of ethyl alcohol?

EXERCISE 5-4

Suppose a closed flask containing liquid water is connected to a vacuum pump and the pressure over the liquid is gradually lowered. If the water temperature is kept at 20°C, at what pressure will the water boil?

EXERCISE 5-5

Answer Exercise 5-4, substituting ethyl alcohol for water.

5-1.3 Solid–Liquid Phase Changes

Solids and liquids are called *condensed phases.* The attractive forces in a condensed phase, either a solid or a liquid, tend to hold the molecules close together. In liquids, molecules are irregularly spaced and randomly oriented. In a crystalline solid, the molecules occupy regular positions, resulting in additional stability (relative to the liquid).

The difference between the energy of a substance in liquid form and its energy in solid form is usually much smaller than the difference between the energies of the liquid and gaseous forms. For example, consider the heat of melting a mole of ice,

$$H_2O(\text{solid}) \longrightarrow H_2O(\text{liquid}) \qquad (2)$$

or, in abbreviation,

$$H_2O(s) \longrightarrow H_2O(l) \qquad (2)$$

The heat accompanying the phase change (*2*) is 1.44 kcal/mole. This is much less than the molar heat of vaporization of water, 10 kcal/mole. Table 5-II contrasts the melting points and the heats of melting per mole (the **molar heat of melting,** or the **molar heat of fusion**) of the same pure substances listed in Table 5-I.

Once again, we find an extreme range among the properties of these substances. The molar heats of melting vary from 0.080 kcal/mole for

Table 5-II. **THE MELTING POINTS AND HEATS OF MELTING OF SOME PURE SUBSTANCES**

SUBSTANCE	PHASE CHANGE (solid) ⟶ (liquid)	MELTING POINT °K	MELTING POINT °C	MOLAR HEAT OF MELTING (kcal/mole)
neon	$Ne(s) \longrightarrow Ne(l)$	24.6	−248.4	0.080
chlorine	$Cl_2(s) \longrightarrow Cl_2(l)$	172	−101	1.53
water	$H_2O(s) \longrightarrow H_2O(l)$	273	0	1.44
sodium	$Na(s) \longrightarrow Na(l)$	371	98	0.63
sodium chloride	$NaCl(s) \longrightarrow NaCl(l)$	1081	808	6.8
copper	$Cu(s) \longrightarrow Cu(l)$	1356	1083	3.11

neon to 6.8 kcal/mole for the substance sodium chloride—a change by a factor of 85. There are very great differences in the forces that bind these solids. Since these differences affect properties other than melting point and heat of melting, they are important to a chemist.

Fig. 5-3. **Melting of ice.**

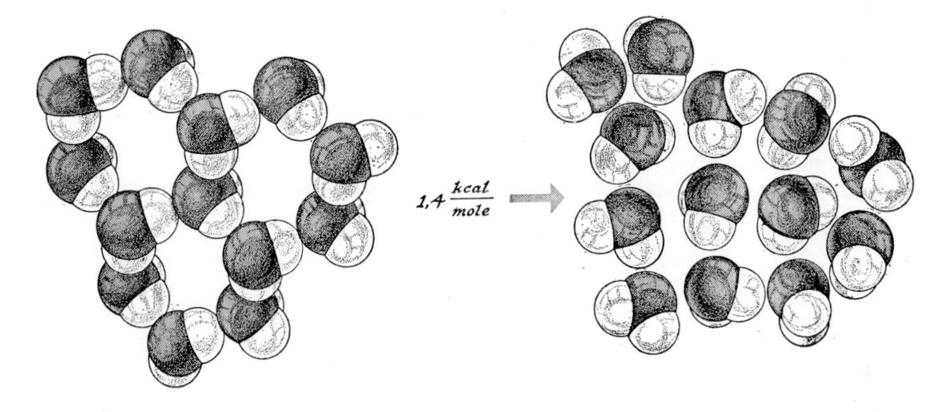

5-2 SOLUTIONS

Sodium chloride, sugar, ethyl alcohol, and water are four pure substances. Each is characterized by definite properties, such as vapor pressure, melting point, boiling point, density. Suppose we mix some of these pure substances. Sodium chloride dissolves when placed in contact with water. The solid disappears, becoming part of the liquid. Likewise, sugar in contact with water dissolves. When ethyl alcohol is added to water, the two pure substances mix to give a liquid similar in appearance to the original liquids. The salt-water mixture, the sugar-water mixture, and the alcohol-water mixtures are called solutions. Solutions differ from pure substances in that their properties vary, depending upon the relative amounts of the constituents. The behavior of solutions during phase changes is dramatically different from that just described for pure substances. These differences provide, at once, reason for making a distinction between pure substances and solutions and, as well, a basis for deciding whether a given material is a pure substance or a solution.

5-2.1 Differentiating Between Pure Substances and Solutions

The earth has many unlike parts—it is *heterogeneous*. Some of the parts are uniform throughout; that is, they are *homogeneous*. Familiar examples of heterogeneous materials are granite (which consists of various minerals suspended in another mineral), oil and vinegar salad dressing (which consists of droplets of oil suspended in aqueous acetic acid), and black smoke (which consists of particles of soot suspended in air). Examples of homogeneous materials are diamond, fresh water, salt water, and clear air. Heterogeneous materials are hard to describe and classify but we can describe homogeneous materials rather precisely.

Both pure substances and solutions are homogenous. *A homogeneous material that contains only one substance is called a* **pure substance.** **A solution** *is a homogeneous material that contains more than one substance.*

We have used the terms gas phase, liquid phase, and solid phase. A *phase* is a homogeneous part of a system—a part which is uniform and alike throughout. A *system* in turn, is any region and the material in it that we wish to consider. It may include only one, or more than one, phase.

Suppose we compare two liquid samples, one of distilled water, and one of salt water. Each sample is a homogeneous system consisting of a single phase. However, one of the liquids is a pure substance whereas the other is a solution. We cannot tell, merely by visual observation, which of these clear liquids is the pure substance and which is the solution. True, there are differences—for example, the salt water has a greater density than the pure water—but even this property does not indicate which is the pure substance.

Let us compare the behavior of these two systems during a phase change. Consider, first, how water acts when it is frozen or vaporized. Pure water freezes at a fixed temperature, 0°C. If we freeze half of a water sample to ice, remove the ice, melt it in another container, and compare the separate samples, we find that the two fractions of the original sample are indistinguishable. In a similar way, if we boil a sample of water until half of it has changed to steam, condense the steam to water in a different vessel, and then compare the separate samples, we find that the fractions of the original sample are indistinguishable. *Such behavior on boiling (condensing) or freezing (melting) characterizes pure substances.* Solutions behave differently.

Suppose we boil off part of a sample of salt water. The temperature of the liquid rises, as shown in curve *b* of Figure 5-4, until boiling

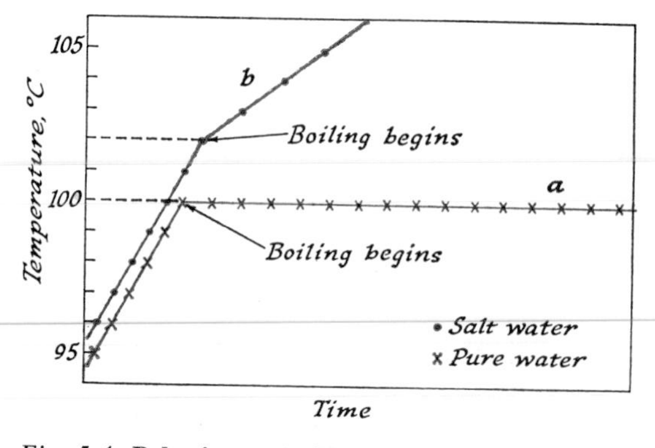

Fig. 5-4. **Behavior on boiling.** (*a*) *A pure substance.* (*b*) *A solution.*

begins. Already, a difference from the behavior of pure water can be noted, shown in curve *a* of Figure 5-4: the boiling point of the salt solution is higher. As boiling continues, the temperature of the pure water remains constant whereas the temperature of the salt solution rises. As the boiling point goes up, the remaining liquid becomes saltier. If we collect the steam from the salt solution and condense it in a separate vessel, we find that the resulting liquid behaves like pure water rather than like the solution from which it came. If we boil off all the water, solid salt remains behind. Thus, by ***distilling***—that is, *evaporating and recondensing in a separate vessel*—we can separate a pure liquid from a solution; and, by ***crystallizing***—that is, *forming a crystalline solid*—we can obtain a pure solid from a solution. Chemists call the pure liquid obtained by distilling and the pure solid obtained by crystallizing, the *components* of the solution. In our

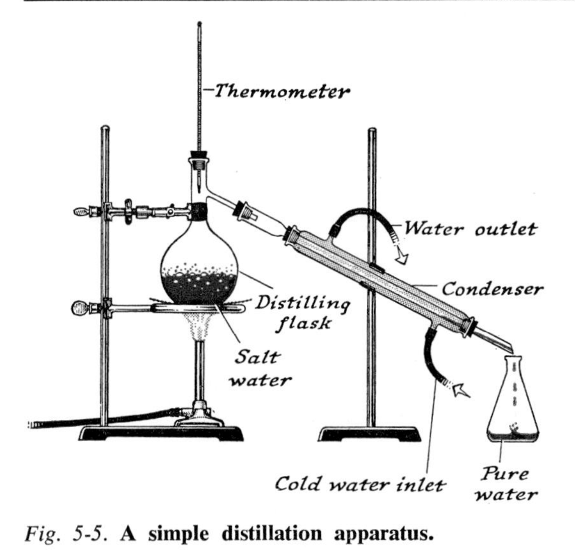

Fig. 5-5. **A simple distillation apparatus.**

experiment shown in Figure 5-4, the components are salt and water.

Pure sodium chloride, like pure water, has a definite melting (freezing) temperature (at a given pressure). Separating operations—such as distilling or freezing—do not separate the salt into components. The composition of the salt, whether expressed in relative numbers of sodium and chlorine atoms or in the relative weights of these atoms, is fixed and is represented by the formula NaCl. Sodium chloride, like water, is an example of a pure substance.

On the other hand, operations such as distilling or freezing usually tend to separate solutions into the pure substances that were the components of the solution. The nearer alike the components are, the harder it is to separate them from the solution, but even in difficult cases, a variety of methods in succession usually brings about a separation. In nature, solutions are much more common than pure substances, and heterogeneous systems are more common than solutions. When we want pure substances, we often must prepare them from solutions through successive phase changes.

We are all familiar with liquid solutions. Gas and solid solutions also exist. We shall consider them briefly and then return to liquid solutions, the most important from a chemist's point of view.

5-2.2 Gaseous Solutions

All gas mixtures are homogeneous; hence all gas mixtures are solutions. Air is an example. There is only one phase—the gas phase—and all the molecules, regardless of the source, behave as gas molecules. The molecules themselves may have come from gaseous substances, liquid substances, or solid substances. Whatever the source of the constituents, this gaseous solution, air, is a single, homogeneous phase. As with other solutions, the constituents of air are separated by phase changes.

5-2.3 Solid Solutions

Solid solutions are more rare. Crystals are stable because of the regularity of the positioning of the atoms. A foreign atom interferes with this regularity and hence with the crystal stability. Therefore, as a crystal forms, it tends to exclude foreign atoms. That is why crystallization provides a good method for purification.

But in metals it is relatively common for solid solutions to form. The atoms of one element may enter the crystal of another element if their atoms are of similar size. Gold and copper form such solid solutions. The gold atoms can replace copper atoms in the copper crystal and, in the same way, copper atoms can replace gold atoms in the gold crystal. Such solid solutions are called alloys. Some solid metals dissolve hydrogen or carbon atoms—steel is iron containing a small amount of dissolved carbon.

Solid solutions, alloys in particular, will be considered again in Chapter 17.

5-2.4 Liquid Solutions

In your laboratory work you will deal mostly with liquid solutions. Liquid solutions can be made by mixing two liquids (for example, alcohol and water), by dissolving a gas in a liquid (for example, carbon dioxide and water), or by dissolving a solid in a liquid (for example, sugar and water). The result is a homogeneous system containing more than one substance—a solution. In such a liquid, each component is diluted by the other component. In salt water, the salt dilutes the water and, of course, the water dilutes the salt. This solution is only partly made up of water molecules and it is found that the vapor pressure of the solution is correspondingly lower than the vapor pressure of pure water. Whereas water must be heated to 100°C to raise the vapor pressure to 760 mm, it is necessary to heat a salt solution above 100°C to reach this vapor pressure. Therefore, the boiling point of salt water is above the boiling point of pure water. The amount the boiling point is raised depends upon the relative amounts of water and salt. The more salt that is added, the higher is the boiling point.

In a similar way, a lower temperature is required to crystallize ice from salt water or from an alcohol-water solution than from pure water. "Antifreeze" substances added to an automobile radiator act on this principle. They dilute the water in the radiator and lower the temperature at which ice can crystallize from the solution. Again, the amount the freezing temperature is lowered depends upon the relative amounts of water and antifreeze compound.

In general, the properties of a solution depend upon the relative amounts of the components. It is important to be able to specify quantitatively what is present in a solution, that is, to specify its *composition*. There are many ways to do this, but one method will suffice for our purposes.

5-2.5 Expressing the Composition of Solutions

The components of a solution are the pure substances that are mixed to form the solution. If there are two components, one is sometimes called the *solvent* and the other the *solute*. These are merely terms of convenience. Since both must intermingle to form the final solution, we cannot make any important distinction between them. When chemists make a liquid solution from a pure liquid and a solid, they usually call the liquid component the solvent.

To indicate the composition of a particular solution we must show the relative amounts as well as the kind of componen s. These relative amounts chemists call **concentrations.** Chemists use different ways of expressing concentration for various purposes. We shall use only one in this course.

Chemists often indicate the concentration of a substance in water solution in terms of the number of moles of the substance dissolved per liter of solution. This is called the **molar concentration.** A one-molar solution (1 *M*) *contains one mole of the solute per liter of total solution*, a two-molar solution (2 *M*) contains two moles of solute per liter, and a 0.1-molar solution (0.1 *M*) contains one-tenth mole of solute per liter. Notice that the concentration of water is not specified, though we must add definite amounts of water to make the solutions.

We can make a 1 *M* solution of sodium chloride by weighing out one mole of the salt. From the formula, $NaCl$, we know that one mole weighs 58.5 grams (23.0 grams + 35.5 grams). We dissolve this salt in some water in a 1 liter volumetric flask—a flask holding just 1 liter when filled exactly to an etched mark. After the salt dissolves, more water is added until the water level reaches the etched mark to make the volume exactly 1 liter. Equally well, we can prepare a 1 *M* sodium chloride solution using a 100 ml volumetric flask. Then the final volume of the solution will be 0.100 liter and we need only one-tenth mole of salt. In this case, we weigh out 5.85 grams of salt, place it in the flask, dissolve it, and add water to the 100 ml mark.

5-2.6 Solubility

When solid is added to a liquid, solid begins to dissolve and the concentration of dissolved material begins to rise. After all of the solid has dissolved, the concentration remains constant, fixed by the amount of dissolved solid and the volume of the solution. If more solid is now added, the concentration will rise further. Finally, however, the addition of more solid no longer raises the concentration of dissolved material. When a fixed amount of liquid has dissolved all of the solid that it can, the concentration reached is called the **solubility** of that solid. A solution in contact with excess solid is said to be **saturated.**

The solubilities of solids in liquids vary widely.

For example, sodium chloride continues to dissolve in water at 20°C until the concentration is about six moles per liter. The solubility of NaCl in water is 6 *M* at 20°C. In contrast, only a minute amount of sodium chloride dissolves in ethyl alcohol at 20°C. This solubility is 0.009 *M*. Even in a single liquid, solubilities differ over wide limits. The solids calcium chloride, $CaCl_2$, and silver nitrate, $AgNO_3$, have solubilities in water exceeding one mole per liter. The solid called silver chloride, AgCl, has a solubility in water of only 10^{-5} mole per liter.

Because of this range of solubilities, the word *soluble* does not have a precise meaning. There is an upper limit to the solubility of even the most soluble solid, and even the least soluble solid furnishes a few dissolved particles per liter of solution. If a compound has a solubility of more than one-tenth mole per liter (0.1 *M*), chemists usually say it is *soluble*. When the solubility lies below 0.1 *M* (10^{-1} *M*), chemists usually say the compound is *slightly soluble*. Compounds with solubility below about 10^{-3} *M* are sometimes said to be *very slightly soluble*, and if the solubility is so low as to be of no interest, the solid is said to have *negligible solubility*. We use glass containers for pure water because glass has a negligible solubility in water.

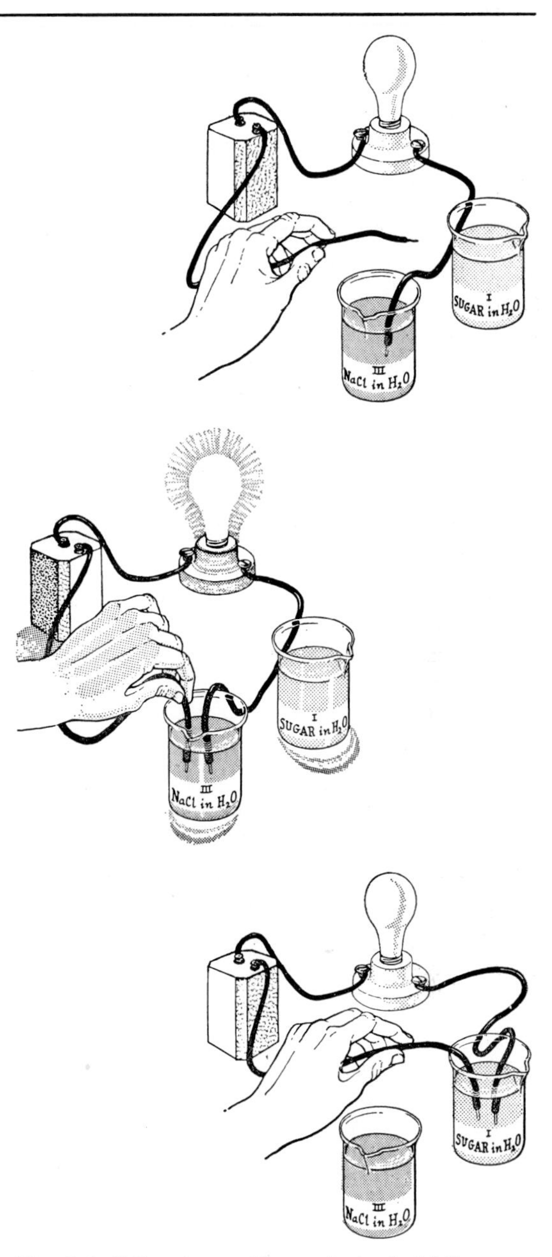

Fig. 5-6. **Salt water readily conducts electricity;** *sugar solution does not.*

5-2.7 Variations Among the Properties of Solutions

Though many solutions are colorless and closely resemble pure water in appearance, the differences among solutions are great. This can be demonstrated with the five pure substances, sodium chloride (salt), iodine, sugar, ethyl alcohol, and water. Two of these substances, ethyl alcohol and water, are liquids at room temperature. Let's investigate the properties of the solutions these two substances form.

First we can investigate, qualitatively, the extent to which the solids dissolve in the liquids. By adding a small piece of each solid to a milliliter of liquid, we easily discover that sugar dissolves both in water and ethyl alcohol, sodium chloride dissolves readily in water but not in ethyl alcohol, and iodine does not dissolve much in water but dissolves readily in ethyl alcohol. Thus we see that the solvent properties of the two liquids are quite distinctive, at least as far as sugar, salt, and iodine are concerned.

The experiment just described gives us four solutions containing a substantial amount of solute:

I	II	III	IV
Sugar in water	Sugar in ethyl alcohol	Sodium chloride in water	Iodine in ethyl alcohol

Of these four solutions, IV is readily distinguished. This solution has a dark brown color. The other three, I, II, and III, are colorless. They can be easily distinguished by taste but chemists have safer and more meaningful ways of distinguishing them. These solutions differ markedly in their ability to conduct an electric current. The two sugar solutions, I and II, have virtually the same conductivity properties as the pure liquids —they do not conduct electric current readily. Solution III conducts electric current much more readily than does pure water.

Thus we find great variation among solutions. Iodine dissolves in ethyl alcohol, coloring the liquid brown, but does not dissolve readily in water. Sodium chloride does not dissolve readily in ethyl alcohol but does dissolve in water, forming a solution that conducts electric current. Sugar dissolves readily both in ethyl alcohol and in water, but neither solution conducts electric current. These differences are very important to the chemist, and variations in electrical conductivity are among the most important. We shall investigate electrical conductivity further but, first, we need to explore the electrical nature of matter.

5-3 ELECTRICAL NATURE OF MATTER

We have mentioned the electrical conductivity of solutions as a means of distinguishing solutions. The interest of a chemist in the electrical nature of matter goes far deeper than this. We shall find that an understanding of electrical behavior furnishes a key to the explanation of chemical properties. We shall find that electrical effects aid us in predicting molecular formulas, in explaining chemical reactions, and in understanding energy changes that accompany them.

5-3.1 Electrical Phenomena

Try to name several electrical phenomena that you have often observed. Before you read on, see if you can name five. Does your list include the following?

(1) The attraction of a comb for your hair on a dry day.
(2) The flash of a bolt of lightning.
(3) The shock you get if you touch a bare wire in a radio set.
(4) The heat generated by an electric current passing through the heating element of an electric stove.
(5) The light emitted by the filament of a light bulb as electric current is passed through it.
(6) The magnetic field generated by a current passing through a coil of wire.
(7) The work done by an electric motor when electric current passes through its coils.
(8) The emission of "radio waves" by the antenna of a radio or television station.

We see electrical devices all around us—furnishing power, light, means of communication—influencing every facet of life. What does it mean to say that an electric current "passes through" a coil of wire? What is an electric current? To begin answering these questions, we must explore an electrometer, a device for detecting and measuring electric charge.

5-3.2 Detection of Electric Charge

Figure 5-7 shows a simple electrometer. It consists of two spheres of very light weight, each coated with a thin film of metal. The spheres are suspended near each other by fine metal threads in a closed box to exclude air drafts. Each suspending thread is connected to a brass terminal. Next to the box is a "battery"—a collection of electrochemical cells. There are two terminal posts on the battery. We shall call these posts P_1 and P_2. If post P_1 is connected by a copper wire

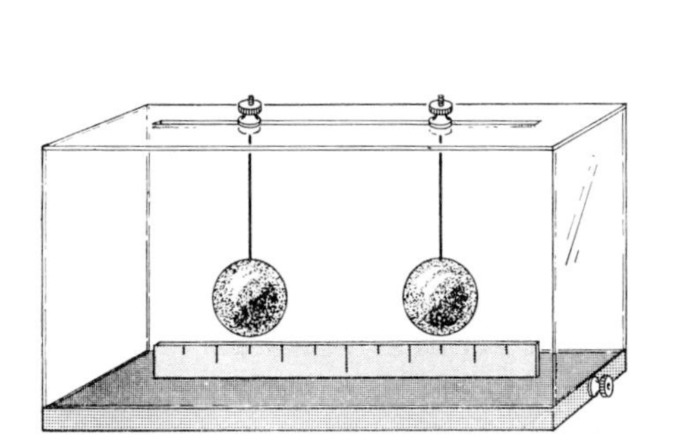

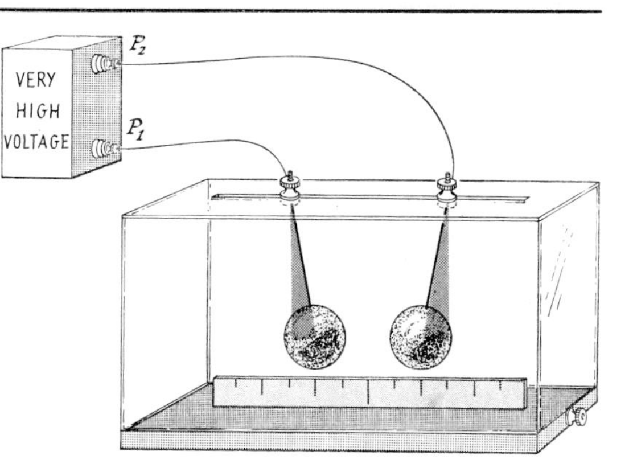

Fig. 5-7. **A simple electrometer.**

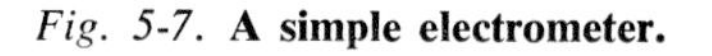

to the left terminal of the electrometer and post P_2 is connected to the right terminal, we observe that the two spheres move toward each other. Evidently the wires have transmitted to the spheres the property of exerting force on each other—an attractive force. The force is still present when the air in the electrometer is removed with a vacuum pump. The spheres react to each other "across space." They feel "force at a distance."

If now the wires are disconnected, the attractive force remains. However, if the two electrometer terminals are connected by a copper wire the spheres return to their original positions and hang vertically again. The attraction is lost.

We see that the battery transfers to the electrometer spheres the property of attracting each other. It is natural to imagine that something has been transferred from the battery to the spheres. This "something" is called electric charge. The movement of this electric charge from the battery through the metal threads to the spheres is called an electric current. This electric charge is lost when the two electrometer terminals are connected by a copper wire.

We can learn more about electric charge by another use of the electrometer. Connect one wire from the battery (say, post P_1) to the base of the electrometer and the other wire (from post P_2) to both terminals, as in Figure 5-8.

This time the two spheres move apart—they repel each other! When both spheres are given electric charge from the battery post labeled P_2, they repel instead of attract. In Figure 5-7 we saw that when one sphere received charge from post P_1 and the second sphere received charge from post P_2, the two spheres attracted. There must be at least two kinds of charge!

Now let us reverse the wires so that both spheres are charged from battery post P_1. This time, P_2 is connected to the base of the electrometer. Again we observe that the spheres move apart. *Whenever both spheres are connected to the same battery post, the two spheres repel each other.*

This represents a large gain in our knowledge of electric charge. One kind of charge comes from battery post P_1. Until we have reason to do otherwise, we shall call this kind of charge C_1. The other kind of charge comes from battery post P_2; we shall call this kind of charge C_2. The spheres attract or repel each other when they carry charge according to the following pattern:

C_1 attracts C_2	unlike charges attract	
C_1 repels C_1	like charges repel	(3)
C_2 repels C_2	like charges repel	

Fig. 5-8. **The electrometer** *with both spheres connected to battery post* P_2.

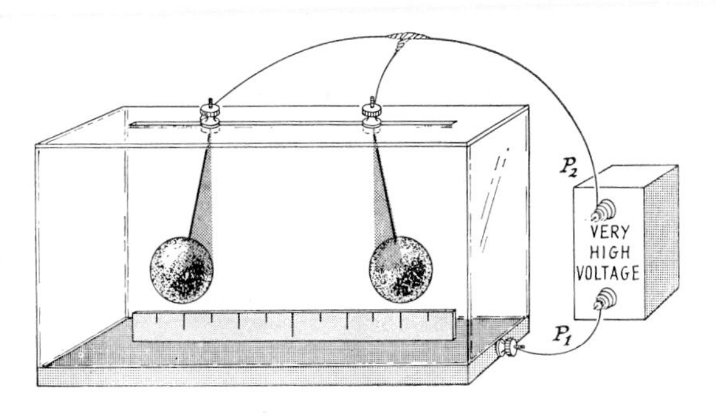

We have one more observation to symbolize. When the spheres were given different charges (as in Figure 5-7), the charges could be removed by connecting the two terminals by a copper wire. Then the spheres lost all attraction for each other. We interpret this behavior to mean that either C_1 or C_2 (or both) has moved through the wire so as to join the other kind of charge. When C_1 and C_2 are united, no charge remains. Symbolically we can say,

$$C_1 + C_2 = \text{no charge} \qquad (4)$$

Our accumulated evidence shows that there are at least two kinds of electric charge, which we have symbolized C_1 and C_2. These two kinds of charge possess the properties (*3*) and (*4*). We may wonder if there are other kinds of charge. This is answered by looking into other ways of producing electric charges. Chemists can assemble a variety of types of electrochemical cells which show the same electrometer behavior just described. Some frictional processes leave charges on the two surfaces rubbed together. The attraction of a comb for your hair is caused by charges left on the comb as it rubbed against your hair. Many decades ago the properties of electric charge were investigated as they are produced by rubbing a hard rubber rod with cat fur. The hard rubber is found to carry charge C_1, and the cat fur carries charge C_2. If a glass rod is rubbed with silk, the glass rod is left with charge C_2 and the silk with charge C_1.

No matter how electric charge is produced, we always find these same two types, C_1 and C_2, and *only these two*. Any method of producing C_1 also produces an equivalent amount of C_2. We conclude *there are two and only two types of electric charge.*

5-3.3 The Effect of Distance

Figure 5-9 shows two electrometers which differ in the spacing of the spheres. Though the charges on the spheres come from the same battery, there is more deflection of the spheres when they are closely spaced (left) than when they are widely spaced. When the spheres are closer together, the deflection is larger. Hence, we conclude that the force of attraction varies with distance and is stronger when the charges are close to each other. Careful quantitative studies show that the force is inversely proportional to the square of the distance r between the two spheres:

$$\text{(Electric force) is proportional to } \frac{1}{r^2} \qquad (5)$$

where

r = distance between centers of the two spheres.

5-3.4 The Electron-Proton Model

These new facts about electrical phenomena can be incorporated into our particle model of the structure of matter if we again allow some

Fig. 5-9. **Contrast of deflections** *in two electrometers with different distances between spheres.*

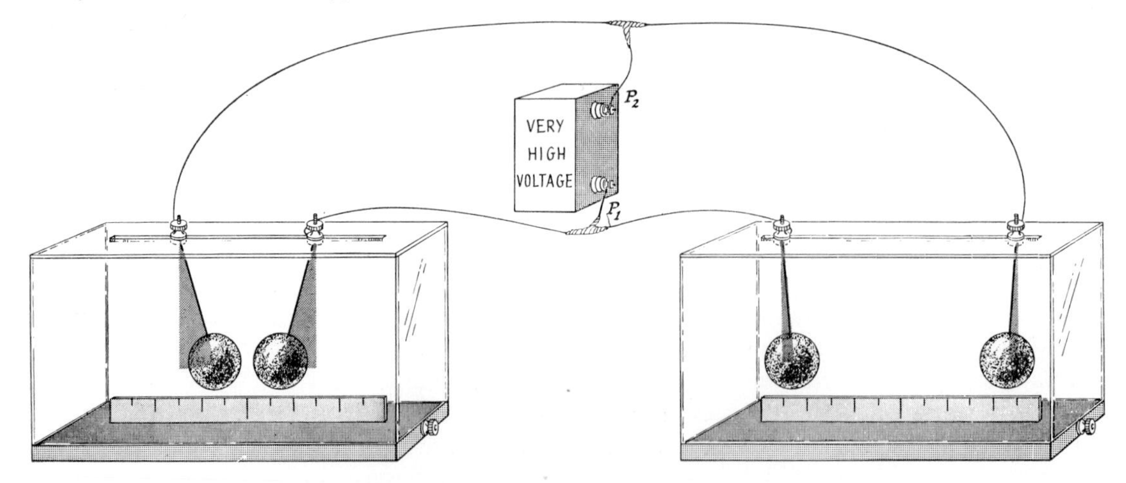

growth of the model. The new idea is that *matter is made up of particles which carry the property called electric charge.* To be specific, we propose that in atoms there are two kinds of particles that carry unit charge, one which carries one portion, or unit, of charge C_1 and one which carries one portion, or unit, of charge C_2. These particles are called *electrons* and *protons.*

Proposal: Matter includes particles, each of which carries a unit of electric charge.
Electrons: Each electron carries one unit of charge C_1.
Protons: Each proton carries one unit of charge C_2.

These particles exert force at a distance on each other in accordance with the electrical behavior we have observed.

Since:

C_1 repels C_1,	electrons repel electrons;
C_2 repels C_2,	protons repel protons;
C_1 attracts C_2,	electrons attract protons;
$C_1 + C_2 =$ no charge,	one electron + one proton = no charge;
or,	one unit C_1 + one unit C_2 = no charge.

The atomic model now can cope with the facts we have learned about electrical behavior. If a piece of matter (such as one of the electrometer spheres) has the same number of electrons and protons, there are just as many units of charge of type C_1 as of type C_2. Since $C_1 + C_2 =$ no charge, the sphere will have no charge. *A body with no net charge* (with equal numbers of protons and electrons) *is said to be* **electrically neutral.** If we remove some of the electrons from the sphere, it will then have an excess of protons, hence a net charge of type C_2. If we add an excess of electrons to the sphere, it will have a net charge of type C_1. *The amount of net charge is the difference between the amount of charge C_1 and charge C_2.*

It is a mathematical convenience if we express the net charge in terms of algebraic symbols. Henceforth we shall identify the type of charge called C_1 as "negative charge" and the type called C_2 as "positive charge." Notice the advantages.

The combination of 5 units of C_1 and 3 units of C_2 leaves a net of 2 units of C_1. This now can be expressed

$$5(-1) + 3(+1) = -5 + 3 = -2$$

EXERCISE 5-6

Suppose ten protons and eleven electrons are brought together. These charges, grouped together, have the same net charge as how many electrons? Remember that one proton plus one electron gives no charge.

EXERCISE 5-7

Write an algebraic expression to obtain the result of Exercise 5-6, using numbers with algebraic signs to represent charges.

5-3.5 Electric Force: A Fundamental Property of Matter

We have learned that a battery can transfer to the spheres of an electrometer a property called electric charge. When this happens, the spheres exert force on each other. The discussion brings up two "wondering why" questions. The first is, "Why do the electric charges appear?" What caused the battery to transfer to the electrometer the property called electric charge? We shall examine this question carefully later in the course because the subject of the operation of an electrochemical cell is extremely important in chemistry. It is the topic of an entire chapter in this book (Chapter 12). For the moment, all we can say is that the electric charge *did* come from the battery of electrochemical cells, thus indicating that the matter within the cells contains electric charges.

The second question probes deeper: "Why do the two electrometer spheres, when charged, exert force on each other?" What is our explanation of this phenomenon? We say that the spheres have an excess of electrons (or protons) and these electrons (or protons) exert force on each other. This does not really explain electric force at a distance. We are left with the equivalent questions, "Why do two electrons (or two protons) repel each other? Why does an electron attract a proton?" Without an answer, we say, "It is a *fundamental property* of matter that it can acquire electric charge and, when it does so, it exerts force on other charged bodies." Such a statement may be taken as a definition of a "fundamental property"—a property which is generally observed but for which diligent search has failed to yield a useful model. Without an explanation of a property, we call the property "fundamental." It is a curious fact that after a property has resisted explanation for quite a time and it becomes classified as a fundamental property, an explanation no longer seems to be necessary.

EXERCISE 5-8

There was a time when atoms were said to be fundamental particles of which matter is composed. Now we describe the structure of the atom in terms of the fundamental particles we have just named, protons and electrons, plus another kind of particle called a neutron. Why are atoms no longer said to be fundamental particles? Do you expect neutrons, protons, and electrons always to be called fundamental particles?

5-4 ELECTRICAL PROPERTIES OF CONDENSED PHASES

Now we are ready to investigate behavior of condensed phases that shows evidence of the presence and movement of electric charge. We have already referred to one of the most important examples—the movement of electric current through water solutions.

5-4.1 The Electrical Conductivity of Water Solutions

The movement of electric charge is called an electric current. Hence when we say electric current flows through a salt solution, we mean there is a movement of electric charge through the solution. We shall be concerned here with the manner in which this charge moves.

Water is a very poor conductor of electricity. Yet when sodium chloride dissolves in water, the solution conducts readily. The dissolved sodium chloride must be responsible. How does the dissolved salt permit charge to move through the liquid? One possibility is that when salt dissolves in water, particles with electric charge are produced. The movement of these charged particles through the solution accounts for the current. Salt has the formula, NaCl—for every sodium atom there is one chlorine atom. Chemists have decided that when sodium chloride dissolves in water, the charged species present are chlorine atoms, each carrying the negative charge of one electron, and sodium atoms, each carrying the positive charge of one proton. We symbolize a chlorine atom with a negative charge as Cl^-. A sodium atom with a positive charge is symbolized Na^+. *Atoms or molecules that carry electric charge are called* ***ions***.

With these symbols, we can write the equation for the reaction that occurs when sodium chloride dissolves in water

$$NaCl(\text{solid}) + \text{water} \longrightarrow Na^+(\text{in water}) + Cl^-(\text{in water}) \quad (6)$$

Equation (*6*) shows that when water dissolves solid sodium chloride, Na^+ ions (sodium ions) and Cl^- ions (chloride ions) are present in the solution. Chemists usually abbreviate this equation as much as is consistent with retaining essential information. On the left of equation (*6*), the term "water" is usually not written since its presence is implied by the symbols on the right.

$$NaCl(\text{solid}) \longrightarrow Na^+(\text{in water}) + Cl^-(\text{in water}) \quad (6)$$

We have already seen that NaCl(solid) is usually written NaCl(*s*). There is a similar abbreviation for "in water." To represent this, the expression "in water" is replaced by the term "aqueous," commonly abbreviated "*aq*". * Thus equation (*6*), showing the reaction of sodium chloride dissolving in water to form a conducting solution, is usually written in the form:

$$NaCl(s) \longrightarrow Na^{+}(aq) + Cl^{-}(aq) \qquad (6)$$

Now we have a model of a salt solution that aids us in discussing electrical conductivity. The solid dissolves, forming the charged particles $Na^{+}(aq)$ and $Cl^{-}(aq)$, which can move about in the solution independently. An electric current can pass through the solution by means of the movement of these ions. The $Cl^{-}(aq)$ ions move in one direction, causing negative charge to move that way. The $Na^{+}(aq)$ ions move in the opposite direction, causing positive charge to move this way. These movements carry charge through the solution and current flows.

Sugar dissolves in water, but the resulting solution conducts electric current no better than does pure water. We conclude that when sugar dissolves, no charged particles result: no ions are formed. Sugar must be quite different from sodium chloride.

Calcium chloride, $CaCl_2$, is another crystalline solid that dissolves readily in water. The resulting solution conducts electric current, as does the sodium chloride solution. Calcium chloride is, in this regard, like sodium chloride and unlike sugar. The equation for the reaction is

$$CaCl_2(s) \longrightarrow Ca^{+2}(aq) + 2Cl^{-}(aq) \qquad (7)$$

Equation (*7*) shows that when calcium chloride dissolves, ions are present—$Ca^{+2}(aq)$ and $Cl^{-}(aq)$ ions. In this case, each calcium ion has the positive charge of two protons. Therefore it has twice the positive charge held by a sodium ion, $Na^{+}(aq)$. The chloride ion that forms, $Cl^{-}(aq)$, is the same negative ion that is present in the sodium chloride solution, though it comes from the calcium chloride solid instead of the sodium chloride. Because both $CaCl_2(s)$ and $NaCl(s)$ dissolve in water to form aqueous ions, they are considered to be similar.

Silver nitrate, $AgNO_3$, is a third solid substance that dissolves in water to give a conducting solution. The reaction is

$$AgNO_3(s) \longrightarrow Ag^{+}(aq) + NO_3^{-}(aq) \qquad (8)$$

This time the ions formed are silver ions, $Ag^{+}(aq)$, and nitrate ions, $NO_3^{-}(aq)$. The aqueous silver ion is a silver atom with the positive charge of a proton; it carries the same charge as does an aqueous sodium ion. The aqueous nitrate ion carries the negative charge of an electron—the same charge carried by the aqueous chloride ion. This time, however, the negative charge is carried by four atoms, a nitrogen and three oxygen atoms, that remain together. Since this group, NO_3^{-}, remains together and acts as a unit, it has a distinctive name, nitrate ion.

These three solids, sodium chloride, calcium chloride, and silver nitrate are similar, hence they are classified together. They all dissolve in water to form aqueous ions and give conducting solutions. These solids are called **ionic solids.**

The ease with which an aqueous salt solution conducts electric current is determined by how much salt is dissolved in the water, as well as by the fact that ions are formed. A solution containing 0.1 moles per liter conducts much more readily than a solution containing 0.01 moles per liter. Thus the conductivity is determined by the concentration of ions, as well as by their presence.

Silver chloride is a solid that shows this effect. This solid does not dissolve readily in water. When solid silver chloride is placed in water, very little solid enters the solution and there is only a very slight increase in the conductivity of the solution. Yet there is a real and measurable increase—ions *are* formed. Careful measurements show that even though silver chloride is much less soluble in water than sodium chloride, it is like sodium chloride in that all the solid that does dissolve forms aqueous ions. The reaction is

$$AgCl(s) \longrightarrow Ag^{+}(aq) + Cl^{-}(aq) \qquad (9)$$

Silver chloride, like sodium chloride, is an ionic solid.

* The adjective, aqueous, comes from the Latin name for water, *aqua*.

5-4.2 Precipitation Reactions in Aqueous Solutions

Though both silver nitrate and sodium chloride have high solubility in water, silver chloride is only slightly soluble. What will happen if we mix a solution of silver nitrate and sodium chloride? Then, we will have a solution that includes the species present in a solution of silver chloride, $Ag^+(aq)$ and $Cl^-(aq)$, but now they are present at high concentration! The $Ag^+(aq)$ came from reaction (*8*) and the $Cl^-(aq)$ came from reaction (*6*) and their concentrations far exceed the solubility of silver chloride. The result is that solid will be formed. *The formation of solid from a solution is called* ***precipitation*****:**

$$Ag^+(aq) + Cl^-(aq) \longrightarrow AgCl(s) \qquad (10)$$

Notice that reaction (*10*) indicates the change that takes place when silver nitrate solutions and sodium chloride solutions are mixed. We could have written a more complete equation:

$$Ag^+(aq) + NO_3^-(aq) + Na^+(aq) + Cl^-(aq) \longrightarrow AgCl(s) + NO_3^-(aq) + Na^+(aq)$$

However, the two ions $NO_3^-(aq)$ and $Na^+(aq)$ do not play an active role in the reaction, nor do they influence the reaction that does occur [reaction (*10*)]. Consequently, they are *not* included in the equation for the reaction. *The balanced chemical equation should show only species which actually participate in the reaction.* These species are called the **predominant reacting species.**

Equations (*6*), (*7*), (*8*), (*9*), and (*10*) involve charged species, ions. When we considered how to balance equations for chemical reactions (Section 3-2.1), we dealt with reactions involving electrically neutral particles. We were guided by the rule that atoms are conserved. This principle is still applicable to reactions involving ions. *In addition*, we must consider the charge balance. A chemical reaction does not produce or consume electric charge. Consequently, the sum of the electric charges among the reactants must be the same as the sum of the electric charges among the products. In reaction (*7*), calcium chloride dissolves to give aqueous Ca^{+2} and Cl^- ions. The balanced equation tells us that the neutral solid calcium chloride dissolves to give one Ca^{+2} ion for every two Cl^- ions. Summing these electric charges,

$$\begin{bmatrix}\text{charge on}\\ CaCl_2\text{ solid}\end{bmatrix} = \begin{bmatrix}\text{charge on}\\ Ca^{+2}\text{ ion}\end{bmatrix} + 2\begin{bmatrix}\text{charge on}\\ Cl^-\text{ ion}\end{bmatrix}$$

$$0 = (2+) + 2\ (1-)$$

$$0 = 0 \qquad (11)$$

In a balanced equation for a chemical reaction, charge is conserved.

EXERCISE 5-9

Balance the equations for the reactions given below. For each balanced equation, sum up the charges of the reactants and compare to the sum of the charges of the products.

(a) $PbCl_2(s) \longrightarrow Pb^{+2}(aq) + Cl^-(aq)$
(b) $K_2Cr_2O_7(s) \longrightarrow K^+(aq) + Cr_2O_7^{-2}(aq)$
(c) $Cr_2O_7^{-2}(aq) + H_2O \longrightarrow CrO_4^{-2}(aq) + H^+(aq)$

5-4.3 The Electrical Conductivity of Solids

We have, in this chapter, encountered a number of properties of solids. In Table 5-II, we found that melting points and heats of melting of different solids vary widely. To melt a mole of solid neon requires only 80 calories of heat, whereas a mole of solid copper requires over 3000 calories. Some solids dissolve in water to form conducting solutions (as does sodium chloride), others dissolve in water but no conductivity results (as with sugar). Some solids dissolve in ethyl alcohol but not in water (iodine, for example). Solids also range in appearance. There is little resemblance between a transparent piece of glass and a lustrous piece of aluminum foil, nor between a lump of coal and a clear crystal of sodium chloride.

The great variations among solids make it desirable to find useful classification schemes. Though this topic is taken up much later in the course (Chapter 17), a beginning is provided by a look at the electrical conductivity of solids.

The high electrical conductivity of a substance like copper or silver is familiar to all. Conductivity measurements on many other solids show that all of the substances that conduct electricity as readily as copper and silver are of similar appearance. These good conductors could almost all be classified visually as metals. *The most distinctive property of metallic substances is high electrical conductivity.*

When we study a solid that does not have the characteristic lustrous appearance of a metal, we find that the conductivity is extremely low. This includes the solids we have called ionic solids: sodium chloride, sodium nitrate, silver nitrate, and silver chloride. It includes, as well, the *molecular crystals*, such as ice. This solid, shown in Figure 5-3, is made up of molecules (such as exist in the gas phase) regularly packed in an orderly array. These poor conductors differ widely from the metals in almost every property. Thus electrical conductivity furnishes the key to one of the most fundamental classification schemes for substances.

5-4.4 Ionic Solids

Referring to Tables 5-I and 5-II, we find that both sodium chloride and copper have extremely high melting and boiling points. These two solids have little else in common. Sodium chloride has none of the other properties that identify a metal. It has no luster, rather, it forms a transparent crystal. It does not conduct electricity nor is it a good heat conductor. The kind of forces holding this crystal together must be quite different from those in metals.

The sodium chloride crystal contains an equal number of sodium atoms and chlorine atoms, but they are not present as molecules. On the basis of much experimental evidence, chemists have concluded that sodium chloride crystals are built up of sodium ions, Na^+, and chloride ions, Cl^-, rather than of neutral atoms or molecules. The numbers of Na^+ and Cl^- ions must be equal because the entire crystal is electrically neutral. Nevertheless, there is electrical attraction between these oppositely charged particles. This attraction between positive and negative ions accounts for the binding in an ionic solid.

To represent the composition of such a solid, we use the formula NaCl. However, this formula does not indicate that molecules of sodium chloride are present—it is *not* a molecular formula. Since the formula NaCl shows the composition only it is called an **empirical formula.**

Figure 5-10 shows a representation of the arrangement of the ions in the sodium chloride crystal. The ions are arranged in layers. A layer in the interior of a crystal has a similar layer lying in front of it and a similar layer lying behind it. These layers are displaced so that a Cl^- ion lies in front of each Na^+ ion and a Cl^- ion lies behind each Na^+ ion. Thus, each ion is surrounded by six oppositely charged ions. We call this arrangement the *sodium chloride arrangement* or *sodium chloride lattice*. Because of the proximity of the oppositely charged ions in this arrangement, it is strongly bonded and the melting point of such a crystal is high.

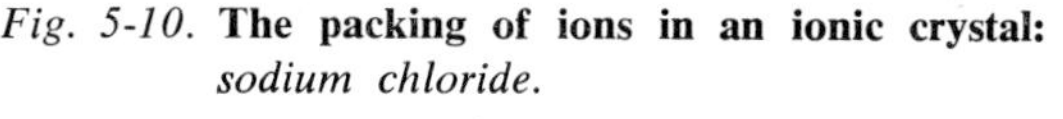

Fig. 5-10. **The packing of ions in an ionic crystal:** *sodium chloride.*

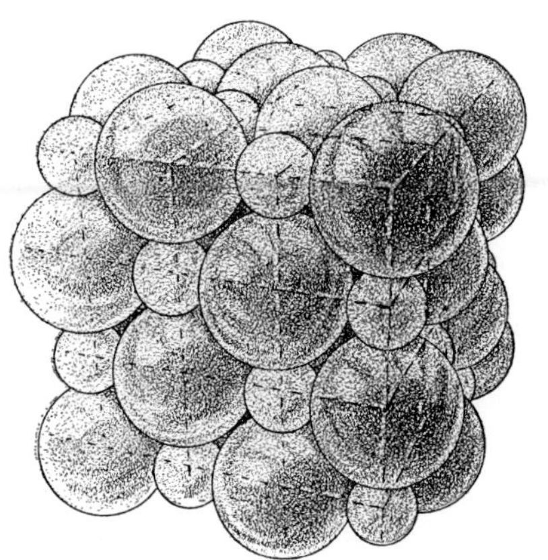

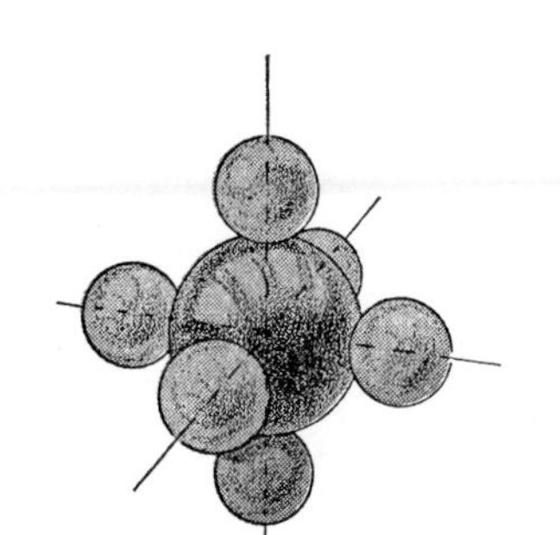

When an ionic solid like sodium chloride is melted, the molten salt conducts electric current. The conductivity is like that of an aqueous salt solution: Na^+ and Cl^- ions are present. The extremely high melting temperature (808°C) shows that a large amount of energy is needed to tear apart the regular NaCl crystalline arrangement to free the ions so they can move.

In contrast, solid sodium chloride dissolves readily in water at room temperature and *without* a large heat effect. This can only mean that the water interacts strongly with the ions—so strongly that aqueous ions are about as stable as are ions in the crystal. In fact, water interacts so strongly with ions that some molecular crystals dissolve in water to form conducting solutions. For example, solid hydrogen chloride, HCl(*s*), is a molecular crystal similar to the ice crystal. The solid is made up of HCl molecules, not of ions like the ionic sodium chloride. Yet HCl(*s*) dissolves in water to form a conducting solution containing hydrogen ions, $H^+(aq)$, and chloride ions, $Cl^-(aq)$. Thus we cannot safely interpret the conductivity of an aqueous solution to mean that the solid dissolved was an ionic solid. We can, however, state the opposite: When an ionic solid dissolves in water, a conducting solution is obtained.

QUESTIONS AND PROBLEMS

1. A liquid is heated at its boiling point. Although energy is used to heat the liquid, its temperature does not rise. Explain.

2. What is the maximum amount of heat that you can lose as one gram of water evaporates from your skin?

3. Note in Table 5-I the correlation between the normal boiling point and heat of vaporization of a number of liquids. Suggest possible reasons for this regularity.

4. Which would likely cause the more severe burn, one gram of $H_2O(g)$ at 100°C or one gram of $H_2O(l)$ at 100°C?

5. Liquids used in rocket fuels are passed over the outer wall of the combustion chamber before being fed into the chamber itself. What advantages does this system offer?

6. Which of the following will require more energy?

   (a) Changing a mole of liquid water into gaseous water.
   (b) Decomposing, by electrolysis, one mole of water.

   Explain.

7. Pick the liquid having the higher vapor pressure from each of the following pairs. Assume all substances are at room temperature.

   (a) Mercury, water.
   (b) Gasoline, motor oil.
   (c) A perfume, honey.

8. Explain why the boiling point of water is lower in Denver, Colorado (altitude, 5,280 feet), than in Boston, Massachusetts (at sea level).

9. Both carbon tetrachloride, CCl_4 (used in dry cleaning and in some fire extinguishers) and mercury, Hg, are liquids whose vapors are poisonous to breathe. If CCl_4 is spilled, the danger can be removed merely by airing the room overnight but if mercury is spilled, it is necessary to pick up the liquid droplets with a "vacuum cleaner" device. Explain.

10. Because of its excellent heat conductivity, liquid sodium has been proposed as a cooling liquid for use in nuclear power plants.

    (a) Over what temperature range could sodium be used in a cooling system built to operate at one atmosphere pressure or lower?
    (b) How much heat would be absorbed per kilogram of sodium to melt the solid when the cooling system is put in operation?
    (c) How much heat would be absorbed per kilogram of sodium if the temperature rose too high and the sodium vaporized?

    Use the data in Tables 5-I (p. 67) and 5-II (p. 69).

11. Water is a commonly used cooling agent in power plants. Repeat Problem 10 considering one kilogram of water instead of sodium. Contrast the results for these two coolants.

12. How much heat must be removed to freeze an ice tray full of water at 0°C if the ice tray holds 500 grams of water?

13. List three heterogeneous materials not given in Section 5-2.1.

14. List three homogeneous materials not given in Section 5-2.1.

15. Which of the following statements about sea water is FALSE?

    (a) It boils at a higher temperature than pure water.
    (b) It melts at a lower temperature than pure water.
    (c) The boiling point rises as the liquid boils away.
    (d) The melting point falls as the liquid freezes.
    (e) The density is the same as that of pure water.

16. How many grams of methanol, CH_3OH, must be added to 2.00 moles of H_2O to make a solution containing equal numbers of H_2O and CH_3OH molecules? How many molecules (of all kinds) does the resulting solution contain?

17. How many grams of ammonium chloride, NH_4Cl, are present in 0.30 liter of a 0.40 M NH_4Cl solution?

    Answer. 6.4 grams.

18. Write directions for preparing the following aqueous solutions:

    (a) 1.0 liter of 1.0 M lead nitrate, $Pb(NO_3)_2$, solution.
    (b) 2.0 liters of 0.50 M ammonium chloride, NH_4Cl, solution.
    (c) 0.50 liter of 2.0 M potassium chromate, K_2CrO_4, solution.

19. How many liters of a 0.250 M K_2CrO_4 solution contain 38.8 grams of K_2CrO_4?

20. List three properties of a solution you would expect to vary as the concentration of the solute varies.

21. Give two forces other than electric that are felt at a distance.

22. What would you expect to observe if one electrometer sphere were charged by your hair and the other by the comb used to comb your hair?

23. Why do scientists claim there are only two kinds of electric charge?

24. It is known that electric charges attract or repel each other with a force that is inversely proportional to the square of the distance between them. If two spheres like those in the electrometer (Figure 5-7) are negatively charged, what would be the change in the force of repulsion if the distance between them were increased to four times the original distance?

25. Why do two electrically neutral objects with mass attract each other?

26. Each of the following ionic solids dissolves in water to form conducting solutions. Write equations for each reaction.

    (a) potassium chloride, KCl
    (b) sodium nitrate, $NaNO_3$
    (c) calcium bromide, $CaBr_2$

    Answer. $CaBr_2(s) \longrightarrow Ca^{+2}(aq) + 2Br^-(aq)$.

    (d) lithium iodide, LiI

27. A chloride of iron called ferric chloride, $FeCl_3$, dissolves in water to form a conducting solution containing ferric ions, Fe^{+3}, and chloride ions, Cl^-.

    (a) Write the equation for this reaction.
    (b) If 0.10 mole of $FeCl_3$ is dissolved in 1.0 liter of water, what is the concentration of ferric ion and of chloride ion?

    Answer. Concentration of Fe^{+3} = 0.10 M.
    Concentration of Cl^- = 0.30 M.

28. The salt ammonium sulfate, $(NH_4)_2SO_4$, dissolves in water to form a conducting solution containing ammonium ions, NH_4^+, and sulfate ions, SO_4^{-2}.

    (a) Write the balanced equation for the reaction when this ionic solid dissolves in water.
    (b) Verify the conservation of charge by comparing the charge of the reactant to the sum of the charges of the products.
    (c) Suppose 1.32 grams of ammonium sulfate is dissolved in water and diluted to 0.500 liter. Calculate the concentrations of $NH_4^+(aq)$ and $SO_4^{-2}(aq)$.

29. 1.00 liter of solution contains 0.100 mole of ferric chloride, $FeCl_3$, and 0.100 mole of ammonium chloride, NH_4Cl. Calculate the concentrations of Fe^{+3}, Cl^-, and NH_4^+ ions.

    Answer. Concentration Fe^{+3} = 0.100 *M*,
    Concentration NH_4^+ = 0.100 *M*,
    Concentration Cl^- = 0.400 *M*.

30. In Experiment 10 you mixed lead nitrate and sodium iodide. Write an equation for the reaction that occurred. Show only the predominant reacting species.

31. Write equations for the reactions between aqueous bromide ions and:

    (a) aqueous lead ions,
    (b) aqueous silver ions.

    Both lead bromide, $PbBr_2$, and silver bromide, AgBr, are only slightly soluble.

32. When solutions of barium chloride, $BaCl_2$, and potassium chromate, K_2CrO_4, are mixed, the following reaction occurs:

$$2K^+(aq) + CrO_4^{-2}(aq) + Ba^{+2}(aq) + 2Cl^-(aq) \longrightarrow BaCrO_4(s) + 2K^+(aq) + 2Cl^-(aq)$$

    (a) Show how charge is conserved.
    (b) Rewrite the equation showing predominant reacting species only.
    (c) Suppose 1.00 liter of 0.500 *M* $BaCl_2$ is mixed with 1.00 liter of 0.200 *M* K_2CrO_4. Assuming $BaCrO_4$ has negligible solubility, calculate the concentrations of all ions present when precipitation stops.

    Answer. Concentration K^+ = 0.200 *M*,
    Concentration Cl^- = 0.500 *M*,
    Concentration CrO_4^{-2} = negligible,
    Concentration Ba^{+2} = 0.150 *M*.

CHAPTER 

Structure of the Atom and the Periodic Table

The eighth element, starting from a given one, is a kind of repetition of the first, like the eighth note of an octave in music.

J. A. R. NEWLANDS, 1864

We have already learned that nature has great variety. Around us we find gases, liquids, and solids. To liquefy air, we must cool it to about −180°C, far colder than the coldest winter. To liquefy rock, we must heat it to temperatures above 1000°C, the climate found in an active volcano. When we examined chemical reactivity, we found even more variety. A candle burns quietly and slowly, once lit, though it does not react appreciably until lit. Iron also reacts with oxygen very slowly (it rusts), though not as slowly as we might like. Hydrogen, by contrast, reacts explosively with oxygen when it is ignited. In contrast to the slow reactions of paraffin wax and iron with oxygen, and the instantaneous reaction of hydrogen with oxygen, helium gas will *never* react with oxygen.

Turning to the atomic view of matter, we find more than a hundred different elements. Each of these elements has a kind of atom that is somehow different from all of the others. With these 100 elements, chemists have prepared about one and one-half million different compounds, each having its own special properties. Each year about 100,000 new compounds are reported. Again we must deal with great variety.

We have already remarked in Chapter 1 that "the mere cataloguing of observations is not science." We could never cope with this great variety in nature if we did not make use of its regularities in organizing our knowledge. The fact that chemists have been able to synthesize more than a million compounds shows that they have been successful in this organization. Their success stems in large part from the regularities embodied in the *periodic table*.

The periodic table groups elements with similar chemistry. It is of great value just as a correlating device. It is even more powerful when coupled with an understanding of the structure of atoms. So, it is appropriate to consider this topic before examining the relationships that establish the periodic table.

6-1 STRUCTURE OF THE ATOM

Scientists have developed a highly sophisticated view of the structure of the atom. The currently accepted model is called "the nuclear atom." We shall present it without trying to show immediately all of the experimental evidence that led to this particular model. Rest assured, though, that every feature of the nuclear atom picture rests upon experimental evidence, as we shall see in Chapter 14.

6-1.1 A Model: The Nuclear Atom

An atom contains electrons and protons. Since mass is associated with all matter, it is natural to assume that atoms, which form matter, have mass. And since any sample of matter occupies a certain volume, we can also assume that each atom has volume. Almost all the mass of the atom is concentrated in a region that is much smaller than the total volume of the atom. This region is called the nucleus of the atom. *The rest of the volume of the atom is occupied by electrons.*

The nucleus carries a positive electric charge. The element hydrogen has the lightest atoms, and the nuclei of these atoms have the smallest positive charge anyone has observed. Every atom of hydrogen has one proton in its nucleus. The charge on the nucleus of an atom of hydrogen is that of a single proton, 1+ unit of charge. All other nuclei have positive charges that are exactly an integer times the proton charge; a nucleus may have 2+ units, 3+ units, 4+ units, and so on. Each nucleus contains a definite number of protons and the charge on the nucleus is fixed by this number. *All atoms of a particular element have the same nuclear charge.* All hydrogen atoms have a nuclear charge of 1+; all helium atoms have a nuclear charge of 2+; all lithium atoms have a nuclear charge of 3+; and so on. We shall see that *the nuclear charge determines the chemistry of an atom.*

Since the nucleus has positive charge, it attracts electrons (each with negative charge). If a nucleus attracts the number of electrons just equal to the nuclear charge, an electrically neutral atom is formed. Consider a nucleus containing two protons, a helium nucleus. When the helium atom has two electrons as well (2− charge), an electrically neutral helium atom results:

$$\begin{array}{ccccc} \text{2 protons} & + & \text{2 electrons} & = & \text{no charge} \\ (2+) & + & (2-) & = & 0 \end{array}$$

Electrons can be removed from or added to a neutral atom, giving it a net charge. This is how ions are formed. Thus,

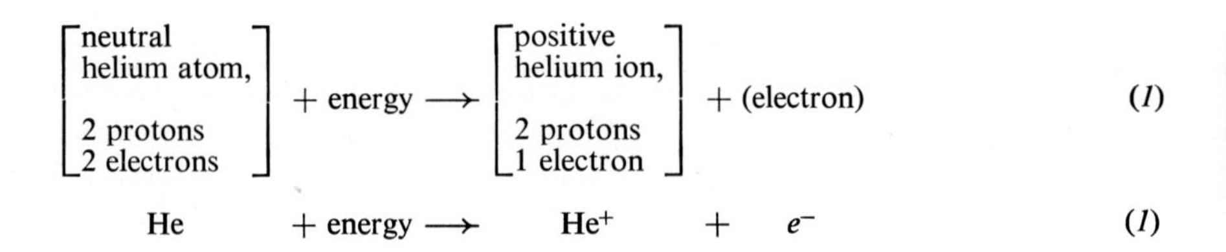

and, as another example,

$$\begin{bmatrix} \text{neutral} \\ \text{fluorine atom,} \\ \\ \text{9 protons} \\ \text{9 electrons} \end{bmatrix} + \text{energy} \longrightarrow \begin{bmatrix} \text{positive} \\ \text{fluorine ion,} \\ \\ \text{9 protons} \\ \text{8 electrons} \end{bmatrix} + \text{(electron)} \qquad (2)$$

$$\text{F} \quad + \text{energy} \longrightarrow \quad \text{F}^+ \quad + \quad e^- \qquad (2)$$

Since a positive charge attracts a negative charge, it is difficult to take the electron away from the positive helium or fluorine nucleus. Scientists say "work must be done" or "energy is required" to form a positive ion from a neutral atom, as in (*1*) or (*2*). "Work" and "energy" are synonymous here and they indicate that an external agent must exert force on the electron to make it leave the neutral atom. In analogy, the attraction between electron and nucleus is like a stretched rubber band connecting the two particles. Continued force applied to the two particles can result in the rubber band being stretched until it breaks, releasing the two particles, but at the expense of work.

Some neutral atoms can *gain* electrons, forming negative ions. Thus a neutral fluorine atom can *add* an electron to form a negative ion, F^-. This change, for fluorine atoms, does not require the input of energy; it *releases* energy:

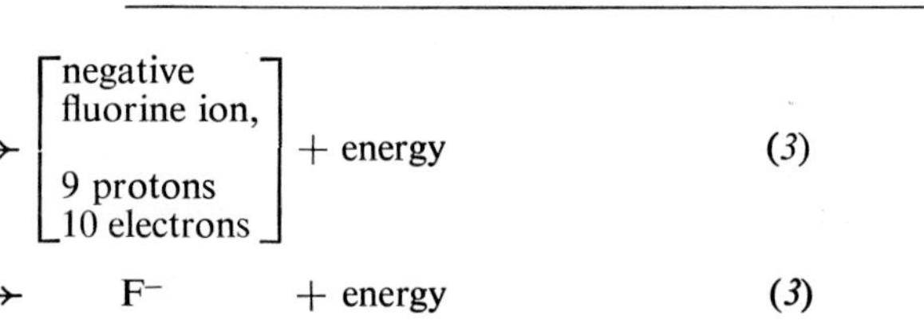

$$\begin{bmatrix}\text{neutral} \\ \text{fluorine atom,} \\ \\ \text{9 protons} \\ \text{9 electrons}\end{bmatrix} + (\text{electron}) \longrightarrow \begin{bmatrix}\text{negative} \\ \text{fluorine ion,} \\ \\ \text{9 protons} \\ \text{10 electrons}\end{bmatrix} + \text{energy} \qquad (3)$$

$$F \quad + \quad e^- \quad \longrightarrow \quad F^- \quad + \text{energy} \qquad (3)$$

6-1.2 The Mass of an Atom and Its Parts

Protons are in the nucleus and electrons surround it. Most of the mass of he atom is in the nucleus. These two statements imply that an electron weighs far less than a proton; this is the case. Experiments have been performed in which individual electrons and protons have been weighed (they are described in Chapter 14). These experiments show that the mass of the electron is smaller than that of a proton by a factor of $\frac{1}{1840}$.

This means that most of the mass of the atom must be furnished by the nucleus. However, the mass of the nucleus is not determined by the number of protons alone. For example, a helium nucleus has two protons and a hydrogen nucleus has one proton. Yet a helium atom is measured to be four times heavier than a hydrogen atom. What can be the composition of the helium nucleus? A partial answer to this problem was obtained when a third particle, the neutron, was discovered. The neutron carries no charge; it is a neutral particle. Its mass is almost identical to the mass of the proton. Thus the nucleus of the helium atom must consist of two neutrons and two protons. Then its charge will be 2+ but its mass will be four times the mass of the hydrogen atom.

Now our nuclear model suffices. We can build up the atoms for all elements. Each atom has a nucleus consisting of protons and neutrons. The protons are responsible for all of the nuclear charge and part of the mass. The neutrons are responsible for the rest of the mass of the nucleus. The neutron plays a role in binding the nucleus together, apparently adding attractive forces which predominate over the electrical repulsions among the protons.*

Around the nucleus are enough electrons to make the atom as a whole, electrically neutral.

Table 6-I

CHARGE AND MASS OF SOME FUNDAMENTAL PARTICLES

PARTICLE	CHARGE (relative to the electron charge)	APPROXIMATE MASS (relative to the mass of a proton)
Electron	1−	$\frac{1}{1840}$
Proton	1+	1
Neutron	0	1

The charge and mass of each of the three fundamental particles we have discussed are shown in Table 6-I.

* This model still does not explain what forces hold the nucleus together in spite of the proton repulsions. We do know that the helium nucleus is stable—it can exist indefinitely—but the model does not explain *why* it is stable. Thus we use models because they help us to explain many important facts, even though they do not explain all the facts.

6-1.3 The Sizes of Atoms

How large is an atom? We cannot answer this question for an isolated atom. We can, however, devise experiments in which we can find how closely the nucleus of one atom can approach the nucleus of another atom. As atoms approach, they are held apart by the repulsion of the positively charged nuclei. The electrons of the two atoms also repel one another but they are attracted by the nuclei. The closeness of approach of two nuclei will depend upon a balance between the repulsive and attractive forces. It also depends upon the energy of motion of the atoms as they approach one another. If we think of atoms as spheres, we find that their diameters vary from 0.000 000 01 to 0.000 000 05 cm (from 1×10^{-8} to 5×10^{-8} cm). Nuclei are much smaller. A typical nuclear diameter is 10^{-13} cm, about 1/100,000 the atom diameter.

Suppose we take Yankee Stadium (seating capacity, 67,000) as a model for the atom. To keep the proper scale, the nucleus would be about the size of a flea! For the hydrogen atom, the flea would represent one proton. He would be located at the center of the stadium, somewhere behind second base. The one electron present in the neutral atom would wander hither and yon, occupying *all* the rest of the stadium. For the helium atom, the nucleus would be replaced by a cluster of four fleas—representing two protons and two neutrons. The two electrons of the neutral helium atom would now share the ample space of the huge stadium. This situation is made even more difficult to picture by the fact that the fleas, occupying a minute volume in center field, represent almost all of the mass that is carried in the stadium.

Fig. 6-1. **The nucleus is much smaller than the atom.** *The nucleus occupies about the same relative volume in the atom as does a flea in Yankee Stadium.*

6-1.4 Atomic Number

What do the atoms of one element have in common to distinguish them from the atoms of all other elements? Each hydrogen nucleus bears a charge of 1+. Each neutral hydrogen atom has one electron, charge 1−, situated in the relatively large volume of the atom outside the nucleus. Each helium atom has a nucleus whose charge is 2+ and each neutral atom has two electrons around the nucleus. The element lithium has atoms heavier than hydrogen or helium, and each lithium atom has a nucleus whose charge is 3+. Three electrons around the nucleus are required to form a neutral lithium atom.

Thus, each of the chemical elements consists of atoms whose nuclei contain a particular number of protons, hence a particular nuclear charge. *The number of protons in the nucleus is called the*

atomic number. Of course, all atomic numbers are whole numbers. Thus, oxygen with atomic number 8, has eight protons in the nucleus (nuclear charge 8+). A neutral oxygen atom must have eight electrons (each with charge 1−) as well.

The atomic number of each of the elements is listed in the table on the inside of the back cover of this book. You will find there that each element has a distinctive name, symbol, and atomic number. A given element can be identified by any of these. For example, helium can be called by its name, helium, by its symbol, He, or by its atomic number, the element of atomic number 2.

In the periodic table we shall see that the elements have been listed in the order of increasing atomic number. An example of the periodic table is on the wall of your classroom and there is a copy on the inside of the cover at the front of this book. In each box in the table the number above the chemical symbol for the element is the atomic number.

6-1.5 Mass Number and Isotopes

All of the atoms of an element have the same nuclear charge. Do all of the atoms of an element have the same mass? Almost all hydrogen atoms do have the same mass—the sum of the mass of one proton and the mass of one electron. For these atoms the nucleus consists of a single proton. However, a small fraction of the hydrogen atoms 0.016% of them) have nuclei whose mass is approximately twice as great as the mass of the proton. (Compare with the helium nucleus.) To explain the mass of these hydrogen atoms, we conclude that each of their nuclei consists of a neutron (charge zero, mass 1) and a proton (charge 1+, mass 1). This kind of hydrogen atom is called *hydrogen-2;* another name commonly

Table 6-II. **VITAL STATISTICS OF SOME COMMON ISOTOPES**

NAME OF ISOTOPE	ABUNDANCE IN NATURE	ATOMIC NUMBER	MASS NUMBER	NUCLEUS COMPOSITION*		NUCLEUS MASS	NUCLEUS CHARGE	NUMBER OF ELECTRONS IN NEUTRAL ATOM
hydrogen-1	99.984%	1	1	1*p*		1	1+	1
hydrogen-2	0.016	1	2	1*p*,	1*n*	2	1+	1
helium-3	1.34×10^{-4}	2	3	2*p*,	1*n*	3	2+	2
helium-4	100	2	4	2*p*,	2*n*	4	2+	2
lithium-6	7.40	3	6	3*p*,	3*n*	6	3+	3
lithium-7	92.6	3	7	3*p*,	4*n*	7	3+	3
beryllium-9	100	4	9	4*p*,	5*n*	9	4+	4
boron-10	18.83	5	10	5*p*,	5*n*	10	5+	5
boron-11	81.17	5	11	5*p*,	6*n*	11	5+	5
carbon-12	98.892	6	12	6*p*,	6*n*	12	6+	6
carbon-13	1.108	6	13	6*p*,	7*n*	13	6+	6
nitrogen-14	99.64	7	14	7*p*,	7*n*	14	7+	7
nitrogen-15	0.36	7	15	7*p*,	8*n*	15	7+	7
oxygen-16	99.76	8	16	8*p*,	8*n*	16	8+	8
oxygen-17	0.04	8	17	8*p*,	9*n*	17	8+	8
oxygen-18	0.20	8	18	8*p*,	10*n*	18	8+	8
fluorine-19	100	9	19	9*p*,	10*n*	19	9+	9
chlorine-35	75.4	17	35	17*p*,	18*n*	35	17+	17
chlorine-37	24.6	17	37	17*p*,	20*n*	37	17+	17
gold-197	100	79	197	79*p*,	118*n*	197	79+	79
uranium-235	0.71	92	235	92*p*,	143*n*	235	92+	92
uranium-238	99.28	92	238	92*p*,	146*n*	238	92+	92

* *p* = proton, *n* = neutron.

used is *deuterium*. The two kinds of hydrogen atoms (having the same atomic number but different masses) are called **isotopes.** An isotope is identified by specifying, first, which element it is (usually by the symbol or name of the element) and, second, the sum of the number of protons and the number of neutrons. *The number of protons plus the number of neutrons in a nucleus is called the mass number of the nucleus.* The mass number is, of course, always an integer.

Reviewing:

Atomic number = number of protons in the nucleus (fixes nuclear charge)

Mass number = number of protons and neutrons in the nucleus (fixes nuclear mass)

Most of the chemical elements consist of mixtures of isotopes. Oxygen, atomic number 8, has three stable isotopes. The kind having mass number 16 is most abundant. About 99.76% of the oxygen atoms consist of this isotope. Only 0.04% of the oxygen atoms have mass number 17 and about 0.20% have mass number 18. The nucleus of an oxygen-16 atom consists of eight protons and eight neutrons—charge 8+, mass 16. The nucleus of an oxygen-17 atom consists of eight protons and nine neutrons—charge 8+, mass 17. The nucleus of an oxygen-18 atom consists of eight protons and ten neutrons—charge 8+, mass 18. Table 6-II summarizes the data on the atomic structure of a few common isotopes.

The nuclear charge and the electrons it attracts primarily determine the ways in which atoms behave toward other atoms. Mass differences cause only minor chemical effects. Since the isotopes of an element have the same nuclear charge and the same number of electrons per neutral atom, they react in the same ways. Thus we can speak of the chemistry of oxygen without specifying which one of the three stable isotopes is reacting. Only the most precise measurements will indicate the very slight chemical differences among them.

6-2 THE SIMPLEST CHEMICAL FAMILY—THE INERT GASES

We are now ready to consider why the elements are arranged as they are in the periodic table. We shall examine the known elements to discover the significance and usefulness of this table—a table so important it is printed on the inside cover of this and almost every other general chemistry textbook.

When the elements are arranged in order of increasing atomic number, each successive element has different chemistry from its neighbors. On the other hand, there are striking likenesses among some of the elements. For example, of the one-hundred or so known elements, only about a dozen are gases at normal conditions of temperature and pressure. Of these, six elements are so remarkably similar in chemistry that they are conveniently considered together. Further, they are identified by a family name, "the inert gases."

If we look at the properties of other elements, we find that this recurrence of properties is usual. It is convenient to group all of the elements according to their chemistry into families or groups. The elements in a particular group have similar chemistry. Knowledge about one element in a group then aids in understanding the chemistry of other elements in that group. In the periodic table each group appears in a vertical

Fig. 6-2. **The inert gas elements**—*a chemical family.*

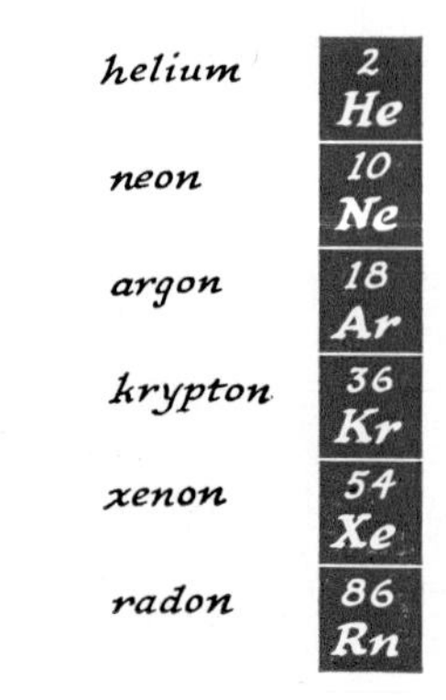

column. The key to this arrangement lies in the elements called the inert gases. It is one of nature's quirks that these gases provide the key to the organization of our chemical knowledge because these elements are distinctive in their almost total *absence of chemical reactivity*. The first of the inert gases is helium.

6-2.1 Helium

Helium, the second element in the periodic table, has atomic number 2. This means its nucleus contains two protons and has a 2+ charge. The neutral atom, then, contains two electrons. There are two stable isotopes, helium-4 and helium-3, but the helium found in nature is almost pure helium-4. Helium is found in certain natural gas fields and is separated as a by-product. Sources of helium are rare and most of the world supply is produced in the United States, mainly in Texas and Kansas.

Helium is a monatomic gas and, as yet, *no stable compounds of helium have been found.* The attractive forces between the atoms of helium are unusually weak, as shown by the normal boiling point. To liquefy helium, it must be cooled to −268.9°C or 4.2°K. No other element or compound has a boiling point as low. Helium has another distinction which reflects these weak forces: it is the only substance known which cannot be solidified at any temperature unless it is subjected to pressure. Helium becomes solid at 1.1°K at a pressure of 26 atmospheres.

Chemical reactions tend to occur among atoms to form more stable arrangements of the atoms. Helium takes part in no chemical reactions. *Helium atoms must be particularly stable.*

6-2.2 Neon, Argon, Krypton, Xenon, Radon

Among all of the hundred or so known elements, there are only five with chemistry resembling that of helium. We have remarked that these elements; *neon, argon, krypton, xenon,* and *radon,* together with *helium,* are known as the **inert gases.** It is only since 1962 that chemists have been able to prepare *any compounds at all* involving these elements. The few compounds that have been prepared are extremely reactive, decomposing readily to restore the inert gas to the elementary state. Everything that is known of the chemistry of the inert gas elements indicates that *the atoms of the inert gases must be particularly stable.*

The physical properties of the inert gases are shown in Table 6-III. There is much information contained in this table, and we shall examine it in parts.

BOILING POINT

The inert gas elements are all gases at room temperature. Helium has the lowest known boiling point, 4.2°K. Neon has the third lowest known boiling point, 27.2°K. (Hydrogen, H_2, has the second lowest known boiling point, 20.4°K.) The boiling point of argon is still quite low, 87.3°K, but not so low as to distinguish it from a number of diatomic molecules. (Compare the boiling points of nitrogen, N_2, b.p. = 77.4°K; fluorine, F_2, b.p. = 85.2°K; oxygen, O_2, b.p. = 90.2°K.) Krypton, xenon, and radon have successively higher boiling points. Apparently, as the atomic number goes up, the boiling point goes up. Figure 6-3 shows the boiling point trend of the inert gases in a plot of boiling point against atomic number. Since the atomic number gives

Table 6-III. **SOME PROPERTIES OF THE INERT GAS ELEMENTS**

PROPERTY	He	Ne	Ar	Kr	Xe	Rn
Atomic number	2	10	18	36	54	86
Atomic weight	4.00	20.2	39.9	83.7	131	222
Boiling point (°K)	4.2	27.2	87.3	120	165	211
Melting point (°K)	—	24.6	83.9	116	161	202
Atomic volume, liquid (ml/mole of atoms)	31.8	16.8	28.5	32.2	42.9	50.5

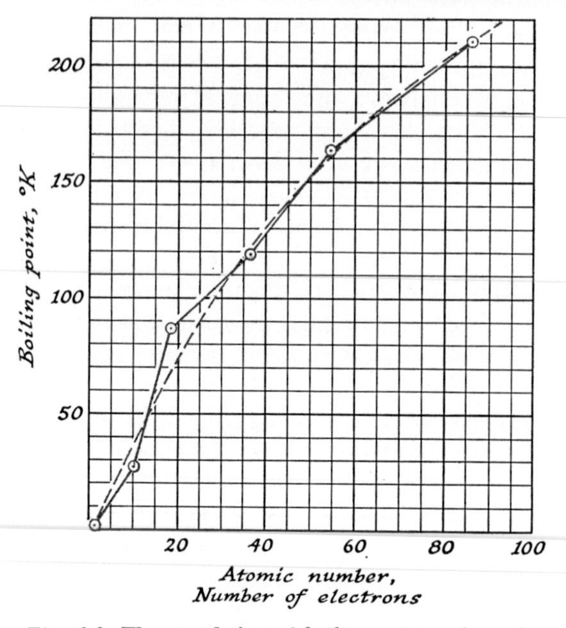

Fig. 6-3. **The correlation** *of boiling point and number of electrons per atom for the inert gases.*

the number of protons in the nucleus, it also gives the number of electrons held by each atom. We can interpret a higher boiling point to mean that more energy must be added to disrupt the liquid state. Hence, the weak attractive forces which cause the inert gases to liquefy increase as the number of electrons per atom increases.

MELTING POINT

At temperatures only slightly below the liquefaction temperatures, the liquids freeze. The solids are all simple crystals in which the atoms are close-packed in a regular lattice arrangement. The narrow temperature range over which any one of these liquids can exist suggests that the forces holding the crystal together are very much like the forces in the liquid.

ATOMIC VOLUME

We picture the atoms in a liquid and in a solid as being packed rather tightly. The packing is random in the liquid and regular in the solid. With this picture, we can deduce from the volume per mole of atoms the volume to be assigned to a single atom. Consulting Table 6-III, we find that helium is distinctive in its atomic volume (indeed, as it is in every other property listed in this table). Otherwise, we see that volume per mole of atoms increases with atomic number, that is, with the number of electrons around the nucleus. Notice that the volume per mole of atoms of neon (16.8 ml) is only slightly less than that of water (18 ml). The water molecule occupies slightly more space than does the neon atom.

PHYSICAL PROPERTIES: SUMMARY

We have compared the physical properties of the inert gases among themselves and with the corresponding properties of a few simple diatomic molecules. We find that the forces between the atoms are indeed weak though comparable to the forces between some stable molecules. The physical properties, however, do not distinguish this group of elements. *The uniqueness of the inert gases relates to compound formation;* when compared with the other elements, *the inert gases are seen to have a specially small tendency to form stable compounds.*

6-2.3 Number of Electrons and Stability

The inert character of stable compounds makes them of special interest to us. We might make a

Fig. 6-4. **Trends in the physical properties of the inert gases.**

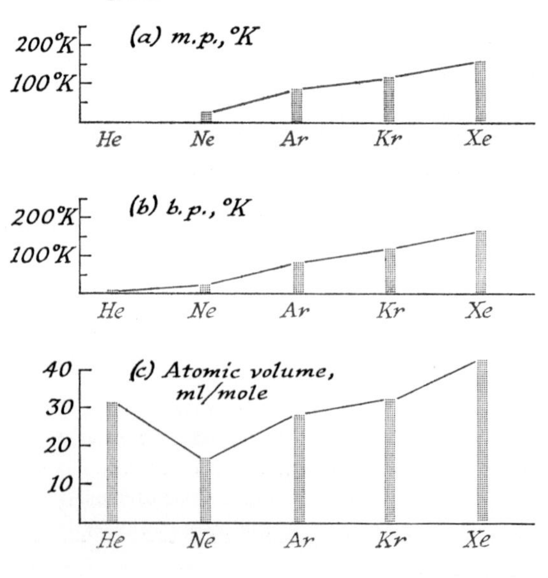

list of the number of electrons possessed by each inert gas. (Remember the lost child organizing his information?) We shall find these numbers to be of particular value in all of our future study of chemistry.

In Table 6-IV there is much food for thought. First, each of the especially stable atoms has an even number of electrons. Next we see that there seems to be some regularity in the differences between the number of electrons possessed by a given inert gas and that of its predecessor. The first two differences are eight and the second two differences are eighteen. If there were another inert gas, would it have $86 + 32 = 118$ electrons? What is the special significance of the numbers 8, 18, and 32? Is it true for any other element that the electron arrangements of helium, neon, argon, and so on are especially stable? We shall see that this is true, not just for some elements, but for *all* elements.

Table 6-IV

THE ELECTRON POPULATIONS OF THE INERT GASES

INERT GAS	TOTAL NUMBER OF ELECTRONS	CHANGE IN NUMBER OF ELECTRONS
helium	2	
neon	10	$10 - 2 = 8$
argon	18	$18 - 10 = 8$
krypton	36	$36 - 18 = 18$
xenon	54	$54 - 36 = 18$
radon	86	$86 - 54 = 32$

6-2.4 Sodium Chloride—Atoms Trying to Be Inert Gas Atoms

Sodium chloride is a compound of an element, chlorine, which just precedes an inert gas, and an element, sodium, which just follows an inert gas. Sodium has atomic number (nuclear charge) 11, one larger than neon, which has atomic number 10. Hence, the neutral atom of sodium has one more electron than the number held by the especially stable neon atom. Chlorine has atomic number (nuclear charge) 17, one below that of argon, 18. The neutral atom of chlorine has one less than the number of electrons held by the especially stable argon atom. We find sodium and chlorine combined in a one-to-one ratio in the very stable compound called sodium chloride, $NaCl(s)$.

We have already discussed the structure of solid sodium chloride in Chapter 5. We said there, "On the basis of much experimental evidence, chemists have concluded that *sodium chloride crystals are built up of charged particles* rather than of neutral atoms." The discussion went on to identify the ions in the lattice as Cl^- ions packed tightly around Na^+ ions (see Figure 5-10 on page 81). But how many electrons does a chlorine ion, Cl^-, have? By gaining an electron, the chlorine now has 18 electrons, exactly the same number of electrons as does argon, the adjacent inert gas. In a similar way, but by losing an electron, the sodium has contrived to reach the 10 electron population of its adjacent inert gas, neon. The atoms reached these inert gas-like electron arrangements through compound formation and the resulting compound thereby acquired some of the unique stability of the inert gases.

With this in mind, let us explore the chemistry of all of the elements immediately adjacent to the inert gases. These two vertical columns of the periodic table are called the *alkalies* and the *halogens.*

6-3 THE ALKALIES

The six elements adjacent to and following the six inert gases are *lithium, sodium, potassium, rubidium, cesium,* and *francium.* These elements have similar chemistries and are called the **alkalies** (or, the **alkali metals**). Figure 6-5 shows that these elements are neighbors to the inert gases. Their chemistries can be understood in terms of the special stabilities of the 1+ ions

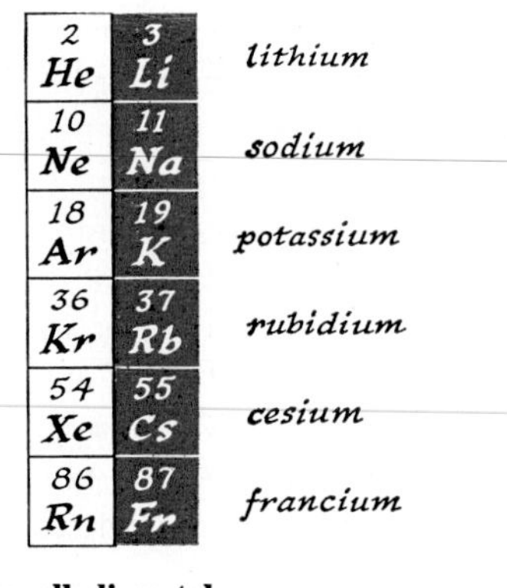

Fig. 6-5. **The alkali metals.**

Fig. 6-6. **Potassium is soft;** *it can be cut with a knife.*

which have the inert gas electron arrangements.

All the alkalies are metals in their elementary states. When the metal surfaces are clean, they have a bright, silvery luster. The metals are excellent conductors of electricity and heat. They are soft and malleable, and have low melting points (compared with almost all other elementary metals).

In Chapter 5 we identified metals by their high electrical conductivity. Now we can explain why they conduct electric current so well. It is because there are some electrons present in the crystal lattice that are extremely mobile. These "conduction electrons" move throughout the metallic crystal without specific attachment to particular atoms. The alkali elements form metals because of the ease of freeing one electron per atom to provide a reservoir of conduction electrons. The ease of freeing these conduction electrons derives from the stability of the residual, inert gas-like atoms.

6-3.1 Physical Properties of the Alkali Elements

Table 6-V lists the same properties for the alkali metals that were listed in Table 6-III for the inert gases.

BOILING POINT AND MELTING POINT

All the alkali metals are solids at room temperature, though cesium melts just above room temperature. Notice that both melting points and boiling points decrease as atomic number increases (the opposite of the inert gas behavior). Figure 6-7a and 6-7b contrast the trends in the alkali metal melting and boiling points to the opposite trends in the inert gases. Notice also the extremely wide temperature range over which the alkali liquids are stable. Sodium, for example, melts at 371°K and boils at 1162°K,

Table 6-V. **SOME PROPERTIES OF THE ALKALI ELEMENTS**

PROPERTY	LITHIUM	SODIUM	POTASSIUM	RUBIDIUM	CESIUM
Atomic number	3	11	19	37	55
Atomic weight	6.94	23.0	39.1	85.4	133
Boiling point (°K)	1599	1162	1030	952	963
(°C)	1326	889	757	679	690
Melting point (°K)	453	371	336.4	311.8	301.7
(°C)	180	98	63.4	38.8	28.7
Atomic volume, solid (ml/mole of atoms)	13.0	23.7	45.4	55.8	70.0
Density of solid at 20°C	0.535	0.971	0.862	1.53	1.90

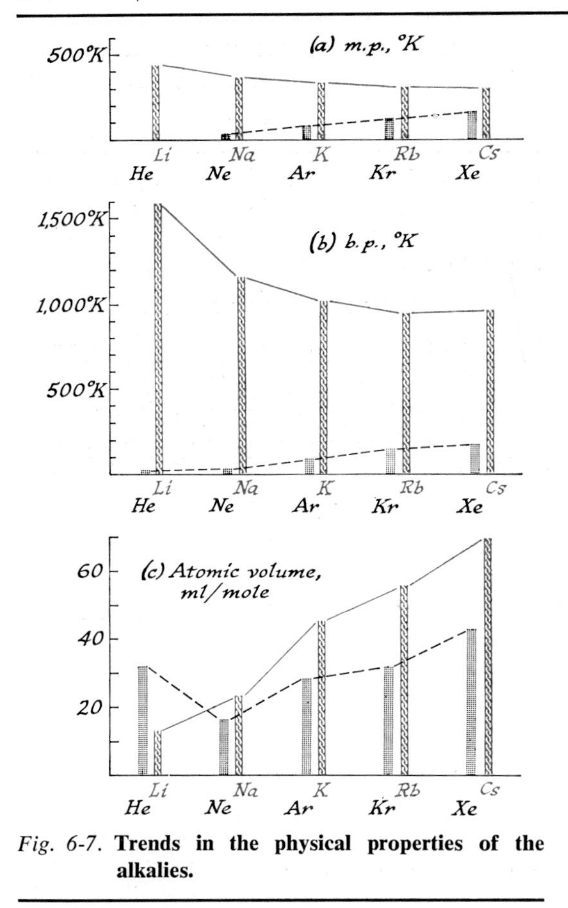

Fig. 6-7. **Trends in the physical properties of the alkalies.**

almost 800° higher. Contrast neon, which melts at 24.6°K and boils only 2.6° higher, at 27.2°K. How different are the alkali metals from the inert gases, only one atomic number removed!

ATOMIC VOLUME

The atomic volumes of the alkali metals increase with atomic number, as do those of the inert gases. Notice, however, that the volume occupied by an alkali atom is somewhat larger than that of the adjacent inert gas (with the exception of the lithium and helium—helium is the cause of this anomaly). The sodium atom in sodium metal occupies 30% more volume than does neon. Cesium occupies close to twice the volume of xenon.

6-3.2 Chemistry of the Alkali Metals

The alkali metals are exact opposites of the inert gases in chemical reactivity. These metals react vigorously when in contact with oxygen and chlorine, and even with such a placid reagent as water. Let us investigate some of these reactions.

REACTIONS OF THE ALKALI METALS WITH CHLORINE

When chlorine gas is brought into contact with sodium metal, sodium chloride is formed:

$$Na(s) + \tfrac{1}{2}Cl_2(g) \longrightarrow NaCl(s) + \text{energy} \quad (4)$$

Sodium chloride has a lattice built up of sodium ions, Na^+, and chloride ions, Cl^-. Therefore, the reaction *(4)* involves the transfer of electrons from sodium atoms to chlorine atoms. The resulting ions attract each other because they have opposite electric charges. Of course, it takes some energy to remove an electron from a sodium atom to form a sodium ion. However, it takes only a moderate amount of energy because the Na^+ ion that is formed has the electron population of an inert gas, neon. This electron, removed from a sodium atom, is then added to a chlorine atom to form a chloride ion, Cl^-. This reaction *releases* a small amount of energy—the Cl^- ion also has the electron population of an inert gas, argon. At a rather low net cost of energy, an electron can be removed from a sodium atom and transferred to a chlorine atom, giving Na^+ and Cl^- ions. Now these two ions can move toward each other with a large reduction in total energy. The stability of the sodium chloride crystal can be said to depend upon the electrical attraction of the oppositely charged ions. The crystal is said to be held together by **ionic bonds.**

This chemistry is characteristic of all of the alkali metals. Each of them reacts with chlorine gas in a similar way:

$$Li(s) + \tfrac{1}{2}Cl_2(g) \longrightarrow LiCl(s) + \text{energy} \quad (5)$$
$$Na(s) + \tfrac{1}{2}Cl_2(g) \longrightarrow NaCl(s) + \text{energy} \quad (6)$$
$$K(s) + \tfrac{1}{2}Cl_2(g) \longrightarrow KCl(s) + \text{energy} \quad (7)$$
$$Rb(s) + \tfrac{1}{2}Cl_2(g) \longrightarrow RbCl(s) + \text{energy} \quad (8)$$
$$Cs(s) + \tfrac{1}{2}Cl_2(g) \longrightarrow CsCl(s) + \text{energy} \quad (9)$$

In every case, the alkali metal reacts to form a stable, ionic solid in which the alkali is present as an inert gas-like ion. The product is, in each case, a crystalline substance with high solubility in water.

REACTIONS OF THE ALKALI METALS WITH WATER

Sodium metal reacts vigorously with water to form hydrogen gas and an aqueous solution of sodium hydroxide, NaOH:

$$2Na(s) + 2H_2O \longrightarrow 2Na^+(aq) + 2OH^-(aq) + H_2(g) + \text{energy} \quad (10)$$

Energy is liberated and the reaction often takes place so rapidly that the temperature rises and the hydrogen, mixing with air, explodes. Thus, sodium metal is dangerous and must be handled with caution. This chemistry is also characteristic of all of the alkali metals.

EXERCISE 6-1

Write the equations for the reactions between water and: lithium, potassium, rubidium, cesium.

Again, we see that the alkali metals display likeness in their reactions with water. Furthermore, the reaction products always include an aqueous ion of the alkali element in which one electron has been removed, giving a 1+ ion.

SUMMARY OF CHEMISTRY OF THE ALKALI ELEMENTS

The alkali metals are extremely reactive. Thus, there is a dramatic change in chemistry as we pass from the inert gases to the next column in the periodic table. The chemistry of the alkali metals is interesting and often spectacular. Thus, these metals react with chlorine, water, and oxygen, always forming a +1 ion that is stable in contact with most substances. The chemistry of these +1 ions, on the other hand, is drab, reflecting the stabilities of the inert gas electron arrangements that they have acquired.

6-4 THE HALOGENS

Now let's slide to the left in the periodic table and consider the column of elements *fluorine*, *chlorine*, *bromine*, *iodine*, and *astatine*. Each of these elements has one less electron than does its neighboring inert gas. These elements are called the **halogens.** (The discussion that follows does not include astatine because this halogen is very rare.)

Fig. 6-8. **The halogens.**

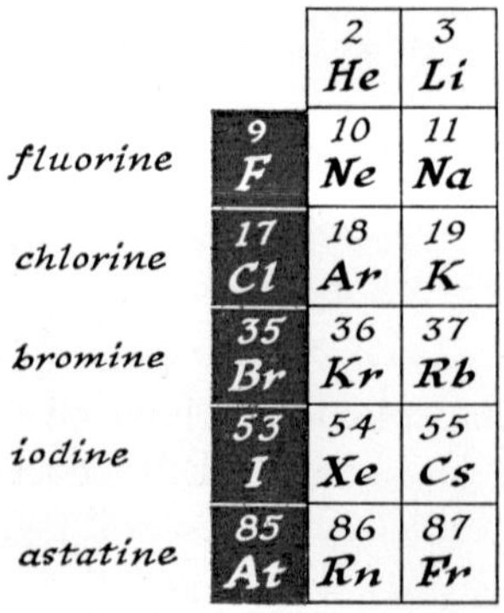

		2 He	3 Li
fluorine	9 F	10 Ne	11 Na
chlorine	17 Cl	18 Ar	19 K
bromine	35 Br	36 Kr	37 Rb
iodine	53 I	54 Xe	55 Cs
astatine	85 At	86 Rn	87 Fr

6-4.1 Physical Properties of the Halogens

Table 6-VI lists some properties of the halogens. In the elemental state, the halogens form stable diatomic molecules. This stability is indicated by the fact that it takes extremely high temperatures to disrupt halogen molecules to form the monatomic species. For example, it is known that the chlorine near the surface of the sun, at a temperature near 6000°C, is present as a gas consisting of single chlorine atoms. At more normal temperatures, chlorine atoms react with each other to form molecules:

$$2Cl(g) \longrightarrow Cl_2(g) \quad (11)$$

Then, no further reactions among chlorine molecules occur.

Apparently the diatomic molecules of the halogens already have achieved some of the stability characteristic of the inert gas electron arrangement. How is this possible? How could one chlorine atom satisfy its need for one more electron (so it can reach the argon stability) by

Table 6-VI. **SOME PROPERTIES OF THE HALOGENS**

PROPERTY	FLUORINE	CHLORINE	BROMINE	IODINE	ASTATINE
Atomic number	9	17	35	53	85
Atomic weight	19.0	35.5	79.9	127	
Molecular formula	F_2	Cl_2	Br_2	I_2	
Boiling point (°K)	85	238.9	331.8	457	
(°C)	−188	−34.1	58.8	184	
Melting point (°K)	55	172	265.7	387	
(°C)	−218	−101	−7.3	114	
Atomic volume, solid (ml/mole of atoms)	14.6	18.7	23.5	25.7	

combining with another chlorine atom, an atom with a similar need? We answer this question by suggesting that the two atoms *share* two electrons, each atom contributing one electron. If the two atoms huddle close together and place this communal pair of electrons between them, each atom acts more as though it had the special stability of the inert gas. This results in the formation of a stable aggregate of atoms, a molecule. Its formula is Cl_2. The same argument can be made to explain the diatomic molecular formulas of the other halogens. Because each of these molecules is bonded by a *shared pair* of electrons, the bond is called a **covalent bond.**

BOILING POINT AND MELTING POINT

We have already made a comparison between the physical properties of some of the halogens and those of inert gases. The comparison suggests that after the formation of a diatomic molecule, the bonding capacity of the two halogen atoms is "used up." The only attractive interactions remaining between two such diatomic molecules are of the extremely weak variety that account for the liquefaction of the inert gases. Thus, the melting points rise as atomic number increases (remember that the alkali metals did the reverse) and there is but a narrow temperature range over which the liquid is stable. Fluorine and chlorine are gases under normal conditions, bromine is a liquid, and iodine is a solid. These differences in physical state result from the accident that normal conditions fall where they do. On a planet with a "normal" temperature of 25°K, all of the halogens would be solids whereas neon would be a liquid, helium a gas, and argon, krypton, and xenon would be solids.

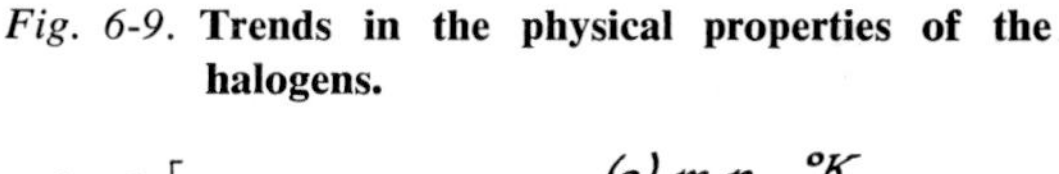

Fig. 6-9. **Trends in the physical properties of the halogens.**

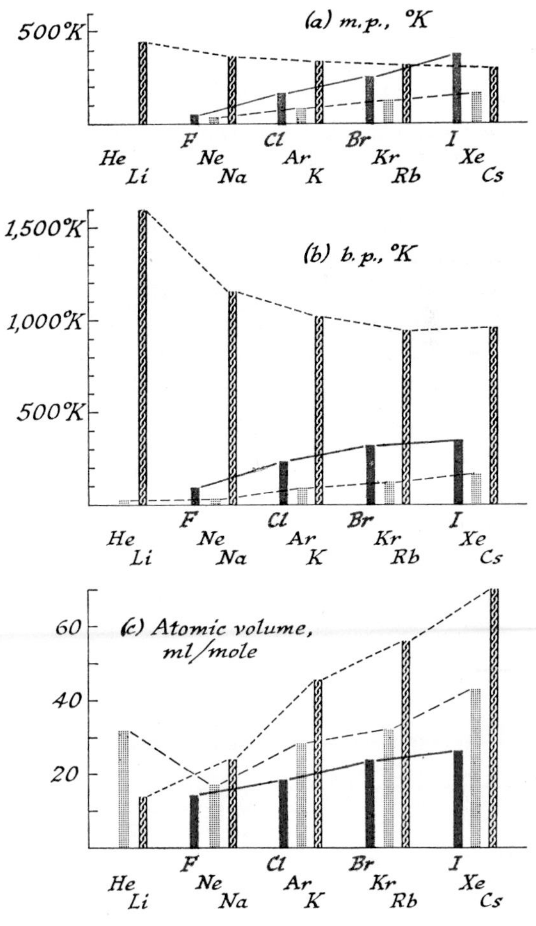

ATOMIC VOLUMES

Here we find a continuation of the trend displayed by the inert gases and alkali metals. Compare the atomic volumes of the three adjacent elements in the solid state:

fluorine 14.6 ml	neon 20.2 ml	sodium 23.7 ml
chlorine 18.7 ml	argon 24.2 ml	potassium 45.4 ml
bromine 23.5 ml	krypton 41.9 ml	rubidium 55.8 ml

In each set, the atomic volumes increase going from halogen to inert gas to alkali metal, as shown graphically in Figure 6-9c. Figure 6-10 shows models constructed on the same scale to show the relative sizes of atoms indicated by the atomic volumes and by the packing of the ions in the ionic solids.

Figure 6-10. **Models (to scale)** *of halogen atoms, inert gas atoms, and alkali atoms.*

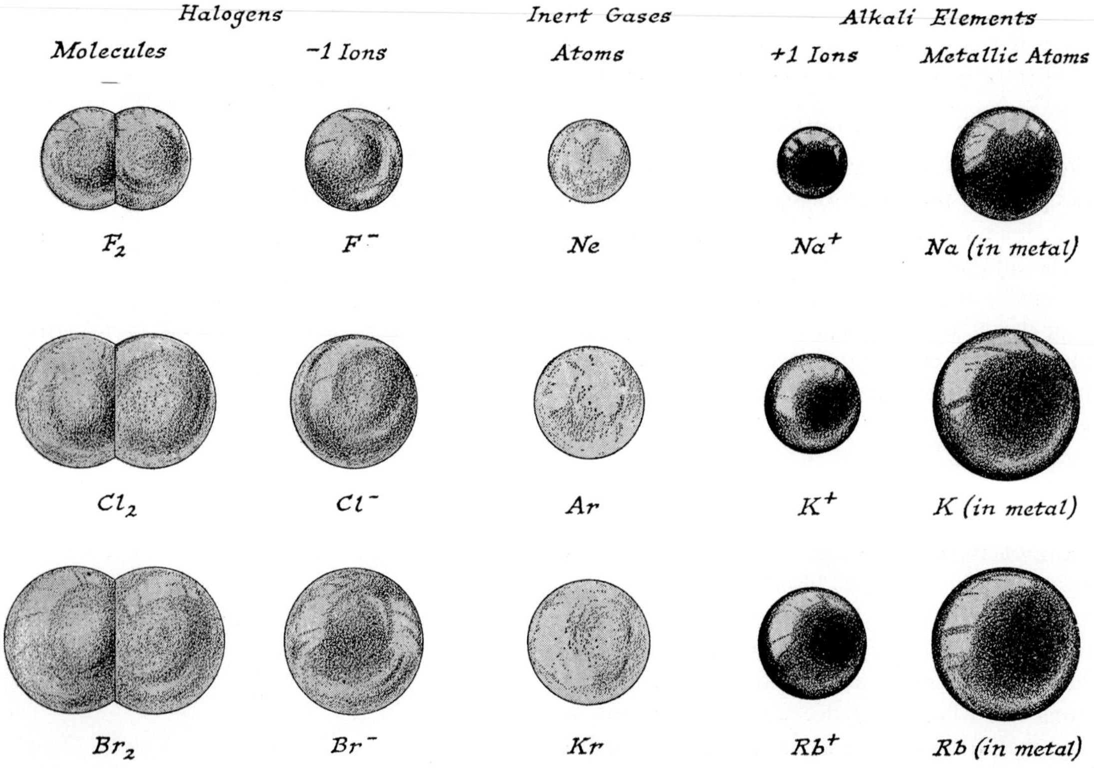

6-4.2 Chemistry of the Halogens

The reactions of the alkali metals with chlorine were used to display the similarities of the alkali metals. In a similar way, the reactions of the halogens with one of the alkali metals, say sodium, show similarity within this group. The reactions that occur are as follows.

$$Na(s) + \tfrac{1}{2}F_2(g) \longrightarrow NaF(s) + \text{energy} \quad (12)$$
$$Na(s) + \tfrac{1}{2}Cl_2(g) \longrightarrow NaCl(s) + \text{energy} \quad (13)$$
$$Na(s) + \tfrac{1}{2}Br_2(g) \longrightarrow NaBr(s) + \text{energy} \quad (14)$$
$$Na(s) + \tfrac{1}{2}I_2(g) \longrightarrow NaI(s) + \text{energy} \quad (15)$$

These reactions all proceed readily and they produce ionic solids with the general empirical formula, NaX. Each of these solids has a crystal structure made up of positively charged sodium atoms and negatively charged halogen atoms. These negative ions, F^-, Cl^-, Br^-, and I^-, are called **halide ions.** The stabilities of these ions can be related to the stabilities of the corresponding inert gas electron arrangement.

The halogens also react with hydrogen gas to form the hydrogen halides:

$$H_2(g) + F_2(g) \longrightarrow 2HF(g) \quad \text{hydrogen fluoride} \quad (16)$$

$$H_2(g) + Cl_2(g) \longrightarrow 2HCl(g) \quad \text{hydrogen chloride} \quad (17)$$

$$H_2(g) + Br_2(g) \longrightarrow 2HBr(g) \quad \text{hydrogen bromide} \quad (18)$$

$$H_2(g) + I_2(g) \longrightarrow 2HI(g) \quad \text{hydrogen iodide} \quad (19)$$

None of these reactions, (*16*), (*17*), (*18*), or (*19*), proceeds readily at room temperatures. This is because the bonds holding the atoms together in the hydrogen and in the halogen molecules must be broken if new bonds are to form between hydrogen atoms and halogen atoms. The breaking of the bonds is favored by high temperatures, however, and, once started, these reactions tend to proceed rapidly or even explosively.

6-4.3 Chemistry of the Halide Ions

Because of the stabilities of halides, most elements form stable halide compounds. Thus calcium forms the compounds CaF_2, $CaCl_2$, $CaBr_2$, and CaI_2, all ionic solids. In each crystal, the calcium ion carries a $+2$ charge, and each of the halide ions carries a -1 charge. The empirical formulas are all of the type CaX_2.

The alkali halides are relatively unreactive substances. They all display high solubility in water and quite low solubility in ethyl alcohol.

We have seen in Experiment 8 that silver chloride has low solubility in water. This is also true for silver bromide and silver iodide. In fact, these low solubilities provide a sensitive test for the presence of chloride ions, bromide ions, and iodide ions in aqueous solutions. If silver nitrate solution, which contains silver ions, $Ag^+(aq)$, and nitrate ions, $NO_3^-(aq)$, is added to a solution containing $I^-(aq)$ ions, a yellow precipitate of $AgI(s)$ forms:

$$Ag^+(aq) + I^-(aq) \longrightarrow \underset{\text{yellow}}{AgI(s)} \quad (20)$$

If Br^- ions are present, the reaction is

$$Ag^+(aq) + Br^-(aq) \longrightarrow \underset{\text{light yellow}}{AgBr(s)} \quad (21)$$

and if Cl^- ions are present, the reaction is

$$Ag^+(aq) + Cl^-(aq) \longrightarrow \underset{\text{white}}{AgCl(s)} \quad (22)$$

Silver fluoride is soluble. Therefore, no precipitate forms as Ag^+ ions are added to a solution of F^- ions.

All the hydrogen halides are gaseous at room temperature but hydrogen fluoride liquefies at 19.9°C and 1 atmosphere pressure. The most important chemistry of the hydrogen halides relates to their aqueous solutions. All of the hydrogen halides dissolve in water to give solutions that conduct electric current, suggesting that ions are present. The reactions may be written:

$$HF(g) + \text{water} \longrightarrow H^+(aq) + F^-(aq) \quad (23)$$

$$HCl(g) + \text{water} \longrightarrow H^+(aq) + Cl^-(aq) \quad (24)$$

$$HBr(g) + \text{water} \longrightarrow H^+(aq) + Br^-(aq) \quad (25)$$

$$HI(g) + \text{water} \longrightarrow H^+(aq) + I^-(aq) \quad (26)$$

These solutions have similar properties and are called acid solutions. The common species in the solutions is the aqueous hydrogen ion, $H^+(aq)$, and the properties of aqueous acid solutions are attributed to this ion. We shall investigate these solutions in Chapter 11.

6-5 HYDROGEN—A FAMILY BY ITSELF

Perhaps you are wondering why the element hydrogen was not included among the halogens. It is, after all, an element with one less electron than its neighboring inert gas, helium. On the other hand, the hydrogen atom has but one electron and, in a sense, it is like an alkali metal. The removal of one electron from an alkali metal atom leaves a specially stable electron population—that of an inert gas. The removal of one electron from a hydrogen atom leaves it with *no* electrons, which also turns out to be specially stable. We shall see both of these influences in the chemistry of hydrogen. This element forms a family by itself, one having some similarities to the halogens and some similarities to the alkalies.

6-5.1 Physical Properties

Hydrogen is a diatomic gas at normal conditions. Its melting point is 15.9°K and its normal boiling point is 20.4°K. This is the second lowest boiling point of any element. Table 6-VII lists these properties.

Table 6-VII

SOME PROPERTIES OF HYDROGEN

Atomic number	1
Atomic weight	1.008
Molecular formula	H_2
Boiling point (°K)	20.4
(°C)	−252.8
Melting point (°K)	14.0
(°C)	−259.2
Atomic volume, solid (ml/mole of atoms)	13.1

We see that the physical properties of hydrogen are like those of the halogens. Hydrogen is a diatomic gas, like the halogens, rather than a metal, like the alkalies. Its melting point is very low and it has a narrow temperature range over which the liquid is stable. However, the family relationships among the elements are based upon their chemistry, so we must investigate the reactions of hydrogen before classifying this unique element.

6-5.2 Some Chemistry of Hydrogen

One of the most distinctive reactions characterizing *both* the alkalies and the halogens is their reaction with each other. The example we have discussed most is the reaction between sodium and chlorine to give sodium chloride. Sodium chloride is an ionic solid that dissolves in water to give positively charged sodium ions, $Na^+(aq)$, and negatively charged chloride ions, $Cl^-(aq)$:

$$Na(s) + \tfrac{1}{2}Cl_2(g) \longrightarrow NaCl(s) \quad (27)$$

$$NaCl(s) + \text{water} \longrightarrow Na^+(aq) + Cl^-(aq) \quad (28)$$

Does hydrogen react like sodium or chlorine in reaction (*27*)? Experiments show that hydrogen can take *either* position in reaction (*27*):

$$Na(s) + \tfrac{1}{2}H_2(g) \longrightarrow NaH(s) \quad (29)$$

$$\tfrac{1}{2}H_2(g) + \tfrac{1}{2}Cl_2(g) \longrightarrow HCl(g) \quad (30)$$

$$HCl(g) + \text{water} \longrightarrow H^+(aq) + Cl^-(aq) \quad (31)$$

The compound sodium hydride, formed in reaction (*29*), is a crystalline compound with physical properties similar to those of sodium chloride. The chemical properties are very different, however. Whereas sodium burns readily in chlorine, it reacts with hydrogen only on heating to about 300°C. While sodium chloride is a stable substance that dissolves in water to form $Na^+(aq)$ and $Cl^-(aq)$, the alkali hydrides burn in air and some of them ignite spontaneously. In contact with water, a vigorous reaction occurs, releasing hydrogen:

$$NaH(s) + H_2O \longrightarrow H_2(g) + Na^+(aq) + OH^-(aq) \quad (32)$$

Thus, in reaction (*29*), hydrogen reacts with sodium like a halogen [as in reaction (*27*)], but the product, sodium hydride, is very different in its chemistry from sodium chloride.

Reaction (*30*) shows hydrogen acting like an alkali metal. Though the product, hydrogen chloride, is not an ionic solid like sodium chloride, it does dissolve in water to give aqueous ions. The formation of $H^+(aq)$ and $Cl^-(aq)$ is strikingly like the analogous formation of $Na^+(aq)$ and $Cl^-(aq)$ by sodium chloride. In fact, the tendency of hydrogen to form a positively charged ion in water, $H^+(aq)$, and the absence of any evidence for a negatively charged ion in water, $H^-(aq)$, is one of the most significant differences between hydrogen and the halogens.

An overall view of the chemistry of hydrogen requires that it be classified alone—as a separate

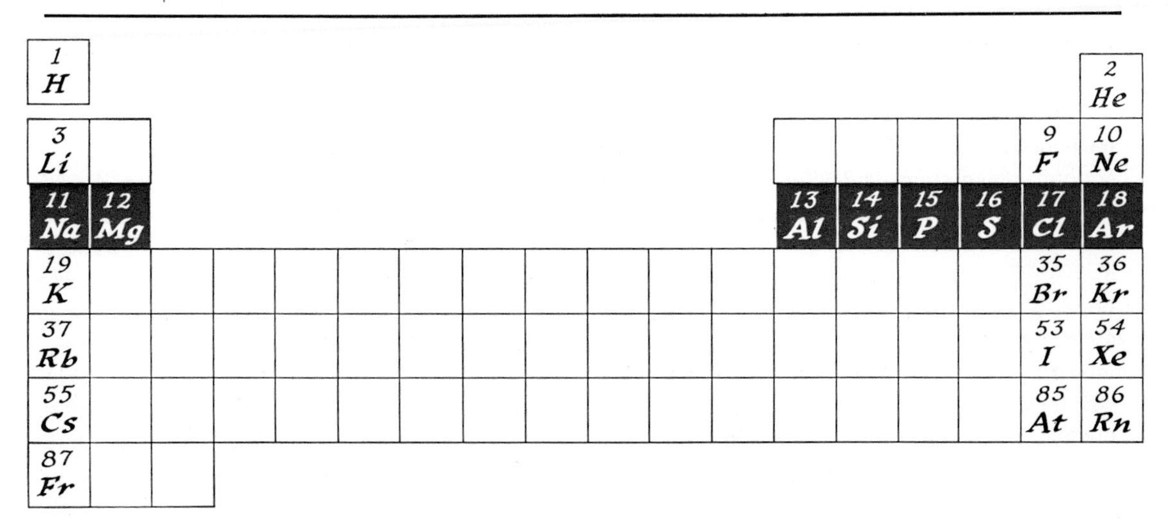

Fig. 6-11. **The placement of the third-row elements** *in the periodic table.*

chemical family. There are some important similarities to halogens—as we have seen, it is a stable diatomic gas—but its chemistry is more like that of the alkalies. Therefore, hydrogen is usually shown on the left side of the periodic table with the alkalies but separated from them to indicate its distinctive character.

6-6 THE THIRD-ROW ELEMENTS

The chemistry of the elements we have examined thus far in this chapter is dominated by the special stabilities of the inert gas electron populations. We can expect to see this same factor at work in the chemistry of the elements in other parts of the periodic table. We shall now take an excursion across a horizontal row of the periodic table to see the trend in chemistry as we pass from element to element. We shall consider the third row, which contains the elements *sodium*, *magnesium*, *aluminum*, *silicon*, *phosphorus*, *sulfur*, *chlorine*, and *argon*.

Table 6-VIII. **SOME PROPERTIES OF THE ELEMENTS OF THE THIRD ROW OF THE PERIODIC TABLE**

PROPERTY	Na	Mg	Al	Si	P	S	Cl	Ar
Atomic number	11	12	13	14	15	16	17	18
Atomic weight	23.0	24.3	27.0	28.1	31.0	32.1	35.5	39.9
Molecular formula	metal	metal	metal	network solid	P_4	S_8	Cl_2	Ar
Boiling point (°K)	1162	1393	2600	2628	553	718	238.9	87
(°C)	889	1120	2327	2355	280	445	−34.1	−186
Melting point (°K)	371	923	933	1683	317.2	392	172	84
(°C)	98	650	660	1410	44.2	119	−101	−189
Atomic volume, solid (ml/mole of atoms)	23.7	14.0	9.99	12.1	16.9	15.6	18.7	24.2

6-6.1 Physical Properties of the Third-row Elements

Table 6-VIII presents some properties of the elements we are considering. The first three, sodium, magnesium, and aluminum, are metallic. The melting points and boiling points are high and increase as we go from element to element. This trend reflects stronger and stronger bonding and it is paralleled by a decrease in the atomic volume.

The fourth element, silicon, forms a solid in which each silicon atom is bonded to four neighboring silicon atoms placed equidistant from each other. (This arrangement places the four neighbors at the four corners of a regular tetrahedron.) This arrangement generates a three-dimensional network, hence is called a **network solid.** Typically such a network solid has a high melting point and a high boiling point.

The remaining four elements form molecular solids. The atoms of white phosphorus, sulfur, and chlorine are strongly bonded into small molecules (formulas, P_4, S_8, and Cl_2, respectively) but only weak attractions exist between the molecules. The properties are all appropriate to this description. Of course there is no simple trend in the properties since the molecular units are so different.

6-6.2 Compounds of the Third-Row Elements

To see the trend in chemistry as we move across the periodic table, we will consider three types of compounds: the hydrides, the chlorides, and the oxides.

HYDRIDES

The hydrides are the compounds formed with hydrogen. Hot, molten sodium metal reacts with hydrogen gas to form a solid, salt-like hydride having the empirical formula, NaH. The ions in the salt are thought to be the Na^+ ion and the H^- ion. The H^- ion may exist in the solid or its melt but it does not exist in a water solution. Magnesium forms a similar salt-like hydride having the empirical formula MgH_2. Apparently two electrons per neutral magnesium atom are removed to form the Mg^{+2} ion (with the stable, neon-like electron arrangement).

Aluminum forms a hydride which seems to be more molecular in character than it is salt-like. The empirical formula is, however, AlH_3. The remaining elements form hydrides which are definitely molecular, gaseous compounds. The formulas are, respectively, SiH_4, PH_3, H_2S, and HCl. These formulas are shown in Table 6-IX, together with the ratio of number of atoms of hydrogen per atom of the element (H/M). We see the regularity of the trend in combining capacity of the elements (as reflected in the ratio, H/M). On the left side, the elements sodium and magnesium (and, to some extent, aluminum) achieve the neon-like electron arrangement by surrendering electrons to hydrogen atoms, one electron to each hydrogen atom. Note that each hydrogen atom thereby achieves the helium-like electron structure. From silicon onward, the

Table 6-IX. THE FORMULAS OF SOME COMPOUNDS OF THE THIRD-ROW ELEMENTS

	Na	Mg	Al	Si	P	S	Cl	Ar
Hydrides								
Formula	NaH	MgH_2	AlH_3	SiH_4	PH_3	H_2S	HCl	—
H/M	1	2	3	4	3	2	1	0
Chlorides								
Formula	NaCl	$MgCl_2$	Al_2Cl_6	$SiCl_4$	PCl_5, PCl_3	S_2Cl_2	Cl_2	—
Cl/M	1	2	3	4	5, 3	1	1	0
Oxides								
Formula	Na_2O	MgO	Al_2O_3	SiO_2	P_4O_{10}	SO_3	Cl_2O_7, Cl_2O	—
2(O/M)	1	2	3	4	5	6	7, 1	0

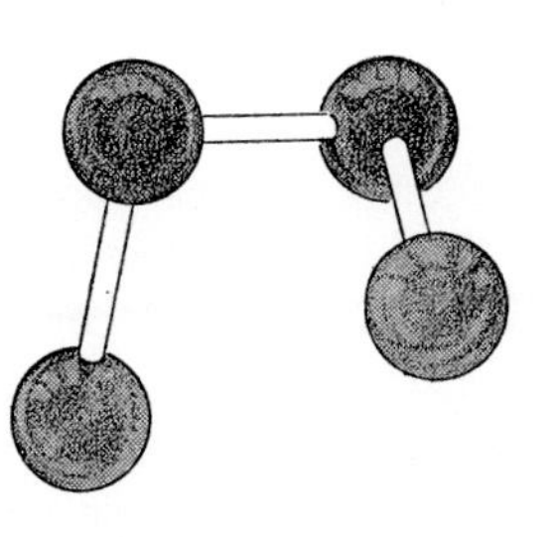

Fig. 6-12. **The structure of S_2Cl_2.**

molecular compounds indicate that the elements are achieving the next higher inert gas electron arrangement (argon) by sharing electrons with hydrogen atoms. For these compounds, the bonding to the hydrogen atoms resembles more the covalent bonding we find in Cl_2 than the ionic bonding we find in sodium chloride. Whatever the nature of the bond, however, the combining capacity of each element is influenced by the tendency to form the next lower or next higher inert gas electron population.

CHLORIDES AND OXIDES

We can find the same sort of trends in the combining capacity by examining the chlorides and the oxides.* Again, we find the bonding can be understood in terms of electron transfer or electron sharing, in either case, to reach an inert gas electron arrangement. There is only one apparent anomaly—the combining capacity of sulfur in the compound S_2Cl_2. This is not a real anomaly, however, for the structure of this compound reveals the expected combining power of 2. The structure is shown in Figure 6-12. The atoms in this molecule are arranged so that each sulfur atom reaches the argon-like electron arrangement by sharing one pair of electrons with a chlorine atom and one pair of electrons with another sulfur atom.

SUMMARY

The simple trend in the molecular formulas shown by the third-row elements demonstrates the importance of the inert gas electron populations. The usefulness of the regularities is evident. Merely from the positions of two atoms in the periodic table, it is possible to predict the most likely molecular formulas. In Chapters 16 and 17 we shall see that the properties of a substance can often be predicted from its molecular formula. Thus, we shall use the periodic table continuously throughout the course as an aid in correlating and in predicting the properties of substances.

6-7 THE PERIODIC TABLE

The power of the periodic table is evident in the chemistry we have viewed. By arranging the atoms in the array shown on the inside of the front cover, we simplify the problem of understanding the variety of chemistry found in nature. The elements grouped in a vertical column have pronounced similarities. General statements can be made about their chemistries and the compounds they form. Furthermore, the formulas of these compounds and the nature of the bonds that hold them together can be understood in terms of the special stabilities of the inert gases.

The periodicity of chemical properties was dis-

* It must be remarked that some stable compounds have been omitted (for example, Na_2O_2, SO_2). The compounds listed are stable and display the trends in bonding capacity.

covered about one-hundred years ago. J. W. Döbereiner (a German chemist) in 1828 recognized similarities among certain elements (chlorine, bromine, and iodine; lithium, sodium, and potassium; etc.) and he grouped them as "triads." (Remember, "Cylindrical objects burn?") J. A. R. Newlands (an English chemist) in 1864 was ridiculed for proposing a "law of octaves" which foresaw the differences of eight that we noted in Table 6-IV. Simultaneously, Lothar Meyer (a German chemist and physicist) proposed a periodic table similar to that of Newlands. Independently and in this same year (the time was ripe for the next step, "Wooden objects burn") D. I. Mendeleev (a Russian) framed the periodic table in more complete form. He even predicted both the existence and properties of elements not then known. The subsequent discovery of these elements and corroboration of their properties solidified the acceptance of the periodic table. It remains, one-hundred years later, the most important single correlation of chemistry. It permits us to deal with the great variety we find in nature.

QUESTIONS AND PROBLEMS

1. For which of the following processes will energy be absorbed?

   (a) Separating an electron from an electron.
   (b) Separating an electron from a proton.
   (c) Separating a proton from a proton.
   (d) Removing an electron from a neutral atom.

2. Which of the following statements is FALSE? The atoms of oxygen differ from the atoms of every other element in the following ways:

   (a) the nuclei of oxygen atoms have a different number of protons than the nuclei of any other element;
   (b) atoms of oxygen have a higher ratio of neutrons to protons than the atoms of any other element;
   (c) neutral atoms of oxygen have a different number of electrons than neutral atoms of any other element;
   (d) atoms of oxygen have different chemical behavior than do atoms of any other element.

3. For *every* atom, less energy is needed to remove one electron from the neutral atom than is needed to remove another electron from the resulting ion. Explain.

4. List the number and kind of fundamental particles found in a neutral lithium atom that has a nucleus with a nuclear charge three times that of a hydrogen nucleus and with seven times the mass.

5. The nucleus of an aluminum atom has a diameter of about 2×10^{-13} cm. The atom has an average diameter of about 3×10^{-8} cm. Calculate the ratio of the diameters.

6. Suppose a copper atom is thought of as occupying a sphere 2.6×10^{-8} cm in diameter. If a spherical model of the copper atom is made with a 5.2 cm diameter, how much of an enlargement is this?

7. Suppose an atom is likened to bees flying around their beehive. The beehive would be compared to the nucleus and the bees roving about the countryside would be compared to the electrons of the atom.

   (a) If the radius of the beehive is 25 cm, what would be the average radius of the flight of the bees to maintain proper scale with the atom? Express your answer in kilometers.
   (b) At any instant, where is the concentration of bees apt to be highest?
   (c) Describe qualitatively the distribution of bees around the hive as a function of direction and of distance.

8. Helium, as found in nature, consists of two isotopes. Most of the atoms have a mass number 4 but a few have a mass number 3. For each isotope, indicate the:

   (a) atomic number;
   (b) number of protons;
   (c) number of neutrons;
   (d) mass number;
   (e) nuclear charge.

9. Fill in the blanks of the following table.

ELEMENT	ATOMIC NO.	PARTICLES PER ATOM: PROTONS	ELECTRONS	NEUTRONS	MASS NUMBER
aluminum (Al)	13				27
beryllium (Be)		4			9
bismuth (Bi)	83				209
calcium (Ca)			20	20	
carbon (C)		6		6	
fluorine (F)			9		19
phosphorus (P)	15			16	
iodine (I)			53		127

10. How do isotopes of one element differ from each other? How are they the same?

11. How much would 0.754 mole of chlorine-35 atoms weigh? How much would 0.246 mole of chlorine-37 atoms weigh? What is the weight of a mole of "average" atoms in a mixture of the above samples? What is the atomic weight of the naturally occurring mixture of these two isotopes of chlorine?

12. What is the significance of the trends in the boiling points and melting points of the inert gases in terms of attractions among the atoms?

13. Why is argon used in many electric light lamps?

14. Calculate the ratio of the number of electrons in a neutral xenon atom to the number in a neutral neon atom. Compare this number to the ratio of the atomic volumes of these two elements. On the basis of these two ratios, discuss the effects of electron-electron repulsions and electron-nuclear attractions on atomic size.

15. The molar heats of vaporization of the inert gases (in kcal/mole) are: He, 0.020; Ne, 0.405; Ar, 1.59; Kr, 2.16; Xe, 3.02; Rn, 3.92. Using the data in Table 6-III, (p. 91), plot the boiling points (vertical axis) against the heats of vaporization (horizontal axis). Suggest a generalization based upon a simple curve passing near the plotted points. Write an equation for the straight line passing through the origin (that is, through zero) and through the point for radon.

16. Lithium forms the following compounds: lithium oxide, Li_2O; lithium hydroxide, LiOH; lithium sulfide, Li_2S. Name and write the formulas of the corresponding sodium and potassium compounds.

17. An alkali element produces ions having the same electron population as atoms of the preceding inert gas. In what ways do these ions differ from the inert gases? In what ways are they alike?

18. There is a large difference between the energy needed to remove an electron from a neutral, gaseous sodium atom and a neutral, gaseous neon atom:

$$Na(g) + 118.4 \text{ kcal} \longrightarrow Na^+(g) + e^-$$
$$Ne(g) + 497.0 \text{ kcal} \longrightarrow Ne^+(g) + e^-$$

Explain how these energies are consistent with the proposal that the electron arrangements of the inert gases are specially stable.

19. Refer to the halogen column in the periodic table. How many electrons must each halogen atom gain to have an electron population equal that of an atom of the adjacent inert gas? What property does this population impart to each ion?

20. How do the trends in physical properties for the halogens compare with those for the inert gases? Compare boiling points, melting points, and atomic volumes.

21. Use your knowledge of the usefulness of the periodic table to fill in the blank spaces in Table 6-VI, p. 97, under Astatine. List some chemical reactions expected for astatine.

22. Chlorine is commonly used as a germicide in swimming pools. When chlorine dissolves in water, it reacts to form hypochlorous acid, HOCl, as follows:

$$Cl_2 + H_2O \rightleftarrows HOCl(aq) + H^+(aq) + Cl^-(aq)$$

Predict what happens when bromine, Br_2, dissolves in water. Write the equation for the reaction.

23. Zinc metal dissolves in a solution of gaseous chlorine in water as follows:

$$Zn(s) + Cl_2(aq) \longrightarrow Zn^{+2}(aq) + 2Cl^-(aq)$$

Zinc does not dissolve in a solution of gaseous hydrogen in water but it *does* dissolve in an aqueous solution of hydrogen chloride:

$$Zn(s) + 2H^+(aq) + 2Cl^-(aq) \longrightarrow Zn^{+2}(aq) + H_2(g) + 2Cl^-(aq)$$

Recognizing that zinc metal must release electrons to form $Zn^{+2}(aq)$, explain how these reactions demonstrate that gaseous hydrogen does not behave like a halogen.

24. Write the molecular formulas of the hydrogen compounds of the second-row elements, Li, Be, B, C, N, O, F, Ne. Indicate, for each compound, the H/M ratio.

25. Indicate the electron rearrangement (gain or loss) in each kind of atom assuming it attains inert gas-like electron structure in the following reactions.

(a) $2Rb + Br_2 \longrightarrow 2RbBr$
(b) $2Cs + I_2 \longrightarrow 2CsI$
(c) $Mg + S \longrightarrow MgS$
(d) $2Ba + O_2 \longrightarrow 2BaO$

26. Which of the following is NOT a correct formula for a substance at normal laboratory conditions?

(a) $H_2S(g)$
(b) $CaCl_2(s)$
(c) $He(g)$
(d) $NaNe(s)$
(e) $Al_2O_3(s)$

27. Magnesium metal burns in air, emitting enough light to be useful as a flare, and forming clouds of white smoke. Write the equation for the reaction. What is the composition of the smoke?

28. Use the formulas for magnesium oxide, MgO, and magnesium chloride, $MgCl_2$, together with the periodic table to decide that magnesium ions have the same number of electrons as each of the following, EXCEPT

(a) neon atoms, Ne;
(b) sodium ions, Na^+;
(c) fluoride ions, F^-;
(d) oxide ions, O^{-2};
(e) calcium ions, Ca^{+2}.

29. Sodium metal reacts with water to form sodium ions, Na^+, hydroxide ions, OH^-, and hydrogen gas, H_2, as follows:

$$2Na(s) + 2H_2O \longrightarrow 2Na^+(aq) + 2OH^-(aq) + H_2(g)$$

Assuming calcium metal reacts in a similar way, write the equation for the analogous reaction between calcium and water. Remember that calcium is in the second column of the periodic table and sodium is in the first.

30. Use Table 6-IX, p. 102, and the periodic table to write possible formulas for the following compounds:

(a) a hydride of barium, element 56;
(b) a chloride of germanium, element 32;
(c) an oxide of indium, element 49;
(d) an oxide of cesium, element 55;
(e) a fluoride of tin, element 50.

31. All of the isotopes of the element with atomic number 87 are radioactive. Hence, it is not found in nature. Yet, prior to its preparation by nuclear bombardment, chemists were confident they knew the chemical reactions this element would show. Explain. What predictions about this element would you make?

DMITRI MENDELEEV, 1834–1907

Element 101 is named Mendelevium in honor of the great Russian chemist, Dmitri Mendeleev. The youngest of seventeen children, he was born in Tobolska where his grandfather published the first newspaper in Siberia and his father was the high school principal. Dmitri received his early education from a political exile, but when his father died, his mother traveled west in search of better educational opportunities for Dmitri.

At the University of St. Petersburg (now Leningrad), he distinguished himself in science and mathematics and earned the doctorate with a thesis on a subject that remains of current interest, "The Union of Alcohol and Water." Subsequent studies in France and Germany permitted him to attend the 1858 Karlsruhe (Germany) conference at which Avogadro's Hypothesis was heatedly debated. Later, he visited the oil fields of Pennsylvania to see the first oil well. Upon his return to Russia, he developed a new commercial distillation process.

He became a professor of chemistry at St. Petersburg when only 32. Searching for regularities, he arranged the elements by their properties. This organization led him to propose the periodic table and use it to predict the existence and properties of a number of additional elements. When some of those that were foretold in 1869 were actually discovered a few years later, Mendeleev was hailed as a prophet.

This inspiring teacher and tireless experimenter was so deeply concerned over social issues that he resigned his professorship rather than obey an order to cease interfering with affairs of government. He made enemies by supporting liberal movements and even defied the Czar's wishes by refusing to cut his hair. Nevertheless, he won the appointment as Director of the Bureau of Weights and Measures.

When Mendeleev first published his chart, there were 63 elements known. One year after his death, there were 86. The rapidity of this increase was made possible by the most important generalization of chemistry, the periodic table.

CHAPTER 7

Energy Effects in Chemical Reactions

Although a typical chemical reaction · · · may appear far removed from the working of an engine, the same fundamental principles of heat and work apply to both.

F. T. WALL, 1958

Chemical reactions form the heart of chemistry. And there is no more important aspect of chemical reactions than the energy effects that are caused. You will realize this if you let your thoughts wander between the warmth the little child in the fable derived from the combustion of wood and the celestial joy ride an astronaut receives from the reactions of his rocket fuels. How much energy is involved in a chemical reaction? How do we find this out? Where does this energy come from? We shall investigate these questions in this chapter.

7-1 HEAT AND CHEMICAL REACTIONS

At a temperature of 600°C, steam passed over hot coal (coal is mostly carbon) reacts to give carbon monoxide and hydrogen:

$$H_2O(g) + C(s) \longrightarrow CO(g) + H_2(g) \qquad (1)$$

This reaction is quite useful because the mixture of product gases, called "water gas," is an excellent industrial fuel. In the commercial preparation of water gas, the chemical engineer must allow for the absorption of heat during the reaction. In fact, he must periodically turn off the steam and reheat the coal to keep the reaction going. To aid the engineer, we might measure the amount of heat absorbed by the system and write it into the chemical reaction. Such a measurement shows that 31.4 kcal of heat are *absorbed* per mole of carbon reacted. Since the heat is used up (as are the reactants), we can rewrite reaction (*1*) and show the heat on the left side of the equation:

$$H_2O(g) + C(s) + 31.4 \text{ kcal} \longrightarrow CO(g) + H_2(g) \qquad (1a)$$

Now we might think of the mechanical engineer who is designing a boiler to be heated with water gas fuel. He is interested in burning the water gas. This involves two chemical reactions of combustion:

$$CO(g) + \tfrac{1}{2}O_2(g) \longrightarrow CO_2(g) \quad (2)$$

and

$$H_2(g) + \tfrac{1}{2}O_2(g) \longrightarrow H_2O(g) \quad (3)$$

These reactions release heat, and our mechanical engineer wishes to know how much. Again, we might help by measuring these amounts of heat and adding the information to reactions (*2*) and (*3*). Since heat is *produced* by the reaction (as is a chemical product), we should place it on the right side of the equation. Experiments show:

$$CO(g) + \tfrac{1}{2}O_2(g) \longrightarrow CO_2(g) + 67.6 \text{ kcal} \quad (2a)$$

$$H_2(g) + \tfrac{1}{2}O_2(g) \longrightarrow H_2O(g) + 57.8 \text{ kcal} \quad (3a)$$

Now let us talk with the business manager. He thinks in terms of gains and losses. He is likely to observe that the consumption of coal and water to generate water gas is followed by the combustion of the water gas to form carbon dioxide and water. Without knowing much chemistry, he can see that what is finally accomplished is the combustion of coal to form carbon dioxide. The overall reaction is obtained by adding reactions (*1*), (*2*), and (*3*):

$$H_2O(g) + C(s) \longrightarrow CO(g) + H_2(g) \quad (1)$$

$$CO(g) + \tfrac{1}{2}O_2(g) \longrightarrow CO_2(g) \quad (2)$$

$$H_2(g) + \tfrac{1}{2}O_2(g) \longrightarrow H_2O(g) \quad (3)$$

Overall reaction: $C(s) + O_2(g) \longrightarrow CO_2(g)$ (*4*)

The business manager is frugal so he asks, "Why not burn the coal directly and save the cost ot manufacturing the water gas?" The mechanical engineer is practical so he asks, "How much heat will the boiler receive if I use coal instead of water gas?" The chemical engineer goes to the laboratory to find the answers by measuring the heat released per mole of carbon burned in reaction (*4*). The laboratory result shows that reaction (*4*) releases 94.0 kcal/mole:

$$C(s) + O_2(g) \longrightarrow CO_2(g) + 94.0 \text{ kcal} \quad (4a)$$

The chemical engineer now can answer all of the questions. If one mole of carbon is burned directly, 94.0 kcal of heat are released for the mechanical engineer. The same amount of coal converted into water gas releases the sum of the heats evolved in reactions (*2a*) and (*3a*):

$$67.6 + 57.8 = 125.4 \text{ kcal}$$

The mechanical engineer has a better fuel in water gas than in coal.

With new respect, the business manager might now ask the chemical engineer, "Where did this extra heat come from?" "Did we get something for nothing?" The answer is, "No." The water gas releases more heat per mole of carbon because the chemical engineer put that amount of heat in during reaction (*1a*). The business manager's ledger is shown in Table 7-I.

Table 7-I

HEAT EFFECTS IN THE MANUFACTURE AND USE OF WATER GAS

	DEBIT	CREDIT
Reaction (*1a*): heat absorbed	31.4 kcal	—
Reaction (*2a*): heat released	—	67.6 kcal
Reaction (*3a*): heat released	—	57.8 kcal
Overall reaction: (*1a*) + (*2a*) + (*3a*) = (*4a*)	31.4 kcal absorbed	125.4 kcal released
		125.4
		− 31.4
		94.0 net
Experiment reaction (*4a*): Heat released		94.0 kcal

7-1.1 Heat Content of a Substance

The example just given shows that the 31.4 kcal absorbed during reaction (*1a*) was "stored" in the water gas. Furthermore, the amount of energy "stored" is definite, not alterable at the demand of the business manager or the whim of the chemical engineer. How much energy is stored depends upon the reactants and products in reaction. We must add a fixed amount of energy (as heat) to coal and steam to make a specified amount of carbon monoxide and hydrogen. This heat is retained by the CO and H_2 molecules, as shown in Table 7-I. We might say that reaction (*1a*) increases the "heat content" of the atoms of the reactants by rearranging them to form the products. Apparently a mole of each molecular substance has a characteristic "heat content" just as it has a characteristic mass. This ***heat content*** *measures the energy stored in a substance during its formation.* The heat effect in a chemical reaction measures the difference between the heat contents of the products and the heat contents of the reactants. If more energy is stored in the reactants than in the products, then heat will be released during reaction. Conversely, heat will be absorbed if more energy is stored in the products than in the reactants.

This idea—that each molecular substance has a characteristic heat content—provides a good explanation of the heat effects found in chemical reactions. Chemists symbolize heat content by H. Since the heat effect in a reaction is the *difference* between the H's of the products and the H's of the reactants, the heat of reaction is called ΔH, the Greek letter Δ (delta) signifying "difference."

We can see what ΔH means in terms of an example. Consider reaction (*1*):

$$H_2O(g) + C(s) \longrightarrow CO(g) + H_2(g) \qquad (1)$$

A heat content, H, is assigned to each substance. Then ΔH for reaction (*1*) is the difference:

$$\Delta H = \begin{pmatrix}\text{heat content}\\ \text{of products}\end{pmatrix} - \begin{pmatrix}\text{heat content}\\ \text{of reactants}\end{pmatrix}$$

$$= [H_{CO} + H_{H_2}] - [H_{H_2O} + H_C]$$

$$= H_{CO} + H_{H_2} - H_{H_2O} - H_C$$

Since the reaction consumes heat, the heat content of the products is higher and ΔH will be positive. We can express this by writing

$$H_2O(g) + C(s) \longrightarrow CO(g) + H_2(g) \qquad \Delta H = +31.4 \text{ kcal} \quad (1b)$$

Reaction (*1b*) is exactly equivalent to reaction (*1a*):

$$H_2O(g) + C(s) + 31.4 \text{ kcal} \longrightarrow CO(g) + H_2(g) \quad (1a)$$

Let's try this on reaction (*2*):

$$\Delta H = \begin{pmatrix}\text{heat content}\\ \text{of products}\end{pmatrix} - \begin{pmatrix}\text{heat content}\\ \text{of reactants}\end{pmatrix}$$

$$= H_{CO_2} - [H_{CO} + H_{\frac{1}{2}O_2}]$$

$$= H_{CO_2} - H_{CO} - H_{\frac{1}{2}O_2}$$

In this reaction, heat is evolved; hence, the heat content of the products is *below* that of the reactants. Therefore, ΔH must be negative:

$$CO(g) + \tfrac{1}{2}O_2(g) \longrightarrow CO_2(g) \qquad \Delta H = -67.6 \text{ kcal} \quad (2b)$$

This has exactly the same meaning as

$$CO(g) + \tfrac{1}{2}O_2(g) \longrightarrow CO_2(g) + 67.6 \text{ kcal} \quad (2a)$$

We see that the sign of ΔH is sensible. It is positive when heat content is rising (by heat absorption) and it is negative when heat content is dropping (by heat evolution). This is shown diagrammatically in Figure 7-1.

Fig. 7-1. **Heat content change** *during a reaction.*

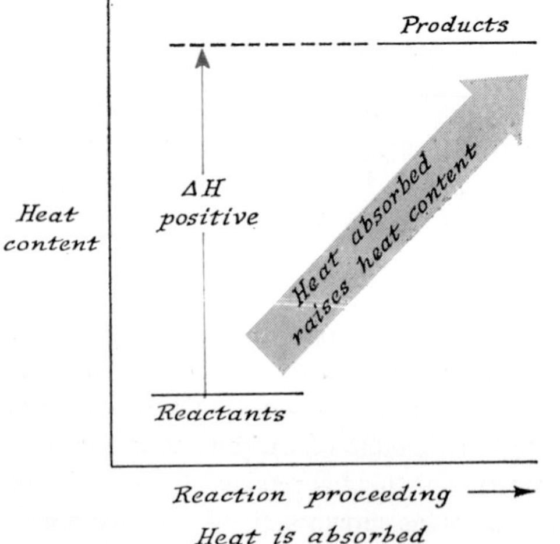

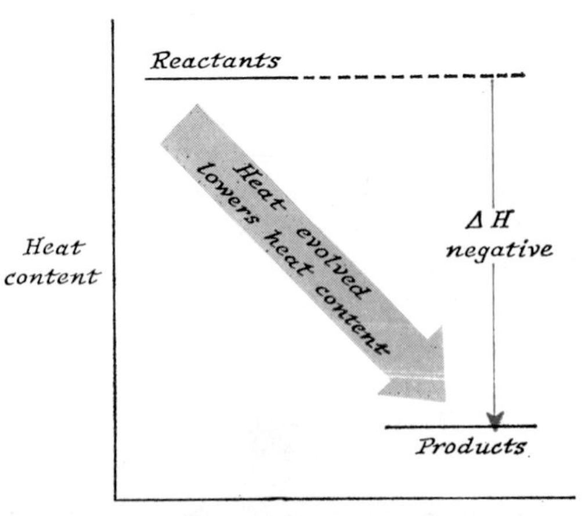

7-1.2 Additivity of Reaction Heats

Let us return to the debit and credit balance we found in our water gas fuel problem. In terms of ΔH, the heat effects are as follows:

$$H_2O(g) + C(s) \longrightarrow CO(g) + H_2(g) \qquad \Delta H_1 = +31.4 \text{ kcal} \quad (1b)$$

$$CO(g) + \tfrac{1}{2}O_2(g) \longrightarrow CO_2(g) \qquad \Delta H_2 = -67.6 \text{ kcal} \quad (2b)$$

$$H_2(g) + \tfrac{1}{2}O_2(g) \longrightarrow H_2O(g) \qquad \Delta H_3 = -57.8 \text{ kcal} \quad (3b)$$

Overall reaction (1) + (2) + (3) = (4):

$$C(s) + O_2(g) \longrightarrow CO_2(g) \qquad \Delta H_4 = -94.0 \text{ kcal} \quad (4b)$$

We discover that not only is reaction (4) equal to the sum of reactions (1b) + (2b) + (3b) in terms of atoms, but also that

$$\begin{aligned} \Delta H_4 &= \Delta H_1 + \Delta H_2 + \Delta H_3 \\ &= 31.4 + (-67.6) + (-57.8) \\ &= 31.4 - 67.6 - 57.8 \\ &= -94.0 \text{ kcal} \end{aligned}$$

We see that *when a reaction can be expressed as the algebraic sum of a sequence of two or more other reactions, then the heat of the reaction is the algebraic sum of the heats of these reactions.* This generalization has been found to be applicable to every reaction that has been tested. Because the generalization has been so widely tested, it is called a law—the **Law of Additivity of Reaction Heats.***

7-1.3 The Measurement of Reaction Heat

The measurement of reaction heats is called ***calorimetry***—a name obviously related to the unit of heat, the calorie. You already have some experience in calorimetry. In Experiment 5 you measured the heat of combustion of a candle and the heat of solidification of paraffin. Then in Experiment 13 you measured the heat evolved when NaOH reacted with HCl. The device you used was a simple calorimeter.

* This generalization was first proposed in the year 1840 by G. H. Hess on the basis of his experimental measurements of reaction heats. It is sometimes called Hess's Law of Constant Heat Summation.

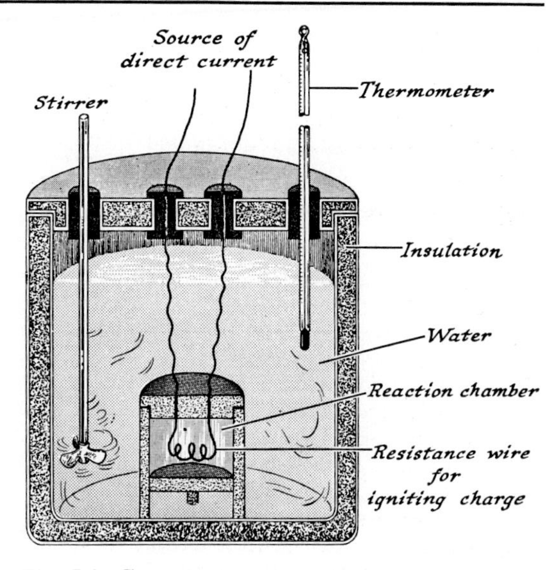

Fig. 7-2. **General plan of a calorimeter.**

Calorimeters vary in details and are adapted to the particular reaction being studied. Figure 7-2 shows the general plan of a calorimeter that might be used in measuring the heat evolved during a combustion reaction. It might be applied to the combustion of a candle to yield a much more reliable answer than can be obtained by the crude technique of Experiment 5.

Essentially, the device consists of an insulated vessel containing a known weight of water. A weighed amount of the substance to be burned and an excess of oxygen are introduced under pressure into a reaction chamber placed in the vessel. The reaction mixture is ignited by means of an electrical resistance wire sealed into the reaction chamber. The heat produced by the reaction changes the temperature of the water, which is stirred to keep its temperature uniform. From this temperature change and from the amount of heat required to raise the temperature of the calorimeter and its contents by one degree, the heat of combustion per mole of substance burned can be calculated. By using a large amount of water, the actual change in temperature is kept small. This is desirable in that it keeps the final temperature of the products of reaction fairly close to the initial temperature of the reactants.

EXERCISE 7-1

Suppose reactants are mixed in a calorimeter at 25°C and the reaction heat causes the temperature of the products and calorimeter to rise to 35°C. The resultant determination of ΔH applies to what temperature? Explain why it is desirable to keep the final temperature close to the initial temperature in a calorimetric measurement.

In the combustion reaction as carried out in the calorimeter of Figure 7-2, the *volume* of the system is kept constant and pressure may change because the reaction chamber is sealed. In the laboratory experiments you have conducted, you kept the pressure constant by leaving the system open to the surroundings. In such an experiment, the volume may change. There is a small difference between these two types of measurements. The difference arises from the energy used when a system expands against the pressure of the atmosphere. In a constant volume calorimeter, there is no such expansion; hence, this contribution to the reaction heat is not present. Experiments show that this difference is usually small. However, *the symbol ΔH represents the heat effect that accompanies a chemical reaction carried out at constant pressure*—the condition we usually have when the reaction occurs in an open beaker.

7-1.4 Predicting the Heat of a Reaction

Chemists have measured the heats of many reactions. With the measured values, many unmeasured reaction heats can be predicted by applying the Law of Additivity of Reaction Heats. Consequently, a compilation of known reaction heats is extremely useful. Table 7-II is such a compilation.

Suppose we are interested in the heat of combustion of nitric oxide, NO:

$$NO(g) + \tfrac{1}{2}O_2(g) \longrightarrow NO_2(g) \qquad \Delta H_5 \quad (5)$$

Since reaction (5) can be obtained by combining two reactions in Table 7-II, we can predict ΔH_5. In Table 7-II we find

$$\tfrac{1}{2}N_2(g) + \tfrac{1}{2}O_2(g) \longrightarrow NO(g) \qquad \Delta H_6 = +21.6 \text{ kcal/mole NO} \quad (6)$$

$$\tfrac{1}{2}N_2(g) + O_2(g) \longrightarrow NO_2(g) \qquad \Delta H_7 = +8.1 \text{ kcal/mole } NO_2 \quad (7)$$

Now we wish to obtain reaction (5) by combining reactions (6) and (7). Since NO is a reactant in reaction (5), we need the reverse of reaction (6). We obtain the heat of the reverse reaction merely by changing the algebraic sign of ΔH_6. If 21.6 kcal of heat are absorbed when one mole of NO is formed, then 21.6 kcal of heat will be released when one mole of NO is decomposed in the reverse reaction:

$$NO(g) \longrightarrow \tfrac{1}{2}N_2(g) + \tfrac{1}{2}O_2(g) \qquad \Delta H_8 = -21.6 \text{ kcal/mole NO} \quad (8)$$

Now we can add reactions (7) and (8) to obtain reaction (5):

Table 7-II. **HEATS OF REACTION BETWEEN ELEMENTS, t = 25°C, p = 1 atm**

ELEMENTS		COMPOUND FORMULA	COMPOUND NAME	HEAT OF REACTION (kcal/mole of product)
$H_2(g) + \tfrac{1}{2}O_2(g)$	$\longrightarrow$	$H_2O(g)$	water vapor	−57.8
$H_2(g) + \tfrac{1}{2}O_2(g)$	$\longrightarrow$	$H_2O(l)$	water	−68.3
$S(s) + O_2(g)$	$\longrightarrow$	$SO_2(g)$	sulfur dioxide	−71.0
$H_2(g) + S(s) + 2O_2(g)$	$\longrightarrow$	$H_2SO_4(l)$	sulfuric acid	−194
$\tfrac{1}{2}N_2(g) + \tfrac{1}{2}O_2(g)$	$\longrightarrow$	$NO(g)$	nitric oxide	+21.6
$\tfrac{1}{2}N_2(g) + O_2(g)$	$\longrightarrow$	$NO_2(g)$	nitrogen dioxide	+8.1
$\tfrac{1}{2}N_2(g) + \tfrac{3}{2}H_2(g)$	$\longrightarrow$	$NH_3(g)$	ammonia	−11.0
$C(s) + \tfrac{1}{2}O_2(g)$	$\longrightarrow$	$CO(g)$	carbon monoxide	−26.4
$C(s) + O_2(g)$	$\longrightarrow$	$CO_2(g)$	carbon dioxide	−94.0
$2C(s) + 3H_2(g)$	$\longrightarrow$	$C_2H_6(g)$	ethane	−20.2
$3C(s) + 4H_2(g)$	$\longrightarrow$	$C_3H_8(g)$	propane	−24.8
$\tfrac{1}{2}H_2(g) + \tfrac{1}{2}I_2(g)$	$\longrightarrow$	$HI(g)$	hydrogen iodide	+6.2

$$NO(g) \longrightarrow \tfrac{1}{2}N_2(g) + \tfrac{1}{2}O_2(g) \qquad \Delta H_8 = -21.6 \text{ kcal/mole NO} \quad (8)$$

$$\tfrac{1}{2}N_2(g) + O_2(g) \longrightarrow NO_2(g) \qquad \Delta H_7 = +8.1 \text{ kcal/mole } NO_2 \quad (7)$$

Overall reaction:

$$NO(g) + O_2(g) + \tfrac{1}{2}N_2(g) \longrightarrow NO_2(g) + \tfrac{1}{2}O_2(g) + \tfrac{1}{2}N_2(g) \qquad \Delta H_5 = -21.6 + 8.1$$

or

$$NO(g) + \tfrac{1}{2}O_2(g) \longrightarrow NO_2(g) \qquad \Delta H_5 = -13.5 \text{ kcal/mole NO} \quad (5)$$

EXERCISE 7-2

Predict the heat of the reaction

$$CO(g) + \tfrac{1}{2}O_2(g) \longrightarrow CO_2(g)$$

from two reactions listed in Table 7-II. Compare your result with ΔH_{2b} given in Section 7-1.2.

The usefulness of Table 7-II is obvious from these examples. Many, many reaction heats can be predicted—in fact, the heat of any reaction that can be obtained by adding two or more of the reactions in the table. Furthermore, there is an easy way to decide whether the table contains the necessary information for a particular example. A given reaction can be obtained by adding reactions in Table 7-II, provided every *compound* in the reaction is included in the table. The elements participating in the reactions automatically will appear in proper amount.

Consider a more complicated example—the oxidation of ammonia, NH_3:

$$NH_3(g) + \tfrac{7}{4}O_2(g) \longrightarrow NO_2(g) + \tfrac{3}{2}H_2O(g) \qquad \Delta H_9 \quad (9)$$

In reaction (9) we find three compounds, $NH_3(g)$, $NO_2(g)$, and $H_2O(g)$. These are all found in Table 7-II. Consequently, we are able to calculate ΔH_9.

EXERCISE 7-3

Convince yourself that reaction (9) and also $\Delta H_9 = -67.6$ kcal can be obtained by carrying out the indicated summation:

Subtract $\tfrac{1}{2}N_2(g) + \tfrac{3}{2}H_2(g) \longrightarrow NH_3(g)$
Subtract $\Delta H = -11.0$ kcal

Add $\tfrac{1}{2}N_2(g) + O_2(g) \longrightarrow NO_2(g)$
Add $\Delta H = +8.1$ kcal

Add $\tfrac{3}{2}$ times $H_2(g) + \tfrac{1}{2}O_2(g) \longrightarrow H_2O(g)$
Add $\tfrac{3}{2}$ times $\Delta H = -57.8$ kcal

Thus, when we wish to predict the heat of some reaction, it takes but a moment to decide whether the compilation of Table 7-II includes the necessary reactions. If every *compound* in the reaction of interest is in the "Compound" column of Table 7-II, then the prediction can be made. This is a convenience provided by a compilation of heats of reaction between elements and explains why these reaction heats are the ones chemists tabulate. Of course, the list in Table 7-II includes only a small fraction of the known values. Many more reaction heats are tabulated in handbooks; they are listed in indexes under Heat of Formation.

7-2 THE LAW OF CONSERVATION OF ENERGY

We have seen how chemists measure the heat of a reaction. Using a compilation of measured values, we can predict the energy changes of many reactions that have not been measured. Thus, the rule of Additivity of Reaction Heats is a very useful and reliable generalization. It makes us wonder "Why should it be so?" The explanation, as usual, is found by connecting the behavior of a chemical system to the behavior of other systems that are better understood.

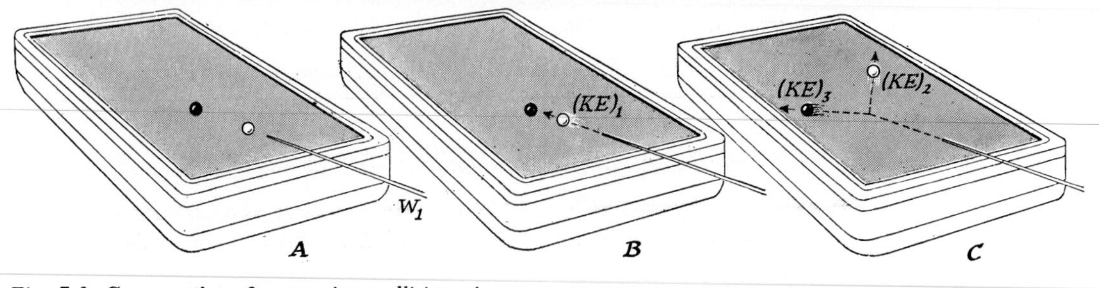

Fig. 7-3. **Conservation of energy** *in a collision of billiard balls.*

7-2.1 Conservation of Energy in a Billiard Ball Collision

Figure 7-3 shows an experimental study of the collision of hard spheres. The experimenter imparts energy of motion to the white ball (see Figure 7-3A, 7-3B). He does so by doing work by striking the ball with the end of a cylindrical stick (a cue). The amount of *energy of motion* (**kinetic energy**) received by the ball is fixed by the amount of work done. If the ball is struck softly (little work being done), it moves slowly. If the ball is struck hard (much work being done), it moves rapidly. The kinetic energy of the white ball appears because work was done—the amount of work, W_1, determines and equals the amount of kinetic energy, $(KE)_1$. In symbols,

$$W_1 = (KE)_1 \qquad (10)$$

Suppose the direction of motion of the white ball causes it to contact the motionless red ball. A collision occurs. Figure 7-3C shows the result. The white ball has a lower kinetic energy, $(KE)_2$, but now the red ball is moving! The red ball now has energy of motion—let's call it $(KE)_3$. Measurements show that velocities are such that the kinetic energy gained by the red ball is equal to the kinetic energy lost by the white ball. In symbols,

$$(KE)_1 = (KE)_2 + (KE)_3 \qquad (11)$$

Let us review this experiment. First, an amount of work was performed, W_1, with the cylindrical stick. The amount of kinetic energy received by the white ball, $(KE)_1$, was exactly fixed by W_1. After the white ball collided with the red ball, the sum of the energies of the two balls, $(KE)_2 + (KE)_3$, is exactly equal to $(KE)_1$. We can write

$$W_1 = (KE)_1 = (KE)_2 + (KE)_3 \qquad (12)$$

If we recognize work as a form of energy, we may say that *energy was conserved* during the experiment. Every quantity in (*12*) is known—we can measure the work done as we do it and we can learn the magnitudes of $(KE)_1$, $(KE)_2$, and $(KE)_3$ through velocity and mass measurements. Many such experiments are performed every day and the results are always in agreement. In a billiard parlor, *energy is conserved.*

7-2.2 Conservation of Energy in a Stretched Rubber Band

Figure 7-4 shows an experimental study of the stretching of a rubber band. The rubber band is

Fig. 7-4. **Conservation of energy** *in a stretched rubber band.*

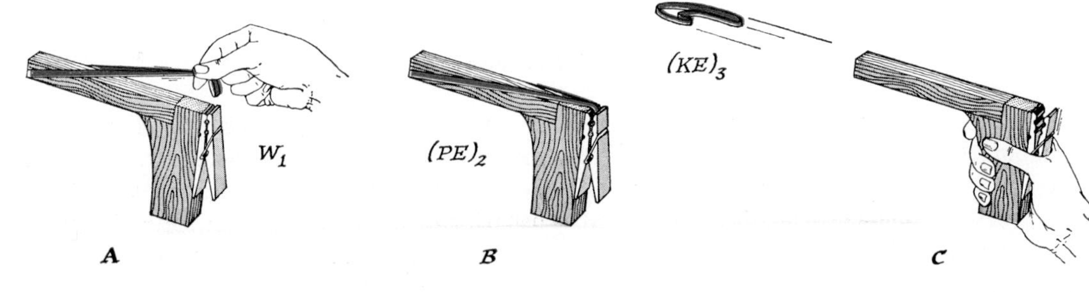

stretched and hooked over the end of the testing device, as shown in Figure 7-4A. Work must be done to stretch it—let's call it W_1. In Figure 7-4C the rubber band has been released and has returned to its initial length, but it now has energy of motion, $(KE)_3$. How much energy? The kinetic energy the rubber band has depends upon how much work was done in stretching it. W_1 fixes $(KE)_3$. We may write

$$W_1 = (KE)_3 \qquad (13)$$

Again we recognize work as a form of energy and, since $W_1 = (KE)_3$, the overall result is that energy was conserved. But was energy conserved in Figure 7-4B? If so, where is the energy? The work W_1 has already been done. Though the rubber band is stretched, there is no outward manifestation of energy. The rubber band is motionless, so it has received no kinetic energy. Yet, we know (from previous experiments) that the rubber band will receive the energy $(KE)_3$ at any instant that we choose to release it. The initial and final result are reminiscent of the billiard balls: energy is conserved. It would be convenient to say that energy is conserved in Figure 7-4B, as well. So, we invent a new form of energy—*energy of position*, or **potential energy.** We say that as the work W_1 was performed it was stored in the rubber band as potential energy.

Now we can review this experiment. In symbols we have

$$\underset{\text{work}}{W_1} = \underset{\text{potential energy}}{(PE)_2} = \underset{\text{kinetic energy}}{(KE)_3} \qquad (14)$$

The amount of work performed fixed W_1. Measurements of mass and velocity of the rubber band tell us, experimentally, the magnitude of $(KE)_3$. How do we know $(PE)_2$? How are we sure that $(PE)_2$ is equal to W_1 and to $(KE)_3$? The evidence we have is that we put an amount of energy into the system and can recover all of it later at will. It is natural to say the energy is stored in the meantime. Then we can say that the rubber band is just like the billiard ball collision: *energy is conserved* at all times.

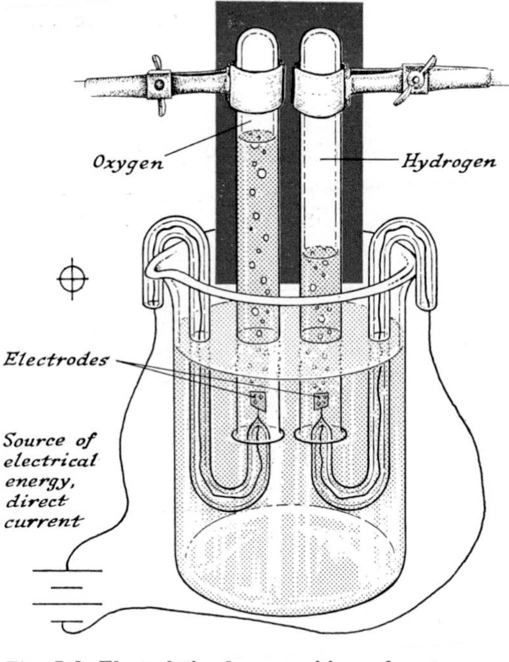

Fig. 7-5. **Electrolytic decomposition of water.**

7-2.3 Conservation of Energy in a Chemical Reaction

Figure 7-5 shows an apparatus in which an electric current can be passed through water. As remarked in Section 3-1.2, the electric current causes a decomposition of water. As work is done (electrical work), hydrogen gas and oxygen gas are produced. Measurements of the electric current and voltage show that 68.3 kcal of electrical work, W_1, must be done to decompose one mole of water. The equation for the reaction is

$$\underset{\text{electrical work}}{68.3 \text{ kcal}} + H_2O(l) \longrightarrow H_2(g) + \tfrac{1}{2}O_2(g) \qquad (15)$$

Now suppose we measure the heat of reaction of hydrogen and oxygen in a calorimeter like that shown in Figure 7-2. This experiment has been performed many times; 68.3 kcal of heat, Q_3, are produced for every mole of water formed. The equation for this reaction is

$$H_2(g) + \tfrac{1}{2}O_2(g) \longrightarrow H_2O(l) + 68.3 \text{ kcal heat} \qquad (16)$$

We have a situation just like that of the rubber band. We put a readily measured amount of energy, W_1, into the system and, at any time later that we choose, this energy can be recovered as heat, Q_3:

$$W_1 = Q_3 \tag{17}$$

The overall result is that energy is conserved.

Figure 7-7 shows this schematically in a diagram like that of Figure 7-1. If the heat content of two moles of water is represented by a line on this diagram, then the energy of two moles of hydrogen plus one mole of oxygen should be represented by a line 136.6 kcal higher. Now the diagram indicates that when water is decomposed, energy must be put in to raise the heat content enough to form the products. When hydrogen burns in oxygen, the heat content drops and energy is released.

Again we may ask: Where was the energy put in reaction (*15*) before we carry out reaction (*16*)? The rubber band experiment guides us.

It is easier to explain why $W_1 = Q_3$ if we say that the energy W_1 was stored in the chemical substances $H_2(g)$ and $O_2(g)$. We assign to these (and all other) substances the capacity to store energy and we call it **heat content.** This permits us to say that energy is conserved at all times during a chemical reaction as it is in billiard ball collisions and in stretched rubber bands.

7-2.4 The Basis for the Law of Conservation of Energy

We see that there are many forms of energy. We have talked about work as a form of energy and

Fig. 7-6. **Conservation of energy** *in a chemical reaction.*

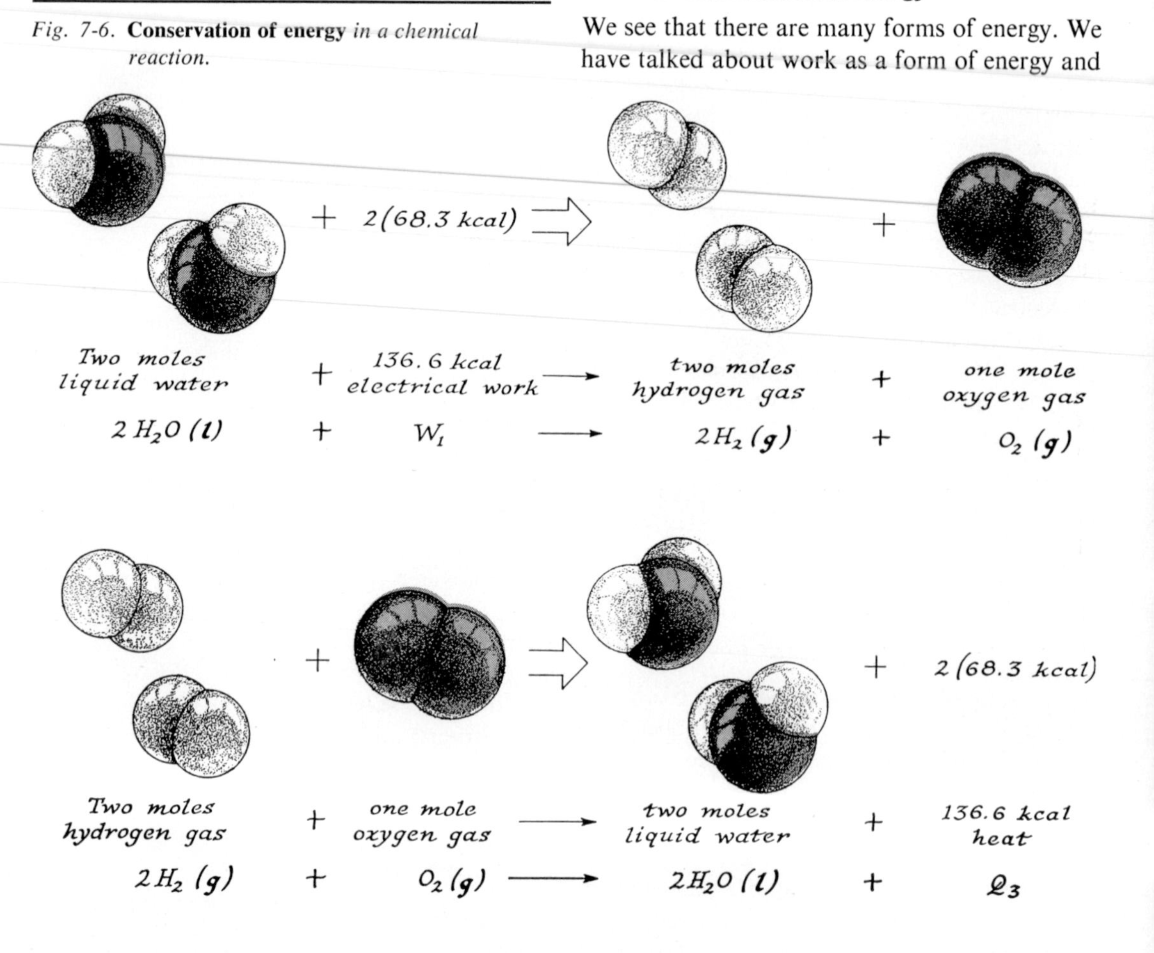

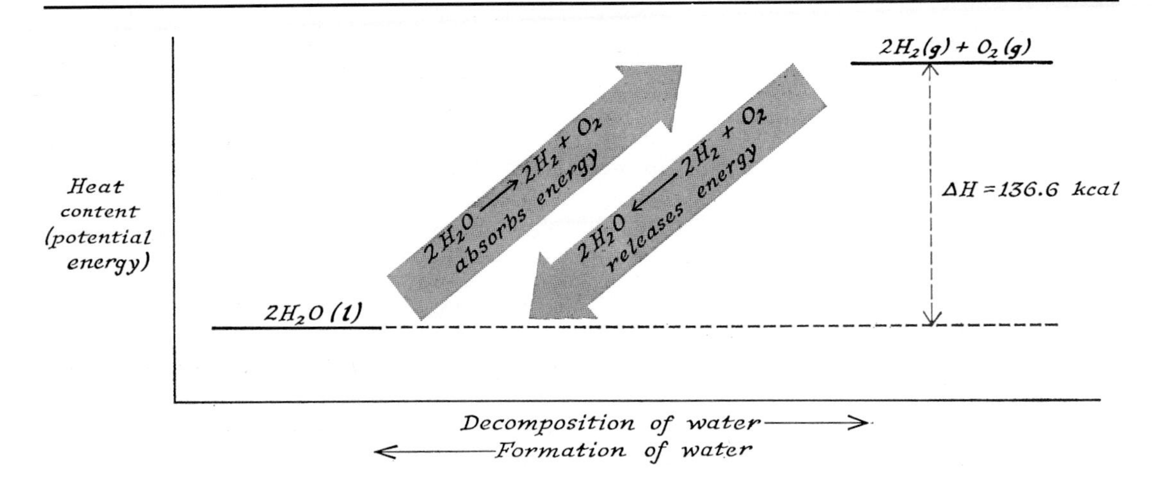

Fig. 7-7. **The energy change** *in a chemical reaction.*

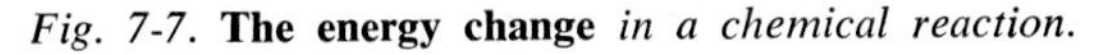

referred to two kinds of work, muscular work and electrical work. Kinetic energy and heat are other readily measured forms of energy. We have added two additional forms that are only indirectly detectable—the potential energy of the stretched rubber band and the heat content of chemical substances. With these two added forms, we can write:

Energy is always conserved.

This is called the **Law of Conservation of Energy.** It says that in every experiment so far performed, energy was conserved provided all of the different forms of energy are taken into account. Because the number of such experiments is extremely large and varied in type, the law gives a reliable basis for predicting. Don't be upset that the law is true only because we added forms of energy to account for energy not directly measurable. This law is like any other law—its usefulness justifies it. As long as the several forms of energy give us a model that is always consistent with experiments, the law remains useful.

In specific reference to the heat effects in chemical reactions, hundreds of different reactions have been studied calorimetrically. The results are always in accord with the Law of Additivity of Reaction Heats. If we assign a characteristic heat content to each chemical substance, then all of these experiments support the Law of Conservation of Energy. Since the Law of Conservation of Energy is consistent with so many different reactions, it can be safely assumed to apply to a reaction which hasn't been studied before.

The term "law" is usually applied to the older scientific generalizations. Modern scientists do not apply the term to new generalizations because they realize that all "laws of nature" are human statements—human generalizations—and are subject to revision. For example, later in this chapter you will find that matter and energy are one and the same. At that time, you will see that the two conservation laws, the Conservation of Matter and the Conservation of Energy, are really but one law, the Conservation of Matter (which is Energy). Yet in any chemical process the mass equivalent of the reaction heat is negligible. Under these conditions, the Law of Conservation of Matter and the Law of Conservation of Energy can be considered as independent statements. In this form the conservation laws are very useful, even though the statements we have made about them do not apply under all conditions.

7-3 THE ENERGY STORED IN A MOLECULE

In the discussion of Sections 7-2.1 to 7-2.4, we found it useful to talk about different "forms" of energy. Two of these are heat and heat content. Heat content is sometimes called "chemical energy" because its magnitude is intimately tied up with chemical composition. These are macroscopic* manifestations of energy. Two other macroscopic manifestations of energy are kinetic energy (of a thrown baseball, for example) and potential energy (of a baseball at the top of the flight of a high foul ball, for example). Thus, we need to identify several "forms" of energy when discussing the macroscopic properties of substances: heat, heat content (chemical energy), kinetic energy, potential energy, electrical energy, and mechanical work. The presence and amount of each energy form is determined by methods uniquely applicable to that form. We determine the quantity of heat released in a calorimeter by measuring a temperature rise with a thermometer. We measure the kinetic energy of a baseball with a watch and a meter stick. You would learn little about the kinetic energy of a baseball by throwing it at a thermometer and nothing about water temperature by wearing your wrist watch in the shower.

When we turn to the molecular scale, however, we discover that all of these macroscopic forms of energy can be discussed in terms of the two kinds of energy we assigned to the baseball, kinetic and potential energies. We can "explain" all macroscopic forms of energy with a microscopic model involving only the energy of motion and energy of position of the atomic and molecular particles. The explanation has the special advantage given in Section 1-1.3 (pp. 5–8). All of our experience and knowledge about the properties of moving baseballs (and billiard balls and rubber bands and automobiles and pendulums and gyroscopes) can be used in clarifying the nature of heat, heat content, electrical energy, etc. To see this, we must consider how chemists discuss the energy held by a molecule.

* **Macroscopic** means on a large scale—the opposite of microscopic. In general, it is used to indicate weighable and visible amounts.

7-3.1 The Energy of a Molecule

Let us picture a molecule in terms of a model consisting of balls of proper relative masses hooked together by springs. The springs represent the bonds between the atoms. We can start the springs vibrating and then toss the entire assembly through space in an end-over-end motion. There are now three kinds of kinetic energy associated with our model, as pictured in Figure 7-8.

This model applies quite well to a molecule in the gaseous state, but in the liquid state, and (even more so) in the solid state, all these motions are restricted. In these phases the chief kinetic energy manifestation is a back-and-forth motion of the molecule about a fixed point.

Fig. 7-8. **Types of motion of a molecule of carbon dioxide, CO_2. A.** *Translational motion; the molecule moves from place to place.* **B.** *Rotational motion; the molecule rotates about its center of mass.* **C.** *Vibrational motion; the atoms move alternately toward and away from the center of mass.*

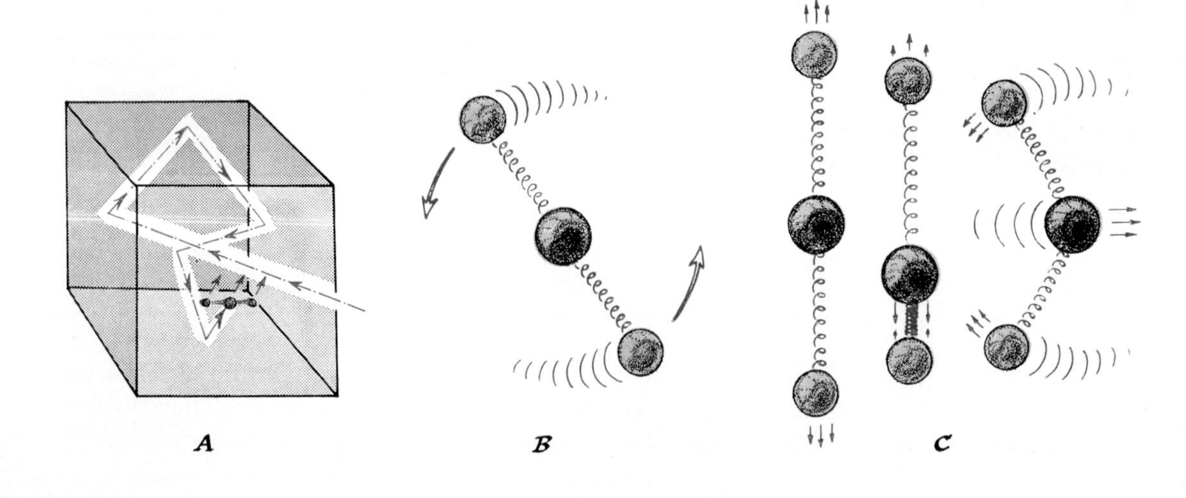

In addition to the various kinds of kinetic energy listed in Figure 7-8, there is potential energy related to the forces which act between molecules. These forces are attractive, having a very small average value in the gaseous state in which the molecules are far apart. The forces are, on the average, larger in the liquid state and are still larger in the solid state.

Next, there is present, within the molecule, chemical energy which is related to the forces which hold the atoms together in the molecule. This is referred to as chemical bond energy.

In addition, each atom has energy, some associated with the electrons and some with the nucleus. The electrons in the atom possess kinetic energy and, because of their attraction to the nucleus and repulsion from each other, they also possess potential energy. The algebraic sum of these kinetic and potential energies represents the energy necessary to pull an electron away from an atom.

Finally, there is present within the nucleus of each atom a store of energy. This energy is related to the forces holding the nuclear particles together. Since each nucleus remains intact and apparently uninfluenced through chemical reactions, this nuclear energy does not change. Hence, the nuclear contribution to the molecular heat content does not usually concern a chemist.

The sum of all of these forms of molecular energy makes up the molecular heat content. If we add together the molecular heat content of 6.02×10^{23} molecules of a given kind, we obtain the molar heat content of that substance.

7-3.2 Energy Changes on Warming

Having this view of the make-up of the heat content of a substance, we can now visualize the effects brought on by warming the substance. If the temperature is low at first, the substance will be a solid. Warming the solid increases the kinetic energy of the back-and-forth motions of the molecules about their regular crystal positions. As the temperature rises, these motions disturb the regularity of the crystal more and more. Too much of this random movement destroys the lattice completely. At the temperature above which the kinetic energy of the particles causes so much random movement that the lattice is no longer stable, a phase change occurs: the solid melts.

In the liquid each molecule has considerably more freedom of movement, particularly for translation and rotation. Warming the liquid enhances the amount of molecular movement. As always, kinetic energy provides a randomizing effect, tending to carry the molecules everywhere in the container. As the energy of motion rises (with the rising temperature), more of the molecules are able to move away from the liquid region where the potential energy is a minimum. Another phase change occurs: the liquid vaporizes.

If, now, we continue warming the substance sufficiently, we will reach a point at which the kinetic energies in vibration, rotation, and translation become comparable to chemical bond energies. Then molecules begin to disintegrate. This is the reason that only the very simplest molecules—diatomic molecules—are found in the Sun. There the temperature is so high (6000°K at the surface) that more complex molecules cannot survive.

Finally, if we continue the heating still further, we will ultimately reach a temperature at which the kinetic energies are large enough to disrupt the nuclei. Then, "nuclear reactions" begin. The conditions in some stars are considered to be suitable for rapid nuclear reactions.

To conclude this study, let's consider the magnitudes of the energy effects. Phase changes usually involve energies of several kilocalories per mole. Chemical reactions usually involve energies of 50 to several hundred kilocalories per mole. Thus, we see that the energies involved in chemical reactions are usually 10 to 100 times larger than those involved in phase changes.

EXERCISE 7-4

Show that the ratio of the molar heat of formation of water from the elements (a chemical reaction) to the molar heat of the fusion of water (a phase change) is of the order of 50.

7-4 THE ENERGY STORED IN A NUCLEUS

Now let us consider nuclear changes. The fact that nuclei do remain intact during chemical reactions suggests that much larger energies are required for nuclear changes. Experimentally, this proves to be true. Nuclear reactions usually involve energy changes over a million times greater than those we find in chemical reactions. This enormous factor accounts for the current interest in nuclear reactions as a source of energy.

One such nuclear reaction is represented by the equation

$$^{235}_{92}U + ^{1}_{0}n \longrightarrow ^{141}_{56}Ba + ^{92}_{36}Kr + 3^{1}_{0}n + \text{energy} \quad (18)$$

Before we examine the details of this rather strange looking equation, let us focus our attention on the "+ energy" term. The numerical value is of the order of 4.5×10^9 kcal/mole of uranium. Look at that figure again and compare it to the molar heat of combustion of carbon. Roughly, what is the ratio of these two energies? It is 10^7, or 10 million!

Now let us examine the reaction in more detail. Forget momentarily the subscripts and superscripts. Recall from Chapter 6 that the neutron (n) is one of the fundamental particles visualized as present in nuclei. What has happened?

$$U + n \longrightarrow Ba + Kr + 3n \quad (19)$$

Instead of producing new kinds of substances by combination of atoms, the element uranium has combined with a neutron and as a result has split into two other elements—barium and krypton—plus three more neutrons. Atoms of a given element are characterized by their atomic number, the number of units of positive charge on the nucleus. For one element to change into another element the nucleus must be altered. In our example the uranium nucleus, as a result of reacting with a neutron, splits or fissions into two other nuclei and releases, in addition, neutrons.*

How we get neutrons and how we get them to react with uranium nuclei is not essential to our present discussion.

A glance at the periodic table will show that the subscripts we have attached to our symbols are the atomic numbers of the elements designated by the symbols—92 for U, 56 for Ba, 36 for Kr. The zero subscript attached to the neutron denotes the lack of charge on this particle. If we look at the subscripts,

$$_{92}U + _{0}n \longrightarrow _{56}Ba + _{36}Kr + 3_{0}n \quad (20)$$

we notice that their sum on each side of the equation is identical:

$$92 + 0 = 56 + 36 + (3 \times 0)$$

This identity is but another way of expressing the law of conservation of charge.

In the model of nuclear structure you were given in Chapter 6, the nucleus was pictured as being built up of protons and neutrons. These two kinds of particles are given the general name **nucleon.** The ***mass number*** *of a nucleus is equal to the number of nucleons present.* The superscripts in our equation are mass numbers:

$$^{235}U + ^{1}n \longrightarrow ^{141}Ba + ^{92}Kr + 3^{1}n \quad (21)$$

Apparently the mass numbers are also conserved:

$$235 + 1 = 141 + 92 + (3 \times 1)$$

We may rephrase this in the form of a rule: *The total number of nucleons is unchanged during nuclear reactions.*

EXERCISE 7-5

According to the model of Chapter 6, how many nucleons would be present in a uranium nucleus of mass number 235? How many protons are pictured as being present? How many neutrons?

---

* Perhaps you have already recognized our nuclear reaction as a **fission** reaction. It is of the type of reaction used in an atomic pile, the energy source of a nuclear power plant. The example we have selected is only one of the ways the uranium nucleus can divide. Lanthanum and bromine nuclei are also produced, cerium and selenium, and so on, each pair of fission products being such that the sum of their atomic numbers is always 92.

Actually, then, by our symbol $^{235}_{92}U$ we are representing not an atom, but a nucleus. Our equation is written in terms of nuclei and particles associated with them. This nuclear equation tells us nothing about what compound of uranium was bombarded with neutrons or what compound of barium is formed. We are summarizing only the *nuclear changes*. During the nuclear change there is much disruption of other atoms because of the tremendous amounts of energy liberated. We do not know in detail what happens but eventually we return to electrically neutral substances (chemical compounds) and the neutrons are consumed by other nuclei.

It is easy to determine that there is an electron balance between the reactants and products:

Nucleus:	$_{92}U$	$_0n$	$_{56}Ba$	$_{36}Kr$	$_0n$
Electrons associated with nucleus:	92	0	56	36	0

$$92 = 56 + 36$$

7-4.1 Exact Mass Relationships

Although the mass numbers of the proton and neutron are both one, the masses of these fundamental particles are not identical. The mass of one mole of protons is 1.00762 grams and that of one mole of neutrons is 1.00893 grams. Further investigation would show that the experimentally measured mass of the nucleus of any given isotope is not the exact sum of the masses of protons and neutrons confined in the nucleus according to our model. For example, the mass of the nucleus of the uranium isotope of mass number 235 is less than the exact sum of the masses of 92 protons and 143 neutrons.

One of the consequences of the special theory of relativity formulated in 1905 by the great German theoretical physicist, Albert Einstein, was that we came to realize that mass and energy are one and the same. Although this was a very radical notion at the time Einstein first presented his theory, the equation relating mass to energy is probably already familiar to you. The formula $E = mc^2$ has become almost a part of common idiom since the successful application of nuclear energy became a part of modern technology in the mid 1940's. In this equation c is the speed of light, 3.00×10^{10} cm/second. Apparently a small value of mass (m) is equivalent to a tremendous amount of energy since the proportionality constant (c^2) relating mass to energy is numerically 9.00×10^{20}.

We can use this idea of the relation of mass to energy in several ways. The mass of a ^{235}U nucleus is less than the sum of the masses of the 92 protons and 143 neutrons postulated to be in it. The difference in mass represents the binding energy which holds the nucleons together in the nucleus. Here we have used the concept of expressing the nuclear binding energy in terms of the implied decrease in mass. We can do the same, if we wish, for a chemical reaction. Again let us return to the molar heat of combustion of carbon, roughly 10^2 kcal:

$$C(s) + O_2(g) \longrightarrow CO_2(g) \quad \Delta H = -94 \text{ kcal} \qquad (22)$$

The mass change associated with an energy change of 10^2 kcal* is of the order of 5×10^{-9} grams. This is a quantity far too small to be detected on any balance capable of weighing the 12 grams of carbon and 32 grams of oxygen consumed in the reaction. Since the chemical "mass defects" are too small to measure, we do not use this terminology in chemistry.

If we wish to gain some idea of the alteration of mass in a nuclear change, we cannot use the fission reaction because the exact masses of the nuclei involved are not known. Let us look at another type of reaction of possible importance in the production of nuclear energy:

$$^2_1H + ^3_1H \longrightarrow ^4_2He + ^1_0n \qquad (23)$$

This reaction is called **fusion** since nuclei are combining to form a heavier nucleus. The energy associated with this change is 4.05×10^7 kcal/mole of ^2_1H nuclei.

Let us do a little bookkeeping with the exact masses of these nuclei. Actually we will simplify a bit and use the exact masses of the atoms. This will make no difference. The masses of the atoms differ from the nuclear masses by the masses of the number of electrons in each atom. We have shown that electrons are conserved in nuclear changes. Exact masses of atoms (that is, exact masses of each isotopic species and not the chemical atomic weights shown on the inside back cover) are readily available. For our hydrogen-helium reaction we have

Reactants:	^2H	2.01471 g/mole	
	^3H	3.01707	
		5.03178	
Products:	^4He	4.00390	
	^1n	1.00893	
		5.01283	
Reactants:			5.03178
Products:			5.01283
Mass Difference:			0.01895 g/mole

Compare this mass difference of about 0.02 g/mole with one of about 5×10^{-9} g/mole for the combustion of carbon.

In closing, let us remind ourselves of the difference between nuclear and chemical reactions. In nuclear reactions, changes in the nuclei take place. In chemical reactions, the nuclei remain intact and the changes are explainable in terms of the electrons outside the nucleus.

* Before using the $E = mc^2$ relation to calculate the amount of mass associated with this energy change, you must pay attention to the relation of various energy units.

QUESTIONS AND PROBLEMS

1. Given

$$3C(s) + 2Fe_2O_3(s) + 110.8\ \text{kcal} \longrightarrow 4Fe(s) + 3CO_2(g)$$

Rewrite the equation using one mole of carbon and use the ΔH notation.

2. Given

$$\tfrac{1}{2}H_2(g) + \tfrac{1}{2}Br_2(l) \longrightarrow HBr(g) \qquad \Delta H = -8.60\ \text{kcal/mole HBr}$$

Rewrite the equation for one mole of hydrogen gas and include the heat effect as a term in the equation.

3. Which of the following reactions are endothermic?

(a) $H_2(g) + \tfrac{1}{2}O_2(g) \longrightarrow H_2O(g) \qquad \Delta H = -57.8\ \text{kcal}$

(b) $\tfrac{1}{2}N_2(g) + \tfrac{1}{2}O_2(g) \longrightarrow NO(g) \qquad \Delta H = +21.6\ \text{kcal}$

(c) $\tfrac{1}{2}N_2(g) + O_2(g) + 8.1\ \text{kcal} \longrightarrow NO_2(g)$

(d) $\tfrac{1}{2}N_2(g) + \tfrac{3}{2}H_2(g) \longrightarrow NH_3(g) + 11.0\ \text{kcal}$

(e) $NH_3(g) \longrightarrow \tfrac{1}{2}N_2(g) + \tfrac{3}{2}H_2(g) \qquad \Delta H = +11.0\ \text{kcal}$

4. What is the minimum energy required to synthesize one mole of nitric oxide, NO, from the elements?

5. How much energy is liberated when 0.100 mole of H_2 (at 25°C and 1 atmosphere) is combined with enough $O_2(g)$ to make liquid water at 25°C and 1 atmosphere?

6. How much energy is consumed in the decomposition of 5.0 grams of $H_2O(l)$ at 25°C and 1 atmosphere into its gaseous elements at 25°C and 1 atmosphere?

7. Using Table 7-II, calculate the heat of burning ethane in oxygen to give CO_2 and water vapor.

Answer. $\Delta H = -341$ kcal/mole C_2H_6.

8. Given

$$C(\text{diamond}) + O_2(g) \longrightarrow CO_2(g) \qquad \Delta H = -94.50\ \text{kcal}$$

$$C(\text{graphite}) + O_2(g) \longrightarrow CO_2(g) \qquad \Delta H = -94.05\ \text{kcal}$$

Find ΔH for the manufacture of diamond from graphite.

$$C(\text{graphite}) \longrightarrow C(\text{diamond})$$

Is heat absorbed or evolved as graphite is converted to diamond?

9. To change the temperature of a particular calorimeter and the water it contains by one degree requires 1550 calories. The complete combustion of 1.40 grams of ethylene gas, $C_2H_4(g)$, in the calorimeter causes a temperature rise of 10.7 degrees. Find the heat of combustion per mole of ethylene.

10. The "thermite reaction" is spectacular and highly exothermic. It involves the reaction between Fe_2O_3, ferric oxide, and metallic aluminum. The reaction produces white-hot, molten iron in a few seconds. Given:

$$2Al + \tfrac{3}{2}O_2 \longrightarrow Al_2O_3 \qquad \Delta H_1 = -400\ \text{kcal/mole}$$

$$2Fe + \tfrac{3}{2}O_2 \longrightarrow Fe_2O_3 \qquad \Delta H_2 = -200\ \text{kcal/mole}$$

Determine the amount of heat liberated in the reaction of 1 mole of Fe_2O_3 with Al.

Answer. $\Delta H = -200$ kcal/mole Fe_2O_3.

11. How much energy is released in the manufacture of 1.00 kg of iron by the "thermite reaction" mentioned in Problem 10?

12. How many grams of water could be heated from 0°C to 100°C by the heat liberated per mole of aluminum oxide formed by the "thermite reaction," as described in Problem 10?

13. Which would be the better fuel on the basis of the heat released per mole burned, nitric oxide, NO, or ammonia, NH_3? Assume the products are $NO_2(g)$ and $H_2O(g)$.

14. What is the minimum energy required to synthesize sulfur dioxide from sulfuric acid?

$$H_2SO_4(l) \longrightarrow SO_2(g) + H_2O(g) + \tfrac{1}{2}O_2(g)$$

Answer. $\Delta H = +65$ kcal/mole $SO_2(g)$.

15. Why is the Law of Conservation of Energy considered to be valid?

16. What do you think would happen in scientific circles if a clearcut, well-verified exception was

found to the Law of Conservation of Energy as stated in the text?

17. Is energy conserved when a ball of mud is dropped from your hand to the ground? Explain your answer.

18. What becomes of the energy supplied to water molecules as they are heated in a closed container from 25°C to 35°C?

19. Outline the events and associated energy changes that occur on the molecular level when steam at 150°C and 1 atmosphere pressure loses energy continually until it finally becomes ice at −10°C.

20. The heat of combustion of methane, CH_4, is −210 kcal/mole:

$$CH_4(g) + 2O_2 \longrightarrow CO_2 + 2H_2O \qquad \Delta H = -210 \text{ kcal}$$

Discuss why this fuel is better than water gas if the comparison is based on one mole of carbon atoms.

21. In a nuclear reaction of the type called "nuclear fusion," two nuclei come together to form a larger nucleus. For example, deuterium nuclei, ${}^2_1H$, and tritium nuclei, ${}^3_1H$, can "fuse" to form helium nuclei, ${}^4_2He$, and a neutron:

$$ {}^2_1H + {}^3_1H \longrightarrow {}^4_2He + {}^1_0n \qquad \Delta H = -4.05 \times 10^7 \text{ kcal}$$

How many grams of hydrogen would have to be burned (to gaseous water) to liberate the same amount of heat as liberated by fusion of one mole of ${}^2_1H$ nuclei? Express the answer in tons (1 ton = 9.07×10^5 g).

22. Which of the following reactions is most likely to have a heat effect of −505 kcal? Which would be -1.7×10^6 kcal? Which would be +7.2 kcal?

(a) $UF_6(l) \longrightarrow UF_6(g)$ $\quad \Delta H = ?$

(b) $U(s) + 3F_2(g) \longrightarrow UF_6$ $\quad \Delta H = ?$

(c) ${}^{238}_{92}U + {}^1_0n \longrightarrow {}^{239}_{92}U$ $\quad \Delta H = ?$

23. Fission of uranium gives a variety of fission products, including praseodymium, Pr. If the process by which praseodymium is formed gives ${}^{147}_{59}Pr$ and three neutrons, what is the other nuclear product?

$$ {}^{235}_{92}U + {}^1_0n \longrightarrow {}^{147}_{59}Pr + ? + 3{}^1_0n$$

CHAPTER 8

The Rates of Chemical Reactions

. . . a molecular system . . . [passes] . . . from one state of equilibrium to another . . . by means of all possible intermediate paths, but the path most economical of energy will be more often travelled.

HENRY EYRING, 1945

A candle remains in contact with air indefinitely without observable reaction but it reacts when given a start with a lighted match. A mixture of household gas and air in a closed room remains indefinitely without reacting but it may explode violently if so much as a glowing cigarette is brought into the room. A piece of iron reacts quite slowly with air (it rusts) but a piece of white phosphorus bursts into flame when it is exposed to air. These are all reactions with oxygen from the air but they have extremely different time behaviors. *Reactions proceed at different rates.* We care about reaction rate because we must understand how rapidly a reaction proceeds and what factors determine its rate in order to bring the reaction under control.*

Let us see what the expression "the rate of a reaction" means in terms of an example—the reaction between carbon monoxide gas, CO, and nitrogen dioxide, NO_2. Chemical tests show that the products are carbon dioxide, CO_2, and nitric oxide, NO. The equation for the reaction is

$$CO + NO_2 \longrightarrow CO_2 + NO \qquad (1)$$

Suppose we prepare a mixture of carbon monoxide and nitrogen dioxide and then heat it to 200°C. When the gas is heated, we observe a gradual disappearance of the reddish brown color of NO_2. Reaction is taking place. We can find its time behavior by measuring the change in color during a measured time interval. Since the other gases are colorless, this color change indicates the number of moles of NO_2 that have reacted during the time interval. The quotient of these two, moles reacted divided by the time interval, is called the rate of the reaction:

$$\text{Rate} = \frac{\text{quantity } NO_2 \text{ consumed}}{\text{time interval}} = \text{quantity } NO_2 \text{ consumed per unit time}$$

We can express the rate of reaction (*1*) in terms of the rate of consumption of either CO or NO_2. Equally well, we can express the time behavior

* The study of reaction rates is called **chemical kinetics.**

of the reaction in terms of the appearance of either product, CO_2 or NO. Which is used depends upon convenience of measurement. If the experimenter prefers to measure the production of carbon dioxide, he would express the rate in the form

$$\text{Rate} = \frac{\text{quantity } CO_2 \text{ produced}}{\text{time interval}}$$

$$= \text{quantity } CO_2 \text{ produced per unit time}$$

The quantity consumed or produced is conveniently expressed in partial pressure units if the substance is a gas. Concentration units are convenient if the reactant or product is in solution. The time measurement is also expressed in whatever units fit the reaction: microseconds for the explosion of household gas and oxygen, seconds or minutes for the burning of a candle, days for the rusting of iron, months for the rotting of wood.

8-1 FACTORS AFFECTING REACTION RATES

In the laboratory you have observed the reaction of ferrous ion, $Fe^{+2}(aq)$, with permanganate ion, $MnO_4^-(aq)$, and also the reaction of oxalate ion, $C_2O_4^{-2}(aq)$, with permanganate ion, $MnO_4^-(aq)$. These studies show that *the rate of a reaction depends upon the nature of the reacting substances.* In Experiment 14, the reaction between IO_3^- and HSO_3^- shows that *the rate of a reaction depends upon concentrations of reactants and on the temperature.* Let us examine these factors one at a time.

8-1.1 The Nature of the Reactants

Compare the following three reactions, all of which occur in water solutions:

$$5C_2O_4^{-2}(aq) + 2MnO_4^-(aq) + 16H^+(aq) \longrightarrow 10CO_2(g) + 2Mn^{+2}(aq) + 8H_2O \quad \text{slow} \quad (2)$$

$$5Fe^{+2}(aq) + MnO_4^-(aq) + 8H^+(aq) \longrightarrow 5Fe^{+3}(aq) + Mn^{+2}(aq) + 4H_2O \quad \text{very fast} \quad (3)$$

$$Fe^{+2}(aq) + Ce^{+4}(aq) \longrightarrow Fe^{+3}(aq) + Ce^{+3}(aq) \quad \text{very fast} \quad (4)$$

Both ferrous ion, $Fe^{+2}(aq)$ and oxalate ion, $C_2O_4^{-2}(aq)$, have the capability of decolorizing a solution containing permanganate ion at room temperature. Yet, there is a great contrast in the time required for the decoloration. The difference lies in specific characteristics of $Fe^{+2}(aq)$ and $C_2O_4^{-2}(aq)$. On the other hand, $Fe^{+2}(aq)$ is also changed to $Fe^{+3}(aq)$ by reacting either with $MnO_4^-(aq)$ or with ceric ion, $Ce^{+4}(aq)$. One of these reactions is simple and the other involves many molecules. Yet, these are both rapid reactions.

Here are two reactions that take place in the gas phase.

$$2NO + O_2 \longrightarrow 2NO_2 \quad \text{moderate at 20°C} \quad (5)$$

$$CH_4 + 2O_2 \longrightarrow CO_2 + 2H_2O \quad \text{extremely slow at 20°C} \quad (6)$$

The oxidation of nitric oxide, NO, is a reaction involved in smog production. It is moderately rapid at normal temperatures. The oxidation of methane, CH_4 (household gas), however, occurs so slowly at room temperature that we may say that, for all practical purposes it doesn't react at all. Again, the difference in the reaction rates must depend upon specific characteristics of the reactants, NO and CH_4.

The determination of the molecular characteristics which are important in rate behavior is an interesting frontier of chemistry. It seems that chemical reactions which involve the breaking of several chemical bonds and the formation of new chemical bonds tend to proceed slowly at room temperature. Reaction (*2*) is of this type—there are many bonds which must be broken in the five $C_2O_4^{-2}$ ions and the two MnO_4^- ions to form the $10CO_2$ and $2Mn^{+2}$. This reaction might be expected to proceed slowly (as it does). Reaction (6) also involves breaking of bonds and forming of new bonds and it is slow at room temperature. In contrast, reaction (*3*) is very rapid though it involves breaking of chemical bonds and it might be expected to proceed slowly.

We see that we cannot be certain of a prediction that a reaction might be slow. Reaction (*4*) apparently does not require bond breaking or bond formation. It can be expected to be rapid (as it is). A prediction of this type is usually reliable. Reaction (*5*) requires breaking of but one bond and the formation of two. It has a moderate reaction rate, rapid at high pressures and slow at low pressures.

These and other examples lead to the following rules:

(1) Reactions that do not involve bond rearrangements are usually rapid at room temperature.
(2) Reactions in which bonds are broken tend to be slow at room temperature.

We can say little more about how the nature of the reactants determines the reaction rate until we consider in detail how some reactions take place. For the time being, it will suffice to observe that this is an active field of study and much remains to be learned.

EXERCISE 8-1

Are any of the following three reactions likely to be extremely rapid at room temperature? Are any likely to be extremely slow at room temperature? Explain.

(a) $Cr^{+2}(aq) + Fe^{+3}(aq) \longrightarrow Cr^{+3}(aq) + Fe^{+2}(aq)$

(b) $3Fe^{+2}(aq) + NO_3^-(aq) + 4H^+(aq) \longrightarrow 3Fe^{+3}(aq) + NO(g) + 2H_2O$

(c) $\underset{\text{liquid gasoline}}{C_8H_{18}} + 12\frac{1}{2}O_2(g) \longrightarrow 8CO_2(g) + 9H_2O(g)$

8-1.2 Effect of Concentration: Collision Theory

Henceforth we shall concentrate our attention on one reaction at a time. The nature of the reactants will be held constant while the other factors that affect rates are considered. The first of these factors is concentration.

Chemists have learned that, for many reactions, raising the concentration of a reactant increases the reaction rate. Not infrequently, though, there will be no effect. In this section we shall consider how a rate increase with rising reactant concentration is explained. In Section 8-1.3 we shall explore why some reactions proceed at a rate independent of the concentration of one or more reactants. Both explanations are based upon a model of the way chemical reactions take place on the molecular scale.

In the molecular view of matter, it is natural to assume that two molecules must come close together in order to react. Therefore, we postulate that chemical reactions depend upon collisions between the reacting particles—atoms, molecules, or ions. This model of reaction rate behavior is called the **collision theory.** It provides a successful basis for understanding the effect of concentration. Just as an increase of the number of cars in motion on a highway leads to a higher rate of formation of dented fenders, increasing the number of particles in a given volume gives more frequent molecular collisions. The higher frequency of collisions results in a higher rate of reaction.

Consider a homogeneous system—one in which all components are in the same phase. According to the collision theory, we can expect that increasing the concentration of one or more reactants will result in an increase in the rate of the reaction. Lowering the concentration has the opposite effect. This is exactly the behavior found in the reaction between $HSO_3^-(aq)$ and $IO_3^-(aq)$ when the concentrations are varied by adding or removing reactants or solvent (Experiment 14). In gases, also homogeneous systems, the concentration of an individual reactant can be raised by admitting more of that substance into the mixture. The concentrations of all gaseous components can be raised simultaneously by decreasing the volume occupied by the mixture. Decreasing the volume by compressing the gas raises the concentration of all reactants, hence increases the rates of reactions taking place. Increasing the volume by expanding the gas has the opposite effect on concentrations, hence decreases reaction rates.

In a heterogeneous reaction system, the com-

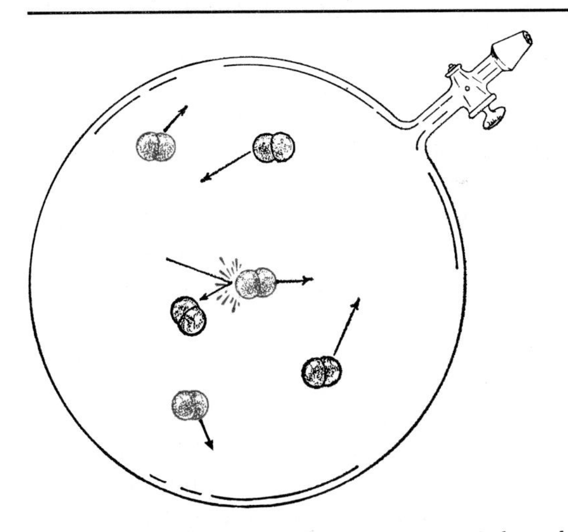

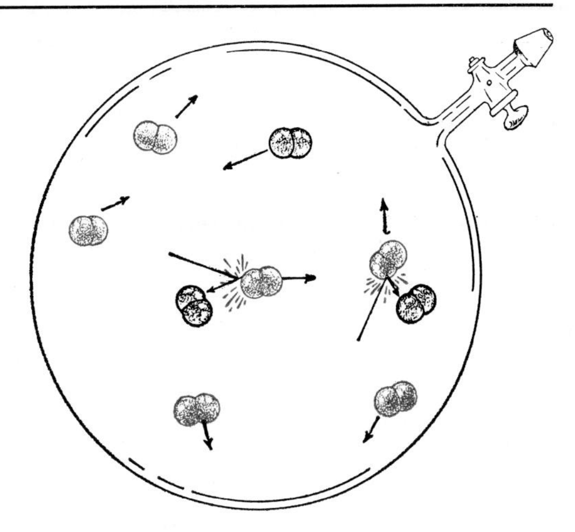

Fig. 8-1. **The number of collisions per second** *depends upon concentration.*

ponents are in two or more different phases. As an example, consider

$$\text{wood (solid)} + \text{oxygen (gas)} \xrightarrow{\text{burning}} \text{carbon dioxide (gas)} + \text{water (gas)}$$

In a system of this sort, the rate of the reaction depends upon the amount of interface between the phases, or, in other words, the area of contact between them. For example, a log burns in air at a relatively slow rate. If the amount of exposed surface of the wood is increased by reducing the log to splinters, the burning is much more rapid. If, further, the wood is reduced to fine sawdust and the latter is suspended in a current of air, the combustion takes place explosively. Where one of the reactants is a gas, such as in the above example, the concentration of the gas is also a factor. A piece of wood burns much more rapidly in pure oxygen than it does in ordinary air, in which the oxygen makes up only about 20% of the mixture.

We see that the collision theory provides a good explanation of reaction rate behavior. It is quite reasonable that the reaction rate should depend upon collisions among the reactant molecules. In fact, it is so reasonable that we are left wondering why the concentrations of some reactants in some reactions do not affect the rate. The explanation is found in the detailed steps by which the reaction takes place, the reaction mechanism.

8-1.3 Reaction Mechanism

As has been proposed, in order for a chemical reaction to occur, particles must collide. The particles may be atoms, molecules, or ions. As a result of collisions, there can be rearrangements of atoms, electrons, and chemical bonds, with the resultant production of new species. As an example, let us take another look at the reaction between Fe^{+2} and MnO_4^- in acid solution:

$$5Fe^{+2}(aq) + MnO_4^-(aq) + 8H^+(aq) \longrightarrow 5Fe^{+3}(aq) + Mn^{+2}(aq) + 4H_2O \quad (7)$$

The equation indicates that one MnO_4^- ion, five Fe^{+2} ions, and eight H^+ ions (a total of fourteen ions) must react with each other. If this reaction were to take place in a single step, these fourteen ions would have to collide with each other *simultaneously*. The probability of such an event occurring is extremely small—so small that a reaction which depended upon such a collision would proceed at a rate immeasurably slow. Since the reaction occurs at an easily measured rate, it must proceed by some sequence of steps, none of which involves such an improbable collision.

As a matter of fact, chemists regard the collision of even four molecules as an extremely improbable event if they are at low concentration or if they are in the gas phase. We conclude that a complex chemical reaction which proceeds at a measurable rate probably takes place in a series of simpler steps. The series of reaction steps is called the **reaction mechanism.**

Consider the oxidation of gaseous hydrogen bromide, HBr, a reaction that is reasonably rapid in the temperature range from 400 to 600°C:

$$4HBr(g) + O_2(g) \longrightarrow 2H_2O(g) + 2Br_2(g) \quad (8)$$

By the collision theory, we expect that increasing the partial pressure (and thus, the concentration) of either the HBr or O_2 will speed up the reaction. Experiments show this is the case. Quantitative studies of the rate of reaction (*8*) at various pressures and with various mixtures show that oxygen and hydrogen bromide are equally effective in changing the reaction rate. However, this result raises a question. Since reaction (*8*) requires four molecules of HBr for every one molecule of O_2, why does a change in the HBr pressure have just the same effect as an equal change in the O_2 pressure?

The explanation is found by considering the details of the process by which reaction (*8*) occurs. The overall reaction brings together five molecules, four of HBr and one of O_2. However, the chance that five gaseous molecules will collide simultaneously is practically zero. The reaction must occur in a sequence of simpler steps.

All of the studies of reaction (*8*) are explained by the following series of reactions:

$$\begin{aligned} HBr + O_2 &\longrightarrow HOOBr && \text{slow} \quad (9) \\ HOOBr + HBr &\longrightarrow 2HOBr && \text{fast} \quad (10) \\ HOBr + HBr &\longrightarrow H_2O + Br_2 && \text{fast} \quad (11) \end{aligned}$$

First, observe that adding reactions (*9*) and (*10*) plus twice reaction (*11*) gives the overall reaction (*8*). Next, we see that each step in the sequence requires only two molecules to collide. Finally, the proposal that reaction (*9*) is slow whereas reactions (*10*) and (*11*) are fast explains why HBr and O_2 have the same effect on the reaction rate.

Reaction (*9*) is a "bottle-neck" in the oxidation of hydrogen bromide. As fast as HOOBr is formed by this slow reaction it is consumed in the rapid reaction (*10*). But no matter how rapid reactions (*10*) and (*11*) are, they can produce H_2O and Br_2 only as fast as the slowest reaction in the sequence. Hence, the factors that determine the rate of reaction (*9*) determine the rate of the overall process.

The sequence of reactions (*9*), (*10*), and (*11*) is called the reaction mechanism of the overall reaction (*8*). Because it is the slowest reaction in the mechanism, reaction (*9*) is the step that fixes the rate. *The slowest reaction in a reaction mechanism is called the* ***rate determining step.***

There are two features of this example that are rather common. First, none of the steps in the reaction mechanism requires the collision of more than two particles. *Most chemical reactions proceed by sequences of steps, each involving only two-particle collisions.* Second, the overall or net reaction does *not* show the mechanism. In general, *the mechanism of a reaction cannot be deduced from the net equation for the reaction*; the various steps by which atoms are rearranged and recombined must be determined through experiment.

EXERCISE 8-2

Imagine five people working together to wash a stack of very greasy dishes. The first two clear the table and hand the dishes to the third person who washes them and hands them on. The last two persons dry and stack them. Which step is likely to be the rate determining step? In the light of your answer, discuss how the rate of the overall process would be affected if a sixth person joined the group (a) as a table clearer; (b) as a second dishwasher; (c) as a dish dryer.

8-1.4 The Quantitative Effect of Concentration

The reaction mechanism is deduced from quantitative studies of the dependence of the rate upon the concentrations or pressures of the various reactants. To interpret such studies, we need to develop our collision theory model.

Consider the reaction between gaseous hydrogen, H_2, and gaseous iodine, I_2:

$$H_2(g) + I_2(g) \longrightarrow 2HI(g) \quad (12)$$

Each time a molecule of H_2 collides with an iodine molecule, reaction may occur. The frequency of these encounters, for a particular H_2 molecule, is determined by how many I_2 molecules are present. Doubling the number of I_2 molecules per unit volume would just double the collisions. Tripling the number of I_2 molecules per unit volume would triple the collisions. Since the iodine partial pressure fixes the iodine concentration, the rate of the reaction is proportional to the iodine partial pressure:

$$\text{(rate) is proportional to} \begin{bmatrix} \text{iodine} \\ \text{partial} \\ \text{pressure} \end{bmatrix} \tag{13}$$

In the same way, a particular iodine molecule must find a hydrogen molecule to react. The rate of the reaction is proportional to the partial pressure of the hydrogen:

$$\text{(rate) is proportional to} \begin{bmatrix} \text{hydrogen} \\ \text{partial} \\ \text{pressure} \end{bmatrix} \tag{14}$$

In view of (*13*) and (*14*), the rate must be proportional to the *product* of the partial pressure of iodine and hydrogen:

(rate) is proportional to

$$\begin{bmatrix} \text{hydrogen} \\ \text{partial} \\ \text{pressure} \end{bmatrix} \times \begin{bmatrix} \text{iodine} \\ \text{partial} \\ \text{pressure} \end{bmatrix} \tag{15}$$

In symbols, we can write

$$\text{rate} = k[p_{H_2}] \times [p_{I_2}] \tag{16}$$

8-1.5 Effect of Temperature: Collision Theory

In Experiments 12 and 14, you discovered that temperature has a marked effect upon the rate of chemical reactions. Thus, raising the temperature speeded up the reaction between IO_3^- and HSO_3^-. That is the same effect, qualitatively, that is observed in the reaction of a candle with air. The match "lighted" the candle by raising its temperature (at the wick). Once started, the reaction of combustion releases enough heat to keep the temperature high, thus keeping the reaction going at a reasonable rate. Raising the temperature speeded up the reaction. The same type of explanation applies to the explosion of a kitchen full of household gas and air when a cigarette is brought into the room. Around the glowing tip of the cigarette the gas temperature is raised. At this locale, the reaction speeds up, liberating heat. This heat warms the nearby region even more and the reaction goes somewhat faster. This acceleration continues until finally (in a millisecond or so) it reaches an explosive rate—the most rapid reaction permitted by the collisional properties of the gas. Raising the temperature started it all by speeding up the reaction.

In all of these reactions (and in almost all others), increasing the temperature has a very pronounced effect, always speeding up the reaction. Two questions come to mind. "Why does a temperature rise speed up a reaction?" and "Why does a temperature rise have such a large effect?" To answer these questions, we return to our collision theory.

From what we know about molecular sizes, we can calculate that a particular CH_4 molecule collides with an oxygen molecule about once every one-thousandth of a microsecond (10^{-9} seconds) in a mixture of household gas (methane, formula CH_4) and air under normal conditions. This means that every second this methane molecule encounters 10^9 oxygen molecules! Yet the reaction does not proceed noticeably. We can conclude either that most of the collisions are ineffective or that the collision theory is not a good explanation. We shall see that the former is the case—we can understand why most collisions might be ineffective in terms of ideas that are consistent with the collision theory.

Chemists have learned that chemical reactions occur when collisions occur but *only when the collision involves more than a certain amount of energy*. We can understand this by returning to our analogy of cars bumping each other on a highway. In a line of heavy traffic one frequently receives gentle bumps from the car in front or the car behind. No damage is done to the cars—only to tempers. But occasionally a high speed collision occurs. If this occurs with enough energy, a bumper may be knocked off a car and a fender may be dented. It is the high energy collisions which cause the auto damage and it is high energy molecular collisions which cause the "molecular damage" that we call a chemical reaction. Just as a certain amount of energy is required to break loose a bumper, a certain amount of energy is required to cause a chemical reaction. In either instance, if there is more than this "threshold energy," the reaction can occur and if there is less, it cannot.

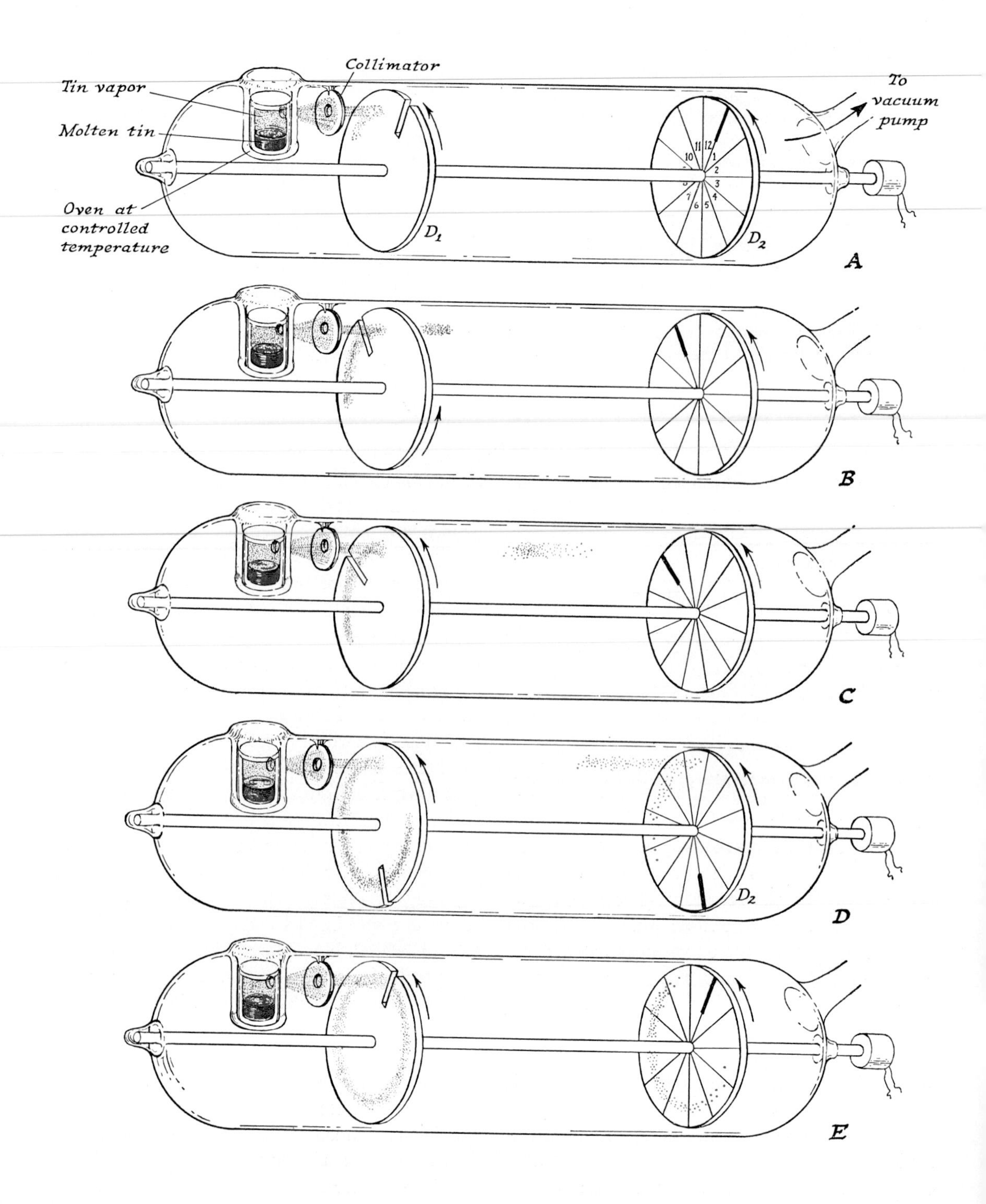

Fig. 8-2. **Rotating disc** *for measurement of atomic velocities.*

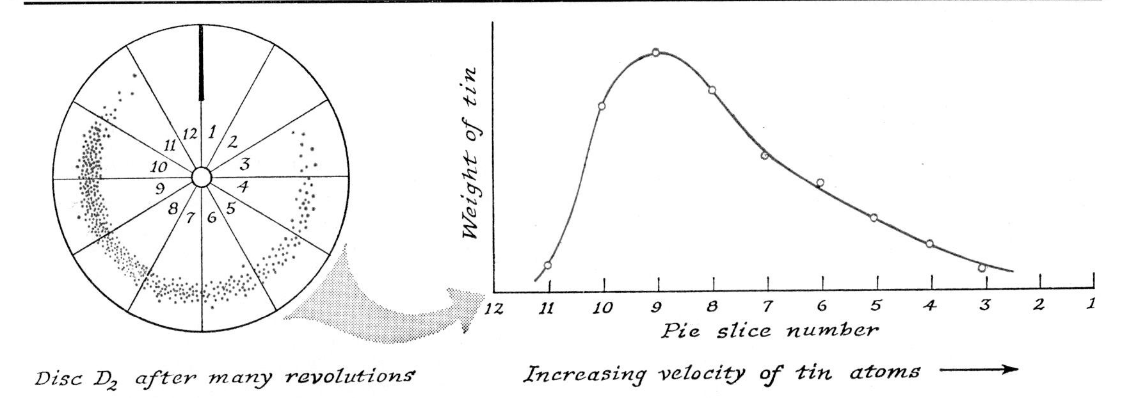

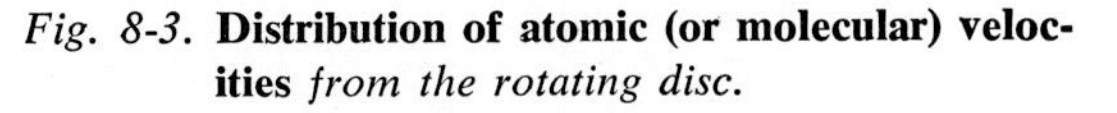

Fig. 8-3. **Distribution of atomic (or molecular) velocities** *from the rotating disc.*

DISTRIBUTION OF KINETIC ENERGIES

This discussion of threshold energy causes us to wonder what energies are possessed by molecules at a given temperature. We have already compared the molecules of a gas with billiard balls rebounding on a billiard table. When billiard balls bounce around, colliding with each other, some of them move rapidly and some slowly. Do molecules behave this way? Experiment provides the answer.

Figure 8-2 shows a device for measuring the distribution of atomic or molecular velocities. It consists of two discs, D_1 and D_2, rotating rapidly on a common axle. They rotate in a vacuum chamber in front of an oven containing molten tin and held at a controlled temperature. Vapor streams out of the small opening in the oven and strikes the rotating disc, D_1. When the disc has rotated to the position shown in Figure 8-1B, a small amount of gas has passed through the slot in disc D_1. A short time later, shown in Figure 8-1C, the atoms of tin have traveled part of the way toward the second rotating disc. The fastest moving atoms have traveled farther than the others—they are leading the way. The slowest moving atoms are beginning to lag behind. Still later, Figure 8-1D, the atoms have spread out in space even more, and the fastest atoms have already reached the second rotating disc. Now as the atoms reach disc D_2, they condense on it. The position where a given atom condenses on disc D_2 depends upon how long that atom took to travel from D_1 to D_2 and how fast D_2 is rotating.

As the slotted disc D_1 lets through burst after burst of tin atoms, a layer of tin builds up on the surface of disc D_2. The pattern of this layer is determined by the distribution of velocities of the atoms escaping from the oven at temperature T. Figure 8-3 shows the disc D_2 divided into sections, like slices of a pie. The fastest moving atoms are condensed on pie slices 3, 4, and 5. The slowest moving atoms are condensed on the pie slices 10 and 11. If the disc is cut up and each slice weighed, the amount of tin can be determined. A plot of the weight of tin against the pie slice number indicates the distribution of atomic velocities.

Clearly, the plot of Figure 8-3 contains information about the distribution of kinetic energies. From the rate of rotation of the discs and the distance between them we can calculate the velocity an atom must have to condense on a particular pie slice. From the atomic mass and its velocity, we learn the atom's kinetic energy. Figure 8-4 shows the result. At a temperature T_1

Fig. 8-4. **Effect of temperature** *on atomic (or molecular) kinetic energy distribution.*

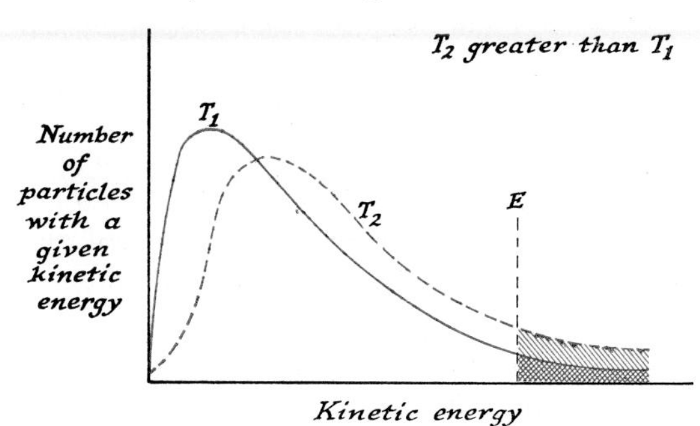

a few atoms have very low kinetic energies and some have very high kinetic energies. Most of them have intermediate kinetic energies, as shown by the solid curve. At a higher temperature, T_2, the energy distribution is altered to that shown as a dashed curve. As can be seen, increasing the temperature causes a general shift of the distribution toward one with more molecules having high kinetic energies. Moreover, in going from T_1 to T_2 there is a large increase in the number of molecules having kinetic energies above a certain value.

THRESHOLD ENERGY AND REACTION RATE

We can apply these curves to our reaction rate problem. Suppose a reaction can proceed only if two molecules collide with kinetic energy exceeding a certain threshold energy, E. Figure 8-4 shows us a typical situation. At T_1 the vertically shaded area is proportional to the number of the molecules which possess this energy or more. Since only a small number of molecules have as much as this energy, few collisions are effective, and the reaction is slow. But if we raise the temperature to T_2, the number of molecules with energy E or greater is raised in proportion to the diagonally shaded area. Only a small temperature change is needed to make a large change in the area out on the tail of the energy distribution curve. Consequently, the reaction rate is very sensitive to change in temperature.

This argument is based upon the "typical" situation in which E is well out on the tail of the curve. Suppose it is not; suppose E is near the maximum of the curve at T_1, or is even to the left of it. Then a large number of the molecules have the requisite energy, even at the lower temperature, T_1. Since collisions occur so rapidly (remember, one every 10^{-9} second or so), the reaction is over in a blink of the eye. This reaction would be called "instantaneous." The circumstances shown in Figure 8-4 are "typical" only of a slow reaction.

It should be remarked that raising the temperature also increases the rate by increasing the frequency of the collisions. This is, however, a very small effect compared with that caused by the increase in the number of molecules with sufficient energy to cause reaction.

8-2 THE ROLE OF ENERGY IN REACTION RATES

Now that we have this evidence about the molecular velocity distribution, we can see how temperature changes the fraction of the molecular collisions involving an energy exceeding the threshold energy E. Now our understanding of the role of energy in fixing reaction rates can be expanded.

8-2.1 Activation Energy

Suppose you wished to make an automobile trip from Los Angeles to San Francisco. This trip requires that the Tehachapi Mountains be crossed. As shown in Figure 8-5, this can be accomplished by taking the highway through the Tejon Pass. Of course, the high altitude of this pass, 4200 feet, makes this part of the trip the slowest and the biggest test of your automobile. This is the point in the trip where radiator fluids boil and engine troubles develop. This is the point where the older cars turn back and limp home. Yet, this is the lowest pass in these mountains, hence the most favorable route. If your car will surmount a 4200 foot pass, it will undoubtedly make the trip to San Francisco.

Chemical reactions are similar. As molecules collide and reaction takes place, the atoms must take up, momentarily, bonding arrangements that are less stable than either reactants or products. These high energy molecular arrangements are like the mountain pass—they place an energy barrier between reactants and products. Only if the colliding molecules have enough energy to surmount the barrier imposed by the unstable arrangements can reaction take place. This barrier determines the "threshold energy" or minimum energy necessary to permit a reaction to occur. It is called the **activation energy.**

This barrier can be shown graphically by amplifying Figure 7-1 (p. 110), which showed the relative energies of reactants and products. In Figure 8-6 we see that the diagram becomes the equivalent of the road map in our automobile trip analogy. This diagram applies to the reaction between carbon monoxide, CO, and nitrogen dioxide, NO_2, to produce carbon dioxide,

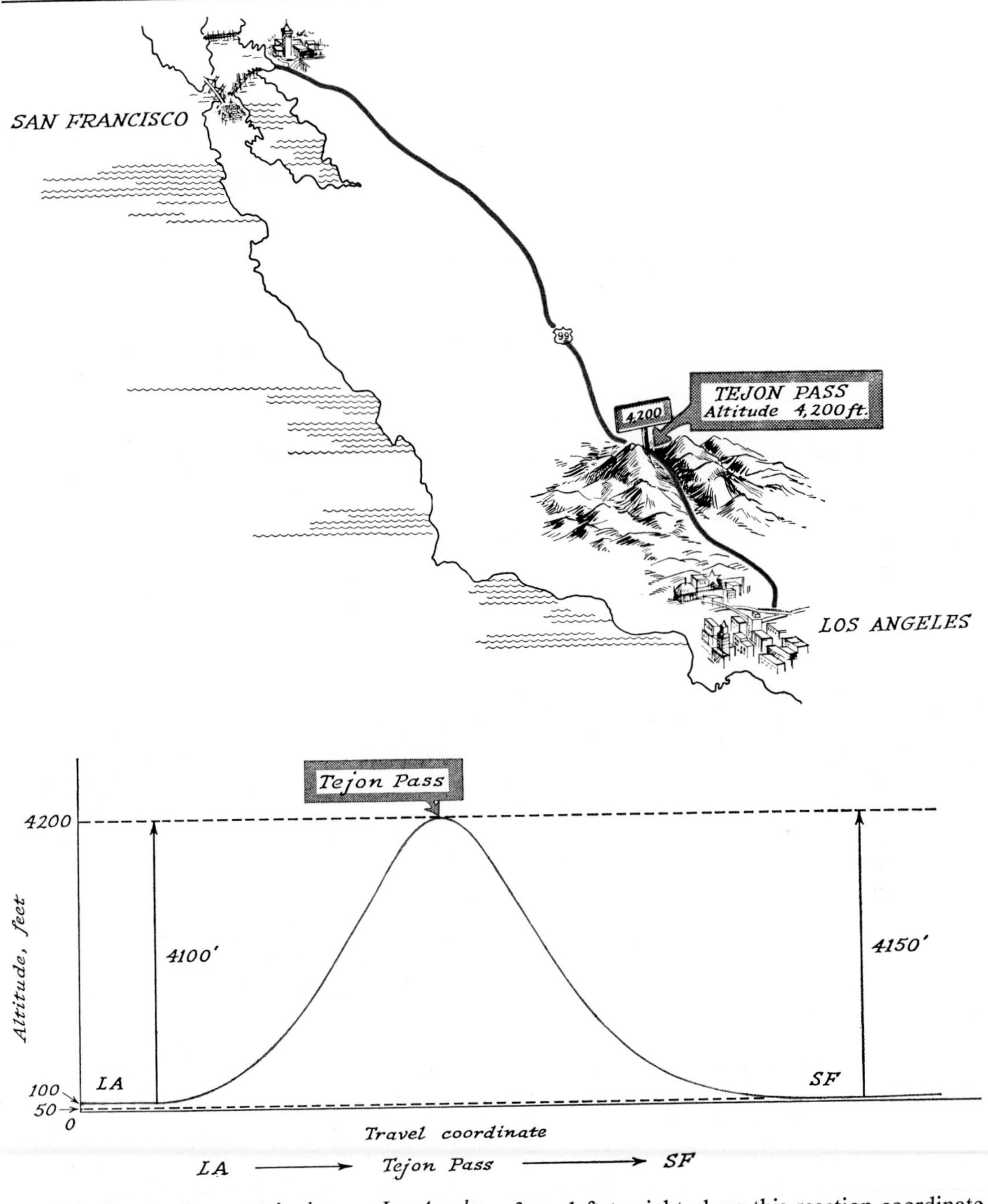

Fig. 8-5. **Crossing the mountains** *between Los Angeles and San Francisco.*

CO_2, and nitric oxide, NO. The horizontal axis of the diagram, called the *reaction coordinate*, shows the progress of the reaction. Proceeding from left to right along this reaction coordinate signifies the CO and NO_2 molecules approaching each other, colliding, and going through intermediate processes of reaction which result in the formation of CO_2 and NO, and, finally, the separation of the latter two molecules. The vertical axis represents the total potential energy of the

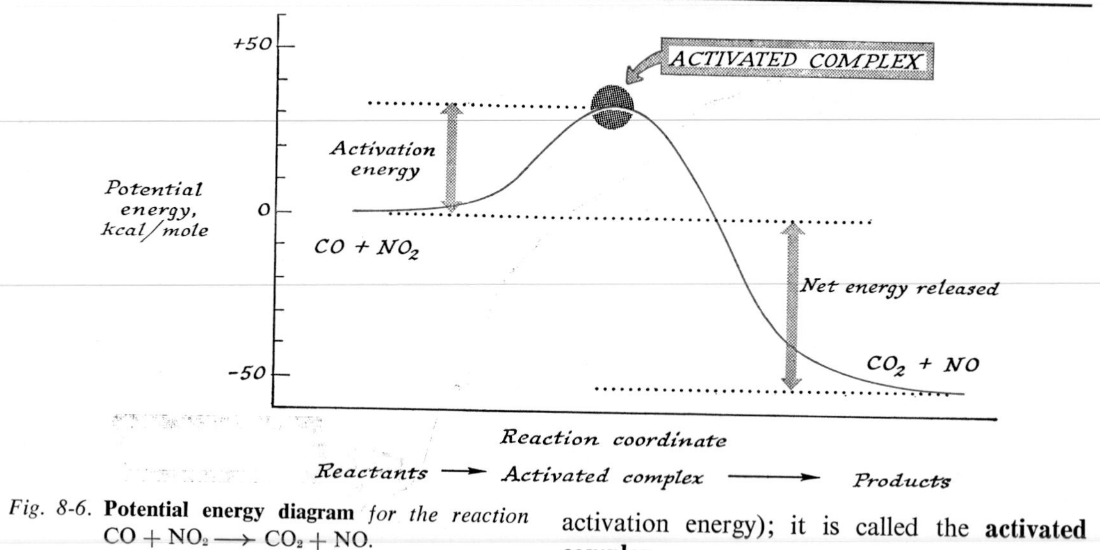

Fig. 8-6. **Potential energy diagram** *for the reaction* $CO + NO_2 \longrightarrow CO_2 + NO$.

system. Thus, the curve provides a history of the potential energy change during a collision which results in a reaction. The energy required to surmount the potential energy "barrier" to reaction usually is provided by the kinetic energies of the colliding particles, as fixed by the temperature.

Let us move from left to right along this curve and describe the events which occur. Along the flat region at the left, CO and NO_2 are approaching each other. In this region, they possess kinetic energy and their total potential energy shows no change. The beginning of the rise in the curve signifies that the two molecules have come sufficiently close to have an effect on each other. During this approach, the molecules slow down as their kinetic energies furnish the potential energy to climb the curve. If they have sufficient kinetic energy, they can ascend the left side of the "barrier" all the way up to the summit. Attaining this point is interpreted as follows: CO and NO_2 had sufficient kinetic energy to overcome the mutually repulsive forces of their nuclei and negative electron clouds and thus come very close to each other. Here at the summit the molecular cluster is unstable with respect to either the forward reaction (to give CO_2 and NO) or the reverse reaction (to restore the molecules of CO and NO_2). This transitory arrangement is of key importance (its potential energy fixes the activation energy); it is called the **activated complex.**

Now there are two possibilities: (1) the activated complex may separate into the two original CO and NO_2 molecules, which would then retrace their former path on the curve, or, (2) the activated complex may separate into CO_2 and NO molecules. The latter possibility is represented by moving down the right side of the "barrier." In the flat region at the right, CO_2 and NO have separated beyond the point of having any effect on each other and the potential energy of the activated complex has become kinetic energy again.

In the event that the CO and NO_2 molecules do not have sufficient energy to attain the summit, they reach a point only part way up the left side of the "barrier." Then, repelling one another, they separate again, going downhill to the left.

We have labeled the difference between the high potential energy at the activated complex and the lower energy of the reactants as the activation energy. The ***activation energy*** *is the energy necessary to transform the reactants into the activated complex.* This may involve weakening or breaking bonds, forcing reactants close together in opposition to repulsive forces, or storing energy in a vibrating molecule so that it reacts on collision. Increasing the temperature affects reaction rate by increasing the number of molecular collisions that involve sufficient energy to form this activated complex. The magnitude of the activation energy for a reaction can be determined by measuring experimentally the change in reaction rate with temperature.

8-2.2 Heat of Reaction

We can deduce the heat of reaction from Figure 8-6. In our example, the reactants are at a higher total energy than are the products. This means that in the course of the reaction there will be a net *release* of energy. This reaction is *exothermic*. Figure 8-6 shows that the reaction releases 54 kcal of heat per mole of carbon monoxide consumed. Notice that the height of the energy barrier between reactants and products has no effect on the net heat release. We must put in an amount of energy equal to the activation energy to get to the top of the barrier but we get it all back on the way down the other side.

Now let us consider the reverse reaction. We need not draw another reaction diagram, since Figure 8-6 will suffice. Now we are interested in the reaction between CO_2 and NO to produce CO and NO_2:

$$CO + NO_2 \longleftarrow CO_2 + NO \qquad (17)$$

This reaction begins at the lower energy appropriate to the chemical stability of $CO_2 + NO$ (at the right side of Figure 8-6) and ends at the higher energy appropriate to the chemical stability of $CO + NO_2$ (at the left side of Figure 8-6). The difference in energy, the heat of this reaction, is just equal to that of the reverse reaction but is opposite in sign. This reaction *absorbs* 54 kcal of heat per mole of carbon monoxide produced. It is *endothermic*.

Figure 8-6 contains one other very interesting piece of information concerning the rate of the reverse reaction, (*17*). This reaction rate is controlled by the energy barrier confronting the colliding molecules of CO_2 and NO. We see from the diagram that the activation energy for this reaction is higher than that for the reaction we studied earlier. Further, it is higher by exactly the heat of reaction. We conclude that the reaction between CO_2 and NO will be slower, at any given temperature, than the reverse reaction between CO and NO_2, if the rates are compared at the same partial pressures.

The relationship between activation energies for the forward and reverse reactions can be expressed mathematically. The activation energy is denoted by the symbol $\Delta H^{\ddagger}$ (read "delta-*H*-cross") and the heat of the reaction by ΔH. Hence we may write:

$$\Delta H_R^{\ddagger} = \Delta H_L^{\ddagger} + \Delta H \qquad (18)$$

$\Delta H_R^{\ddagger}$ = activation energy for reaction proceeding to right (always endothermic)
$\Delta H_L^{\ddagger}$ = activation energy for reaction proceeding to left (always endothermic)
ΔH = heat absorbed during reaction proceeding left to right (either endothermic or exothermic).

The heat of reaction, ΔH, is positive if heat is absorbed as the reaction proceeds, left to right. It is negative if heat is evolved. In our example, $CO + NO_2 \longrightarrow CO_2 + NO$,

$$\Delta H_R^{\ddagger} = +32 \text{ kcal/mole}$$
$$\Delta H_L^{\ddagger} = +86 \text{ kcal/mole}$$
$$\Delta H = -54 \text{ kcal/mole}$$

We see that

$$(32) = (86) + (-54)$$

which is in accordance with equation (*18*). This relationship is important because it implies that we need only *two* of the three quantities, $\Delta H_R^{\ddagger}$, $\Delta H_L^{\ddagger}$, and ΔH and then we can calculate the third. For example, if we measure the heat of the reaction, ΔH, and also measure the rate of the reaction between CO and NO_2 in order to determine $\Delta H_R^{\ddagger}$, then we can calculate $\Delta H_L^{\ddagger}$ by equation (*18*). From $\Delta H_L^{\ddagger}$ we learn something about the rate of reaction of $CO_2 + NO$.

8-2.3 Action of Catalysts

Many reactions proceed quite slowly when the reactants are mixed alone but can be made to take place much more rapidly by the introduction of other substances. These latter substances, called **catalysts,** are not used up in the reaction. The process of increasing the rate of a reaction through the use of a catalyst is referred to as **catalysis.** You have seen at least one example of catalytic action, the effect of Mn^{+2}*(aq)* in speeding up the reaction between $C_2O_4^{-2}$*(aq)* and MnO_4^-*(aq)*.

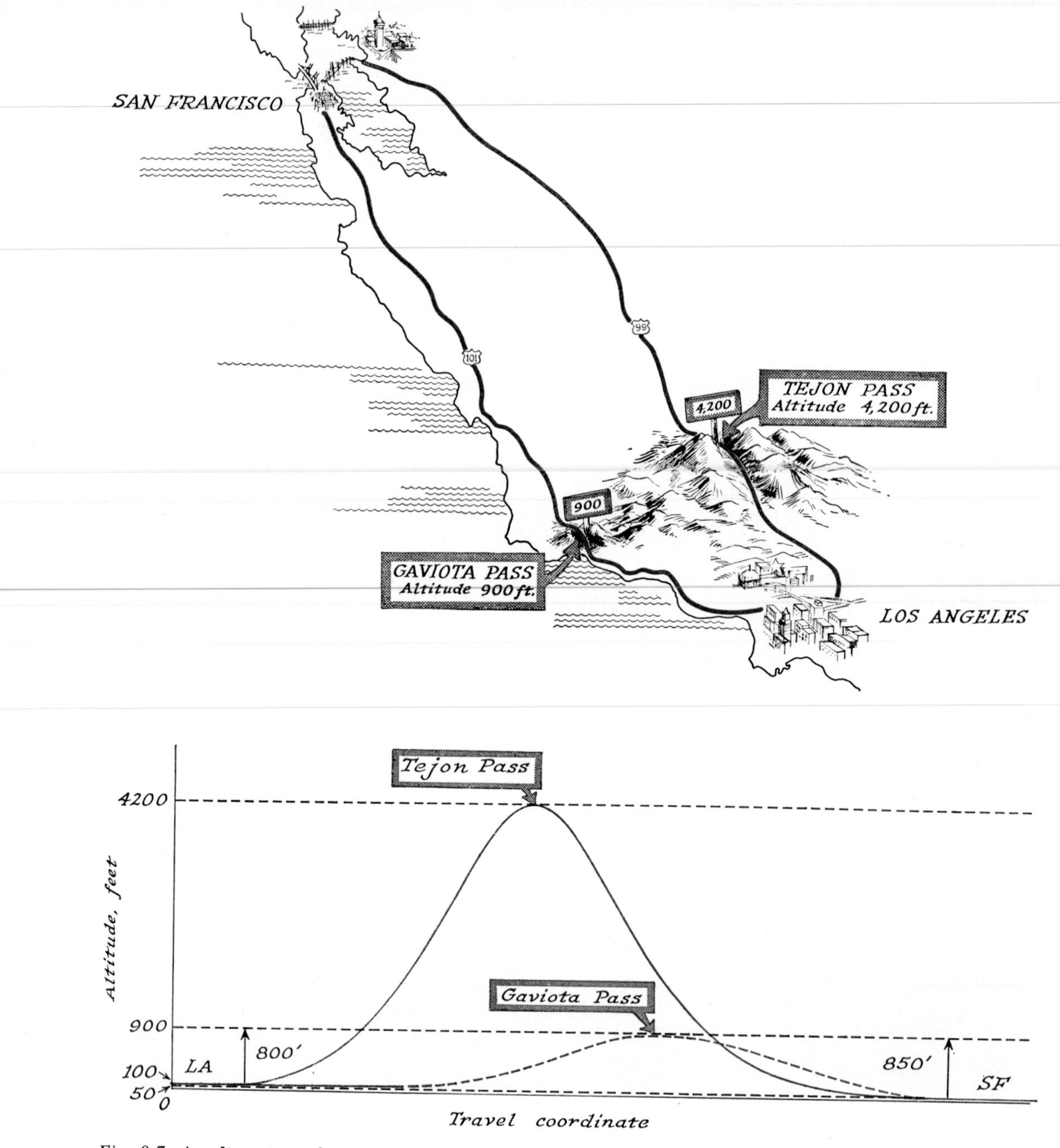

Fig. 8-7. **An alternate, easier route** *through the mountains between Los Angeles and San Francisco.*

The action of a catalyst can be explained in terms of our mountain pass analogy. In Figure 8-5 we see a formidable mountain pass obstructing travel between Los Angeles and San Francisco. Many people who would like to take this trip cannot because their automobile isn't up to this much of a climb. For this reason, an alternate route was improved. The coast route, shown in Figure 8-7 is somewhat longer but it is easier because the highest pass, Gaviota Pass, is only

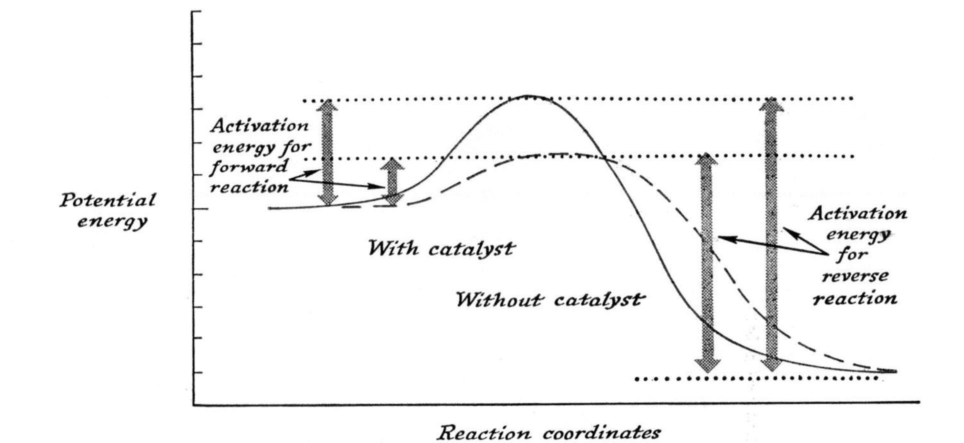

Fig. 8-8. **Effect of a catalyst** *on a reaction and its reverse.*

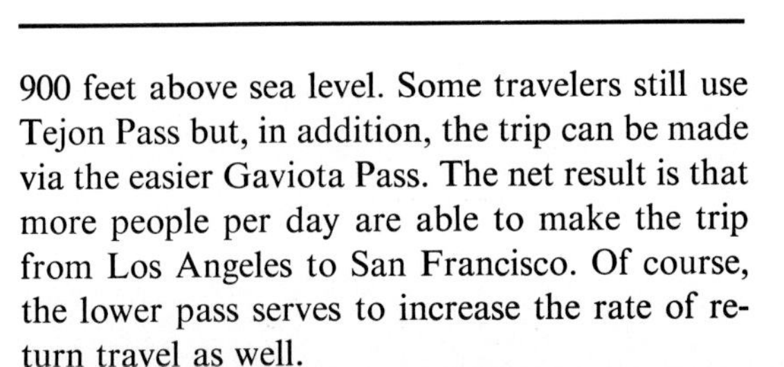

900 feet above sea level. Some travelers still use Tejon Pass but, in addition, the trip can be made via the easier Gaviota Pass. The net result is that more people per day are able to make the trip from Los Angeles to San Francisco. Of course, the lower pass serves to increase the rate of return travel as well.

Figure 8-8 shows the analogous situation for a chemical reaction. The solid curve shows the activation energy barrier which must be surmounted for reaction to take place. When a catalyst is added, a new reaction path is provided with a different activation energy barrier, as suggested by the dashed curve. This new reaction path corresponds to a new reaction mechanism that permits the reaction to occur via a different activated complex. Hence, more particles can get over the new, lower energy barrier and the rate of the reaction is increased. Note that the activation energy for the reverse reaction is lowered exactly the same amount as for the forward reaction. This accounts for the experimental fact that a catalyst for a reaction has an equal effect on the reverse reaction; that is, both reactions are speeded up by the same factor. If a catalyst doubles the rate in one direction, it also doubles the rate in the reverse direction.

8-2.4 Examples of Catalysts

In all cases of catalysis, the catalyst acts by inserting intermediate steps in a reaction—steps that would not occur without the catalyst. The catalyst itself must be regenerated in a subsequent step. (An added substance which is permanently used by reaction is a reactant, not a catalyst.) An example is provided by the catalytic action of acid on the decomposition of formic acid, HCOOH. A model of formic acid is shown in Figure 8-9. The carbon atom has attached to it a hydrogen atom, an oxygen atom, and an OH group.

Figure 8-10 shows how this molecule might decompose. If the hydrogen atom attached to carbon migrates over to the OH group, the carbon-oxygen bond can break to give a molecule of water and a molecule of carbon monoxide. This migration, shown in the center drawing, requires a large amount of energy. This means there is a high activation energy. Hence, the reaction occurs very slowly:

$$HCOOH \longrightarrow H_2O + CO \qquad (19)$$

Fig. 8-9. **A model of formic acid.**

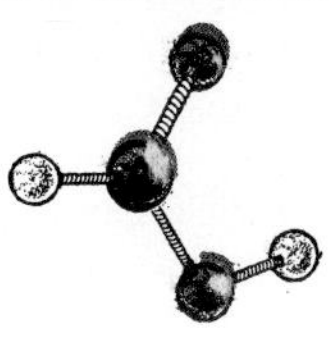

Ball-and-Spring Model

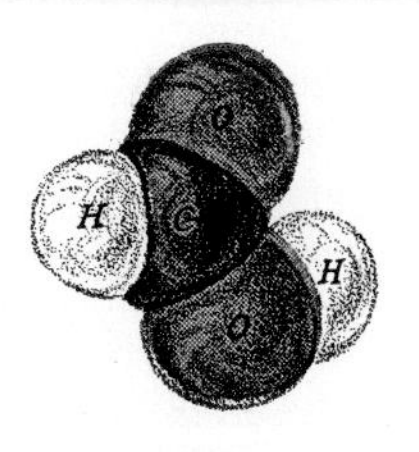

Space-Filling Model

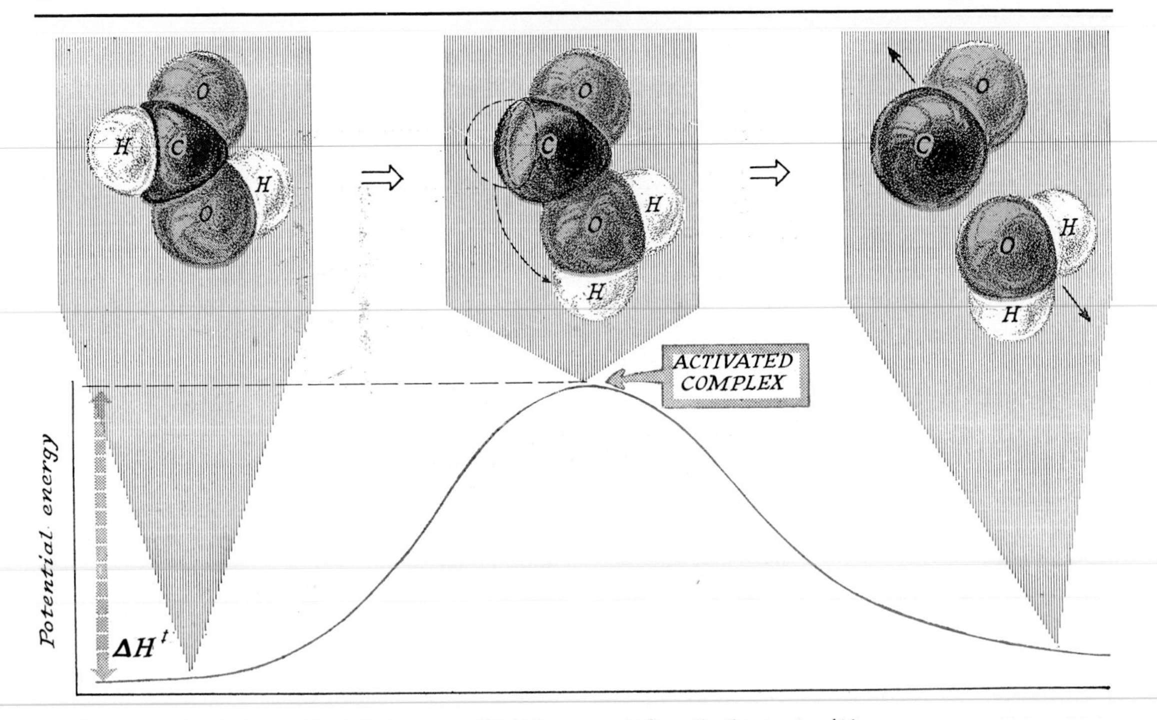

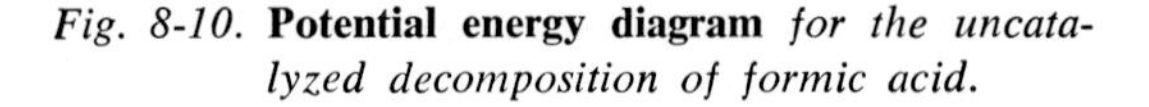

Fig. 8-10. **Potential energy diagram** *for the uncatalyzed decomposition of formic acid.*

If sulfuric acid, H_2SO_4, is added to an aqueous solution of formic acid, carbon monoxide bubbles out rapidly. This also occurs if phosphoric acid, H_3PO_4, is added instead. The common factor is that both of these acids release hydrogen ions, H^+. Yet, careful analysis shows that the concentration of hydrogen ion is constant during the rapid decomposition of formic acid. Evidently, hydrogen ion acts as a catalyst in the decomposition of formic acid.

Chemists have a rather clear picture of how H^+ catalyzes reaction (*19*). The availability of H^+ in the solution makes a new reaction path available. The new reaction mechanism begins with the addition of a hydrogen ion to formic acid, as shown in Figure 8-11. Thus, the catalyst is consumed at first, forming a new species, $(HCOOH_2)^+$. In this species one of the carbon-oxygen bonds is weakened. With only a small expenditure of energy, the next reaction shown in Figure 8-11 can occur, producing $(HCO)^+$ and H_2O. Finally, $(HCO)^+$ decomposes to produce carbon monoxide, CO, and H^+. This last reaction of the sequence regenerates the catalyst, H^+.

Each of the steps in this new reaction mechanism is governed by the same principles that govern a simple reaction. Each reaction has an activation energy. The overall reaction has a potential energy diagram that is merely a composite of the simple energy curves of the succeeding steps.

The highest energy required in this new reaction path is only 18 kcal, much lower than the activation energy shown in Figure 8-10 for the uncatalyzed reaction. Hence the rate of decomposition is much faster when acid is present.

Notice that catalytic action does not *cause* the reaction. A catalyst speeds up a reaction that might take place in its absence but at a much lower rate.

In some cases, the catalyst is a solid substance on whose surface a reactant molecule can be held (adsorbed) in a position favorable for reaction until a molecule of another reactant reaches the same point on the solid. Metals such as iron, nickel, platinum and palladium seem to act in this way in reactions involving gases. There is evidence that in some cases of surface adsorption, bonds of reactant particles are weakened or actually broken, thus aiding reaction with another reactant particle.

A very large number of catalysts, called enzymes, are found in living tissues. Among the best known examples of these are the digestive enzymes, such as the ptyalin in saliva and the pepsin in gastric juice. A common function of these two enzymes is to hasten the breakdown of large molecules, such as starch and protein, into simpler molecules which can be utilized by body cells. In addition to the relatively small number of digestive enzymes, there are many other enzymes involved in biochemical processes. Enzymes are considered again in Chapter 24.

The specific methods by which catalysts work are not clearly understood in most cases. Finding a catalyst suitable for a given reaction usually requires a long period

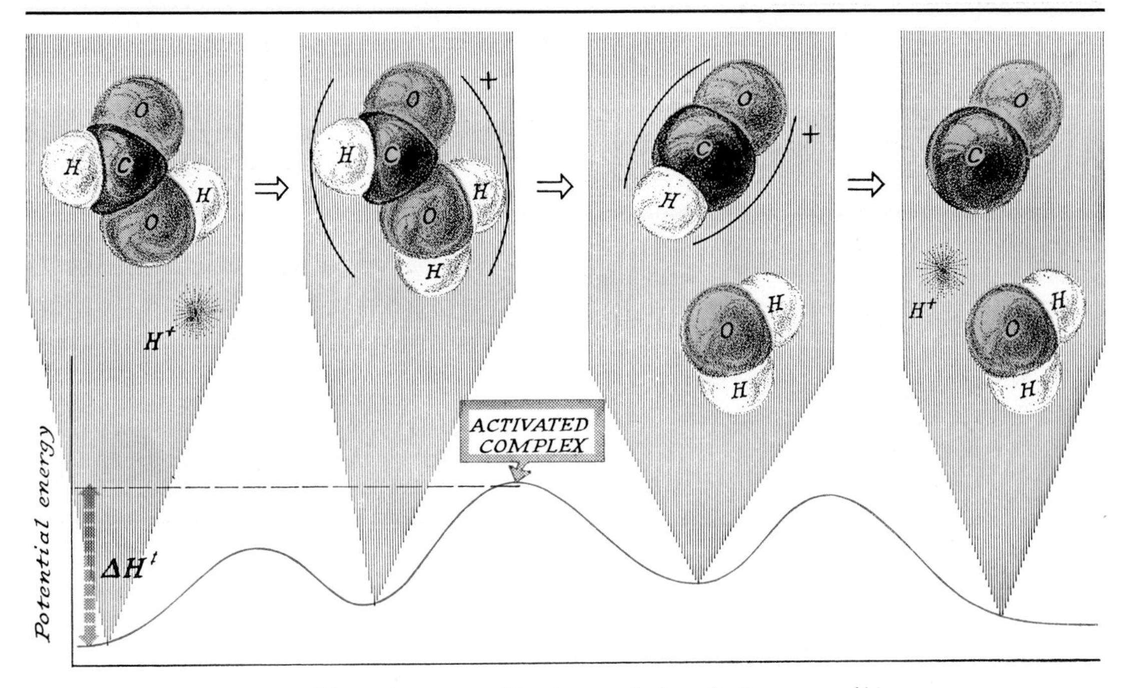

Fig. 8-11. **Potential energy diagram** *of the catalyzed decomposition of formic acid.*

of laboratory experimentation. Yet, we can look forward to the time when a catalyst can be tailor-made to fit a particular need. This exciting prospect accounts for the great activity on this chemical frontier.

QUESTIONS AND PROBLEMS

1. The rate of movement of an automobile can be expressed in the units miles per hour. In what units would you discuss the rate of:

   (a) movement of movie film through a projector;
   (b) rotation of a motor shaft;
   (c) gain of altitude;
   (d) consumption of milk by a family;
   (e) production of automobiles by an auto assembly plant.

2. Pick the member of each pair having the greater reaction rate. Assume similar conditions within each pair.

   (a) Iron rusting or copper tarnishing.
   (b) Wax burning or paper burning.
   (c) Evaporation of gasoline or evaporation of water.

3. Describe two homogeneous and two heterogeneous systems that are not described in the text.

4. Explain why there is danger of explosion where a large amount of dry, powdered, combustible material is produced.

5. Explain (at the molecular level) why an increase in concentration of a reactant may cause an increase in rate of reaction.

6. Consider two gases A and B in a container at room temperature. What effect will the following changes have on the rate of the reaction between these gases?

   (a) The pressure is doubled.
   (b) The number of molecules of gas A is doubled.
   (c) The temperature is decreased at constant volume.

7. In an important industrial process for producing ammonia (the Haber Process) the overall reaction is

$$N_2(g) + 3H_2(g) \longrightarrow 2NH_3(g) + 24{,}000 \text{ calories}$$

A yield of approximately 98% can be obtained at 200°C and 1000 atmospheres of pressure. The process makes use of a catalyst which is usually finely divided, mixed iron oxides containing small amounts of potassium oxide, K_2O, and aluminum oxide, Al_2O_3.

(a) Is this reaction exothermic or endothermic?
(b) Suggest a reason for the fact that this reaction is generally carried out at a temperature of 500°C and 350 atmospheres in spite of the fact that the yield under these circumstances is only about 30%.
(c) What is the ΔH for the reaction in kilocalories per mole of $NH_3(g)$?
(d) How many grams of hydrogen must react to form 1.60 moles of ammonia?

8. Give several ways by which the rate of combustion in a candle flame might be increased. State why the rate would be increased.

9. State three methods by which the pressure of a gaseous system can be increased.

10. Do you expect the reaction

$$C_2H_4(g) + 3O_2(g) \longrightarrow 2CO_2(g) + 2H_2O(g)$$

to represent the mechanism by which ethylene, C_2H_4, burns? Why?

11. A group of students is preparing a ten-page directory. The pages have been printed and are stacked in ten piles, page by page. The pages must be: (1) assembled in order, (2) straightened, and (3) stapled in sets. If three students work together, each performing a different operation, which might be the rate-controlling step? What would be the effect on the overall rate if the first step were changed by ten helpers joining the individual assembling the sheets? What if these ten helpers joined the student working on the second step? The third step?

12. Describe the life and death of an ordinary, empty water glass. Utilize the concept "threshold energy."

13. An increase in temperature of 10°C rarely doubles the kinetic energy of particles and hence the number of collisions is not doubled. Yet, this temperature increase may be enough to double the rate of a slow reaction. How can this be explained?

14. In a collision of particles, what is the primary factor that determines whether a reaction will occur?

15. In Figure 8-6, why is kinetic energy decreasing as NO_2 and CO go up the left side of the barrier and why is kinetic energy increasing as CO_2 and NO go down the right side? Explain in terms of conservation of energy and also in terms of what is occurring to the various particles in relation to each other.

16. Phosphorus, P_4, exposed to air burns spontaneously to give P_4O_{10}; the ΔH of this reaction is -712 kcal/mole P_4.

(a) Draw an energy diagram for the net reaction, explaining the critical parts of the curve.
(b) How much heat is produced when 12.4 grams of phosphorus burn?

17. Considering that so little energy is required to convert graphite to diamond (recall Problem 8, Chapter 7), how do you account for the great difficulty found in the industrial process for accomplishing this?

18. Why does a burning match light a candle?

19. Draw an energy diagram for the reaction

$$C(s) + O_2(g) \longrightarrow CO_2(g)$$

(a) when the C is in large chunks of coal.
(b) Is the curve changed if very fine carbon powder is used?

20. Sketch a potential energy diagram which might represent an endothermic reaction. (Label parts of curve representing activated complex, activation energy, net energy absorbed.)

21. Why is it difficult to "hardboil" an egg at the top of Pike's Peak? Is it also difficult to cook scrambled eggs there? Explain.

22. Give two factors that would increase the rate of a reaction and explain why these do increase the rate.

HENRY EYRING, 1901 –

Henry Eyring is one of the most active and honored chemists of our time. His advancement of the theory of reaction rates benefits practically every field of chemistry and chemical technology. His 300 published papers and five books range through chemistry, physics, metallurgy, and biology.

Henry was born in Chihuahua, Mexico, on a cattle ranch. The Eyring family was forced to abandon this home eleven years later under threat by the revolutionist Salazar. With other displaced American colonists, the family moved to Texas, then to Arizona. By distinguishing himself in high school, Henry Eyring earned a scholarship to the University of Arizona. Years later, this university gave him their Distinguished Alumnus Award, proud that he had earned a B.S. in mining engineering and an M.S. in metallurgy.

After graduation he was highly successful but not satisfied as an engineer in a flotation mill. He turned to the University of California where he received the Ph.D. in physical chemistry and where he felt the inspiration of the great G. N. Lewis. Two years of teaching at the University of Wisconsin, a year of study in Germany, and a year as lecturer at the University of California won him a faculty position at Princeton University. In 1946 he went to the University of Utah as Chairman of the Chemistry Department and Dean of the Graduate School, carrying with him world recognition in chemistry.

Henry Eyring's research has been original and frequently unorthodox. He was one of the first chemists to apply quantum mechanics in chemistry. He unleashed a revolution in the treatment of reaction rates by use of detailed thermodynamic reasoning. Having formulated the idea of the activated complex, Eyring proceeded to find a myriad of fruitful applications—to viscous flow of liquids, to diffusion in liquids, to conductance, to adsorption, to catalysis.

Anyone who hears Eyring speak on chemistry leaves convinced that clarifying chemical behavior is exhilarating fun. Eyring's enjoyment of science is as obvious as is the importance of his fundamental contributions. There remains the noteworthy facet of this great scientist that he is deeply religious and gives generously of his time and energy to his church. He shows deep concern about the political, social, and ethical implications of science and is always ready to discuss them. Thus it can be said that, in the broadest sense, Henry Eyring acts as a catalyst of men's minds.

CHAPTER 9

Equilibrium in Chemical Reactions

. . . by . . . equilibrium, we mean a state in which the properties of a system, as experimentally measured, would suffer no further observable change even after the lapse of an indefinite period of time. It is not intimated that the individual particles are unchanging.

G. N. LEWIS *and* M. RANDALL, 1923

In Chapter 8 we discussed the rate of the reaction between CO and NO_2,

$$CO(g) + NO_2(g) \longrightarrow CO_2(g) + NO(g) \quad (1)$$

Then, later in that chapter, we turned to the question of the rate of the reaction that is the reverse of (*1*),

$$CO_2(g) + NO(g) \longrightarrow CO(g) + NO_2(g) \quad (2)$$

Indeed! Can't this reaction make up its mind? If we mix $CO(g)$ and $NO_2(g)$, reaction (*1*) begins. But as soon as it does, $CO_2(g)$ and $NO(g)$ are formed. As these products accumulate, reaction (*2*) becomes possible, undoing reaction (*1*). Which wins out?

By direct observation of the reddish-brown color of NO_2 we can see the progress of reaction (*1*). The NO_2 is consumed at first, but after a time the color stops changing. When changes no longer occur in a reacting chemical system, we say the system has reached a **state of equilibrium.** The equilibrium situation raises many interesting questions. How do we recognize equilibrium? What are the molecules doing at the state of equilibrium? What factors change the state of equilibrium? What is the composition of the gas mixture at equilibrium? In this chapter we shall seek answers to these questions.

9-1 QUALITATIVE ASPECTS OF EQUILIBRIUM

We have encountered equilibrium before—in our consideration of phase changes. In Section 5-1.2 we considered the liquid-gas equilibrium that fixes the vapor pressure of a liquid, and in Sec-

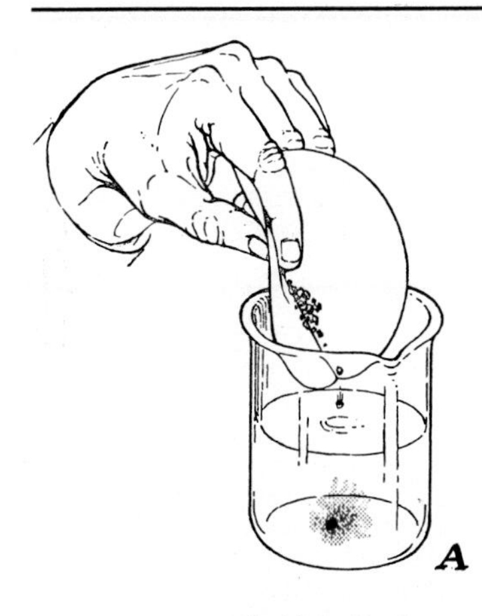

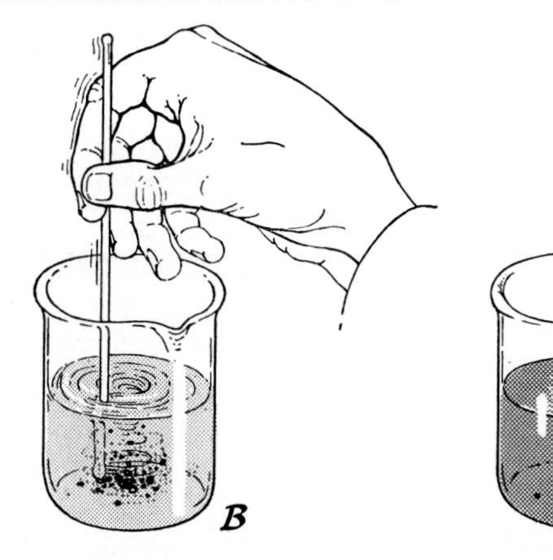

C

Solid begins to dissolve

Solid still dissolving, color deepening

No more changes, equilibrium exists

Fig. 9-1. **Iodine dissolving in an alcohol-water mixture.** *Equilibrium is recognized by constant color of the solution.*

tion 5-2.4 we considered the solid-liquid equilibrium that fixes the solubility of a solid in a liquid. With this as background, let us consider the first question about equilibrium: How do we recognize it?

9-1.1 Recognizing Equilibrium

Figure 9-1 shows the addition of solid iodine to a mixture of water and alcohol. At first the liquid is colorless but very quickly a reddish color appears near the solid. Stirring the liquid causes swirls of the reddish color to move out—solid iodine is dissolving to become part of the liquid. Changes are evident: the liquid takes on an increasing color and the pieces of solid iodine diminish in size as time passes. Finally, however, the color stops changing (see Figure 9-1). Solid is still present but the pieces of iodine no longer diminish in size. Since we can detect no more evidence of change, we say that the system is at equilibrium. *Equilibrium is characterized by constancy of macroscopic properties.**

Calcium carbonate, $CaCO_3$, decomposes upon heating to form carbon dioxide gas, CO_2, and calcium oxide (lime), CaO:†

$$CaCO_3(s) \xrightarrow[\text{temp.}]{\text{high}} CaO(s) + CO_2(g) \qquad (3)$$

Figure 9-2 shows the result of heating solid $CaCO_3$, initially under a vacuum, to 800°C (part A). Decomposition begins according to reaction (*3*) and the gas pressure rises (part B). The pressure continues to rise until it reaches 190 mm (part C). Thereafter, no further change is evident. Since we can detect no more evidence of change, we say that the system is at equilibrium. *Equilibrium is characterized by constancy of macroscopic properties.*

Though a system at equilibrium is constant in properties, constancy is not the only requirement. Consider a laboratory burner flame. There is a well-defined structure to the flame—an inner cone surrounded by a luminous region whose appearance does not change. A temperature measurement made at a particular place in the flame shows that the temperature at that spot is constant. At another place in the flame the temperature might be different but, again, it would be constant, not changing with time. A measurement of the gas flow rate shows a constant movement of gas into the flame. Yet a laboratory burner flame is *not* at equilibrium be-

* Remember that the word *macroscopic* was defined in Chapter 7. It means *a large amount of material—enough to see and weigh.*

† Reaction (*3*) is used for the manufacture of millions of tons of lime every year in the United States, for use, principally, in plaster.

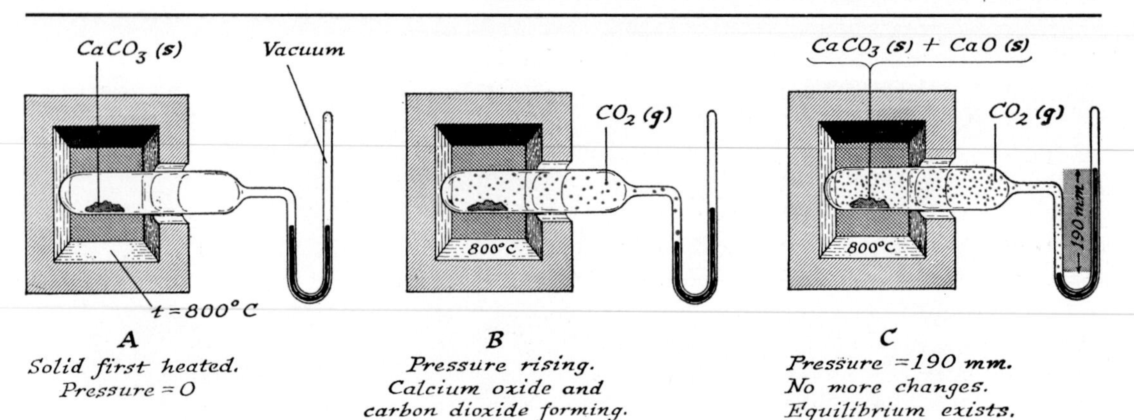

Fig. 9-2. **The thermal decomposition of calcium carbonate,** $CaCO_3(s) \rightleftarrows CaO(s) + CO_2(g)$.

cause chemical change is occurring. Methane, CH_4, and oxygen, O_2, are continuously fed into the flame and carbon dioxide, CO_2, and water, H_2O, are continuously leaving. Substances are entering and leaving at all times. Such a system is called an **open system.** Furthermore, the temperature is not uniform throughout the system. Equilibrium can exist only in a **closed system**—*a system containing a constant amount of matter with all of this matter at the same temperature.* The laboratory burner flame is called a **steady state** to indicate that some of its properties are constant but equilibrium does not exist.

Now we can give a complete statement about recognizing equilibrium: **equilibrium is recognized by the constancy of macroscopic properties in a closed system at a uniform temperature.**

EXERCISE 9-1

Which of the following systems constitute steady state situations, and which are at equilibrium? For each, a constant property is indicated.

(a) An open pan of water is boiling on a stove. The temperature of the water is constant.
(b) A balloon contains air and a few drops of water. The pressure in the balloon is constant.
(c) An ant-hill follows its daily life. The population of the ant-hill is constant.

9-1.2 The Dynamic Nature of Equilibrium

The constancy of properties at equilibrium refers to macroscopic measurements. Now we will consider what the equilibrium is like on the molecular level, as chemists picture it.

SOLUBILITY

Figure 9-1C shows a system at equilibrium. Solid iodine has dissolved in an alcohol-water mixture until the solution is saturated. Then no more solid dissolves and the color of the solution remains constant.

Among the molecules, however, business is going on as usual. Iodine dissolves by the detachment of surface layer molecules from the iodine crystals. The rate at which this process occurs is fixed by the stability of the crystal (tending to hold the molecules in the surface layer) and the temperature (the thermal agitation tending to dislodge the molecules from their lattice positions). As the dissolving continues, the concentration of iodine molecules in the solution increases.

Occasionally a molecule moving about in the solution encounters the surface of an iodine crystal and lodges there. This addition to the crystal is called **precipitation,** or **crystallization,** and it occurs more and more often as the concentration of iodine in solution rises.

Here we have two opposing processes. At a given temperature, molecules leave the surface of the crystal at a constant rate, tending to increase the concentration in solution. On the other hand, dissolved molecules are continually striking the surface and precipitating, tending to decrease the concentration of molecules in solution. When enough material has dissolved so that the rate of return of molecules to the surface of the solid is just equal to the rate at which they are leaving the surface, no more net change will occur. Even though molecules are continually dissolving and others are precipitating, as long as these two processes are in balance, the amount of iodine dissolved per unit volume of solution will be constant. This macroscopic property, the solubility, is now constant: the system is in solubility equilibrium. But chemists interpret this constancy as a balance between two opposing processes which continue at equilibrium. **At equilibrium, microscopic processes continue but in a balance that yields no macroscopic changes.**

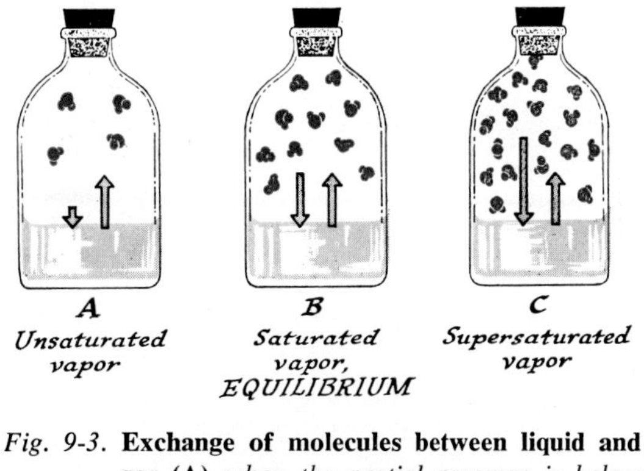

Fig. 9-3. **Exchange of molecules between liquid and gas (A)** *when the partial pressure is below the vapor pressure;* **(B)** *at equilibrium;* **(C)** *when the partial pressure is above the vapor pressure.*

VAPOR PRESSURE

Consideration of the dissolving of iodine in an alcohol-water mixture on the molecular level reveals the dynamic nature of the equilibrium state. The same type of argument is applicable to vapor pressure.

We have already noted that if we place liquid water in a flask at 20°C and seal the flask, some water molecules leave the liquid and enter the gas phase. The partial pressure rises as more and more water molecules become part of the gas. Finally, however, the pressure stops rising and the partial pressure of water becomes constant. This partial pressure is the vapor pressure and equilibrium now exists.

Yet, it is reasonable to suppose that water molecules from the liquid are still evaporating, even at equilibrium. Molecules in the liquid have no way of "knowing" that the partial pressure of the vapor is equal to the vapor pressure. In the gas phase, the randomly moving molecules continue to strike the surface of the liquid and condense. Equilibrium corresponds to a perfect balance between this continuing evaporation and condensation. Then no net changes can be detected.*

* When the partial pressure of the water equals the vapor pressure, the gas above the liquid is said to be **saturated.** The word "saturated" has the same meaning as it did relative to solubility: the gas phase contains as much water vapor as it can hold at equilibrium.

Figure 9-3 shows this schematically. If the partial pressure of the vapor is less than the equilibrium value (as in Figure 9-3A), the rate of evaporation exceeds the rate of condensation until the partial pressure of the vapor equals the equilibrium vapor pressure. If we inject an excess of vapor into the bottle (as in Figure 9-3C), condensation will proceed faster than evaporation until the excess of vapor has condensed. The equilibrium vapor pressure corresponds to that concentration of water vapor at which condensation and evaporation occur at exactly the same rate (as in Figure 9-3B). *At equilibrium, microscopic processes continue but in a balance that yields no macroscopic changes.*

CHEMICAL REACTIONS

Let us examine a chemical reaction to see if these same conditions apply. Suppose we fill two identical bulbs to equal pressures of nitrogen dioxide. Now immerse the first bulb (bulb *A*) in an ice bath and the second bulb (bulb *B*) in boiling water, as in Figure 9-4. The gas in bulb *A* at 0°C is almost colorless; the gas in bulb *B* at 100°C is reddish-brown. The predominant molecular species in the cold bulb must be different from that in the hot bulb. A variety of experiments shows that the cold bulb contains mostly N_2O_4 molecules. These same experiments show that the hot bulb contains mostly NO_2 molecules. The N_2O_4 molecules absorb no visible light, so

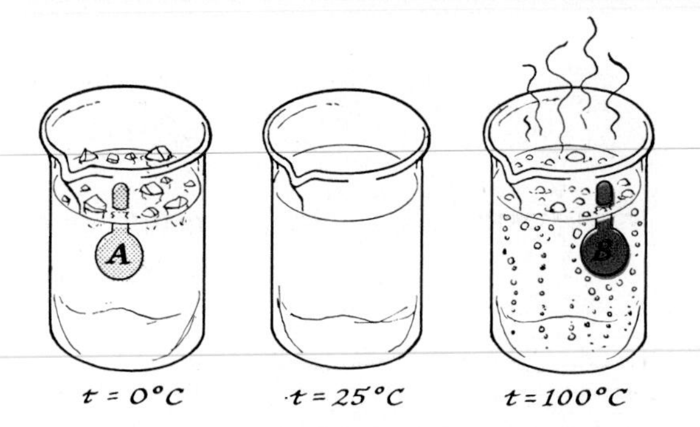

Fig. 9-4. **Nitrogen dioxide gas at different temperatures.** *Bulb A: At 0°C: N_2O_4 (almost colorless). Bulb B: At 100°C: NO_2 (reddish-brown).*

the cold gas is almost colorless. The NO_2 molecules do absorb some visible light, so the hot gas is reddish-brown.

Now let us transfer these two bulbs to a bath at room temperature, as shown in Figure 9-5. Immediately the color begins to intensify in bulb *A*. The color shows that a chemical change has occurred, forming NO_2 molecules from N_2O_4:

$$\text{In bulb } A \qquad N_2O_4(g) \longrightarrow 2NO_2(g) \qquad (4)$$

At the same time, the color in bulb *B* begins to pale, showing that a chemical change has occurred in this bulb as well, forming N_2O_4 molecules from NO_2:

$$\text{In bulb } B \qquad 2NO_2(g) \longrightarrow N_2O_4(g) \qquad (5)$$

In each bulb the colors continue to change, bulb *A* becoming darker and bulb *B* becoming more

Fig. 9-5. **Nitrogen dioxide gas at room temperature.** *Bulb A and bulb B after transfer to water bath at 25°C.*

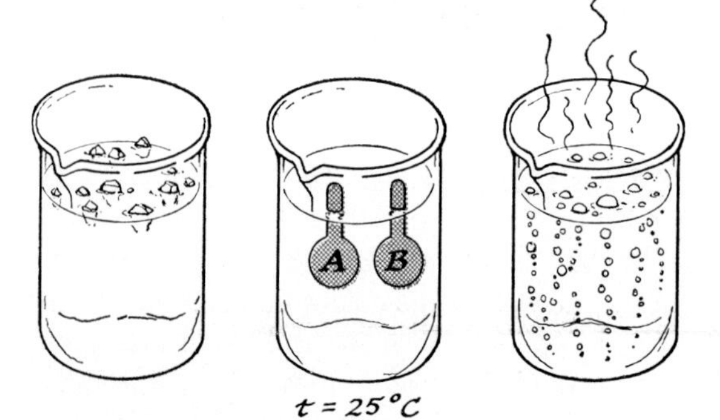

pale. Finally, as the two bulbs approach the same temperature, the colors stop changing. A close examination shows that the two colors are now the same!

By direct visual observation we can watch the contents of these two bulbs approach the constancy of macroscopic properties (in this case, color) that indicates equilibrium. In bulb *A* equilibrium was approached by the dissociation of N_2O_4, reaction (*4*); in bulb *B* it was approached by the opposite reaction, reaction (*5*). Here it is clear why the color of each bulb stopped changing at the particular hue characteristic of the equilibrium state at 25°C. The reaction between NO_2 and N_2O_4 can proceed in both directions:

$$N_2O_4(g) \longrightarrow 2NO_2(g) \qquad (4)$$
$$N_2O_4(g) \longleftarrow 2NO_2(g) \qquad (5)$$

Since N_2O_4 molecules can dissociate in bulb *A*, they must also be able to dissociate in bulb *B*. Surely an N_2O_4 molecule doesn't act differently in bulb *A* (at 25°C) than it does in bulb *B* (at 25°C). The same sort of statements must apply to the combination of two NO_2 molecules. If the reaction occurs in bulb *B*, then it must also occur in bulb *A*. The net change we see (by observing the changing NO_2 color) represents, then, the *difference* in the rate of production of NO_2 by reaction (*4*) and the rate of loss of NO_2 by reaction (*5*). Changes will cease when these two rates are exactly equal. If we approach equilibrium from a lower temperature (which favors N_2O_4), then reaction (*4*) predominates at first. But as more and more NO_2 is produced, reaction (*5*) becomes faster and faster. When reaction (*5*) becomes just as fast as reaction (*4*), then equilibrium has been reached: macroscopic properties no longer change even though both reactions still proceed in a state of balance. At this time we replace the single arrow in reaction (*4*) by a double arrow ($\rightleftarrows$) or an equals sign (=) to show that equilibrium prevails:*

$$N_2O_4(g) \rightleftarrows 2NO_2(g) \qquad (6)$$

or

$$N_2O_4(g) = 2NO_2(g) \qquad (6)$$

* These alternative notations, $\rightleftarrows$ and =, are used by chemists interchangeably in equations for chemical reactions. Both notations will be seen.

In bulb B we approached equilibrium from a higher temperature (which favors NO_2); then reaction (5) predominated at first. Using the same sort of argument we applied to bulb A, we see that as time progresses, reaction (4) becomes more and more rapid (as N_2O_4 is produced) and reaction (5) becomes slower (as NO_2 is used up). Finally, when the rates become equal, equilibrium is reached and the equilibrium expression (6) is applicable in bulb B.

For chemical reactions, just as for phase changes, *at equilibrium, microscopic processes continue but in a balance which gives no macroscopic changes.*

9-1.3 The State of Equilibrium

It is most important to note that in our description of the equilibrium state we have not implied that at equilibrium the number of moles of N_2O_4 remaining is the same as the number of moles of NO_2 produced. *Equation (6) gives us no information concerning the fraction of the nitrogen dioxide present as NO_2 at equilibrium.* This is easily verified by heating the water surrounding bulbs A and B about 10°. The colors of the gases in both bulbs change to a new equilibrium color (one corresponding to the presence of more NO_2). Yet the same expression is applicable:

$$N_2O_4(g) \rightleftharpoons 2NO_2(g) \qquad (6)$$

What does equation (6) tell us, then? First, it tells us that equilibrium prevails (the $\rightleftharpoons$ sign tells us that). Next, it tells us that there are two types of molecules present, N_2O_4 molecules and NO_2 molecules. Finally, it tells us that during the approach to equilibrium, *two* molecules of NO_2 are produced (or consumed) for every *one* molecule of N_2O_4 dissociated (or formed). It does *not* tell us whether at equilibrium there will be much or little NO_2 compared with the amount of N_2O_4.

To emphasize this point, let us consider another familiar reaction,

$$H_2O(g) \rightleftharpoons H_2(g) + \tfrac{1}{2}O_2(g) \qquad (7)$$

Until we are given the necessary information we have no idea how complete the decomposition of water is at equilibrium. All we know is that for every mole of water which decomposes we will obtain 1 mole of hydrogen and $\frac{1}{2}$ mole of oxygen.

It has been determined that in a closed vessel at 2273°K and a total pressure equal to one atmosphere at the time equilibrium is attained, 0.6% of the water has dissociated. If we started with one mole of water, $0.6\% = 0.6 \times \frac{1}{100} = 0.006$ mole would decompose. There would be left $1 - 0.006 = 0.994$ mole of undecomposed water. There would be formed 0.006 mole of H_2 and 0.003 mole of oxygen. We can summarize this as follows:

	$H_2O(g)$ $\rightleftharpoons$	$H_2(g)$ +	$\frac{1}{2}O_2(g)$
Initial moles	1	0	0
Moles present at equilibrium	0.994	0.006	0.003

In other words, if we start with water, not much of it has decomposed when the equilibrium state is attained at 2273°K.

Now what about approaching the equilibrium state by starting with hydrogen and oxygen? Let us start with 1 mole of hydrogen and $\frac{1}{2}$ mole of oxygen and allow the reaction to attain equilibrium at 2273°K and a total pressure equal to one atmosphere. At equilibrium we find present 0.994 mole of water, 0.006 mole of H_2, and 0.003 mole of O_2. This can be summarized as follows:

	$H_2(g)$ +	$\frac{1}{2}O_2(g)$ $\rightleftharpoons$	$H_2O(g)$
Initial moles	1	0.5	0
Moles present at equilibrium	0.006	0.003	0.994

If we start with hydrogen and oxygen, equilibrium is attained after most of the hydrogen and oxygen have united to form water. More important, though, the partial pressures at equilibrium are the same as those obtained beginning with pure H_2O. The equilibrium pressures are fixed by the temperature, the composition, and the total pressure; they do not depend upon the direction from which equilibrium is approached. The balanced equation does not indicate the concentrations (or partial pressures) at equilibrium.

9-1.4 Altering the State of Equilibrium

We have seen that, qualitatively, the state of equilibrium for a system is characterized by the relative amounts of products and reactants present. With reference to the decomposition of water, any change in conditions which would cause more than 0.6% of the water to dissociate at equilibrium would be said to change the state of equilibrium for the reaction,

$$H_2O(g) \rightleftharpoons H_2(g) + \tfrac{1}{2}O_2(g) \qquad (7)$$

in favor of the formation of more hydrogen and oxygen.

What conditions might alter the equilibrium state? Concentration and temperature! These are factors that affect the rate of reaction. Equilibrium is attained when the rates of opposing reactions become equal. Any condition that changes the rate of one of the reactions involved in the equilibrium may affect the conditions at equilibrium.

CONCENTRATION

Consider the reaction you encountered in the laboratory—that between ferric ion (Fe^{+3}) and thiocyanate ion (SCN^-):

$$Fe^{+3}(aq) + SCN^-(aq) \rightleftharpoons FeSCN^{+2}(aq) \qquad (8)$$

Again we have visual evidence of concentration at equilibrium since the intensity of the color is fixed by the concentration of the $FeSCN^{+2}$ ion. The addition of either more ferric ion [by adding a soluble salt such as ferric nitrate, $Fe(NO_3)_3$] or more thiocyanate ion (by adding, say, sodium thiocyanate) changes the concentration of one of the reactants in (8). Immediately the color of the solution darkens, showing that there is an increase in the amount of the colored ion, $FeSCN^{+2}$. *The equilibrium concentrations are affected if the concentrations of reactants (or products) are altered.*

Fig. 9-6. **Equilibrium conditions** *are affected by the reactant concentrations.*

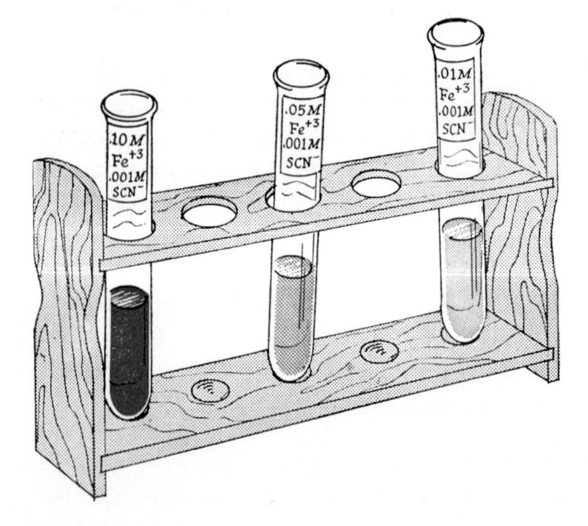

TEMPERATURE

We have already considered an example of the change of equilibrium concentrations as the temperature is altered. The relative amounts of NO_2 and N_2O_4 are readily and obviously affected by a temperature change. *The equilibrium concentrations are affected if the temperature is altered.*

CATALYSTS

Catalysts increase the rate of reactions. It is found experimentally that *addition of a catalyst to a system at equilibrium does* **not** *alter the equilibrium state*. Hence it must be true that any catalyst has the same effect on the rates of the forward and reverse reactions. You will recall that the effect of a catalyst on reaction rates can be discussed in terms of lowering the activation energy. This lowering is effective in increasing the rate in both directions, forward and reverse. Thus, a catalyst produces no net change in the equilibrium concentrations even though the system may reach equilibrium much more rapidly than it did without the catalyst.

9-1.5 Attainment of Equilibrium

The equilibrium state is not always attained in chemical reactions. Consider reaction (7):

$$H_2O(g) \rightleftharpoons H_2(g) + \tfrac{1}{2}O_2(g) \qquad \Delta H = +57.8 \text{ kcal} \qquad (7)$$

A large amount of heat is absorbed in this reaction, 57.8 kcal/mole of water decomposed. If the temperature is lowered, the state of equilibrium is even more favorable to the production of water at room temperature than it is at 2273°K. Yet a mixture of hydrogen and oxygen can remain at room temperature for a long period without apparent reaction. Equilibrium is not attained in this system because the rate of the reaction between hydrogen and oxygen at room temperature is too low. This explanation is easily verified by speeding up the reaction slightly. If a mixture of H_2 and O_2 is disturbed with a small spark, reaction begins and it enthusiastically (and explosively) continues until most of the gases have been converted to water.

This distinction between the conditions in a chemical system at equilibrium and the rate at which these conditions are attained is very important in chemistry. By arguments that we shall consider a chemist can decide with confidence whether equilibrium favors reactants or products or neither. He cannot predict, however, how rapidly the system will approach the equilibrium conditions. That is a matter of reaction rates, and the chemist must perform separate experiments to learn whether a given rate is rapid or not.

9-1.6 Predicting New Equilibrium Concentrations: Le Chatelier's Principle

We are not satisfied with the conclusion that this change or that change affects the equilibrium concentrations. We would also like to predict the *direction* of the effect (does it favor products or reactants?) and the *magnitude* of the effect (how much does it favor products or reactants?). The first desire, to know the qualitative effects, is answered by a generalization first proposed by a French chemist, Henry Louis Le Chatelier, and now called Le Chatelier's Principle.

Le Chatelier sought regularities among a large amount of experimental data concerning equilibria. To summarize the regularities he found, he made this generalization: **If an equilibrium system is subjected to a change, processes occur that tend to counteract partially the imposed change.** This generalization has been found to be applicable to such a large number of systems that it is now called a principle. Let us see how it applies to our examples.

CONCENTRATION AND LE CHATELIER'S PRINCIPLE

If a soluble thiocyanate salt is added to an equilibrium solution containing both $Fe^{+3}(aq)$ and $SCN^-(aq)$, the color of the complex ion increases:

$$Fe^{+3}(aq) + SCN^-(aq) \rightleftharpoons FeSCN^{+2}(aq) \quad (8)$$

A new state of equilibrium is then attained in which more $FeSCN^{+2}$ is present than was there before the addition of SCN^-. Increasing the concentration of SCN^- has increased the concentration of the $FeSCN^{+2}$ ion. This is in accord with Le Chatelier's Principle. The change imposed on the system was an increase in the concentration of SCN^-. This change can be counteracted in part by some Fe^{+3} and SCN^- ions reacting to form more $FeSCN^{+2}$. The same argument applies to an addition of ferric ion from a soluble ferric salt. In each case, the formation of $FeSCN^{+2}$ uses up a portion of the added reactant, partially counteracting the change.

PRESSURE AND LE CHATELIER'S PRINCIPLE

Instead of altering the concentration of one individual component in an equilibrium system, we can alter the concentration of all gaseous components by changing the pressure at which the system is confined. Let us start with the system represented by equation (7) and double the total pressure. The system now occupies a much smaller volume than it did previously. The total number of moles present per unit volume is greater than it was under the original equilibrium conditions. This change can be counteracted in part if some hydrogen and oxygen combine to form gaseous water. Then the total number of moles present is reduced ($1\frac{1}{2}$ moles unite to form 1 mole). Hence, we can predict that increasing the concentration of all components by increasing the pressure will shift the state of equilibrium in favor of the formation of gaseous water. This is in accord with experiment.

A change in total pressure does not always shift equilibrium. The first reaction mentioned in this chapter exemplifies this:

$$CO(g) + NO_2(g) \rightleftharpoons CO_2(g) + NO(g) \quad (9)$$

If we increase the pressure on a mixture of these four gases at equilibrium, the gases are compressed to a smaller volume. Once again the concentrations are all increased. What does Le Chatelier's Principle tell us here? If the equilibrium state is altered to favor products, some CO and NO_2 molecules react to form an exactly equal number of molecules of CO_2 and NO. Since there is no change in the total number of moles, the proposed change in the equilibrium does *not* partially reduce the pressure change. Le Chatelier's Principle tells us that processes occur so as to "counteract partially the imposed change." Here neither a change favoring the reactants nor a change favoring products will "counteract" the imposed pressure change. Hence, Le Chatelier's Principle leads us to expect no change of the equilibrium state for reaction (*9*) when the pressure is altered. Experimentally we find that no change is observed. The equilibrium state is not affected by a pressure change for any equilibrium gas mixture where the number of reactant molecules is the same as the number of product molecules in the balanced reaction.

EXERCISE 9-2

Does Le Chatelier's Principle predict a change of equilibrium concentrations for the following reactions if the gas mixture is compressed? If so, does the change favor reactants or products?

(a) $N_2O_4(g) \rightleftharpoons 2NO_2(g)$
(b) $H_2(g) + I_2(g) \rightleftharpoons 2HI(g)$
(c) $N_2(g) + 3H_2(g) \rightleftharpoons 2NH_3(g)$

TEMPERATURE AND LE CHATELIER'S PRINCIPLE

Let us add to reaction (*4*) the information that the decomposition of N_2O_4 is endothermic:

$$N_2O_4(g) \rightleftharpoons 2NO_2(g) \qquad \Delta H = +14.1 \text{ kcal} \quad (4)$$

Our experimental observations indicated that warming a bulb containing NO_2 and N_2O_4 caused a shift of the equilibrium state in favor of the formation of NO_2 (the reddish-brown color deepened). It is easy to see that this is in accord with Le Chatelier's Principle. A rise in temperature is caused by an input of heat. At the higher temperature, the equilibrium is changed to form more NO_2. The formation of NO_2 absorbs a portion of the heat that caused the temperature rise.

Raising the temperature of liquid water raises its vapor pressure. This is in accord with Le Chatelier's Principle since heat is absorbed as the liquid vaporizes. This absorption of heat, which accompanies the change to the new equilibrium conditions, partially counteracts the temperature rise which caused the change.

9-1.7 Application of Equilibrium Principles: The Haber Process

Knowledge of chemical principles pays rewards in technological progress. Control of chemical reactions is the key. The large scale commercial production of nitrogen compounds provides a practical example of the beneficial application of Le Chatelier's Principle.

The most difficult step in the process for the conversion of the inert nitrogen of the atmosphere into important commercial compounds such as fertilizers and explosives involves the reaction

$$N_2(g) + 3H_2(g) \rightleftharpoons 2NH_3(g) + 22 \text{ kcal} \quad (10a)$$

or

$$N_2(g) + 3H_2(g) \rightleftharpoons 2NH_3(g) \qquad \Delta H = -22 \text{ kcal} \quad (10b)$$

Can we predict the optimum conditions for a high yield of NH_3? Should the system be allowed to attain equilibrium at a low or a high temperature? Application of Le Chatelier's Principle suggests that the lower the temperature the more the equilibrium state will favor the production of NH_3. Should we use a low or a high pressure? The production of NH_3 represents a decrease in total moles present from 4 to 2. Again Le Chatelier's Principle suggests use of pressure to increase concentration. But what about practicality? At low temperatures reaction rates are slow. Therefore a compromise is necessary. Low temperature is required for a desirable equilibrium state and high temperature is necessary for a satisfactory rate. The compromise used industrially involves an intermediate temperature around 500°C and even then the success of the process depends upon the presence of a suitable catalyst to achieve a reasonable reaction rate.

With regard to pressure, another compromise is needed. It is expensive to build high pressure equipment. A pressure of about 350 atmospheres is actually used. Under these conditions, 350 atmospheres and 500°C, only about 30% of the reactants are converted to NH_3. The NH_3 is removed from the mixture by liquefying it under conditions at which N_2 and H_2 remain as gases. The unreacted N_2 and H_2 are then recycled until the total percent conversion to ammonia is very high.

Prior to World War I the principal sources of nitrogen compounds were some nitrate deposits in Chile. Fritz Haber, a German chemist, successfully developed the process we have just described, thus allowing chemists to use the almost unlimited supply of nitrogen in the atmosphere as a source of nitrogen compounds.

9-2 QUANTITATIVE ASPECTS OF EQUILIBRIUM

Le Chatelier's Principle permits the chemist to make qualitative predictions about the equilibrium state. Despite the usefulness of such predictions, they represent far less than we wish to know. It is a help to know that raising the pressure will favor production of NH_3 in reaction (*10a*). But *how much* will the pressure change favor NH_3 production? Will the yield change by a factor of ten or by one-tenth of a percent? To control a reaction, we need quantitative information about equilibrium. Experiments show that quantitative predictions are possible and they can be explained in terms of our view of equilibrium on the molecular level.

9-2.1 The Equilibrium Constant

By means of colorimetric determination in the laboratory you measured the concentration of $FeSCN^{+2}$, which we shall designate [$FeSCN^{+2}$],* in solutions containing ferric and thiocyanate ions, Fe^{+3} and SCN^-. The reaction is

$$Fe^{+3}(aq) + SCN^-(aq) \rightleftharpoons FeSCN^{+2}(aq) \quad (8)$$

From [$FeSCN^{+2}$] and the initial values of [Fe^{+3}] and [SCN^-] you calculated the values of [Fe^{+3}] and [SCN^-] at equilibrium. You then made calculations for various combinations of these values. Many experiments just like these show that the ratio

$$\frac{[FeSCN^{+2}]}{[Fe^{+3}][SCN^-]} \quad (11)$$

comes closest to being a fixed value. Note that this ratio is the quotient of the equilibrium concentration of the single substance produced in the reaction divided by the product of the equilibrium concentrations of the reactants.

Colorimetric analysis based on visual estimation is not very exact. Some more accurate data on the H_2, I_2, HI system at equilibrium are shown in Table 9-I. The reaction is

$$2HI(g) \rightleftharpoons H_2(g) + I_2(g) \quad (12)$$

The data have been expressed in concentrations, although pressure units are more usual for a reaction involving gases.

EXERCISE 9-3

For the last two experiments in Table 9-I, numbers 4 and 5, why is [H_2] = [I_2]? For experiment 1, what were the initial concentrations of H_2 and I_2 before the reaction occurred to form HI?

Let's work with these data. Heartened by your results from your own laboratory data, let us compute the value of the ratio

$$\frac{[H_2][I_2]}{[HI]} \quad (13)$$

* Hereafter, we shall regularly use the square brackets notation, [ ], to indicate concentration. Thus, we read [Fe^{+3}] as "ferric ion concentration."

Table 9-I. **EQUILIBRIUM CONCENTRATION AT 698.6°K OF HYDROGEN, IODINE, AND HYDROGEN IODIDE**

EXPT. NO.	$[H_2]$ (moles/liter)	$[I_2]$ (moles/liter)	$[HI]$ (moles/liter)
1	1.8313×10^{-3}	3.1292×10^{-3}	17.671×10^{-3}
2	2.9070×10^{-3}	1.7069×10^{-3}	16.482×10^{-3}
3	4.5647×10^{-3}	0.7378×10^{-3}	13.544×10^{-3} *
4	0.4789×10^{-3}	0.4789×10^{-3}	3.531×10^{-3}
5	1.1409×10^{-3}	1.1409×10^{-3}	8.410×10^{-3}

* Values above the line were obtained by heating hydrogen and iodine together; values below the line, by heating pure hydrogen iodide.

We obtain the numbers in Table 9-II. In view of the precision of the data from which these ratios are derived, the ratios are far from constant. Now let us try the ratio

$$\frac{[H_2][I_2]}{[HI]^2} \qquad (14)$$

These calculations are summarized in Table 9-III.

The results are most encouraging and imply that with a fair degree of accuracy we can write

$$\frac{[H_2][I_2]}{[HI]^2} = \text{a constant}$$

$$= 1.835 \times 10^{-2} \text{ at } 698.6°K \qquad (15)$$

Look at this ratio in terms of reaction (*12*):

$$2HI(g) \rightleftharpoons H_2(g) + I_2(g) \qquad (12)$$

The ratio (*15*) is the product of the equilibrium concentrations of the substances produced in the reaction, $[H_2] \times [I_2]$, divided by the *square* of the concentration of the reacting substance, $[HI]^2$. In this ratio, the power to which we raise the concentration of each substance is equal to its coefficient in reaction (*12*).

9-2.2 The Law of Chemical Equilibrium

Let us summarize what we have learned. For the reaction

$$Fe^{+3}(aq) + SCN^-(aq) \rightleftharpoons FeSCN^{+2}(aq) \qquad (8)$$

we found that the concentrations of the molecules involved have a simple relationship.

$$\frac{[FeSCN^{+2}]}{[Fe^{+3}][SCN^-]} = \text{a constant} \qquad (16)$$

Then we considered precise equilibrium data for the reaction

$$2HI(g) \rightleftharpoons H_2(g) + I_2(g) \qquad (12)$$

Table 9-II

VALUES OF $\frac{[H_2][I_2]}{[HI]}$ FOR DATA OF TABLE 9-I

EXPT. NO.	$\frac{[H_2][I_2]}{[HI]}$
1	32.429×10^{-5}
2	30.105×10^{-5}
3	24.866×10^{-5}
4	6.495×10^{-5}
5	15.477×10^{-5}

Table 9-III

VALUES OF $\frac{[H_2][I_2]}{[HI]^2}$ FOR DATA OF TABLE 9-I

EXPT. NO.	$\frac{[H_2][I_2]}{[HI]^2}$
1	1.8351×10^{-2}
2	1.8265×10^{-2}
3	1.8359×10^{-2}
4	1.8390×10^{-2}
5	1.8403×10^{-2}
Average	1.835×10^{-2}

The concentrations of the molecules appearing in reaction (*12*) were found to have a simple relationship,

$$\frac{[H_2][I_2]}{[HI]^2} = \text{a constant} \qquad (14)$$

In each of our simple relationships, (*16*) and (*14*), the concentrations of the products appear in the numerator. In each relationship the concentrations of reactants appear in the denominator. In reaction (*12*), two molecules of hydrogen iodide react. This influences expression (*14*) because it is necessary to square the concentration of hydrogen iodide, [HI], in order to obtain a constant ratio.

These observations and many others like them lead to the generalization known as the Law of Chemical Equilibrium. For a reaction

$$aA + bB \rightleftharpoons eE + fF \qquad (17)$$

when equilibrium exists, there will be a simple relation between the concentrations of products, $[E]$ and $[F]$, and the concentrations of reactants, $[A]$ and $[B]$:

$$\frac{[E]^e[F]^f}{[A]^a[B]^b} = K = \text{a constant at constant temperature} \qquad (18)$$

In this generalized equation, (*18*), we see that again the numerator is the product of the equilibrium concentrations of the substances formed, each raised to the power equal to the number of moles of that substance in the chemical equation. The denominator is again the product of the equilibrium concentrations of the reacting substances, each raised to a power equal to the number of moles of the substance in the chemical equation. The quotient of these two remains constant. The constant K is called the equilibrium constant. This generalization is one of the most useful in all of chemistry. From the equation for any chemical reaction one can immediately write an expression, in terms of the concentrations of reactants and products, that will be constant at any given temperature. If this constant is measured (by measuring all of the concentrations in a particular equilibrium solution), then it can be used in calculations for any other equilibrium solution at that same temperature.

In Table 9-IV are listed some reactions along with the equilibrium law relation of concentrations and the numerical values of the equilibrium constants. First, let's verify the forms of the equilibrium law relation among the concentra-

Table 9-IV. **SOME EQUILIBRIUM CONSTANTS**

REACTION	EQUILIBRIUM LAW RELATION	K AT STATED TEMP.
$Cu(s) + 2Ag^+(aq) \rightleftharpoons Cu^{+2}(aq) + 2Ag(s)$	$K = \frac{[Cu^{+2}]}{[Ag^+]^2}$	2×10^{15} at 25°C
$Ag^+(aq) + 2NH_3(aq) \rightleftharpoons Ag(NH_3)_2^+(aq)$	$K = \frac{[Ag(NH_3)_2^+]}{[Ag^+][NH_3]^2}$	1.7×10^7 at 25°C
$N_2O_4(g) \rightleftharpoons 2NO_2(g)$	$K = \frac{[NO_2]^2}{[N_2O_4]}$	0.87 at 55°C
$2HI(g) \rightleftharpoons H_2(g) + I_2(g)$	$K = \frac{[H_2][I_2]}{[HI]^2}$	0.018 at 423°C
$HSO_4^-(aq) \rightleftharpoons H^+(aq) + SO_4^{-2}(aq)$	$K = \frac{[H^+][SO_4^{-2}]}{[HSO_4^-]}$	0.013 at 25°C
$CH_3COOH(aq) \rightleftharpoons H^+(aq) + CH_3COO^-(aq)$	$K = \frac{[H^+][CH_3COO^-]}{[CH_3COOH]}$	1.8×10^{-5} at 25°C
$AgCl(s) \rightleftharpoons Ag^+(aq) + Cl^-(aq)$	$K = [Ag^+][Cl^-]$	1.7×10^{-10} at 25°C
$H_2O \rightleftharpoons H^+(aq) + OH^-(aq)$	$K = [H^+][OH^-]$	10^{-14} at 25°C
$AgI(s) \rightleftharpoons Ag^+(aq) + I^-(aq)$	$K = [Ag^+][I^-]$	10^{-16} at 25°C

tions. The very first has an unexpected form. For the reaction,

$$Cu(s) + 2Ag^+(aq) = Cu^{+2}(aq) + 2Ag(s) \quad (19)$$

you do not find

$$\frac{[Cu^{+2}][Ag]^2}{[Ag^+]^2[Cu]} = K \quad (20)$$

but rather, you find

$$\frac{[Cu^{+2}]}{[Ag^+]^2} = K \quad (21)$$

This is because the concentrations of solid copper and solid silver are incorporated into the equilibrium constant. The concentration of solid copper is fixed by the density of the metal—it cannot be altered either by the chemist or by the progress of the reaction. The same is true of the concentration of solid silver. Since neither of these concentrations varies, no matter how much solid is added, there is no need to write them each time an equilibrium calculation is made. Equation (*21*) will suffice.

EXERCISE 9-4

If we assign the equilibrium constant K' to expression (*20*) and K to expression (*21*),

$$K' = \frac{[Cu^{+2}][Ag]^2}{[Ag^+]^2[Cu]} \qquad K = \frac{[Cu^{+2}]}{[Ag^+]^2}$$

show that

$$K = K' \frac{[Cu]}{[Ag]^2}$$

Another K of unexpected form applies to the reaction

$$H_2O \rightleftharpoons H^+(aq) + OH^-(aq) \quad (22)$$

For this reaction we might have written

$$\frac{[H^+][OH^-]}{[H_2O]} = K \quad (23)$$

Instead, Table 9-IV lists expression (*24*) as

$$[H^+][OH^-] = K \quad (24)$$

The concentration of water, $[H_2O]$, does not appear in the denominator of expression (*24*). This is usually done in treating aqueous reactions that consume or produce water. It is justified because the variation in the concentration of water during reaction is so slight in dilute aqueous solutions. We can treat $[H_2O]$ as a concentration that does not vary. Hence, it can be incorporated in the equilibrium constant.

EXERCISE 9-5

Water has a density of one gram per milliliter. Calculate the concentration of water (expressed in moles per liter) in pure water. Now calculate the concentration of water in 0.10 M aqueous solution of acetic acid, CH_3COOH, assuming each molecule of CH_3COOH occupies the same volume as one molecule of H_2O.

In summary, the concentrations of solids and the concentrations of solvent (usually water) can be and usually are incorporated in the equilibrium constant, so they do not appear in the equilibrium law relation.

Now look at the numerical values of the equilibrium constants. The K's listed range from 10^{+15} to 10^{-16}, so we see there is a wide variation. We want to acquire a sense of the relation between the size of the equilibrium constant and the state of equilibrium. A large value of K must mean that at equilibrium there are much larger concentrations present of products than of reactants. Remember that the numerator of our equilibrium expression contains the concentrations of the products of the reaction. The value of 2×10^{15} for the K for reaction (*19*) certainly indicates that if a reaction is initiated by placing metallic copper in a solution containing Ag^+ (for example, in silver nitrate solution), when equilibrium is finally reached, the concentration of Cu^{+2} ion, $[Cu^{+2}]$, is very much greater than the square of the silver ion concentration, $[Ag^+]^2$.

A small value of K for a given reaction implies that very little of the products have to be formed from the reactants before equilibrium is attained. The value of $K = 10^{-16}$ for the reaction

$$AgI(s) \rightleftharpoons Ag^+(aq) + I^-(aq) \quad (25)$$

indicates that very little solid AgI can dissolve

before equilibrium is attained. Silver iodide has extremely low solubility. Conversely, if 0.1 *M* solutions of KI and $AgNO_3$ are mixed, the values of $[Ag^+]$ and $[I^-]$ are large and the equilibrium state cannot be reached until the $[Ag^+]$ and $[I^-]$ have been greatly reduced by the precipitation of AgI.

9-2.3 The Law of Chemical Equilibrium Derived from Rates of Opposing Reactions

Chemists picture equilibrium as a dynamic balance between opposing reactions. An understanding of the Law of Chemical Equilibrium can be built upon this basis.

Consider the oxidation of nitric oxide, NO, to nitrogen dioxide, NO_2:

$$2NO(g) + O_2(g) \longrightarrow 2NO_2(g) \qquad (26)$$

The reaction to the right, (*R*), proceeds with a rate that is found experimentally to depend upon the concentrations of the reactants as follows:

$$(\text{rate})_R = k_R[NO]^2[O_2] \qquad (27)$$

The reverse reaction to the left, (*L*), has been studied as well. The rate of this reaction

$$2NO(g) + O_2(g) \longleftarrow 2NO_2(g) \qquad (28)$$

depends upon the concentrations as follows:

$$(\text{rate})_L = k_L[NO_2]^2 \qquad (29)$$

Expressions (*27*) and (*29*) show how the rates of reaction (*26*) and its reverse, reaction (*28*), depend upon the concentrations. Now we can apply our microscopic view of the equilibrium state. Chemical changes will cease (on the macroscopic scale) when the rate of reaction (*26*) is exactly equal to that of reaction (*28*). When this is so, we can equate expressions (*27*) and (*29*):

$$(\text{rate})_R = (\text{rate})_L \qquad (30)$$

or

$$k_R[NO]^2[O_2] = k_L[NO_2]^2 \qquad (31)$$

By algebraic rearrangement, equation (*31*) can be written

$$\frac{k_R}{k_L} = \frac{[NO_2]^2}{[NO]^2[O_2]} \qquad (32)$$

Since both k_R and k_L are constants at a given temperature, their ratio is constant. Hence (*32*) is the equilibrium law expression for the equilibrium

$$2NO(g) + O_2(g) \rightleftharpoons 2NO_2(g) \qquad (33)$$

and the equilibrium constant

$$K = \frac{k_R}{k_L} \qquad (33)$$

Thus we see that the experimental rate laws for this reaction and its reverse lead to the equilibrium law. In every reaction that has been sufficiently studied, this same result is obtained. We are led to have confidence in the molecular view of equilibrium as a dynamic balance between opposing reactions.

9-2.4 The Factors Which Determine Equilibrium

We have learned much about equilibrium. It is characterized by constancy of macroscopic properties but with molecular processes continuing in a state of dynamic balance. At equilibrium we can conclude that every reaction that takes place does so at the same reaction rate as its reverse reaction.

We have gone further and discovered that the equilibrium conditions imply a constant relationship among the concentrations of reactants and products. This relationship is called the Law of Chemical Equilibrium. Using this law, we can express the conditions at equilibrium in terms of a number *K*, called the equilibrium constant.

Despite this detailed familiarity with equilibrium, there is one facet we have not considered at all. What determines the equilibrium constant? Why does one reaction favor reactants and another reaction favor products? What factors cause sodium chloride to have a large solubility in water and silver chloride to have a low solubility? Why does equilibrium favor the reaction of oxygen with iron to form Fe_2O_3 (rust) but not the reaction of oxygen with gold? As scientists, we cannot resist wondering what factors determine the conditions at equilibrium.

This is the activity of science we called "Wondering why" (Section 1-1.3)—we are searching for explanation. An explanation is a likeness which connects the system under study with a model system which is well understood. We might begin by considering Figure 9-7. Here we see a golf bag thrown into the rear of a station wagon. Unfortunately, the ball pocket is open and all of the golf balls have spilled out onto the floor of the station wagon. Because the floor has a step in it, the golf balls on the upper level possess some potential energy (energy of position). The golf balls tend to roll to the lower level spontaneously, as shown in Figure 9-8. As a golf ball does this, its potential energy becomes kinetic energy (energy of motion). Finally, this

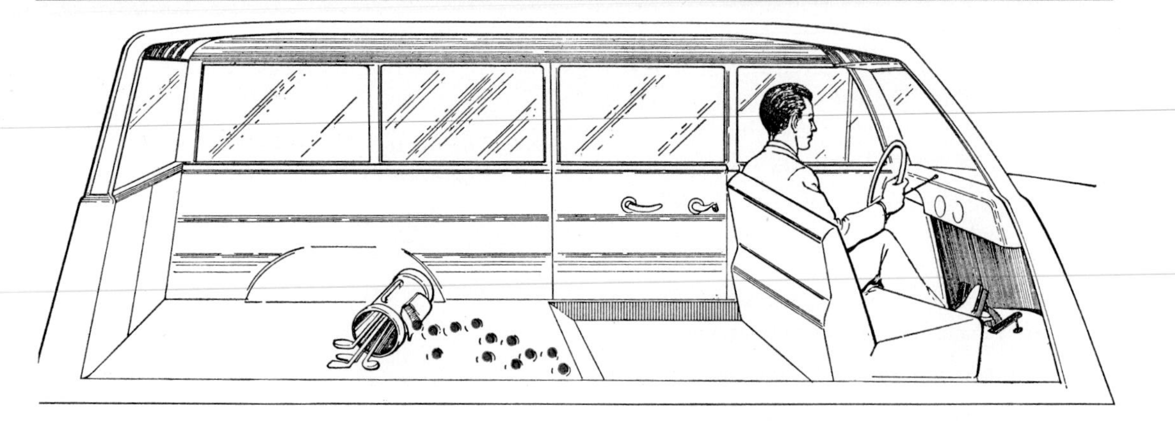

Fig. 9-7. **Golf balls rolling on the floor of a station wagon.**

kinetic energy is dissipated into heat. Now the golf balls lie at rest at the lower floor level.

This situation has some similarities to the chemical change in a spontaneous, exothermic reaction. The reactants of high heat content react spontaneously to form products of lower heat content. As each molecular reaction occurs, the excess heat content becomes kinetic energy. The product molecules separate from each other with high kinetic energy. As they collide with other molecules, this energy is dissipated into heat. Figure 9-8 shows this through a heat content diagram for the chemical reaction.

(1) There are two states of each system:

	Initial State		Final State
Golf balls:	on upper level	⟶	on lower level
Reaction:	reactants	⟶	products

Fig. 9-8. **Comparison of a chemical reaction** *to golf balls rolling downhill.*

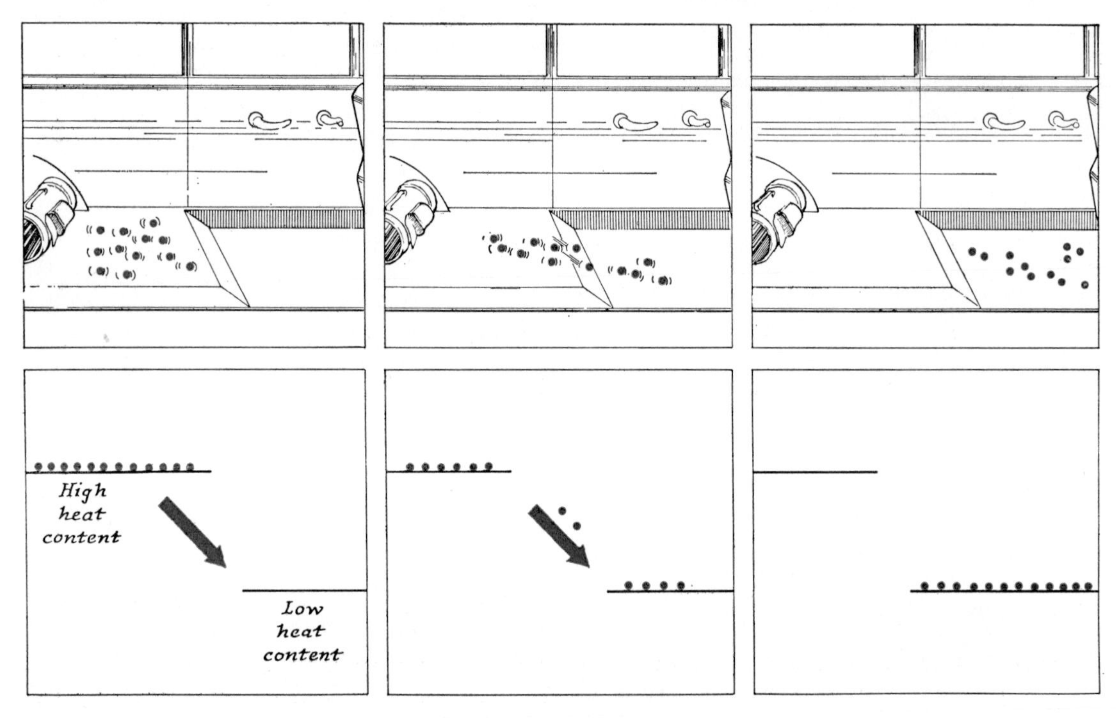

(2) The potential energy of the initial state is higher than the potential energy of the final state:

	Initial State		Final State
Golf balls:	high potential energy	⟶	low potential energy
Reaction:	high heat content	⟶	low heat content

(3) As the change from initial state to final state proceeds, the form of the energy changes:

	Initial State		Final State
Golf balls:	potential energy	⟶	kinetic energy, and then heat
Reaction:	heat content	⟶	molecular kinetic energy, and then heat

(4) The changes from initial to final state proceed spontaneously toward lowest potential energy, the direction corresponding to "rolling downhill":

	Initial State ⟶ Final State
Golf balls:	spontaneous
Reaction:	spontaneous

Having established these similarities, we might offer a possible generalization:

Since: golf balls always roll downhill spontaneously

Perhaps: reactions always proceed spontaneously in the direction toward minimum energy.

This proposal leads us to expect that a reaction will tend to proceed spontaneously if the products have lower energy than the reactants. This expectation is in accord with experience with many reactions, especially for those which release a large amount of heat.

There are two basic and serious difficulties with our proposed explanation. (Remember "Cylindrical objects burn"?)

(1) *Some endothermic reactions proceed spontaneously.* One example is the evaporation of a liquid. Water evaporates spontaneously but it absorbs heat as it does so. It is not "rolling downhill" energetically. When ammonium chloride dissolves in water, the solution becomes cooler. Again heat is absorbed —yet the ammonium chloride goes ahead and dissolves.

(2) Another difficulty is that *spontaneous chemical reactions do not go to completion.* Even if a spontaneous reaction is exothermic, it proceeds only till it reaches equilibrium. But in our golf ball analogy, "equilibrium" is reached when all of the golf balls are on the lower level. Our analogy would lead us to expect that an exothermic reaction would proceed until *all* of the reactants are converted to products, not to a dynamic equilibrium.

Because of these failures, we need to alter our proposed explanation. We must seek a new analogy that gives a better correspondence with the behavior of chemical reactions. How should we alter our golf ball analogy to bring it into better accord with experimental facts? Here is a possible view.

Consider how the golf ball situation shown in Figure 9-8 will change when the station wagon is driven over a bumpy road. Now the golf balls are shaken and jostled about; they roll around and collide with each other. Every now and then one of the golf balls even accumulates enough energy (through collisions) to return to the upper level of the station wagon floor. Of course, any golf ball that is bounced up tends to roll back down to the lower level a little later. As this bumpy ride continues, a state is reached in which golf balls are being jostled up to the higher level at the same rate they are rolling back down to the lower level. Then "equilibrium" exists. Some of the golf balls are on the lower level and some on the upper level. Since the rate of rolling up equals the rate of rolling down, a dynamic balance exists.

This analogy solves the problems of the simpler "golf balls roll downhill" picture. The bumpy road model contains a new feature that gives a basis for expecting "reaction" in the

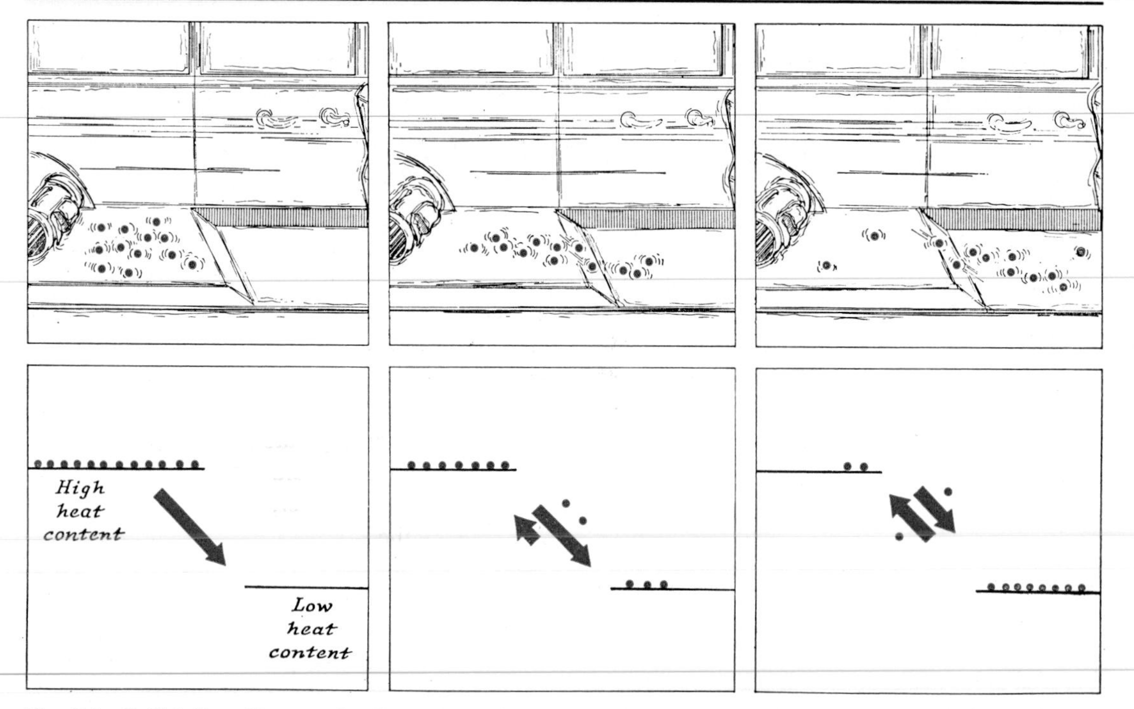

Fig. 9-9. **Golf balls rolling on the floor of a station wagon** *driving on a bumpy road.*

endothermic direction. Some golf balls roll *uphill* if they are shaken hard enough. The tendency to roll back down will always keep them coming back to the lower level and, finally, equilibrium will be reached when the rate of rolling down equals the rate of jostling up.

What happens if the road becomes smoother? The "jostling up" reaction is less favored—the equilibrium conditions change in favor of the golf balls at the lower level.

Now turn to the chemical reaction. What feature in a reacting chemical system corresponds to the jostling of the bumpy road in our analogy? It is the *temperature*. At any temperature except absolute zero there is a constant random jostling of the molecules. Some molecules have low kinetic energies, some have high kinetic energies—we looked at the distribution of the energies in Chapter 8 (see Figure 8-4, page 131). Some of the molecules will occasionally accumulate enough energy to "roll uphill" to less stable molecular forms. On the one hand, molecular changes will take place in the direction of minimum energy. On the other hand, the molecular changes will finally reach a dynamic equilibrium when the random jostlings or energy transfers at the temperature of the system are restoring molecules to the molecular forms of higher energy at the same rate as they are "rolling downhill" to the lower energy forms.

Now we have an analogy that does aid us in understanding chemical reactions and equilibrium. We can see the following features of chemical reactions:

(1) Chemical reactions proceed spontaneously to approach the equilibrium state.
(2) *One factor that fixes the equilibrium state is the energy.* **Equilibrium tends to favor the state of the lowest energy.**
(3) *The other factor that fixes the equilibrium state is the randomness implied by the temperature.* **Equilibrium tends to favor the state of greatest randomness.**
(4) **The equilibrium state is a compromise between these two factors, minimum energy and maximum randomness.** At very low temperatures, energy tends to be the more important factor. Then equilibrium favors the molecular substances with the

lowest heat content. At very high temperatures, randomness becomes more important. Then equilibrium favors a random distribution among reactants and products without regard for energy differences.

Our analogy can be stretched one point further. We might ask whether the relative area of the upper floor level compared with that of the lower floor level has any bearing on the distribution of golf balls. After all, if the upper level area is small, as in Figure 9-7, few golf balls are likely to remain there. Contrast Figure 9-10, in which the golf bag has been removed. Now the upper floor level has much more area. Golf balls which reach the upper level will now have a great deal of space. They can roll around longer before returning to the lower level. The effect of extending the upper level will be to increase the fraction of the golf balls that occupy the upper level at "equilibrium."

This extension of the analogy increases its value in considering chemical reactions. The simplest example is probably the vaporization of a liquid. It is true that the molecules have lower energy when they cluster together tightly in the liquid state. On the other hand, the gaseous state provides a broad upper level. Every molecule which vaporizes has an amount of space available to it much larger than it had in the crowded liquid. This "available space" factor, accompanied by the random jostlings of temperature to overcome the potential energy difference, aids vaporization.

Now let us review what happens as we warm a solid substance from a very low temperature to a very high temperature. As the temperature is raised, small energy differences become unimportant. Thus, if the temperature of the solid is raised too much, the lower energy of the regular solid becomes unimportant compared with the random thermal energies. The solid melts, surrendering this energy stability in return for the randomness of the liquid state. If the temperature is raised still further, the energies of attraction among the molecules become unimportant compared with the random thermal energies. Then the liquid vaporizes, surrendering the lower potential energy afforded by the molecules remaining close together in favor of the still higher randomness of the gaseous state. If we raise the temperature still further, the energies that hold molecules together begin to become unimportant compared with random thermal energies. Finally, at extremely high energies, molecules no longer exist—all is chaos. This is the chemical situation within the Sun. Since at such high energies chemical reactions become unimportant, all chemists on the Sun are out of work. We'd better return to room temperature to apply our knowledge of equilibrium to chemical systems within our interest.

Fig. 9-10. **Golf balls rolling on the floor of a station wagon.** *The effect of extending the upper level.*

QUESTIONS AND PROBLEMS

1. Which of the following are equilibrium situations and which are steady state situations?

   (a) A playing basketball team and the bench of reserves. The number of players "in the game" and the number "on the bench" are constant.
   (b) The liquid mercury and mercury vapor in a thermometer. Temperature is constant.
   (c) Grand Coulee dam and the lake behind it. Water level is constant, though a river flows into the lake.
   (d) A well-fed lion in a cage. The lion's weight is constant.

   Answer. (a) and (b) are equilibrium situations.

2. Which of the following are equilibrium situations and which are steady state situations?

   (a) A block of wood floating on water.
   (b) During the noon hour, the water fountain constantly has a line of ten persons.
   (c) When a capillary tube is dipped in water, water rises in the capillary (because of surface tension) to a height h and remains constant there.
   (d) The capillary system of (c) considered over such a long period that evaporation of water out the end of the capillary cannot be neglected.
   (e) At a particular point in the reaction chamber of a jet engine, the gas composition (fuel, air, and products) is constant.

3. What, specifically, is "equal" in a chemical reaction that has attained a state of equilibrium?

4. One drop of water may or may not establish a state of vapor pressure equilibrium when placed in a closed bottle. Explain.

5. Why are chemical equilibria referred to as "dynamic"?

6. What do the following experiments (done at 25°C) show about the state of equilibrium?

   (a) One liter of water is added a few milliliters at a time, to a kilogram of salt which only partly dissolves.
   (b) A large salt shaker containing one kilogram of salt is gradually emptied into one liter of water. The same amount of solid dissolves as in (a).

7. The following chemical equation represents the reaction between hydrogen and chlorine to form hydrogen chloride:

$$H_2(g) + Cl_2(g) \rightleftharpoons 2HCl(g) + 44.0 \text{ kcal}$$

   (a) List four important pieces of information conveyed by this equation.
   (b) What are three important areas of interest concerning this reaction for which no information is indicated?

8. How does a catalyst affect the equilibrium conditions of a chemical system?

9. In any discussion of chemical equilibrium why are concentrations always expressed in moles, rather than in grams, per unit volume?

10. If the phase change represented by

$$\text{Heat} + H_2O(l) \rightleftharpoons H_2O(g)$$

   has reached equilibrium in a closed system:

   (a) What will be the effect of a reduction of volume, thus increasing the pressure?
   (b) What will be the effect of an increase in temperature?
   (c) What will be the effect of injecting some steam into the closed system, thus raising the pressure?

11. Methanol (methyl alcohol) is made according to the following net equation:

$$CO(g) + 2H_2(g) \rightleftharpoons CH_3OH(g) + \text{heat}$$

   Predict the effect on equilibrium concentrations of an increase in: (a) Temperature. (b) Pressure.

   Answer. (a) Decreases CH_3OH.
   (b) Increases CH_3OH.

12. Consider the reaction:

$$4HCl(g) + O_2(g) \rightleftharpoons 2H_2O(g) + 2Cl_2(g) + 27 \text{ kcal}$$

What effect would the following changes have on the equilibrium concentration of Cl_2? Give your reasons for each answer.

(a) Increasing the temperature of the reaction vessel.
(b) Decreasing the total pressure.
(c) Increasing the concentration of O_2.
(d) Increasing the volume of the reaction chamber.
(e) Adding a catalyst.

13. Write the equation for the dissociation of $HI(g)$ into its elements.

(a) Will HI dissociate to a greater or a lesser extent as the temperature is increased? $\Delta H = -6.2$ kcal/mole $HI(g)$.
(b) How many grams of iodine will result if at equilibrium 0.050 mole of HI has dissociated?

14. Consider two separate closed systems, each at equilibrium:

(a) HI and the elements from which it is formed,
(b) H_2S and the elements from which it is formed.

What would happen in each if the total pressure were increased? Assume conditions are such that all reactants and products are gases.

15. Each of the following systems has come to equilibrium. What would be the effect on the equilibrium concentration (increase, decrease, no change) of each substance in the system when the listed reagent is added?

REACTION	ADDED REAGENT
(a) $C_2H_6(g) \rightleftharpoons H_2(g) + C_2H_4(g)$	$H_2(g)$
(b) $Cu^{+2}(aq) + 4NH_3(g) \rightleftharpoons Cu(NH_3)_4^{+2}(aq)$	$CuSO_4(s)$
(c) $Ag^+(aq) + Cl^-(aq) \rightleftharpoons AgCl(s)$	$AgCl(s)$
(d) $PbSO_4(s) + H^+(aq) \rightleftharpoons Pb^{+2}(aq) + HSO_4^-(aq)$	$Pb(NO_3)_2$ solution
(e) $CO(g) + \frac{1}{2}O_2(g) \rightleftharpoons CO_2(g)$ + heat	heat

Answer. (a) C_2H_6 (increase), H_2 (increase), C_2H_4 (decrease).

16. Nitric oxide, NO, releases 13.5 kcal/mole when it reacts with oxygen to give nitrogen dioxide. Write the equation for this reaction and predict the effect of (i) raising the temperature, and of (ii) increasing the concentration of NO (at a fixed temperature) on:

(a) the equilibrium concentrations;
(b) the numerical value of the equilibrium constant;
(c) the speed of formation of NO_2.

17. Given:

$$SO_2(g) + \tfrac{1}{2}O_2(g) \rightleftharpoons SO_3(g) + 23 \text{ kcal}$$

(a) For this reaction discuss the conditions that favor a high equilibrium concentration of SO_3.
(b) How many grams of oxygen gas are needed to form 1.00 gram of SO_3?

Answer. 0.200 gram of O_2.

18. Given:

$$CaCO_3(s) \rightleftharpoons CaO(s) + CO_2(g)$$

(closed system)

At a fixed temperature, what effect would adding more $CaCO_3$ have on the concentration of CO_2 in the region above the solid phase? Explain.

19. Given:

$$H_2(g) + I_2(g) \rightleftharpoons 2HI(g) \text{ (closed system)}$$

At 450°C, $K = 50.0$ for the above reaction. What is the equilibrium constant of the reverse reaction at 450°C?

20. Write the expression indicating the equilibrium law relations for the following reactions.

(a) $N_2(g) + 3H_2(g) \rightleftharpoons 2NH_3(g)$
(b) $CO(g) + NO_2(g) \rightleftharpoons CO_2(g) + NO(g)$
(c) $Zn(s) + 2Ag^+(aq) \rightleftharpoons Zn^{+2}(aq) + 2Ag(s)$
(d) $PbI_2(s) \rightleftharpoons Pb^{+2}(aq) + 2I^-(aq)$
(e) $CN^-(aq) + H_2O(l) \rightleftharpoons HCN(aq) + OH^-(aq)$

21. Equilibrium constants are given for several systems below. In which case does the reaction as written occur to the greatest extent?

REACTION	K
(a) $CH_3COOH(aq) \rightleftarrows H^+(aq) + CH_3COO^-(aq)$	1.8×10^{-5}
(b) $CdS(s) \rightleftarrows Cd^{+2}(aq) + S^{-2}(aq)$	7.1×10^{-28}
(c) $H^+(aq) + HS^-(aq) \rightleftarrows H_2S(aq)$	1×10^7

22. In the reaction

$$2HI(g) \rightleftarrows H_2(g) + I_2(g)$$

at 448°C the partial pressures of the gas at equilibrium are as follows:

$[HI] = 4 \times 10^{-3}$ atm;
$[H_2] = 7.5 \times 10^{-3}$ atm;
$[I_2] = 4.3 \times 10^{-5}$ atm.

What is the equilibrium constant for this reaction?

23. Reactants A and B are mixed, each at a concentration of 0.80 mole/liter. They react slowly, producing C and D: $A + B \rightleftarrows C + D$. When equilibrium is reached, the concentration of C is measured and found to be 0.60 mole/liter. Calculate the value of the equilibrium constant.

Answer. $K = 9.0$.

24. The water gas reaction

$$CO_2(g) + H_2(g) \rightleftarrows CO(g) + H_2O(g)$$

was carried out at 900°C with the following results.

TRIAL NO.	PARTIAL PRESSURE, ATM AT EQUILIBRIUM			
	CO	H_2O	CO_2	H_2
1	0.352	0.352	0.648	0.148
2	0.266	0.266	0.234	0.234
3	0.186	0.686	0.314	0.314

(a) Write the equilibrium constant expression.
(b) Verify that the expression in (a) is a constant, using the data given.

25. Select from each of the following pairs the more random system.

(a) A brand new deck of cards arranged according to suit and number.
(a′) The same deck of cards after shuffling.

(b) A box full of sugar cubes.
(b′) Sugar cubes thrown on the floor.

(c) A hay stack.
(c′) Stacked fire wood.

26. For each of the following reactions, state: (i) whether tendency toward minimum energy favors reactants or products, (ii) whether tendency toward maximum randomness favors reactants or products.

(a) $H_2O(l) \rightleftarrows H_2O(s)$ $\Delta H = -1.4$ kcal
(b) $H_2O(l) \rightleftarrows H_2O(g)$ $\Delta H = +10$ kcal
(c) $CaCO_3(s) + 43\text{ kcal} \rightleftarrows CaO(s) + CO_2(g)$
(d) $I_2(s) + 1.6\text{ kcal} \rightleftarrows I_2$ (in alcohol)
(e) $4Fe(s) + 3O_2(g) \rightleftarrows 2Fe_2O_3(s) + 400\text{ kcal}$

CHAPTER 

Solubility Equilibria

. . . solubility . . . depends fundamentally upon the ease with which . . . two molecular species are able to mix, and if the two species display a certain hostility toward mixing, . . . saturation [is] attained at smaller concentration . . .

JOEL H. HILDEBRAND, 1936

The principles of equilibrium have wide applicability and great utility. For example, they aid us in understanding and controlling the solubility of solids and gases in liquids. We shall consider, first, the solubility of a molecular solid in a liquid, then the solubility of a gas in a liquid. The usefulness of equilibrium principles is even greater in treating the solubilities of ionic solids in water. Much of this chapter will be devoted to aqueous solutions of ionic solids.

10-1 SOLUBILITY: A CASE OF EQUILIBRIUM

The starting point in any quantitative equilibrium calculation is the Equilibrium Law. For a generalized reaction,

$$aA + bB + \cdots \rightleftharpoons eE + fF + \cdots \qquad (1)$$

equilibrium exists when the concentrations satisfy the relation

$$K = \text{a constant} = \frac{[E]^e[F]^f \cdots}{[A]^a[B]^b \cdots} \qquad (2)$$

First, we shall apply expression (*2*) to the solution system of solid iodine dissolving in liquid ethyl alcohol.

10-1.1 The Solubility of Iodine in Ethyl Alcohol

As a solid dissolves in a liquid, atoms or molecules leave the solid and become part of the liquid. These atoms or molecules may carry no charge (then they are electrically neutral) or they may be ions. The iodine-alcohol system is of the former kind. As iodine dissolves, neutral molecules of iodine, I_2, leave the regular crystal lattice and these molecules become part of the liquid phase. At equilibrium, excess solid must remain and a fixed concentration of iodine is present in solution. This fixed concentration is called the solubility.

For this system at equilibrium, the reaction is

$$I_2(\text{solid}) = I_2(\text{alcohol solution}) \qquad (3)$$

The Equilibrium Law applied to this reaction gives

$$K = \text{a constant} = [\text{concentration } I_2 \text{ in alcohol}]$$
$$K = [I_2] \qquad (4)$$

10-1.2 The Dynamic Nature of Solubility Equilibrium

The simple form of the equilibrium expression (*4*) follows directly from the dynamic nature of the solubility equilibrium. There must be a dynamic balance between the rate that iodine molecules leave the crystal and the rate that iodine molecules return to the crystal. To understand this dynamic balance, we must consider the factors that determine these two rates.

RATE OF DISSOLVING

One of the factors that influences the rate of dissolving of solid is the area, A, of the crystal surface that contacts the liquid. If many crystals (with large A) are dissolving simultaneously, the rate of dissolving is faster than if only a few crystals (with small A) are in the solvent. The rate of dissolving is proportional to this liquid-solid surface area, A.

A molecule of iodine is more stable in the crystal than in the solution. The potential energy must rise as a molecule leaves the crystal and the principles that govern rates of reaction are operative. Presumably there is an activated complex for the process. The rate at which molecules leave a square centimeter of surface, passing over the energy barrier, is determined by the height of the barrier and by the temperature. We can call this rate k_d. Changing the temperature does not affect the activated complex, but the molecular energy distribution is altered (see Figure 8-3, p. 131). Hence, k_d is a function of temperature. These two factors determine completely the rate of dissolving:

$$(\text{rate of dissolving}) = \begin{pmatrix}\text{surface}\\ \text{area}\end{pmatrix} \times \begin{pmatrix}\text{rate molecules leave}\\ \text{a square centimeter}\\ \text{of crystal surface}\end{pmatrix}$$
$$= \quad (A) \quad \times \quad (k_d)$$
$$(\text{rate of dissolving}) = A \times k_d \qquad (5)$$

RATE OF PRECIPITATION

The rate of precipitation is the rate at which molecules return to the surface and fit into the crystal lattice. To do this, the molecules in solution must first strike the crystal surface. Again, the more surface, the more frequently will dissolved molecules encounter a piece of crystal. The rate of precipitation is proportional to A.

In addition, the rate that molecules strike the surface depends upon how many molecules there are per unit volume of solution. As the concentration rises, more and more molecules strike the surface per unit time. The rate of precipitation is proportional to the iodine concentration, $[I_2]$.

The last factor is, again, the rate that molecules can pass over the energy barrier—the activated complex for precipitation. Again there is a rate constant, k_p, that is determined by temperature and the height of the energy barrier to precipitation.

We have, then, three factors that determine the rate of precipitation:

$$(\text{rate of precipitation}) = \begin{pmatrix}\text{surface}\\ \text{area}\end{pmatrix}\begin{pmatrix}\text{concentration of}\\ \text{dissolved } I_2\end{pmatrix}\begin{pmatrix}\text{rate dissolved molecules pass}\\ \text{over activation energy barrier}\end{pmatrix}$$
$$= \quad (A) \quad \times \quad [I_2] \quad \times \quad (k_p)$$
$$(\text{rate of precipitation}) = A \times k_p \times [I_2] \qquad (6)$$

THE DYNAMIC NATURE OF EQUILIBRIUM

At equilibrium, we can equate (*5*) and (*6*):

$$\text{(rate of dissolving)} = \text{(rate of precipitation)}$$

$$A \times k_d = A \times k_p \times [I_2] \qquad (7)$$

The area of contact, A, appears both on the left and on the right of expression (*7*). Hence, it cancels out. Dividing both sides of (*7*) by k_p, we obtain

$$\frac{k_d}{k_p} = [I_2] \qquad (8)$$

Since k_d and k_p each depend upon temperature, their ratio depends upon temperature. Otherwise, however, each is constant. We can write

$$K = [I_2] \qquad (4)$$

where

$$K = \frac{k_d}{k_p}$$

Thus, by expressing the dynamic balance between the rates of dissolving and precipitation, we obtain (*4*). The concentration of I_2 at equilibrium is a constant, fixed by the temperature. This constant equals the solubility.

10-1.3 The Factors That Fix Solubility of a Solid

All of the discussion we have just applied to the dissolving of iodine in ethyl alcohol applies equally well to the dissolving of iodine in carbon

Fig. 10-1. **Solubility equilibrium is dynamic.**

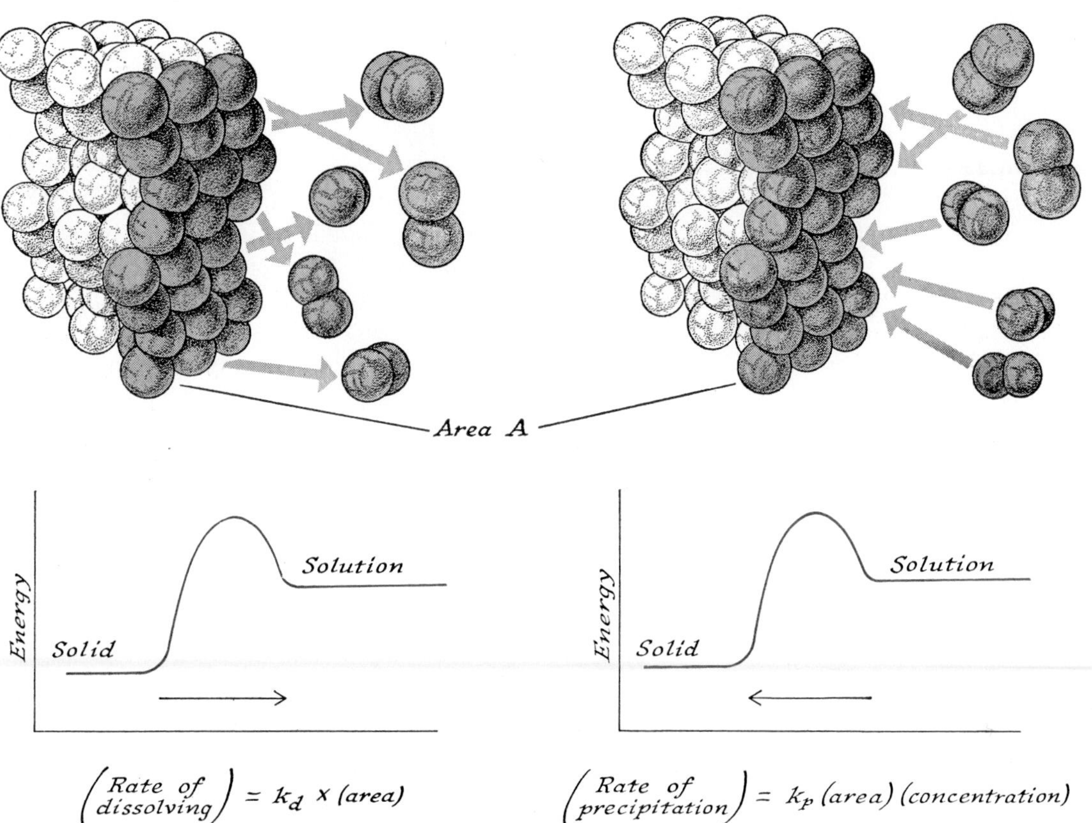

tetrachloride, CCl_4. Iodine at room temperature dissolves in carbon tetrachloride at a certain rate that, at equilibrium, exactly equals the rate of precipitation. Again we reach the simple equilibrium expression

$$K = \frac{k_d}{k_p} = [I_2] \qquad (4)$$

Despite this qualitative similarity, the solubility of iodine in CCl_4 is very different from its solubility in alcohol. One liter of alcohol dissolves 0.84 mole of iodine, whereas one liter of CCl_4 dissolves only 0.12 mole:

$$K_{\text{alcohol}} = 0.84 \text{ mole/liter} \qquad (9)$$

$$K_{CCl_4} = 0.12 \text{ mole/liter} \qquad (10)$$

Why are these constants so different? To see why, we must turn to the two factors that control every equilibrium, tendency toward minimum energy and tendency toward maximum randomness.

THE EFFECT OF RANDOMNESS

In either solvent, alcohol or carbon tetrachloride, the dissolving process destroys the regular crystal lattice of iodine and forms the disordered solution. The dissolving process *increases* randomness. The tendency toward maximum randomness tends to cause solids to dissolve.

THE EFFECT OF ENERGY

Experiment shows that heat is absorbed as iodine dissolves. The regular, ideally packed iodine crystal gives an iodine molecule a lower potential energy than does the random and loosely packed solvent environment. We see that the second factor, tendency toward minimum energy, favors precipitation and growth of the crystal.

Now we see the opposing factors at equilibrium. To increase randomness, solid tends to dissolve. To lower energy, solid tends to precipitate. Equilibrium is reached when the concentration is such that these two tendencies are equal.

How much the energy factor favors the crystal depends upon the change in heat content as a mole of solid dissolves. This change is called the heat of solution. The heats of solution of iodine in these two solvents have been measured; they are as follows:

$$I_2(s) + 1.6 \text{ kcal} \rightleftharpoons I_2(\text{in alcohol}) \qquad (11)$$

$$I_2(s) + 5.8 \text{ kcal} \rightleftharpoons I_2(\text{in } CCl_4) \qquad (12)$$

We see that there is a much greater energy rise when a mole of I_2 dissolves in CCl_4 than when a mole of I_2 dissolves in alcohol. Thus the energy factor (favoring the crystal) that opposes the randomness factor (favoring solution) is much

Fig. 10-2. **A large energy difference** *between solid and solution lowers the solubility.*

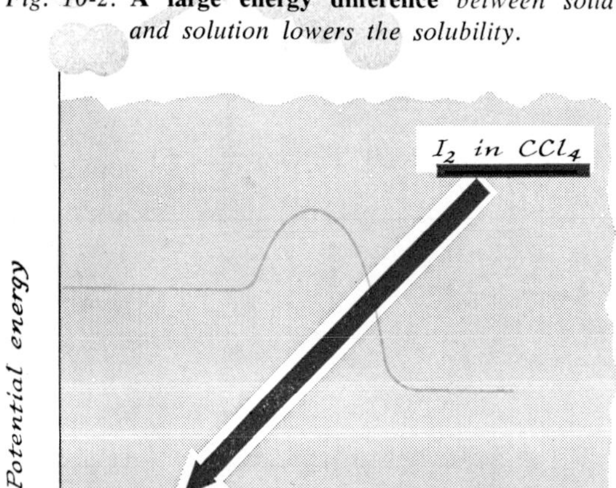

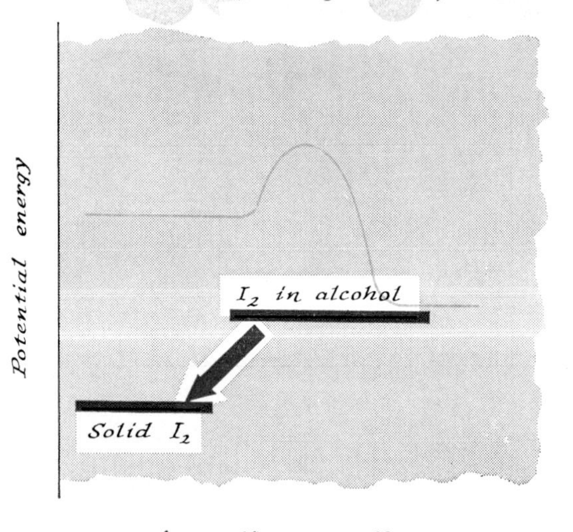

larger for CCl_4 than for alcohol. The solubility of iodine in CCl_4 is not as high as it is in alcohol.

This energy rise establishes, for this case, the "hostility" toward mixing referred to in the quotation at the beginning of the chapter. The larger the "hostility," as measured by heat absorbed on mixing, the lower will be the solubility.

THE EFFECT OF TEMPERATURE

Raising the temperature always tends to favor the more random state. For these solvents, this means that more solid will dissolve, since the liquid solution is more random than the solid. The solubility of iodine *increases* as temperature is raised, both in alcohol and in carbon tetrachloride.

EXERCISE 10-1

The heat of solution of iodine in benzene is +4.2 kcal/mole (heat is absorbed). Assuming the increase in randomness is the same when iodine dissolves in liquid benzene as it is in ethyl alcohol and in CCl_4, justify the prediction that the solubility of I_2 in benzene is higher than in CCl_4 but lower than in alcohol.

10-1.4 Solubility of a Gas in a Liquid

Gases, too, dissolve in liquids. Let us apply our understanding of equilibrium to this type of system.

THE EFFECT OF RANDOMNESS

The gaseous state is more random than the liquid state since the molecules move freely through a much larger space as a gas. Hence randomness *decreases* as a gas dissolves in a liquid. In this case, unlike solids, the tendency toward maximum randomness favors the gas phase and opposes the dissolving process.

THE EFFECT OF ENERGY

In a gas the molecules are far apart and they interact very weakly. As a gas molecule enters the liquid, it comes close to the solvent molecules and they attract each other, lowering the potential energy. Again we find a contrast to the behavior of solids. When a gas dissolves in a liquid, *heat is evolved.* The tendency toward minimum energy favors the dissolving process.

Thus we see that the equilibrium solubility of a gas again involves a balance between randomness and energy as it does for a solid, but the effects are opposite. For a gas, the tendency toward maximum randomness favors the gas phase, opposing dissolving. The tendency toward minimum energy favors the liquid state, hence favors dissolving.

As an example, consider the solubilities of the two gases, oxygen, O_2, and nitrous oxide, N_2O, in water. The heats of solution have been measured and are as follows:

$$O_2(g) \rightleftharpoons O_2(aq) + 3.0 \text{ kcal/mole } O_2 \qquad (13)$$

$$N_2O(g) \rightleftharpoons N_2O(aq) + 4.8 \text{ kcal/mole } N_2O \qquad (14)$$

Assuming the randomness factor is about the same, the gas with the larger heat effect (favoring dissolving) should have the higher solubility. The measured solubilities at one atmosphere pressure and 20°C of oxygen and nitrous oxide in water are, respectively, O_2, 1.4×10^{-3} mole/liter and N_2O, 27×10^{-3} mole/liter, consistent with our prediction.

THE EFFECT OF TEMPERATURE

Raising the temperature always tends to favor the more random state. This means that less gas will dissolve, since the gas is more random than the liquid. The solubility of a gas *decreases* as temperature is raised.

EXERCISE 10-2

From the heat of solution of chlorine in water, −6.0 kcal/mole (heat evolved), how do you expect the solubility of chlorine at one atmosphere pressure and 20°C to compare with that of oxygen and of nitrous oxide, N_2O?

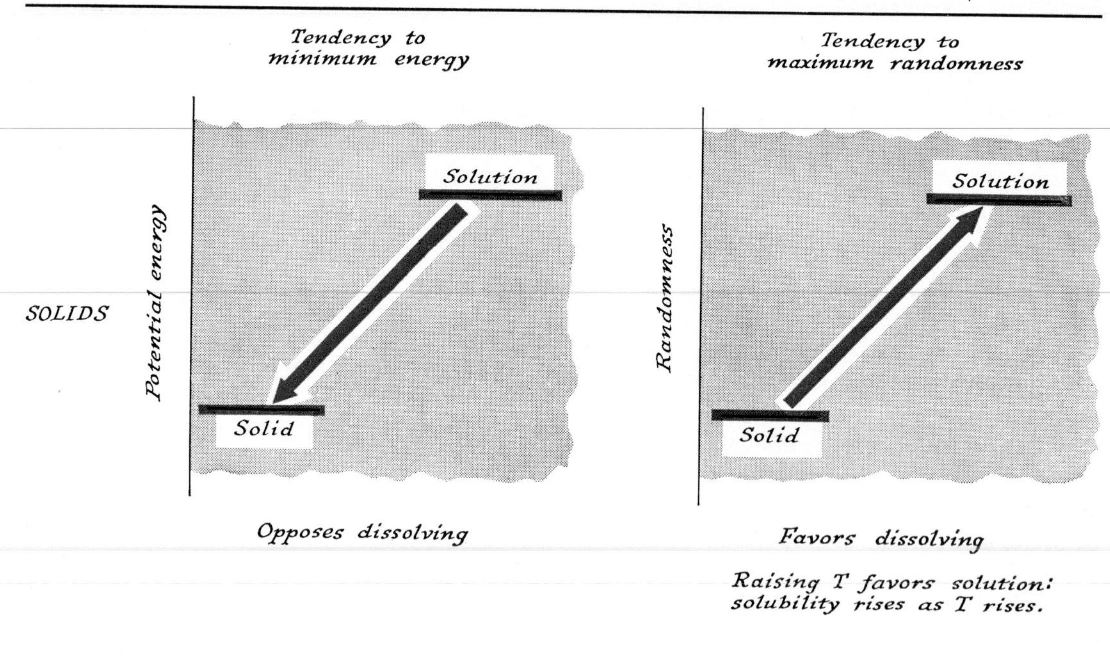

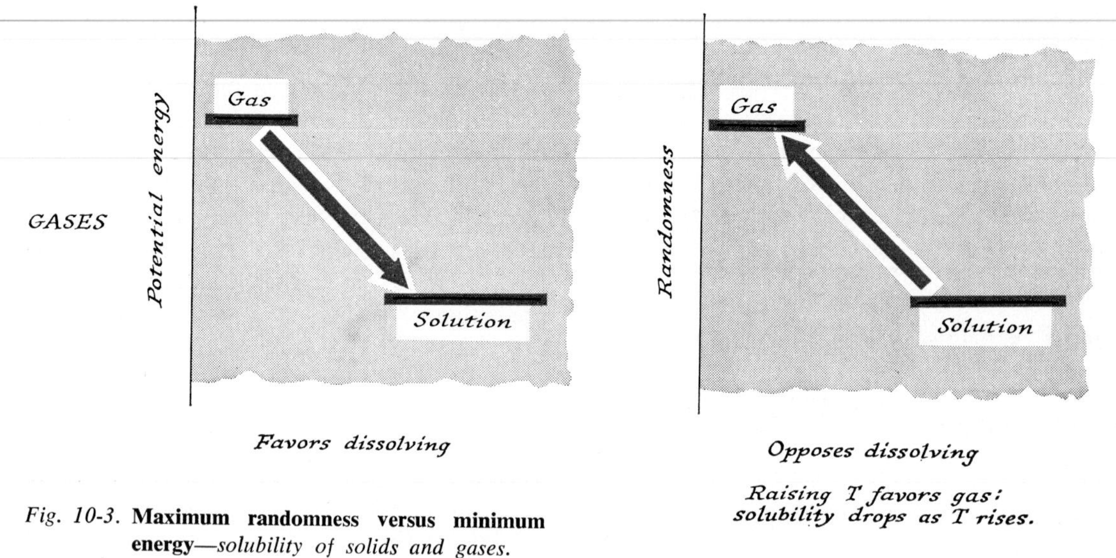

Fig. 10-3. **Maximum randomness versus minimum energy**—*solubility of solids and gases.*

10-2 AQUEOUS SOLUTIONS

Expression (*4*) is applicable to the solubility equilibria of some substances in water, but not of all. Contrast, for example, water solutions of sugar, salt, and hydrochloric acid. Sugar forms a molecular solid and, as it dissolves in water, the sugar molecules remain intact. These molecules leave the crystal and become a part of the liquid. This is exactly the situation we described for iodine in alcohol, and expression (*4*) is applicable to aqueous sugar solutions. But sodium chloride, NaCl, behaves quite differently. As salt dissolves, positively charged sodium ions and negatively charged chloride ions enter the solution and these ions behave quite independently:

$$NaCl(s) \longrightarrow Na^+(aq) + Cl^-(aq) \qquad (15)$$

Hydrochloric acid, HCl, is similar. This substance is a gas at normal conditions. At very low temperatures it condenses to a molecular solid. When HCl dissolves in water, positively charged hydrogen ions and negatively charged chloride ions are found in the solution. As with sodium chloride, a conducting solution containing ions is formed:

$$HCl(g) \longrightarrow H^+(aq) + Cl^-(aq) \qquad (16)$$

Substances like $NaCl(s)$ and $HCl(g)$ that dissolve in water to form conducting solutions are called **electrolytes.** The conduction involves movement of the ions through the solution, positive ions moving in one direction and negative ions in the other. This shows that the positive and negative ions behave independently. In view of this independence of the ions, solubility behavior in an electrolyte solution is more complicated than that given by expression (*4*). We shall find that equilibrium principles are correspondingly more important.

10-2.1 Types of Compounds That Are Electrolytes

The ions in an electrolyte solution can arise in two major ways. They may already be present in the pure compound, as in ionic solids. When such a solid is placed in water, the ions separate and move throughout the solution.* However, some compounds that form ions in water are *not* considered to contain ions when pure, whether in the solid, liquid, or gas phase. Hydrochloric acid, HCl, and sulfuric acid, H_2SO_4, are good examples of the second type of compound. They form molecular liquids (or solids, if cold enough). But in water they form ions: HCl gives hydrogen ion, $H^+(aq)$, and chloride ion, $Cl^-(aq)$; H_2SO_4 gives hydrogen ion, $H^+(aq)$, hydrogen sulfate or bisulfate ion, HSO_4^-, and sulfate ion, SO_4^{-2}:

$$HCl(g) + \text{water} \longrightarrow H^+(aq) + Cl^-(aq) \qquad (16)$$

$$H_2SO_4(l) + \text{water} \longrightarrow H^+(aq) + HSO_4^-(aq) \qquad (17)$$

$$HSO_4^-(aq) \rightleftarrows H^+(aq) + SO_4^{-2}(aq) \qquad (18)$$

However they are formed, and from whatever source, aqueous ions are individual species with properties not possessed by the materials from which they came. Furthermore, the properties of a particular kind of ion are independent of the source. Chloride ions from sodium chloride, $NaCl(s)$, have the same properties as chloride ions in an aqueous solution of hydrochloric acid, HCl. In a mixture of the two, all of the chloride ions act alike; none "remembers" whether it entered the solution from an ionic NaCl lattice or from a gaseous HCl molecule.

Since the properties of an ionic solution (that is, a solution containing ions) differ in important ways from those of nonconducting solutions, it is important to be able to predict which substances are likely to form ionic solutions in water. The periodic table guides us.

In Chapter 6 we saw that the chemistry of sodium can be understood in terms of the special stability of the inert gas electron population of neon. An electron can be pulled away from a sodium atom relatively easily to form a sodium ion, Na^+. Chlorine, on the other hand, readily accepts an electron to form chloride ion, Cl^-, achieving the inert gas population of argon. When sodium and chlorine react, the product, sodium chloride, is an ionic solid, made up of Na^+ ions and Cl^- ions packed in a regular lattice. Sodium chloride dissolves in water to give $Na^+(aq)$ and $Cl^-(aq)$ ions. Sodium chloride is an electrolyte; it forms a conducting solution in water.

This example illustrates the guiding principles. Sodium is a metal—electrons can be pulled away from sodium relatively easily to form positive ions. Chlorine is a nonmetal—it tends to accept electrons readily to form negative ions. When a metallic element reacts with a nonmetallic element, the resulting compound usually forms a conducting solution when dissolved in water.

* Liquids that form conducting solutions are called **ionizing solvents.** A few other compounds (ammonia, NH_3, sulfur dioxide, SO_2, sulfuric acid, H_2SO_4, etc.) are "ionizing solvents" but water is by far the most important. We will discuss water exclusively but the same ideas apply to the other solvents in which ions form.

Positive ions that form soluble compounds with ALL anions

H^+																	He
Li^+	Be^{+2}											B	CO_3^{-2}	NH_4^+ / NO_3^-	OH^-	F^-	Ne
Na^+	Mg^{+2}											Al^{+3}	Si	PO_4^{-3}	S^{-2} / SO_4^{-2}	Cl^-	Ar
K^+	Ca^{+2}	Sc^{+3}	Ti	V	Cr^{+3}	Mn^{+2}	Fe^{+2} / Fe^{+3}	Co^{+2}	Ni^{+2}	Cu^+ / Cu^{+2}	Zn^{+2}	Ga^{+3}	Ge	As^{+3} / As^{+5}	Se	Br^-	Kr
Rb^+	Sr^{+2}	Y^{+3}	Zr	Nb	Mo	Tc	Ru	Rh	Pd	Ag^+	Cd^{+2}	In^{+3}	Sn^{+2} / Sn^{+4}	Sb^{+3} / Sb^{+5}	Te	I^-	Xe
Cs^+	Ba^{+2}	La^{+3} & R.E.	Hf	Ta	W	Re	Os	Ir	Pt	Au	Hg_2^{+2} / Hg^{+2}	Tl^+	Pb^{+2}	Bi^{+3}	Po	At^-	Rn
Fr^+	Ra^{+2}																

Fig. 10-4. **Almost all compounds** *of the alkalies, hydrogen ion, and ammonium ion are soluble in water.*

The metals are found toward the left side of the periodic table and the nonmetals are at the right side. *A compound containing elements from the opposite sides of the periodic table can be expected to form a conducting solution when dissolved in water.* Notice from our examples that hydrogen reacts with nonmetals to form compounds that give conducting solutions in water. In this sense, hydrogen acts like a metallic element.

EXERCISE 10-3

Using the periodic table as a guide, predict which of the following compounds form ionic solutions in water: silicon carbide, SiC; magnesium bromide, $MgBr_2$; carbon tetrabromide, CBr_4; chromic chloride, $CrCl_3$.

10-2.2 A Qualitative View of Aqueous Solubilities

Hereafter in this chapter we shall be concerned exclusively with substances that form ionic solutions in water. Since each substance is electrically neutral before it dissolves, it must form ions of positive charge and, as well, ions of negative charge. *Ions with positive charges are called* ***cations****. Ions with negative charges are called* ***anions***. A conducting solution is electrically neutral; it contains both anions and cations.

First, let us consider substances with high solubility. As was stated on p. 73, chemists consider a substance to be soluble if it dissolves to a concentration exceeding one-tenth of a mole per liter (0.1 *M*) at room temperature. Using this meaning of the word soluble, we can say that some cations (positive ions) form soluble compounds with *all* anions (negative ions). These cations are the hydrogen ion, $H^+(aq)$, ammonium ion, NH_4^+, and the alkali ions, Li^+, Na^+, K^+, Rb^+, Cs^+, Fr^+. Figure 10-4 shows the placement of these ions in the periodic table.

The same sort of remark can be made about two anions (negative ions). Almost all compounds involving nitrate ion, NO_3^-, and acetate ion, CH_3COO^-, are soluble in water.*

Other anions (negative ions) form compounds of high solubility in water with some metal cations (positive ions) and compounds of low solubilities with others. Figure 10-5 indicates, for five anions, the metal ions that form compounds of *low* solubilities. Figure 10-5A refers to chlorides, bromides, and iodides, Cl^-, Br^-, and I^-; Figure 10-5B refers to sulfates, SO_4^{-2}; Figure 10-5C refers to sulfides, S^{-2}. Notice the difference between Figure 10-4 and Figure 10-5. The cross-

* There are a few compounds of alkalies, nitrate, and acetate that have low solubilities, but most of them are quite complex in composition. For example, sodium uranyl acetate, $NaUO_2(CH_3COO)_3$ has low solubility. Silver acetate is an exception but its solubility is moderate.

hatching in Figure 10-4 identifies metal ions that form *soluble* compounds. Figure 10-5 identifies the ones with *low* solubility.

Figure 10-6 continues this pictorial presentation of solubilities. Figure 10-6A shows the positive ions that form hydroxides of low solubility. Figure 10-6B shows the positive ions that have low solubility when combined with phosphate ion, PO_4^{-3}, carbonate ion, CO_3^{-2}, and sulfite ion, SO_3^{-2}.

These figures furnish a handy summary of solubility behavior. We see from Figure 10-5A that few chlorides have low solubilities. The few that do contain cations of metals clustered toward the right side of the periodic table (silver ion, Ag^+, cuprous ion, Cu^+, mercurous ion, Hg_2^{+2}, and lead ion, Pb^{+2}) but they do not fall in a single column. This irregularity is not unusual in solubility behavior and is seen again in Figures 10-5B and 10-6A. In these two figures, the elements in the second column (the alkaline earths) show a trend in behavior. Thus, beryllium and magnesium ions (Be^{+2} and Mg^{+2}) form soluble sulfates. The others—calcium, strontium, barium, and radium ions (Ca^{+2}, Sr^{+2}, Ba^{+2}, and Ra^{+2})—form sulfates with low solubilities. Just the opposite behavior is seen in Figure 10-6A for the compounds of these same elements with hydroxide ion, OH^-. As for the elements in the middle of the periodic table, we see that they form compounds of low solubilities with the ions: sulfide, S^{-2}, hydroxide, OH^-, phosphate, PO_4^{-3}, carbonate, CO_3^{-2}, and sulfite, SO_3^{-2}.

The information contained in Figures 10-4, 10-5, and 10-6 is collected for reference in Table 10-I.

Table 10-I. **SOLUBILITY OF COMMON COMPOUNDS IN WATER**

NEGATIVE IONS (Anions)	+	POSITIVE IONS (Cations)	FORM	COMPOUNDS WITH SOLUBILITY:
All		Alkali ions Li^+, Na^+, K^+, Rb^+, Cs^+, Fr^+		Soluble
All		Hydrogen ion, $H^+(aq)$		Soluble
All		Ammonium ion, NH_4^+		Soluble
Nitrate, NO_3^-		All		Soluble
Acetate, CH_3COO^-		All		Soluble
Chloride, Cl^-; Bromide, Br^-; Iodide, I^-		Ag^+, Pb^{+2}, Hg_2^{+2}, Cu^+		Low Solubility
		All others		Soluble
Sulfate, SO_4^{-2}		Ba^{+2}, Sr^{+2}, Pb^{+2}		Low Solubility
		All others		Soluble
Sulfide, S^{-2}		Alkali ions, $H^+(aq)$, NH_4^+, Be^{+2},		Soluble
		Mg^{+2}, Ca^{+2}, Sr^{+2}, Ba^{+2}		Soluble
		All others		Low Solubility
Hydroxide, OH^-		Alkali ions, $H^+(aq)$, NH_4^+, Sr^{+2}, Ba^{+2}		Soluble
		All others		Low Solubility
Phosphate, PO_4^{-3}; Carbonate, CO_3^{-2}; Sulfite, SO_3^{-2}		Alkali ions, $H^+(aq)$, NH_4^+		Soluble
		All others		Low Solubility

A

Positive ions that form compounds of LOW solubility with Cl^-, Br^-, I^-

H^+																	He
Li^+	Be^{+2}											B	CO_3^{-2}	NH_4^+/NO_3^-	OH^-	F^-	Ne
Na^+	Mg^{+2}											Al^{+3}	Si	PO_4^{-3}	S^{-2}/SO_4^{-2}	Cl^-	Ar
K^+	Ca^{+2}	Sc^{+3}	Ti	V	Cr^{+3}	Mn^{+2}	Fe^{+2}/Fe^{+3}	Co^{+2}	Ni^{+2}	Cu^+/Cu^{+2}	Zn^{+2}	Ga^{+3}	Ge	As^{+3}/As^{+5}	Se	Br^-	Kr
Rb^+	Sr^{+2}	Y^{+3}	Zr	Nb	Mo	Tc	Ru	Rh	Pd	Ag^+	Cd^{+2}	In^{+3}	Sn^{+2}/Sn^{+4}	Sb^{+3}/Sb^{+5}	Te	I^-	Xe
Cs^+	Ba^{+2}	La^{+3} & R.E.	Hf	Ta	W	Re	Os	Ir	Pt	Au	Hg_2^{+2}/Hg^{+2}	Tl^+	Pb^{+2}	Bi^{+3}	Po	At^-	Rn
Fr^+	Ra^{+2}																

B

Positive ions that form compounds of LOW solubility with SO_4^{-2}

H^+																	He
Li^+	Be^{+2}											B	CO_3^{-2}	NH_4^+/NO_3^-	OH^-	F^-	Ne
Na^+	Mg^{+2}											Al^{+3}	Si	PO_4^{-3}	S^{-2}/SO_4^{-2}	Cl^-	Ar
K^+	Ca^{+2}	Sc^{+3}	Ti	V	Cr^{+3}	Mn^{+2}	Fe^{+2}/Fe^{+3}	Co^{+2}	Ni^{+2}	Cu^+/Cu^{+2}	Zn^{+2}	Ga^{+3}	Ge	As^{+3}/As^{+5}	Se	Br^-	Kr
Rb^+	Sr^{+2}	Y^{+3}	Zr	Nb	Mo	Tc	Ru	Rh	Pd	Ag^+	Cd^{+2}	In^{+3}	Sn^{+2}/Sn^{+4}	Sb^{+3}/Sb^{+5}	Te	I^-	Xe
Cs^+	Ba^{+2}	La^{+3} & R.E.	Hf	Ta	W	Re	Os	Ir	Pt	Au	Hg_2^{+2}/Hg^{+2}	Tl^+	Pb^{+2}	Bi^{+3}	Po	At^-	Rn
Fr^+	Ra^{+2}																

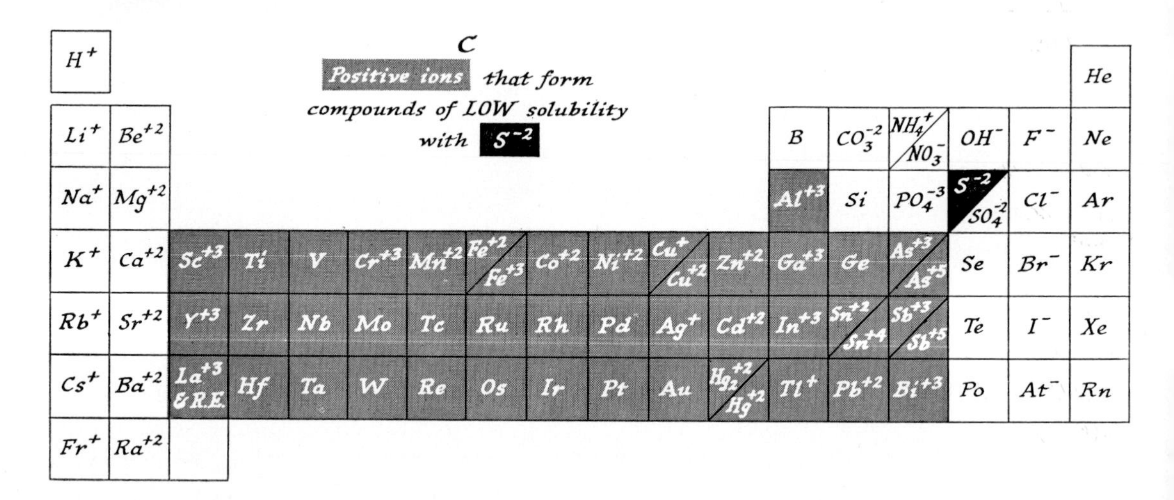

Fig. 10-5. **Positive ions** *forming compounds of low solubilities with various anions.*

A

Positive ions that form compounds of LOW solubility with hydroxide ion, OH^-

H^+																	He
Li^+	Be^{+2}											B	CO_3^{-2}	NH_4^+ / NO_3^-	OH^-	F^-	Ne
Na^+	Mg^{+2}											Al^{+3}	Si	PO_4^{-3}	S^{-2} / SO_4^{-2}	Cl^-	Ar
K^+	Ca^{+2}	Sc^{+3}	Ti	V	Cr^{+3}	Mn^{+2}	Fe^{+2} / Fe^{+3}	Co^{+2}	Ni^{+2}	Cu^+ / Cu^{+2}	Zn^{+2}	Ga^{+3}	Ge	As^{+3} / As^{+5}	Se	Br^-	Kr
Rb^+	Sr^{+2}	Y^{+3}	Zr	Nb	Mo	Tc	Ru	Rh	Pd	Ag^+	Cd^{+2}	In^{+3}	Sn^{+2} / Sn^{+4}	Sb^{+3} / Sb^{+5}	Te	I^-	Xe
Cs^+	Ba^{+2}	La^{+3} & R.E.	Hf	Ta	W	Re	Os	Ir	Pt	Au	Hg_2^{+2} / Hg^{+2}	Tl^+	Pb^{+2}	Bi^{+3}	Po	At^-	Rn
Fr^+	Ra^{+2}																

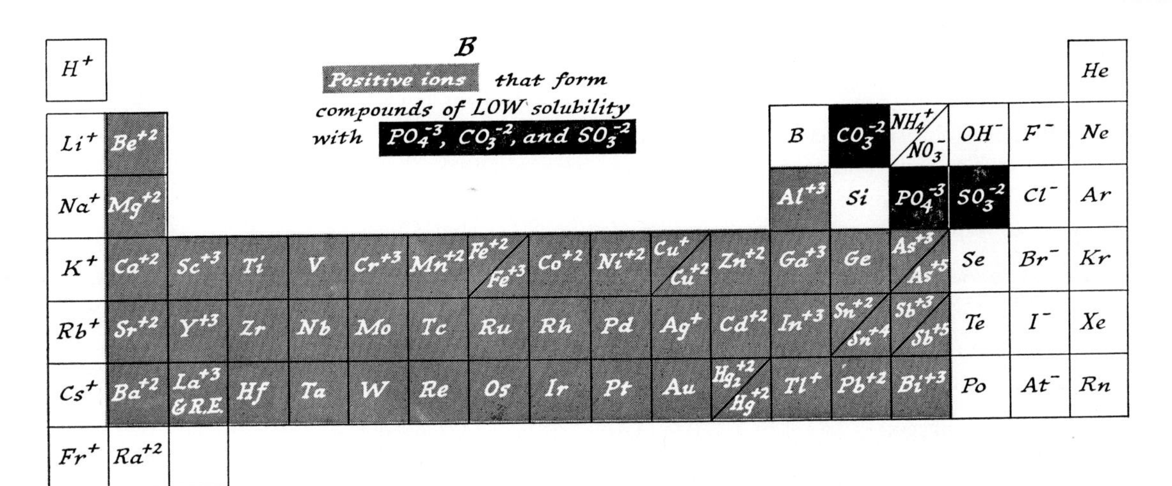

Fig. 10-6. **More positive ions** *forming compounds of low solubilities with various anions.*

EXERCISE 10-4

Use Figures 10-4, 10-5, and 10-6 to decide the solubility of each of the compounds listed below. Write "sol" if the compound is soluble and "low" if it has low solubility.

$Mg(NO_3)_2$, $MgCl_2$, $MgSO_4$, $Mg(OH)_2$, $MgCO_3$,
$Ca(NO_3)_2$, $CaCl_2$, $CaSO_4$, $Ca(OH)_2$, $CaCO_3$,
$Sr(NO_3)_2$, $SrCl_2$, $SrSO_4$, $Sr(OH)_2$, $SrCO_3$.

EXERCISE 10-5

Decide the formula of each of the following compounds and indicate those with low solubility in water. Silver carbonate; aluminum chloride; aluminum hydroxide; cuprous chloride (the chloride of Cu^+); cupric chloride (the chloride of Cu^{+2}); ammonium bromide.

10-2.3 The Equilibrium Law

Table 10-I and the schematic presentations of Figures 10-4, 10-5, and 10-6 are useful for a quick and qualitative view of solubilities. But chemists are not satisfied with the statement that a substance has low solubility. We must know *how much* of the substance dissolves. We must be able to treat solubility in a quantitative fashion. Fortunately this can be done with the aid of the principles of equilibrium, developed in Chapter 9.

As mentioned earlier, the quantitative concentration relationship that exists at equilibrium is shown in the Equilibrium Law Relation:

$$K = \frac{[E]^e[F]^f \cdots}{[A]^a[B]^b \cdots} \qquad (2)$$

Expression (*2*) applies to a solubility equilibrium, provided we write the chemical reaction to show the important molecular species present. In Section 10-1 we considered the solubility of iodine in alcohol. Since iodine dissolves to give a solution containing molecules of iodine, the concentration of iodine itself fixed the solubility. The situation is quite different for substances that dissolve to form ions. When silver chloride dissolves in water, no molecules of silver chloride, AgCl, seem to be present. Instead, silver ions, Ag^+, and chloride ions, Cl^-, are found in the solution. The concentrations of these species, Ag^+ and Cl^-, are the ones which fix the equilibrium solubility. The counterpart of equation (*1*) will be

$$AgCl(s) \rightleftharpoons Ag^+(aq) + Cl^-(aq) \qquad (19)$$

and equilibrium will exist when the concentrations are in agreement with the expression

$$K = \text{a constant} = [Ag^+] \times [Cl^-] \qquad (20)$$

Just as in expression (*4*), the "concentration" of the solid (silver chloride) does not appear in the equilibrium expression (*20*); it does not vary.

To consider a more complicated example, consider the application of expression (*2*) to the solubility of lead chloride, $PbCl_2$:

$$PbCl_2(s) \rightleftharpoons Pb^{+2}(aq) + 2Cl^-(aq) \qquad (21)$$

At equilibrium,

$$K = \text{a constant} = [Pb^{+2}] \times [Cl^-]^2 \qquad (22)$$

Solubility equilibrium constants, such as (*20*) and (*22*), are given a special name—the **solubility product.** It is symbolized K_{sp}. A low value of K_{sp} means the concentrations of ions are low at equilibrium. Hence the solubility must be low. Table 10-II lists solubility products for some common compounds.

EXERCISE 10-6

Write the equation for the dissolving of calcium sulfate, $CaSO_4$, and the solubility product expression.

EXERCISE 10-7

Write the equation for the dissolving of silver chromate, Ag_2CrO_4, and the solubility product expression. Silver chromate dissolves to give Ag^+ and CrO_4^{-2} ions.

10-2.4 Calculation of the Solubility of Cuprous Chloride in Water

The solubility product is learned from measurements of the solubility. In turn, it can be used as a basis for calculations of solubility. Suppose we wish to know how much cuprous chloride, CuCl, will dissolve in one liter of water. We begin by writing the balanced equation for the reaction:

$$CuCl(s) \rightleftharpoons Cu^+(aq) + Cl^-(aq) \qquad (23)$$

From this equation, we can write the equilibrium expression:

$$K_{sp} = [Cu^+][Cl^-] \qquad (24)$$

Now the numerical value of K_{sp} is found in Table 10-II:

$$K_{sp} = 3.2 \times 10^{-7} = [Cu^+][Cl^-] \qquad (25)$$

Table 10-II

SOME SOLUBILITY PRODUCTS AT ROOM TEMPERATURE

TlCl	1.9×10^{-4}	$SrCrO_4$	3.6×10^{-5}
CuCl	3.2×10^{-7}	$BaCrO_4$	8.5×10^{-11}
AgCl	1.7×10^{-10}	$PbCrO_4$	2×10^{-16}
TlBr	3.6×10^{-6}	$CaSO_4$	2.4×10^{-5}
CuBr	5.9×10^{-9}	$SrSO_4$	7.6×10^{-7}
AgBr	5.0×10^{-13}	$PbSO_4$	1.3×10^{-8}
		$BaSO_4$	1.5×10^{-9}
TlI	8.9×10^{-8}	$RaSO_4$	4×10^{-11}
CuI	1.1×10^{-12}		
AgI	8.5×10^{-17}	$AgBrO_3$	5.4×10^{-5}
		$AgIO_3$	3.1×10^{-8}

Expression (*25*) indicates that cuprous chloride dissolves, according to reaction (*23*), until the molar concentrations of cuprous ion and chloride ion rise enough to make their product equal to 3.2×10^{-7}.

Now is the time to dust off the algebra and put it to work. Suppose we designate the solubility of cuprous chloride in water by a symbol, *s*. This symbol *s* equals the number of moles of solid cuprous chloride that dissolve in one liter of water. Remembering equation (*23*), we see that *s* moles of solid cuprous chloride will produce *s* moles of cuprous ion, Cu^+, and *s* moles of chloride ion, Cl^-. Hence these concentrations must be equal, as shown below.

$$[Cu^+] = [Cl^-] = s \text{ moles/liter} = \text{moles CuCl dissolved} \quad (26)$$

Substituting (*26*) into (*25*), we have

$$K_{sp} = 3.2 \times 10^{-7} = (s) \times (s) = s^2$$
$$s^2 = 3.2 \times 10^{-7} = 32 \times 10^{-8}$$
$$s = \sqrt{32 \times 10^{-8}} = 5.7 \times 10^{-4} \text{ mole/liter} = 0.00057\ M \quad (27)$$

EXERCISE 10-8

Calculate the solubility, in moles per liter, of calcium sulfate in water, using the solubility product given in Table 10-II.

10-2.5 Will a Precipitate Form?

When two solutions are mixed, a precipitate may form. For example, suppose solutions of calcium chloride, $CaCl_2$, and sodium sulfate, Na_2SO_4, are mixed. The mixture contains both calcium ions, Ca^{+2}, and sulfate ions, SO_4^{-2}, so solid calcium sulfate may form. The solubility product permits us to predict with confidence whether it will or not.

Let us consider two cases to show how the prediction is made:

(1) Equal volumes of 0.02 *M* $CaCl_2$ and 0.0004 *M* Na_2SO_4 are mixed.
(2) Equal volumes of 0.08 *M* $CaCl_2$ and 0.02 *M* Na_2SO_4 are mixed.

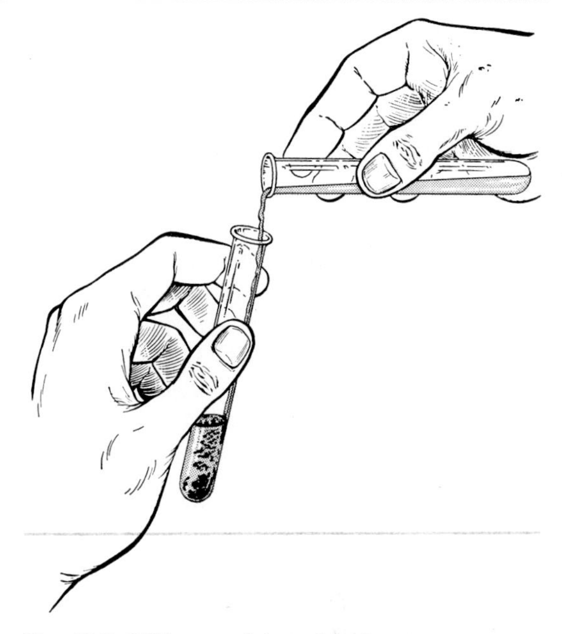

Fig. 10-7. **Will a precipitate form?**

Will a precipitate form in either case? In both?

The first step is to write the balanced equation for the reaction of calcium sulfate dissolving in water and then use the Equilibrium Law:

$$CaSO_4(s) \rightleftharpoons Ca^{+2}(aq) + SO_4^{-2}(aq) \quad (28)$$

$$K_{sp} = [Ca^{+2}][SO_4^{-2}] \quad (29)$$

The next step is to find the concentration of each ion in the final mixture. After we mix equal volumes, each ion is present in twice as much solution so the concentration is only half as great as before mixing. Therefore, in case 1,

$$[Ca^{+2}] = \frac{0.02\ M}{2} = 0.01\ M = 1 \times 10^{-2}\ M$$

$$[SO_4^{-2}] = \frac{0.0004\ M}{2} = 0.0002\ M = 2 \times 10^{-4}\ M$$

A trial value of the ion product must be compared to K_{sp}:

$$[Ca^{+2}] \times [SO_4^{-2}] = (1 \times 10^{-2}) \times (2 \times 10^{-4}) = 2 \times 10^{-6}$$

The trial product, 2×10^{-6}, is *less* than $K_{sp} = 2.4 \times 10^{-5}$ so *precipitation will not occur in the mixture of case 1*. In case 2,

$$[Ca^{+2}] = \frac{0.08\ M}{2} = 0.04\ M = 4 \times 10^{-2}\ M$$

$$[SO_4^{-2}] = \frac{0.02\ M}{2} = 0.01\ M = 1 \times 10^{-2}\ M$$

Again we calculate a trial value of the ion product,

$$[Ca^{+2}] \times [SO_4^{-2}] = (4 \times 10^{-2}) \times (1 \times 10^{-2}) = 4 \times 10^{-4}$$

This time the trial product, 4×10^{-4}, is *greater* than $K_{sp} = 2.4 \times 10^{-5}$ so a *precipitate does form.* Solid calcium sulfate, $CaSO_4$, will continue to form, lowering the concentrations $[Ca^{+2}]$ and $[SO_4^{-2}]$ until they are low enough that the ion product equals K_{sp}. Then equilibrium exists and no more precipitation occurs.

EXERCISE 10-9

A 50 ml volume of 0.04 *M* $Ca(NO_3)_2$ solution is added to 150 ml of 0.008 *M* $(NH_4)_2SO_4$ solution. Show that a trial value of the calcium sulfate ion product is 6×10^{-5}. Will a precipitate form?

10-2.6 Precipitations Used for Separations

A chemist is often interested in separating substances in a solution mixture. Such a problem is solved by applying equilibrium considerations.

Suppose we have a solution known to contain both lead nitrate, $Pb(NO_3)_2$, and magnesium nitrate, $Mg(NO_3)_2$. The lead and magnesium can be separated by removing from the solution all of the lead ion, Pb^{+2}, as a solid lead compound. We must avoid, of course, precipitation of any magnesium compound. Consulting Figure 10-5 or Table 10-I, we see that lead ion and sulfate ion form a compound with low solubility. If enough sodium sulfate, Na_2SO_4, is added, lead sulfate, $PbSO_4$, will precipitate. Since Figure 10-5B and Table 10-I indicate that magnesium sulfate, $MgSO_4$, is soluble, there will be no precipitate of $MgSO_4$. The solid can be removed from the liquid by filtration and the desired separation has been obtained.

EXERCISE 10-10

Use Figures 10-5 and 10-6 or Table 10-I to decide which of the following soluble salts would permit a separation of magnesium and lead through a precipitation reaction: sodium iodide, NaI; sodium sulfide, Na_2S; sodium carbonate, Na_2CO_3.

QUESTIONS AND PROBLEMS

1. Sugar is added to a cup of coffee until no more sugar will dissolve. Does addition of another spoonful of sugar increase the rate at which the sugar molecules leave the crystal phase and enter the liquid phase? Will the sweetness of the liquid be increased by this addition? Explain.

2. In view of the discussion of the factors that determine the rate of dissolving (Section 10-1.2), propose two methods for increasing the rate at which sugar dissolves in water.

3. When a solid evaporates directly (without melting), the process is called **sublimation.** Evaporation of "dry ice" (solid CO_2) is a familiar example. Two other substances that sublime are FCN and ICN:

   $$FCN(s) \rightleftharpoons FCN(g) \qquad \Delta H = +5.7 \text{ kcal}$$
   $$ICN(s) \rightleftharpoons ICN(g) \qquad \Delta H = +14.2 \text{ kcal}$$

   (a) In sublimation, does the tendency toward maximum randomness favor solid or gas?
   (b) In sublimation, does the tendency toward minimum energy favor solid or gas?
   (c) The vapor pressure of solid FCN is 760 mm at 201°K. In view of part *b*, would you expect solid ICN to have a lower or higher vapor pressure than solid FCN at this same temperature, 201°K?

4. Liquid chloroform, $CHCl_3$, and liquid acetone, CH_3COCH_3, dissolve in each other in all proportions (they are said to be **miscible**).

   (a) When pure $CHCl_3$ is mixed with pure acetone, is randomness increased or decreased?

(b) Does the tendency toward maximum randomness favor reactants or product in the reaction:

$$CHCl_3(l) + CH_3COCH_3(l) \longrightarrow 1{:}1 \text{ solution} \qquad \Delta H = -495 \text{ cal}$$

(c) Considering the sign of ΔH shown in part *b*, does the tendency toward minimum energy favor reactants or product?

(d) In view of your answers to parts *b* and *c*, discuss the experimental fact that these two liquids are miscible.

5. Which of the following substances can be expected to dissolve in the indicated solvent to form, primarily, ions? Which would form molecules?

(a) sucrose in water.
(b) RbBr in water.
(c) $CHCl_3$ in water.
(d) $CsNO_3$ in water.
(e) HNO_3 in water.
(f) S_8 in carbon disulfide, CS_2.
(g) ICl in ethyl alcohol.

6. Which of the substances listed in Problem 5 would be called electrolytes?

7. Assume the following compounds dissolve in water to form separate, mobile ions in solution. Write the formulas and names for the ions that can be expected.

(a) HI
(b) $CaCl_2$
(c) Na_2CO_3
(d) $Ba(OH)_2$
(e) KNO_3
(f) NH_4Cl

8. Write the equation for the reaction that occurs when each of the following electrolytes is dissolved in water.

(a) lithium hydroxide (solid).
(b) nitric acid (liquid).
(c) potassium sulfate (solid).
(d) sodium nitrate (solid).
(e) ammonium iodide (solid).
(f) potassium carbonate (solid).

Answer. (a) $LiOH(s) \longrightarrow Li^+(aq) + OH^-(aq)$.

9. What would you expect to happen if equal volumes of 0.1 *M* $MgSO_4$ and 0.1 *M* $ZnCl_2$ were mixed together?

10. Predict what would happen if equal volumes of 0.2 *M* Na_2SO_3 and 0.2 *M* $MgSO_4$ were mixed. If a reaction takes place, write the net ionic equation.

11. Using Figures 10-4 to 10-6 (or Table 10-I), make a statement about the solubilities of the compounds containing the following ions.

Anion	Cations
(a) carbonate, CO_3^{-2}	alkali ions, Li^+, Na^+, K^+, Rb^+, Cs^+
(b) carbonate, CO_3^{-2}	alkaline earth ions, Be^{+2}, Mg^{+2}, Ca^{+2}, Sr^{+2}, Ba^{+2}
(c) sulfide, S^{-2}	alkaline earth ions, Be^{+2}, Mg^{+2}, Ca^{+2}, Sr^{+2}, Ba^{+2}
(d) hydroxide, OH^-	the cations of the fourth row of the periodic table
(e) chloride, Cl^-	the cations of the fifth row of the periodic table

Answer. (a) All alkali carbonates are soluble.
(b) All alkaline earth carbonates have low solubilities.

12. Write the empirical formulas for each of the following compounds and indicate which have low solubilities.

(a) silver sulfide.
(b) potassium sulfide.
(c) ammonium sulfide.
(d) nickel sulfide.
(e) ferrous sulfide (Fe^{+2}).
(f) ferric sulfide (Fe^{+3}).

13. Write net ionic equations for any reactions that will occur upon mixing equal volumes of 0.2 *M* solutions of the following pairs of compounds.

(a) silver nitrate and ammonium bromide.
(b) $SrBr_2$ and $NaNO_3$.
(c) sodium hydroxide and aluminum chloride.
(d) NaI and $Pb(NO_3)_2$.
(e) barium chloride and sodium sulfate.

Answer. (a) $Ag^+(aq) + Br^-(aq) \longrightarrow AgBr(s)$.

14. What ions could be present in a solution if samples of it gave:

(a) A precipitate when either $Cl^-(aq)$ or $SO_4^{-2}(aq)$ is added?
(b) A precipitate when $Cl^-(aq)$ is added but none when $SO_4^{-2}(aq)$ is added?
(c) A precipitate when $SO_4^{-2}(aq)$ is added but none when $Cl^-(aq)$ is added?

15. What cations from the fourth row of the periodic table could be present in a solution with the following behavior.

(a) No precipitate is formed with hydroxide ion.
(b) A precipitate forms with hydroxide ion and with sulfate ion.
(c) A precipitate forms with hydroxide ion and with sulfide ion.
(d) A precipitate forms with carbonate ion, none with sulfide ion.

16. The solubility of silver chloride is so low that all but a negligible amount of it is precipitated when excess sodium chloride solution is added to silver nitrate solution. What would be the weight of the precipitate formed when 100 ml of 0.5 M NaCl is added to 50.0 ml of 0.100 M $AgNO_3$?

Answer. 0.715 gram.

17. Write the solubility product expression for each of the following reactions.

(a) $BaSO_4(s) \rightleftharpoons Ba^{+2}(aq) + SO_4^{-2}(aq)$
(b) $Zn(OH)_2(s) \rightleftharpoons Zn^{+2}(aq) + 2OH^-(aq)$
(c) $Ca_3(PO_4)_2(s) \rightleftharpoons 3Ca^{+2}(aq) + 2PO_4^{-3}(aq)$

18. Write the solubility product expression applicable to the solubility of each of the following substances in water.

(a) calcium carbonate.
(b) silver sulfide.
(c) aluminum hydroxide.

19. The solubility product of AgCl is 1.4×10^{-4} at 100°C. Calculate the solubility of silver chloride in boiling water.

20. Experiments show that 0.0059 gram of $SrCO_3$ will dissolve in 1.0 liter of water at 25°C. What is K_{sp} for $SrCO_3$?

Answer. 1.6×10^{-9}.

21. How many milligrams of silver bromide dissolve in 20 liters of water? (Use the data given in Table 10-II.)

22. To one liter of 0.001 M H_2SO_4 is added 0.002 mole of solid $Pb(NO_3)_2$. As the lead nitrate dissolves, will lead sulfate precipitate?

23. Suppose 10 ml of 1.0 M $AgNO_3$ is diluted to one liter with tap water. If the chloride concentration in the tap water is about 10^{-5} M, will a precipitate form?

24. The test described in Problem 23 does *not* give a precipitate if the laboratory distilled water is used. What is the maximum chloride concentration that could be present?

25. Will a precipitate exist at equilibrium if $\frac{1}{2}$ liter of a 2×10^{-3} M $AlCl_3$ solution and $\frac{1}{2}$ liter of a 4×10^{-2} M solution of sodium hydroxide are mixed and diluted to 10^3 liters with water at room temperature? ($K_{sp} = 5 \times 10^{-33}$.)

26. Use Figures 10-5 and 10-6 or Table 10-I to decide which of the following soluble substances would permit a separation of aqueous magnesium and barium ions. For those that are effective, write the equation for the reaction that occurs.

(a) ammonium carbonate.
(b) sodium bromide.
(c) potassium sulfate.
(d) sodium hydroxide.

27. To a solution containing 0.1 M of each of the ions Ag^+, Cu^+, Fe^{+2}, and Ca^{+2} is added 2 M NaBr solution, giving precipitate *A*. After filtration, a sulfide solution is added to the solution and a black precipitate forms, precipitate *B*. This precipitate is removed by filtration and 2 M sodium carbonate solution is added, giving precipitate *C*. What is the composition of each precipitate, *A*, *B*, and *C*?

CHAPTER 11

Aqueous Acids and Bases

The acids, · · · are compounded of two substances · · · the one constitutes acidity, and is common to all acids, · · · the other is peculiar to each acid, and distinguishes it from the rest · · ·

A. LAVOISIER, 1789

In Chapter 10 we used the principles of equilibrium to help us understand solubility in liquids. In such a system constituents in solution reach the dynamic balance of equilibrium with another phase, a solid or a gas. Equilibrium can also exist among two or more constituents present in the same solution. One of the examples already encountered (in Chapter 9 and in Experiment 15) is

$$Fe^{+3}(aq) + SCN^{-}(aq) \rightleftharpoons FeSCN^{+2}(aq) \quad (1)$$

for which experiment showed

$$K = \frac{[FeSCN^{+2}]}{[Fe^{+3}][SCN^{-}]} \quad (2)$$

In reaction (*1*), all of the molecular species involved in the equilibrium are in the solution as dissolved species. Though the equilibrium relationship that exists among the concentrations is a little more complicated than in the solubility product expressions, the guiding principles are the same.

In this chapter we shall explore some more equilibria like (*1*), in which all of the important species are dissolved. We will consider mainly those equilibria in which one of the ions is $H^{+}(aq)$. This type of equilibrium furnishes one of the most important classes of chemical reactions of all those that occur in water.

11-1 ELECTROLYTES—STRONG OR WEAK

In Section 10-2 we considered *electrolytes*, substances that dissolve in water to give solutions containing ions. Thus far, we have considered electrolytes such as $HCl(g)$ and $NaOH(s)$:

$$HCl(g) + \text{water} \longrightarrow H^{+}(aq) + Cl^{-}(aq) \quad (3)$$

$$NaOH(s) + \text{water} \longrightarrow Na^{+}(aq) + OH^{-}(aq) \quad (4)$$

According to reaction (*3*), when HCl gas dissolves in water, *all* of the HCl molecules break up, or *dissociate*, into ions, $H^{+}(aq)$ and $Cl^{-}(aq)$. There is no experimental evidence for the presence of molecules of HCl in aqueous hydrochloric acid solutions. In a similar way, there is no evidence for the presence of molecules of NaOH in aqueous solution—apparently the sodium hydroxide crystal breaks up completely into sodium ions, $Na^+(aq)$ and hydroxide ions, $OH^-(aq)$. *A substance that dissolves and exclusively gives ions is called a* **strong electrolyte.**

Not all substances that form conducting solutions break up, or dissociate, so completely. For example, vinegar is just an aqueous solution of acetic acid. Such a solution conducts electric current, showing that ions are present:

$$CH_3COOH \rightleftharpoons H^+(aq) + CH_3COO^-(aq) \quad (5)$$

The conductivity of a 0.1 *M* acetic acid solution is much lower, however, than that of a 0.1 *M* hydrogen chloride solution. This and other experiments show that only a small fraction of the dissolved acetic acid, CH_3COOH, has formed ions. Such *a substance that dissolves and dissociates to ions only to a limited extent is called a* **weak electrolyte.**

Fig. 11-1. **A strong electrolyte solution conducts better than a weak electrolyte solution.**

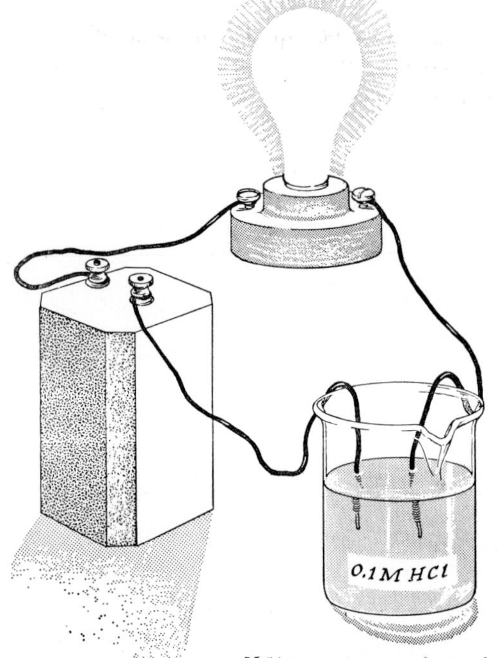

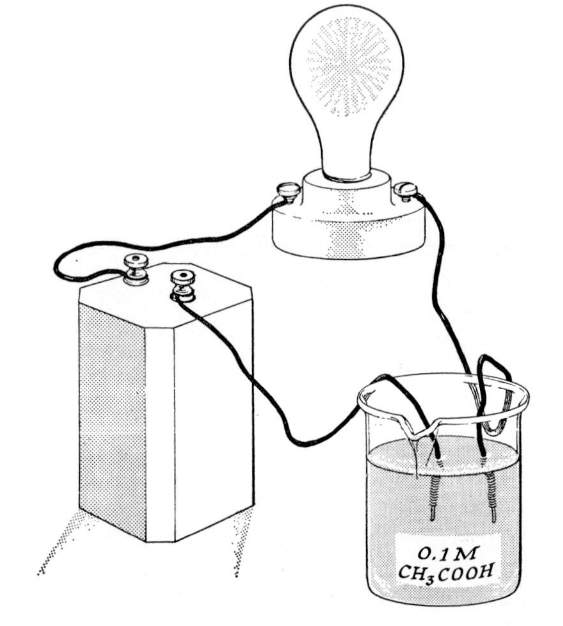

Careful measurements show that water is, itself, a weak electrolyte. We shall consider it first.

11-1.1 Water: A Weak Electrolyte

Pure water does not conduct electric current readily. Yet an extremely sensitive meter shows that even the purest water has a tiny conductivity. To conduct electric current, water must dissociate to a very small extent, forming ions. The ions prove to be hydrogen ion, $H^+(aq)$, and hydroxide ion, $OH^-(aq)$:

$$H_2O(l) \rightleftharpoons H^+(aq) + OH^-(aq) \quad (6)$$

This is an equilibrium involving three species in the liquid phase, H_2O, $H^+(aq)$, and $OH^-(aq)$. The equilibrium law can be written

$$K = \frac{[H^+][OH^-]}{[H_2O]} \quad (7)$$

However, as we have remarked in Section 9-2.2, the concentration of H_2O in water is so large, 55.5 *M*, and so few ions are formed that its concentration is virtually constant. Consequently, expression (7) is usually simplified by incorporating the factor 55.5 in the constant. We shall do this, and label the constant K_w to indicate that it includes the factor $[H_2O]$:

$$K_w = [H^+][OH^-] \quad (8)$$

$$K_w = [H_2O] \times K = 55.5 \times K \quad (9)$$

The magnitude of K_w is given in Table 9-IV at 25°C,

$$K_w = 1.00 \times 10^{-14} \quad (10)$$

Having this value of K_w, we can calculate the ion concentrations, using the same methods we applied to solubility.

In reaction (*6*) there is one $H^+(aq)$ ion formed for every $OH^-(aq)$ ion. Hence, in pure water where the only source of ions is reaction (*6*), the concentrations of $H^+(aq)$ and $OH^-(aq)$ must be equal. That is,

$$\text{in pure water, } [OH^-] = [H^+] \quad (11)$$

Substituting (*11*) into the equilibrium expression, we can calculate the concentrations of the two types of ions:

$$K_w = [H^+] \times [OH^-] = [H^+] \times [H^+] = [H^+]^2$$

or,

$$[H^+] = \sqrt{K_w} = \sqrt{1.00 \times 10^{-14}}$$

$$[H^+] = 1.00 \times 10^{-7}\ M \quad (12)$$

Because of (*11*), we can also write

$$[OH^-] = [H^+] = 1.00 \times 10^{-7}\ M \quad (13)$$

Now we can explain the low conductivity of pure water. Though water dissociates into ions, $H^+(aq)$ and $OH^-(aq)$, it does so only to a very slight extent. At equilibrium, the ion concentrations are only 10^{-7} *M*. *Water is a weak electrolyte.*

11-1.2 The Change of K_w with Temperature

Experiments show that reaction (*6*) absorbs energy:

$$H_2O(l) + 13.68 \text{ kcal} \longrightarrow H^+(aq) + OH^-(aq) \quad (14)$$

Even though only a minute fraction of the water present actually is dissociated at equilibrium, if we measure the energy effect and divide by the number of moles, we find that it takes 13.68 kcal per mole of water broken into ions.

This heat effect can be used in predicting how K_w changes with temperature. Le Chatelier's Principle indicates that increased temperature will shift the equilibrium condition toward larger concentrations of the ions (so as to absorb heat). Hence, K_w is expected to increase. This agrees with the experimental values for K_w, given for various temperatures in Table 11-I.

Table 11-I

VALUES OF K_w AT VARIOUS TEMPERATURES

TEMPERATURE (°C)	K_w
0	0.114×10^{-14}
10	0.295×10^{-14}
20	0.676×10^{-14}
25	1.00×10^{-14}
60	9.55×10^{-14}

11-1.3 The Special Roles of $H^+(aq)$ and $OH^-(aq)$ in Water

The equilibrium reaction (*6*) gives the two ions $H^+(aq)$ and $OH^-(aq)$ special roles in aqueous solutions:

$$H_2O(l) \rightleftharpoons H^+(aq) + OH^-(aq) \quad (6)$$

In pure water, where the only source of ions is reaction (*6*), the concentrations of $H^+(aq)$ and $OH^-(aq)$ must be equal. But what if we add some HCl to the solution? We have already noted that HCl is a strong electrolyte, dissolving to give the ions $H^+(aq)$ and $Cl^-(aq)$. Thus, hydrogen chloride adds $H^+(aq)$ but not $OH^-(aq)$ to the solution. The concentrations $[H^+]$ and $[OH^-]$ are no longer equal. However, they are still found to be "tied together" by the equilibrium relationship

$$K_w = [H^+][OH^-] \quad (8)$$

or,

$$[OH^-] = \frac{K_w}{[H^+]} \quad (15)$$

Expression (*15*) shows that as the concentration of H^+ rises (for example, as we add HCl to water), the concentration of $[OH^-]$ must decrease.

Suppose, on the other hand, that we add sodium hydroxide, NaOH, to pure water. Sodium hydroxide is also a strong electrolyte, adding

OH^- ions to the solution. Now we can rewrite (*8*) in the form

$$[H^+] = \frac{K_w}{[OH^-]} \qquad (16)$$

and we see that raising the concentration $[OH^-]$ lowers the concentration $[H^+]$.

In this way the two ions $H^+(aq)$ and $OH^-(aq)$ are connected in the chemistry of water. When $[H^+]$ is high, $[OH^-]$ must be low. When $[OH^-]$ is high, $[H^+]$ must be low.

Let's consider an example. Suppose we dissolve 0.10 mole of hydrogen chloride in 1.0 liter of water. Since HCl is a strong electrolyte, 0.10 mole of HCl forms 0.10 mole of $H^+(aq)$ ions and 0.10 mole of $Cl^-(aq)$ ions in the one liter volume. The concentrations of $H^+(aq)$ and $Cl^-(aq)$ due to the HCl are equal:*

$$[H^+] = [Cl^-] = 0.10\ M \qquad (17)$$

If the concentration $[H^+]$ is 0.10 *M*, expression (*15*) allows us to calculate the concentration $[OH^-]$:

$$[OH^-] = \frac{K_w}{[H^+]} = \frac{1.00 \times 10^{-14}}{0.10} = \frac{1.00 \times 10^{-14}}{1.0 \times 10^{-1}}$$

$$[OH^-] = 1.0 \times 10^{-13} \qquad (18)$$

We see that adding the 0.10 mole of HCl to 1.0 liter of water lowered the OH^- concentration from 10^{-7} *M* (its value in pure water) to 10^{-13} *M*, a change by a factor of a million!

EXERCISE 11-1

Show that the addition of 0.010 mole of solid NaOH to 1.0 liter of water reduces the concentration of $H^+(aq)$ to 1.0×10^{-12} *M*.

EXERCISE 11-2

Suppose that 3.65 grams of HCl are dissolved in 10.0 liters of water. What is the value of $[H^+]$? Use expression (*15*) to show that $[OH^-] = 1.00 \times 10^{-12}$ *M*.

Thus we see that the concentrations of the two ions, $H^+(aq)$ and $OH^-(aq)$, always remain related through the equilibrium relation (*8*). In pure water they are equal, $[H^+] = [OH^-]$. In a solution of HCl, in which $[H^+]$ is high, $[OH^-]$ must be low. In a solution of NaOH, in which $[OH^-]$ is high, $[H^+]$ must be low. Furthermore, as shown in the calculation (*18*) and Exercise 11-1, with rather low concentrations of HCl and NaOH, the concentration of $H^+(aq)$ [or $OH^-(aq)$] can be varied from 0.1 *M* to 10^{-13} *M*, by the immense factor of 10^{12} (a million million).

This ease with which we can control and vary the concentrations of $H^+(aq)$ and $OH^-(aq)$ would be only a curiosity but for one fact. *The ions $H^+(aq)$ and $OH^-(aq)$ take part in many important reactions that occur in aqueous solution.* Thus, if $H^+(aq)$ is a reactant or a product in a reaction, the variation of the concentration of hydrogen ion by a factor of 10^{12} can have an enormous effect. At equilibrium such a change causes reaction to occur, altering the concentrations of all of the other reactants and products until the equilibrium law relation again equals the equilibrium constant. Furthermore, there are many reactions for which either the hydrogen ion or the hydroxide ion is a catalyst. An example was discussed in Chapter 8, the catalysis of the decomposition of formic acid by sulfuric acid. Formic acid is reasonably stable until the hydrogen ion concentration is raised, then the rate of the decomposition reaction becomes very rapid.

In these ways, by affecting equilibrium conditions and by changing reaction rates, the concentrations of $H^+(aq)$ and $OH^-(aq)$ give us immense leverage in controlling the chemistry of aqueous solutions.

EXERCISE 11-3

The color of a solution of potassium chromate, K_2CrO_4, changes to the color of a solution of potassium dichromate, $K_2Cr_2O_7$, when a few drops of HCl solution are added. Write the balanced equation for the reaction between $CrO_4^{-2}(aq)$ and $H^+(aq)$ to produce $Cr_2O_7^{-2}$ and explain the color change on the basis of Le Chatelier's Principle.

---

* There is a small additional concentration of $H^+(aq)$, owing to the dissociation of water, but it will be quite negligible compared with the 0.10 *M* concentration provided by the HCl.

11-2 EXPERIMENTAL INTRODUCTION TO ACIDS AND BASES

The significant influence that $H^+(aq)$ and $OH^-(aq)$ exert on aqueous solution chemistry was recognized long ago. Consequently chemists have long identified by the class names "acids" and "bases" those substances that change the concentrations of these two ions. We shall see, later in this chapter, that the meanings of the terms "acid" and "base" are evolving and have become more general in modern usage. This is a natural and desirable development as chemists attempt to relate the chemistry of aqueous solutions to the chemistry that occurs in other solvents and in other phases. For the moment, however, we shall keep our attention focused on aqueous solutions and consider how a chemist recognizes acids and bases.

11-2.1 Properties of Aqueous Solutions of Acids

Consider the following compounds:

HCl : hydrochloric acid (or, hydrogen chloride),
HNO_3 : nitric acid,
CH_3COOH : acetic acid,
H_2SO_4 : sulfuric acid,
H_3PO_4 : phosphoric acid.

Each of these five compounds is called an acid. They all share this name because they have the following important properties in common.

Hydrogen Containing. Each of these compounds contains hydrogen.

Fig. 11-2. **Some familiar acids.**

Electrical Conductivity. Each of these compounds dissolves in water to form solutions that conduct electricity. Ions are present in these aqueous solutions.

Liberation of Hydrogen Gas. The aqueous solutions of each compound produce hydrogen gas, H_2, if zinc metal is added.

Color of Litmus. Litmus, a dye, is red in color when placed in these aqueous solutions.

Taste. The dilute, aqueous solutions of each compound are sour tasting.*

Because the aqueous solutions of these compounds have these properties in common, the compounds are conveniently classed together and identified as acids. In fact, these properties constitute the simplest definition of an acid. They provide a basis for deciding whether some other compound should be classified as an acid.

What is the common factor that makes these different substances behave in the same ways? In water they all form conducting solutions; we conclude that they all form ions in water. Each substance contains hydrogen and each reacts with zinc metal to produce hydrogen gas. Perhaps all of these aqueous solutions contain the *same* ion and this ion accounts for the formation of $H_2(g)$. It is reasonable to propose that the common ion is $H^+(aq)$. We postulate: *a substance has the properties of an* ***acid*** *if it can release hydrogen ions.*

11-2.2 Properties of Aqueous Solutions of Bases

Consider the following compounds:

$NaOH$: sodium hydroxide,
KOH : potassium hydroxide,
$Mg(OH)_2$: magnesium hydroxide,
Na_2CO_3 : sodium carbonate,
NH_3 : ammonia.

* Many chemicals, including some acids and some bases, are poisons. Therefore chemists rarely use taste as a voluntary means of deciding if an unidentified solution is acidic or basic.

Each of these five compounds is called a base. They all share this name because they have the following important properties in common.

Electrical Conductivity. Like acids, these compounds dissolve in water to form conducting solutions. Ions are present in an aqueous solution of a base.

Reaction with Acids. When one of these compounds is added to an acid solution, it destroys the identifying properties of the acid solution—all but electrical conductivity.

Color of Litmus. The dye, litmus, is blue in color when placed in an aqueous solution of any of these compounds.

Taste. The dilute aqueous solutions taste bitter.*

Feel. The aqueous solutions feel "slippery."

Again, these properties constitute the simplest definition of a base. These properties provide a basis for deciding whether some other compound should be classified as a base.

11-2.3 An Explanation of the Properties of Bases

Using the same argument we used for acids, we can seek a common factor that accounts for the similarities of bases. Because of the electrical conductivity, we might seek an ion. Because of the ability to counteract the properties of acids, we ought to seek an ion which can remove the hydrogen ion, $H^+(aq)$, since hydrogen ion accounts for the properties of acids.

Sodium hydroxide, NaOH, when dissolved in water, gives a solution with the properties of a base. The hydroxides of many elements—those from the left side of the periodic table—behave in the same way. Perhaps they dissolve to form ions of the sort

$$NaOH(s) \rightleftharpoons Na^+(aq) + OH^-(aq) \quad (19)$$

$$KOH(s) \rightleftharpoons K^+(aq) + OH^-(aq) \quad (20)$$

$$Mg(OH)_2(s) \rightleftharpoons Mg^{+2}(aq) + 2OH^-(aq) \quad (21)$$

$$Ca(OH)_2(s) \rightleftharpoons Ca^{+2}(aq) + 2OH^-(aq) \quad (22)$$

The hydroxide ion, $OH^-(aq)$, could react with hydrogen ion to account for the second property of bases, the removal of acid properties:

$$OH^-(aq) + H^+(aq) \rightleftharpoons H_2O \quad (23)$$

The similarities among the hydroxides are obvious. Let's compare sodium carbonate and ammonia. Sodium carbonate, Na_2CO_3, dissolves in water to give a solution with the properties that identify a base. Quantitative studies of the solubilities of carbonates show that carbonate ion, CO_3^{-2}, can react with water. The reactions are

$$Na_2CO_3(s) \rightleftharpoons 2Na^+(aq) + CO_3^{-2}(aq) \quad (24)$$

$$CO_3^{-2}(aq) + H_2O \rightleftharpoons HCO_3^-(aq) + OH^-(aq) \quad (25)$$

Reaction (*25*) indicates that the presence of carbonate ion in water increases the hydroxide ion, OH^-, concentration. This is a constituent that is present in the solutions of NaOH, KOH, $Mg(OH)_2$, and $Ca(OH)_2$.

Reaction (*25*) also provides a basis for understanding the removal of acid properties. If CO_3^{-2} readily forms bicarbonate ion,† HCO_3^-, then reaction (*26*) is likely:

$$CO_3^{-2}(aq) + H^+(aq) \rightleftharpoons HCO_3^-(aq) \quad (26)$$

We see that the existence of the stable bicarbonate ion, $HCO_3^-(aq)$ produces the chemical species, $OH^-(aq)$ in common with solutions of the hydroxides. We can postulate that $OH^-(aq)$ accounts for the slippery "feel" and bitter taste of the basic solutions. The stability of bicarbonate ion also explains the removal of acid properties through reaction (*26*).

The fifth substance listed as a base is ammonia, NH_3. Ammonia readily forms ammonium ion, NH_4^+. Ammonia can react with water,

$$NH_3(aq) + H_2O \rightleftharpoons NH_4^+(aq) + OH^-(aq) \quad (27)$$

and with hydrogen ion,

$$NH_3(aq) + H^+(aq) \rightleftharpoons NH_4^+(aq) \quad (28)$$

Once again, the formation of NH_4^+ accounts for ammonia having the properties of a base. Reaction (*27*) produces hydroxide ion, which, by our postulate, accounts for the taste and "feel" properties of solutions of bases. Reaction (*28*) shows

* If you have forgotten the danger of tasting chemicals, reread the preceding footnote.

†Bicarbonate ion, HCO_3^-, is also called "hydrogen carbonate" or "monohydrogen carbonate."

how ammonia can act to destroy the acid properties of a solution containing hydrogen ions, $H^+(aq)$.

Investigation of the reactions of other compounds that have the properties of a base shows that each compound can produce hydroxide ions in water. The $OH^-(aq)$ ions may be produced directly (as when solid NaOH dissolves in water) or through reaction with water (as when Na_2CO_3 and NH_3 dissolve in water):

$$NaOH(s) \rightleftharpoons Na^+(aq) + OH^-(aq) \quad (19)$$

$$CO_3^{-2}(aq) + H_2O \rightleftharpoons HCO_3^-(aq) + OH^-(aq) \quad (25)$$

$$NH_3(aq) + H_2O \rightleftharpoons NH_4^+(aq) + OH^-(aq) \quad (27)$$

Furthermore, *any substance that can produce hydroxide ions in water also can combine with hydrogen ions:*

$$OH^-(aq) + H^+(aq) \rightleftharpoons H_2O \quad (23)$$

$$CO_3^{-2}(aq) + H^+(aq) \rightleftharpoons HCO_3^-(aq) \quad (26)$$

$$NH_3(aq) + H^+(aq) \rightleftharpoons NH_4^+(aq) \quad (28)$$

Since production of $OH^-(aq)$ and reaction with $H^+(aq)$ go hand in hand when we are dealing with aqueous solutions, a base can be described *either* as a substance that produces $OH^-(aq)$ *or* as a substance that can react with $H^+(aq)$. In solvents other than water, the latter description is more generally useful. Therefore, we postulate: *a substance has the properties of a* **base** *if it can combine with hydrogen ions.*

11-2.4 Acids and Bases: Summary

Let us repeat our two definitions and explanations.

Fig. 11-3. **Some familiar bases.**

Sodium carbonate *Ammonia* *Third*

DEFINITIONS

An *acid* is a hydrogen-containing substance that has the following properties when dissolved in water:

it is an electrical conductor;
it reacts with Zn to give $H_2(g)$;
it makes litmus red;
it tastes sour.

A *base* is a substance that has the following properties when dissolved in water:

it is an electrical conductor;
it reacts with an acid, removing the acidic properties;
it makes litmus blue;
it tastes bitter;
it feels slippery.

EXPLANATIONS

A substance is an acid if it can release hydrogen ions, $H^+(aq)$.

A substance is a base if it can react with hydrogen ions, $H^+(aq)$.

11-2.5 The Nature of H⁺(aq)

Both of our explanations of the properties of acids and bases involve the hydrogen ion, $H^+(aq)$. This species has great importance in the chemistry of aqueous solutions, so we shall consider what is known about it.

Before considering what a chemist means by the symbols $H^+(aq)$, we must discuss more generally the interaction of ions with water. Lithium chloride provides a good example. Lithium chloride dissolves in water spontaneously at 25°C, forming a conducting solution. At equilibrium, it has a high solubility:

$$LiCl(s) \rightleftharpoons Li^+(aq) + Cl^-(aq) \quad t = 25°C \quad (29)$$

Lithium chloride melts spontaneously above 613°C and forms a liquid that conducts electricity:

$$LiCl(s) \rightleftharpoons Li^+(l) + Cl^-(l) \quad t = 613°C \quad (30)$$

Equilibrium in any reaction is determined by a compromise between tendency toward minimum energy ("golf balls roll downhill") and tendency toward maximum randomness. Reaction (*29*) and reaction (*30*) both involve increase in randomness since the regular solid lattice dissolves or melts to become part of a disordered liquid state. Both reactions produce ions. But reaction (*29*) proceeds readily at 25°C, whereas reaction (*30*) does not

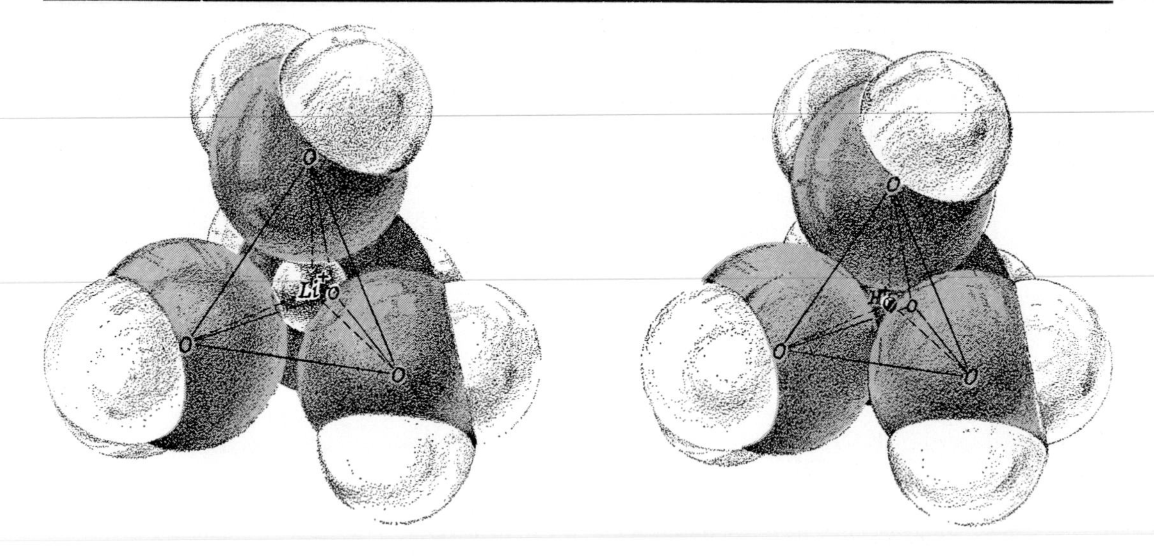

Fig. 11-4. **Possible tetrahedral * arrangements** *of water molecules around Li^+ and H^+ ions.*

occur until the solid is quite hot, 613°C. The difference must be in the special stabilities of Li^+ and Cl^- ions in water. The high melting point of lithium chloride shows that the crystal is very stable. The high solubility of lithium chloride in water can be explained only by saying that $Li^+(aq)$ and $Cl^-(aq)$ must also be very stable. This means that water must interact strongly with these ions.

A similar situation exists for hydrochloric acid, HCl. This gaseous compound dissolves readily in water at 25°C:

$$HCl(g) \rightleftharpoons H^+(aq) + Cl^-(aq) \qquad t = 25°C \quad (31)$$

The HCl molecule is a stable one—it must be heated to a few thousand degrees before the atoms will separate. Even then, neutral atoms are obtained and still higher temperatures are needed before gaseous ions are obtained.

$$HCl(g) \rightleftharpoons H(g) + Cl(g) \qquad t \text{ very high} \quad (32)$$

The high temperature required to separate the two atoms of a molecule of HCl shows that HCl is very stable. Again, we can explain the solubility of HCl in water by saying $H^+(aq)$ and $Cl^-(aq)$ must also be very stable. Water must interact strongly with these ions.

This is why we have been symbolizing these aqueous ions as $Li^+(aq)$, $H^+(aq)$, and $Cl^-(aq)$. The notation (aq) reminds us that the ions interact strongly with the solvent. The symbol is purposely vague because, in most cases, chemists are not certain of the arrangement of the water molecules around a given ion. For $Li^+(aq)$ and $H^+(aq)$, there may be simple packing of four water molecules around each ion, as shown in Figure 11-4. This figure suggests similarity between $Li^+(aq)$ and $H^+(aq)$.

However, most chemists feel there is quite a difference between these two ions. After all, H^+ is unique—the proton has no electrons. Many chemists suggest that the proton attaches itself strongly to one molecule of water, forming a new molecular species, $H_3O^+(aq)$. Figure 11-5

Fig. 11-5. **A model of hydronium ion, H_3O^+.**

* The word tetrahedral means four-sided. If the oxygen atoms of the four water molecules are connected by lines, the lines form a four-sided figure.

shows a model of this proposed ion—called hydronium ion—surrounded by solvent. Notice that the three hydrogen atoms are pictured as equivalent. There is a pleasing similarity to the formation of the well-established ammonium ion, NH_4^+, with its four equivalent hydrogen atoms:

$$H_2O + H^+ \rightleftharpoons H_3O^+(aq) \quad (33)$$

$$NH_3(aq) + H^+ \rightleftharpoons NH_4^+(aq) \quad (34)$$

Still other chemists feel there are probably several arrangements of water molecules around the proton. In addition to H_3O^+, there may be molecules such as $H_5O_2^+$, $H_7O_3^+$, $H_9O_4^+$, etc.

Unfortunately, the experimental data do not provide a definite answer to the nature of $H^+(aq)$. The hydronium ion, shown in Figure 11-5, does exist in certain crystal structures.* Spectroscopic studies† indicate that several species are present in water. Thermal and electrical conductivities of aqueous acid solutions have been interpreted to indicate the presence of a molecular unit, $H_9O_4^+$.

Figure 11-4 shows the hydrogen ion surrounded by four oxygen atoms, each part of a water molecule. We could write the formula for this arrangement as $H^+ \cdot 4H_2O$ or $H_9O_4^+$. But Figure 11-6 shows how an H_3O^+ ion can serve as the basis for another structure with the formula $H_3O^+ \cdot 3H_2O$, again giving $H_9O_4^+$. Still, there seems to be no compelling experimental basis for preference.

So we are faced with at least three plausible structures for the species $H^+(aq)$, as shown in Figures 11-4, 11-5, and 11-6. Under these circumstances, convenience in discussing the experimental properties of $H^+(aq)$ governs our choice.

So far in this chapter, we have referred to the aqueous hydrogen ion as $H^+(aq)$. Later in the chapter we will consider a more general theory of acids and bases. Then it will be more convenient to designate the aqueous hydrogen ion as $H_3O^+(aq)$. The reason will be clear—such usage aids us in seeing regularity in the behavior of a

* Hydronium ion, H_3O^+, is a structural unit in solid perchloric acid hydrate, $HClO_4 \cdot H_2O$, as shown by nuclear magnetic resonance studies.

† Spectroscopy refers to the study of the absorption of light—in this case, by aqueous solutions of acids such as HCl.

Fig. 11-6. **A possible model for $H_9O_4^+$ based upon H_3O^+.**

larger class of acids and bases. That is sufficient basis for the use of any theory.

11-2.6 Acid–Base Titrations

We have noted that the two concentrations $[H^+]$ and $[OH^-]$ are "tied together." If 0.100 mole of HCl is dissolved in 0.100 liter of water to raise $[H^+]$ to 1.00 *M*, then $[OH^-]$ goes down to 1.00×10^{-14} *M* and the product $[H^+] \times [OH^-]$ remains equal to $K_w = 1.00 \times 10^{-14}$. Perhaps you are wondering what happens to the hydroxide ions to reduce their concentration from 1.00×10^{-7} *M* (their concentration in pure water) to the new value, 1.00×10^{-14} *M*. The answer is that they are consumed through reaction with the added $H^+(aq)$:

$$OH^-(aq) + H^+(aq) \longrightarrow H_2O \qquad (35)$$

This is in accord with Le Chatelier's Principle. Addition of HCl to water raises $[H^+]$. By Le Chatelier's Principle, processes take place that tend to counteract partially the imposed change. Reaction with $OH^-(aq)$ does tend to counteract the raised concentration of $H^+(aq)$.

In the case we have considered, the actual amount by which $[H^+]$ is reduced is extremely small. To reduce $[OH^-]$ from 1.00×10^{-7} *M* to 1.00×10^{-14} *M* (by a factor of 10^7), reaction (*35*) must consume about 10^{-7} mole of hydroxide ion for every liter of solution. Since one mole of $OH^-(aq)$ reacts with one mole of $H^+(aq)$, the amount of $H^+(aq)$ required is also 10^{-7} mole for every liter of solution. Subtracting 10^{-7} mole/liter from a concentration near 1 mole/liter causes such a small change in $[H^+]$ that it need not be considered in calculations (such as in Exercises 11-1 and 11-2).

HCl AND NaOH IN THE SAME SOLUTION: EXCESS HCl

Suppose that to the 0.100 liter of 1.00 *M* HCl solution we add 0.090 mole of solid sodium hydroxide. Now we have added *both* $H^+(aq)$ and $OH^-(aq)$ in high concentration *to the same solution*. What will happen?

Immediately after the sodium hydroxide dissolves, the concentrations of $H^+(aq)$ and $OH^-(aq)$ do not satisfy the equilibrium expression. Their product far exceeds 1.00×10^{-14}:

$$\text{initial } [H^+] = 1.00\ M$$

$$\text{initial } [OH^-] = \frac{0.090 \text{ mole}}{0.100 \text{ liter}} = 0.90\ M$$

$$\text{initial product, } [H^+] \times [OH^-] = 9.01 \times 10^{-1}$$
$$(\text{far exceeding } 1.00 \times 10^{-14})$$

Again Le Chatelier's Principle tells us qualitatively what will occur and the equilibrium expression tells us quantitatively. If we add $OH^-(aq)$, a change will take place that tends to counteract partially the resulting increase in $[OH^-]$. This occurs through the reaction between $OH^-(aq)$ and $H^+(aq)$, consuming both ions and reducing the value of $[H^+] \times [OH^-]$. Reaction continues until this product reaches the equilibrium value, $K_w = 1.00 \times 10^{-14}$.

Since K_w is so small, the reaction consumes almost all of one of the constituents (H^+ or OH^-) if the other is present in excess. In our example, $[H^+]$ initially exceeds $[OH^-]$ by 0.10 mole/liter:

$$\begin{array}{ccccc} \text{initial } [H^+] & - & \text{initial } [OH^-] & = & \text{excess } [H^+] \\ 1.00\ M & - & 0.90\ M & = & 0.10\ M \end{array}$$

In Section 11-1.3 we calculated that if the concentration $[H^+] = 0.100$ *M*, then at equilibrium, $[OH^-] = 1.00 \times 10^{-13}$ *M*. Thus the rather small excess of hydrogen ion, 0.10 *M*, is sufficient to guarantee that the reaction between $H^+(aq)$ and $OH^-(aq)$ consumes most of the 0.90 *M* OH^-, reducing $[OH^-]$ to 1.0×10^{-13} *M*.

EXERCISE 11-4

Suppose that 0.099 mole of solid NaOH is added to 0.100 liter of 1.00 *M* HCl.

(a) How many *more* moles of HCl are present in the solution than moles of NaOH?
(b) From the excess number of moles and the volume, calculate the concentration of excess $H^+(aq)$.
(c) Calculate the excess concentration of $H^+(aq)$ from the difference between the initial concentrations of HCl and NaOH.

(d) Calculate the concentration of $OH^-(aq)$ at equilibrium (see your calculations for Exercise 11-2).

HCl AND NaOH IN THE SAME SOLUTION: EXCESS NaOH

Returning to our original 0.100 liter of 1.00 M HCl, let us now consider the addition of 0.101 mole of solid NaOH. Again we have added both $H^+(aq)$ and $OH^-(aq)$ to the same solution, and the concentrations immediately after mixing do not satisfy the equilibrium expression:

$$\text{initial } [H^+] = 1.00\ M$$

$$\text{initial } [OH^-] = \frac{0.101 \text{ mole}}{0.100 \text{ liter}} = 1.01\ M$$

$$\text{initial product, } [H^+] \times [OH^-] = 1.01 \quad (\text{far exceeding } 1.00 \times 10^{-14})$$

This solution contains excess hydroxide ion. Since $OH^-(aq)$ is in excess, almost all of the $H^+(aq)$ will be consumed, forming water:

$$\begin{array}{ccccc} \text{initial } [OH^-] & - & \text{initial } [H^+] & = & \text{excess } [OH^-] \\ 1.01\ M & - & 1.00\ M & = & 0.01\ M \end{array}$$

In Exercise 11-1 we calculated the equilibrium concentration $[H^+]$ in a solution containing $[OH^-] = 0.01\ M$. The result is $[H^+] = 1.0 \times 10^{-12}\ M$.

HCl AND NaOH IN THE SAME SOLUTION: NO EXCESS OF EITHER

In each example used in this section, a number of moles of NaOH was added to 0.100 liter of 1.00 M HCl with one or the other constituent in excess. Reaction between $H^+(aq)$ and $OH^-(aq)$ consumes almost all of the constituent not in excess. Let us now consider the case in which there is excess of *neither* HCl nor NaOH.

Suppose we add 0.100 mole of NaOH to 0.100 liter of 1.00 M HCl. The initial values of $[H^+]$ and $[OH^-]$ are equal and their product far exceeds 1.00×10^{-14}

$$\text{initial } [H^+] = 1.00\ M$$

$$\text{initial } [OH^-] = \frac{0.100 \text{ mole}}{0.100 \text{ liter}} = 1.00\ M$$

$$\text{initial product, } [H^+] \times [OH^-] = 1.00 \quad (\text{far exceeding } 1.00 \times 10^{-14})$$

Reaction between $H^+(aq)$ and $OH^-(aq)$ must occur, forming water:

$$OH^-(aq) + H^+(aq) \rightleftharpoons H_2O$$

Since one mole of $OH^-(aq)$ consumes one mole of $H^+(aq)$, the concentrations $[H^+]$ and $[OH^-]$ remain equal as reaction (*35*) proceeds. When equilibrium is reached, they will still be equal. This is exactly the situation in pure water. As we saw in Section 11-1.2,

$$[H^+] = [OH^-] = \sqrt{K_w} = 1.00 \times 10^{-7}\ M$$

A solution containing exactly equivalent amounts of acid and base is neither acidic nor basic. Such a solution is called a **neutral** solution.*

PROGRESSIVE ADDITION OF NaOH TO HCl: A TITRATION

Now we have considered the progressive addition of more and more NaOH to a fixed amount of HCl solution. The results are compiled in Table 11-II.

Table 11-II shows that the concentration of hydrogen ions changes drastically as the amount of NaOH nears the equivalent amount of HCl. There is a change of $[H^+]$ by a factor of 10^{10} as the initial concentration of $OH^-(aq)$ is changed from 0.99 to 1.01. Thus the concentration changes by a huge factor near the point at which the amounts of acid and base are equivalent. Because of this, *the progressive addition of a base to an acid*, a **titration,** furnishes a sensitive means of comparing the concentrations of an acid and a base solution.

An acid–base titration is carried out by adding carefully measured amounts of a base solution to a known volume of the acid solution. The acid solution contains some substance that provides visual evidence of the magnitude of $[H^+]$. The dye litmus is such a substance. As mentioned in Sections 11-2.1 and 11-2.2, litmus is red in solutions containing excess $[H^+]$. Litmus is blue in

---

* This use of the word "neutral" for a solution with equal amounts of H^+ and OH^- has its disadvantages because the same word is used in reference to electrical neutrality. Aqueous solutions are *always* electrically neutral whether there is an excess of either H^+ or OH^- or an excess of neither.

Table 11-II. **THE CONCENTRATIONS $[H^+]$ AND $[OH^-]$ IN SOLUTIONS CONTAINING BOTH HCl AND NaOH**

INITIAL CONC. H^+	INITIAL CONC. OH^-	EXCESS CONC. H^+ OR OH^-	CALC. $[H^+]$	CALC. $[OH^-]$
1.00 *M*	none	1.00 *M* H^+	1.00 *M*	1.0×10^{-14} *M*
1.00	0.90 *M*	0.10 *M* H^+	1.0×10^{-1}	1.0×10^{-13}
1.00	0.99	0.01 *M* H^+	1.0×10^{-2}	1.0×10^{-12}
1.00	1.00	none	1.0×10^{-7}	1.0×10^{-7}
1.00	1.01	0.01 *M* OH^-	1.0×10^{-12}	1.00×10^{-2}

solutions in which $[H^+]$ is less than $[OH^-]$. A small amount of litmus in the solution will tint the solution red until the point in the titration at which the number of moles of $OH^-(aq)$ becomes equal to the number of moles of $H^+(aq)$. Then, the slightest addition of more $OH^-(aq)$ causes a drastic reduction in $[H^+]$ and the solution color turns to blue. A dye whose color is sensitive to the change of $[H^+]$ (such as litmus) is called an **acid–base indicator.**

11-2.7 pH

For compact expression of $H^+(aq)$ concentrations, chemists use a quantity, *p*H, defined by the equation

$$pH = -\log_{10} [H^+]$$

Since $[H^+] = 10^{-7}$ *M* in a neutral solution at 25°C, it follows that for such a solution

$$pH = -\log_{10} [10^{-7}] = -(-7) = +7$$

This result helps us to understand the symbol *p*H, first defined by a Danish chemist, Sørenson. He used the *p* to stand for the Danish word *potenz* (power) and H to stand for hydrogen. After a change of sign, *p*H is the power of ten needed to express the hydrogen ion concentration in moles per liter. In acidic solutions, *p*H is less than 7 ($pH < 7$); and in basic solutions, *p*H is greater than 7 ($pH > 7$). Table 11-III expresses the results of Table 11-II in terms of *p*H.

Table 11-III

CONCENTRATIONS OF $H^+(aq)$ AND $OH^-(aq)$ EXPRESSED IN TERMS OF pH

ACIDITY OR BASICITY	$[H^+]$	*p*H
acidic	1.00	0
acidic	10^{-1}	1
acidic	10^{-2}	2
neutral	10^{-7}	7
basic	10^{-13}	13

11-3 STRENGTHS OF ACIDS

Earlier in this chapter, strong and weak electrolytes were distinguished in terms of the degree to which the dissolved material forms ions. As a particular case, such distinctions can be made in terms of acids, furnishing a quantitative basis for defining the strength of an acid.

11-3.1 Weak Acids

We have contrasted the electrical conductivities of 0.1 *M* aqueous solutions of hydrochloric acid and acetic acid (see Figure 11-1). In water, hydrochloric acid dissociates completely to ions; HCl is a strong electrolyte. Because one of the ions released is $H^+(aq)$, HCl is also called a **strong acid.** Acetic acid, on the other hand, dissociates to ions only to a slight extent; acetic acid is a weak electrolyte. Because one of the ions released is $H^+(aq)$, acetic acid is called a **weak acid.**

These qualitative ideas can be expressed more usefully in terms of the principles of equilibrium.

For example, let us contrast the behavior of two weak acids, acetic acid and hydrofluoric acid:

$$CH_3COOH(aq) \rightleftharpoons H^+(aq) + CH_3COO^-(aq) \quad (36)$$

$$HF(aq) \rightleftharpoons H^+(aq) + F^-(aq) \quad (37)$$

Measurements of the electrical conductivities of 0.10 M solutions of these two acids show that there are more ions present in the HF solution than in the acetic acid solution. We can conclude that acetic acid is a weaker acid than HF. This information is conveyed quantitatively in terms of the equilibrium constants for reactions (*36*) and (*37*):

$$K_{CH_3COOH} = \frac{[H^+][CH_3COO^-]}{[CH_3COOH]} = 1.8 \times 10^{-5} \quad (38)$$

$$K_{HF} = \frac{[H^+][F^-]}{[HF]} = 6.7 \times 10^{-4} \quad (39)$$

Since K_{HF} is a larger number than K_{CH_3COOH}, hydrofluoric acid dissociates in water to a larger extent than does acetic acid. Though HF is a weak acid (only partially dissociated), it is a stronger acid than is acetic acid.

We can express these ideas in terms of a general acid, HB. The acidic nature of HB is connected to its ability to release hydrogen ions,

$$HB(aq) \rightleftharpoons H^+(aq) + B^-(aq) \quad (40)$$

The equilibrium constant for reaction (*40*) measures quantitatively the ease with which HB releases $H^+(aq)$ ions,

$$K_A = \frac{[H^+][B^-]}{[HB]} \quad (41)$$

Tables of K_A furnish a quantitative measure of acid strengths with which we can compare different acids and predict their properties. Several values of K_A are given in Table 11-IV.

We see in Table 11-IV that the equilibrium view of acid strengths suggests that we regard water itself as a weak acid. It can release hydrogen ions and the extent to which it does so is indicated in its equilibrium constant, just as for the other acids. We shall see that this type of comparison, stimulated by our equilibrium considerations, leads us to a valuable generalization of the acid–base concept.

EXERCISE 11-5

Which of the following acids is the strongest acid and which the weakest?

nitrous acid, HNO_2; $K_{HNO_2} = 5.1 \times 10^{-4}$
sulfurous acid, H_2SO_3; $K_{H_2SO_3} = 1.7 \times 10^{-2}$
phosphoric acid, H_3PO_4; $K_{H_3PO_4} = 7.1 \times 10^{-3}$

Table 11-IV. **RELATIVE STRENGTHS OF ACIDS IN AQUEOUS SOLUTION AT ROOM TEMPERATURE**

$$K_A = \frac{[H^+][B^-]}{[HB]}$$

ACID	STRENGTH	REACTION	K_A
HCl	very strong	$HCl(g) \longrightarrow H^+(aq) + Cl^-(aq)$	very large
HNO_3	↓	$HNO_3(g) \longrightarrow H^+(aq) + NO_3^-(aq)$	very large
H_2SO_4	very strong	$H_2SO_4 \longrightarrow H^+(aq) + HSO_4^-(aq)$	large
HSO_4^-	strong	$HSO_4^-(aq) \rightleftharpoons H^+(aq) + SO_4^{-2}(aq)$	1.3×10^{-2}
HF	weak	$HF(aq) \rightleftharpoons H^+(aq) + F^-(aq)$	6.7×10^{-4}
CH_3COOH	↓	$CH_3COOH(aq) \rightleftharpoons H^+(aq) + CH_3COO^-(aq)$	1.8×10^{-5}
H_2CO_3 ($CO_2 + H_2O$)	↓	$H_2CO_3(aq) \rightleftharpoons H^+(aq) + HCO_3^-(aq)$	4.4×10^{-7}
H_2S	weak	$H_2S(aq) \rightleftharpoons H^+(aq) + HS^-(aq)$	1.0×10^{-7}
NH_4^+	↓	$NH_4^+(aq) \rightleftharpoons H^+(aq) + NH_3(aq)$	5.7×10^{-10}
HCO_3^-	↓	$HCO_3^-(aq) \rightleftharpoons H^+(aq) + CO_3^{-2}(aq)$	4.7×10^{-11}
H_2O	very weak	$H_2O(aq) \rightleftharpoons H^+(aq) + OH^-(aq)$	1.8×10^{-16} *

* The equilibrium constant, K_A, for water equals $\frac{K_w}{[H_2O]} = \frac{1.00 \times 10^{-14}}{55.5}$. See Section 11-1.1.

11-3.2 Equilibrium Calculations of Acidity

The acidity of a solution has pronounced effects on many chemical reactions. It is therefore important to be able to learn and control the hydrogen ion concentration. This control is obtained through application of the Equilibrium Law. Common types of calculation, based on this law, are those needed to determine K_A from experimental data and those using K_A to find $[H^+]$. We will illustrate both of these types, using benzoic acid, C_6H_5COOH, as an example.

DETERMINATION OF K_A

To apply the Equilibrium Law to acid solutions, a chemist must know the numerical value of the equilibrium constant, K_A. Experiments which provide this information require the measurement of hydrogen ion concentration. Acid-sensitive dyes, such as litmus, offer the easiest estimate of $[H^+]$.

A typical example is as follows. Benzoic acid, C_6H_5COOH, is a solid substance with only moderate solubility in water. The aqueous solutions conduct electric current and have the other properties of an acid listed in Section 11-2.1. We can describe this behavior with reaction (*42*) leading to the equilibrium relation (*43*):

$$C_6H_5COOH(aq) \rightleftharpoons H^+(aq) + C_6H_5COO^-(aq) \quad (42)$$

$$K_A = \frac{[H^+][C_6H_5COO^-]}{[C_6H_5COOH]} \quad (43)$$

The following experiment was performed to determine the equilibrium constant in (*43*). A 1.22 gram sample of benzoic acid was dissolved in 1.00 liter of water at 25°C. With dyes whose color is sensitive to acidity (indicators) the concentration of $H^+(aq)$ was estimated to be $8 \times 10^{-4}\ M$.

To make use of these data, we must first express all quantities in terms of moles. The molecular weight of benzoic acid, C_6H_5COOH, is 122.1 grams/mole. Hence,

$$1.22 \text{ g } C_6H_5COOH = \frac{1.22 \text{ g}}{122.1 \text{ g/mole}}$$
$$= 0.0100 \text{ mole } C_6H_5COOH$$

Now we can calculate the concentration of benzoic acid if we assume very little of it has reacted to form $H^+(aq)$ according to reaction (*42*):

$$[C_6H_5COOH] = \frac{\text{moles}}{\text{volume}} = \frac{0.0100 \text{ mole}}{1.00 \text{ liter}}$$

$$[C_6H_5COOH] = 0.0100\ M = 1.00 \times 10^{-2}\ M \quad (44)$$

and, by measurement,

$$[H^+] = 8 \times 10^{-4}\ M \quad (45)$$

(Notice that $[H^+]$ is less than 10% of $[C_6H_5COOH]$, verifying our assumption that very little of the acid reacted.) Now we know two of the concentrations in expression (*43*) and, to complete the calculation, we must know the concentration of benzoate ion, $[C_6H_5COO^-]$. Since the benzoic acid was dissolved in pure water, the only source of $C_6H_5COO^-$ is reaction (*42*). This is also the source of hydrogen ion, $H^+(aq)$. Since these two ions are both produced only by reaction (*42*), their concentrations must be equal in this solution. That is,

$$[H^+] = [C_6H_5COO^-]$$

Thus, if $[H^+] = 8 \times 10^{-4}$, then, also

$$[C_6H_5COO^-] = 8 \times 10^{-4} \quad (46)$$

Now we can complete the calculation by substituting (*44*), (*45*), and (*46*) into (*43*):

$$K_A = \frac{[H^+][C_6H_5COO^-]}{[C_6H_5COOH]} = \frac{[8 \times 10^{-4}] \times [8 \times 10^{-4}]}{[1.0 \times 10^{-2}]}$$
$$= \frac{64 \times 10^{-8}}{1.0 \times 10^{-2}}$$
$$K_A = 64 \times 10^{-6} = 6.4 \times 10^{-5}$$

CALCULATION OF $[H^+]$

Having established experimentally the numerical value of K_A, we can use it in calculations of equilibrium concentrations.

As an example, suppose a chemist needs to know the hydrogen ion concentration in a solution containing both 0.010 M benzoic acid, C_6H_5COOH, and 0.030 M sodium benzoate, C_6H_5COONa. Of course, he could go to the laboratory and proceed to investigate the colors of indicator dyes placed in the solution. However, it is easier to calculate the value of $[H^+]$, using the accurate value of K_A listed in Appendix 2.

Sodium benzoate is a strong electrolyte; its aqueous solutions contain sodium ions, $Na^+(aq)$, and benzoate ions, $C_6H_5COO^-(aq)$. Hence the equilibrium involved is the same as before:

$$C_6H_5COOH(aq) \rightleftharpoons H^+(aq) + C_6H_5COO^-(aq) \quad (42)$$

At equilibrium, the concentrations must be in accord with the equilibrium expression. That is,

$$K_A = \frac{[H^+][C_6H_5COO^-]}{[C_6H_5COOH]} = 6.6 \times 10^{-5} \qquad (43)$$

First, let us assume that very little $[H^+]$ is formed through dissociation of benzoic acid. This assumption implies that the concentrations of benzoate ion and benzoic acid are very little affected by reaction (*42*). Assuming this, we see that two of the concentrations in (*39*) are already specified:

$$[C_6H_5COOH] = 0.010\ M$$
$$[C_6H_5COO^-] = 0.030\ M$$

$$K_A = \frac{[H^+][C_6H_5COO^-]}{[C_6H_5COOH]} = \frac{[H^+](0.030)}{(0.010)} \qquad (47)$$

Rearranging (*47*), we obtain

$$[H^+] = K_A \times \frac{(0.010)}{(0.030)} = 6.6 \times 10^{-5} \times \frac{0.010}{0.030}$$

$$[H^+] = 2.2 \times 10^{-5}$$

The calculation cannot be considered complete until we check the assumption. Was it reasonable to assume that the concentrations of benzoate ion and benzoic acid were not changed by reaction (*42*)? To decide, we compare the magnitude of $[H^+]$, $2.2 \times 10^{-5}\ M$, to the benzoate ion and benzoic acid concentrations. We find that $[C_6H_5COOH] = 0.010\ M$ is about 500 times larger than the concentration change necessary to form $2.2 \times 10^{-5}\ M$ H^+. The same argument applies to $[C_6H_5COO^-]$. The assumption is valid.

11-3.3 Competition for H^+ Among Weak Acids

We have explained the properties of acids in terms of their abilities to release hydrogen ions, $H^+(aq)$. Thus acetic acid is a weak acid because of the slight extent to which reaction (*48*) releases $H^+(aq)$:

$$CH_3COOH(aq) \rightleftharpoons H^+(aq) + CH_3COO^-(aq) \qquad (48)$$

We have explained the properties of bases in terms of their abilities to *react* with hydrogen ion. Thus ammonia is a base because it can react as in (*49*):

$$NH_3(aq) + H^+(aq) \rightleftharpoons NH_4^+(aq) \qquad (49)$$

Now consider the result of mixing aqueous solutions of acetic acid and ammonia. The reaction that occurs can be compared to a sequence of reactions,

$$CH_3COOH(aq) \rightleftharpoons H^+(aq) + CH_3COO^-(aq)$$
$$NH_3(aq) + H^+(aq) \rightleftharpoons NH_4^+(aq)$$

Net reaction

$$CH_3COOH(aq) + NH_3(aq) \rightleftharpoons CH_3COO^-(aq) + NH_4^+(aq) \qquad (50)$$

Practically, the result of reactions (*48*) and (*49*) is reaction (*50*). In reaction (*50*), we see that acetic acid acts as an acid in the same sense that it does in (*48*). In either case, it releases hydrogen ions. In (*48*) acetic acid releases hydrogen ions and forms $H^+(aq)$ and in (*50*) it releases hydrogen ions to NH_3 and forms NH_4^+. In the same way, ammonia acts as a base in (*50*) by reacting with the hydrogen ion released by acetic acid. So reaction (*50*) is an acid–base reaction, though the net reaction does not show $H^+(aq)$ explicitly.

Now by taking one more step we can view acid–base reaction in a broader sense. Suppose we mix aqueous solutions of ammonium chloride, NH_4Cl, and sodium acetate, CH_3COONa. A sniff indicates ammonia has been formed. Reaction occurs,

$$NH_4^+(aq) + CH_3COO^-(aq) \rightleftharpoons CH_3COOH(aq) + NH_3(aq) \qquad (51)$$

Reaction (*51*) is just the reverse of reaction (*50*). Inspection of this reaction reveals that *reaction (51), too, is an acid–base reaction!* Once again there is an acid that releases H^+—it is NH_4^+—and a base that accepts H^+—the base is CH_3COO^-. Once again the net effect of the reaction is transfer of a hydrogen ion from one species to another. We see that the acid–base reaction between acetic acid and ammonia gave two products, one an acid, NH_4^+, and one a base, CH_3COO^-. A little thought will convince you that *every* acid–base reaction does so. The transfer of a hydrogen ion from an acid to a base necessarily implies that it might be handed back. The reaction of handing it back, the reverse reaction, is just as much a hydrogen ion transfer, hence an acid–base reaction, as is the original transfer.

Notice that we are now referring to reactions in which a hydrogen ion is transferred from an acid to a base without specifically involving the aqueous species $H^+(aq)$. A hydrogen ion, H^+, is nothing more than a proton. Consequently we can frame a more general view of acid–base reactions in terms of *proton transfer*. The main value of this view is that it is applicable to a wider range of chemical systems, including non-aqueous systems.

We generalize our view of the acid–base type of reaction as follows. In our example, reaction (*50*),

$$\underset{\text{an acid}}{CH_3COOH} + \underset{\text{a base}}{NH_3} \rightleftarrows \underset{\text{an acid}}{NH_4^+} + \underset{\text{a base}}{CH_3COO^-} \quad (50)$$

The acetic acid reacts as an acid, giving up its proton, to form acetate, CH_3COO^-, a substance that can act as a base. We can write (*50*) in a general form:

$$HB_1 + B_2 \rightleftarrows HB_2 + B_1 \quad (52)$$

$$\text{Acid}_1 + \text{Base}_2 \rightleftarrows \text{Acid}_2 + \text{Base}_1 \quad (53)$$

We see that *an acid and a base react, through proton transfer, to form another acid and another base.**

We can use this more general view to discuss the strengths of acids. In our generalized acid–base reaction (*52*), the proton transfer implies the chemical bond in HB_1 must be broken and the chemical bond in HB_2 must be formed. If the HB_1 bond is easily broken, then HB_1 will be a strong acid. Then equilibrium will tend to favor a proton transfer from HB_1 to some other base, B_2. If, on the other hand, the HB_1 bond is extremely stable, then this substance will be a weak acid. Equilibrium will tend to favor a proton transfer from some other acid, HB_2, to base B_1, forming the stable HB_1 bond.

11-3.4 Hydronium Ion in the Proton Transfer Theory of Acids

In the proton transfer view of acid–base reactions, an acid and a base react to form another acid and another base. Let us see how this theory encompasses the elementary reaction between $H^+(aq)$ and $OH^-(aq)$ and the reaction of dissociation of acetic acid, reactions (*54*) and (*55*):

$$H^+(aq) + OH^-(aq) \rightleftarrows H_2O \quad (54)$$

$$CH_3COOH(aq) \rightleftarrows H^+(aq) + CH_3COO^-(aq) \quad (55)$$

It does so by making a specific assumption about the nature of the species $H^+(aq)$. It is considered to have the molecular formula $H_3O^+(aq)$. Thus, when HCl dissolves in water, the reaction is written

$$HCl(g) + H_2O \rightleftarrows H_3O^+(aq) + Cl^-(aq) \quad (56)$$

instead of

$$HCl(g) \rightleftarrows H^+(aq) + Cl^-(aq) \quad (57)$$

Whenever $H^+(aq)$ might appear in an equation for a reaction, it is replaced by the hydronium ion, H_3O^+, and a molecule of water is added to the other side of the equation. We write (*55*) in the form

$$CH_3COOH(aq) + H_2O \rightleftarrows H_3O(aq) + CH_3COO^-(aq) \quad (58)$$

Now the dissociation of acetic acid can be regarded as an acid–base reaction. The acid CH_3COOH transfers a proton to the base H_2O forming the acid H_3O^+ and the base CH_3COO^-.

The reaction (*54*) now takes the form

$$H_3O^+(aq) + OH^-(aq) \rightleftarrows H_2O + H_2O \quad (59)$$

In (*59*) the acid H_3O^+ transfers a proton to the base OH^-, forming an acid, H_2O, and a base, H_2O. We see that within the proton transfer theory, the molecule H_2O must be assigned the properties of an acid and, as well, those of a base.

This designation of the species $H^+(aq)$ in terms of hydronium ion, H_3O^+, is not necessitated by experimental evidence that proves the unique existence of this molecule, H_3O^+, in dilute aqueous solutions (see Section 11-2.5). Nevertheless, the convenience of this assumption, as an aid in correlating acid–base behavior, amply justifies its use.

11-3.5 Contrast of Acid–Base Definitions

In this chapter we first identified acids and bases in aqueous solution by investigating the proper-

* This more general view of acids and bases is named the Brønsted-Lowry theory after the two scientists who proposed it, J. N. Brønsted and T. M. Lowry.

ties possessed by acid solutions and base solutions. By so doing, we *defined* an acid in terms of the properties of solutions of acids. Now we are *explaining* the behavior of an acid in terms of the process of proton transfer. Because this explanation fits a large number of experimental facts well and conveniently, it has come into common use. If a chemist is asked what substances are acids, he is liable to refer to the explanation rather than to the identifying properties. At this point, he has shifted to a new definition. Let us compare these two definitions.

Definition 1. An *acid* is a substance that has the properties listed below when dissolved in water:

it is an electrical conductor;
it reacts with Zn to give $H_2(g)$;
it makes litmus red;
it tastes sour.

Definition 2. An *acid* is a substance that can release protons.

The first definition is of the type called an "operational definition." To understand this term, consider the intended meaning of the word "definition." According to the dictionary, "definition" means "a statement of what a thing is." With a definition to help, it is possible to sort the universe into two piles—one pile containing those objects that fit the definition, and another pile containing those that do not. The "statement" gives criteria by which this sorting can be carried out. An "operational definition" is, then, a definition that lists, as criteria, measurements or observations (that is, "operations") by which you could decide whether a given object is "in" or "out."

The second definition is a "conceptual definition." It defines the class in terms of an explanation of why the class has its properties. To see the difference—and the relative merit—of the two kinds of definition, let us consider an analogy.

Contrast the following pair of definitions.

(1) A "star" is an athlete who regularly scores an unusually large number of points or who, in clever defensive acts, repetitiously and advantageously contributes to the welfare of his team.
(2) A "star" is an athlete with unusual muscular coordination.

The first definition is "operational." It explicitly states criteria for deciding whether a player is a "star." He is one who "regularly scores an unusually large number of points" or who "in clever defensive acts, repetitiously and. . . ." To decide if a player is a "star," you count his scoring or, alternatively, number the outstanding defensive plays he makes.

The second definition might be called a "conceptual definition." It offers an explanation of *why* the athlete enjoys special success in his sport. It is one step removed from things that show on the scoreboard.

The first definition gives no clue why one player is a "star" and another player is not. Its value is that it is practical, useful, and surely correct. It does not matter how awkward a basketball player is; if he scores 30 points per game he is assured of a considerable amount of public acclaim.

The value of the second (conceptual) definition is that it contains more information about "stardom." If accurate, it has the deeper significance. It might help the basketball coach more in developing the optimum characteristics of his squad. It permits him to *predict* athletic skill in advance of the first game.

Returning to our two definitions of an acid, the first, the operational definition, gives clearcut instructions on how to decide whether a given substance is an acid. Dissolve it in water and see if it has certain properties. The second (conceptual) definition, however, has the deeper significance since it includes our knowledge of why an acid has these particular properties. It provides a basis for finding hidden likenesses between acid–base reactions in water and other reactions in other solvents. Each type of definition has its merit; neither is *the* definition.

QUESTIONS AND PROBLEMS

1. What is the concentration of $H^+(aq)$ in an aqueous solution in which $[OH^-] = 1.0 \times 10^{-3}$ *M*?

2. 100 ml of the HCl solution described in Exercise 11-2 (p. 182) is diluted with water to 1.00 liter. What is the concentration of $H^+(aq)$? What is $[OH^-]$ in this solution?

3. Vinegar, lemon juice, and curdled milk, all taste sour. What other properties would you expect them to have in common?

4. Give the name and formula of three hydrogen containing compounds that are not classified as acids. State for each compound one or more properties common to acids that it does not possess.

5. As a solution of barium hydroxide is mixed with a solution of sulfuric acid, a white precipitate forms and the electrical conductivity decreases markedly. Write equations for the reactions that occur and account for the conductivity change.

6. An eyedropper is calibrated by counting the number of drops required to deliver 1.0 ml. Twenty drops are required.

   (a) What is the volume of one drop?
   (b) Suppose one such drop of 0.20 *M* HCl is added to 100 ml of water. What is $[H^+]$?
   (c) By what factor did $[H^+]$ change when the one drop was added?

   Answer. (c) $\times$ 1000.

7. Suppose drops (from the same eyedropper) of 0.10 *M* NaOH are added, one at a time, to the 100 ml of HCl in Problem 6b.

   (a) What will be $[H^+]$ after one drop is added?
   (b) What will be $[H^+]$ after two drops are added?
   (c) What will be $[H^+]$ after three drops are added?

8. Calculate $[H^+]$ and $[OH^-]$ in a solution made by mixing 50.0 ml 0.200 *M* HCl and 49.0 ml 0.200 *M* NaOH.

   Answer. $[OH^-] = 5 \times 10^{-12}$ *M*.

9. Calculate $[H^+]$ and $[OH^-]$ in a solution made by mixing 50.0 ml 0.200 *M* HCl and 49.9 ml 0.200 *M* NaOH.

10. How much more 0.200 *M* NaOH solution need be added to the solution in Problem 9 to change $[H^+]$ to 10^{-7} *M*?

11. An acid is a substance HB that can form $H^+(aq)$ in the equilibrium:

$$HB(aq) \rightleftharpoons H^+(aq) + B^-(aq)$$

   (a) Does equilibrium favor reactants or products for a strong acid?
   (b) Does equilibrium favor reactants or products for a very weak acid?
   (c) If acid HB_1 is a stronger acid than acid HB_2, is K_1 a larger or smaller number than K_2?

$$K_1 = \frac{[H^+][B_1^-]}{[HB_1]} \qquad K_2 = \frac{[H^+][B_2^-]}{[HB_2]}$$

12. (a) Which of the following acids is the strongest and which is the weakest?

   ammonium ion, NH_4^+ (in an NH_4Cl solution);
   bisulfate ion, HSO_4^- (in a $KHSO_4$ solution);
   hydrogen sulfide, H_2S.

   (b) If 0.1 *M* solutions are made of NH_4Cl, $KHSO_4$, and H_2S, in which will $[H^+]$ be highest and in which will it be lowest?

13. (a) Nitric acid is a very strong acid. What is $[H^+]$ in a 0.050 *M* HNO_3 solution?
   (b) Hydrogen peroxide, H_2O_2, is a very weak acid. What is $[H_2O_2]$ in a 0.050 *M* H_2O_2 solution?

14. From a study of Appendix 2, what generalization can you make concerning acids which contain more than one atom of hydrogen in their molecules or ions?

15. A 0.25 *M* solution of benzoic acid (symbolize it HB) is found to have a hydrogen ion concentration $[H^+] = 4 \times 10^{-3}$ *M*.

   (a) Assuming the simple reaction $HB(aq) \rightleftharpoons H^+(aq) + B^-(aq)$, calculate K_A for benzoic acid.
   (b) Compare the values of $[HB]$, $[H^+]$, $[B^-]$, and K_A used in this problem to the corresponding quantities in the benzoic acid calculation presented in the text, Section 11-3.2.

16. If 23 grams of formic acid, HCOOH, are dissolved in 10.0 liters of water at 20°C, the $[H^+]$ is found to be 3.0×10^{-3} *M*. Calculate K_A.

17. A chemist dissolved 25 grams of CH_3COOH in enough water to make one liter of solution. What is the concentration of this acetic acid solution? What is the concentration of $H^+(aq)$? Assume a negligible change in $[CH_3COOH]$ because of dissociation to $H^+(aq)$.

18. When sodium acetate, CH_3COONa, is added to an aqueous solution of hydrogen fluoride, HF, a reaction occurs in which the weak acid HF loses H^+.

    (a) Write the equation for the reaction.
    (b) What weak acid is competing with HF for H^+?

    Answer. (b) CH_3COOH, acetic acid.

19. (a) Write the equation for the reaction that shows the acid–base reaction between hydrogen sulfide, H_2S, and carbonate ion, CO_3^{-2}.
    (b) What are the two acids competing for H^+?
    (c) From the values of K_A for these two acids (see Table 11-IV), predict whether the equilibrium favors reactants or products.

    Answer. (c) Products.

20. Write the equations for the reaction between each of the following acid–base pairs. For each reaction, predict whether reactants or products are favored (using the values of K_A given in Appendix 2).

    (a) $HNO_2(aq) + NH_3(aq) \rightleftharpoons$
    (b) $NH_4^+(aq) + F^-(aq) \rightleftharpoons$
    (c) $C_6H_5COOH(aq) + CH_3COO^-(aq) \rightleftharpoons$

    Answer. (a) $HNO_2(aq) + NH_3(aq) \rightleftharpoons NO_2^-(aq) + NH_4^+(aq)$. Products, $NO_2^-(aq)$ and $NH_4^+(aq)$ favored.

21. Write the equations for the reactions between each of the following acid–base pairs. For each reaction, predict whether reactants or products are favored.

    (a) $H_2SO_3(aq) + HCO_3^-(aq) \rightleftharpoons$
    (b) $H_2CO_3(aq) + SO_3^{-2}(aq) \rightleftharpoons$
    (c) $H_2SO_3(aq) + SO_3^{-2}(aq) \rightleftharpoons$

22. If the *p*H of a solution is 5, what is $[H^+]$? Is the solution acidic or basic?

23. What is $[H^+]$ in a solution of $pH = 8$? Is the solution acidic or basic? What is $[OH^-]$ in the same solution?

24. Devise an operational and also a conceptual definition of a gas.

SVANTE AUGUST ARRHENIUS 1859–1927

During a full life this great Swedish chemist met practically all the important men of science of his day and won their affection as well as their highest regard. He is said to have had a genius for friendship. Nevertheless, his early career was filled with a battle for acceptance.

At 22 Arrhenius had performed many experiments concerned with the passage of electricity through aqueous solutions and he decided to continue this work in preparation for his doctorate. For two years he collected voluminous data on hundreds of solutions and concentrations while working in the laboratory of the University of Upsala. He then formulated a carefully considered hypothesis that aqueous solutions contain charged species, ions. This was a revolutionary suggestion and his professors found it so different from their own ideas that they only grudgingly granted his degree.

Undiscouraged, Arrhenius sent copies of his thesis to other scientists. Although few took his radical idea seriously, the great German scientist, Ostwald, became so excited that he traveled to Sweden to meet Arrhenius. Encouraged by this support, Arrhenius traveled and studied in Germany and Holland. Finally, in 1889, his paper "On the Dissociation of Substances in Aqueous Solutions" was published.

He was invited to come to Leipzig as a professor at the University but chose to return to Sweden as a lecturer and teacher at a high school in Stockholm. His theory was still not generally accepted, and those who did not agree with it dubbed its proponents "the wild horde of the Ionians." Even Arrhenius' assignment as professor at Stockholm in 1893 was questioned until a storm of protest came from scientists in Germany. Within two years of this appointment he was elected President of the University and was named a Nobel laureate, the award being only the third such in chemistry. Arrhenius finally received the acclaim he had so long deserved. He was offered the coveted position of professor of chemistry at Berlin but the King of Sweden founded the Nobel Institute for Physical Chemistry and, in 1905, Arrhenius became its director. He continued as a tireless experimenter and an extremely versatile scientist until his death in 1927.

Arrhenius' success in science must be credited not only to his brilliance as a scientist but also to his conviction in his views. His understanding of the electrical properties of aqueous solutions was so far ahead of contemporary thought that it would have been ignored but for his confidence in the usefulness of his theory and his refusal to abandon it. It is fitting tribute that the ionic model of aqueous solutions has changed permanently the face of inorganic chemistry.

CHAPTER 

Oxidation-Reduction Reactions

Chemical thermodynamics enables one to state what may happen when two substances react.

WENDELL M. LATIMER, 1953

We have now made use of the principles of equilibrium in two general types of reactions. In Chapter 10 we considered reactions involving a solid and a solution: dissolving and precipitation. In Chapter 11 we turned to reactions occurring entirely in solution and involving proton transfer. Now we shall take a more general view of equilibrium in aqueous solutions, a view provided by an investigation of the chemistry of an electrochemical cell.

12-1 ELECTROCHEMICAL CELLS

Electrochemical cells are familiar—a flashlight operates on current drawn from electrochemical cells called dry cells, and automobiles are started with the aid of a battery, a set of electrochemical cells in tandem. The last time you changed the dry cells in a flashlight because the old ones were "dead," did you wonder what had happened inside those cells? Why does electric current flow from a new dry cell but not from one that has been used many hours? We shall see that this is an important question in chemistry. By studying the chemical reactions that occur in an electrochemical cell we discover a basis for predicting whether equilibrium in a chemical reaction favors reactants or products. The reactions are of the type called oxidation-reduction reactions, which is the subject of this chapter.

12-1.1 The Chemistry of an Electrochemical Cell

Let's begin our investigation of an electrochemical cell by assembling one. Fill a beaker with a dilute solution of silver nitrate (about 0.1 *M* will do) and another beaker with dilute copper sulfate. Put a silver rod in the $AgNO_3$ solution and a copper rod in the $CuSO_4$ solution. With a wire, connect the silver rod to one terminal of an

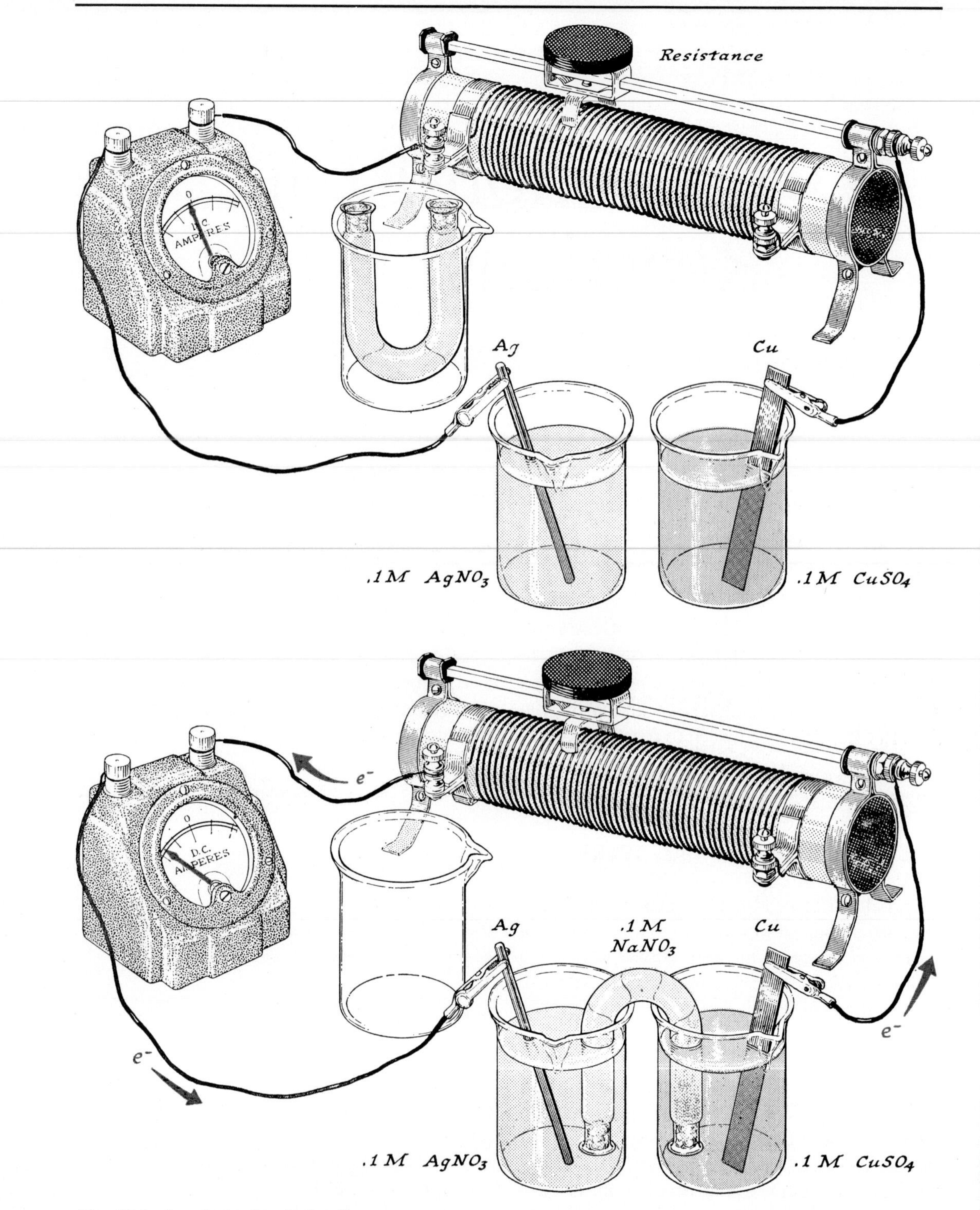

Fig. 12-1. **An electrochemical cell.**

ammeter to measure the electric current. Connect the other terminal of the ammeter through a wire resistance, *R*, to the copper rod. Finally, connect the two solutions to complete the electric circuit. Figure 12-1 shows suitable equipment. These two drawings show how a connection can be made between the two solutions to complete the electric circuit. A glass tube containing a sodium nitrate solution furnishes an electrical path. It is called a salt bridge.

As soon as the last connection is made, things start to happen. The ammeter needle deflects—electric current is moving through the meter and the wire resistance, *R*. The direction of current flow is that of electrons moving from the copper rod to the silver rod. The resistance becomes warm—the cell is doing work as it forces electrons through *R*. In the beakers, the copper rod dissolves and the silver rod grows. As time goes by, the ammeter shows less and less current flow until, finally, there is none.

Now let's be more quantitative. Let's repeat the experiment, weighing the metal rods before and after the test. The weighing shows that during the test the copper rod has become 0.635 gram lighter and the silver rod has become 2.16 grams heavier. Chemical reaction has occurred and, as any good chemist will do, we immediately ask, "How many moles of copper and silver are involved?"

$$\text{Moles Cu dissolved} = \frac{\text{wt Cu dissolved}}{\text{atomic wt Cu}} = \frac{0.635 \text{ g}}{63.5 \text{ g/mole}} = 0.0100 \text{ mole}$$

$$\text{Moles Ag deposited} = \frac{\text{wt Ag deposited}}{\text{atomic wt Ag}} = \frac{2.16 \text{ g}}{108 \text{ g/mole}} = 0.0200 \text{ mole}$$

We see that there is a simple relationship between the weight of copper dissolved and the weight of silver deposited. One mole of copper dissolves in the right beaker for every two moles of silver deposited in the left beaker. Copper ions, $Cu^{+2}(aq)$, are formed in the right beaker from the neutral copper metal atoms. This means atoms of copper release electrons into the copper rod. These electrons move into the wire, through the resistance, and through the ammeter. They arrive at the silver rod in the left beaker, where silver metal is formed from silver ions, $Ag^{+}(aq)$. Here, the positive silver ions draw electrons from the silver rod to become neutral silver metal atoms. Summarizing these processes, we have:

In the right beaker, $Cu(s) \longrightarrow Cu^{+2}(aq) + 2e^-$ (*1*)

In the left beaker, $2Ag^{+}(aq) + 2e^- \longrightarrow 2Ag(s)$ (*2*)

Overall reaction, $Cu(s) + 2Ag^{+}(aq) \longrightarrow 2Ag(s) + Cu^{+2}(aq)$ (*3*)

The overall reaction describes what goes on in the entire electrochemical cell. In half of the cell, the right beaker, reaction (*1*) occurs. In the other half of the cell, the left beaker, reaction (*2*) occurs. Hence, reactions (*1*) and (*2*) are called *half-cell reactions* or *half-reactions*.

There are several interesting features about these half-reactions:

(1) *The two half-reactions are written separately.* In our electrochemical cell the half-reactions occur in separate beakers. As the name implies, there must be two such reactions.
(2) *Electrons are shown as part of the reaction.* Our ammeter shows that electrons are involved. They flow when the reaction starts, and do not flow when the reaction stops. The meter also indicates that the electrons leave the copper rod, pass through the wire, and enter the silver rod.
(3) *New chemical species are produced in each half of the cell.* The copper rod is converted to copper ions (the rod loses weight) and the silver ions are changed to metal (the silver rod gains weight). The new species can be explained in terms of gain of electrons (by silver) and loss of electrons (by copper).
(4) *The half-reactions, when combined, express the overall, or net, reaction.*

The net reaction (*3*) is obtained by combining (*1*) and (*2*) so as to cause the exact balancing of electrons lost by copper atoms, in (*1*), and electrons gained by silver ion, in (*2*). This cancellation is necessary because electrical measurements show that the electrochemical cell operates without accumulation or consumption of electric charge. The reaction mixture always remains electrically neutral. *The number of electrons lost equals the number of electrons gained.*

We see that the overall chemical reaction that occurs in an electrochemical cell is conveniently described in terms of two types of half-reactions. In one, electrons are lost; in the other, they are gained. To distinguish these half-reactions we need two identifying names.

The half-reaction in which electrons are lost is called ***oxidation.***

Oxidation $Cu(s) \longrightarrow Cu^{+2}(aq) + 2e^-$ (*1*), (*4*)*

The half-reaction in which electrons are gained is called ***reduction.***

Reduction $2Ag^{+}(aq) + 2e^- \longrightarrow 2Ag(s)$ (*2*), (*5*)

The overall reaction is called an ***oxidation-reduction reaction.***

Oxidation-reduction reaction

$$Cu(s) + 2Ag^{+}(aq) \longrightarrow Cu^{+2}(aq) + 2Ag(s) \quad (3), (6)$$

It is often convenient and usually informative to treat oxidation-reduction in terms of half-reactions. When it *is* convenient, *oxidation* is involved in the half-reaction showing *loss of electrons*, and *reduction* is involved in the half-reaction showing *gain of electrons*.

12-1.2 Oxidation-Reduction Reactions in a Beaker

These ideas, developed for an electrochemical cell, have great importance in chemistry because they are also applicable to chemical reactions that occur in a single beaker. Without an electric circuit or an opportunity for electric current to flow, the chemical changes that occur in a cell can be duplicated in a single solution. It is reasonable to apply the same explanation.

* In this and later chapters, each equation is assigned a consecutive number (given on the right). If the equation has occurred earlier in the chapter, the earlier number is given as well (on the left).

COPPER OXIDIZED BY $Ag^{+}(aq)$ IN A BEAKER

We can easily demonstrate that reaction (*3*) can occur even when the half-cells are not separated. You did this in Experiment 7. A copper wire immersed in $AgNO_3$ solution caused copper to dissolve [the blue color of $Cu^{+2}(aq)$ appeared] and metallic silver was precipitated. The ratio of (Cu dissolved) to (Ag formed) was the same as that in our cell, hence the net reaction was the same.

EXERCISE 12-1

Compare the mole ratio Ag/Cu derived from your own data for Experiment 7 to the electrochemical data given in Section 12-1.1.

The moles of silver deposited per mole of copper dissolved are the same whether reaction (*3*) is carried out in an electrochemical cell or in a single beaker, as in Experiment 7. If, in the cell, electrons are transferred from copper metal (forming Cu^{+2}) to silver ion (forming metallic silver), then electrons must have been transferred from copper metal to silver ion in Experiment 7.

Fig. 12-2. **Oxidation-reduction reactions** *can occur in a beaker.*

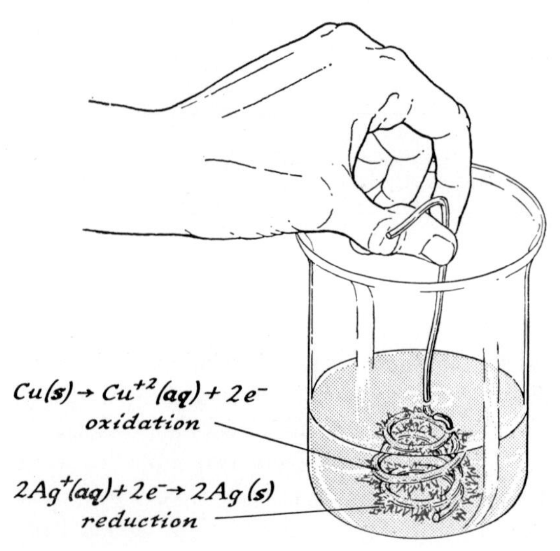

Thus, Experiment 7 involved the same oxidation-reduction reaction but *the electron transfer must have occurred locally between individual copper atoms* (in the metal) and *individual silver ions* (in the solution near the metal surface). This local transfer replaces the wire "middleman" in the cell, which carries electrons from one beaker (where they are released by copper) to the other (where they are accepted by silver ions).

ZINC OXIDIZED BY $H^+(aq)$ IN A BEAKER

Many oxidation-reduction reactions (nicknamed "redox" reactions) take place in aqueous solution. One of these was mentioned in Section 11-2.1 when we characterized acids:

$$Zn(s) + 2H^+(aq) \longrightarrow Zn^{+2}(aq) + H_2(g) \quad (7)$$

Each zinc atom loses two electrons in changing to a zinc ion, therefore zinc is oxidized. Each hydrogen ion gains an electron, changing to a hydrogen atom, therefore hydrogen is reduced. (After reduction, two hydrogen atoms combine to form molecular H_2.) As before, reaction (*7*) can be separated into two half-reactions:

$$Zn(s) \longrightarrow Zn^{+2}(aq) + 2e^- \quad (8)$$

$$2H^+(aq) + 2e^- \longrightarrow H_2(g)^* \quad (9)$$

Net reaction

$$Zn(s) + 2H^+ \longrightarrow Zn^{+2} + H_2(g) \quad (7), (10)$$

Thus, the reaction by which a metal dissolves in an acid is conveniently discussed in terms of oxidation and reduction involving electron transfer. The reaction can be divided into half-reactions to show the electron gain (by H^+ ions) and the electron loss (by metal atoms).

Not all metals react with aqueous acids. Among the common metals, magnesium, aluminum, iron, and nickel liberate H_2 as zinc does. Other metals, including copper, mercury, silver, and gold, do not produce measurable amounts of hydrogen even though we make sure that the equilibrium state has been attained. With these metals, hydrogen is not produced and it is surely not just because of slow reactions. Apparently some metals release electrons to H^+ [as zinc does in reaction (*10*)] and others do not.

* From this point on in this chapter we will consider only aqueous solutions, hence we will not specify *(aq)* for each ion.

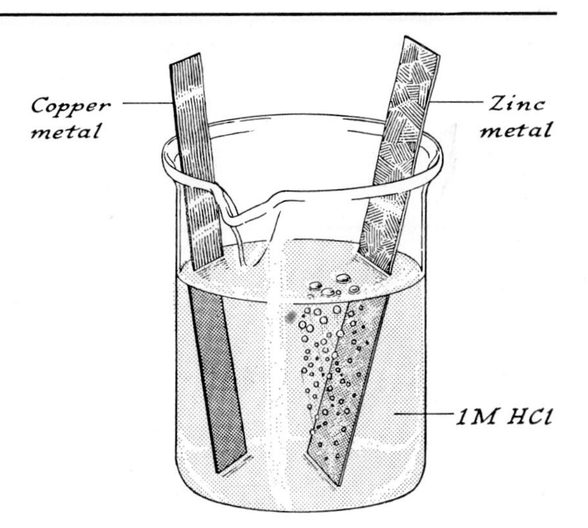

Fig. 12-3. **Some metals** *release electrons to* H^+ *and others do not.*

ZINC OXIDIZED BY $Cu^{+2}(aq)$ IN A BEAKER

As a third oxidation-reduction example, suppose a strip of metallic zinc is placed in a solution of copper nitrate, $Cu(NO_3)_2$. The strip becomes coated with reddish metallic copper and the bluish color of the solution disappears. The presence of zinc ion, Zn^{+2}, among the products can be shown when the Cu^{+2} color is gone. Then if hydrogen sulfide gas is passed into the mixture, white zinc sulfide, ZnS, can be seen. The reaction between metallic zinc and the aqueous copper nitrate is

$$Zn(s) + Cu^{+2} \longrightarrow Zn^{+2} + Cu(s) \quad (11)$$

Zinc has lost electrons in reaction (*11*) to form Zn^{+2}:

$$Zn(s) \longrightarrow Zn^{+2} + 2e^- \quad (8), (12)$$

Zinc is oxidized. If zinc is oxidized, releasing electrons, something must be reduced, accepting these electrons. Copper ion is reduced:

$$Cu^{+2} + 2e^- \longrightarrow Cu(s) \quad (13)$$

This time, copper ion *gains* electrons from the zinc, in contrast to the behavior in Experiment 7, where copper metal *lost* electrons to silver.

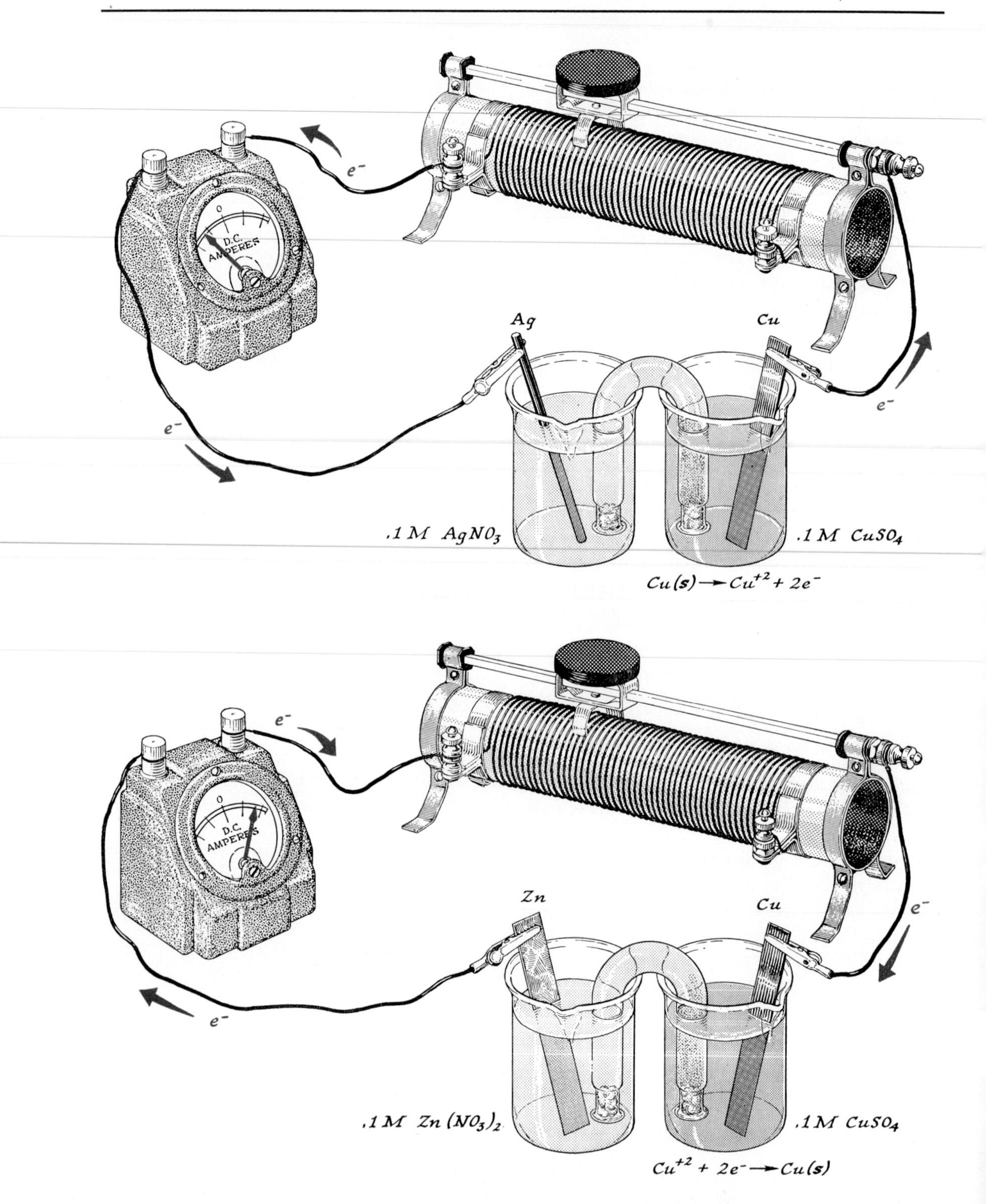

Fig. 12-4. **Two electrochemical cells involving copper:** *with silver, copper is oxidized; with zinc,* Cu^{+2} *is reduced.*

What about the state of equilibrium for the reaction represented by equation (*11*)? Let us place a strip of metallic copper in a zinc sulfate solution. No visible reaction occurs and attempts to detect the presence of cupric ion by adding H_2S to produce the black color of cupric sulfide, CuS, fail. Cupric sulfide has such low solubility that this is an extremely sensitive test, yet the amount of Cu^{+2} formed cannot be detected. Apparently the state of equilibrium for the reaction (*11*) greatly favors the products over the reactants.

12-1.3 Competition for Electrons

These reactions can be viewed as a competition between two kinds of atoms (or molecules) for electrons. Equilibrium is attained when this competition reaches a balance between opposing reactions. In the case of reaction (*3*), copper metal reacting with silver nitrate solution, the Cu(*s*) releases electrons and Ag^+ accepts them so readily that equilibrium greatly favors the products, Cu^{+2} and Ag(*s*). Since randomness tends to favor neither reactants nor products, the equilibrium must favor products because the energy is lowered as the electrons are transferred. If we regard reaction (*3*) as a competition between silver and copper for electrons, stability favors silver over copper.

The same sort of competition for electrons is involved in reaction (*11*), in which Zn(*s*) releases electrons and Cu^{+2} accepts them. This time the competition for electrons is such that equilibrium favors Zn^{+2} and Cu(*s*). By way of contrast, compare the reaction of metallic cobalt placed in a nickel sulfate solution. A reaction occurs,

$$Co(s) + Ni^{+2} \rightleftharpoons Co^{+2} + Ni(s) \qquad (14)$$

At equilibrium, chemical tests show that both Ni^{+2} and Co^{+2} are present at moderate concentrations. In this case, neither reactants (Co and Ni^{+2}) nor products (Co^{+2} and Ni) are greatly favored.

This competition for electrons is reminiscent of the competition for protons among acids and bases. The similarity suggests that we might develop a table in which metals and their ions are listed by tendency to release electrons just as we did in Table 11-IV (p. 191) in which the acid strength indicates tendency for an acid to release H^+.

We can already make some comparisons. We might begin by listing some of the half-reactions we have encountered in this chapter. We shall write them to show the release of electrons and then arrange them in order of their tendency to do so. First we considered reaction (*3*) and discovered that copper releases electrons to silver ion. Therefore, we shall write our first two half-reactions in the order

$$Cu(s) \longrightarrow Cu^{+2} + 2e^- \qquad (1), (15)$$

$$Ag(s) \longrightarrow Ag^+ + e^- \qquad (16)$$

Listing the Cu–Cu^{+2} half-reaction first indicates that it releases electrons more readily than does the Ag–Ag^+ half-reaction.

Now consider reaction (*11*). Since zinc releases electrons to copper ion, we know that we must add it to our list at the top:

$$Zn(s) \longrightarrow Zn^{+2} + 2e^- \qquad (8), (17)$$

$$Cu(s) \longrightarrow Cu^{+2} + 2e^- \qquad (1), (18)$$

$$Ag(s) \longrightarrow Ag^+ + e^- \qquad (16), (19)$$

Listing the Zn–Zn^{+2} half-reaction first tells us that it releases electrons more readily than does the Cu–Cu^{+2} half-reaction. But if this is true, then the Zn–Zn^{+2} half-reaction must also release electrons more readily than does the Ag–Ag^+ half-reaction. *Our list leads us to expect that zinc metal will release electrons to silver ion, reacting to produce zinc ion and silver metal.*

We should test this proposal! We dip a piece of zinc metal in a solution of silver nitrate. The result confirms our expectation; zinc metal dissolves and bright crystals of metallic silver appear.

Our data allow us to make one more addition to the list. By reaction (*7*), zinc reacts with H^+ to give Zn^{+2} and $H_2(g)$. The half-reaction H_2–$2H^+$ must be placed below the Zn–Zn^{+2} half-reaction. How far below? To answer that, remember that copper does not react with H^+ to produce H_2. This indicates that the half-reaction

H_2–$2H^+$ releases electrons more readily than does the half-reaction Cu–Cu^{+2}. Now we can expand our list to that given in Table 12-I.

Table 12-I

SOME HALF-REACTIONS LISTED IN ORDER OF TENDENCY TO RELEASE ELECTRONS

$Zn(s) \longrightarrow Zn^{+2} + 2e^-$	(*8*), (*20*)
$H_2(g) \longrightarrow 2H^+ + 2e^-$	(*21*)
$Cu(s) \longrightarrow Cu^{+2} + 2e^-$	(*1*), (*22*)
$Ag(s) \longrightarrow Ag^+ + e^-$	(*16*), (*23*)

EXERCISE 12-2

From the statement in the text that nickel metal reacts with H^+ to give $H_2(g)$ and the additional information that zinc metal reacts readily with nickel sulfate solution, decide where to add the half-reaction Ni–Ni^{+2} in our list.

The value of this list is obvious. Any half-reaction can be combined with the reverse of another half-reaction (in the proportion for which electrons gained is equal to electrons lost) to give a possible chemical reaction. Our list permits us to predict whether equilibrium favors reactants or products. We would like to expand our list and to make it more quantitative. Electrochemical cells help us do this.

12-1.4 Operation of an Electrochemical Cell

Now let's take a more detailed look into the electrochemical cell. Figure 12-5 shows a cross-section of a cell that uses the same chemical reaction as that depicted in Figure 12-1. The only difference is that the two solutions are connected differently. In Figure 12-1 a tube containing a solution of an electrolyte (such as KNO_3) provides a conducting path. In Figure 12-5 the silver nitrate is placed in a porous porcelain cup. Since the silver nitrate and copper sulfate solutions can seep through the porous cup, they provide their own connection to each other.

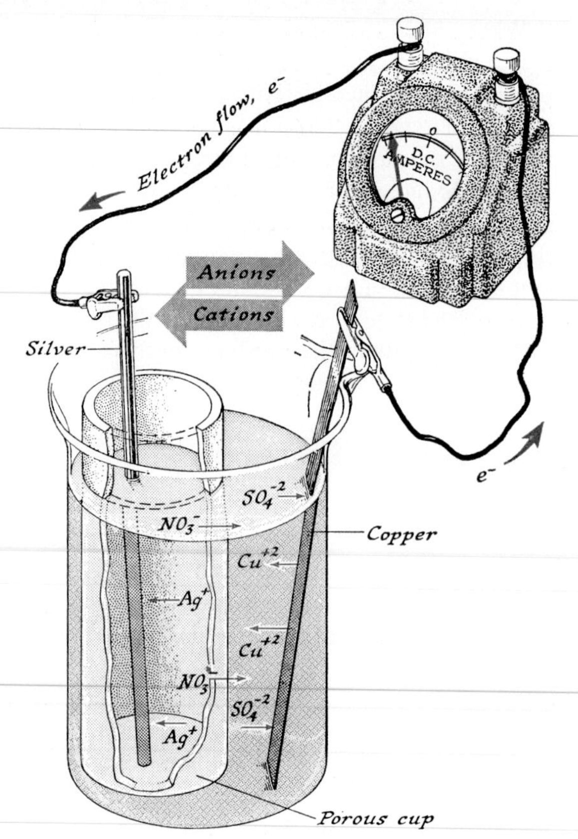

Fig. 12-5. **The operation of an electrochemical cell.**

Before examining the processes in a cell, we should name the parts of a cell and clear away some language matters. The electrons enter and leave the cell through electrical conductors—the copper rod and the silver rod in Figure 12-5—called electrodes. At one electrode, the copper electrode, electrons are released and oxidation occurs. The electrode where oxidation occurs is called the *anode*. At the other electrode, the silver electrode, electrons are gained and reduction occurs. The electrode where reduction occurs is called the *cathode*.

As electrons leave the cell from the anode (electrons are released where oxidation occurs), positively charged Cu^{+2} ions are produced. Negative charge is leaving (by means of the electron movement) and positive charge is produced (the Cu^{+2} ions) in this half of the cell. How is electrical neutrality maintained? It must be maintained by the movement of ions through the solution. Negative ions drift toward the anode and positive ions move away. It is because negative ions in a cell always drift toward the anode that *negative ions are called* ***anions*** (pronounced an′ions). Since positive ions drift away from the anode and toward the cathode, *positive ions are called* ***cations*** (pronounced cat ions).

Here is our electrochemical glossary:

Electrodes: The conductors at which reactions occur in an electrochemical cell.
Anode: The electrode at which oxidation occurs.
Cathode: The electrode at which reduction occurs.
Anion: Negatively charged ion.
Cation: Positively charged ion.

With the verbal matters out of the way, let's take an electrical tour through the cell shown in Figure 12-5. We'll start at the surface of the copper rod and follow the process around the entire circuit and back to the copper rod. Let us begin with a particular copper atom that loses two electrons:

$$Cu(s) \longrightarrow Cu^{+2} + 2e^- \qquad (1), (24)$$

The Cu^{+2} ion drifts away into the solution but the electrons remain in the copper rod. They move up through the copper anode, through the wire, and enter the silver cathode. At the surface of this rod, the electrons encounter Ag^+ ions in the solution. The electrons react with Ag^+ ions to give neutral silver atoms which remain on the rod as silver metal:

$$2Ag^+ + 2e^- \longrightarrow 2Ag(s) \qquad (2), (25)$$

Now there is an excess of positive charge in the solution near the copper anode and a deficiency of positive charge in the solution near the silver cathode. Two negative charges have been moved from the anode half-cell through the wire to the cathode half-cell. This charge movement causes all of the negative ions (anions) in the solution (SO_4^{-2} and NO_3^-) to start drifting toward the anode. All of the positive ions (cations) start drifting toward the cathode. When the movement of all of these ions amounts to the charge movement of two negative charges from the cathode porous cup to the anode beaker, our tour of the cell is completed. Electrical neutrality has been restored and the net reaction is

$$Cu(s) + 2Ag^+ \longrightarrow Cu^{+2} + 2Ag(s) \qquad (3), (26)$$

12-2 ELECTRON TRANSFER AND PREDICTING REACTIONS

The usefulness of Table 12-I is clear. Qualitative predictions of reactions can be made with the aid of the ordered list of half-reactions. Think how the value of the list would be magnified if we had a *quantitative* measure of electron losing tendencies. The voltages of electrochemical cells furnish such a quantitative measure.

12-2.1 Electron Losing Tendency

The circuit shown in Figure 12-1 includes a wire resistance coil, *R*. As the current flows through *R*, heat is generated. The cell is doing electrical work in forcing the electron current through *R*. Again we apply the Law of Conservation of Energy. As the electrons leave *R*, they must have lower potential energy than they had when they entered. As they fall to the lower potential energy, the energy change appears as heat. This potential energy change is measured by voltage. Just as lowering a mass from a higher altitude decreases its potential energy, moving an electric charge to a lower voltage lowers its potential energy.

So the voltage of an electrochemical cell measures its capacity for doing electrical work. Different cells show different voltages. To see the importance of this voltage, consider the experiment shown in Figure 12-6. In Figure 12-6A we have a cell based upon reaction (*27*):

$$Zn(s) + Ni^{+2} \longrightarrow Zn^{+2} + Ni(s) \qquad (27)$$

If the concentrations in the cell are 1 *M*, the voltage in the cell is 0.5 volt. We will call this

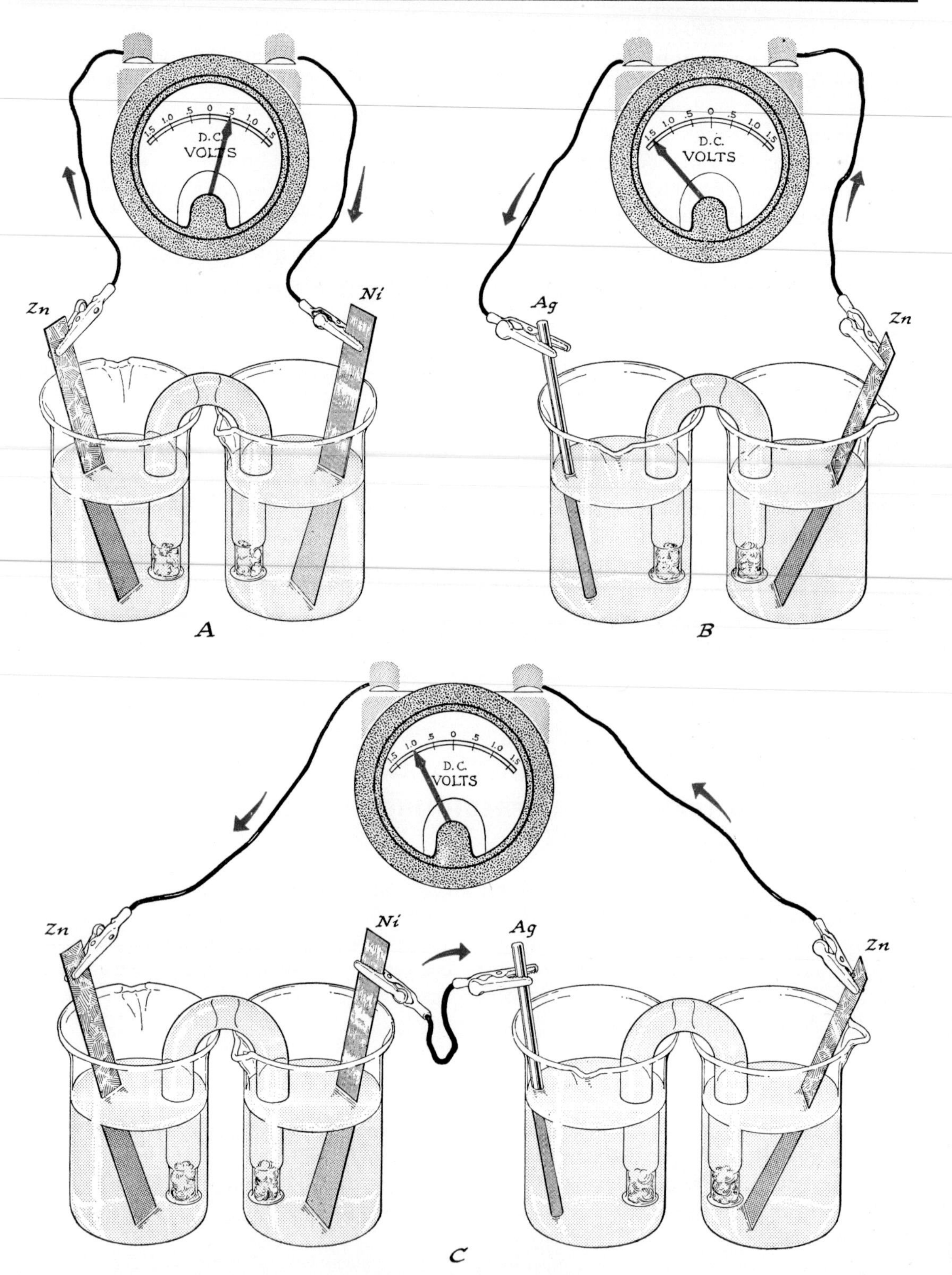

Fig. 12-6. **Two cells in opposition.** *In which direction will current flow?*

voltage $E°$ (Zn–Ni^{+2}). As the cell is shown, electrons move clockwise through the meter. On the right in Figure 12-6B is a cell based upon reaction (*28*). The voltage of this cell, $E°$ (Zn–Ag^+), is 1.5 volts for 1 M concentrations.

$$Zn(s) + 2Ag^+ \longrightarrow Zn^{+2} + 2Ag(s) \qquad (28)$$

The electrical hookup in this cell causes electrons to move counterclockwise through the meter.

Then, in Figure 12-6C, the two cells are reconnected in opposition to each other. The zinc–nickel cell pushes the electrons clockwise and the silver–zinc cell pushes the electrons counterclockwise. In each cell, the zinc electrode has a tendency to dissolve, releasing electrons into the external circuit. But electrons cannot flow in both directions. Which cell will win out? Experiment shows that electron current flows counterclockwise from the zinc electrode of the Zn–Ag^+ cell into the zinc electrode of the Zn–Ni^{+2} cell. We can explain this intuitively by saying the stronger (1.5 volts) cell has overpowered the weaker (0.5 volt) cell. The Zn–Ag^+ cell proceeds to react in the sense it would if it were alone in the circuit. The other cell is forced to react in the direction opposite to that it would take spontaneously. The reaction that generates the higher voltage prevails. *The reaction which generates the higher voltage has the greater tendency to proceed* and the observed voltage measures the tendency.

We can learn one more valuable lesson from these two cells working against each other. The measured voltage of our double cell (Figure 12-6) is 1.0 volt. Since the voltage for each cell is a measure of the tendency to send out electrons, we could calculate the net voltage (1.0 volt) by combining the individual cell voltages. In this case we must subtract the two because they are hooked up to oppose each other. So we find

$$\begin{aligned} E^\circ_{net} &= E^\circ(\text{Zn–Ag}^+) - E^\circ(\text{Zn–Ni}^{+2}) \\ &= 1.5 - 0.5 \\ &= 1.0 \text{ volt} \end{aligned}$$

It is interesting to write an overall reaction for these two cells *when they are operating as in Figure 12-6:*

Zn–Ag^+ cell:

$$Zn(s) \longrightarrow Zn^{+2} + 2e^- \qquad (8), (29)$$

$$2Ag^+ + 2e^- \longrightarrow 2Ag(s) \qquad (2), (30)$$

Zn–Ni^{+2} cell:

$$Zn^{+2} + 2e^- \longrightarrow Zn(s) \qquad (8), (31)$$

$$Ni(s) \longrightarrow Ni^{+2} + 2e^- \qquad (32)$$

Net reaction in both cells:

(*29*) + (*30*) + (*31*) + (*32*)

$$Ni(s) + 2Ag^+ \longrightarrow Ni^{+2} + 2Ag(s) \qquad (33)$$

But the net reaction is just the reaction that occurs in a Ni–Ag^+ cell! What is the voltage of such a cell? Experiment shows that such a cell has a voltage of 1.0 volt. We find our double cell (Figure 12-6) has a voltage identical to that of a Ni–Ag^+ cell alone. The tendency of zinc to release electrons through reaction (*29*) must have influenced the voltage of the Zn–Ag^+ cell. The same tendency of zinc to release electrons must also have influenced the voltage of the Zn–Ni^{+2} cell. When these cells are put in opposition, the tendency of zinc to release electrons exactly cancels.

This shows that the voltage of a given cell may be thought of as being made up of two parts, one part characteristic of one of the half-reactions and one part characteristic of the other half-reaction. Chemists call these two parts "half-cell potentials," a term that emphasizes the relation between voltage and potential energy. The half-cell potentials are symbolized $E°$.

Thus we write for the Zn–Ag^+ cell

$$E^\circ_1 = E^\circ(\text{Zn–Zn}^{+2}) - E^\circ(\text{Ag–Ag}^+) = 1.5 \text{ volts} \qquad (34)$$

and for the Zn–Ni^{+2} cell

$$E^\circ_2 = E^\circ(\text{Zn–Zn}^{+2}) - E^\circ(\text{Ni–Ni}^{+2}) = 0.5 \text{ volt} \qquad (35)$$

If we place these batteries in opposition, the net voltage will be

$$\begin{aligned} E^\circ_3 = E^\circ_1 - E^\circ_2 &= [E^\circ(\text{Zn–Zn}^{+2}) - E^\circ(\text{Ag–Ag}^+)] \\ &\quad - [E^\circ(\text{Zn–Zn}^{+2}) - E^\circ(\text{Ni–Ni}^{+2})] \end{aligned}$$

$$E^\circ_3 = E^\circ(\text{Ni–Ni}^{+2}) - E^\circ(\text{Ag–Ag}^+) = 1.0 \text{ volt} \qquad (36)$$

and also

$$E^\circ_3 = E^\circ_1 - E^\circ_2 = (1.5) - (0.5) = 1.0 \text{ volt}$$

We see that:

(a) the value $E°$(Zn–Zn^{+2}) cancels in (*36*);

(b) (*36*) is just the expected sum of half-reactions for a Ni–Ag cell;

(c) the calculated difference between E_1° and E_2° is just 1.0 volt, the same as measured for a Ni–Ag^+ cell.

MEASURING HALF-CELL POTENTIALS

We would like to measure the contribution each half-reaction makes to the voltage of a cell. Yet *every* cell involves *two* half-reactions and *every* cell voltage measures a *difference* between their half-cell potentials. We can never isolate one half-reaction to measure its E°. An easy escape is to assign an arbitrary value to the potential of some selected half-reaction. Then we can combine all other half-reactions in turn with this reference half-reaction and find values for them relative to our reference. The handiest arbitrary value to assign is zero and chemists have decided to give it to the half-reaction

$$H_2(g) \longrightarrow 2H^+ + 2e^- \qquad E^\circ = 0.00 \cdots \text{ volt} \quad (21), (37)$$

We must control concentration during these measurements of E°, since the voltage of a cell changes as concentrations change. For example, in the laboratory we studied a cell based on reaction (*38*):

$$Zn(s) + Cu^{+2} \longrightarrow Zn^{+2} + Cu(s) \quad (11), (38)$$

We found that the voltmeter readings were different for different concentrations. Were these readings in agreement with predictions we would make on the basis of equilibrium principles? Increasing the concentration of Cu^{+2} ion in reaction (*38*) should increase the tendency to form the products (Le Chatelier's Principle). Experimentally, we find an increase in voltage when more solid copper sulfate is dissolved in the solution around the copper electrode. Conversely, decreasing the concentration of the copper ion should decrease the tendency to form the products. Again in agreement, voltage readings decrease when the concentration of Cu^{+2} ion is reduced by precipitation of CuS. The voltage shows the tendency for reaction to occur.

What happens to any cell or battery as it operates? The voltage decreases until, finally, it reaches zero. Then we say the cell is dead. Equilibrium has been attained and the reaction that has been producing the energy has the same tendency to proceed as does the reverse reaction. Again the voltage measures the net tendency for reaction to occur. At equilibrium there is a balance between forward and reverse reactions, hence there is *no* net tendency for further reaction either way. Therefore, the voltage of a cell at equilibrium is zero.

Since concentration variations have measurable effects on the cell voltage, a measured voltage cannot be interpreted unless the cell concentrations are specified. Because of this, chemists introduce the idea of "standard-state." The standard state for gases is taken as a pressure of one atmosphere at 25°C; the standard state for ions is taken as a concentration of 1 *M*; and the standard state of pure substances is taken as the pure substances themselves as they exist at 25°C. The half-cell potential associated with a half-reaction taking place between substances in their standard states is called E° (the superscript zero means standard state). We can rewrite equation (*37*) to include the specifications of the standard states:

$$H_2(g, 1\text{ atm}) \longrightarrow 2H^+(aq, 1\ M) + 2e^- \qquad E^\circ = 0.00 \cdots \text{ volt (assigned)} \quad (39)$$

Now if we combine a Zn–Zn^{+2} half-cell in its standard state with a H_2–$2H^+$ half-cell in its standard state, the voltage (potential) we measure (0.76 volt) is the value assigned to the half-reaction:

$$Zn(s) \longrightarrow Zn^{+2}(aq, 1\ M) + 2e^- \qquad E^\circ = +0.76 \text{ volt} \quad (40)$$

Similarly, if we combine a Cu–Cu^{+2} half-cell in its standard state with a standard H_2–$2H^+$ half-cell, the voltage (potential) we measure (0.34 volt) is the value assigned to the half-reaction:

$$Cu(s) \longrightarrow Cu^{+2}(aq, 1\ M) + 2e^- \qquad E^\circ = -0.34 \text{ volt} \quad (41)$$

Chemists have determined a large number of these half-cell potentials. The magnitude of the voltage is a quantitative measure of the tendency of that half-reaction to release electrons in comparison to the H_2–$2H^+$ half-reaction. If the sign is *positive*, the half-reaction has *greater* tendency to release electrons than does the H_2–$2H^+$ half-

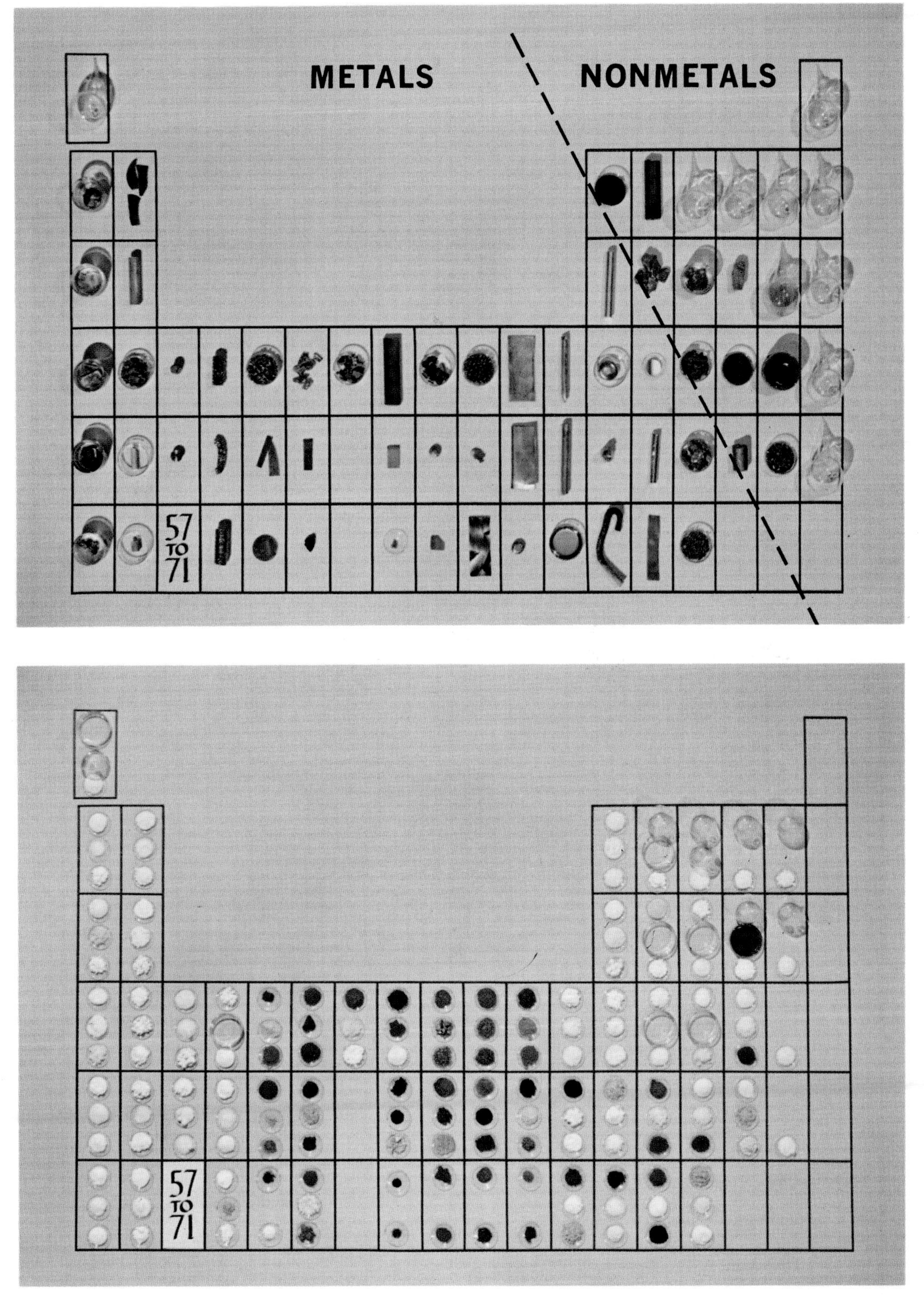

Plate 1. (*Above*) **The elements in the periodic table.** (*Below*) **Compounds in the periodic table.**

OXIDES
CHLORIDES
MISCELLANEOUS

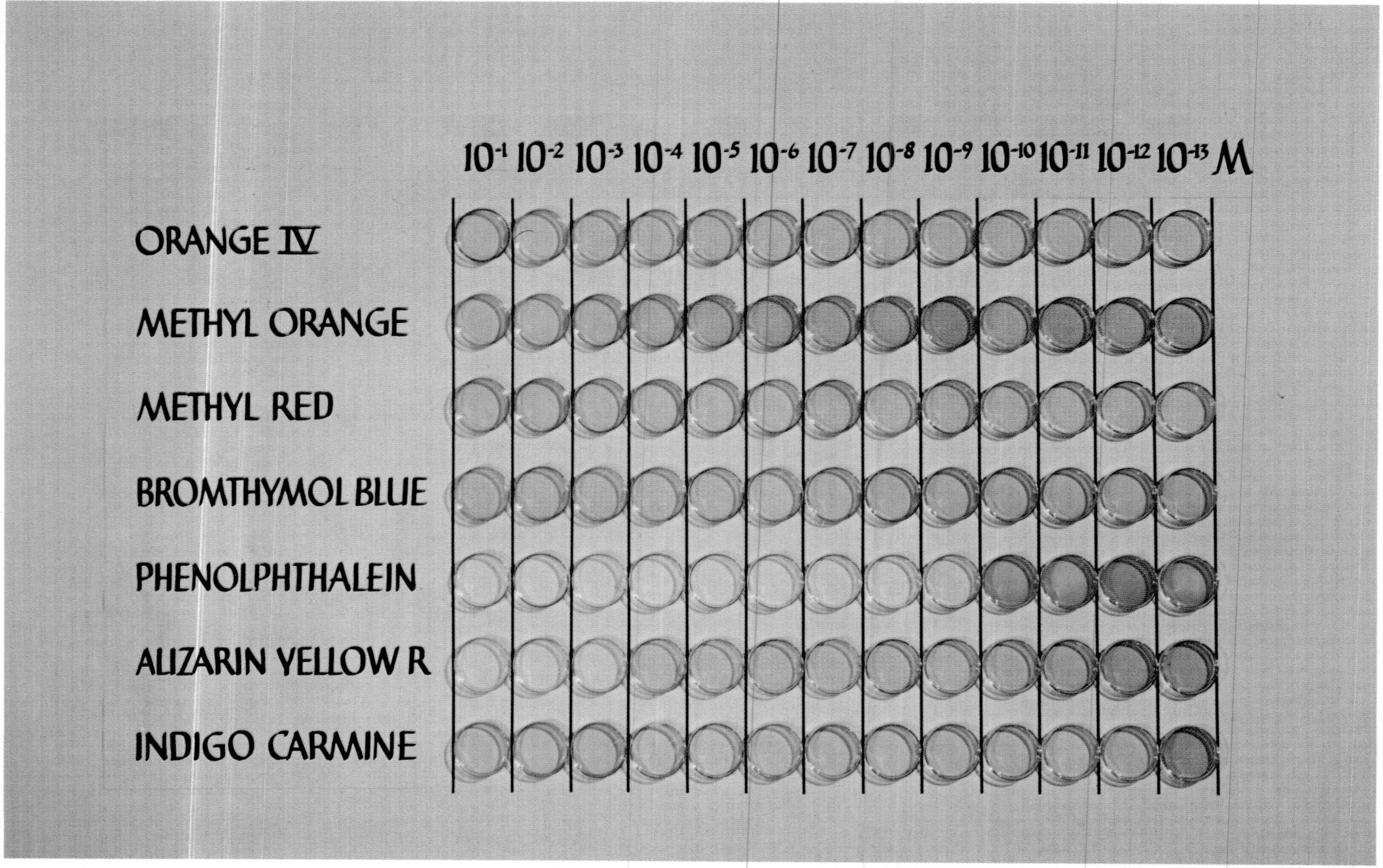

10^{-1} 10^{-2} 10^{-3} 10^{-4} 10^{-5} 10^{-6} 10^{-7} 10^{-8} 10^{-9} 10^{-10} 10^{-11} 10^{-12} 10^{-13} M

ORANGE IV

METHYL ORANGE

METHYL RED

BROMTHYMOL BLUE

PHENOLPHTHALEIN

ALIZARIN YELLOW R

INDIGO CARMINE

Plate II. **Some acid-base indicator colors and the hydrogen ion concentration.**

reaction. If the sign is *negative*, the half-reaction has *less* tendency.

TABLE OF HALF-CELL POTENTIALS

Table 12-II gives the values of the standard oxidation potentials for a number of half-reactions. A more complete table is given in Appendix 3. We have not added the information "1 *M*" for each ion since this is implied by the symbol $E°$. For the same reason, 25°C and 1 atmosphere pressure of gases are understood.

Table 12-II

SELECTED STANDARD OXIDATION POTENTIALS FOR HALF-REACTIONS*

HALF-REACTION Reduced state		Oxidized state	$E°$ (volts)
$Zn(s)$	$\longrightarrow$	$2e^- + Zn^{+2}$	+0.76
$Co(s)$	$\longrightarrow$	$2e^- + Co^{+2}$	+0.28
$Ni(s)$	$\longrightarrow$	$2e^- + Ni^{+2}$	+0.25
$H_2(g)$	$\longrightarrow$	$2e^- + 2H^+$	0.000
$Cu(s)$	$\longrightarrow$	$2e^- + Cu^{+2}$	−0.34
$2I^-$	$\longrightarrow$	$2e^- + I_2(s)$	−0.53
$Ag(s)$	$\longrightarrow$	$e^- + Ag^+$	−0.80
$2Br^-$	$\longrightarrow$	$2e^- + Br_2(l)$	−1.06
$Mn^{+2} + 2H_2O$	$\longrightarrow$	$2e^- + MnO_2 + 4H^+$	−1.28
$2Cl^-$	$\longrightarrow$	$2e^- + Cl_2(g)$	−1.36
$Mn^{+2} + 4H_2O$	$\longrightarrow$	$5e^- + MnO_4^- + 8H^+$	−1.52

* A more complete table is given in Appendix 3.

Let's examine this table to see if it agrees with our laboratory experience. Table 12-I summarized some of these results in a qualitative way. Extracting these four half-reactions from Table 12-II, we find

$$Zn(s) \longrightarrow Zn^{+2} + 2e^- \qquad E° = +0.76 \text{ volt} \qquad (8), (42)$$

$$H_2(g) \longrightarrow 2H^+ + 2e^- \qquad E° = 0.000 \text{ volt} \qquad (21), (43)$$

$$Cu(s) \longrightarrow Cu^{+2} + 2e^- \qquad E° = -0.34 \text{ volt} \qquad (1), (44)$$

$$Ag(s) \longrightarrow Ag^+ + e^- \qquad E° = -0.80 \text{ volt} \qquad (16), (45)$$

The half-reactions, listed in order of decreasing half-cell potentials, are in the same order as in Table 12-I, which was dictated by laboratory experience.

EXERCISE 12-3

In Exercise 12-2 you placed the Ni–Ni^{+2} half-reaction into Table 12-I. Check your placement by examining the half-cell potential of this half-reaction in Table 12-II.

Another way to verify the usefulness of Table 12-II is to compare its voltage predictions to one we have measured. For example, we found a value of approximately 0.5 volt for a cell based on reaction (46):

$$Zn(s) + Ni^{+2} \longrightarrow Zn^{+2} + Ni(s) \qquad (27), (46)$$

This cell involves the following two half-reactions:

$$Zn(s) \longrightarrow Zn^{+2} + 2e^- \qquad E° = +0.76 \text{ volt} \qquad (8), (47)$$

$$Ni(s) \longrightarrow Ni^{+2} + 2e^- \qquad E° = +0.25 \text{ volt} \qquad (32), (48)$$

By the values of $E°$, we conclude that the Zn–Zn^{+2} half-reaction has the greater tendency to release electrons. Hence it will tend to transfer electrons to nickel, forcing half-reaction (48) in the reverse direction. Our net reaction will be obtained by *subtracting* half-reaction (48) from (47):

$$Zn(s) \longrightarrow Zn^{+2} + 2e^- \qquad E_1° = +0.76 \text{ volt} \qquad (47), (49)$$

$$\text{minus} \quad Ni(s) \longrightarrow Ni^{+2} + 2e^- \qquad E_2° = +0.25 \text{ volt} \qquad (48), (50)$$

$$E° = E_1° - E_2°$$

$$E° = (+0.76) - (+0.25)$$

$$\text{Net reaction} \quad Zn(s) + Ni^{+2} \longrightarrow Zn^{+2} + Ni(s) \qquad E° = +0.51 \text{ volt} \qquad (51)$$

Table 12-II predicts the cell will operate so as to dissolve metallic zinc and deposit metallic nickel, and its voltage will be +0.51 volt. This is exactly what occurs in such a cell. Predicting is fun—let's try it again! Another cell we studied is based on reaction (*52*):

$$Zn(s) + 2Ag^+ \longrightarrow Zn^{+2} + 2Ag(s) \quad (28), (52)$$

The two half-reactions involved are

$$Zn(s) \longrightarrow Zn^{+2} + 2e^- \qquad E^\circ = +0.76 \text{ volt} \quad (8), (53)$$

$$Ag(s) \longrightarrow Ag^+ + e^- \qquad E^\circ = -0.80 \text{ volt} \quad (16), (54)$$

By the half-cell potentials, we conclude the Zn–Zn^{+2} half-reaction has the greater tendency to release electrons. It will tend to transfer an electron to silver ion, forcing (*54*) in the reverse direction. Hence we obtain the net reaction by subtracting (*54*) from (*53*). But remember that this subtraction must be in the proportion that causes no net gain or loss of electrons. If two electrons are lost per atom of zinc oxidized in (*53*), then we must double half-reaction (*54*) so that two electrons will be consumed.

	$Zn(s) \longrightarrow Zn^{+2} + 2e^-$	$E_1^\circ = +0.76$ volt	(*53*), (*55*)
minus	$2Ag(s) \longrightarrow 2Ag^+ + 2e^-$	$E_2^\circ = -0.80$ volt	(*54*), (*56*)
		$E^\circ = E_1^\circ - E_2^\circ$	
		$= (+0.76) - (-0.80)$	
Net reaction	$Zn(s) + 2Ag^+ \longrightarrow Zn^{+2} + 2Ag(s)$	$E^\circ = +1.56$ volts	(*28*), (*57*)

Our conclusions are again in agreement with experiment. The cell operates so as to dissolve zinc metal and precipitate silver metal. The voltage is indeed about 1.5 volts. Finally, experiment shows that one mole of zinc does react with two moles of silver ion, as required by the balance of electrons.

Notice that we did *not* double E_2° for (*54*) in obtaining (*57*). The voltage of a half-reaction does not depend upon how many moles we consider. Thus:

$$Ag(s) \longrightarrow Ag^+ + e^- \qquad E_2^\circ = -0.80 \text{ volt} \quad (16), (58)$$

$$2Ag(s) \longrightarrow 2Ag^+ + 2e^- \qquad E_2^\circ = -0.80 \text{ volt} \quad (16), (59)$$

You might wonder what we would have learned if we had assumed that either of these two cells operates with the reverse reaction. Suppose we had proposed a cell based on oxidation of nickel and reduction of zinc:

	$Ni(s) \longrightarrow Ni^{+2} + 2e^-$	$E_1^\circ = +0.25$ volt	(*32*), (*60*)
minus	$Zn(s) \longrightarrow Zn^{+2} + 2e^-$	$E_2^\circ = +0.76$ volt	(*8*), (*61*)
		$E^\circ = E_1^\circ - E_2^\circ$	
		$= (+0.25) - (+0.76)$	
Net reaction	$Ni(s) + Zn^{+2} \longrightarrow Ni^{+2} + Zn(s)$	$E^\circ = -0.51$ volt	(*62*)

Our assumption concerning the chemistry leads us to the reverse reaction of that we obtained earlier—reaction (*51*)—and to an equal voltage but with opposite sign. The significance of the negative voltage (−0.51 volt) is that equilibrium in the reaction favors *reactants, not products.* We obtain the same prediction we did before—since the voltage is negative, the reaction will tend to operate a cell in the reverse direction—to dissolve zinc metal and to precipitate nickel metal. The reaction will occur in the direction written (consuming nickel and precipitating zinc) only if the cell is "overpowered" by an opposing cell of higher voltage than 0.51 volt (as was done in the experiment pictured in Figure 12-6).

12-2.2 Predicting Reactions from Table 12-II

The ideas we have developed for reactions occurring in electrochemical cells are also applicable to reactions that occur in a beaker. Therefore, *chemists use half-cell potentials to predict what chemical reactions can occur spontaneously.*

If a chemist wishes to know whether zinc can be oxidized if it is placed in contact with a solution of nickel sulfate, the values of E° help him decide. The half-cell potential for $Zn-Zn^{+2}$ is +0.76 volt, which is greater than that for $Ni-Ni^{+2}$ (which is +0.25 volt). The difference, +0.51 volt, is positive, indicating that zinc has a greater tendency to lose electrons than does nickel. Therefore, zinc can transfer electrons to Ni^{+2}. The chemist predicts: zinc will react with Ni^{+2}, zinc being oxidized and nickel being reduced.

AN EXAMPLE: COPPER AND SILVER

Suppose the question is whether silver will be oxidized if it is immersed in copper sulfate. The half-cell potential for $Ag-Ag^+$ is −0.80 volt and that for $Cu-Cu^{+2}$ is −0.34 volt. The first value, −0.80 volt, is more negative than the second, −0.34 volt. The difference, then, is still negative: $-0.80 - (-0.34) = -0.46$ volt. The negative answer shows that $Ag-Ag^+$ has *less* tendency to lose electrons than does $Cu-Cu^{+2}$. The reaction will *not* tend to proceed spontaneously. Silver will not be oxidized to an appreciable extent in copper sulfate.

EXERCISE 12-4

Use the values of E° to predict whether cobalt metal will tend to dissolve in a 1 M solution of acid, H^+. Now predict whether cobalt metal will tend to dissolve in a 1 M solution of zinc sulfate (reacting with Zn^{+2}).

GENERALIZING ON PREDICTIONS WITH E°

We can generalize now on the use of Table 12-II. A substance on the left in Table 12-II reacts by losing electrons. A substance on the right reacts by gaining electrons. We may draw the following conclusions:

1. An oxidation-reduction reaction must involve a substance from the left column of Table 12-II (something which can be oxidized), and it must involve a substance from the right column (something which can be reduced).
2. A substance in the left column of Table 12-II tends to react spontaneously with any substance in the right column which is *lower* in the Table.

Applying these rules, we would predict: copper metal could be oxidized to Cu^{+2} by $Br_2(l)$ or $MnO_2(s)$ but not by Ni^{+2} or Zn^{+2}. Of course, copper metal cannot be oxidized by either zinc metal or mercury metal because neither zinc metal nor mercury metal can accept electrons to be reduced (as far as we know from Table 12-II).

EXERCISE 12-5

Use Table 12-II to decide which substances in the following list tend to oxidize bromide ion, Br^-: $Cl_2(g)$, H^+, Ni^{+2}, MnO_4^-.

EXERCISE 12-6

Use Table 12-II to decide which substances in the following list tend to reduce $Br_2(l)$: Cl^-, $H_2(g)$, $Ni(s)$, Mn^{+2}.

PREDICTIONS AND THE EFFECT OF CONCENTRATIONS

All of our predictions have been based upon the values of E° that apply to standard conditions. Yet we often wish to carry out a reaction at other than standard conditions. Our prediction then must be adjusted in accord with Le Chatelier's Principle as we change conditions from standard conditions to others of interest to us.

For example, by comparing E°'s, we predicted that zinc would dissolve in nickel sulfate. These E°'s show that zinc metal will dissolve if zinc ion and nickel ion are both present at 1 M concentration:

$$Zn(s) + Ni^{+2} \longrightarrow Zn^{+2} + Ni(s) \quad (27), (63)$$

In our case, however, no zinc ion is present at all. How does this affect our prediction? By

Le Chatelier's Principle, the removal of Zn^{+2} tends to shift equilibrium toward the products. Therefore, removing Zn^{+2} *increases* the tendency for reaction (*63*) to occur. Our prediction of reaction is still valid.

This will not always be the case, however. Consider the question: "Will silver metal dissolve in 1 *M* H^+?" According to Table 12-II,

$$2Ag(s) \longrightarrow 2Ag^+(1\ M) + 2e^- \qquad E^\circ = -0.80 \text{ volt}$$
$$2H^+(1\ M) + 2e^- \longrightarrow H_2(g) \qquad E^\circ = 0.00 \text{ volt}$$
$$\text{Net reaction} \quad 2Ag(s) + 2H^+(1\ M) \longrightarrow 2Ag^+(1\ M) + H_2(g) \qquad E^\circ = -0.80 \text{ volt} \qquad (64)$$

The negative voltage shows that the state of equilibrium favors the reactants more than the products for the reaction as written. For standard conditions, the reaction will not tend to occur spontaneously. However, if we place Ag(*s*) in 1 *M* H^+, the Ag^+ concentration is not 1 *M*—it is zero. By Le Chatelier's Principle, this increases the tendency to form products, in opposition to our prediction of no reaction. Some silver will dissolve, though only a minute amount because silver metal releases electrons so reluctantly compared with H_2. It is such a small amount, in fact, that no silver chloride precipitate forms, even though silver chloride has a very low solubility.

That some silver does dissolve to form Ag^+ can be verified experimentally by adding a little KI to the solution. Silver iodide has an even lower solubility than does silver chloride. The experiment shows that the amount of silver that dissolves is sufficient to cause a visible precipitate of AgI but not of AgCl. This places the Ag^+ ion concentration below 10^{-10} *M* but above 10^{-17} *M*. Either of these concentrations is so small that we can consider our prediction for the standard state to be applicable here too—silver metal does not dissolve *appreciably* in 1 *M* HCl. In general, the question of whether a prediction based upon the standard state will apply to other conditions depends upon how large is the magnitude of E°. If E° for the overall reaction is only one- or two-tenths volt (either positive or negative), then deviations from standard conditions may invalidate predictions that do not take into account these deviations.

12-2.3 Reliability of Predictions

There is one more limitation on the reliability of predictions based upon E°'s. To see it, we shall consider the three reactions

$$Cu(s) + 2H^+ \longrightarrow Cu^{+2} + H_2(g) \qquad E^\circ = -0.34 \text{ volt} \qquad (65)$$

$$Fe(s) + 2H^+ \longrightarrow Fe^{+2} + H_2(g) \qquad E^\circ = +0.44 \text{ volt} \qquad (66)$$

$$3Fe(s) + 2NO_3^- + 8H^+ \longrightarrow 3Fe^{+2} + 2NO(g) + 4H_2O \qquad E^\circ = +1.40 \text{ volts} \qquad (67)$$

The three values of E° are easily calculated from half-cell potentials. Then, we can predict with confidence that reaction (*65*) will not occur to an appreciable extent if solid copper is immersed in dilute acid. The negative value of E° (−0.34 volt) indicates that equilibrium in (*65*) strongly favors the reactants, not the products.

Furthermore, we can predict that reactions (*66*) and (*67*) *might* occur. The positive values of E° (+0.44 and +1.40 volts) show that equilibrium strongly favors products in these reactions. Again, experiments are warranted. A piece of iron is immersed in dilute acid—bubbles of hydrogen appear. Reaction (*66*) does occur. A piece of iron is immersed in a one molar nitric acid solution—though bubbles of hydrogen may appear, no nitric oxide, NO, gas appears. Reaction (*67*) between iron and nitrate ion does not immediately occur: the reaction rate is extremely slow. This slow rate could not be predicted from the E°'s.

So the equilibrium predictions based on E°'s do not make all experiments unnecessary. They provide no basis whatsoever for anticipating whether a reaction will be very slow or very fast. Experiments must be performed to learn the reaction rate. The E°'s do, however, provide definite and reliable guidance concerning the equilibrium state, thus making many experiments unnecessary; the multitude of reactions that are foredoomed to failure by equilibrium considerations need not be performed.

12-2.4 E° and the Factors That Determine Equilibrium

We see that $E°$ furnishes a basis for predicting the equilibrium state. In Chapter 9 and in subsequent chapters, equilibrium was treated in terms of two opposing tendencies—toward minimum energy and toward maximum randomness. What is the connection between $E°$ and these two tendencies?

Consider two reactions for which $E°$ shows that products are favored, one an exothermic reaction, and the other an endothermic reaction. For the exothermic reaction, when the reactants are mixed they are driven toward equilibrium in accord with the tendency toward minimum energy. Now contrast the endothermic reaction for which $E°$ shows that equilibrium favors products. When these reactants are mixed, they approach equilibrium *against* the tendency toward minimum energy (since heat is absorbed). This reaction is driven by the tendency toward maximum randomness.

Summarizing, *$E°$ measures quantitatively the difference between the tendency to minimum energy and tendency to maximum randomness under the standard state conditions.*

12-2.5 Oxidation Numbers—An Electron Bookkeeping Device

The reaction between ferric ion, Fe^{+3}, and cuprous ion, Cu^{+}, to produce ferrous ion, Fe^{+2}, and cupric ion, Cu^{+2}, is plainly an oxidation-reduction reaction:

$$Fe^{+3} + Cu^{+} \longrightarrow Fe^{+2} + Cu^{+2} \qquad (68)$$

It is readily separated into two half-reactions showing electron transfer:

Oxidation (loss of electrons)

$$Cu^{+} \longrightarrow Cu^{+2} + e^{-} \qquad (69)$$

Reduction (gain of electrons)

$$Fe^{+3} + e^{-} \longrightarrow Fe^{+2} \qquad (70)$$

Because of the presence of Cu^{+} ion, ferric ion is reduced. Chemists say that Cu^{+} ion acts as a *reducing agent* in this reaction—Cu^{+} ion is the "agent" that caused the reduction of ferric ion. At the same time, Cu^{+} is oxidized because of the presence of ferric ion. Hence, Fe^{+3} is called an *oxidizing agent* in this reaction.

Another reaction by which ferric ion can be reduced involves bisulfite* ion, HSO_3^-. The balanced equation is

$$H_2O + HSO_3^- + 2Fe^{+3} \longrightarrow 2Fe^{+2} + HSO_4^- + 2H^{+} \qquad (71)$$

Again half-reaction (*70*) describes what happens to ferric ion:

$$2Fe^{+3} + 2e^{-} \longrightarrow 2Fe^{+2} \qquad (72)$$

Since two electrons are gained by the two ferric ions in half-reaction (*72*), two electrons must be released by the remaining constituents in (*71*). The other half-reaction can be found by subtracting (*72*) from (*71*) to give,

$$H_2O + HSO_3^- - 2e^{-} \longrightarrow HSO_4^- + 2H^{+}$$

or

$$H_2O + HSO_3^- \longrightarrow HSO_4^- + 2H^{+} + 2e^{-} \qquad (73)$$

The combination of H_2O and HSO_3^- acts as a reducing agent toward Fe^{+3}. Since water solutions of Fe^{+3} are quite stable, HSO_3^- is considered to be the actual reducing agent.

Half-reaction (*73*) differs from the others we have looked at. This is not the same simple situation in which a single atom is oxidized by releasing electrons or is reduced by accepting them. Here a complicated ion, HSO_3^-, furnishes electrons to Fe^{+3} in an intricate process in which an oxygen atom is transferred from H_2O to HSO_3^-, forming HSO_4^- and two hydrogen ions. We cannot assign the electron loss to a particular atom. In this situation, it is convenient to have a bookkeeping device that at least keeps track of the number of electrons so that none is forgotten. The bookkeeping device is called assigning oxidation numbers.

Since we don't know the locations of the electrons held by a molecule such as HSO_3^-, we *assume* that the hydrogen atom has a $+1$ charge, that each oxygen atom has a -2 charge, and that the sulfur atom has all the rest of the electrons in the molecule. Of course, if the charges on all the atoms in HSO_3^- are added together, they must sum to -1, the molecular charge. Since there are three oxygen atoms in HSO_3^-, the algebra looks like this:

* This ion, HSO_3^-, is also called hydrogen sulfite.

$$\begin{pmatrix}\text{charge on}\\ \text{hydrogen}\\ \text{atom}\end{pmatrix} + 3\begin{pmatrix}\text{charge on}\\ \text{oxygen}\\ \text{atom}\end{pmatrix} + \begin{pmatrix}\text{charge on}\\ \text{sulfur}\\ \text{atom}\end{pmatrix} = \begin{pmatrix}\text{molecular}\\ \text{charge}\end{pmatrix}$$

$$\underset{\text{assumed}}{(+1)} + 3\underset{\text{assumed}}{(-2)} + \begin{pmatrix}\text{charge on}\\ \text{sulfur}\\ \text{atom}\end{pmatrix} = (-1)$$

$$\begin{pmatrix}\text{charge on}\\ \text{sulfur}\\ \text{atom}\end{pmatrix} = (+4)$$

This fictitious charge is called the *oxidation number* of sulfur:

$$\text{Oxidation number sulfur} = +4 \text{ in } HSO_3^- \qquad (74)$$

The same process can be applied to HSO_4^-. Again assuming hydrogen has a charge of +1 and each of the four oxygen atoms has a charge of −2, we calculate a fictitious charge on the sulfur atom of +6:

$$\text{Oxidation number sulfur} = +6 \text{ in } HSO_4^- \qquad (75)$$

According to the oxidation number bookkeeping, the two electrons released in the HSO_3^-–HSO_4^- half-reaction (*73*) are associated with the change in oxidation number of sulfur from +4 to +6.

This arbitrary scheme of assigning oxidation numbers turns out to be quite useful, provided we don't forget that the oxidation number is a *fictitious* charge. We can see the usefulness by considering a reaction related to the oxidation of HSO_3^-.

Sulfur forms two oxides, SO_2 (a gas at normal conditions) and SO_3 (a liquid that boils at 44.8°C). Under suitable conditions, SO_2 reacts with oxygen to form SO_3:

$$2SO_2(g) + O_2(g) \longrightarrow 2SO_3(g) \qquad (76)$$

Is this an oxidation-reduction reaction? Historically, it surely is, for the term "oxidation" originally referred specifically to reactions with oxygen. Yet our electron-transfer view of oxidation-reduction reactions provides no help in deciding so. Where in reaction (*76*) is there any evidence of electrons being gained or lost? In such a doubtful case, our oxidation number scheme provides an answer. Applying the same assumptions used in treating the HSO_3^-–HSO_4^- half-reaction, we can calculate the oxidation numbers of sulfur in SO_2 and SO_3.

In SO_2:

$$2\begin{pmatrix}\text{charge on}\\ \text{oxygen}\\ \text{atom}\end{pmatrix} + \begin{pmatrix}\text{charge on}\\ \text{sulfur}\\ \text{atom}\end{pmatrix} = \begin{pmatrix}\text{molecular}\\ \text{charge}\end{pmatrix}$$

$$2\underset{\text{assumed}}{(-2)} + \begin{pmatrix}\text{oxidation}\\ \text{number}\\ \text{sulfur}\end{pmatrix} = (0)$$

$$\text{Oxidation number sulfur} = +4 \text{ in } SO_2 \qquad (77)$$

In SO_3:

$$3\begin{pmatrix}\text{charge on}\\ \text{oxygen}\\ \text{atom}\end{pmatrix} + \begin{pmatrix}\text{charge on}\\ \text{sulfur}\\ \text{atom}\end{pmatrix} = \begin{pmatrix}\text{molecular}\\ \text{charge}\end{pmatrix}$$

$$3\underset{\text{assumed}}{(-2)} + \begin{pmatrix}\text{oxidation}\\ \text{number}\\ \text{sulfur}\end{pmatrix} = (0)$$

$$\text{Oxidation number sulfur} = +6 \text{ in } SO_3 \qquad (78)$$

Thus, in reaction (*76*) the sulfur atom changes oxidation number from +4 to +6, just as it did in the HSO_3^-–HSO_4^- half-reaction. The oxidation number gives a basis for connecting the SO_2–SO_3 change to the oxidation of HSO_3^- to HSO_4^-. Both changes are considered to be examples of oxidation.

Oxidation-reduction reactions occurring in aqueous solutions are conveniently treated in terms of half-reactions showing transfer of electrons. Under more general conditions (gaseous state, other solvents, etc.), it is more convenient to treat oxidation-reduction reactions in terms of oxidation numbers, based upon the arbitrary scheme of assigning charge +1 to a hydrogen atom bound to an unlike atom and charge −2 to an oxygen atom when it is bound to unlike atoms. Generally, then, ***an oxidation-reduction reaction** is one in which oxidation numbers change.*

EXERCISE 12-7

The reactions by which SO_2 and SO_3 dissolve in water are not considered to be oxidation-reduction reactions:

$$SO_2(g) + H_2O \longrightarrow HSO_3^-(aq) + H^+(aq) \quad (79)$$

$$SO_3(g) + H_2O \longrightarrow HSO_4^-(aq) + H^+(aq) \quad (80)$$

Convince yourself that none of the atoms in either *(79)* or *(80)* changes oxidation number.

EXERCISE 12-8

In reaction *(76)* the oxidation number of sulfur changes from +4 to +6. According to this, two electrons are released by each sulfur atom oxidized. Show that these electrons are gained by oxygen if we assume oxygen has oxidation number equal to zero in O_2.

12-3 BALANCING OXIDATION-REDUCTION REACTIONS

When a reaction is properly written, it expresses certain conservation laws. A chemical reaction does not destroy or produce atoms. Therefore there must be the same number and types of atoms among the reactants as among the products. A chemical reaction also does not destroy or produce electric charge. Therefore the sum of the charges appearing on the reactant species must be the same as the sum of the charges appearing on the products. The process by which we write a correct chemical reaction, making sure these two conservation laws are followed, is called *balancing* the chemical reaction.*

Oxidation-reduction reactions must be balanced if correct predictions are to be made. Just as in selecting a route for a trip from San Francisco to New York, there are several ways to reach the desired goal. Which route is best depends to some extent upon the likes and dislikes of the traveler. We will discuss two ways to balance oxidation-reduction reactions—first, using half-reactions and, next, using the oxidation numbers we have just introduced.

12-3.1 Use of Half-Reactions for Balancing Oxidation-Reduction Reactions

Suppose we want to describe by an equation what happens when pure lithium metal is added to a 1 *M* HCl solution. Our first step must be to decide what products will be obtained. This can be determined only by experiment. Often you will know already what the experiment would give. Sometimes what the products will be is not obvious but it is easily learned from a reference book. Sometimes you must do the experiment yourself or proceed on the basis of an assumption.

For lithium metal in 1 *M* HCl, the observed facts are that the metal dissolves spontaneously and a gas bubbles out of the solution. From Appendix 3 we select the two half-reactions (notice that the half-reactions are already "balanced" in both charge and number of atoms):

$$Li(s) \longrightarrow Li^+ + e^- \quad (81)$$

$$H_2(g) \longrightarrow 2H^+ + 2e^- \quad (21), (82)$$

Both of these half-reactions show production of electrons. But we know there must be an electron used for each produced, so one of the equations must be reversed. Experiment shows us it is the second because hydrogen gas is evolved from the solution. The first equation is correct as written. Lithium metal dissolves and is converted to ions. Thus,

$$Li(s) \longrightarrow Li^+ + e^- \quad (81), (83)$$

$$2H^+ + 2e^- \longrightarrow H_2(g) \quad (9), (84)$$

These equations now correctly express the observed facts and they must be properly combined. The first step is to make them indicate the same number of electrons produced and used. We "balance the electrons." By inspection we find that we achieve balance by doubling the first equation.

$$2Li(s) \longrightarrow 2Li^+ + 2e^- \quad (81), (85)$$

$$2H^+ + 2e^- \longrightarrow H_2(g) \quad (9), (86)$$

* Some chemists prefer to call this process *balancing the equation* for the reaction.

Now we add the two equations to get the net reaction,

$$2H^+ + 2Li(s) \longrightarrow 2Li^+ + H_2(g) \quad (87)$$

Note that the electrons cancel—we took care that this would happen because we know electrons are neither consumed nor produced in the net reaction.

As a final check, let us verify the conservation of charge:

$$\begin{array}{lcl} 2H^+ + 2Li(s) & \longrightarrow & 2Li^+ + H_2(g) \\ +2 + 0 & & +2 + 0 \\ +2 & = & +2 \end{array}$$

As a more complex case, suppose we want to write the equation for the reaction that occurs when hydrogen sulfide gas, H_2S, is bubbled into an acidified potassium permanganate solution, $KMnO_4$. When we do this, we observe that the purple color of the MnO_4^- ion disappears and that the resulting mixture is cloudy (sulfur particles). From Appendix 3 we find the two half-reactions

$$H_2S(g) \longrightarrow S(s) + 2H^+ + 2e^- \quad (88)$$

$$Mn^{+2} + 4H_2O \longrightarrow MnO_4^- + 8H^+ + 5e^- \quad (89)$$

Since we know that sulfur is formed, we will use equation (*88*) as it is written. However, the purple MnO_4^- is being changed to almost colorless Mn^{+2} so we will rewrite equation (*89*) as the reverse of what we obtained from Appendix 3:

$$H_2S(g) \longrightarrow S(s) + 2H^+ + 2e^- \quad (88), (90)$$

$$MnO_4^- + 8H^+ + 5e^- \longrightarrow Mn^{+2} + 4H_2O \quad (91)$$

Balancing the electrons is the next step, but this time it is a little more difficult. We see that if we multiply equation (*90*) by 5 and equation (*91*) by 2, there will be 10 electrons in each case. Arithmetic teachers call this finding the least common multiple:

$$5H_2S(g) \longrightarrow 5S(s) + 10H^+ + 10e^- \quad (92)$$

$$2MnO_4^- + 16H^+ + 10e^- \longrightarrow 2Mn^{+2} + 8H_2O \quad (93)$$

$$5H_2S(g) + 2MnO_4^- + 16H^+ \longrightarrow 5S(s) + 10H^+ + 2Mn^{+2} + 8H_2O \quad (94)$$

The electrons cancel out. The atoms and charges are properly balanced, but there is one remaining disturbance. There are H^+ ions included among the reactants as well as the products. Cancelling the excess, we obtain the final, balanced reaction:

$$5H_2S(g) + 2MnO_4^- + 6H^+ \longrightarrow 5S(s) + 2Mn^{+2} + 8H_2O \quad (95)$$

Before leaving the equation, let us check the electric charge balance:

$$\begin{array}{lcl} 5H_2S(g) + 2MnO_4^- + 6H^+ & \longrightarrow & 5S(s) + 2Mn^{+2} + 8H_2O \quad (95) \\ 5(0) + 2(-1) + 6(+1) & & 5(0) + 2(+2) + 8(0) \\ -2 + +6 & & +4 \\ +4 & = & +4 \end{array}$$

12-3.2 Balancing Half-Reactions

When potassium chlorate solution, $KClO_3$, is added to hydrochloric acid, chlorine gas is evolved. Although we can find the half-reaction, $2Cl^- = Cl_2(g) + 2e^-$, in Appendix 3, we find no equation with ClO_3^- ion involved. We can surmise that ClO_3^- is accepting electrons and changing into chlorine. Let us write a partial half-reaction in which we indicate an unknown number of electrons and in which we have conserved only chlorine atoms:

$$ClO_3^- + xe^- \longrightarrow \tfrac{1}{2}Cl_2(g) \quad (96)$$

From experience we note that in acid solution the oxygen in such oxidizing agents as MnO_4^- and $Cr_2O_7^{-2}$ ends up as water and that H^+ is consumed. Let us include this notion by showing $6H^+$ among the reactants and $3H_2O$ among the products:

$$ClO_3^- + 6H^+ + xe^- \longrightarrow \tfrac{1}{2}Cl_2(g) + 3H_2O \quad (97)$$

Finally, we have to remember that charge is conserved. Since the products are neutral molecules, x will have to be 5 in order that the total charge represented among the reactants is zero. Our desired half-reaction is

$$ClO_3^- + 6H^+ + 5e^- \longrightarrow \tfrac{1}{2}Cl_2(g) + 3H_2O \quad (98)$$

Now we can return to working out the equation for the reaction we observed. The least common multiple between 1 and 5 is 5. Writing the half-reactions to involve 5 electrons and adding them, we obtain

$$5Cl^- \longrightarrow \tfrac{5}{2}Cl_2(g) + 5e^- \quad (99)$$

$$ClO_3^- + 6H^+ + 5e^- \longrightarrow \tfrac{1}{2}Cl_2(g) + 3H_2O \quad (100)$$

$$5Cl^- + ClO_3^- + 6H^+ \longrightarrow 3Cl_2(g) + 3H_2O \quad (101)$$

Again demonstrate to yourself that atoms and charge are conserved.

12-3.3 Use of Oxidation Number in Balancing Oxidation-Reduction Reactions

We have already introduced oxidation numbers as a device for assigning a fictitious charge to an atom in a molecule. According to this scheme, oxidation-reduction reactions involve changes of oxidation numbers. Consideration of conservation of charge reveals that there must be a balance between changes of oxidation number. Consequently, oxidation numbers provide just as good a basis for balancing equations as do half-reactions.

ASSIGNING OXIDATION NUMBERS

For the present, we will limit ourselves to molecules containing hydrogen and/or oxygen along with the element to which we wish to assign an oxidation number. The rules we will utilize are as follows:

(1) The oxidation number of a monatomic ion is equal to the charge on the ion.
(2) The oxidation number of any substance in the elementary state is zero.
(3) The oxidation number of hydrogen is taken to be +1 (except in H_2, which is the elementary state).
(4) The oxidation number of oxygen is taken to be −2 (except in O_2; ozone; O_3; and peroxides).
(5) The other oxidation numbers are selected to make the sum of the oxidation numbers equal to the charge on the molecule.
(6) Reactions occur such that the net change of oxidation numbers is zero. (This last rule is really a result of the conservation of charge.)

Do not worry about the exceptions included within parentheses in rules 3 and 4. Your attention will be called to them later when substances involving them are considered.

EXERCISE 12-9

Show that the oxidation number of nitrogen is +5 in each of the two species NO_3^- and N_2O_5.

BALANCING REACTIONS

Just as before, the first step in balancing a reaction must be to decide the products. Again, experiment provides the answer. Let us reconsider one of the same examples we balanced previously by the half-reaction method. For these we already know the products.

In the second example of Section 12-3.1, we find H_2S gas reacts with MnO_4^- to give solid sulfur and Mn^{+2}:

$$MnO_4^- + H_2S(g) \text{ gives } S(s) + Mn^{+2} \quad (102)$$

First, we assign oxidation numbers to each element, using rules 1–5. We find

	MnO_4^- +	$H_2S(g)$	$S(s)$ +	Mn^{+2}
Oxidation number	+7	−2	0	+2

with changes,

$$\text{for manganese } +7 \xrightarrow{-5} +2$$
$$\text{for sulfur } -2 \xrightarrow[+2]{} 0$$

If the gain in oxidation number by sulfur is to equal the loss by manganese, then five atoms of sulfur must react with two atoms of manganese:

$$2MnO_4^- + 5H_2S(g) \text{ gives } 5S(s) + 2Mn^{+2} \quad \text{(not balanced)} \quad (103)$$

$$2(+7) \xrightarrow{2(-5) = -10} 2(+2)$$
$$5(-2) \xrightarrow[5(+2) = +10]{} 5(0)$$

Now we proceed to ensure conservation of oxygen atoms. There are eight oxygen atoms on the left in (*103*), hence we must add eight molecules of H_2O to the right. (The reaction occurs in aqueous solution, so there is plenty of H_2O.)

$$2MnO_4^- + 5H_2S(g) \text{ gives } 5S(s) + 2Mn^{+2} + 8H_2O \text{ (not balanced)} \quad (104)$$

Next we must ensure conservation of hydrogen atoms. On the left, there are 10 hydrogen atoms (in $5H_2S$) and on the right 16 (in $8H_2O$). In aqueous solutions (in neutral or acidic solution) we assume that these six hydrogen atoms needed on the left are provided by H^+:

$$2MnO_4^- + 5H_2S(g) + 6H^+ \longrightarrow 5S(s) + 2Mn^{+2} + 8H_2O \quad (95), (105)$$

The equation is balanced now but experience dictates that a check should *always* be made on the basis of charge balance:

$$2(-1) + 5(0) + 6(+1) \qquad 5(0) + 2(+2) + 8(0)$$
$$-2 \qquad +6 \qquad +4$$
$$+4 = +4$$

Of course, the oxidation number method gives the same balanced equation as the half-reaction method.

12-4 ELECTROLYSIS

So far in this chapter we have dealt with reactions that proceed spontaneously. But the same ideas and names are applied to reactions that we force to take place, against their natural tendency, by supplying energy with an externally applied electric current. Such a process is termed electrolysis or "separation by electricity."

We have dealt with electrolysis before—every time we discussed or measured the electrical conductivity of an electrolyte solution. To see this, let's consider the processes that occur when we cause electric charge to pass through an aqueous solution of hydrogen iodide.

A distinguishing property of ionic solutions is electrical conductivity, just as it is a distinguishing property for metals, but the current-carrying mechanism differs. Electric charge moves through a metal wire, we believe, by means of electron movement. Electrons flow through the wire without changing the metal chemically. In contrast, the movement of electric charge through an aqueous solution of an electrolyte causes significant chemical changes.

Figure 12-7 shows, on the right, the behavior of an aqueous hydrogen iodide solution during conduction. The two carbon rods are connected by wires to the terminals of a 2 volt battery. Electrons flow from the battery through the left carbon rod, entering the solution. An equal number of electrons leave the solution through the right carbon rod to return to the battery. The hydrogen ion, H^+, has the ability to accept an

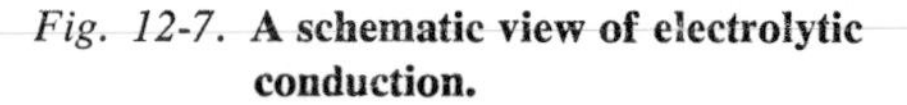

Fig. 12-7. **A schematic view of electrolytic conduction.**

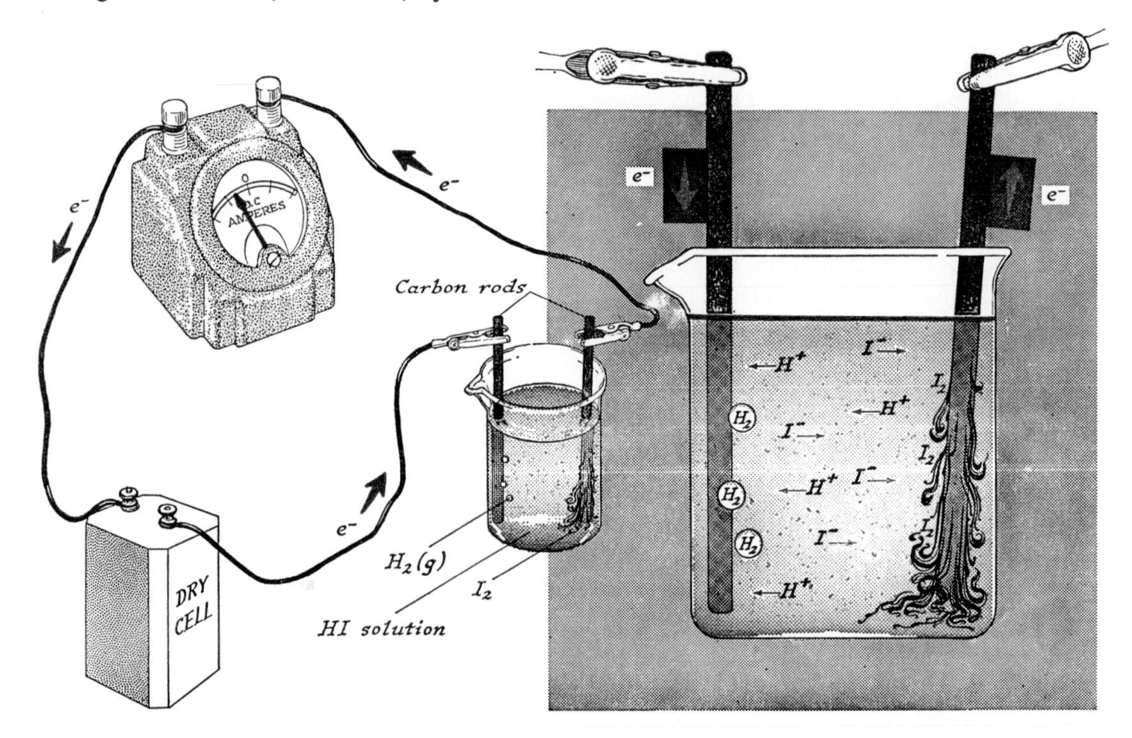

electron at the left electrode, where electrons are in excess. The H^+ ion is changed chemically to a neutral atom. An iodide ion, I^-, has one excess electron that can be released at the right electrode, where electrons are in deficiency. The I^- ion changes chemically to a neutral atom. The net result of these two occurrences is the process we call *electrolysis*.

Let us sum up the process during the movement of one electron through the entire circuit shown on the left in Figure 12-7.

(1) An electron on the left carbon rod is gone; another electron has shown up at the right carbon rod.

(2) One $H^+(aq)$ ion and one $I^-(aq)$ ion are gone; one H atom and one I atom have been formed.

(3) As this process continues to take place, the H^+ ions in the solution at the left tend to be used up and the same occurs for I^- ions in the solution at the right. Since only positive ions are used up at the left, the remaining negative ions are electrically repelled from this region and are attracted toward the solution at the right where positive ions are plentiful. Here, negative ions are used up, so the remaining positive ions are repelled and they are attracted toward the left where negative ions are plentiful. Iodide ions, I^-, move from left to right through the solution, carrying negative charge. At the same time, hydrogen ions, H^+, move from right to left through the solution, carrying positive charge. This drift of the ions through the solution, positive ions in one direction and negative ions in the other, explains the conduction in aqueous solutions.

(4) The battery performs work in forcing current to flow through the solution and in causing chemical changes to occur that would not proceed spontaneously.

The net reaction is

$$2H^+ + 2I^- \longrightarrow H_2(g) + I_2 \qquad (106)$$

Reaction (*106*) is just an oxidation-reduction reaction and it is readily separated into the two half-reactions

$$2H^+ + 2e^- = H_2(g) \qquad E^\circ = \ \ 0.000 \text{ volt}$$
$$2I^- = I_2(g) + 2e^- \qquad E^\circ = -0.53 \text{ volt}$$
$$\text{Overall reaction} \quad 2H^+ + 2I^- = H_2(g) + I_2(g) \qquad E^\circ = -0.53 \text{ volt} \qquad (106), (107)$$

The negative value of $E^\circ = -0.53$ volt tells us that the reaction will not occur spontaneously as written. This voltage tells us further that electrolysis will occur only if a cell with a voltage exceeding -0.53 volt is placed in the external circuit so as to oppose the voltage generated by the cell itself.

EXERCISE 12-10

From Appendix 3, estimate the minimum voltage required to cause electrolysis of 1 *M* HCl, forming $H_2(g)$ and $O_2(g)$, each at 1 atmosphere pressure. Show that at this voltage electrolysis to produce $H_2(g)$ and $Cl_2(g)$ will not occur.

QUESTIONS AND PROBLEMS

1. One method of obtaining copper metal is to let a solution containing Cu^{+2} ions trickle over scrap iron. Write the equations for the two half-reactions involved. Assume the iron becomes Fe^{+2}. Indicate in which half-reaction oxidation is taking place.

2. (a) If a neutral atom becomes positively charged, has it been oxidized or reduced? Write a general equation using *M* for the neutral atom.

   (b) If an ion X^{-1} acquires a -2 charge, has it

been oxidized or reduced? Write a general equation.

3. Aluminum metal reacts with aqueous acidic solutions to liberate hydrogen gas. Write the two half-reactions and the net ionic reaction.

4. When copper is placed in concentrated nitric acid, vigorous bubbling takes place as a brown gas is evolved. The copper disappears and the solution changes from colorless to a greenish-blue. The brown gas is nitrogen dioxide, NO_2, and the solution's color is due to the formation of cupric ion, Cu^{+2}. Using half-reactions from Appendix 3, write the net ionic equation for this reaction.

5. Nickel metal reacts with cupric ions, Cu^{+2}, but not with zinc ions, Zn^{+2}; magnesium metal does react with Zn^{+2}. In each case of reaction, ions of +2 charge are formed. Use these data to expand the table of reactions on p. 206.

6. In acid solution the following are true: H_2S will react with oxygen to give H_2O and sulfur. H_2S will not react in the corresponding reaction with selenium or tellurium. H_2Se will react with sulfur giving H_2S and selenium but it will not react with tellurium. Arrange the hydrides of column VI, H_2O, H_2S, H_2Se, and H_2Te, in order of their tendency to lose electrons to form the elements, O_2, S, Se, and Te.

7. If you wish to replate a silver spoon, would you make it the anode or cathode in a cell? Use half-reactions in your explanation. How many moles of electrons are needed to plate out 1.0 gram of Ag?

8. Figure 12-5 shows electrons leaving the $Cu(s)$ and going to the $Ag(s)$. Experimentally, both half-cells are found to be electrically neutral before current flows and to remain so as the cell operates. Explain this.

9. In the electrolysis of aqueous cupric bromide, $CuBr_2$, 0.500 gram of copper is deposited at one electrode. How many grams of bromine are formed at the other electrode? Write the anode and cathode half-reactions.

Answer. 1.26 grams of $Br_2(l)$

10. Complete the following equations. Determine the net potential of such a cell and decide whether reaction can occur.

(a) $Zn + Ag^+ \longrightarrow$
(b) $Cu + Ag^+ \longrightarrow$
(c) $Sn + Fe^{+2} \longrightarrow$
(d) $Hg + H^+ \longrightarrow$

11. For each of the following,

(i) write the half-reactions;
(ii) determine the net reaction;
(iii) predict whether the reaction can occur giving the basis for your prediction:

(a) $Mg(s) + Sn^{+2} \longrightarrow$
(b) $Mn(s) + Cs^+ \longrightarrow$
(c) $Cu(s) + Cl_2(g) \longrightarrow$
(d) $Zn(s) + Fe^{+2} \longrightarrow$
(e) $Fe(s) + Fe^{+3} \longrightarrow$

12. A half-cell consisting of a palladium rod dipping into a 1 *M* $Pd(NO_3)_2$ solution is connected with a standard hydrogen half-cell. The cell voltage is 0.99 volt and the platinum electrode in the hydrogen half-cell is the anode. Determine $E°$ for the reaction

$$Pd \longrightarrow Pd^{+2} + 2e^-$$

13. Suppose chemists had chosen to call the $2I^- \longrightarrow I_2 + 2e^-$ half-cell potential zero.

(a) What would be $E°$ for $Na \longrightarrow Na^+ + e^-$?
(b) How much would the net potential for the reaction $2Na + I_2 \longrightarrow 2Na^+ + 2I^-$ change?

14. If a piece of copper metal is dipped into a solution containing Cr^{+3} ions, what will happen? Explain, using $E°$s.

15. What would happen if an aluminum spoon is used to stir an $Fe(NO_3)_2$ solution? What would happen if an iron spoon is used to stir an $AlCl_3$ solution?

16. Can 1 *M* $Fe_2(SO_4)_3$ solution be stored in a container made of nickel metal? Explain your answer.

17. Suppose water is added to each of the beakers containing copper sulfate in the two electrochemical cells shown in Figure 12-4 (p. 204). What change will occur in the voltage in each cell? Explain.

18. Determine the oxidation numbers of carbon in the compounds carbon monoxide, CO, carbon dioxide, CO_2, and in diamond.

19. Determine the oxidation number of uranium in each of the known compounds: UO_3, U_3O_8, U_2O_5, UO_2, UO, K_2UO_4, $Mg_2U_2O_7$.

20. By use of half-reactions, give a balanced equation for each of the following reactions:

(a) $H_2O_2 + I^- + H^+$ gives $H_2O + I_2$

(b) $Cr_2O_7^{-2} + Fe^{+2} + H^+$ gives $Cr^{+3} + Fe^{+3} + H_2O$

(c) $Cu + NO_3^- + H^+$ gives $Cu^{+2} + NO + H_2O$

(d) $MnO_4^- + Sn^{+2} + H^+$ gives $Mn^{+2} + Sn^{+4} + H_2O$

21. By use of oxidation numbers, give a balanced equation for each of the following reactions:

(a) $HBr + H_2SO_4$ gives $SO_2 + Br_2 + H_2O$

(b) $NO_3^- + Cl^- + H^+$ gives $NO + Cl_2 + H_2O$

(c) $Zn + NO_3^- + H^+$ gives $Zn^{+2} + NO_2 + H_2O$

(d) BrO^- gives $Br^- + BrO_3^-$

22. Use oxidation numbers to balance the reaction between ferrous ion, Fe^{+2}, and permanganate ion, MnO_4^-, in acid solution to produce ferric ion, Fe^{+3}, and manganous ion, Mn^{+2}.

23. Show the arbitrariness of oxidation numbers by balancing the reaction discussed in Problem 22 with the assumption that the oxidation number of manganese in MnO_4^- is +2. Compare with the result obtained in Problem 22.

24. In order to make $Na(s)$ and $Cl_2(g)$, an electric current is passed through $NaCl(l)$. What does the energy supplied to this reaction do?

CHAPTER 13

Chemical Calculations

The sceptical chemist . . . draws conclusions regarding chemical materials . . . chiefly on the basis of quantitative chemical analysis which . . . is the touchstone of all chemical hypothesis.

G. T. MORGAN, 1930

Chemistry is a quantitative science. This means that a chemist wishes to know more than the qualitative fact that a reaction occurs. He must answer questions beginning "How much . . .?" The quantities may be expressed in grams, volumes, concentrations, percentage composition, or a host of other practical units. Ultimately, however, the understanding of chemistry requires that amounts be related quantitatively to balanced chemical reactions. *The study of the quantitative relationships implied by a chemical reaction is called* ***stoichiometry.***

Stoichiometric calculations are based upon two assumptions. First, we assume that *only a single reaction need be considered* to describe the chemical changes occurring. Second, we assume that *the reaction is complete.* For example, consider the question, How much iron is produced per mole of Fe_2O_3 reacted with aluminum in the following reaction?

$$2Al(s) + Fe_2O_3(s) \rightleftharpoons Al_2O_3(s) + 2Fe(s) \quad (1)$$

We base our calculations upon the two assumptions above, whether they are stated or not. First, we assume this is the only reaction that occurs. Reaction (*2*), for example, is assumed to be unimportant.

$$2Al(s) + 3Fe_2O_3(s) \rightleftharpoons 6FeO(s) + Al_2O_3(s) \quad (2)$$

Sometimes such an assumption is based upon experience, sometimes upon hope. Furthermore, it is assumed that a mole of Fe_2O_3 reacts *completely* according to reaction (*1*). None of the Fe_2O_3 remains unreacted at the finish, either because of equilibrium, because of mechanical losses of some sort, or because insufficient aluminum was added. In practice these conditions are sometimes difficult to obtain.

If these assumptions are valid, however, stoichiometric calculations provide a reliable basis for quantitative predictions. It is important to be able to make these calculations with ease. Fortunately, they all can be made with a single pattern based upon the mole concept.

13-1 A PATTERN FOR STOICHIOMETRIC CALCULATIONS

The equation for a chemical reaction speaks in terms of molecules or of moles. It contains the basis for stoichiometric calculations. However, in the laboratory a chemist measures amounts in such units as grams and milliliters. The first step in any quantitative calculation, then, is to convert the measured amounts to moles. In mole units, the balanced reaction connects quantities of reactants and products. Finally, the result is expressed in the desired units (which may not necessarily be the same as the original units).

Let's put this down schematically. Suppose two substances, *A* and *B*, combine according to a known reaction. We wish to know how much *B* will react with (or, be produced from) a measured quantity of *A*. The solution to this typical problem of stoichiometry consists of three steps.

We shall apply this scheme to a series of types of calculations to show its general applicability. The calculations are all connected with the manufacture of sulfuric acid, H_2SO_4, one of the most important commercial chemicals.

Step I. Amount of *A* in measured units $\xrightarrow[\text{Mol wt of } A]{\text{Convert to moles of } A}$ Moles of *A* (3)

Step II. Moles of *A* $\xrightarrow[\text{Balanced reaction}]{\text{Convert to moles of } B}$ Moles of *B* (4)

Step III. Moles of *B* $\xrightarrow[\text{Mol wt of } B]{\text{Convert to amount of } B}$ Amount of *B* in desired units (5)

13-2 THE MANUFACTURE OF SULFURIC ACID

Many millions of tons of sulfuric acid, H_2SO_4, are produced every year. Its uses are so wide that the amount consumed per year by a country can be taken as a crude index of the technological development of that country. Two manufacturing processes have widespread industrial importance and both will be described. These processes are so highly perfected that the cost of this useful chemical is only about $22.00 per ton!

The chemical reactions appear simple. They begin with pure sulfur (which occurs in natural deposits in the elemental state). First, sulfur is burned to give gaseous sulfur dioxide, SO_2. Next, the SO_2 is further oxidized, catalytically, to sulfur trioxide, SO_3. Finally, addition of water forms sulfuric acid. The reactions are:

$$S_8(s) + 8O_2(g) \rightleftharpoons 8SO_2(g) \quad (6)$$

$$8SO_2(g) + 4O_2(g) \rightleftharpoons 8SO_3(g) \quad \text{(catalyst needed)} \quad (7)$$

$$8SO_3(g) + 8H_2O(l) \rightleftharpoons 8H_2SO_4(l) \quad (8)$$

Overall reaction

$$S_8(s) + 12O_2(g) + 8H_2O(l) \rightleftharpoons 8H_2SO_4(l) \quad (9)$$

Now let's investigate some of the quantitative questions that are connected with this important process.

EXERCISE 13-1

If H_2SO_4 is purchased at a price of $22.00 per ton, how many moles are obtained for a penny? (Note: 1 pound = 453.6 grams.)

13-2.1 Weight–Weight Calculations

A shovelful of sulfur containing 1.00 kg is placed in the hopper to be converted to sulfuric acid. *What weight of H_2SO_4 will be formed?*

This practical question is of a familiar form. We wish to calculate the weight of product resulting from a specified weight of reactant. Our calculational pattern is applicable.

First, a reaction must be assumed. According to the intent of the process, the overall reaction is (*9*):

$$S_8(s) + 12O_2(g) + 8H_2O(l) \rightleftharpoons 8H_2SO_4(l) \quad (9, 10)$$

Our calculation proceeds on the assumption that equation (*10*) is the only reaction that occurs and that the entire kilogram of sulfur is consumed in it.

Step I. We must convert grams of sulfur to moles. The molecular weight of sulfur is needed.

$$\text{Mol wt } S_8 = 8 \times (\text{atomic wt of sulfur})$$
$$= 8 \times (32.1) = 256.8 \text{ g/mole} \quad (11)$$

Now the number of moles of S_8 in 1.00 kg of sulfur is

$$\text{Moles } S_8 = \frac{(\text{wt sulfur})}{(\text{mol wt})} = \frac{(1.00 \times 10^3 \text{ g})}{(256.8 \text{ g/mole})}$$
$$= 3.89 \text{ moles} \quad (12)$$

Step II. The next step is to decide how many moles of H_2SO_4 can be produced from 3.89 moles of S_8. The balanced reaction (*10*) tells us that

one mole of S_8 forms eight moles H_2SO_4

hence

3.89 moles of S_8 form $8 \times (3.89)$ moles H_2SO_4
3.89 moles of S_8 form *31.1 moles H_2SO_4* (*13*)

Step III. The amount of sulfuric acid formed is 31.1 moles. How much does this weigh?

$$\text{Wt } H_2SO_4 = (\text{moles } H_2SO_4)(\text{mol wt of } H_2SO_4)$$
$$= (31.1 \text{ moles})(98.1 \text{ g/mole})$$
$$= 3051 \text{ g}$$
$$\text{Wt } H_2SO_4 = 3.05 \text{ kg}$$

Answer. 1.00 kg of sulfur produces 3.05 kg H_2SO_4 by reaction (*10*). (*14*)

13-2.2 Weight–Gas Volume Calculations

The first step in the manufacture of H_2SO_4 is to burn it to sulfur dioxide. Sulfur burns spontaneously in air, liberating heat.

$$S_8(s) + 8O_2(g) \rightleftharpoons 8SO_2(g)$$
$$\Delta H = -70.96 \text{ kcal/mole } SO_2 \quad (6), (15)$$

An important economic feature of the modern processes is the utilization of this heat in another step in which heat is absorbed.

Of course, a plant designer must anticipate the weights and volumes of the constituents at each stage of the process. Hence, he must be able to answer such a question as, "*What weight of sulfur will burn to produce 100,000 liters of pure SO_2 at 500°C and one atmosphere pressure?*"

Again, we must assume reaction (*15*) can be carried out exclusively and completely. This time the calculation begins with a specified amount of a product and we wish to calculate the corresponding amount of a reactant. Note that we are immediately confronted with a question of significant figures. How many significant figures are intended in the volume, 100,000 liters? In the absence of other information, let's let common sense dictate. Would it be of value to make the calculation to six significant figures? Undoubtedly not. Two, or at most three, significant figures will probably suffice for the purposes of the plant designer. Other conditions, such as the weight of sulfur and the temperature, can't be easily controlled to more than this accuracy. Let's carry three significant figures to be sure we have enough. The volume of SO_2 is, then, 1.00×10^5 liters.*

Step I. We must convert the specified volume of SO_2 into moles. A convenient way to do this is to calculate the volume this gas would occupy under conditions at which we know the volume occupied by one mole of gas. For example, we know that one mole of gas occupies 22.4 liters at 0°C and one atmosphere pressure. Increasing the temperature of a gas at constant

* This volume, 1.00×10^5 liters, is about the volume of a small room—a practical dimension for a reaction chamber.

pressure increases the volume in proportion to the *absolute* temperature:

$$0°C = 273°K$$
$$500°C = 273 + 500 = 773°K$$

1.00 mole of gas occupies 22.4 liters at 273°K, 1 atm

1.00 mole of gas occupies $22.4 \times \frac{773}{273}$ liters at 773°K, 1 atm

1.00 mole of gas occupies 63.4 liters at 773°K, 1 atm

or

63.4 liters SO_2 at 773°K, 1 atm contain 1.00 mole SO_2

hence

1.00×10^5 liters contain

$$\frac{1.00 \times 10^5}{63.4} \times 1.00 \text{ mole } SO_2$$

1.00×10^5 liters contain *1.58×10^3* moles SO_2 at 773°K, 1 atm (*16*)

Step II. Next we decide how many moles of S_8 are needed to produce 1.58×10^3 moles of SO_2. The balanced reaction (*15*) indicates

8 moles SO_2 are produced from 1 mole S_8

1 mole SO_2 is produced from $\frac{1}{8}$ mole S_8

1.58×10^3 moles SO_2 are produced from $\frac{1}{8}(1.58 \times 10^3)$ moles S_8

1.58×10^3 moles SO_2 are produced from 198 moles S_8 (*17*)

Step III. How much do 198 moles of S_8 weigh?

$$\text{Wt } S_8 = (\text{moles } S_8)(\text{mol wt } S_8)$$
$$= (198 \text{ moles})(256.8 \text{ g/mole})$$
$$\text{Wt } S_8 = 50.8 \times 10^3 \text{ g}$$

Answer. 1.00×10^5 liters of SO_2 are produced from 50.8 kilograms of sulfur by reaction (15). (*18*)

13-2.3 Gas Volume–Gas Volume Calculations

After sulfur dioxide is produced by combustion of sulfur, further oxidation is needed in the manufacture of H_2SO_4. The reaction, producing sulfur trioxide, SO_3, is exothermic; heat is released:

$$SO_2(g) + \tfrac{1}{2}O_2(g) \rightleftharpoons SO_3(g) \qquad \Delta H = -23.5 \text{ kcal/mole } SO_3 \quad (7), (19)$$

Yet the reaction is quite slow, even at high temperatures. Evidently the rate is controlled by a high activation energy. In fact, the practical use of reaction (*19*) depends upon the presence of a catalyst to provide a reaction path with a lower activation energy. The two important commercial methods for manufacture of H_2SO_4 differ principally in the choice of catalyst for this step.

The older process is called the *lead chamber* process. It uses a mixture of gaseous oxides of nitrogen—nitric oxide, NO, and nitrogen dioxide, NO_2—as the catalyst. This process has been in use and under development for over 200 years. It is named after the large room-like chambers lined with lead in which the gaseous reactions are carried out. The lead walls react with the acid and become coated with an inert protective coating of lead sulfate.

The newer process uses a solid catalyst for reaction (*19*). Either finely divided platinum or vanadium pentoxide, V_2O_5, is effective. Because catalysis occurs where the gas contacts the surface of the catalyst, this process is called the *contact process.*

EXERCISE 13-2

Reaction (*19*) is carried out at a high temperature (about 500°C in the contact process). How does temperature affect equilibrium, according to Le Chatelier's Principle? In view of your answer, propose an explanation of why the temperature is kept high.

Reaction (*19*) requires the reaction of oxygen from air and sulfur dioxide. *What volume of air, at 500°C and one atmosphere pressure, is needed to react with the 1.00×10^5 liters of SO_2 produced from 50.8 kilograms of sulfur?*

Step I. In considering a chemical reaction between gases, we can apply Avogadro's Hypothesis: Equal volumes of gases contain equal numbers of molecules (at the same pressure and temperature). The volume of the SO_2, 1.00×10^5 liters, is already a measure of the number of moles of SO_2.

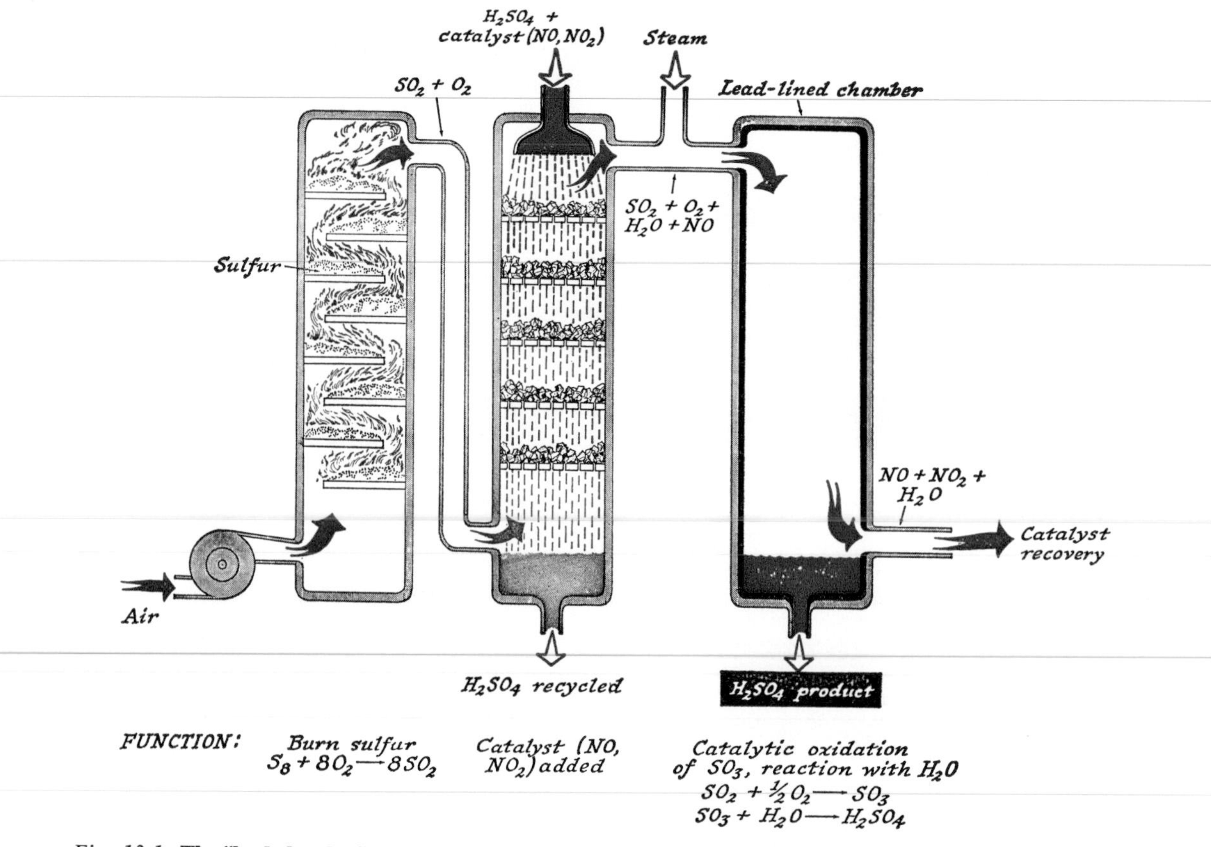

Fig. 13-1. **The "lead chamber" process** *for H_2SO_4 manufacture.*

Step II. By reaction (*19*), 1 mole of SO_2 reacts with $\frac{1}{2}$ mole of O_2. By Avogadro's Hypothesis, 1 liter of SO_2 reacts with $\frac{1}{2}$ liter of O_2 (if they are at the same temperature and pressure). Hence 1.00×10^5 liters of SO_2 react with $\frac{1}{2}(1.00 \times 10^5)$ liters of O_2 if both SO_2 and O_2 are pure and measured at 500°C and one atmosphere.

Answer. We need 0.500×10^5 liters of pure O_2 at 500°C and one atmosphere to oxidize 1.00×10^5 liters of SO_2. (*20*)

Step III. Now we must convert to the desired units. We need the number of moles of oxygen that are present in 0.500×10^5 liters (at 500°C, 1 atm), but we wish to use air instead of pure oxygen. If air contains about 20% oxygen (by volume), then it takes 5 liters of air (at 500°C, 1 atm) to provide the amount of oxygen in 1 liter of pure oxygen (at 500°C, 1 atm). If 0.500×10^5 liters of pure O_2 are needed, then $5(0.500 \times 10^5) = 2.50 \times 10^5$ liters of air are needed.

Answer. 2.50×10^5 liters of air (at 500°C, 1 atm) react with 1.00×10^5 liters of SO_2 (at 500°C, 1 atm). (*21*)

13-2.4 Weight–Liquid Volume Calculations

The last step in the preparation of commercial sulfuric acid is to allow the sulfur trioxide to react with steam:

$$SO_3(g) + H_2O(g) \rightleftharpoons H_2SO_4(l) \quad (9), (22)$$

This results in a concentrated sulfuric acid solution that contains 98% H_2SO_4. It is a viscous, colorless liquid. When it is mixed with water, so much heat is liberated that the operation must be carried out very cautiously. The sulfuric acid is slowly poured into the water, *not the reverse.* The density of this concentrated sulfuric acid solution (98%) is 1.84 grams/ml and its concentration is 18.3 *M*.

One of the important uses of sulfuric acid is that of an oxidizing agent. For example, when heated, it will even dissolve carbon. The reaction is

$$C + 2H_2SO_4 \rightleftharpoons CO_2 + 2H_2O + 2SO_2 \quad (23)$$

EXERCISE 13-3

Verify that reaction (*23*) is an oxidation-reduction reaction and that the oxidation number change of carbon is balanced by the oxidation number change of the sulfur.

How many liters of concentrated sulfuric acid would be consumed in reaction (23) to oxidize 1.00 kg of carbon?

Step I. We wish to oxidize 1.00 kg of carbon. The number of moles of carbon is

$$\text{Moles carbon} = \frac{(\text{wt C})}{(\text{at wt C})} = \frac{1.00 \times 10^3 \text{ g}}{12.01 \text{ g/mole}}$$

$$= 83.2 \text{ moles}$$

Moles carbon = 83.2 (*24*)

Step II. Now we must decide how many moles of sulfuric acid are needed. Reaction (*23*) shows that

1 mole C reacts with 2 moles H_2SO_4
83.2 moles C react with 2(83.2) moles H_2SO_4
83.2 moles C react with 166 moles H_2SO_4 (*25*)

Step III. We need 166 moles of H_2SO_4. What volume of concentrated sulfuric acid (18.3 *M*) is needed?

18.3 moles H_2SO_4 are present in 1.00 liter 98% H_2SO_4

166 moles H_2SO_4 are present in $\frac{166}{18.3}$ liters 98% H_2SO_4

166 moles H_2SO_4 *are present in 9.07 liters*

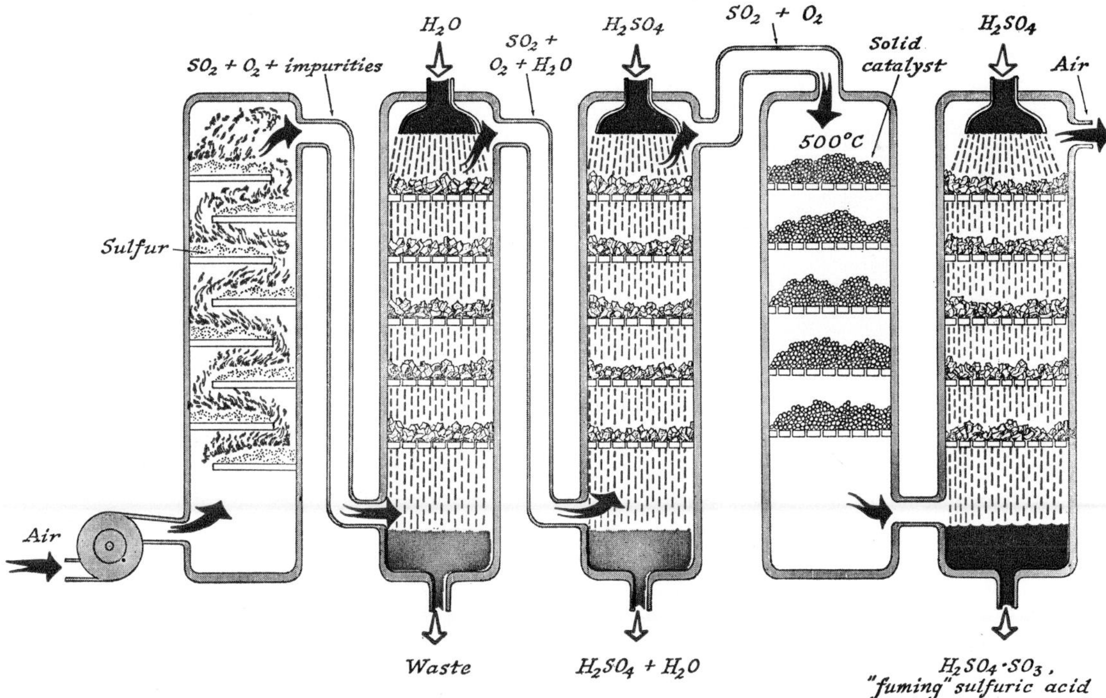

Fig. 13-2. **The "contact" process** *for* H_2SO_4 *manufacture.*

Answer. 1.00 kg of carbon is oxidized by 9.07 liters of concentrated H_2SO_4 in reaction (23). (26)

13-2.5 Liquid Volume–Volume Calculations

A second major use of sulfuric acid of commerce is in reactions with bases. In laboratory use it is diluted to a much lower concentration and can be used as a standard acid. A typical problem would be the titration of a base solution of unknown concentration using a sulfuric acid solution of known concentration. For example, "*What is the concentration of a sodium hydroxide solution if 25.43 ml of the NaOH solution just reacts with 18.51 ml of 0.1250 M H_2SO_4 (to produce a neutral solution)?*"

Step I. We are given the concentration and volume of H_2SO_4 solution. How many moles of H_2SO_4 are present?

$$\text{Moles } H_2SO_4 \text{ present} = (\text{volume})(\text{concentration}) = (18.51 \times 10^{-3} \text{ liter})(0.1250\ M)$$

$$\text{Moles } H_2SO_4 = 2.314 \times 10^{-3} \text{ mole}$$

Step II. We are interested in the reaction between $H^+(aq)$ and $OH^-(aq)$. Sulfuric acid gives 2 moles of $H^+(aq)$ per mole of H_2SO_4 dissolved in water.

$$H_2SO_4(l) \longrightarrow 2H^+(aq) + SO_4^{-2}(aq) \quad (27)$$

Hence

$$\text{Moles } H^+(aq) = 2(\text{moles } H_2SO_4) = 2(2.314 \times 10^{-3})$$

$$\textit{Moles } H^+(aq) = 4.628 \times 10^{-3} \quad (28)$$

Now we are concerned with the acid–base reaction

$$H^+(aq) + OH^-(aq) \rightleftharpoons H_2O \quad (29)$$

By reaction (29),

one mole of $OH^-(aq)$ ion reacts with one mole of $H^+(aq)$ ion. *4.628 × 10^{-3} mole of $OH^-(aq)$ ion reacts with 4.628 × 10^{-3} mole of $H^+(aq)$ ion.* (30)

Step III. We can now calculate the concentration of hydroxide ion. We now know that 4.628×10^{-3} mole of $OH^-(aq)$ is contained in 25.43 ml of sodium hydroxide solution:

$$\text{Concentration hydroxide ion} = \frac{(\text{moles } OH^- \text{ ion})}{(\text{volume})}$$

$$[OH^-] = \frac{(4.628 \times 10^{-3} \text{ mole})}{(25.43 \times 10^{-3} \text{ liter})}$$

$$[OH^-] = 0.1820\ M$$

Answer. The sodium hydroxide solution has a concentration of 0.1820 M. (31)

QUESTIONS AND PROBLEMS

1. In Experiment 7, would the ratio between moles of copper atoms used and moles of silver atoms formed change if silver sulfate, Ag_2SO_4, had been used rather than silver nitrate, $AgNO_3$? Explain.

2. Although sodium carbonate is needed in the manufacture of glass, very little is found in nature. It is made using two very abundant chemicals, calcium carbonate (marble) and sodium chloride (salt). The process involves many steps, but the overall reaction is

$$CaCO_3 + 2NaCl \longrightarrow Na_2CO_3 + CaCl_2$$

(a) How many grams of sodium chloride react with 1.00 kg of calcium carbonate?
(b) How many grams of sodium carbonate are produced?

3. Some catalysts used in gasoline manufacture consist of finely divided platinum supported on an inert solid. Suppose that the platinum is formed by the high temperature reaction between platinum dioxide, PtO_2, and hydrogen gas to form platinum metal and water.

(a) What is the oxidation number of platinum in platinum dioxide?
(b) Is hydrogen an oxidizing or reducing agent in this reaction?
(c) How many grams of hydrogen are needed to produce 1.0 gram of platinum metal?

(d) How many moles of water are produced along with 1.0 gram of Pt?

(e) How many grams of water are produced along with 1.0 gram of Pt?

Answer. (e) 0.18 gram of H_2O

4. Hydrazine, N_2H_4, and hydrogen peroxide, H_2O_2, are used together as a rocket fuel. The products are N_2 and H_2O. How many grams of hydrogen peroxide are needed per 1.00×10^3 grams of hydrazine carried by a rocket?

Answer. 2.12×10^3 grams of H_2O_2

5. Iodine is recovered from iodates in Chile saltpeter by the reaction

$HSO_3^- + IO_3^-$ gives $I_2 + SO_4^{-2} + H^+ + H_2O$

(a) How many grams of sodium iodate, $NaIO_3$, react with 1.00 mole of $KHSO_3$?

(b) How many grams of iodine, I_2, are produced?

6. The hourly energy requirements of an astronaut can be satisfied by the energy released when 34 grams of sucrose are "burned" in his body. How many grams of oxygen would need to be carried in a space capsule to meet this requirement?

sucrose + oxygen gives carbon dioxide + water
$C_{12}H_{22}O_{11} + O_2$ gives $CO_2 + H_2O$

7. The chlorine used to purify your drinking water was possibly made by electrolyzing molten NaCl to produce liquid sodium and gaseous chlorine.

(a) How many grams of sodium chloride are needed to produce 355 grams of chlorine gas?

(b) What volume would this gas occupy at STP?

8. A reaction involved in the production of iron from iron ore is

$Fe_2O_3 + CO$ gives $Fe + CO_2$

$\Delta H = -4.3$ kcal/mole Fe_2O_3

(a) How many grams of CO must react to release 13 kcal?

(b) How many liters of CO(STP) are needed to produce 1.0 kg of Fe?

9. More C_8H_{18}, a hydrocarbon that is useful in gasoline, can be obtained from petroleum if this reaction takes place:

$$C_{16}H_{32}(g) + 2H_2(g) \longrightarrow 2C_8H_{18}(g)$$

(a) How many grams of C_8H_{18} can be made using 224 liters of H_2 at STP?

(b) What pressure conditions favor production of $C_8H_{18}(g)$?

Answer. (a) 1.14×10^3 g C_8H_{18}

10. How many liters of oxygen gas, at STP, will be released by decomposing 14.9 grams of NaOCl to produce $O_2(g) + Cl^-(aq)$ (as in Experiment 14a)?

11. A compound found in kerosene, a mixture of hydrocarbons, is decane, $C_{10}H_{22}$. A stove might burn 1.0 kg of kerosene per hour. Assume kerosene is $C_{10}H_{22}$ and answer the following:

(a) How many liters (STP) of oxygen are needed per hour?

(b) How many liters (STP) of carbon dioxide are produced per hour?

12. How many grams of zinc metal are needed to react with hydrochloric acid to produce enough hydrogen gas to fill an 11.2 liter balloon at STP? What would be the volume of this balloon at 27°C and 680 mm Hg pressure? How many grams of zinc would be needed if sulfuric acid were used?

13. How many liters of air (STP) are needed to burn 2.2 liters (STP) of methane, CH_4, gas in your laboratory burner? How much heat is released? The ΔH for combustion of CH_4 is -210 kcal/mole of CH_4. Assume air is 20% oxygen.

14. In the reaction

$NH_3(g) + O_2(g)$ gives $NO(g) + H_2O(g)$,

if 4.48 liters of ammonia gas measured at STP are used, how many liters of oxygen measured at STP will be needed to react with all the ammonia?

Answer. 5.60 liters of O_2 at STP

15. The following reaction is carried out with all gas volumes measured at the same pressure and temperature:

$C_4H_{10}(g) + O_2(g)$ gives $CO_2(g) + H_2O(g)$

(a) How many liters of oxygen are required to produce 2.0 liters of CO_2?

(b) If 15 liters of oxygen are used, how many liters of butane, C_4H_{10}, will be burned?

(c) If 8.0 liters each of oxygen and butane are mixed, how many liters of CO_2 are produced (assume complete reaction)?

Answer. (a) 3.2 liters of O_2

16. What volume of Cl_2 gas at 37°C and 753 mm could be obtained from 58.4 liters of HCl, also measured at 37°C and 753 mm, if the following reaction could be carried effectively to completion?

$$HCl(g) + O_2(g) \text{ gives } H_2O(g) + Cl_2(g)$$

17. Suppose 105 liters of NH_3 and 285 liters of O_2 are allowed to react until the reaction:

$$NH_3(g) + O_2(g) \text{ gives } H_2O(g) + NO_2(g)$$

is complete. The temperature and pressure are maintained constant at 200°C and 0.30 atmosphere during all volume measurements. What gas and what volume of it measured at the stated conditions remains unreacted?

18. A 6 volt lead storage battery contains 700 grams of pure $H_2SO_4(l)$ dissolved in water.

(a) How many grams of solid sodium carbonate, Na_2CO_3, would be needed to neutralize this acid (giving CO_2 gas and H_2O) if it were spilled?

(b) How many liters of 2.0 *M* Na_2CO_3 solution would be needed?

19. Nitric acid, HNO_3, is made by the process

$$3NO_2(g) + H_2O(l) \rightleftharpoons 2HNO_3(l) + NO(g)$$

Commercial concentrated acid contains 68% by weight HNO_3 in water. The solution is 15 *M*. How many liters of concentrated acid are needed to react with 0.100 kg of copper metal?

$$Cu(s) + H^+(aq) + NO_3^-(aq) \text{ gives } Cu^{+2}(aq) + NO_2(g) + H_2O(l)$$

Answer. 0.42 liter of HNO_3

20. How many grams of silver metal will react with 2.0 liters of 6.0 *M* HNO_3? The reaction is

$$Ag(s) + H^+(aq) + NO_3^-(aq) \text{ gives } Ag^+(aq) + NO(g) + H_2O(l)$$

21. A measured volume, 10.00 liters, of the waste process water from a cotton mill require 23.62 ml of 0.1000 *M* hydrochloric acid to produce a neutral solution. What is the hydroxide ion concentration in the waste?

22. What weight of silver chloride may be obtained from 1.0 liter of 1.0 *M* $AgNO_3$, if 12 ml of 0.15 *M* NaCl are added?

23. How many milliliters of a 0.050 *M* $KMnO_4$ solution are required to oxidize 2.00 grams of $FeSO_4$ in a dilute acid solution?

Answer. 53 ml of $KMnO_4$ solution

CHAPTER 14

Why We Believe in Atoms

From the time of Dalton · · · the history of the atom has been a march of triumph. Wherever the concept of the atom was employed for the interpretation of observational measurements, it supplied lucid explanation; conversely, such success became overwhelming evidence for the existence of the atom.

HANS REICHENBACH, 1951

In Chapter 2 you were introduced to atoms and in Chapter 6 they were described in more detail. You were told that the atom contains charged particles, that it has a nucleus made up of neutrons and protons, and that the nucleus is surrounded by electrons. The atom is incredibly small but the nucleus is even smaller. But also you were told that every theory (including the atomic theory) should be thought about and criticized—the evidence upon which it is based should be examined and understood. It is one thing to ask "Do we believe in atoms?" and quite another to ask "*Why* do we believe in atoms?" In this chapter we shall try to answer this last, harder question.

Let's begin with an unpretentious example that shows how we make such decisions in day-to-day living.

A new tenant is told by his neighbor that the garbage collector comes every Thursday, early in the morning. Later, in answer to a question from his wife about the same matter, the tenant says, "I have been told there is a garbage collector and that he comes early Thursday morning. We shall see if this is true." The tenant, a scientist, accepts the statement of the neighbor (who has had opportunity to make observations on the subject). However, he accepts it *tentatively* until he himself knows the evidence for the conclusion.

After a few weeks, the new tenant has made a number of observations consistent with the existence of a Thursday garbage collector. Most important, the garbage does disappear every Thursday morning. Second, he receives a bill from the city once a month for municipal services. And there are several supplementary observations that are consistent—often he is awakened at 5:00 A.M. on Thursdays by a loud banging and sounds of a truck. Occasionally the banging is accompanied by gay whistling, sometimes by a dog's bark.

The tenant now has many reasons to believe

in the existence of the garbage collector. Yet he has never seen him. Being a curious man and a scientist, he sets his alarm clock one Wednesday night to ring at 5:00 A.M. Looking out the window Thursday morning, his first observation is that it is surprisingly dark out and things are difficult to see. Nevertheless, he discerns a shadowy form pass by, a form that looks like a man carrying a large object.

Seeing is believing! But which of these pieces of evidence really constitutes "seeing" the garbage collector? Which piece of evidence is the basis for "believing" there is a garbage collector? The answer is, *all* of the evidence taken together, constitutes "seeing." And *all* of the evidence taken together, furnishes the basis for accepting the "garbage collector theory of garbage disappearance." The direct vision of a shadowy form at 5:00 A.M. would not constitute "seeing a garbage collector" if the garbage didn't disappear at that time. (The form might have been the paper boy or the milkman.) Neither would the garbage disappearance alone consist of "seeing" the garbage collector. (Perhaps a dog comes by every Thursday and eats the garbage. Remember, a dog's bark was heard!) No, the tenant is convinced there is a garbage collector because the assumption is consistent with so many observations, and it is inconsistent with none. Other possible explanations fit the observations too, but not as well (the tenant has never heard a dog whistle gaily). The garbage collector theory passes the test of a good theory—it is useful in explaining a large number of experimental observations. This was true even before the tenant set eyes on the shadowy form at 5:00 A.M.

Yet we must agree, there are advantages to the "direct vision" type of experiment. Often more detailed information can be obtained this way. Is the garbage collector tall? Does he have a mustache? Could the garbage collector be a woman? This type of information is less easily obtained from other methods of observation. It is worthwhile setting the alarm clock, even after we have become convinced there is a garbage collector.

At the beginning of this course you were a new tenant. You were told that chemists believe in atoms and you were asked to accept this proposal tentatively until you yourself knew the evidence for it. Since that time, we have used the atomic theory continuously in our discussions of chemical phenomena. The atomic theory passes the test of a good theory: it is useful in explaining a large number of experimental observations. We have become convinced there are atoms.

Now we are going to review the types of evidence that form the basis for belief in the atomic theory. We shall include a number of experiments that are close, in concept, to the "direct vision" type. These are particularly convincing and they provide detailed information that is less readily obtained in other ways.

14-1 CHEMICAL EVIDENCE FOR THE ATOMIC THEORY

Let us begin by looking again at the kinds of evidence we already have for the existence of atoms—the evidence from chemistry. We shall consider, in turn, the definite composition of compounds, the simple weight relations among compounds, and the reacting volumes of gases. Each behavior provides experimental support for the atomic theory.

14-1.1 The Law of Definite Composition

Compounds are found to have definite composition, no matter how prepared. For example, 2.016 grams of hydrogen are found combined with 16.00 grams of oxygen in the compound water whether the water is prepared by burning hydrogen in oxygen, by decomposing gaseous nitrous acid, by heating barium chloride dihydrate, or by some other process:

$$H_2(g) + \tfrac{1}{2}O_2(g) \longrightarrow H_2O(g)$$
2.016 g hydrogen/16.00 g oxygen (*1*)

$$2HNO_2(g) \longrightarrow NO(g) + NO_2(g) + H_2O(g)$$
2.016 g hydrogen/16.00 g oxygen (*2*)

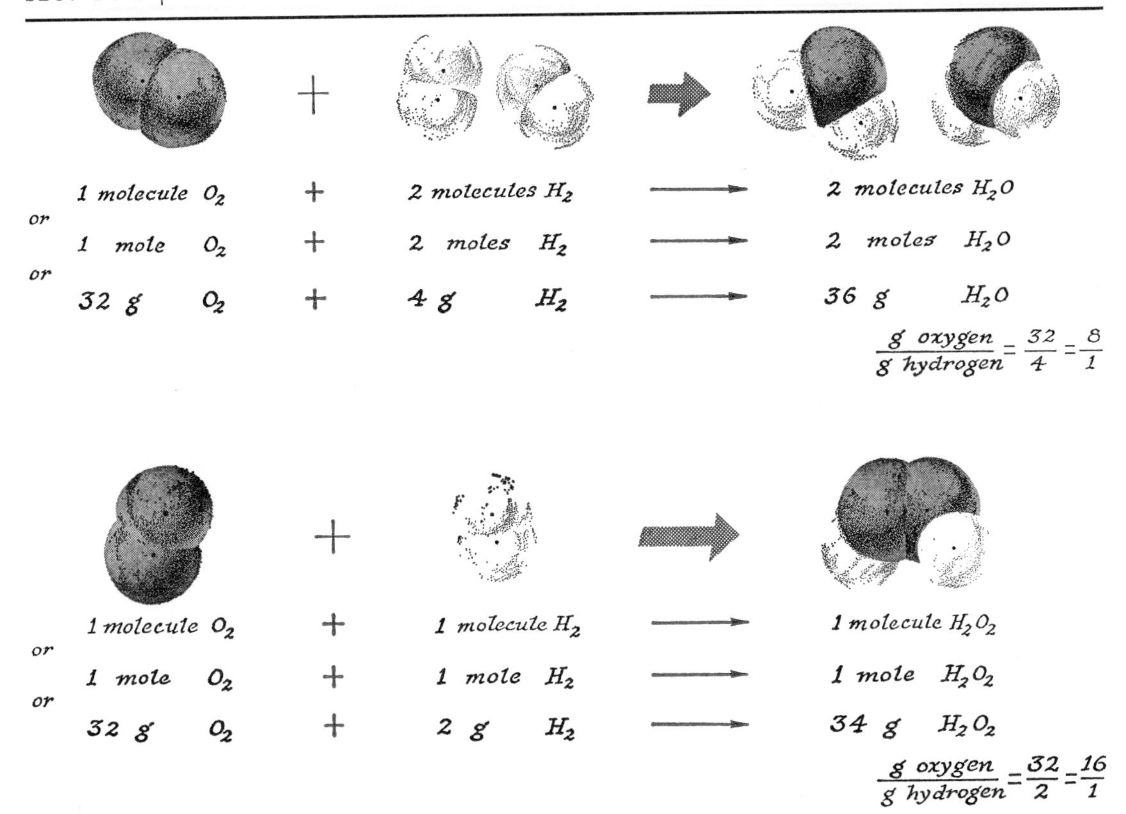

Fig. 14-1. **Simple multiple proportions** *of oxygen to hydrogen in H_2O and H_2O_2.*

$$\tfrac{1}{2}[BaCl_2 \cdot 2H_2O](s) \longrightarrow \tfrac{1}{2}BaCl_2(s) + H_2O(g)$$
$$2.016 \text{ g hydrogen}/16.00 \text{ g oxygen} \quad (3)$$

The atomic theory provides a ready explanation for the definite composition of chemical compounds. It says that compounds are composed of atoms, and every sample of a given compound must contain the same relative number of atoms of each of its elements. Since the atoms of each element have a characteristic weight, the weight composition of a compound is always the same.* Thus, the definite composition of compounds provides experimental support for the atomic theory.

14-1.2 The Law of Simple Multiple Proportions

In many cases, two elements brought together under different conditions can form two or more different compounds. In addition to water, hydrogen and oxygen can form a compound called hydrogen peroxide, H_2O_2, in which the weight of hydrogen is $\frac{1}{16}$ the weight of oxygen in the compound. This weight ratio in water is $\frac{1}{8}$, exactly twice as large. What a simple relationship! The weight ratio is one to sixteen in one compound of hydrogen and oxygen and one to eight in another. Such simple numerical relationships are always found among different compounds of a set of elements. This is explained very clearly within the atomic theory. Each molecule of hydrogen peroxide contains two atoms of hydrogen and two atoms of oxygen. The ratio of the number of hydrogen atoms to oxygen atoms is $2/2 = 1$. In contrast, a molecule of water contains two atoms of hydrogen and only one atom of oxygen. The ratio of hydrogen atoms to oxygen atoms is $2/1 = 2$. Since there are twice as many hydrogen atoms per oxygen atom in water as in hydrogen peroxide, of course the weight ratio of hydrogen to oxygen in water is twice that in hydrogen peroxide.

* This statement applies if the naturally occurring distribution of isotopes is not disturbed.

In general, different compounds of the same two elements have different atomic ratios. Since these atomic ratios are always ratios of integers, 1/1, 1/2, 2/1, 2/3, etc., the weight ratios will be simple multiples of each other. Thus the atomic theory explains the observation that different compounds of the same two elements have relative compositions by weight that are simple multiples of each other.

This success of the atomic theory is not surprising to a historian of science. *The atomic theory was first deduced from the laws of chemical composition.* In the first decade of the nineteenth century, an English scientist named John Dalton wondered why chemical compounds display such simple weight relations. He proposed that perhaps each element consists of discrete particles and perhaps each compound is composed of molecules that can be formed only by a unique combination of these particles. Suddenly many facts of chemistry became understandable in terms of this proposal. The continued success of the atomic theory in correlating a multitude of new observations accounts for its survival. Today, many other types of evidence can be cited to support the atomic postulate, but the laws of chemical composition still provide the cornerstone for our belief in this theory of the structure of matter.

EXERCISE 14-1

Two compounds are known that contain only nitrogen and fluorine. Careful analysis shows that 23.67 grams of compound I contain 19.00 grams of fluorine and that 26.00 grams of compound II contain 19.00 grams of fluorine.

(a) For each compound, calculate the weight of nitrogen combined with 19.00 grams of fluorine.
(b) What is the ratio of the calculated weight of nitrogen in compound II to that in I?
(c) Compound I is NF_3. This compound has one atom of nitrogen per three atoms of fluorine. How many atoms of nitrogen are there per three atoms of fluorine for each of the molecular formulas N_2F_2 and N_2F_4? Compare these atom ratios to the weight ratio obtained in part (b) and convince yourself that compound II could have the formula N_2F_4 but not N_2F_2.

14-1.3 The Law of Combining Volumes

Gases are found to react in simple proportions by volume, and the volume of any gaseous product bears a whole-number ratio to that of any gaseous reactant. Thus, *two* volumes of hydrogen react with exactly *one* volume of oxygen to produce exactly *two* volumes of water vapor (all at the same temperature and pressure). These integer relationships naturally suggest a particle model of matter and, with Avogadro's Hypothesis, are readily explained on the basis of the atomic theory.

Once again it is no surprise that the simple integer volume ratios are readily explained with the atomic theory. The atomic theory was devised for this purpose, as is indicated in Chapter 2.

To summarize, we find that the weight and volume relations that are observed in chemical changes provide an experimental foundation for the atomic theory. All of contemporary chemical thought is based upon the atomic model and, hence, every successful chemical interpretation strengthens our belief in the usefulness of this theory.

14-1.4 Chemical Evidence for the Electrical Nature of Atoms

You have been told that the atomic nucleus bears a positive charge and is surrounded by a number of negatively charged particles called electrons. Also, the nucleus is supposed to contain most of the mass of the atom and to be made of protons and neutrons, each of which has nearly two thousand times the mass of the electron. How do we know that atoms are built this way? How do we know that there is such a particle as an electron? Again, weight relations associated with chemical reactions provide key evidence.

In Chapter 12 we discussed the operation of

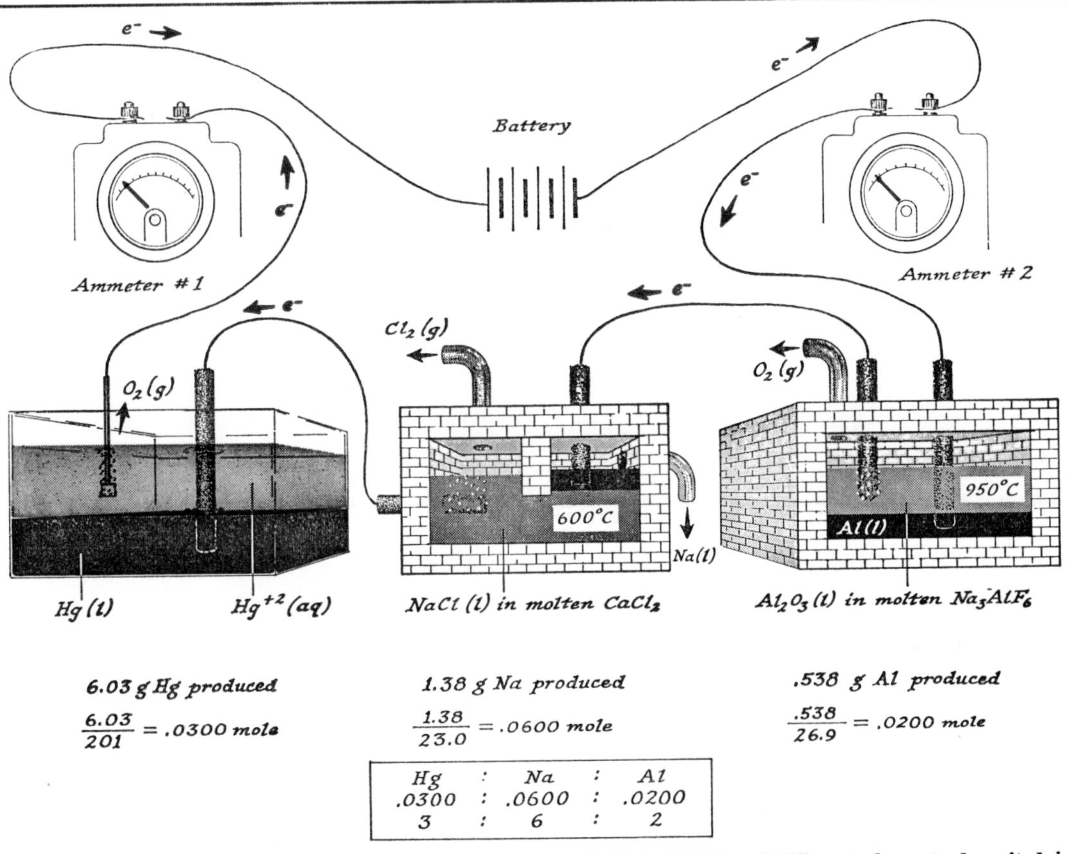

Fig. 14-2. **Weights of different elements deposited** *by a given amount of electricity.*

an electrochemical cell. We successfully interpreted the chemical changes brought about by the movement of electric charge in terms of the atomic theory. To understand the full impact of these experiments on the development of the atomic theory, we must turn back the scientific clock to the views held in the nineteenth century. When Michael Faraday first performed his electrolysis experiments (in the early 1830's), the atomic theory had been proposed but no one had yet suggested the existence of electrons. There was no reason to suspect that electricity consisted of individual units. Faraday observed that the quantity of electricity necessary to deposit a given weight of an element from solutions of its different compounds was always equal to a constant, or some simple multiple of this constant. For example, the amount of electricity that will deposit 6.03 grams of metallic mercury from a solution of mercuric perchlorate, $Hg(ClO_4)_2$, will deposit the same number of grams of mercury from a solution of mercuric nitrate, $Hg(NO_3)_2$. On the other hand, this same amount of electricity will deposit exactly twice as much mercury, $2 \times (6.03) = 12.1$ grams, from a solution of mercurous perchlorate, $Hg_2(ClO_4)_2$. If we restate Faraday's experimental finding in terms of the atomic theory, we see that the number of atoms of mercury deposited by a certain quantity of electricity is a constant or a simple multiple of this constant. Apparently this certain quantity of electricity can "count" atoms. A simple interpretation is that there are "packages" of electricity. During electrolysis, these "packages" are parcelled out, one to an atom, or two to an atom, or three.

The second of Faraday's observations was that the weights of *different* elements that were deposited by the same amount of electricity formed simple whole-number ratios when divided by the atomic weights of these elements. For example, suppose electric current is passed through the three electrolysis cells pictured in Figure 14-2. The two ammeters have the same reading, showing that the current entering the cell at the right is identical to that leaving the cell at the left. Thus, the electric circuit guarantees that the same amount of electricity passes through each of the three cells.

In the first cell the net reaction is the production of metallic mercury and gaseous oxygen through electrolysis of aqueous mercuric nitrate:

$$2Hg^{+2}(aq) + 2H_2O \longrightarrow 2Hg(l) + O_2(g) + 4H^+(aq) \quad (4)$$

After current has passed through the cell for a definite time, the weight of the mercury produced is found to be 6.03 grams.

In the second cell molten sodium chloride is electrolyzed.* The net reaction is

$$NaCl(l) \longrightarrow Na(l) + \tfrac{1}{2}Cl_2(g) \quad (5)$$

The same current that produced 6.03 grams of mercury is found to produce 1.38 grams of molten sodium.

The third cell represents another industrial process, the electrolytic process for manufacturing aluminum. Here, Al_2O_3 is electrolyzed† and the net reaction in the cell is

$$Al_2O_3(l) \longrightarrow 2Al(l) + \tfrac{3}{2}O_2(g) \quad (6)$$

Here we find that the same current that produced 6.03 grams of mercury produces 0.538 gram of aluminum.

Thus after the same amount of electricity is passed through the three cells, the weights of metals produced are found to be

6.03 g Hg
1.38 g Na
0.538 g Al

How are these weights related? Faraday realized that simple numbers result in such a case if each weight is divided by the appropriate atomic weight:

$$\frac{\text{wt Hg}}{\text{at wt Hg}} = \frac{6.03 \text{ g}}{201 \text{ g/mole}} = 0.0300 \text{ mole}$$

$$\frac{\text{wt Na}}{\text{at wt Na}} = \frac{1.38 \text{ g}}{23.0 \text{ g/mole}} = 0.0600 \text{ mole}$$

$$\frac{\text{wt Al}}{\text{at wt Al}} = \frac{0.538 \text{ g}}{26.9 \text{ g/mole}} = 0.0200 \text{ mole}$$

These numbers are simply related to each other as shown by the ratios:

Hg	Na	Al
0.0300 :	0.0600 :	0.0200
3 :	6 :	2

Within the atomic theory, this result means that a certain amount of electricity will deposit a fixed number of atoms, or some simple multiple of this number, *whatever, the element*. Thus, in both of Faraday's experiments, we find that an atom can carry only a fixed quantity of charge, or some simple multiple of this quantity. Therefore, electric charge comes in packages. An atom can carry one package, two packages, possibly three packages, of charge, but not 1.5872 packages. Whatever the package of charge is, it is the same for all atoms. The realization that electric charge comes in packages led to the proposal that electricity is composed of particles. Since atoms carry electric charges, atoms must contain these particles.

EXERCISE 14-2

As current is passed through the cells shown in Figure 14-2, the oxygen produced in the first cell is collected and its volume is compared with the volume of chlorine produced in the center cell (the volumes being compared at identical temperatures and pressures). The volume of chlorine is found to be exactly double that of oxygen. Applying Avogadro's Hypothesis, explain how this result shows that electricity can "count" atoms.

* In practice, calcium chloride must be added to such a cell to lower the melting point of the salt mixture and, even then, the temperature must be high (600°C). This is the commercial method for manufacturing metallic sodium.

† This is the basis for the commercial manufacture of aluminum. Another salt is added as solvent to lower the melting point. A mixture of Al_2O_3 and Na_3AlF_6 (cryolite) can be electrolyzed at 950°C.

14-2 "SEEING" PARTS OF ATOMS

Despite the convincing support for the atomic theory provided by chemical evidence, there is intuitive appeal to evidence that is closer to the "direct vision" type. From such experiments comes a much more detailed view of the atom and its make-up.

14-2.1 "Seeing" Electrons

The Faraday experiments were the original basis for the suggestion that electricity consists of individual charges called electrons. Other experiments involving the passage of electricity through gases provide further evidence that electrons do exist.

Consider the apparatus shown in Figure 14-3. A glass tube is fitted with electrodes so that a potential difference of 10,000 volts can be applied across a space filled with a desired gas at various pressures. Suppose neon, for example, is placed in the tube. With the voltage applied, the gas will begin to conduct electricity when its pressure is reduced to about 0.01 atmosphere. The tube then glows with the familiar color of a "neon sign." If a different gas is used, the color is different, but otherwise, the behavior is about the same. If the pressure is reduced still further to about 10^{-6} atmosphere, the glow from the gas

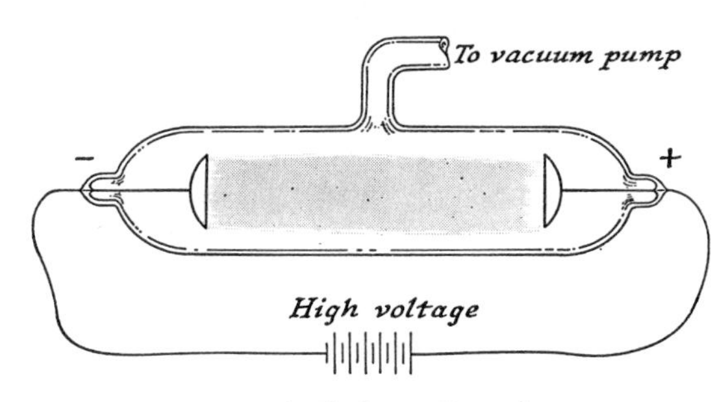

Fig. 14-3. **An electric discharge through a gas.**

disappears but there remains a fluorescent glow from the glass walls of the tube. The apparatus shown in Figure 14-4 demonstrates that the fluorescent glow is either caused by particles or by light rays that travel from the negative electrode and past the positive electrode (in this apparatus, through the triangular hole in the center of the electrode). When the tube operates, a glowing area appears on the glass wall at position *A* directly opposite the hole in the positive electrode and just the same shape as the hole. Because the glowing area is "shadowed" by the positive electrode, it must be caused by rays that travel in straight lines.

Figure 14-5 shows this same apparatus fitted with an auxiliary pair of electrode plates, P_1 and P_2. An electrical

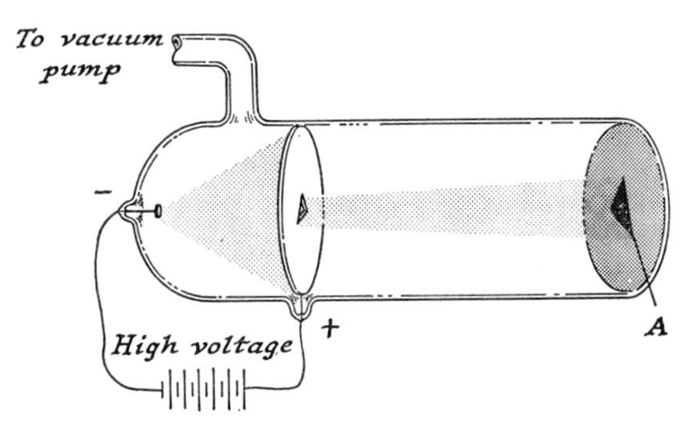

Fig. 14-4. **An electric discharge tube,** *very low pressure. Electrons travel from the negative electrode to the positive electrode; some of them pass through the triangular hole to produce a triangular spot on the fluorescent screen.*

voltage can be applied to plates P_1 and P_2 when switch *S* is closed. When switch S_1 is open, the fluorescence appears at position *A*, just as in Figure 14-4. When S_1 is closed, however, the fluorescent spot moves to position *B*. Apparently the fluorescent spot is caused by particles that are attracted to the positively charged electrode, P_2. Therefore these particles must be negatively charged. Light cannot be deflected in this way, hence the fluorescent glow cannot be caused by light. Such experiments with discharge tubes show that negatively charged particles exist. These particles are now known to be electrons.

This experiment does have some features of a "direct vision" observation. First, the glowing spot is directly visible. Second, it is easy to imagine an invisible stream of particles hurtling through the triangular hole in the electrode to crash against the fluorescent screen in a burst

Fig. 14-5. **An electric discharge tube** *with deflection electrodes.*

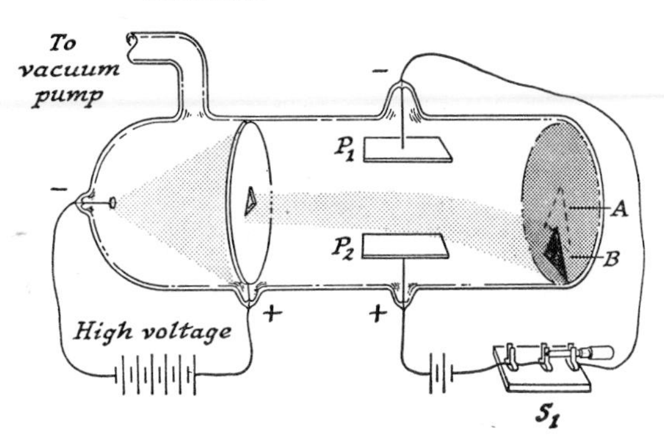

of light. Third, the experiment conveys detailed information about these particles, information difficult to obtain any other way. The electric charge on a particle is clearly evident from the deflection experiments. Accurate measurements of such deflections even lead to a measure of the ratio of electron charge to electron mass.

Yet there is a real difference between this experiment and the "direct vision" of the shadowy form of the garbage collector. We don't see the electron directly; rather we see a burst of light on the fluorescent screen. The light isn't considered to be the electron: the burst of light is caused by molecular damage to the screen—damage resulting from the electron crash. A more apt comparison from our analogy would be our seeing some footprints in the garden. We assume the footprints are caused by the garbage collector. Then from certain properties of the footprints—size, depth, spacing—we form a detailed image of his height, weight, stride. You will find that this is typical of most of the experiments that might be called "seeing" atoms and their components. We see their "footprints"—bursts of light on a screen, marks on a photographic plate, discharges in a Geiger counter, etc. These "footprints" substantiate in every way the atomic theory and furnish detailed information on the nature of atoms

THE RATIO OF ELECTRON CHARGE TO ELECTRON MASS, e/m

By the action of a magnetic field electrons may be made to follow a curved path. Such experiments lead to a determination of the ratio of the electric charge of an electron e, to its mass, m.

Consider the apparatus shown in Figure 14-6. The equipment is similar to that shown in Figure 14-4 except a fluorescent screen within the tube reveals the trajectory of the particles that pass through the slot in the positive electrode. When a magnetic field is added, the electron trajectory is curved. A mathematical analysis of the curvature permits an interpretation of this experiment that leads to a determination of e/m.

Fig. 14-6. **The effect of a magnetic field** *on the electron beam.*

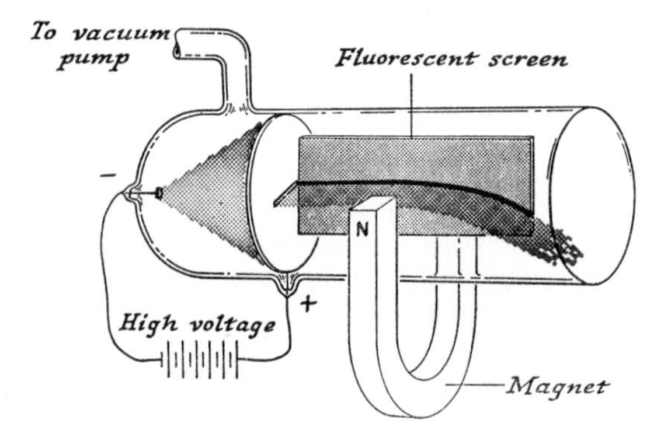

The calculations can be made only if the strength of the magnetic field is known and if the field is uniform. Therefore, apparatus more suitable than that shown in Figure 14-6 is needed. An electric current flowing through a wire coil generates a magnetic field that is easily measured and readily made uniform (by making the coil large compared to the apparatus).

Substituting, then, a large coil for the magnet shown in Figure 14-6, we can proceed with our measurement. The beam of electrons passes through the positive electrode and strikes the far end of the tube, producing a fluorescent spot. When the magnetic field is turned on (by passing current through the coil), the fluorescent spot moves. The spot moves because a charged particle moving in a uniform magnetic field has a path which is an arc of a perfect circle. From the deflection of the spot and the length of the apparatus, the radius of this circular path can be determined. This radius we shall call r.

This radius is useful to us because it is related to the mass, m, charge, e, and velocity, v, of the electron. It is determined also by the strength of the magnetic field, B, as follows:

$$r = \frac{m}{e} \times \frac{v}{B} \qquad (7)$$

This equation shows that the greater the mass or velocity of the particle, the less curved is its path (a small value of r describes a highly curved path). On the other hand, the path of the particle becomes more curved if the magnetic field is made stronger.

We can rearrange equation (7) to the form

$$\frac{e}{m} = \frac{1}{r} \times \frac{v}{B} \qquad (8)$$

Equation (*8*) shows us how to calculate the charge/mass ratio for the electron if r is measured and both v and B are known.

However, the velocity, v, of the electron is still unknown. We must calculate this quantity from the work done on the electron as it was accelerated, moving from the negative electrode to the positive electrode. The work done on the electron is the product of the charge on the electron times the voltage difference, V, between the electrodes:

$$\text{Work done on electron} = e \times V \qquad (9)$$

This work is used to accelerate the electron, giving it kinetic energy. Also, we know that the kinetic energy of the electron can be expressed in terms of its mass and velocity:

$$\text{Kinetic energy of a moving electron} = \tfrac{1}{2}mv^2 \qquad (10)$$

We must equate (*9*) and (*10*): the work done, eV, equals the kinetic energy the electron acquires, $\frac{1}{2}mv^2$:

$$eV = \tfrac{1}{2}mv^2 \qquad (11)$$

The voltage, V, we obtain from a voltmeter reading.

Expressions (*8*) and (*11*) both relate the ratio e/m to

the electron velocity and the measured quantities, r, B, and V. We can calculate e/m if v is eliminated from the two equations. This is an algebraic process that can be done several ways. Here is one way.

Since (*11*) involves v^2, let us square expression (*8*):

$$\frac{e^2}{m^2} = \frac{v^2}{r^2B^2} \qquad (12)$$

Now let us multiply both sides of (*12*) by m/e:

$$\frac{e}{m} = \frac{mv^2}{e} \times \frac{1}{r^2B^2} \qquad (13)$$

Now we can rearrange (*11*) to the form

$$\frac{mv^2}{e} = 2V \qquad (14)$$

And, finally, we can substitute (*14*) into (*13*):

$$\frac{e}{m} = \frac{mv^2}{e} \times \frac{1}{r^2B^2} = (2V) \times \frac{1}{r^2B^2}$$

or

$$\frac{e}{m} = \frac{2V}{r^2B^2} \qquad (15)$$

We measure V (from the voltmeter reading), r (from the deflection of the spot), B (from the current through the magnet coil windings), substitute them into (*15*), and calculate

$$\frac{e}{m} = 1.759 \times 10^8 \frac{\text{coulombs}}{\text{gram}} \qquad (16)$$

THE CHARGE ON THE ELECTRON

Experiments like those described in Figures 14-3 to 14-6 establish that the electron is a negatively charged particle and that it is present in all substances. Further confirmation of the particulate nature of electricity comes from experiments that were conducted by an American physicist, Robert Millikan, in 1906 to determine the charge on the electron. The apparatus used for his experiment is shown schematically in Figure 14-7.

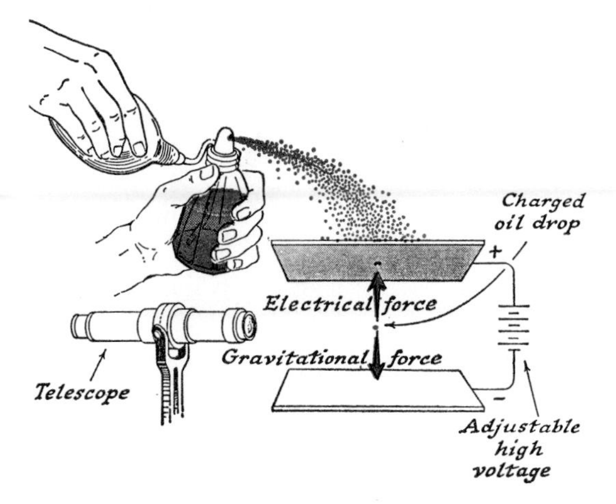

Fig. 14-7. **Millikan's oil-drop apparatus** *for determining the electron charge.*

Tiny droplets of oil or some other liquid are sprayed into the upper part of the apparatus. A few droplets fall through a small hole into the lower chamber. During its production, an oil drop is very likely to become charged by friction. When an oil drop enters the lower chamber, a voltage is applied to the metal plates. If the oil drop is charged, its fall can be completely stopped by adjusting the voltage so that the electrical force on the charged drop is just equal and opposite to the force of gravity.

Millikan made thousands of determinations of the charge on drops of oil, glycerol, and mercury. The charge on the drop was sometimes positive and sometimes negative, but in every case its magnitude was some integral multiple of 1.602×10^{-19} coulomb.* In no case was the charge any less than this. These experiments give a clear demonstration that the fundamental unit of electricity must be a charge of 1.602×10^{-19} coulomb. If the electron carries this fundamental unit of electricity (as we believe it does), the value of the charge on the electron must be 1.602×10^{-19} coulomb:

$$e = 1.602 \times 10^{-19} \text{ coulomb} \qquad (17)$$

We may use this value of the charge on the electron to calculate the mass of an electron. To do so, it is necessary to know the ratio of (electron charge/electron mass) $= e/m$. This ratio is measured with apparatus based on principles displayed in Figures 14-4 and 14-6. Using the result $e/m = 1.759 \times 10^8$ coulombs/g, the mass of an electron is found to be

$$m = \frac{1.602 \times 10^{-19} \text{ coulomb/electron}}{1.759 \times 10^8 \text{ coulomb/g}}$$

$$m = 9.11 \times 10^{-28} \text{ g/electron} \qquad (18)$$

EXERCISE 14-3

Suppose five measurements of oil-drop charges give the values listed below:

4.83×10^{-19} coulomb
3.24×10^{-19}
9.62×10^{-19}
6.44×10^{-19}
4.80×10^{-19}

* The coulomb is a unit of electric charge. Its magnitude can be appraised by its relation to the ampere. One ampere is an electric current of one coulomb of charge passing a point in a wire every second. One mole of electrons has, then, 96,500 coulombs of charge. In a wire carrying 10 amperes, it takes about two and one half hours for one mole of electrons to pass any point.

(a) Divide each charge by the smallest value to investigate the relative magnitudes of these charges.
(b) Assuming each measurement has an uncertainty of $\pm 0.04 \times 10^{-19}$, decide what electron charge is indicated by these experiments alone.

14-2.2 "Seeing" Positive Ions

Experiments can be conducted in which positive ions are detected and their properties measured (charge and mass). These experiments are similar to those we have described for electrons. A gas discharge tube, such as was shown in Figure 14-3, can be used because measurements show that positive ions are present as well as electrons. Whereas electrons are accelerated toward the positive electrode, the positively charged ions are accelerated in the opposite direction, toward the negative electrode. These ions can be removed from the apparatus as a beam in the same way that the electron beam is removed in the apparatus of Figure 14-4. In such fashion, we obtain a beam of positive ions. By deflecting these beams in electric and magnetic fields, the charges and masses of the positive ions can be measured.

The results of experiments of this type show two very important differences from measurements on electrons.

(1) The charge/mass ratio for positive ions changes when the gas in the tube is changed. When the (e/m) measurement is made for electrons, the same value is obtained no matter what gas is introduced.
(2) The charge/mass ratio for positive ions is very much smaller than (e/m) for electrons. These facts are interpreted to mean that the positive ions are ions formed from the gas in the tube. The electric charge is considered to arise from the removal of one or more electrons from an atom or a molecule. Thus the value of the ratio (charge/mass) for positive ions depends upon the gas because each type of atom (or molecule) has a distinctive mass.

"WEIGHING" POSITIVE IONS. THE MASS SPECTROGRAPH

A mass spectrograph is an instrument with which the masses of individual atomic or molecular ions can be measured. One type of mass spectrograph is shown in Figure 14-8. Positive ions are accelerated through a slotted negative electrode and then passed through a uniform magnetic field. The left view in Figure 14-8 shows the apparatus supported between the pole faces of a strong magnet. The right view is an enlargement of the spectrograph with the magnetic field directed vertically through the figure. This is the view of the mass spectro-

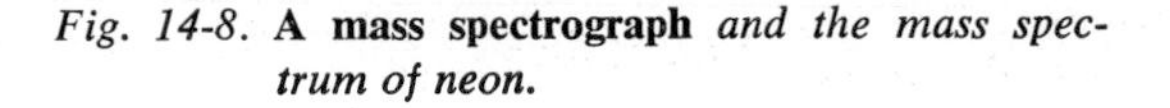

Fig. 14-8. **A mass spectrograph** *and the mass spectrum of neon.*

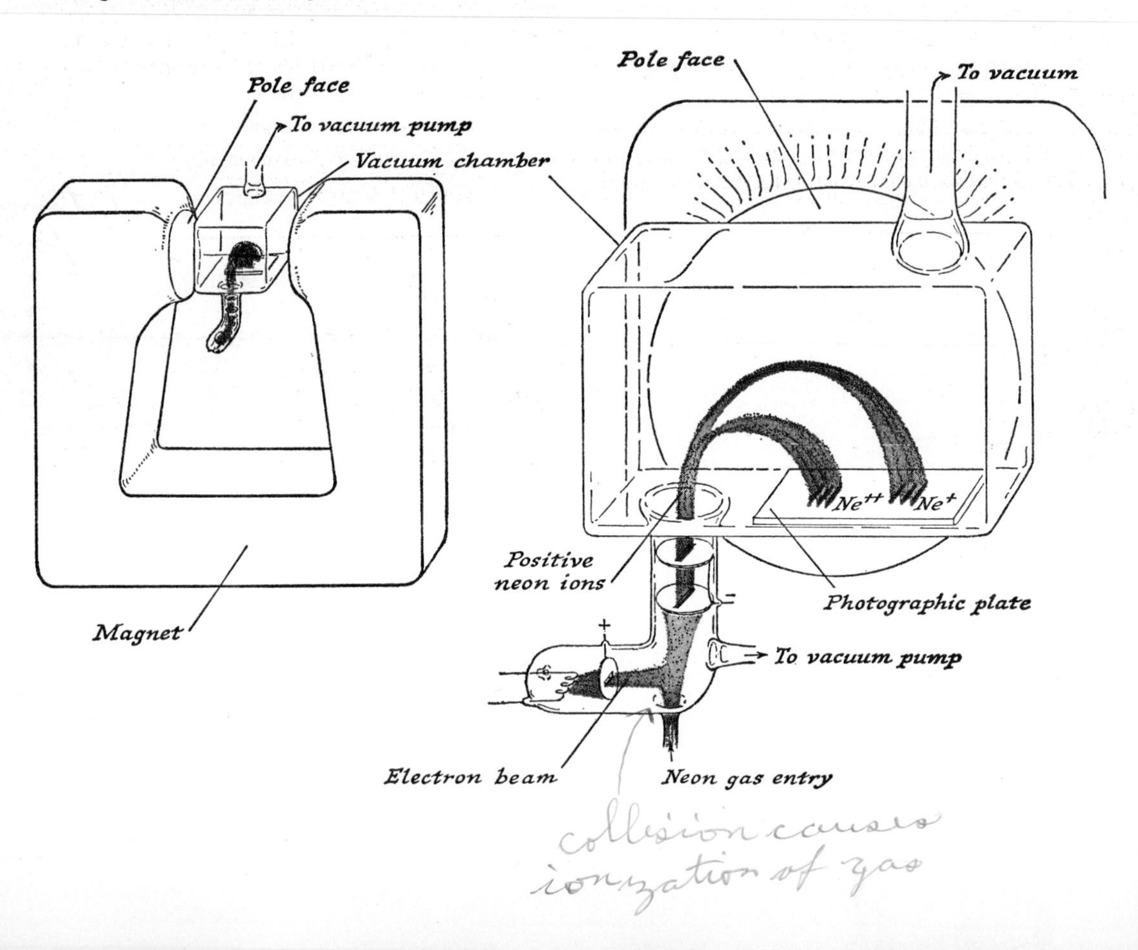

Plate III. **A simple spectrograph and the spectrum of a hot tungsten ribbon.**

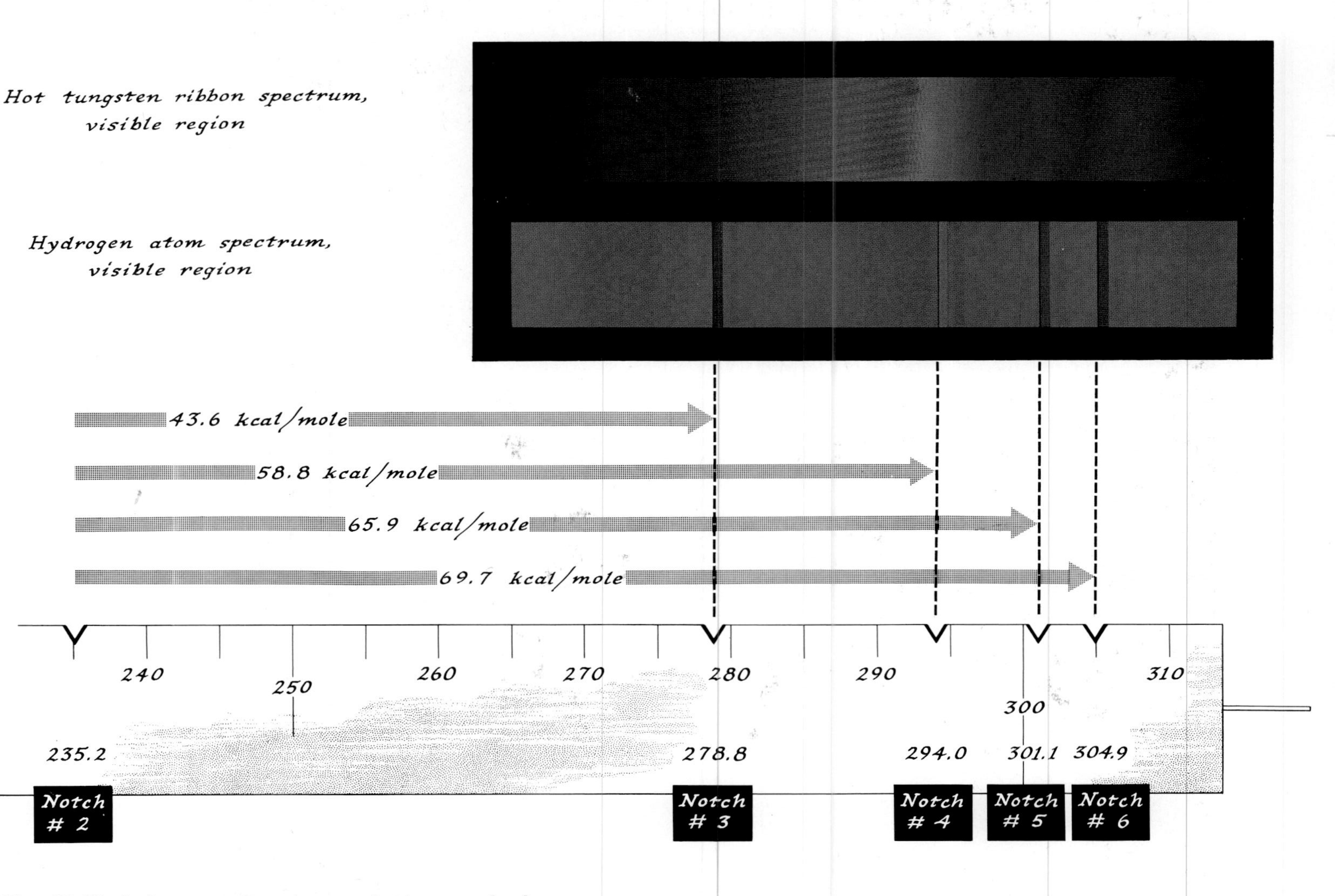

Plate IV. **The hydrogen atom spectrum: a clue to energy levels.**

graph seen by an ant sitting on one pole face of the magnet looking toward the other pole face.

The positive ions can be produced with a glow discharge tube like that shown in Figure 14-3. More usually, however, gaseous atoms or molecules are bombarded with an electron beam as shown in Figure 14-8. If the bombarding electrons have enough energy, they cause positive ion formation when collisions occur with gas molecules. The figure shows neon gas entering at the bottom. The gas passes through the electron beam and some of the atoms collide with electrons to form neon ions. Both Ne^+ and Ne^{+2} ions are formed and they are accelerated by the slotted electrode. As the positive ions enter the magnetic field, they follow a circular path. They have a large radius of curvature if the mass is high, a low radius of curvature if the charge is high. Thus each positive ion follows a circular path fixed by its mass and charge. After circling through an arc of 180°, the ions are collected on a photographic plate. The impact of the ions with the photographic plate causes a reaction that leads to a darkening of the sensitized surface, just as exposure to light does. Such a record shows a spot for each ion at a position fixed by the charge/mass ratio. Measurement of the position of each spot reveals the masses of the ions. The record is called a mass spectrum.

When neon gas is put in the spectrograph, the mass spectrum consists of two widely separated groups of three spots each. The three spots corresponding to large radii are caused by neon ions with a single positive charge, while the three spots corresponding to small radii are caused by doubly charged ions. For each ionic charge there are three slightly separated spots which indicate that neon consists of atoms with three different masses. This shows that ordinary neon consists of three different isotopes. The relative abundances of these isotopes can be determined by measuring the intensity of the spots caused by each of the ion beams.

EXERCISE 14-4

When chlorine, Cl_2, is examined in a mass spectrograph, Cl_2^+, Cl^+, and Cl^{+2} ions are formed. Remembering that there are two isotopes in chlorine, 35 (75%) and 37 (25%), describe qualitatively the appearance of the mass spectrum. Which ion will produce lines at the largest radius? Which at the smallest radius? How many lines will each ion produce?

Figure 14-8 shows a mass spectrograph that uses photographic detection. Nonphotographic detection is also possible. The ions, after being sorted according to mass and charge, can be "counted" by a charge measuring device. The advantage of such a detector is that the result can be presented continuously on a paper chart, thus eliminating the cumbersome and slow photographic process.

THE RATIO OF CHARGE TO POSITIVE ION MASS, e/m

In a mass spectrograph, the factors that determine the trajectory of the ions are the same as those we discussed when we considered the measurement of (e/m) for the electron. In that discussion we derived equation (*19*):

$$\frac{e}{m} = \frac{2V}{r^2B^2} \qquad (19)$$

where e = electron charge,
m = electron mass,
V = accelerating voltage,
B = magnetic field strength.

For positive ions, the charge is e or $2e$ or, in general, some integer, n, times e, the electron charge. We might write a capital M for mass to indicate that the mass of a positive ion is involved. Then we can solve (*19*) for the mass as a function of V, B (experimental conditions we control), n (which will be one, two, three, or some low integer), and r, the radius that we measure on the photographic plate:

$$M = \frac{ner^2B^2}{2V} \qquad (20)$$

With equation (*20*), we can verify quantitatively our identifications of the ionic masses and their charges. The two sets of three spots immediately suggest that one set is caused by three isotopes, each with the same ionic charge, and the second set by the same three isotopes with a different ionic charge. Measurement of the radii shows that one set has a larger radius by just the square root of two. This ratio is consistent with the assignment of +1 to one set of ions and +2 to the other set. Calculations based on the assumption that $n = 1$ for the outer set, together with the accurately measured radii of the paths corresponding to the three spots on the plate give isotopic masses corresponding to 20, 21, and 22 grams/mole, the three stable isotopes of neon. If the true value of n for this outer set of spots had been $n = 2$, the calculated masses would correspond to isotopic masses of 40, 42, and 44 grams/mole. The choice between the two assumptions is usually easily made on the basis of chemical arguments about the possible atoms or molecules present.

EXERCISE 14-5

Suppose a mass spectrograph is used to measure the charge/mass ratio for fluorine ions. Fluorine has only one stable isotope and its atomic weight is 19.0 grams/mole. From the measured charge/mass ratio, 5.08×10^3 coulombs/gram, and the assumption that the ion has one electron charge, calculate the mass of one ion. Repeat the calculation assuming the ion has two electron charges. Now calculate Avogadro's number from the weight of a mole of fluorine ions, using each of your two calculations. Which assumption about ion charge do you prefer? Could the other be correct as well?

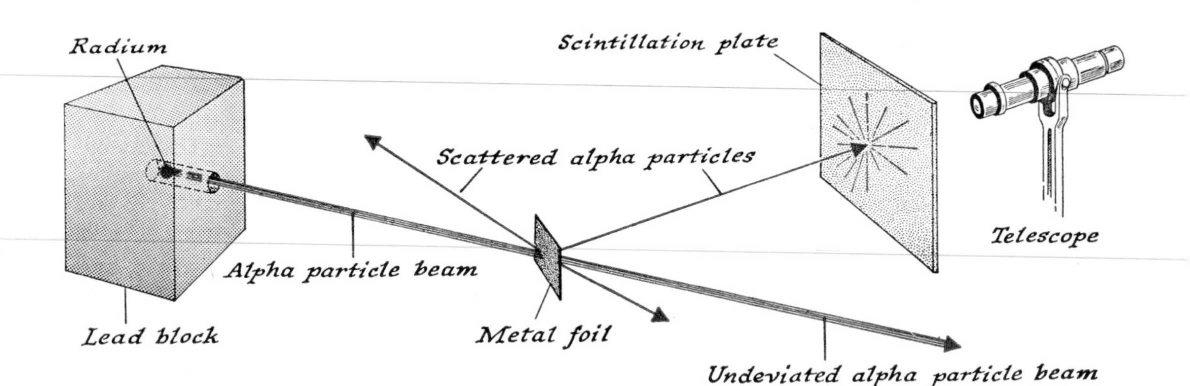

Fig. 14-9. **Rutherford's apparatus** *for observing the scattering of alpha particles by a metal foil. (The entire apparatus is enclosed in a vacuum chamber.)*

14-2.3 "Seeing" the Nucleus: Structure of the Atom

So far we have described experiments that indicate that atoms exist and that indicate atoms are composed of charged particles. We know also that all the positively charged part of the atom is located in a very small but dense region which we call the nucleus. The negatively charged electrons spend most of their time at relatively great distances from the nucleus. The story of how this nuclear model of the atom was first proposed gives a fascinating view of how science progresses.

The first detailed model of the atom, proposed by J. J. Thomson in 1898, was based upon the expectation that the atom was a sphere of positive electricity in which electrons were embedded like plums in a pudding. This picture of the atom was not particularly satisfying because it was not useful in predicting or explaining the chemical properties of the atom. Finally, in 1911, a series of experiments performed in the McGill University laboratory of Ernest Rutherford showed that Thomson's picture of the atom had to be abandoned.

The experiment conducted by Rutherford and his coworkers involved bombarding gold foil with alpha particles, which are doubly charged helium atoms. The apparatus used in their experiment is shown in Figure 14-9. The alpha particles are produced by the radioactive decay of radium, and a narrow beam of these particles emerges from a deep hole in a block of lead. The beam of particles is directed at a thin metal foil, approximately 10,000 atoms thick. The alpha particles are detected by the light they produce when they collide with scintilltaion screens, which are zinc sulfide-covered plates much like the front of the picture tube in a television set. The screen is mounted on an arm in such a way that it can be moved around in a circle whose center is in line with the point where the alpha particles strike the foil. A telescope is mounted behind the screen so that the very small flashes of light produced when individual alpha particles strike the scintillation screen can be detected and counted. The apparatus operates in a vacuum chamber in order that no deflections are caused by the impact of alpha particles upon gaseous molecules.

The first observation made with this apparatus was that apparently all the alpha particles passed through the foil undeflected. Let us see if this result is consistent with the model of the atom proposed by Thomson. You will recall that Thomson's picture of the atom assumed that the positive charge is distributed evenly throughout the entire volume of the atom with the negative electrons embedded in it. Since the electrons weigh so little, the positive part accounts for nearly all of the mass of the atom. Thus the Thomson model pictures the atom as a body of uniform density.

Imagine what our thin metal foil would be like if it were to be made up of Thomson atoms. The physical properties of a solid suggest that the atoms lie very close together, so the metal foil would look something like the diagram shown in Figure 14-10. Of course, the real foil is 10,000 atoms thick. What would happen to the alpha particles if they were shot into a solid of such uniform density? At first we might think that they would be stopped or deflected back upon colliding with the atoms. Since it was observed that the alpha particles went straight through the metal foil, we must reconsider the problem. When we shoot at a paper target with a high-powered rifle, the projectile forces its way through the paper. The alpha particles produced by radium have very high kinetic energy and are very much like bullets from a high-powered rifle. Perhaps the very high kinetic energy allows an alpha particle to force its way right through the atoms of the metal foil. Since a rifle bullet fired into paper passes through undeflected, it seems reasonable to conclude that the alpha particle would also pass through the metal foil undeflected.

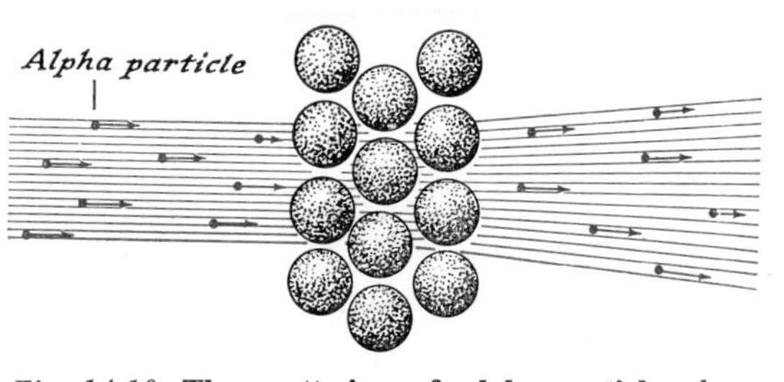

Fig. 14-10. **The scattering of alpha particles** *by a metal foil made of Thomson atoms.*

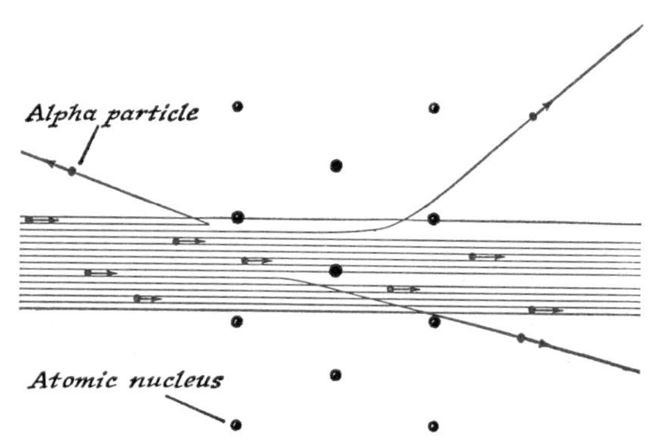

Fig. 14-11. **The scattering of alpha particles** *by a foil made of Rutherford nuclear atoms.*

In summary, in the Thomson model a metal foil is considered to have essentially uniform density. If this is true, there is no way for bombarding alpha particles to be deflected through large angles. At best, the alpha particles might suffer slight deflections from many collisions with many atoms. The model predicts the scattering distribution shown in Figure 14-10.

The first results of Rutherford's experiments seemed to be quite consistent with the Thomson picture of the atom. On more careful examination, an astounding discovery was made. By moving the screen around the metal foil, Rutherford and his co-workers were able to observe that a very few scintillations occurred at many different angles; some of these angles were nearly as large as 180°. It was as if some of the alpha particles had rebounded from a head-on collision with an immovable object. In the words of Rutherford, "It is about as incredible as if you had fired a 15 inch shell at a piece of tissue paper and it came back and hit you." It was impossible to explain the simultaneous observation of large-angle and small-angle deflections by using the Thomson atom.

In order to explain his experimental results, Rutherford designed a new picture of the atom. He proposed that the atom occupies a spherical volume approximately 10^{-8} cm in radius and at the center of each atom there is a nucleus whose radius is about 10^{-12} cm. He further proposed that this nucleus contains most of the mass of the atom, and that it also has a positive charge that is some multiple of the charge on the electron. The region of space outside the nucleus must be occupied by the electrons. We see from Figure 14-11 that Rutherford's picture requires that most of the volume of the atom be a region of very low density.

Using this kind of model of the atom, we can account for the alpha particles that are deflected through both large and small angles. If we allow alpha particles to impinge upon a metal foil composed of atoms based on Rutherford's model, only a few of the particles would be appreciably deflected by the foil. The heavy, fast moving alpha particles can brush past the lighter electrons without being deflected. Since most of the volume of the metal foil is relatively empty space, the greatest number of alpha particles pass through the metal undeflected. It is possible, however, for a few particles to be scattered through very large angles. Since both the alpha particle and the nucleus of the atom are positively charged, they exert a force of repulsion on each other. This force becomes large only when the alpha particle comes quite close to the nucleus. Since the nucleus in question is much heavier than the alpha particle, it can deflect the alpha particle considerably, just as a steel post can deflect a rifle bullet.

Besides providing a qualitative picture of the atom, Rutherford's experiments provided a way of measuring the charge of the nucleus. The force that a nucleus exerts on an alpha particle depends upon the magnitude of the charge on the nucleus. Rutherford showed how to relate the number of alpha particles scattered at any angle to the magnitude of the charge on the nucleus. The first measurements of the nuclear charge by this method were not very accurate, but they did show that different elements have different nuclear charges. By 1920, however, the alpha particle scattering experiments were so refined that they could be used to determine nuclear charge accurately.

14-3 MEASURING DIMENSIONS OF ATOMS AND MOLECULES

There are several ways by which sizes of atoms in molecules and in solids can be estimated. These methods are classified as "spectroscopic" methods because they involve the interaction of light with matter. The measurements show atomic size in the sense that they show how closely the atoms pack together. These packing distances, as measured spectroscopically, have provided the dimensions for the atomic models you have seen.

14-3.1 Light and the Frequency Spectrum

Light can be characterized by its frequency or its wavelength. To understand the meaning of these terms, consider water waves approaching and breaking on a beach. Figure 14-12 shows two measurements we might make, the distance between crests and the time between waves. The distance between crests is called wavelength and it can be expressed in centimeters. The time between waves, τ, indicates how often waves pass a fixed point. Usually the reciprocal, $1/\tau$, is specified. This quantity, with the dimension waves per second, is called frequency and is symbolized ν ("nu"). Light, which is an electromagnetic disturbance traveling through space, has properties much like water waves. The electromagnetic disturbance varies periodically, as does the water disturbance, hence it can be characterized by its frequency. Also, light travels through space with definite distance between the "wave crests" where the electromagnetic disturbance is greatest. This distance is called the wavelength.

Figure 14-13 shows a spectrograph—an instrument that reveals the frequency composition of light. The light entering a narrow slit is focused into a beam by the lens. This beam is passed through the prism. All of the light is refracted (bent) by the angular prism, but different frequencies (colors) are bent through different angles. The result is that the frequency

Fig. 14-12. **Waves can be characterized** *by wavelength or by time between waves.*

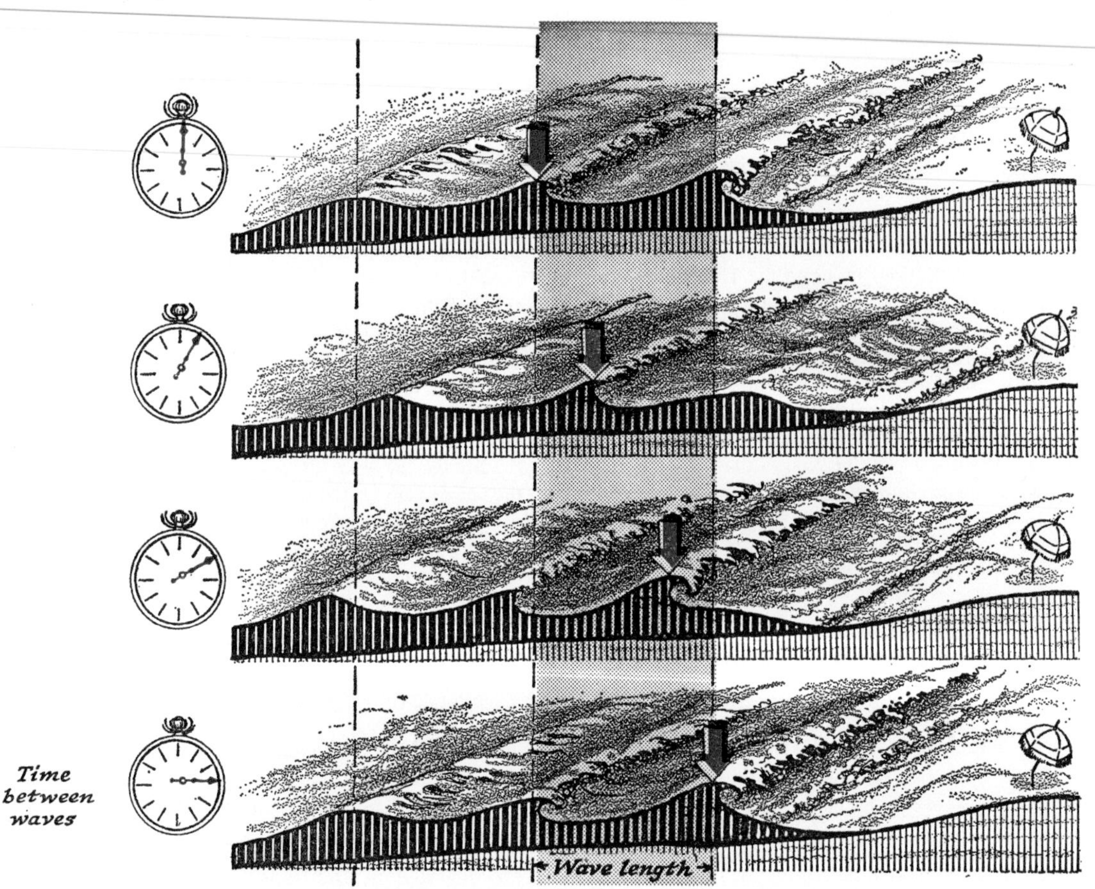

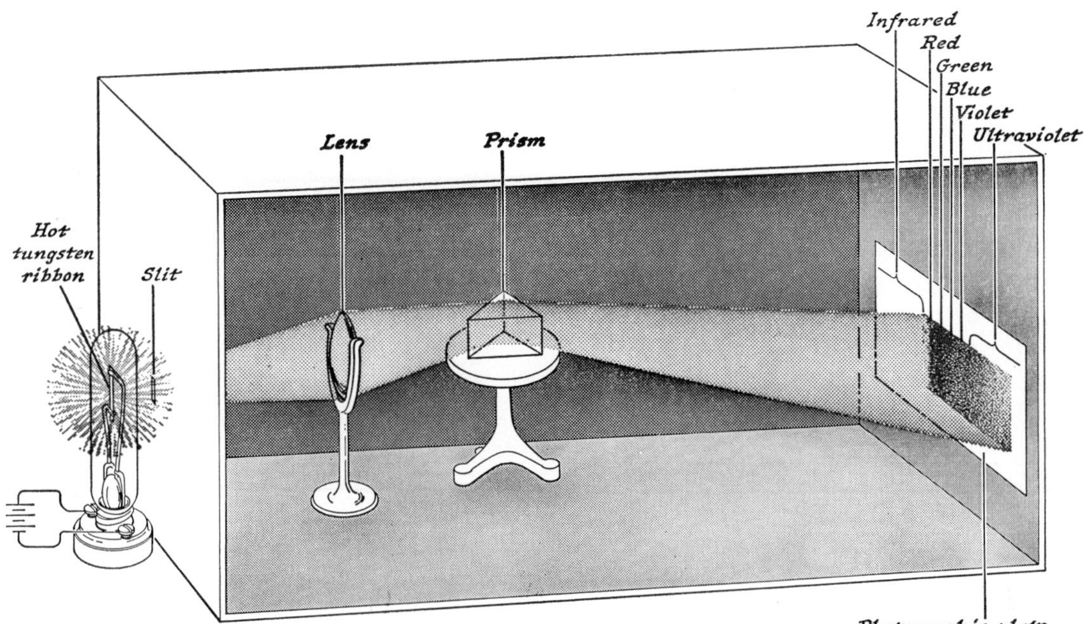

Fig. 14-13. **A simple spectrograph.**

composition of the light entering the slit can be learned from the pattern focused on a photographic film. The light source is a tungsten ribbon heated to a temperature near 1000°C by an electric current.

This separation of light into its component frequencies produces a spectrum. This spectrum is recorded on the photographic film because the darkening of the film (on development) is determined by the light intensity. From the spectrum we learn that different colors correspond to different frequencies. Blue light is found to have a frequency of about 7.5×10^{14} waves per second or, in more usual terminology, $\nu = 7.5 \times 10^{14}$ cycles per second. The red light has a lower frequency; ν is about 4.3×10^{14} cycles per second.

The experiment shown in Figure 14-13 points up another extremely important fact. The photographic film is darkened at larger angles than that at which blue light appears and at smaller angles than that at which red light appears. This implies that the light emitted by the hot ribbon includes frequencies that are not detected by the human eye. The *frequencies lower than the frequency of red light are called* ***infrared*** *frequen-*

Fig. 14-14. **The complete light spectrum.**

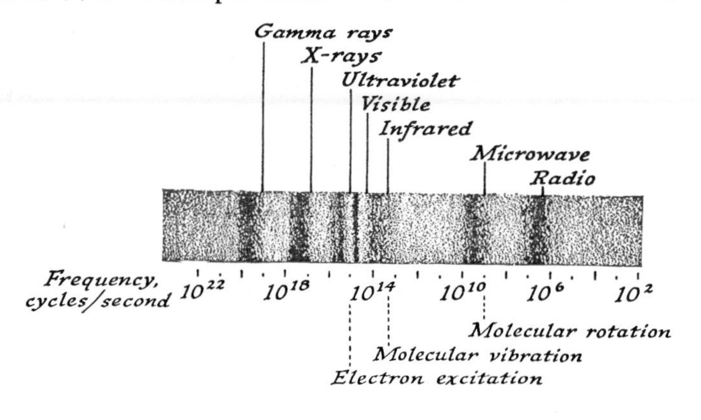

cies. The *frequencies higher than the frequency of violet light are called* **ultraviolet** *frequencies.*

Scientists have now realized that the electromagnetic phenomenon called light extends over an enormous range of frequencies—much wider than the rather narrow region in which the human eye is sensitive. Figure 14-14 shows the range that is commonly studied and the familiar names given to various spectral regions.

There are three spectral regions or ranges of light frequencies that are particularly useful to chemists in learning atomic sizes. We shall discuss each briefly.

14-3.2 X-Ray Diffraction Patterns

X-Rays are light waves of frequencies near 10^{18} cycles per second and wavelengths near 10^{-8} cm. Such light waves, when reflected from the surface of a crystal, give patterns on a photographic film. The pattern is fixed by the spacings of the atoms of the crystal and their spatial arrangement. The pattern is obtained only with X-rays because it results from scattering effects that occur only if the wavelength of the light is close to the atomic separations within the crystal. Therefore, a knowledge of the wavelength of the X-ray light permits an interpretation of the pattern in terms of atomic packing.

Figure 14-15 shows three X-ray diffraction patterns obtained from small crystalline particles of metallic copper, aluminum, and sodium. The qualitative similarity of the patterns given by copper and aluminum shows that they have the same crystal packing. Careful measurements of the spacing of the lines indicate that the atoms in copper, though occupying the same relative positions as the atoms in aluminum, are closer together. In contrast, the pattern of lines produced by sodium does not resemble either of the preceding patterns. Sodium atoms are packed differently in the metallic sodium crystal.

Fig. 14-15. **X-Ray diffraction patterns** *of finely divided metallic copper, aluminum, and sodium.*

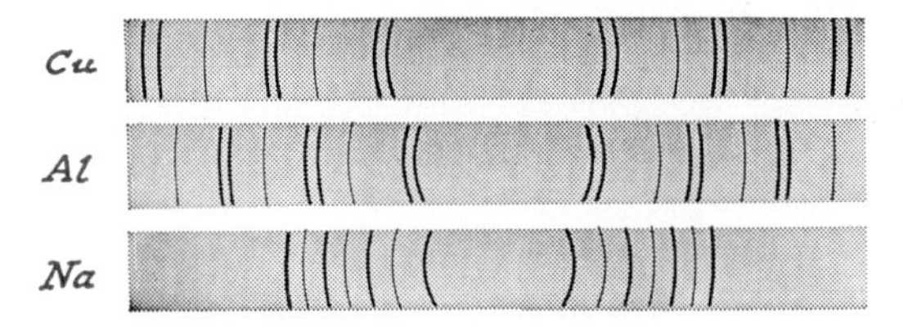

The X-ray diffraction method is applicable to solids and provides such detailed views of crystal geometry as those shown for sodium chloride solid in Figure 5-10, p. 81.

14-3.3 Microwave Spectroscopy: Molecular Rotation

Radio waves are light rays of macroscopic wavelengths (that is, wavelengths of many meters). Using techniques similar to those used in generating radio waves, "microwave" light can be produced with wavelengths in the range 1 mm to 10 cm. In this spectral region, gaseous molecules absorb light because of excitation of rotational movements of the molecules. The frequencies of these rotational motions, shown in Figure 7-8 (p. 118), depend upon the distance of the atoms from the molecular center of gravity.

Microwave spectroscopy is applicable only to gases but it is capable of extremely high accuracy. Interatomic distances and structures have been so measured for many molecules containing only a few atoms.

14-3.4 Infrared Spectroscopy: Molecular Vibration

Light found in the spectrum just beyond the red end of the visible spectrum is called infrared light. The frequencies are in the range 2×10^{13} to 12×10^{13} cycles per second (with wavelengths between 1.5×10^{-3} to 2.5×10^{-4} cm). Molecules absorb light in this spectral region, and analysis of the frequencies shows that the absorptions are associated with the excitation of vibrational motions. These to-and-fro motions of the atoms occur at natural frequencies just like the natural vibrational frequencies of a ball-and-spring model of a molecule. These natural frequencies are fixed by the masses of the atoms, the molecular shape, and the strengths of the chemical bonds that link the atoms together. Again the frequencies absorbed by gaseous molecules provide information about the molecular moments of inertia, the molecular geometry, and the chemical bonds. In addition, infrared study can be extended easily to the liquid and solid state, hence it finds widespread use in chemistry.

Figure 14-16 contrasts the infrared absorption spectra of hydrogen bromide gas, HBr, and deuterium bromide gas, DBr. The horizontal scale shows frequency. For a given frequency, the vertical scale shows the percentage of light of that frequency transmitted by the sample. Thus a reading of 100% means all of the light is transmitted; hence, no light is absorbed at that frequency. Plainly, gaseous HBr absorbs in only one spectral region, that near 7.9×10^{13} cycles per second. This one absorption corresponds to the vibrational excitation of the chemical bond in HBr. There is only one bond, hence only one absorption. The spectrum of gaseous DBr is similar but the absorption occurs near 5.9×10^{13} cycles per second. Of course, the chemistries of DBr and HBr are identical, hence the chemical bond in DBr is identical to that in HBr; nevertheless, the vibration frequencies of these two molecules differ because the atom masses differ. Since the deuterium atom is heavier than the hydrogen atom, it vibrates more slowly.

More complicated molecules, with two or more chemical bonds, have more complicated absorption spectra. However, each molecule has such a characteristic spectrum that the spectrum can be used to detect the presence of that particular molecular substance. Figure 14-17, for example, shows the absorptions shown by liquid carbon tetrachloride, CCl_4, and by liquid carbon disulfide, CS_2. The bottom spectrum is that displayed by liquid CCl_4 containing a small amount of CS_2. The absorptions of CS_2 are evident in the spectrum of the mixture, so the infrared spectrum can be used to detect the impurity and to measure its concentration.

The value of infrared spectra for identifying substances, for verifying purity, and for quantitative analysis rivals their usefulness in learning molecular structure. The infrared spectrum is as important as the melting point for characterizing a pure substance. Thus infrared spectroscopy has become an important addition to the many techniques used by the chemist.

Fig. 14-16. **Infrared absorption spectra** *of gaseous HBr and DBr.*

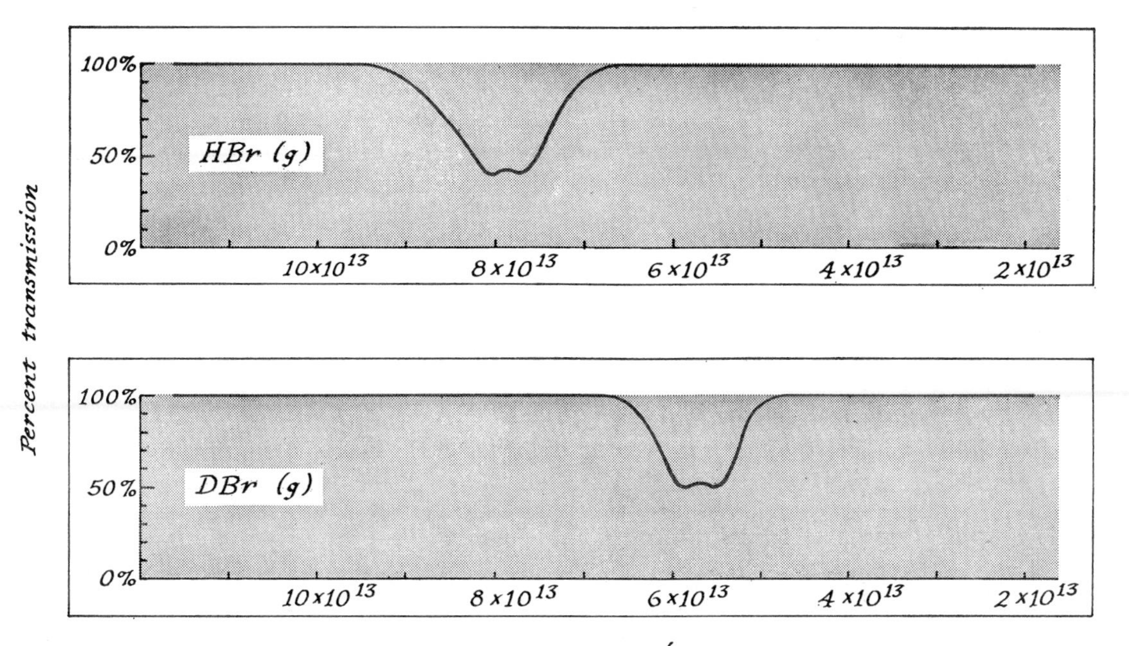

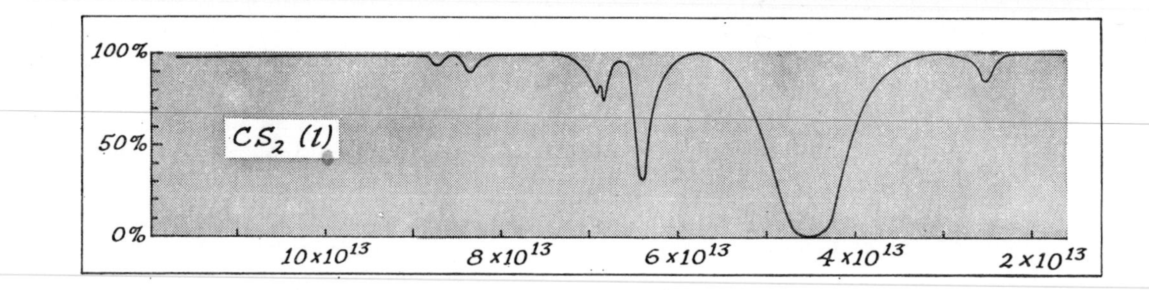

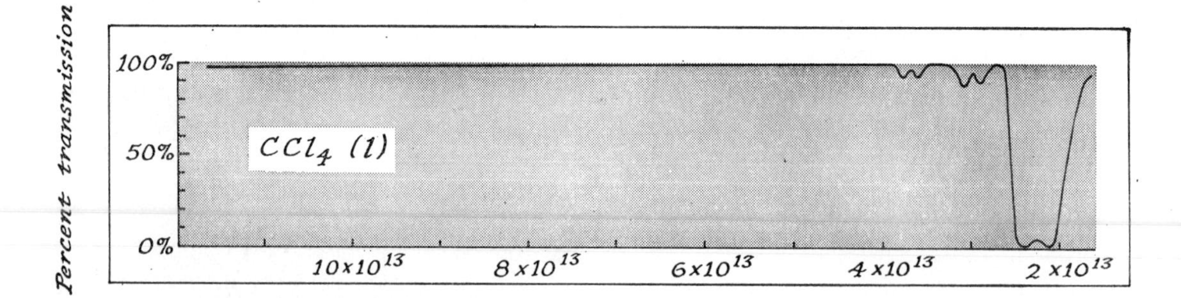

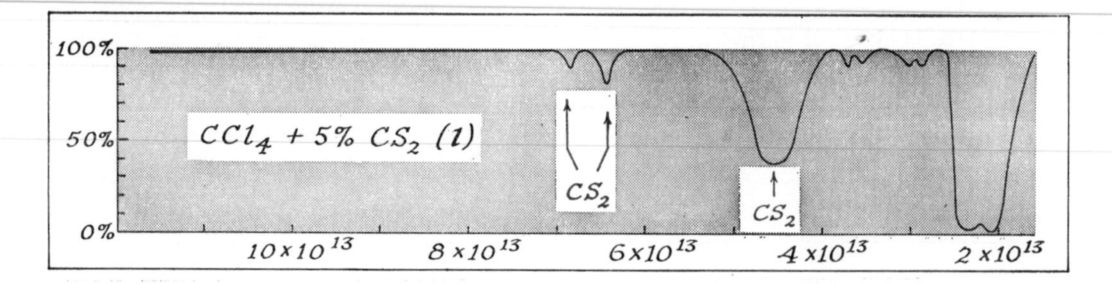

Fig. 14-17. **Infrared absorption spectra** *of liquid carbon tetrachloride, CCl_4, carbon disulfide, CS_2, and a mixture of the two.*

QUESTIONS AND PROBLEMS

1. A compound of carbon and hydrogen is known that contains 1.0 gram of hydrogen for every 3.0 grams of carbon. What is the atomic ratio of hydrogen to carbon in this substance?

2. There are two known compounds containing only tungsten and carbon. One is the very hard alloy, tungsten carbide, used for the edges of cutting tools. Analysis of the two compounds gives, for one, 1.82 grams and, for the other, 3.70 grams of tungsten per 0.12 gram of carbon. Determine the empirical formula of each.

3. John Dalton thought the formula for water was HO (half a century passed before the present formula for water was generally accepted). What relative weights did he then obtain for the weight of oxygen and hydrogen atoms?

4. Nitrogen forms five compounds with oxygen in which 1.00 gram of nitrogen is combined with 0.572, 1.14, 1.73, 2.28, 2.85 grams of oxygen, respectively. Show that the relative weights of the elements in these compounds are in the ratio of small whole numbers. Explain these data using the atomic theory.

5. Using Appendix 3, list two metals that could have given the same number of moles as aluminum did in the experiment shown in Figure 14-2.

6. If n coulombs will deposit 0.119 gram of tin from a solution of $SnSO_4$, how many coulombs are needed to deposit 0.119 gram of tin from a solution of $Sn(SO_4)_2$?

7. Suppose two more cells were attached to the three in Figure 14-2. In one cell, at one of the electrodes copper is being plated from $CuSO_4$ solution and at one of the electrodes in the other cell, bromine, $Br_2(g)$, is being converted to bromide ion, Br^-. How many grams of Cu and Br^- would be formed during the same operation discussed in the figure?

8. Carbon monoxide absorbs light at frequencies near 1.2×10^{11}, near 6.4×10^{13}, and near 1.5×10^{15} cycles per second. It does not absorb at intermediate frequencies.

   (a) Name the spectral regions in which it absorbs (see Figure 14-14).
   (b) Explain why carbon monoxide is colorless.

9. The wavelength and frequency of light are related by the expression $\lambda = c/\nu$, where λ = wavelength in centimeters, ν = frequency in cycles per second, and c = velocity of light = 3.0×10^{10} cm/second. Calculate the wavelength corresponding to each of the three frequencies absorbed by CO (see Problem 8). Express each answer first in centimeters, and then in Angstroms (1 Å = 10^{-8} cm).

   Answer. 1.5×10^{15} cycles/sec: 2.0×10^{-5} cm/cycle = 2.0×10^{3} Å/cycle.

10. The oxygen molecule carries out molecular vibration at a frequency of 2.4×10^{13} cycles/second. If the pressure is such that an oxygen molecule has about 10^9 collisions per second, how many times does the molecule vibrate between collisions?

11. When several oil drops enter the observation chamber of the Millikan apparatus, the voltage is turned on and adjusted. One drop may be made to remain stationary, but some of the others move up while still others continue to fall. Explain these observations.

12. Dust particles may be removed from air by passing the air through an electrical discharge and then between a pair of oppositely charged metal plates. Explain how this removes the dust.

13. How many electrons would be required to weigh one gram? What would be the weight of a "mole" of electrons?

14. About how many molecules would there be in each cubic centimeter of the tube shown in Figure 14-3 when the glow appears? When the glow disappears again because the pressure is too low?

15. Describe the spectrum produced on a photographic plate in a mass spectrograph if a mixture of the isotopes of oxygen (^{16}O, ^{17}O, and ^{18}O) is analyzed. Consider only the record for +1 and +2 ions.

16. Hydroxylamine, NH_2OH, is subjected to electron bombardment. The products are passed through a mass spectrograph. The two pairs of lines formed indicate charge/mass ratios of 0.0625, 0.0588 and 0.1250, 0.1176. How can this be interpreted?

17. Platinum and zinc have the same number of atoms per cubic centimeter. Would thin sheets of these elements differ in the way they scatter alpha particles? Explain.

18. Assume that the nucleus of the fluorine atom is a sphere with a radius of 5×10^{-13} cm. Calculate the density of matter in the fluorine nucleus.

19. An average dimension for the radius of a nucleus is 1×10^{-12} cm and for the radius of an atom is 1×10^{-8} cm. Determine the ratio of atomic volume to nuclear volume.

CHAPTER 15

Electrons and the Periodic Table

It is the behavior and distribution of the electrons around the nucleus that gives the fundamental character of an atom: it must be the same for molecules.

C. A. COULSON, 1951

We have seen that much is known about the structure of the atom. A small nucleus containing protons and neutrons accounts for most of the mass of the atom. Electrons occupy the space around the nucleus like bees around a hive. In the electrically neutral atom, the number of electrons is equal to the number of protons.

Looking back to Chapter 6, we have discovered marvelous regularity among the elements. Of the 100 or so elements, six are unique in their absence of chemical reactivity. Those six elements, the inert gases, provide the key to the most important correlation of chemistry, the periodic table. Not only do these elements furnish the cornerstone for the periodic table but, also, their electron populations seem to play a dominant role in the chemistry of the other elements in the table. An element just preceding an inert gas in the table (one of the halogens) has a strong tendency to acquire an extra electron. The resulting negatively charged ion has, then, the number of electrons possessed by an atom of its inert gas neighbor. In striking contrast, an element just following an inert gas (one of the alkalies) releases electrons quite readily. The resulting positively charged ion has, then, the number of electrons possessed by an atom of its inert gas neighbor. In each type of element, the halogens and the alkalies, the chemistry can be discussed in terms of the tendency of atoms to acquire or release electrons so as to reach the special stability of the inert gases. The importance of this tendency is revealed in the dramatic differences that exist between the chemistry of the halogens and the chemistry of the alkalies.

This special stability associated with the inert gas electron populations was found to pervade the chemistry of every element of the third row of the periodic table (see Section 6-6.2). Each element forms compounds in which it contrives to reach an inert gas electron population. Elements with a few more electrons than an inert gas are apt to donate one or two electrons to some other more needy atom. Elements with a few less electrons than an inert gas are apt to acquire one or two electrons or to negotiate a communal sharing with other atoms. In all cases, the number of electrons transferred or shared is understandable in terms of the inert gas stability.

In this chapter we shall explore our current understanding of this behavior. We are guided in this exploration by a regularity presented in Chapter 6 and reproduced in Table 15-I. The regularity of the differences 2, 8, 8, 18, 18, 32 is a clue of magnificent proportions. The electron populations of the inert gas atoms have startling regularity. This clue has led scientists to a detailed and quantitative understanding of the atomic properties that give rise to the periodic table.

Table 15-I

REGULARITY AMONG THE ELECTRON POPULATIONS OF THE INERT GASES

INERT GAS	ELECTRONS	DIFFERENCES
helium	2	2
neon	10	10 − 2 = 8
argon	18	18 − 10 = 8
krypton	36	36 − 18 = 18
xenon	54	54 − 36 = 18
radon	86	86 − 54 = 32

EXERCISE 15-1

To see that these numbers have regularity, consider the series of numbers 2–8–18–32. (We shall forget, for the time being, that 8 and 18 each appear twice in the series.)

(a) If you were to consider this series incomplete, would you expect the next number (after 32) to be even or odd?
(b) The numbers 2–8–18–32 were obtained by subtracting electron populations (that is, by taking differences). Take differences again and use them to predict the next number beyond 32 in the series.
(c) Divide the numbers 2–8–18–32 by two. Use these numbers as a basis for predicting the next number beyond 32 in the series by another method than taking differences.

15-1 THE HYDROGEN ATOM

Just as the inert gases form the cornerstone of the structure we call the periodic table, the simplest atom, hydrogen, provides the key that unlocks the door of this structure. Atoms of every other element mimic the hydrogen atom. To see that this is so, we must examine the interaction of hydrogen atoms with light. The light emitted (or absorbed) by hydrogen atoms is called the atomic hydrogen spectrum. This spectrum explains the existence of the periodic table.

15-1.1 Light—A Form of Energy

Before we can analyze the spectrum of hydrogen atoms, we must become more familiar with light. In Chapter 14 light was characterized by frequency or wavelength. (Reread Section 14-3.1.) Now we shall consider another property of light—a property less obvious than the "water wave" characteristics. Light is a form of energy.

The statement, "Light is a form of energy," is consistent with quite a bit of common experience. Many of you have used a hand lens to focus light rays on a paper, setting it afire. This experiment is put to work in the huge solar furnaces that achieve temperatures of many thousands of degrees and that melt the most refractory material. The temperature rise in the paper or in the refractory material is caused by the absorption of light. A temperature rise means energy has been absorbed. This energy must have been carried by the light.

We do not need a hand lens to "feel" the energy of light rays. Just remember those lazy afternoons you spent last summer soaking up the warmth of the sun. The afternoon pleasure came from the absorption of the energy by the tissues of your skin. The sunburn pain you suffered that night was caused by the chemical reactions energized by the treacherous light rays. Here we have a personal basis for claiming that light carries energy.

As one other familiar reference, consider photosynthesis. You have undoubtedly heard many times that this is the chemical process by which a plant "stores" the energy of the sun. Much is known about the chemical reactions of photosynthesis and it is indeed true that they result in formation of chemical compounds with higher heat content than the starting substances. These reactions will not occur in the absence of light—the light supplies the energy required to raise the reactants to the higher heat content of the products.

We have all of this familiar experience to build upon, but it is all qualitative. We need a quantitative relationship. *How much* energy is carried by light? The answer is simple in form, but not in concept. Light, too, comes in packages. Each package, called a *photon*, contains an amount of energy determined by the frequency. This statement is contained in the famous equation

$$E = h\nu \qquad (1)$$

The quantity h is called Planck's constant. It is merely a conversion factor that reexpresses frequency, ν, in energy units. The experimental evidence that led to equation (*1*) furnishes a fascinating story—a story that you will hear in your physics class. Our interest is in the application of this equation to the interpretation of the atomic hydrogen spectrum. This spectrum gives a record of the frequencies that are emitted by a hydrogen atom. By equation (*1*), then, the spectrum also gives information about the energies a hydrogen atom can possess.

15-1.2 The Light Emitted by Hydrogen Atoms

Figure 15-1 shows again the spectrograph described in Section 14-3.1. This time the light source is a gas discharge tube such as was shown in Figure 14-3 on p. 239. Just as before, some of the light emitted by the source passes through a narrow slit and is focused to a beam by the lens. Again this light beam is refracted (bent) by the angular prism. The spectrum, or frequency composition of the light, appears on the photographic film.

When hydrogen gas is admitted to the dis-

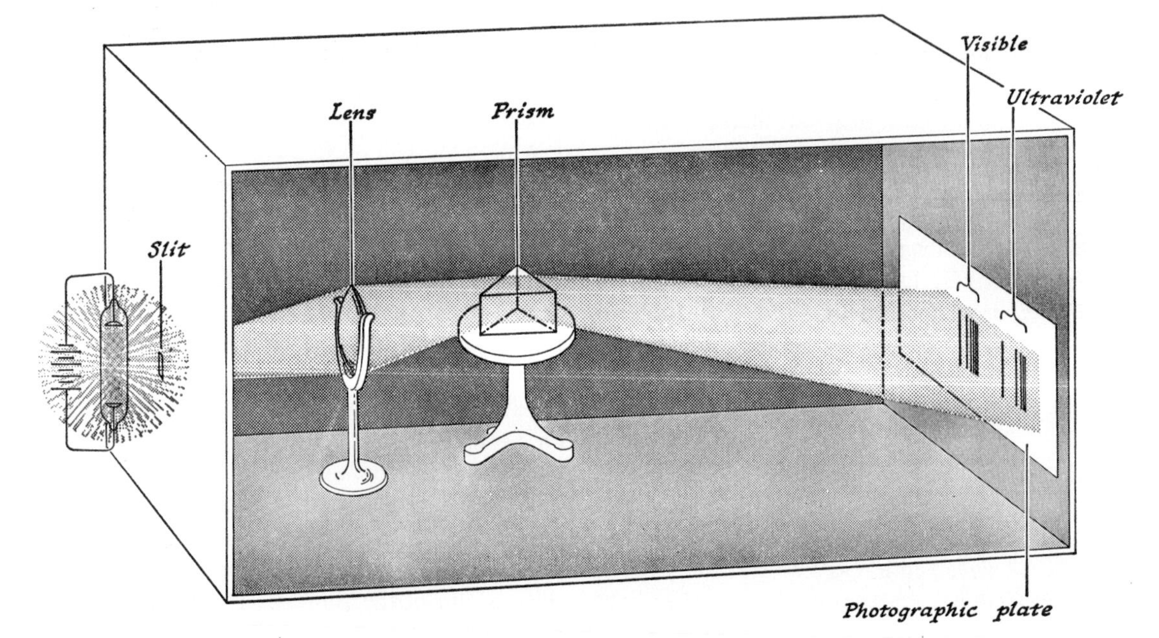

Fig. 15-1. **The spectrum** *of light emitted by a hydrogen discharge tube.*

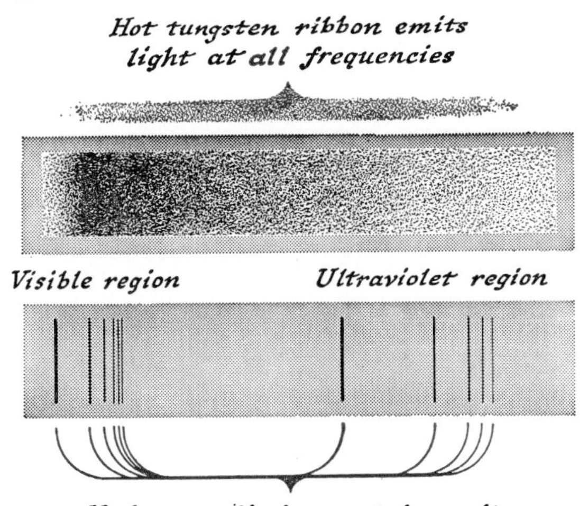

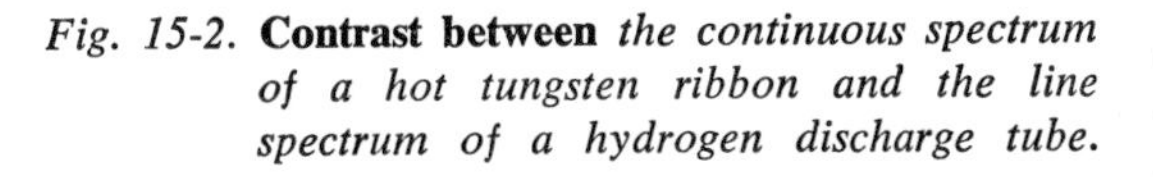

Fig. 15-2. **Contrast between** *the continuous spectrum of a hot tungsten ribbon and the line spectrum of a hydrogen discharge tube.*

charge tube and a high voltage is applied, light is emitted. To the eye, the light appears magenta in color but the spectrograph tells a surprising story! Instead of a fairly continuous darkening across the photographic film (as obtained from the hot tungsten ribbon) the film shows a series of lines! Each line corresponds to a particular frequency emitted by hydrogen atoms. Each space between two lines on the film corresponds to a frequency range in which no light is emitted by the hydrogen atom.

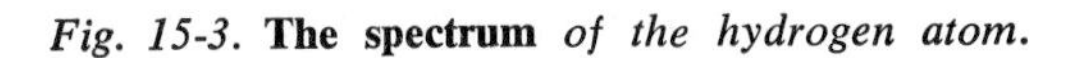

Fig. 15-3. **The spectrum** *of the hydrogen atom.*

	Visible region							Ultraviolet region					
Energy of photons, (kcal/mole)	43.6	58.8	65.9	69.7	72.1	73.6		235.2	278.8	294.0	301.1	304.9	
Frequency, (cycles/sec)	4.567	6.165	6.905	7.307	7.550	7.707	$\times 10^{14}$	2.465	2.922	3.081	3.149	3.196	$\times 10^{15}$
Wavelength, (Å)	6565	4863	4342	4103	3971	3890		1216	1026	973	952	938	

Two peculiarities of the hydrogen discharge spectrum are immediately evident—they were evident to scientists as long ago as 1840. First, hydrogen atoms are very particular about the frequencies they emit. Only special frequencies are observed in the spectrum. Second, the frequencies corresponding to the lines on the film are spaced systematically. There are two groups of lines, one group in the visible part of the light spectrum and another group in the ultraviolet part. Within each group there is a regular decrease in the spacing between successive lines as the frequency increases. The measured frequencies are shown in Figure 15-3.

EXERCISE 15-2

Complete the following table for all of the ultraviolet lines listed in Figure 15-3. Plot the energy spacing against the arbitrary spacing number assigned in the last column to convince yourself that there is regularity in the spacings of these lines. Make the same sort of a table for the lines in the visible group.

GROUP	ENERGY PER MOLE OF PHOTONS	ENERGY SPACING	SPACING NUMBER
Ultraviolet lines	235.2 kcal		
		43.6 kcal	1
	278.8		
			2
	294.0		
			3

To see how these results might be explained, let us translate these spectral lines into energy terms. The hydrogen atom just before light emission has some amount of energy—let us call it E_2. Light of frequency ν is emitted, carrying away energy $h\nu$. The hydrogen atom is left with less energy—let us call it E_1. As is habitual, we assume energy is conserved, so the energy lost by the hydrogen atom must be exactly equal to that carried away by the light:

$$E_2 - E_1 = h\nu \qquad (2)$$

To "explain" why hydrogen atoms emit the line spectrum, we must seek a model with the

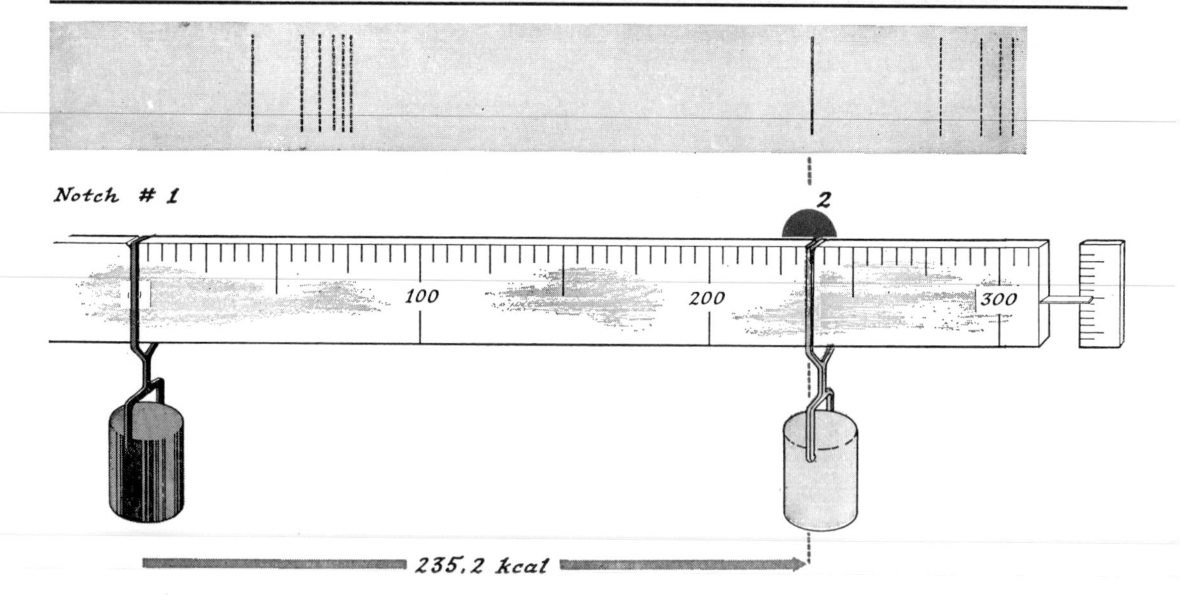

Fig. 15-4. **The notches implied** *by light emission at 235.2 kcal.*

same sort of properties. What sort of a system has the property that its initial energy, E_2, tells it how much energy it can release so as to make the difference $E_2 - E_1$ just exactly one of the special energies, $h\nu$? Fortunately, we have just such a system at hand, though it deals in weights, not energies.

Picture your triple beam balance. The front beam has a sliding hanger that permits you to balance any weight placed on the pan up to the full scale reading on the beam. This front beam might be compared to the hot tungsten ribbon. The beam can be used to balance any small weight change made on the pan: the hot tungsten ribbon emits light of any energy (that is, of any frequency).

But now consider the other two beams—they are quite different. They are notched so that only particular hanger positions can be selected. Each position corresponds to a particular weight and intermediate weights cannot be balanced without the front beam. The notched beam has properties in common with the hydrogen atom. The weight added to the notched beam is fixed by the notches—intermediate weights cannot be measured. In the hydrogen atom the energy is somehow "hung on a notched beam" and the amount of energy it can absorb or release must correspond to the energy difference between two of these "notches."

If we pursue this analogy, we can see how scientists have deduced the energy properties of the hydrogen atom from its line spectrum. We can use the observed energies, listed in Figure 15-3, to construct a notched beam that would "deliver" the observed spectrum. This beam must have a scale calibrated not for weight but for energy. There must be notches on this scale to match the observed lines in the spectrum. We shall find it convenient to begin with the line observed at wavelength 1216 Å or, in energy units, 235.2 kcal/mole of photons. This energy represents the *difference* in energy between two notches. If we number the notches #1 and #2, changing an imaginary hanger from notch #1 to notch #2 delivers 235.2 kcal, as shown in Figure 15-4.

Now consider the other lines in the ultraviolet group. Exercise 15-2 showed that they are related to the one at 235.2 kcal in their systematic spacings. Let's seek regularity among these energies by making a simple assumption. We shall assume that each line in the ultraviolet group corresponds to a movement of the hanger from notch #1 to a new notch. This assumption is arbitrary and will be kept only if it turns out to be helpful. (Remember the lost child and his rule, "Cylindrical objects burn"?)

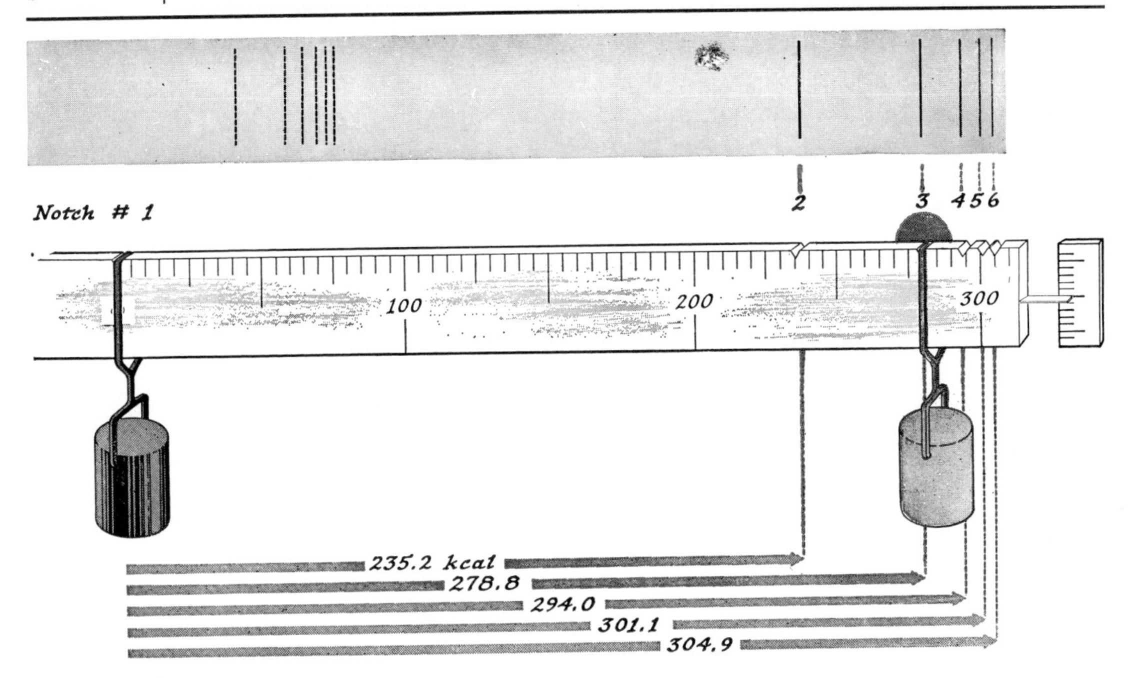

Fig. 15-5. **The notches required by the ultraviolet group of lines if all changes** *begin in notch #1.*

So we can add four more notches to the beam, as shown in Figure 15-5. Notch #3 is cut at a scale position 278.8 kcal relative to notch #1. Notch #4 is at 294.0 kcal, and so on.

The appearance of this notched beam is unfamiliar, to say the least. Can we find any basis for corroboration or contradiction of this interpretation? To seek such evidence, let's base a prediction upon the model. One such prediction concerns the energies that this notched beam would deliver if we investigated movements of the hanger from notch #2 to notch #3. Notch #2 is 235.2 kcal from notch #1, and notch #3 is

Fig. 15-6. **The energy change implied** *by a change from notch #2 to notch #3.*

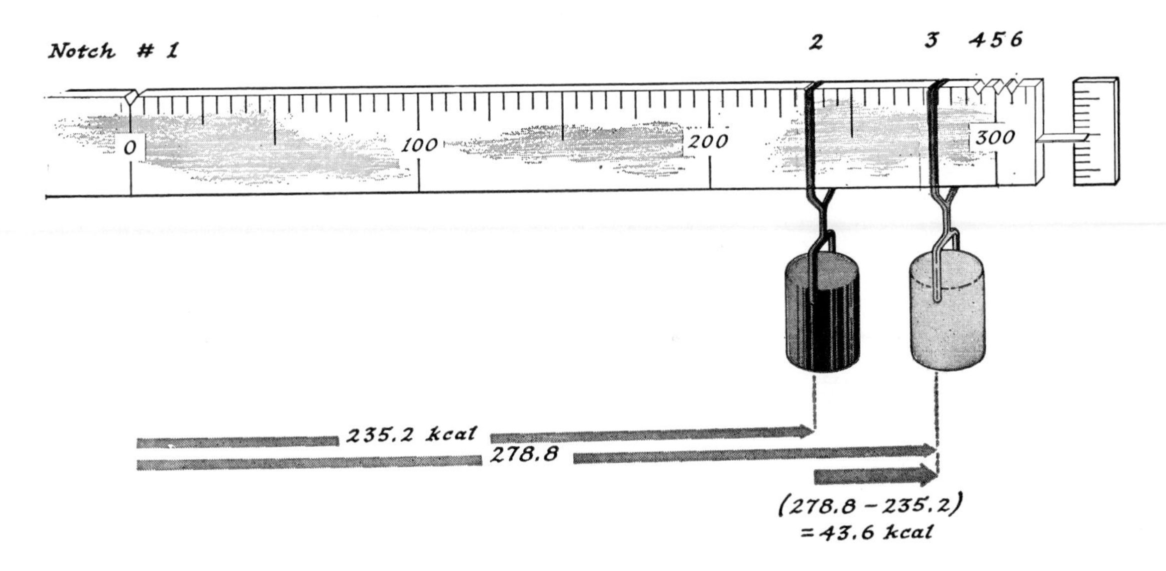

278.8 kcal from notch #1. Therefore, the energy difference going from notch #2 to #3 is the difference between these two numbers:

Energy change from notch #2 to #3
$$= 278.8 - 235.2 = 43.6 \quad (3)$$

Calculation (*3*) indicates that our notched beam model implies light would be emitted with energy at 43.6 kcal as well as at 235.2 kcal and 278.8 kcal. We refer back to the spectrum in Figure 15-3 and, indeed, there *is* light emitted at 43.6 kcal—this is one of the group of lines in the visible region!

With this encouragement, let's calculate the other lines implied by changes beginning in notch #2.

NOTCH CHANGE	ENERGY CHANGE CALCULATED	OBSERVED
#2 → #3	278.8 − 235.2 = 43.6 kcal	43.6 kcal
#2 → #4	294.0 − 235.2 = 58.8	58.8
#2 → #5	301.1 − 235.2 = 65.9	66.0
#2 → #6	304.9 − 235.2 = 69.7	69.8

Plainly, the agreement between calculation and experiment is too good to be accidental. Our notched beam is a useful basis for interpreting the spectrum of the hydrogen atom.

EXERCISE 15-3

Plot the energy of each line of the ultraviolet group against notch number, *n*, using the higher of the two notch numbers assigned to that line in Figure 15-5. For example, plot 235.2 kcal on the vertical axis against 2 on the horizontal axis. Assign to notch #1 the arbitrary value zero and draw a smooth curve through all of the points, including the point for notch #1. Estimate the energy value that would be observed for a notch with very high notch number, as suggested by the curve. (Call this "notch # infinity," $n = \infty$.)

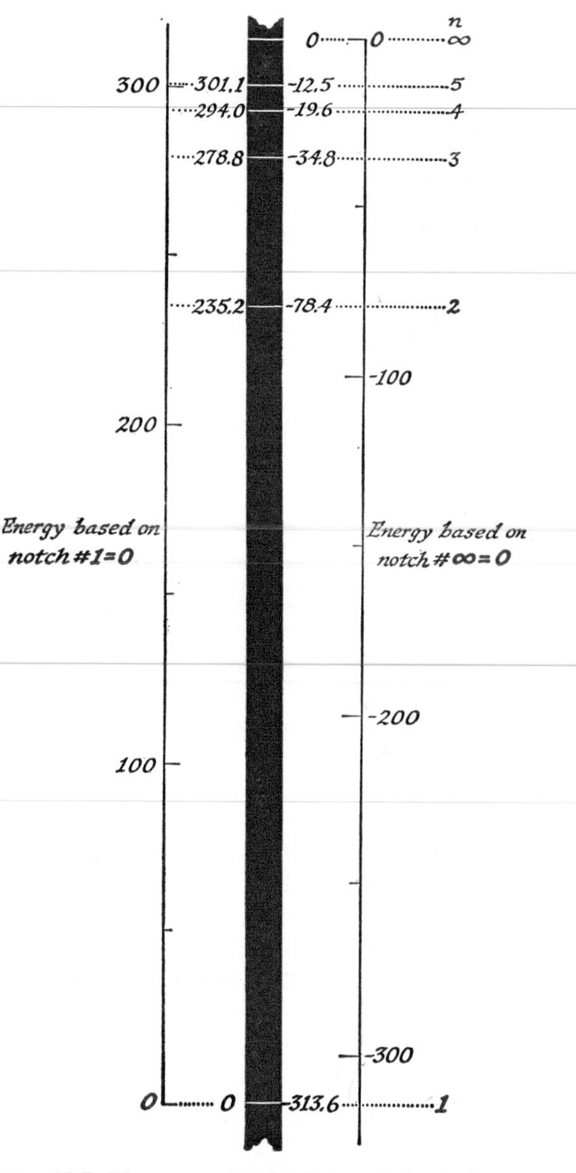

Fig. 15-7. **The energy level scheme** *of the hydrogen atom.*

Historically, the visible emission lines shown in Figure 15-3 were the first atomic hydrogen lines discovered. They were found in the spectrum of the sun by W. H. Wollaston in 1802. In 1862, A. J. Ångstrom announced that there must be hydrogen in the solar atmosphere. These lines were detected first because of the lesser experimental difficulties in the visible spectral region. They are called the "Balmer series" because J. J. Balmer was able to formulate a simple mathematical relation among the frequencies (in 1885). The ultraviolet series shown in Figure 15-3 was actually predicted prior to its discovery (in 1915 by T. Lyman).

Hence our analysis and prediction, selected for logical clarity, reverse the actual chronology. Our prediction of the visible spectrum from the ultraviolet spectrum is more straightforward than was the reverse prediction. Notice that the ultraviolet spectrum utilizes all of the notches involved in producing the visible spectrum. In contrast, none of the visible frequencies involves notch #1, the key to the ultraviolet spectrum.

15-1.3 The Energy Levels of a Hydrogen Atom

Scientists deduced the notched energy scale of the hydrogen atom from the spectrum in just the same way we did.

Of course, more sophisticated language is generally used. For example, the energy possibilities are usually shown on a vertical scale and they are called *energy levels* rather than notches. Figure 15-7 shows the energy level scheme of the hydrogen atom. Each energy level is characterized by an integer, n, the lowest level being given the number 1.

Two energy scales are shown in Figure 15-7. On the left is a scale based upon the energy zero for the $n = 1$ level. On this scale, the levels appear at energies corresponding to our notched beam spacings. The right-hand scale is displaced upward so that the $n = 1$ level corresponds to a negative energy, -313.6 kcal. The zero has been moved upward by this amount so that the zero of the energy scale corresponds to the "notch # infinity" that you estimated in Exercise 15-3. Obviously, the positioning of the energy levels is not affected by this arbitrary change of the zero of energy, so either scale can be used. For reasons that will become evident later in this chapter, the right-hand scale is more convenient and is the one generally used.

EXERCISE 15-4

Calculate the energy change that occurs between notch #1 and notch #2, *using the right-hand scale in Figure 15-7*. (Remember that the energy change is the energy of the final level *minus* the energy of the initial level. Pay careful attention to algebraic signs.) Repeat the calculation for the energy change between notch #2 and #3. Compare your calculations with the numbers shown in Figure 15-6.

Long after this energy level diagram for the hydrogen atom had been established, scientists still pondered its significance. Finally, in the late 1920's, a mathematical scheme was developed that explained the facts. The mathematical scheme is called *quantum mechanics*.

15-1.4 Quantum Mechanics and the Hydrogen Atom

Prior to the development of quantum mechanics, the spectrum of the hydrogen atom posed quite a dilemma. To see the problem, and how it was resolved, let's go back about fifty or sixty years and trace the history of this problem. This is a valuable example because it shows how science advances.

By the year 1912 it was known that the hydrogen atom is composed of a proton and an electron. These two particles are attracted to each other by reason of their electric charges. Physicists felt they should be able to calculate the properties of such a combination. After all, the laws of motion of macroscopic bodies had been studied for centuries. The behavior of electrically charged bodies was also thoroughly understood on the macroscopic scale. Yet scientists could not explain why a hydrogen atom exists, let alone why it would have only particular values of energy. In fact, the laws that had been deduced for macroscopic bodies gave the firm (though incorrect) prediction that the nuclear atom is unstable and the electron should collapse into the nucleus.

At this point a Danish physicist, Niels Bohr, decided to take a fresh start. In effect, he faced the fact that an explanation is a search for likenesses between a system under study and a well-understood model system. An explanation is not good unless the likenesses are strong. Niels Bohr suggested that the mechanical and electrical behavior of macroscopic bodies is not a completely suitable model for the hydrogen atom. He proceeded to seek a new model that did not contradict the known facts.

He began by supposing that the structure of an atom (the arrangement of the electrons around the nucleus) is determined by its energy. To agree with the facts, Bohr proposed that only special atomic structures can exist—he called these special structures "stationary states." Each such state is characterized by a particular energy, and since a set of special atomic structures exists, a corresponding set of energies will be found. Here Bohr departed from the older atomic models (those of classical physics) that permitted structures corresponding to all possible energies.

The most stable state of the atom would be expected to be the one in which the atom has the lowest energy. Bohr reasoned that since we observe that the nuclear atom does exist then it must be a fundamental fact of nature that an atom can exist in its most stable state indefinitely. Even though this fact could not be rationalized (remember, the earlier laws of physics predicted the atom should collapse) it had to be accepted because it was a result of experiments.

Bohr also proposed that although the lowest energy state of the atom is its most stable state, the atom can be excited to its higher allowed energy states (by absorbing light or through a violent collision with other atoms or electrons). The excited atom does not remain in this condition for long; it loses its extra energy by emitting light. Since only certain levels of energy exist, only certain energy changes can occur. The energy change of the atom must be equal to the energy of the light emitted, in accord with equation (*2*), $E_2 - E_1 = h\nu$. Consequently, the frequency of the light emitted by an atom is entirely determined by the values of the allowed energies of its electrons.

These ideas were so revolutionary that they would not have been accepted except for the fact that Bohr was able to propose a way to calculate exactly the energy levels for the hydrogen atom. Within ten years Bohr's calculational methods were completely replaced by better techniques, but his postulate that only certain atomic energy states are possible has been repeatedly shown to be correct.

In this example there is much food for thought concerning the development of science. The wide applicability of the laws of motion and of electromagnetics made it natural for scientists to assume that these same laws, without change, applied to the atom. True, this represented an extrapolation, for the laws were deduced on the macroscopic level. Yet, the same laws that described the motions of the planets also described the motions of tennis balls—why not also the motions of electrons? Many experimental facts said no, but physicists held to the expectation that a way would be found to explain these facts within the framework of the known (and almost sacred) laws. When Bohr finally broke away from the established laws, *he still used them for guidance*, proposing only those changes required by the discordant facts. Perhaps the most ironic part is that Bohr's principal weapon in gaining acceptance for his new attack was his mathematical success in predicting the energy levels of hydrogen, though his model has since been discarded completely. The model he used proved to fit *only* the hydrogen atom and *no other*.

Don't be too eager, though, to scoff at this example. Instead, you may rest assured that some of the theories you will find in this book are waiting to be swept aside as we learn more about nature. The rub—no, the excitement—in the game is that we don't know which theories are fated to go. That remains for some of you to discover!

15-1.5 The Hydrogen Atom and Quantum Numbers

The modern theory of the behavior of matter, called *quantum mechanics*, was developed by several workers in the years 1925–1927. For our purposes the most important result of the quantum mechanical theory is that the motion of an electron is described by the quantum numbers and orbitals. ***Quantum numbers** are integers that identify the stationary states of an atom; the word **orbital** means a spatial description of the motion of an electron* corresponding to a particular stationary state.

THE PRINCIPAL QUANTUM NUMBER, *n*

If we return to our notched beam analogy, as shown in Figure 15-5, we find that we numbered the notches, 1, 2, 3, · · ·. These numbers serve as natural identifying designations. They are the "quantum numbers" of the balance beam.

In Exercises 15-2 and 15-3 you observed that the energy levels of hydrogen vary systematically with the quantum (or notch) number. The smooth curves suggest that the energy could be conveniently expressed mathematically in terms of n, the quantum number. Using the right-hand scale in Figure 15-7, we see that for any value of n, E is always negative. As n becomes larger, the energy rises and approaches zero on the scale. Investigation of the actual energy values shows that the energy levels of Figure 15-7 are *exactly* determined by n according to the relation

$$E_n = -\frac{313.6}{n^2} \text{ kcal/mole} \qquad (4)$$

E_n = energy level with quantum number n
$n = 1, 2, 3, \cdots \infty$

The number n is called the *principal quantum number.*

The mathematical relationship (*4*) is the one Bohr was able to deduce. Current quantum mechanical methods also deduce this relationship, of course, but with a model that is in fundamental discord with the one used by Bohr.

ORBITALS AND THE PRINCIPAL QUANTUM NUMBER, n

Quantum mechanics provides a mathematical framework that leads to expression (*4*). In addition, for the hydrogen atom it tells us a great deal about how the electron moves about the nucleus. It does not, however, tell us an exact path along which the electron moves. All that can be done is to predict the probability of finding an electron at a given point in space. This probability, considered over a period of time, gives an "averaged" picture of how an electron behaves. This description of the electron motion is what we have called an orbital.

Thus, an orbital description of the motion of an electron contains the same information conveyed by the holes made by darts in a dartboard. After the board has been used in many games, the distribution of holes shows how successful earlier players had been in their scoring. There are many holes near the bullseye and, moving away from it, there is a regular decrease in the number of holes per square centimeter of dartboard. At any given distance from the bullseye, the "density" of dart holes (number per square centimeter) is a measure of the probability that the next throw will land there.

We see that the holes in the dartboard tell us *only* the probability that a given throw will land a particular distance from the bullseye. It does *not* tell us the order in which the holes were made in the dartboard. In the case of the electron distribution, the orbital tells us the probability that an experiment designed to locate the electron will find it a particular distance from the nucleus. It does *not* tell us how the electron moves from point to point—its trajectory.

Though quantum mechanics does not tell us the electron trajectory, it does tell us how the orbital changes as n increases. It also indicates that **for each value of n there are n^2 different orbitals.** For the hydrogen atom, the n^2 orbitals for a particular value of n all have the same energy,

$$-\frac{313.6}{n^2}\frac{\text{kcal}}{\text{mole}}$$

s ORBITALS

Consider the lowest energy level of a hydrogen atom, $n = 1$. We have just learned that there are n^2 levels with this energy, and since $n = 1$, there is but one level. It corresponds to an electron distribution that is spherically symmetrical around the nucleus, as shown in Figure 15-8. It is called the 1*s* orbital.* An electron moving in an *s* orbital is called an *s* electron.

The term "spherically symmetric" and the picture of an *s* orbital (Figure 15-8) must be clearly understood. They indicate that if we were to look for the electron somewhere on the surface of a sphere of a particular radius, r_1, which has the nucleus at its center, the probability of finding the electron at any one point on the r_1 sphere is the same as the probability of finding it at any other point on the r_1 sphere. The same would be true at a different radius, r_2, but the probability of finding the electron somewhere on the r_2 sphere would *not* be the same as that on the r_1 sphere. The chance of finding the electron does depend upon the radius of the sphere on

* In the symbol 1*s*, the number 1 tells us that $n = 1$. The letter *s* tells us that the orbital is spherically symmetric. Since the letter *s* has been used for an orbital that is spherically symmetric, we might as well think of it as an abbreviation: *s* = *s*pherical.

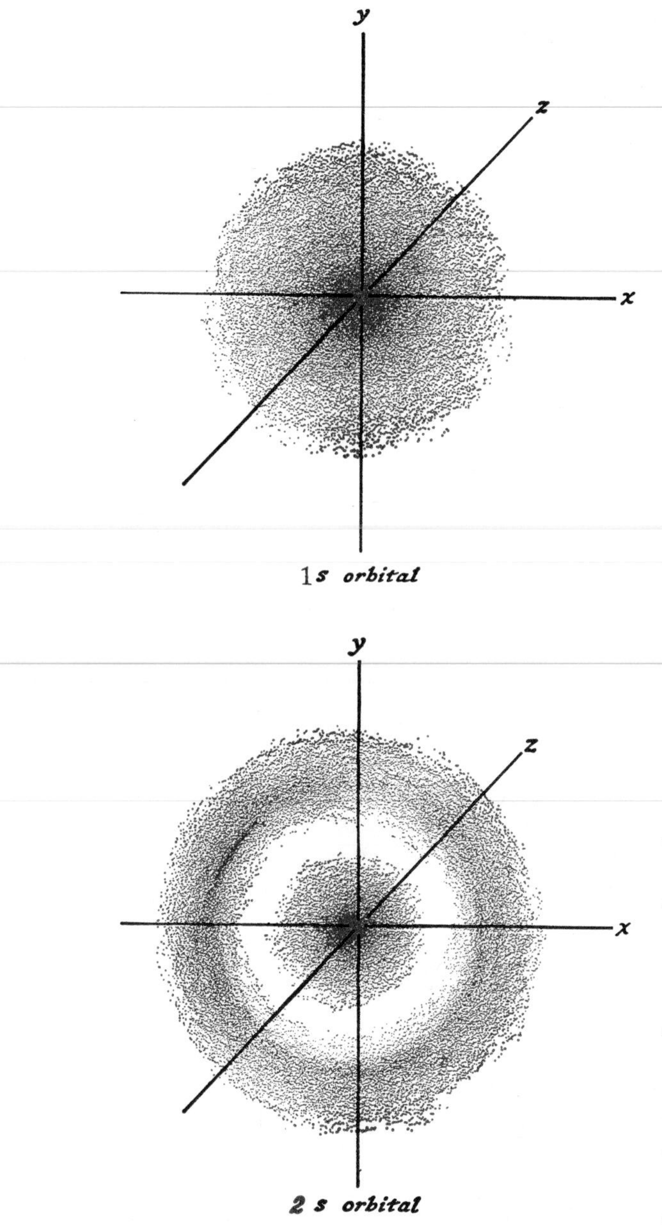

Fig. 15-8. **Atomic orbitals:** *1s and 2s orbitals.*

which we search. A 1*s* electron can be found anywhere from right at the nucleus to a great distance away—but it is most likely to be found approximately 10^{-8} cm from the nucleus.

The next energy level corresponds to $n = 2$. According to our rule, there are $n^2 = 2^2 = 4$ different spatial arrangements with the same energy, $-313.6/4 = -78.4$ kcal/mole. One of them is again spherically symmetric and it is called the 2*s* orbital. Figure 15-8 shows the atomic 2*s* orbital. Here we find the reasonable result that the higher energy of the 2*s* electron permits the electron to spend more time far from the nucleus.

For every value of n, there is one spherically symmetric orbital. As *n* increases, the *ns* orbitals place the electron, on the average, farther and farther from the nucleus.

p ORBITALS

There is only one orbital corresponding to $n = 1$, the 1*s* orbital. For $n = 2$, there are four different spatial arrangements and we have described one of them, the 2*s* orbital. The other three are called 2*p* orbitals. An electron in a *p* orbital behaves in such a way that it is most likely found in either of two regions located on opposite sides of the nucleus. The motion of a *p* electron creates an electron distribution that is shaped somewhat like a dumbbell. We can place the axis of the dumbbell along one of the three perpendicular cartesian coordinate axes. Just as there are three distinct coordinate axes, there are three distinct *p* orbitals, each with its axis perpendicular to the other two. They are sometimes referred to as the p_x, p_y, and p_z orbitals to emphasize their directional character. The p_x orbital is concentrated in the *x* direction—a p_x electron is more apt to be found near the *x* axis than anywhere else. The p_y orbital, on the other hand, is concentrated along the *y* axis. These directional characteristics are useful in explaining the geometrical properties of molecules. Figure 15-9 shows the electron distribution of the 2*p* orbitals.*

Every energy level with n above 1 has three p orbitals. As *n* increases, the *np* orbitals place the electron, on the average, farther and farther from the nucleus, but always with the axial directional properties shown in Figure 15-9.

d AND *f* ORBITALS

At this point we might recast the hydrogen atom energy level diagram to express what we know

* The three *p* orbitals can be considered to extend along the three perpendicular axes, *x*, *y*, and *z*. Hence, *p* might be thought of as an abbreviation: *p* = *p*erpendicular.

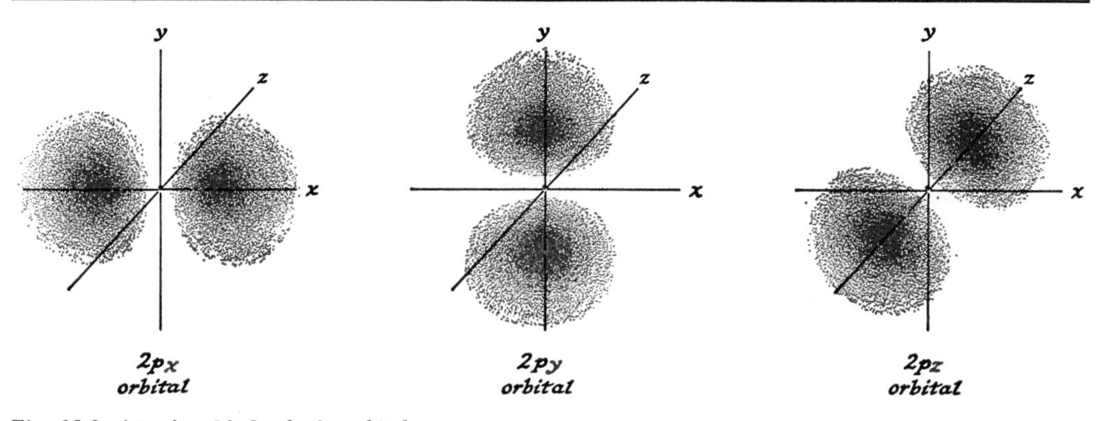

Fig. 15-9. **Atomic orbitals:** *the 2p orbitals.*

about orbitals. Figure 15-10 shows each orbital as a pigeon hole. For a given value of n, there are n^2 total orbitals. For $n = 3$, there are $3^2 = 9$ orbitals, five more than accounted for by one 3*s* orbital and the three 3*p* orbitals. These five orbitals are called *d* orbitals and they have more complicated spatial distribution than do the *p* orbitals.

EXERCISE 15-5

From the information that the numbers of *s*, *p*, and *d* orbitals are 1, 3, and 5, how many of the next higher (*f*) orbitals would you expect? Verify your answer by calculating n^2 for $n = 4$ and comparing to your sum of the numbers of *s*, *p*, *d*, and *f* orbitals.

15-1.6 The Hydrogen Atom and the Periodic Table

At last we are ready to return to the periodic table. At last we are able to begin answering those who are "wondering why" about the special properties of the electron populations in Table 15-I. Let us reproduce Table 15-I together with the numbers of orbitals of the hydrogen atom. The suggestion of a connection is irresistible, as seen in Table 15-II.

The hydrogen atom orbitals give us the numbers 2, 8, 18, and 32—the numbers we find separating the specially stable electron populations of the inert gases. It was necessary to multiply n^2 by two—an important factor that could not have been anticipated. Furthermore, it will be necessary to find an explanation for the occurrence of eight-electron differences both at neon and at argon and eighteen-electron differences both at krypton and at xenon.

Nevertheless, we seem to have a significant start toward explaining the periodic table. We

Table 15-II. **STABLE ELECTRON POPULATIONS AND THE HYDROGEN ATOM**

THE INERT GASES			THE HYDROGEN ATOM		
element	number of electrons	differences	n	number of orbitals	$2 \times n^2$
helium	2	2	$n = 1$	$n^2 = 1$	$2 \times 1 = 2$
neon	10	$10 - 2 = 8$	$n = 2$	$n^2 = 4$	$2 \times 4 = 8$
argon	18	$18 - 10 = 8$	$n = 3$	$n^2 = 9$	$2 \times 9 = 18$
krypton	36	$36 - 18 = 18$	$n = 4$	$n^2 = 16$	$2 \times 16 = 32$
xenon	54	$54 - 36 = 18$			
radon	86	$86 - 54 = 32$			

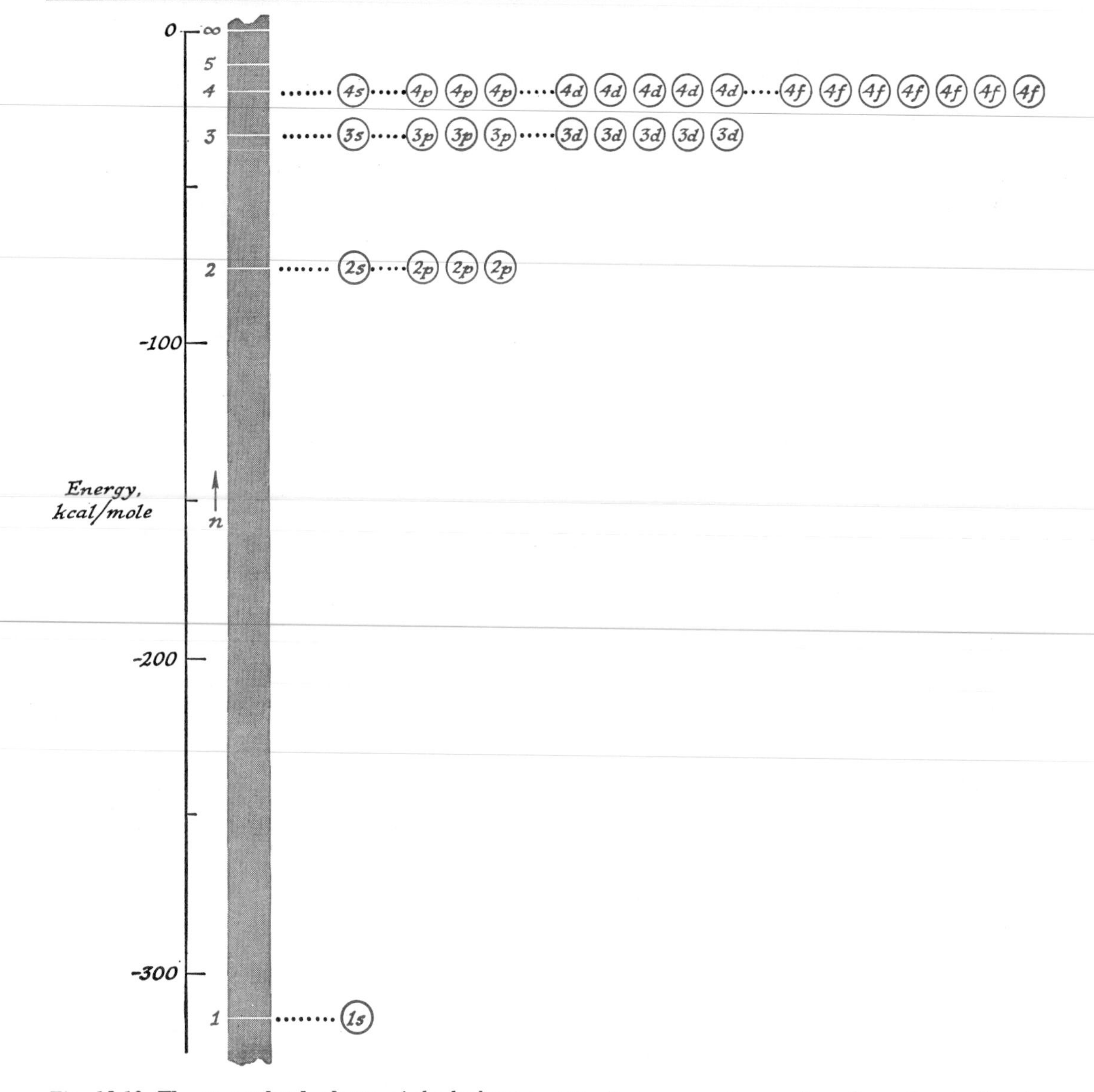

Fig. 15-10. **The energy level scheme** *of the hydrogen atom.*

can understand the chemical trends within the first ten elements in terms of the hydrogen atom orbitals with the following two assumptions:

(1) that atoms of these elements have orbitals and energy levels that are qualitatively like those of the hydrogen atom;
(2) that a single orbital (of any element) can accommodate, at most, two electrons.

The first assumption springs from the similarities noticed in Table 15-II and is well substantiated by a wealth of spectral study of these atoms. The second assumption is stimulated by the factor of two needed in the last column of Table 15-II.

With these two assumptions, we can propose the electronic arrangement of lowest energy for each atom. We do so by mentally placing electrons successively in the empty orbitals of lowest energy. The electron orbital of lowest energy is the 1*s* orbital. The single electron of the hydrogen atom can occupy this orbital. In the helium atom, there are two electrons and the nuclear charge is two. Since each orbital can accommodate two electrons, both of the electrons can go into the 1*s* orbital. The resulting electronic arrangement is described by the notation $1s^2$, which means that there are two electrons in the 1*s* orbital. The notation $1s^2$ is called the *electron configuration.*

Now let us consider what will happen when there are three electrons near a triply charged nucleus, as in the lithium atom. The first two of the three electrons go into the lowest energy orbital, the 1*s* orbital. When this orbital has two electrons in it, it is completely filled, according to our second assumption. Any additional electrons must be placed in orbitals of higher energy. Therefore, the third electron in lithium goes into the 2*s* orbital, and we write the electron configuration as $1s^22s^1$. Despite the nuclear charge of three in the lithium atom, this last electron is rather weakly bound because the 2*s* electron in lithium spends most of its time farther away from the nucleus than do the 1*s* electrons. This electron should be easily removed to give Li^+, which is, indeed, the characteristic behavior of an alkali element.

The beryllium atom has one more electron than does the lithium atom. The fourth electron that enters the beryllium atom can occupy the 2*s* orbital to give a configuration of $1s^22s^2$. The two 2*s* electrons will be most easily removed, tending to form the Be^{+2} ion.

There is no more room in the 2*s* orbital for a fifth electron, which appears when we move on to the boron atom. However, another orbital with principal quantum number 2 is available. A 2*p* orbital accepts the fifth electron, giving the configuration $1s^22s^22p^1$. Continuing this process, we obtain the following configurations:

carbon atom	$1s^2$	$2s^22p_x^12p_y^1$
nitrogen atom	$1s^2$	$2s^22p_x^12p_y^12p_z^1$
oxygen atom	$1s^2$	$2s^22p_x^22p_y^12p_z^1$
fluorine atom	$1s^2$	$2s^22p_x^22p_y^22p_z^1$
neon atom	$1s^2$	$2s^22p_x^22p_y^22p_z^2$

If we proceed to the next element, sodium atom, we are again forced to use an orbital with the next higher quantum number:

sodium atom	$1s^2$	$2s^22p_x^22p_y^22p_z^2$	$3s^1$

Again there is one electron which spends most of its time farther away from the nucleus than any of the others. This one electron could be easily removed to give Na^+, and we return to the type of chemistry shown by lithium.

The hydrogen atom energy levels, together with our two assumptions, have provided a good explanation of some of the properties of the first eleven elements. We shall see that they explain the entire periodic table.

15-2 MANY-ELECTRON ATOMS

All atoms display line spectra. In general these spectra are much more complicated than the atomic hydrogen spectrum shown in Figure 15-3. Nevertheless, these spectra can be interpreted in terms of the concepts we have developed for the hydrogen atom.

15-2.1 Energy Levels of Many-Electron Atoms

Analysis of the spectra of many-electron atoms shows the following similarities to the hydrogen atom case.

(1) All atoms have "stationary states" and can hold only particular values of energy.
(2) The atomic spectra can be understood in terms of transitions between energy levels corresponding to these particular values of energy.
(3) The energy level diagrams resemble the hydrogen atom level diagram except that the n^2 levels with the same value of *n* no longer all have the same energy.

Figure 15-11 shows a schematic energy level diagram of a many-electron atom. Blue patterns

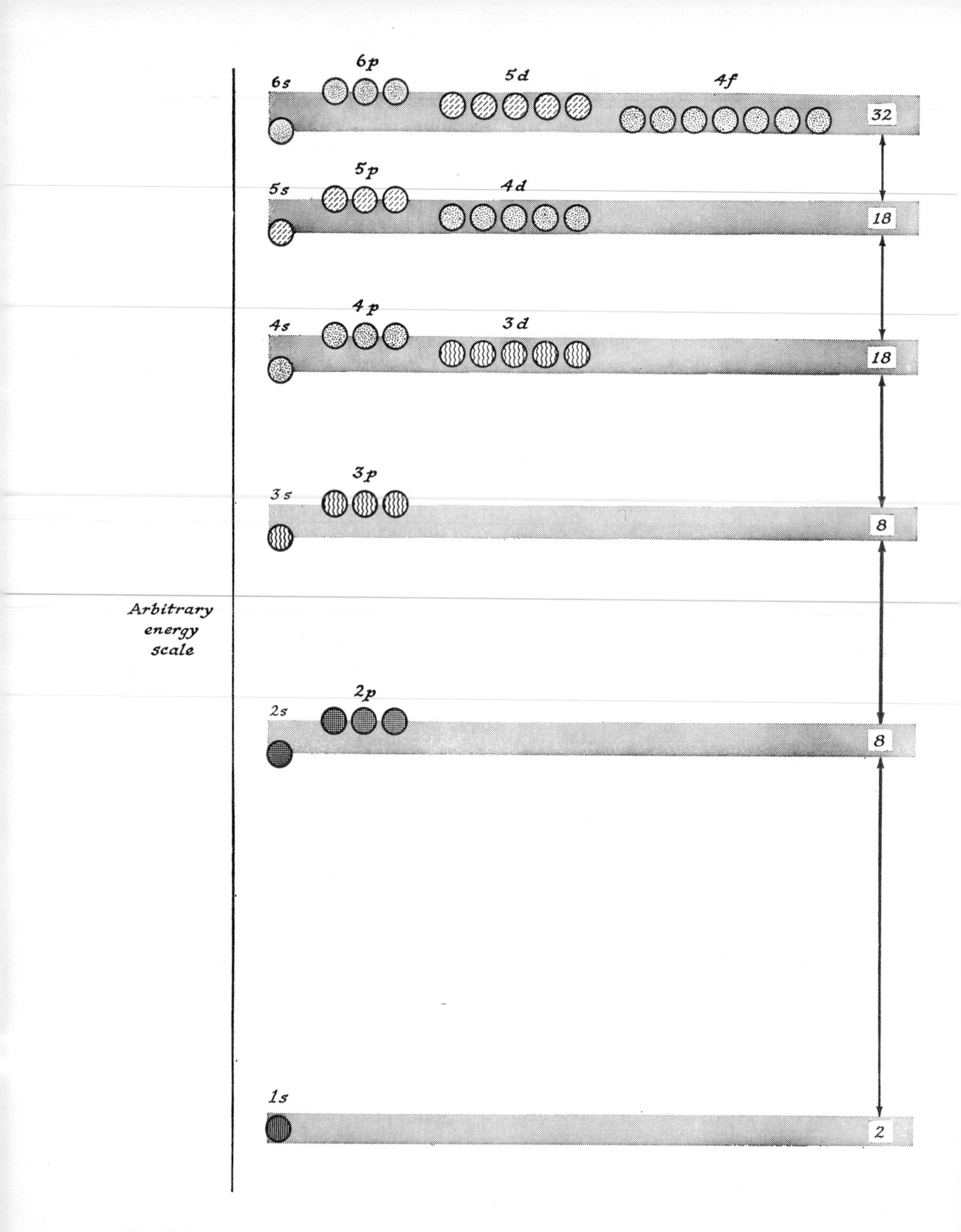

Fig. 15-11. **A schematic energy level diagram** *of a many-electron atom.*

indicate levels with the same hydrogen atom principal quantum number. The effect of placing many electrons into the energy levels of the hydrogen atom is a "tilting" of the diagram. The *p* orbitals are slightly higher in energy than are the *s* orbitals of the same value of *n*. The *d* orbitals and *f* orbitals of this same value of *n* are successively even higher. The result is that the 3*d* orbitals are raised approximately to the energy of the 4*s* and 4*p* orbitals. The 3*s* and 3*p* orbitals are left more or less isolated in energy. They are much higher in energy than are the 2*s* and 2*p* orbitals but much lower than is the cluster of 4*s*, 4*p*, and 3*d* orbitals.

15-2.2 The Periodic Table

Now we can see the development of the entire periodic table. The special stabilities of the inert gases are fixed by the large energy gaps in the energy level diagram, Figure 15-11. The number of orbitals in a cluster, multiplied by two because of our double occupancy assumption, fixes the number of electrons needed to reach the inert gas electron population. The numbers at the right of Figure 15-11 are exactly the numbers we found as differences between inert gas electron populations (see Table 15-I).

It is important to recognize what we have done. We have shown how the energy level diagram of hydrogen was deduced from the atomic hydrogen spectra. We have seen that the energy level diagram of a many-electron atom can be regarded as a tilted hydrogen atom diagram. Such a diagram contains clusters of energy levels with large energy gaps between. These clusters of energy levels provide a basis for explaining the electron populations of the inert gases, provided we assume two electrons can occupy each orbital.

We have not explained why two, but no more than two, electrons can occupy each orbital. This is not known and is accepted because the facts of nature require it. This assumption is called the *Pauli Principle*.

We have given much explanation of why the clusters and energy gaps of the energy level diagram should give rise to a periodicity of chemical properties as we move across the periodic table. We shall seek such an explanation in terms of the energy necessary to remove an electron from an atom, the ionization energy.

15-3 IONIZATION ENERGY AND THE PERIODIC TABLE

The amount of energy required to remove the most loosely bound electron from a gaseous atom is called the ***ionization energy***. We can represent this process by the equation

gaseous atom + energy → gaseous ion + gaseous electron

In terms of our energy level diagram it is the energy necessary to lift an electron from the highest occupied orbital the rest of the way up to the limit corresponding to $n = \infty$. Figure 15-12 shows the ionization process for lithium atom in terms of an energy level diagram. Each orbital is shown as a pigeon hole, designated ○. If this orbital is occupied by one electron, it can be indicated by a diagonal line across the pigeon hole: ⊘. If two electrons occupy the orbital, crossed diagonal lines are shown, one for each electron: ⊗. Of course we assume the Pauli Principle that *only* two electrons can occupy a given orbital. Hence, an orbital shown as ⊗ is

Fig. 15-12. **The ionization of a lithium atom.**

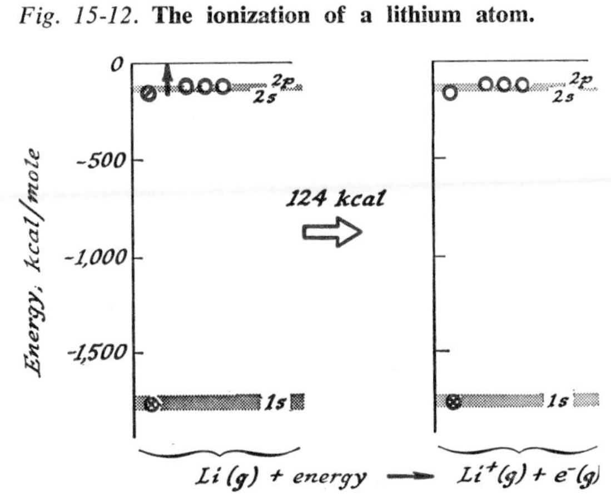

filled; an orbital shown as ⦸ or ⊘ is half-occupied; an orbital shown as ○ is empty.

The term ionization energy is also applied to the removal of the most loosely bound electron from an atom that has already lost one or more electrons (that is, an ion). Hence the "ionization energy" of, say, Mg^+ is the energy of the process

$$Mg^+ (g) + \text{energy} \rightarrow Mg^{+2} (g) + e^-(g)$$

Since the process removes the second electron from a magnesium atom, the ionization energy of Mg^+ is called the "second ionization energy" of magnesium.

Table 15-III

IONIZATION ENERGIES OF THE ELEMENTS

ATOMIC NUMBER	ELEMENT	IONIZATION ENERGY (kcal/mole)
1	H	313.6
2	He	566.7
3	Li	124.3
4	Be	214.9
5	B	191.2
6	C	259.5
7	N	335
8	O	313.8
9	F	401.5
10	Ne	497.0
11	Na	118.4
12	Mg	175.2
13	Al	137.9
14	Si	187.9
15	P	241.7
16	S	238.8
17	Cl	300
18	Ar	363.2
19	K	100.0

15-3.1 Measurement of Ionization Energy

The ionization energy provides a basis for understanding the periodicity of the chemistry of the elements. Owing to the stimulation of Bohr's ideas, many systematic determinations of ionization energies were carried out between 1914 and 1920. The first determinations were done by bombarding an atomic vapor with electrons whose kinetic energy was known accurately. When the kinetic energy of the bombarding electrons is increased to a certain critical value, singly charged positive ions can be detected electrically. These ions result from collisions between atoms and the bombarding electrons that have been given just enough kinetic energy to cause the most weakly bound electron to be ejected from the atom. This critical value was found to be characteristic of the substance being investigated. Table 15-III shows some of the measured ionization energies for some of the lighter elements.*

15-3.2 Trends in Ionization Energies

Examine the figures in the last column of Table 15-III and search for regularities. The most obvious one is the dramatic change in ionization energy between each inert gas and the element following it. This is followed by a slow rise in the ionization energy (another regularity) as we proceed across a row of the periodic table. The best way to see the trend is in a graphical presentation of the data of Table 15-III, as shown in Figure 15-13. We see that the ionization en-

Fig. 15-13. **Ionization energy** *as a function of the atomic number.*

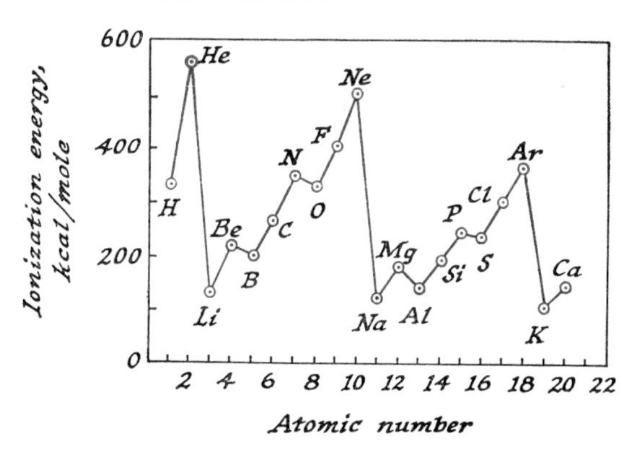

* The ionization energy is given in this book in units of kilocalories per mole, the energy that would be required to remove an electron from each one of a mole of atoms. These units allow an easy comparison between ionization energies and the energy changes that occur in ordinary chemical reactions.

ergy increases more or less regularly across a row of the periodic table, reaching a maximum at the inert gas. As soon as we encounter an alkali metal, we notice that the ionization energy is quite small, and in subsequent elements the general upward trend repeats itself. There is a startling similarity between the regularities we have found in the ionization energies and the periodicity of chemical properties. We shall see that this is not an accident: the trend in chemical behavior as we move across the periodic table can be explained in terms of the trends in the ionization energies.

Let us begin by contrasting the ionization properties of sodium and chlorine:

$$Na(g) \longrightarrow Na^+(g) + e^-(g) \qquad \Delta H = +118.4 \text{ kcal/mole}$$

$$Cl(g) \longrightarrow Cl^+(g) + e^-(g) \qquad \Delta H = +300 \text{ kcal/mole}$$

Since sodium has a low ionization energy, 118.4 kcal/mole, a relatively small amount of energy is required to remove an electron. This is consistent with the chemical evidence that sodium tends to form compounds involving the ion, Na^+. The ease of forming Na^+ ions can be explained in terms of the low ionization energy of the sodium atom. In contrast, it requires a large amount of energy, 300 kcal/mole, to remove an electron from a chlorine atom. It is not surprising, then, that this element shows little tendency to lose electrons in chemical reactions. Instead, chlorine commonly acquires electrons to form negative ions, Cl^-.

Between sodium and chlorine, there is a slow rise in ionization energy. For magnesium and aluminum the ionization energy is still rather low. Hence electrons are readily lost and positive ions can be expected to be important in the chemistry of these elements. As the ionization energy rises, the chemistries of silicon, phosphorus, and sulfur show a trend toward electron sharing. For these elements, an inert gas electron population cannot be reached by losing electrons because the ionization energy required is too high. They seek the inert gas stability by sharing electrons or, for sulfur and chlorine, by acquiring electrons to form negative ions.

These correlations between ionization energy and chemical properties confirm the idea that the electronic structure of an element determines its chemical behavior. In particular, the most weakly bound electrons are of greatest importance in this respect. We shall call *the electrons that are most loosely bound, the* ***valence electrons.***

15-3.3 Ionization Energies and Valence Electrons

It is possible to remove two or more electrons from a many-electron atom. Of course it is always harder to remove the second electron than the first because the second electron to come off leaves an ion with a double positive charge instead of a single positive charge. This gives an additional electrical attraction. Even so, the values of successive ionization energies have great interest to the chemist.

Consider the three elements, sodium, magnesium, and aluminum. For each of these elements we know several ionization energies, corresponding to processes such as the following:

$$Na(g) \longrightarrow Na^+(g) + e^-(g) \qquad E_1 = \text{first ionization energy} \quad (5)$$

$$Na^+(g) \longrightarrow Na^{+2}(g) + e^-(g) \qquad E_2 = \text{second ionization energy} \quad (6)$$

Table 15-IV. **SUCCESSIVE IONIZATION ENERGIES OF Na, Mg, AND Al (KILOCALORIES PER MOLE)**

ELEMENT	ELECTRON CONFIGURATION, NEUTRAL ATOM			E_1	E_2	E_3	E_4
sodium	$1s^2$	$2s^22p^6$	$3s^1$	118	1091	1653	—
magnesium	$1s^2$	$2s^22p^6$	$3s^2$	175	345	1838	2526
aluminum	$1s^2$	$2s^22p^6$	$3s^23p^1$	138	434	656	2767

$$Na^{+2}(g) \longrightarrow Na^{+3}(g) + e^-(g) \qquad E_3 = \text{third ionization energy} \quad (7)$$

$$Na^{+3}(g) \longrightarrow Na^{+4}(g) + e^-(g) \qquad E_4 = \text{fourth ionization energy} \quad (8)$$

The experimental values of these energies are shown in Table 15-IV. Let us begin by comparing sodium and magnesium. For each, the first ionization process removes a 3*s* electron, the most weakly bound. Nevertheless, the ionization energies are somewhat different:

$$Na(g) + 118 \text{ kcal} \longrightarrow Na^+(g) + e^-(g) \quad (9)$$

$$Mg(g) + 175 \text{ kcal} \longrightarrow Mg^+(g) + e^-(g) \quad (10)$$

The difference is caused by the higher nuclear charge of magnesium. Magnesium is element 12, hence it has twelve protons in the nucleus, compared to eleven protons in the nucleus of the sodium atom. Of course the valence electrons are more strongly attracted to the +12 nucleus of Mg than the +11 nucleus of Na.

The second ionization energy, however, reverses the situation:

$$Na^+(g) + 1091 \text{ kcal} \longrightarrow Na^{+2}(g) + e^-(g) \quad (11)$$

$$Mg^+(g) + 345 \text{ kcal} \longrightarrow Mg^{+2}(g) + e^-(g) \quad (12)$$

For sodium, it takes three times as much energy to remove the second electron as it does for magnesium. We can understand the energies in

Fig. 15-14. **Energy level diagram representation** *of the ionization of magnesium and sodium atoms.*

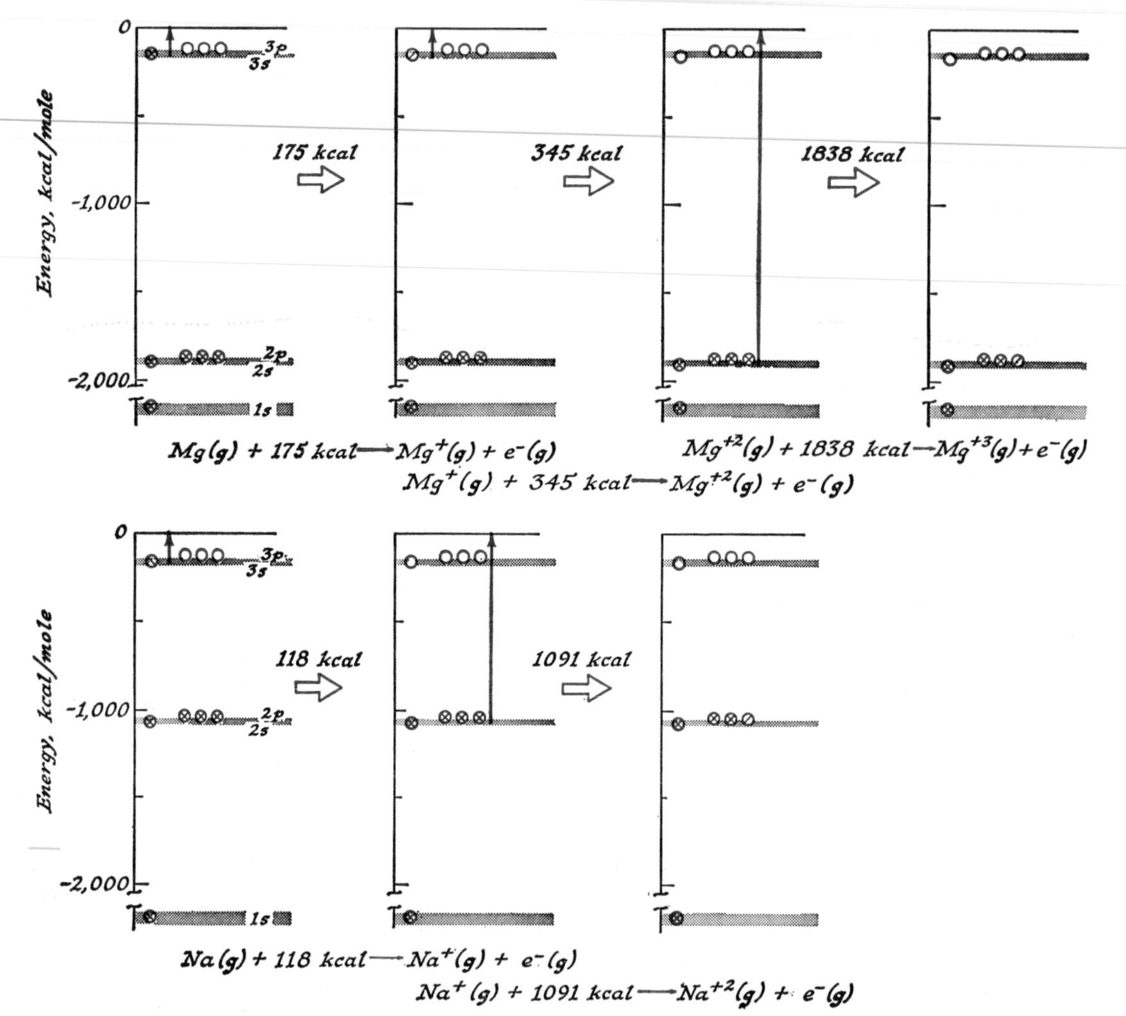

(*11*) and (*12*) in terms of two principles. First, for magnesium the second ionization energy (345 kcal) exceeds the first (175 kcal) because reaction (*12*) takes a 3*s* electron away from the positively charged ion, Mg^+, whereas reaction (*10*) removes a 3*s* electron from an initially neutral atom. This same factor is operative in sodium, of course, but in addition, the second ionization energy of sodium must remove an electron from a 2*p* orbital instead of the 3*s* orbital. This 2*p* orbital is one of a cluster much lower in energy than the 3*s*–3*p* cluster. Figure 15-14 shows that this behavior is readily understood with the aid of the energy level diagram. We conclude that it is difficult to remove two electrons from sodium but it is relatively easy to remove two electrons from magnesium. This is why we say sodium has one valence electron and magnesium has two. Removing more than one electron from sodium or more than two from magnesium is very difficult because orbitals much lower in the energy diagram are involved.

Continuing to aluminum, we see that its first ionization energy is below the first ionization energy of magnesium—despite the fact that aluminum has the higher nuclear charge. We find the explanation, however, in the second column of Table 15-IV. For aluminum we remove a 3*p* electron to produce Al^+ whereas a 3*s* electron is removed from Mg to form Mg^+. A glance at the energy level diagram, Figure 15-11, shows that the 3*p* level is somewhat above the 3*s* level, hence it should be easier to remove the 3*p* electron.

If we continue to remove electrons from aluminum, we discover a very large increase in ionization energy when the fourth electron is removed. Again this is because the fourth electron must be withdrawn from a 2*p* orbital, an orbital much lower on the energy level diagram. We conclude that three electrons, the two 3*s* and the one 3*p*, are more easily removed than the others. Since aluminum has three easily removed electrons, *aluminum is said to have three valence electrons.*

Remembering how we placed electrons in the lowest empty orbitals, two per orbital, we can now generalize concerning the number of valence electrons a given atom possesses. We count the electrons placed in the orbitals that form the highest partially filled cluster of energy levels. These electrons are most easily removed and they determine the chemistry of the atom.

EXERCISE 15-6

Explain why chemists say that boron has three valence electrons and that chlorine has seven. How many valence electrons has fluorine? Oxygen? Nitrogen?

15-3.4 The Fourth Row of the Periodic Table

Turn back to Figure 15-11, the energy level diagram of a many-electron atom, and consider the occupied orbitals of the element potassium. With 19 electrons placed, two at a time, in the orbitals of lowest energy, the electron configuration is

potassium atom $1s^2$ $2s^22p^6$ $3s^23p^6$ $4s^1$

Potassium has one valence electron. It is the first member of the fourth row, the row based on the cluster of orbitals with about the same energy as the 4*s* orbital. There are nine such orbitals, the 4*s* orbital, the three 4*p* orbitals, and the five 3*d* orbitals. Hence the fourth row of the periodic table will differ from the second and third rows. The fourth row, as seen in the periodic table, consists of eighteen elements.

Calcium is the second element of the fourth row. It has two electrons more than the argon inert gas population and these two electrons both occupy the 4*s* orbital:

calcium atom $1s^2$ $2s^22p^6$ $3s^23p^6$ $4s^2$

When we add the next electron to form the element scandium, the orbital of lowest energy that is available is one of the 3*d* orbitals (since the 3*d* orbitals are slightly lower in energy than the 4*p* orbitals). As succeeding electrons are added to form other elements, they enter the 3*d* orbitals until the ten available spaces in these orbitals are filled.

The elements that are formed when the 3d electrons are added are called the ***transition metals*** *or the* ***transition elements.*** Since their chemical properties and electronic structures are unlike those of any of the lighter elements we have discussed, it is reasonable that these transition elements should head a new set of ten columns of the periodic table. The next orbitals in line for occupancy are the 4*p*, and it is not surprising that the chemical properties of the element gallium, which has one 4*p* electron, resemble those of aluminum, which has one 3*p* electron. This row of the periodic table is completed when the 4*p* orbitals are entirely filled. We see that the reason the fourth row of the periodic table contains eighteen elements is that the five 3*d* orbitals have energies that are approximately the same as those of the 4*s* and 4*p* orbitals. The ten extra spaces for electrons provided by the 3*d* orbitals increase the length of the period from eight to eighteen.

Just as the long fourth row of the periodic table arises from filling the 4*s*, 3*d*, and 4*p* orbitals, the fifth row, which also consists of eighteen elements, comes from filling the 5*s*, 4*d*, and 5*p* orbitals. In the sixth row, something new happens. After the 6*s* and the first one of the 5*d* electrons have entered, subsequent electrons go into the 4*f* orbitals. The fact that there are seven 4*f* orbitals means that fourteen electrons can be accommodated in this manner. Filling the 4*f* orbitals gives rise to a series of elements with almost identical chemical properties called the *rare earth elements.* Once the 4*f* orbitals are full, electrons enter the 5*d* and 6*p* orbitals until the sixth period is completed with the inert gas radon.

We see that the rows of the periodic table arise from filling orbitals of approximately the same energy. When all orbitals of similar energy are full (two electrons per orbital), the next electron must be placed in an *s* orbital that has a higher principal quantum number, and a new period of the table starts. We can summarize the relation between the number of elements in each row of the periodic table and the available orbitals of approximately equal energy in Table 15-V.

Table 15-V

THE NUMBER OF ELEMENTS IN EACH ROW OF THE PERIODIC TABLE

ROW OF TABLE	NO. OF ELEMENTS	LOWEST ENERGY ORBITALS AVAILABLE TO BE FILLED
1	2	1*s*
2	8	2*s*, 2*p*
3	8	3*s*, 3*p*
4	18	4*s*, 3*d*, 4*p*
5	18	5*s*, 4*d*, 5*p*
6	32	6*s*, 4*f*, 5*d*, 6*p*
7		7*s*, 5*f*, 6*d*, 7*p*

QUESTIONS AND PROBLEMS

1. Which of the following statements concerning light is FALSE?

   (a) It is a form of energy.
   (b) All photons possess the same amount of energy.
   (c) It cannot be bent by a magnet.
   (d) It includes the part of the spectrum called X-rays.

2. Use the energy level diagram in Figure 15-7 to calculate the energy required to raise the electron in a hydrogen atom from level #1 to level #2; from level #1 to level #3; from level #1 to level #4. Compare these energies with the spectral lines shown in Figure 15-3, p. 255.

3. Your plot in Exercise 15-2 suggested that the energy levels given in Figure 15-7 are systematically related. To explore this relationship further, divide the energy of each level by that of the first level (using the right-hand scale). How are the fractions so obtained related to the numbers of the energy levels?

4. Calculate, using frequency units (cycles per second) and Figure 15-3, the lines predicted by the notched beam due to changes beginning in notch #3. Use the complete light spectrum shown in Figure 14-14 (p. 247) to decide in what spectral region these additional lines were found. (It was this sort of prediction that actually led to the discovery of this set of lines.)

5. According to the quantum mechanical description of the 1*s* orbital of the hydrogen atom, what relation exists between the surface of a sphere centered about the nucleus and the location of an electron?

6. What must be done to a 2*s* electron to make it a 3*s* electron? What happens when a 3*s* electron becomes a 2*s* electron?

7. If the energy difference between two electronic states is 46.12 kcal/mole, what will be the frequency of light emitted when the electron drops from the higher to the lower state?

   Planck's constant
   $$= 9.52 \times 10^{-14} \text{ (kcal sec)/mole}$$

8. Determine the value of E_n for $n = 1, 2, 3, 4$, for a hydrogen atom using the relation $E_n = -313.6/n^2$. For each E_n, indicate how many orbitals have this energy.

9. The quantum mechanical description of the 1*s* orbital is similar in many respects to a description of the holes in a much used dartboard. For example, the "density" of dart holes is constant anywhere on a circle centered about the bullseye, and the "density" of dartholes reaches zero only at a very long distance from the bullseye (effectively, at infinity). What are the corresponding properties of a 1*s* orbital?

   In the light of your answer, point out erroneous features of the following models of a hydrogen atom (both of which were used before quantum mechanics demonstrated their inadequacies).

   (a) A ball of uniform density.
   (b) A "solar system" atom with the electron circling the nucleus at a fixed distance.

10. Name the elements that correspond to each of the following electron configurations

    $1s^2$
    $1s^2 \quad 2s^1$
    $1s^2 \quad 2s^22p^1$
    $1s^2 \quad 2s^22p^3$
    $1s^2 \quad 2s^22p^6 \quad 3s^23p^6 \quad 4s^1$

11. Make a table listing the principal quantum numbers (through three), the types of orbitals, and the number of orbitals of each type.

12. The electron configuration for lithium is $1s^22s^1$ and for beryllium it is $1s^22s^2$. Estimate the approximate ionization energies to remove first one, then a second, electron. Explain your estimates.

13. What trend is observed in the first ionization energy as you move from lithium down the column I metals? On this basis, can you suggest a reason why potassium or cesium might be used in preference to sodium or lithium in photoelectric cells?

14. Consider these two electron populations for neutral atoms:

    A. $1s^2 \quad 2s^22p^6 \quad 3s^1$;
    B. $1s^2 \quad 2s^22p^6 \quad 6s^1$.

    Which of the following is FALSE?

    (a) Energy is required to change *A* to *B*.
    (b) *A* represents a sodium atom.
    (c) *A* and *B* represent different elements.
    (d) Less energy is required to remove one electron from *B* than from *A*.

15. How many valence electrons has carbon? Silicon? Phosphorus? Hydrogen? Write the electron configurations for neutral atoms of each element.

16. The first four ionization energies of boron atoms are as follows:

    $E_1 = 191$ kcal/mole
    $E_2 = 578$
    $E_3 = 872$
    $E_4 = 5962$

    Explain the magnitudes in terms of the electron configurations of boron and deduce the number of valence electrons of boron.

CHAPTER 16

Molecules in the Gas Phase

For the nature of the chemical bond is the problem at the heart of all chemistry.

BRYCE CRAWFORD, JR., 1953

A molecule is a cluster of atoms that persists long enough to have characteristic properties which identify it. The questions we would like to answer are "Why does the cluster of atoms persist?" and "Why does the clustering result in characteristic properties?" In this chapter we will restrict our attention to molecules as they exist in the gas phase. Then, in Chapter 17, we shall consider what additional ideas we need in order to understand the forces which cause the formation of liquids and solids.

16-1 THE COVALENT BOND

When two atoms become fixed together they are said to form a molecule. In "explaining" the properties of a molecule we use models such as two styrofoam spheres glued together, or two wooden spheres held together by a stick, or perhaps by a spring. In each model it is necessary to provide a connection—glue, stick, or spring. It is natural to assume that there is a connection between the atoms in a molecule. This connection is called the chemical bond.

16-1.1 The Hydrogen Molecule

Under normal conditions of temperature and pressure, hydrogen is a gas. By weighing a measured volume of hydrogen gas and applying Avogadro's Hypothesis, we discover that a molecule of hydrogen contains two hydrogen atoms. Only if the temperature is raised to several thousand degrees are the collisions with other molecules sufficiently energetic to knock a hydrogen molecule apart:

$$H_2(g) \rightleftharpoons H(g) + H(g) \qquad \Delta H = 103.4 \text{ kcal} \quad (1)$$

Since energy is absorbed in reaction (*1*), the molecule H_2 is more stable (has a lower energy) than two separated atoms. *This chemical bond (and every chemical bond) forms because the energy is lower when the atoms are near each other.*

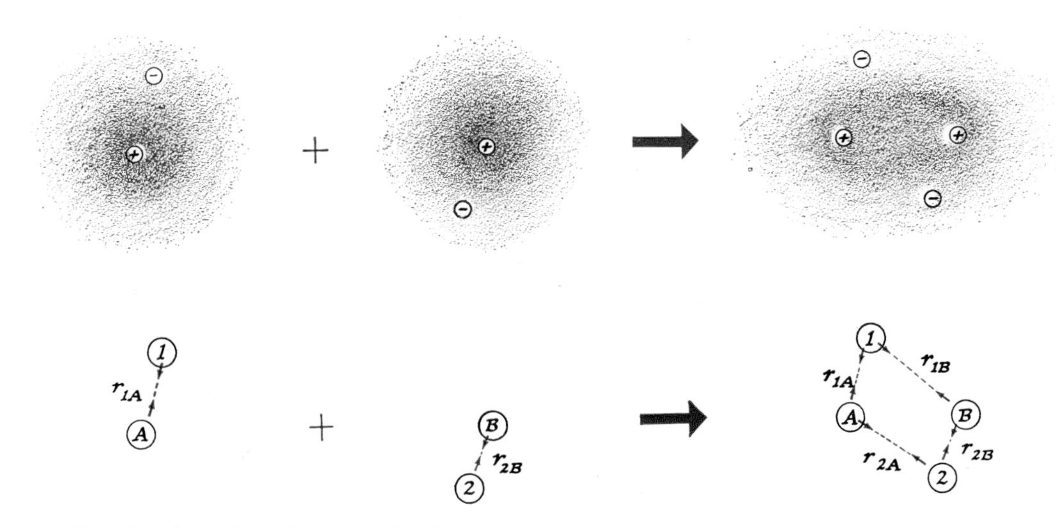

Fig. 16-1. **The formation of a molecule of hydrogen, H_2.**

THE ORIGIN OF THE STABILITY OF THE CHEMICAL BOND

To see why the energy is lower when the atoms are near each other, we must examine interactions among the electric charges of the atoms. Figure 16-1 shows the reverse of reaction (*1*) in a schematic way. Quantum mechanics tells us that the 1*s* orbital of each hydrogen atom has spherical symmetry before reaction. This is suggested by the shading in Figure 16-1. Yet, at any instant, we picture the electron at some particular point, as shown by the negative charge of electron 1 located a distance r_{1A} from nucleus *A*. The energy of hydrogen atom *A* can be explained in terms of the average attraction between electron 1 and nucleus *A*. This is fixed by the average of the distance between the two, r_{1A}. The same is true of hydrogen atom *B*—electron 2 and nucleus *B* attract each other. Now consider the new electrical interactions present after the two atoms have moved close together. Now electron 1 feels the attraction of *both* protons. Electron 2, as well, feels the attraction of *both* protons. This is the "glue" that holds the two atoms together. *The chemical bond in H_2 forms because each of the two electrons is attracted to two protons simultaneously*. This arrangement is energetically more stable than the separated atoms in which

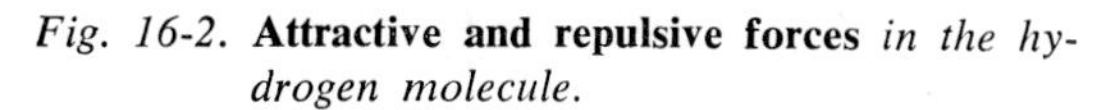

Fig. 16-2. **Attractive and repulsive forces** *in the hydrogen molecule.*

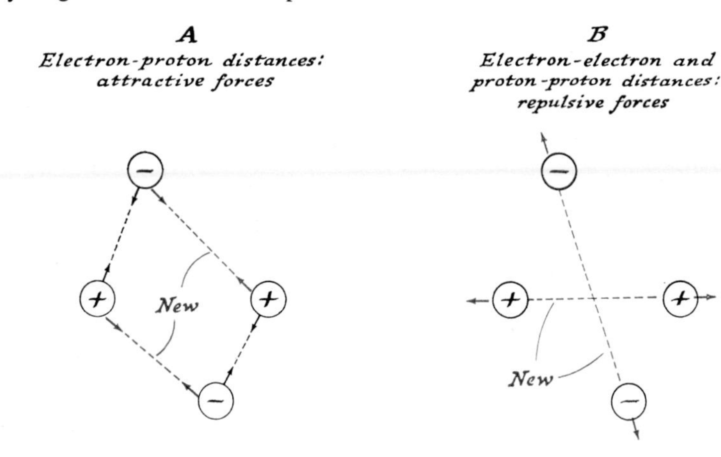

each electron is attracted to only one proton.

Figure 16-2A shows a possible set of the electron–proton distances as they might be seen if it were possible to make an instantaneous photograph. Such distances fix the attractions that cause the chemical bond. But it is well to remember there are also repulsions caused by the approach of the two atoms, as shown in Figure 16-2B. The two electrons repel each other and the two protons do the same. These repulsions tend to push the two atoms apart. Which are more important, the two new attraction terms or the two new repulsion terms illustrated in Figure 16-2? Experiment shows that the new attraction terms dominate—a stable chemical bond is formed. That is not to say the repulsions are not felt. In fact, the proton–proton repulsions prevent the two hydrogen atoms from approaching even closer. The stable bond length in the hydrogen molecule is fixed by a balance between

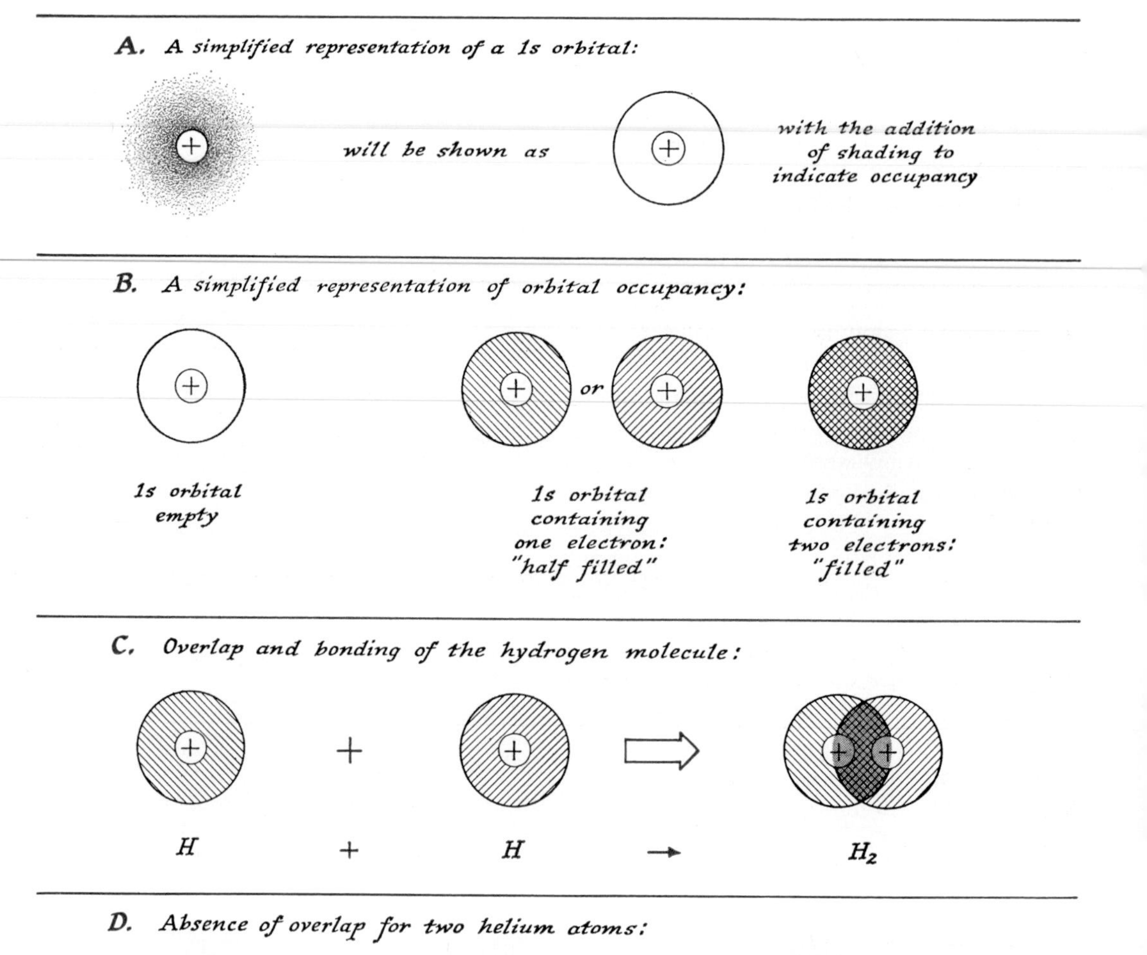

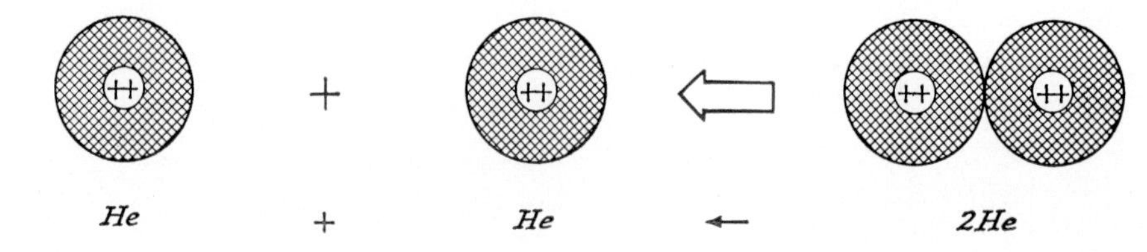

Fig. 16-3. **Schematic representation** *of the interaction between two atoms.*

A

Electron-proton distances: attractive forces

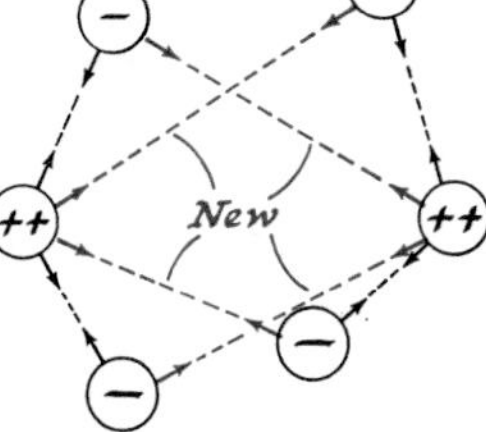

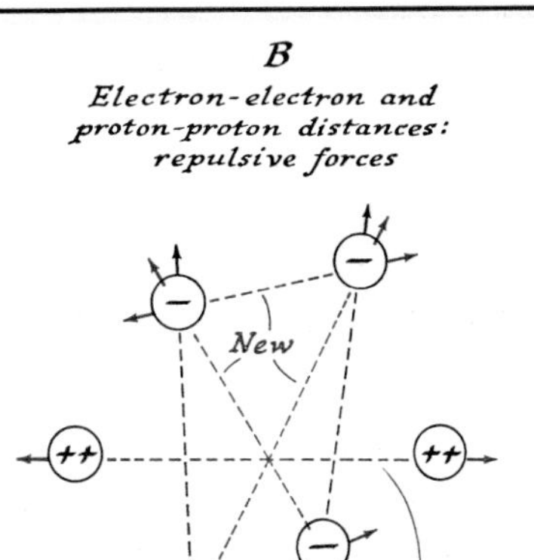

Fig. 16-4. **Attractive and repulsive** *forces when helium atoms approach.*

the forces of attraction (Figure 16-2A) and the forces of repulsion (Figure 16-2B).

We see in Figure 16-2 that bringing two hydrogen atoms together produces two new repulsions and two new attractions. Experiment shows that a chemical bond is formed; the energy of attraction predominates over the energy of repulsion. Why is this so? An explanation is found in the mobility of the electrons. The electrons do not occupy fixed positions but move about through the molecule. They take advantage of this mobility to remain as far from each other as possible. They preferentially occupy positions like those shown in Figure 16-1 in which each electron is closer to both nuclei than to the other electron. The two electrons preferentially move away from positions in which they would be near each other. Thus they are said to "correlate" their motion so as to remain apart, reducing the electron–electron repulsion.

OVERLAP AND THE CHEMICAL BOND

We can simplify our discussion of the chemical bonding in the hydrogen molecule with the aid of Figure 16-3. First, in Figure 16-3A, we picture the electron distribution in cross-section. The electron distribution extends outward far from the nucleus and uniformly in all directions. But the distribution is concentrated near the nucleus so we ought to focus our attention on the center region of the 1*s* orbital. We do so by representing the 1*s* orbital by a circle with radius large enough to contain most of the electron distribution.

An orbital can accommodate either one or two electrons but no more. Figure 16-3B shows a way of differentiating an empty 1*s* orbital, a 1*s* orbital containing one electron, and a 1*s* orbital containing two electrons.

Now, in Figure 16-3C, consider the interaction of two hydrogen atoms. Each atom has a single electron in a 1*s* orbital. As the two hydrogen atoms approach each other, the circles overlap each other. In this region of overlap the two electrons are shared by the two protons (as shown by the crosshatched and shaded area). This sharing, which permits the two electrons to be near both protons a good part of the time, causes the chemical bond. *When a bond arises from equal sharing, it is called a* ***covalent*** *bond.**

16-1.2 Interaction Between Helium Atoms

A measurement of the density of helium gas shows that it is a monatomic gas. Molecules of He_2 do *not* form. What difference between hydrogen atoms and helium atoms accounts for the absence of bonding for helium? The answer to this question also must lie in the attractive and repulsive electrical interactions between two helium atoms when they approach each other. Figure 16-4A shows the attractive forces in one of our hypothetical instantaneous snapshots. There are, of course, four electrons and each is attracted to each nucleus. In Figure 16-4B we see the repulsive forces. Taking score, we find in Figure 16-4A eight attractive interactions, four

* The prefix "co" in "covalent" conveys the notion of "sharing" as it does in the words "coworker," "co-author," etc. The stem of the word, "-valent" refers to "combination."

more than in the separated atoms. In addition, we count seven repulsive interactions, five more than in the separated atoms. Again we appeal to experiment and we learn that the four new attractive terms are *not* sufficient to outweigh the five new repulsions. A chemical bond does not form.

Thus we find that an explanation of the bonding in H_2 and the absence of bonding for He_2 lies in the relative magnitudes of attractive and repulsive terms. Quantum mechanics can be put to work with the aid of advanced and difficult mathematics to calculate these quantities, to tell us which is more important. Unfortunately, solving the mathematics presents such an obstacle that only a handful of the very simplest molecules have been treated with high accuracy. Nevertheless, for some time now chemists have been able to decide whether chemical bonds can form without appealing to a digital computer.

Figure 16-3D shows the simplified representation of the interaction of two helium atoms. This time each helium atom is crosshatched *before* the two atoms approach. This is to indicate there are already *two* electrons in the 1*s* orbital. Our rule of orbital occupancy tells us that the 1*s* orbital can contain *only* two electrons. Consequently, when the second helium atom approaches, its valence orbitals cannot overlap significantly. The helium atom valence electrons *fill* its valence orbitals, preventing it from approaching a second atom close enough to share electrons. The helium atom forms no chemical bonds.*

16-1.3 Representations of Chemical Bonding

We propose, then, that chemical bonds can form if valence electrons can be shared by two atoms using partially filled orbitals. We need a shorthand notation which aids in the use of this rule. Such a shorthand notation is called a representation of the bonding.

* Each helium atom does have, of course, vacant 2*s* and 2*p* orbitals which extend farther out than the filled 1*s* orbital. The electrons of the second helium atom can "overlap" with these vacant orbitals. Since this overlap is at great distance, the resulting attractions are extremely small. This type of interaction presumably accounts for the attractions that cause helium to condense at very low temperatures.

ORBITAL REPRESENTATION OF CHEMICAL BONDING

Our rule about covalent bond formation can be applied quite simply through an orbital representation:

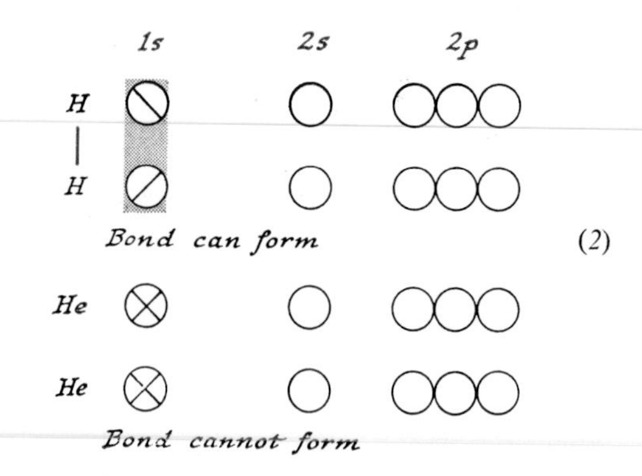

(2)

In this representation there is no need to consider the next higher energy level cluster—the 2*s*, 2*p* orbitals. For hydrogen and helium these are much higher in energy and can give rise only to extremely weak attractions.

ELECTRON DOT REPRESENTATION OF CHEMICAL BONDING

The sharing of electrons can be shown by representing valence electrons as dots placed between the atoms:

$$H\cdot + H\cdot \longrightarrow H:H \qquad (3)$$

We shall use both orbital and electron dot representations to show chemical bonding.

16-1.4 The Bonding of Fluorine

Our explanation of chemical bonding is of value only if it has wide applicability. Let us examine its usefulness in considering the compounds of the second-row elements, beginning with fluorine.

Under normal conditions of temperature and pressure, fluorine is a gas. From gas density experiments we discover that a molecule of fluorine contains two atoms. There is a chemical bond between the two fluorine atoms. Let us see if our expectations agree with these experimental facts.

ORBITAL REPRESENTATION OF THE BONDING OF FLUORINE

A fluorine atom has the orbital occupancy shown below:

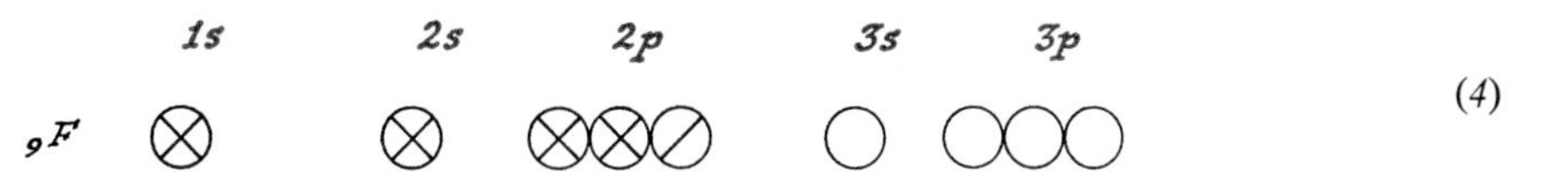

(4)

We see that the neutral fluorine atom has seven valence electrons; that is, seven electrons occupy the outermost partially filled cluster of energy levels. This cluster of energy levels, the valence orbitals, contains one electron less than its capacity permits. Fluorine, then, has the capacity for sharing one electron with some other atom which has similar capacity. If, for example, another fluorine atom approaches, they might share a pair of electrons and form a covalent bond:

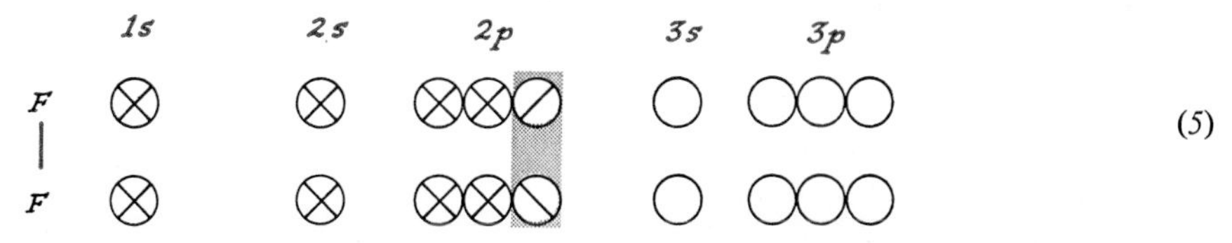

(5)

After the second atom approaches, sharing its electron, each fluorine atom now has "filled" all of its valence orbitals. No additional bonding capacity remains. Hence F_2 does not add a third or fourth atom to form F_3, F_4, etc.

Now consider the possibility of the bonding that might occur if a fluorine atom encounters a hydrogen atom. Again fluorine has an opportunity to share electrons with a second atom having a partially filled valence orbital:

(6)

Fluorine now has, by sharing a pair of electrons with hydrogen, "filled" its valence orbitals. It has no further bonding capacity. The same is true of the hydrogen atom, though for this atom there is but one valence orbital, the 1*s* orbital. Since no partially filled valence orbitals remain for either atom, there is no more bonding capacity and a stable compound is formed, HF.

In each of these cases, F_2 and HF, we find that the fluorine atom forms one bond—in F_2 it is to a second fluorine atom, in HF it is to a hydrogen atom. We describe this single bonding capacity by saying, *fluorine is univalent.*

ELECTRON DOT REPRESENTATION OF THE BONDING OF FLUORINE

In the electron dot method of showing chemical bonds it is necessary to show only the valence electrons. In fluorine there are seven—the pair of electrons in the 1*s* orbital is so tightly bound that it plays little role in the chemistry of fluorine. In this representation, then, we show the reaction between two fluorine atoms as follows:

$$:\underset{..}{\ddot{F}}\cdot + \cdot\underset{..}{\ddot{F}}: \longrightarrow :\underset{..}{\ddot{F}}:\underset{..}{\ddot{F}}: \qquad (7)$$

From (7) we conclude that a covalent bond can form between two fluorine atoms. Furthermore, a census of the number of electrons owned or shared by either of the fluorine atoms shows that the valence orbitals are filled. For example, the fluorine atom on the left feels the electrical attractions of eight electrons near at hand,* as

* Remember, we are omitting from the discussion the tightly bound 1*s* electrons of each fluorine atom.

shown in Figure 16-5. Since eight electrons is just the capacity of the $2s$, $2p$ valence orbitals, each fluorine atom has reached the energetically stable arrangement of an inert gas.

The fluorine atom on the left feels eight electrons *The fluorine atom on the right feels eight electrons*

Fig. 16-5. **The electrons** *near each fluorine atom in* F_2.

Hydrogen fluoride can be represented by the electron dot picture

$$:\ddot{\underset{..}{F}}\cdot + \cdot H \longrightarrow :\ddot{\underset{..}{F}}:H \qquad (8)$$

Again, a census of the number of electrons near each of the atoms shows that this is a stable arrangement. True, the hydrogen atom has, close at hand, only two electrons whereas fluorine has eight. This is energetically desirable, however, because hydrogen has only one valence orbital, the $1s$ orbital. Two electrons just fill this orbital.

The fluorine atom feels eight electrons *The hydrogen atom feels two electrons*

Fig. 16-6. **The electrons** *near each atom in HF.*

We see that the bonding of a fluorine atom to another fluorine atom or to a hydrogen atom can be explained in terms of sharing electrons so as to fill the partially filled valence orbitals. This sharing makes the molecule F_2 (or HF) energetically more stable than the separated atoms would be. *The energy stability results from the shared electrons being attracted simultaneously to both positive nuclei.* A chemical bond results.

ELECTRON AFFINITY OF THE FLUORINE ATOM

Experiment shows that a gaseous fluorine atom can acquire an electron to form a stable ion, $F^-(g)$. We can discuss the energy of formation of this ion in the same way that we treated ionization energies. The first ionization energy of fluorine atom is the energy required to remove an electron from a neutral atom in the gas phase. We shall call this energy E_1. Then the heat of reaction can be written in terms of E_1:

$$F(g) \longrightarrow F^+(g) + e^-(g) \qquad \Delta H = E_1 \qquad (9)$$

The second ionization energy, E_2, refers to reaction (*10*):

$$F^+(g) \longrightarrow F^{+2}(g) + e^-(g) \qquad \Delta H = E_2 \qquad (10)$$

Now we can add a new process with an energy which might logically be called E_0:

$$F^-(g) \longrightarrow F(g) + e^-(g) \qquad \Delta H = E_0 \qquad (11)$$

Comparing equations (*9*), (*10*), and (*11*), we see that E_0 is just the ionization energy of $F^-(g)$. By usual practice, however, the reverse of reaction (*11*) is usually considered. Of course the heat of reaction (*12*) is just the negative of that of reaction (*11*):

$$F(g) + e^-(g) \longrightarrow F^-(g) \qquad \Delta H = -E_0 \qquad (12)$$

The energy change of reaction (*12*) is called the *electron affinity* of the fluorine atom. It is symbolized by E and, as defined here, is a negative quantity if heat is released when the ion is formed:

$$E = -E_0 \qquad (13)$$

Electron affinities are difficult to measure and are known reliably for only a small fraction of the hundred or so elements. The electron affinity of fluorine is one that is known:

$$E = -83 \text{ kcal/mole} \qquad (14)$$

or

$$E_0 = +83 \text{ kcal/mole} \qquad (15)$$

The experimental quantities shown in (*14*) and (*15*) indicate that the F^- ion is more stable than a fluorine atom and an electron. Energetically, a fluorine atom "wants" another electron. It is profitable to express reaction (*12*) in terms of orbital occupancy:

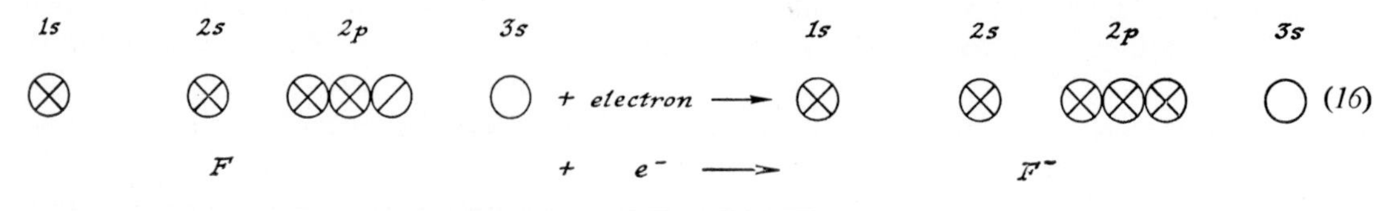

The neutral fluorine atom has seven valence electrons; that is, seven electrons occupy the highest partially filled cluster of energy levels. This cluster of energy levels thus contains one fewer electron than its capacity permits. The electron affinity of fluorine shows that the addition of this last electron is energetically favored. This is in accord with much other experience which shows that there is a special stability to the inert gas electron population.

In view of the electron affinity of a fluorine atom, we can speculate on what would be the result of a collision between two fluorine atoms. Will a reaction occur? The energy is one of the factors which determines the answer. First let us consider a reaction that does *not* occur spontaneously.

$$F(g) + F(g) \rightleftarrows F^+(g) + F^-(g) \qquad (17)$$

Reaction (*17*) can be rewritten in two steps:

	$F(g) \rightleftarrows F^+(g) + e^-(g)$	$\Delta H = E_1 = 401.5$ kcal/mole	(*18*)
	$F(g) + e^-(g) \rightleftarrows F^-(g)$	$\Delta H = -E_0 = -83$	(*19*)
Net	$F(g) + F(g) \rightleftarrows F^+(g) + F^-(g)$	$\Delta H = 318.5$	(*17*)

We see that reaction (*17*) is energetically unfavorable. The stability of F^- is more than outweighed by the difficulty of removing an electron from another fluorine atom.

There is another possible consequence of a collision between two fluorine atoms. The two atoms can remain together to form a molecule. Each atom has a valence electron in a "half-filled" orbital. We can imagine these two atoms orienting so that these "half-filled" orbitals overlap in space. Then the "half-filled" valence orbital of fluorine atom number 1 shares one valence electron of fluorine atom number 2. Thus a part of the electron affinity of fluorine atom number 1 is "satisfied" though it was not necessary to take the electron away from atom number 2. Meanwhile, fluorine atom number 2 is getting the same sort of energy benefit from the valence electron of atom number 1. Each fluorine atom has acquired another electron at least a part of the time. We have gained a part of the stability of reaction (*19*) without paying for reaction (*18*). The most energy that could be released by such an electron sharing would be double the electron affinity of fluorine, $2 \times 83 = 166$ kcal. This takes no account of the work done in bringing the two positive nuclei near each other. Nor can we expect to gain the whole electron affinity under conditions of electron sharing. Yet the energy released when two fluorine atoms form a bond is 36.6 kcal/mole, a reasonable fraction of the maximum possible.

Now we can say why the chemical bond forms between two fluorine atoms. First, the electron affinity of a fluorine atom makes it energetically favorable to acquire one more electron. Two fluorine atoms can realize a part of this energy stability by sharing electrons. *All chemical bonds form because one or more electrons are placed so as to feel electrostatic attraction to two or more positive nuclei simultaneously.*

16-2 BONDING CAPACITY OF THE SECOND-ROW ELEMENTS

In Chapter 6 we saw that the chemical compounds of the third-row elements display a remarkable regularity. Return to Chapter 6 and reread Section 6-6.2. The same simple trend in chemical formulas is discovered in the second row of the periodic table. Now we have a basis for explaining why these trends are found.

16-2.1 The Bonding Capacity of Oxygen Atoms

The neutral oxygen atom has eight electrons. Six of these occupy the 2*s*, 2*p* orbitals and are much more easily removed than the two in the 1*s* orbital. Therefore oxygen has six valence electrons. The 2*s*, 2*p* orbitals are the valence orbitals. They can accommodate the valence electrons in two ways, as follows:

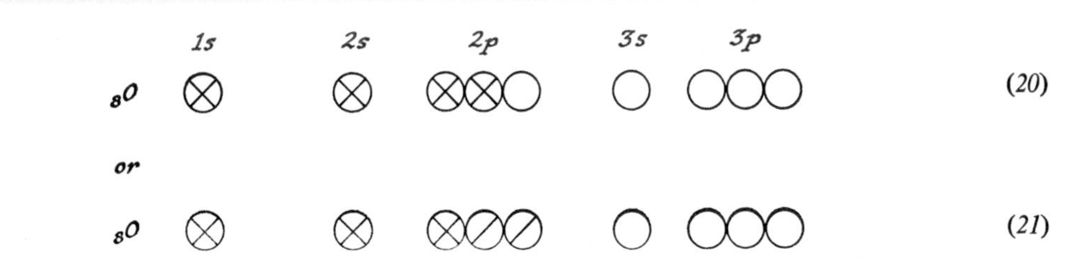

Remember the spatial arrangement of the p orbitals? Each one protrudes along one of the three cartesian axes (as shown in Figure 15-9). If the electrons have the orbital occupancy of (*20*), then two electrons occupy the p orbital protruding along the x axis (p_x) and two electrons occupy the p orbital protruding along the y axis (p_y). In the occupancy of (*21*), two electrons occupy p_x, only one electron occupies p_y, and the last is in p_z. Thus (*21*) differs from (*20*) by the movement of one of the electrons from p_y to a different region of space, p_z. Since electrons repel each other, we can expect that the configuration which keeps the electrons farther apart, (*21*), is the lower in energy. Experiment shows that it is and we shall base our discussion of the bonding of oxygen on orbital occupancy (*21*). However, the occupancy represented by (*20*) also contributes to the chemistry of oxygen atoms.

Suppose a hydrogen atom approaches an oxygen atom in its most stable state, (*21*). Each atom has partially filled valence orbitals. Electron sharing can occur, placing electrons close to two nuclei simultaneously. Hence a stable bond can occur. This is shown in representations (*22*) and (*23*).

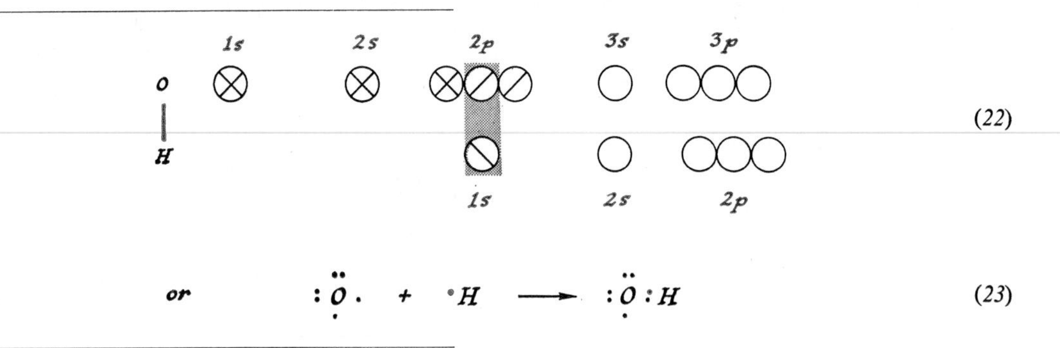

In either representation, (*22*) or (*23*), we see that there is residual bonding capacity remaining in the species OH. In (*22*) the third $2p$ orbital has a single electron but a capacity for two. This means more bonding can occur. In (*23*) a census of the electrons near the oxygen atom indicates there are only seven. The oxygen atom would be more stable if it could add one more electron. With either representation, we conclude that OH should be able to react with another hydrogen atom. See representations (*24*), (*25*).

Now we have the compound H_2O. By either representation, the bonding capacity of oxygen is expended when two bonds are formed. Oxygen is said to be *divalent*, and the compound H_2O is extremely stable. Each of the atoms in H_2O has filled its valence orbitals by electron sharing.

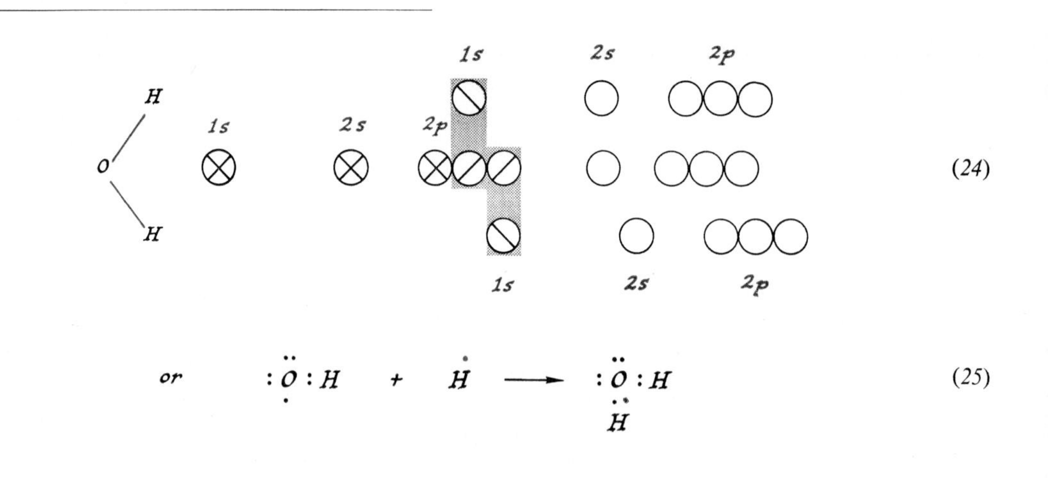

REACTION BETWEEN TWO OH MOLECULES

Though OH is reactive, it is a cluster of atoms with sufficient stability to be identified as a molecule. It is present in a number of high temperature flames, for example. Its chemistry might be expected to be like that of fluorine atoms. Compare the electron dot formulas

$$:\ddot{\underset{\cdot}{F}}: \qquad :\ddot{\underset{\cdot}{O}}:H$$

Since two fluorine atoms react, forming a covalent bond, we can expect two OH molecules to do the same sort of thing

$$:\underset{H}{\ddot{O}}\cdot + \cdot\overset{H}{\ddot{O}}: \longrightarrow :\underset{H}{\ddot{O}}:\overset{H}{\ddot{O}}: \qquad (26)$$

Reaction (*26*) yields the compound H_2O_2. This is the formula of the well-known substance, hydrogen peroxide. By these considerations of chemical bonding, we see that the structure of H_2O_2 must involve an oxygen–oxygen bond:

$$\underset{\displaystyle H}{\overset{|}{O}}\!-\!\overset{\displaystyle H}{\underset{|}{O}} \qquad (27)$$

EXERCISE 16-1

Predict the structure of the compound S_2Cl_2 from the electron dot representation of the atoms. After you have predicted it, turn back to Figure 6-12, p. 103, and check your expectation.

OXYGEN-FLUORINE COMPOUNDS

It is a simple matter to predict that oxygen will form a stable compound with two fluorine atoms, F_2O. The orbital representation is*

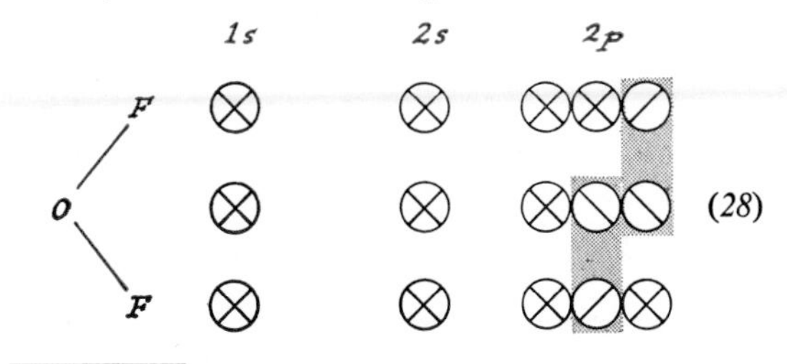

* Henceforth we will omit empty orbitals much higher in energy than the valence orbitals.

The electron dot representation is

$$\begin{matrix} :\ddot{\underset{\cdot\cdot}{O}}:\ddot{\underset{\cdot\cdot}{F}}: \\ :\underset{\cdot\cdot}{F}: \end{matrix} \qquad (29)$$

Again we find that oxygen is divalent.

EXERCISE 16-2

Draw orbital and electron dot representations of each of the following molecules: OF, F_2O_2, HOF, HFO_2. Which of these is apt to be the most reactive?

16-2.2 The Bonding Capacity of Nitrogen Atoms

For the same reason we discussed for oxygen atoms, the nitrogen atom is most stable when it has the maximum number of partially filled valence orbitals. This keeps the electrons as far apart as possible. The most stable state of the nitrogen atom is as follows:

It is now straightforward to predict that nitrogen will form a stable hydrogen compound with formula NH_3. *Nitrogen is trivalent.* A similar compound, NF_3, will be formed with fluorine. The electron dot formulas are

$$\underset{\text{ammonia}}{\begin{matrix} H \\ :\underset{\cdot\cdot}{N}:H \\ H \end{matrix}} \quad \text{and} \quad \underset{\text{nitrogen trifluoride}}{\begin{matrix} :\ddot{F}: \\ :\underset{\cdot\cdot}{\ddot{N}}:\ddot{\underset{\cdot\cdot}{F}}: \\ :\underset{\cdot\cdot}{F}: \end{matrix}} \qquad \begin{matrix} (31) \\ (32) \end{matrix}$$

EXERCISE 16-3

The molecule NH_2 has residual, unused bonding capacity and is extremely reactive. The molecule N_2H_4 (hydrazine) is much more stable. Draw an electron dot representation of the bonding of hydrazine. Draw its structural formula (show which atoms are bonded to each other).

16-2.3 The Bonding Capacity of Carbon Atoms

There are a number of orbital occupancies that we might consider for the carbon atom:

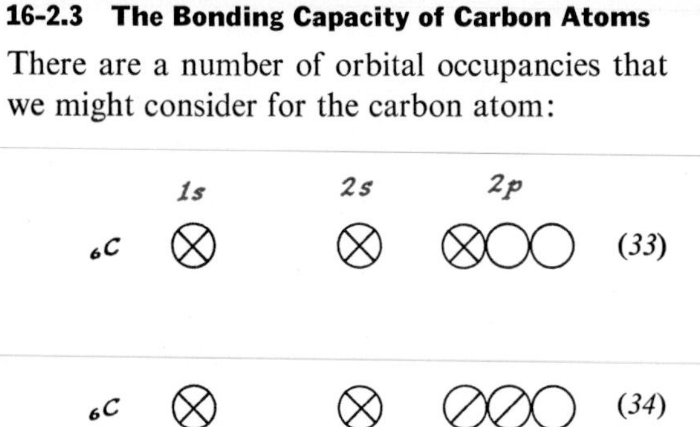

First let us compare (*33*) and (*34*). By our conventional argument that electrons repel each other, configuration (*34*) should be more stable than (*33*). The second, (*34*), places one electron in each of the p_x and p_y orbitals whereas (*33*) places two electrons in the same orbital, p_x. It is an experimental fact that (*34*) is more stable than (*33*).

Now we can predict the chemistry of the carbon atom in this state. It should be divalent, forming compounds CH_2 and CF_2. Let us consider one of these, say CH_2.

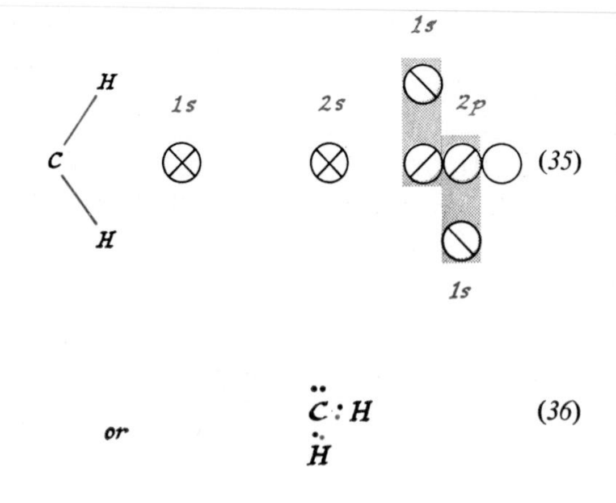

Here is a situation we haven't met before. After using the two available partially filled orbitals to form covalent bonds with hydrogen atoms, there remains a vacant valence orbital. In the electron dot formulation (*36*) we see that the carbon atom finds itself near only six electrons in CH_2. The valence orbitals will accommodate eight electrons. *Because one valence orbital is completely vacant, we can expect CH_2 to be reactive.*

In fact, both CH_2 and CF_2 are considered to be stable* but extremely reactive molecules. Though there is reaction mechanism evidence verifying the existence of each species, it is not possible to prepare either substance pure. This great reactivity shows that energy considerations favor the use of all four of the valence orbitals if possible. This argument leads us to consider a third orbital occupancy:

1s 2s 2p

$_6C$ (37)

With orbital occupancy (*37*), a carbon atom has four half-filled valence orbitals. True, (*37*) is somewhat less stable than (*34*) because an electron was raised from the 2*s* energy level to the slightly higher 2*p* energy level. This process is called "promoting" the electron. On the other hand, the promotional energy is not very large, and in return for it the carbon atom acquires the capacity to form four covalent bonds. Each covalent bond increases stability, more than compensating for the energy investment in promoting one of the 2*s* electrons. With orbital occupancy (*37*), carbon can share pairs of electrons with, for example, four hydrogen atoms or four fluorine atoms. Hence, *carbon is tetravalent:*

methane (38)

:F:
:F:C:F:
:F:

carbon tetrafluoride (39)

---

* The molecule CH_2 is stable in the sense that it does not spontaneously break into smaller fragments. It is reactive because other molecules formed from this group of atoms have much lower energy.

EXERCISE 16-4

Draw electron dot formulas for the molecules CH_3, CF_3, CHF_3, CH_2F_2, CH_3F. Which will be extremely reactive?

EXERCISE 16-5

Draw an electron dot and a structural formula for the molecule C_2H_6 (ethane) which forms if two CH_3 molecules are brought together. Explain why C_2H_6 is much less reactive than CH_3.

16-2.4 The Bonding Capacity of Boron Atoms

The boron atom presents the same sort of option in orbital occupancy as does carbon:

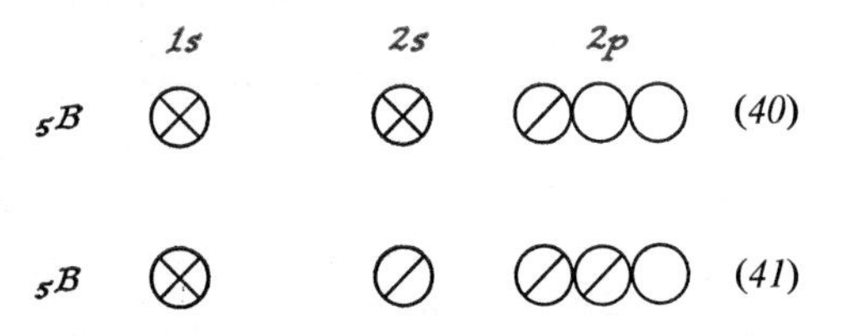

The electron configuration (*41*) is somewhat higher in energy than (*40*). It is necessary to promote a 2*s* electron to the 2*p* state to obtain (*41*). In return, however, the boron atom gains bonding capacity. Whereas a boron atom can form only one covalent bond in configuration (*40*), it can form three in configuration (*41*). Since each bond lowers the energy, the chemistry of boron is fixed by the electron configuration (*41*).

Now we can expect that boron will be trivalent. We predict that there should be molecules such as BH_3 and BF_3

```
  H
  ..
  B : H        (42)
  ..
  H

  ..
: F :
  ..  ..
  B : F :      (43)
  ..  ..
: F :
  ..
```

We cannot help noticing, however, that there remains a completely vacant valence orbital. For example, for BH_3 the orbital representation would be as shown in (*44*).

The last vacant 2*p* orbital is reminiscent of the configuration found for CH_2 [see (*35*)]. Since CH_2 is very reactive, presumably BH_3 will be the

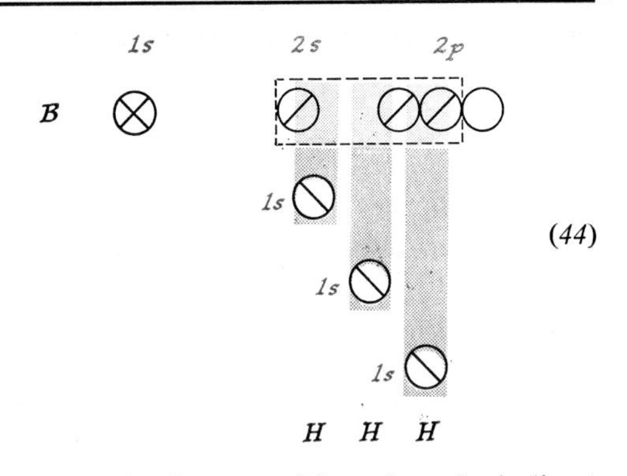

same. Such is the case. There is only indirect evidence establishing the existence of BH_3. Instead boron forms a series of unusual compounds with hydrogen, the simplest of which is called diborane, B_2H_6.

You might wish to predict the structure of diborane (which is now known) but do not be discouraged if you are not able to. Its structure, once elucidated, came as quite a surprise to even the most sophisticated chemists. The explanation of the structure is, even today, composed of a large proportion of words and a small proportion of understanding.

Boron is an obliging element. On the one hand it conforms to our expectation that the unused valence orbital will affect the bonding capacity of boron, as shown by the reactivity of BH_3. On the other hand boron conforms to our expectation that electron configuration (*41*) will make boron trivalent. The compound BF_3 is a stable, gaseous compound and, in contrast to BH_3, is readily prepared in a pure form. The explanation of how the fluorine atoms are able to satisfy in part the bonding capacity of the vacant 2*p* orbital (though hydrogen atoms cannot) must wait until a later chemistry course. For our interest here, BF_3 is the most stable boron-fluorine compound and it demonstrates the trivalent bonding capacity of boron.

16-2.5 The Bonding Capacity of Beryllium Atoms

The beryllium atom, like boron and carbon, can promote an electron in order to form more chemical bonds:

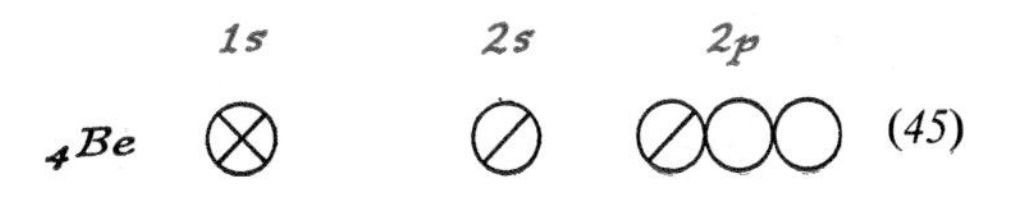

Therefore we should expect in the gaseous state to find molecules such as BeH_2 and BeF_2. These molecules have been detected. On the other hand, beryllium has the trouble boron has, only in a double dose. It has *two* vacant valence orbitals. As a result, BeH_2 and BeF_2 molecules, as such, are obtained only at extremely high temperatures (say, above 1000°K). At lower temperatures these vacant valence orbitals cause a condensation to a solid in which these orbitals can participate in bonding. We shall discuss these solids in the next chapter.

16-2.6 The Bonding Capacity of Lithium Atoms

There is little new to be said about the bonding capacity of a lithium atom. With just one valence electron, it should form gaseous molecules LiH and LiF. Because of the vacant valence orbitals, these substances will be expected only at extremely high temperatures. These expectations are in accord with the facts, as shown in Table 16-I, which summarizes the formulas and the melting and boiling points of the stable fluorides of the second-row elements. In each case, the formula given in the table is the actual molecular formula of the species found in the gas phase.

16-2.7 Valence

Very often the word valence is used in discussing the nature of chemical bonding. Unfortunately this word has been used as a noun to mean a number of different things. Sometimes valence has been used to mean the charge on an ion, sometimes it has meant the total number of atoms to which a particular atom will bond, and at other times, the word valence has been used to mean oxidation number. Perhaps the most widely accepted definition of the word is that it is the number of hydrogen atoms with which an atom can combine, or release, in a chemical reaction. It is clear that a word with so many meanings might confuse a discussion of chemical bonding. For this reason we have avoided and will continue to avoid using the word as a noun in this book.

We have, however, made a careful definition of the term "valence electrons" ("the electrons that are most loosely bound"; see p. 269). We have also used carefully the term "valence orbitals" to mean the entire cluster of orbitals of about the same energy as those which are occupied by the valence electrons. In both of these uses, the word valence is used as an adjective.

Table 16-I. THE FLUORIDES OF THE ELEMENTS IN THE SECOND ROW OF THE PERIODIC TABLE

	Li	Be	B	C	N	O	F
Formula	LiF	BeF_2	BF_3	CF_4	NF_3	F_2O	F_2
Melting point (°K)	1143	1073	146	89	56	49	50
Boiling point (°K)	1949	—	172	145	153	128	85

16-3 TREND IN BOND TYPE AMONG THE FIRST-ROW FLUORIDES

We have termed the chemical bond in the hydrogen molecule, H_2, a covalent bond. This indicates that electrons are shared so that they are simultaneously and, on the average, equally near two nuclei. This makes the system more stable and a chemical bond results.

All chemical bonds occur because electrons can be placed simultaneously near two nuclei. Yet it is often true that the electron-sharing which permits this is not exactly equal sharing. Sometimes the electrons, though close to both nuclei, tend to distribute nearer to one nucleus than to the other. We can see why by contrasting the chemical bonding in gaseous fluorine, F_2, and in gaseous lithium fluoride, LiF.

16-3.1 The Bonding in Gaseous Lithium Fluoride

We have already treated the bonding in an F_2 molecule. Since neither fluorine atom can pull an electron entirely away from the other, they compromise by sharing a pair of electrons equally. How does the chemical bonding in the lithium fluoride molecule compare?

As we have mentioned earlier, lithium has one valence electron, hence can share a pair of electrons with one fluorine atom:

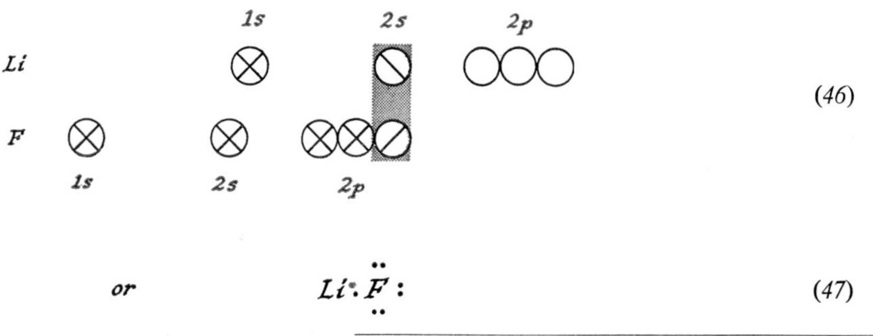

Thus we can expect a stable molecular species, LiF. The term "stable" again means that energy is required to disrupt the molecule. The chemical bond lowers the energy because the bonding electron pair feels simultaneously both the lithium nucleus and the fluorine nucleus. That is not to say, however, that the electrons are shared equally. After all, the lithium and fluorine atoms attract the electrons differently. This is shown by the ionization energies of these two atoms:

$$F(g) \longrightarrow F^+(g) + e^-(g) \qquad \Delta H = 401.5 \text{ kcal/mole} \quad (48)$$

$$Li(g) \longrightarrow Li^+(g) + e^-(g) \qquad \Delta H = 124.3 \text{ kcal/mole} \quad (49)$$

Clearly the fluorine atom holds electrons much more strongly than does the lithium atom. As a result, the electron pair in the lithium fluoride bond is more strongly attracted to the fluorine atom than to the lithium. The energy is lower if the electrons spill toward the fluorine atom. *When the bonding electrons move closer to one of the two atoms, the bond is said to have* ***ionic character.***

In the most extreme situation, the bonding electrons move so close to one of the atoms that this atom has virtually the electron distribution of the negative ion. This is the case in gaseous LiF. In an electron dot representation, we might show

$$\text{Li} \;\; :\ddot{\underset{\cdot\cdot}{\text{F}}}: \qquad (50)$$

or

$$Li^+ \;\; F^- \qquad (51)$$

When a formula like (*50*) or (*51*) provides a useful basis for discussing the properties of a molecule, the bond in that molecule is said to be an *ionic bond.*

CONTRAST OF COVALENT AND IONIC BONDS

The fluorine molecule is held together by the energy gain resulting from placing a bonding pair of electrons near both fluorine nuclei simultaneously. The electrons move about in the molecule in such a manner that, on the average, they are distributed symmetrically between and around the identical fluorine nuclei. This symmetrical distribution is reasonable, since the two fluorine nuclei attract the bonding electrons equally. The lithium fluoride molecule also is held together by the energy gain resulting from placing a bonding pair of electrons near the lithium and fluorine atoms simultaneously. In this case, however, the electrons move in such a way as to remain closer to the fluorine than the

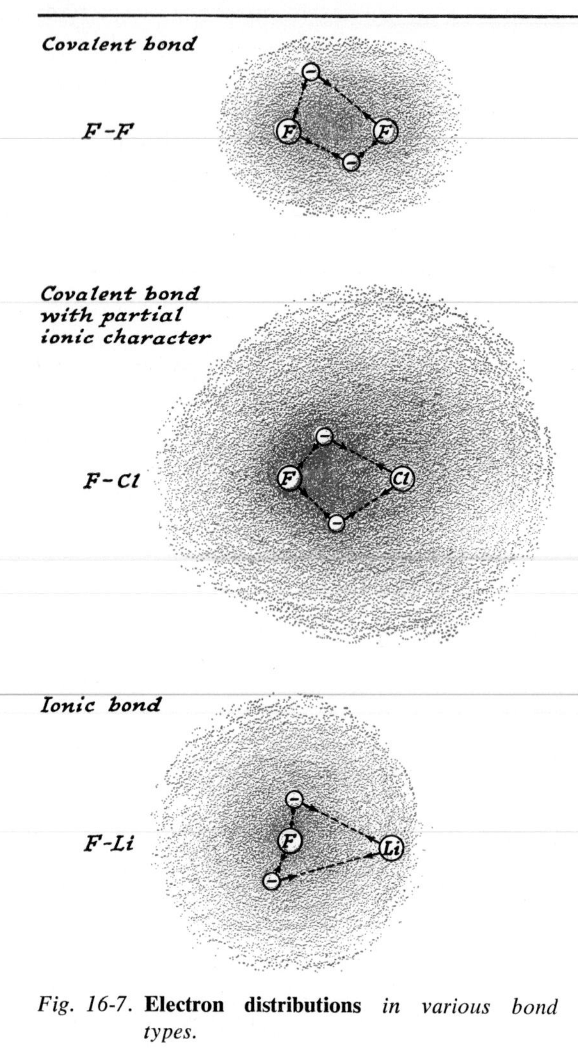

Fig. 16-7. **Electron distributions** *in various bond types.*

lithium atom. Fluorine attracts the bonding electrons more strongly than does lithium.

We see again that there is but one principle which causes a chemical bond between two atoms: *all chemical bonds form because electrons are placed simultaneously near two positive nuclei.* The term *covalent bond* indicates that the most stable distribution of the electrons (as far as energy is concerned) is symmetrical between the two atoms. When the bonding electrons are somewhat closer to one of the atoms than the other, the bond is said to have *ionic character.* The term *ionic bond* indicates the electrons are displaced so much toward one atom that it is a good approximation to represent the bonded atoms as a pair of ions near each other. Figure 16-7 shows schematically how the electron distributions are pictured in covalent, partially ionic, and ionic bonds. The figure also shows how the electrons might look in an instantaneous snapshot. In each type of bond, the electron–nucleus attractions account for the energy stability of the molecule.

THE ELECTRIC DIPOLE OF THE IONIC BOND

The spilling of negative electric charge toward one of the atoms in an ionic bond causes a charge separation. This can be represented crudely as in the last drawing in Figure 16-8. The molecule is electrically positive at the lithium end and electrically negative at the fluorine end. It is said to possess an *electric dipole.* The molecule is then called a *polar molecule.* The forces between molecules possessing electric dipoles are much stronger than those between nonpolar molecules. We shall see in Chapter 17 that these forces, too, involve the same electrical interactions we have discussed here.

The last representation of Figure 16-8 is commonly used and it is the simplest way of showing a bond dipole. The arrow means that the negative charge is mainly at one end of the bond. The directional property of the arrow implies that the force this molecule exerts on another molecule depends upon the direction of approach of the second molecule.

16-3.2 Ionic Character in Bonds to Fluorine

We can expect the effects just discussed to be at work in the bonds fluorine forms with other ele-

Fig. 16-8. **Representations of the electric dipole** *of gaseous lithium fluoride.*

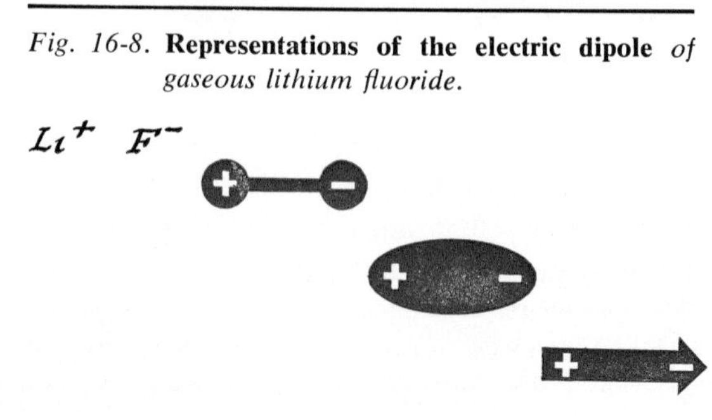

ments. The ionization energies of the elements give us a rough clue to the electron–nuclear attractions. Table 16-II compares the ionization energies of each element of the second row with that of fluorine. The last column describes the type of chemical bond.

The trend in bond type shown in Table 16-II has important influence on the trend in properties of the fluorine compounds. The trend arises because of the increasing difference between ionization energies of the two bonded atoms.

Table 16-II. **BOND TYPES IN SOME FLUORINE COMPOUNDS**

COMPOUND	BOND	IONIZATION ENERGIES (kcal/mole): ELEMENT BONDED TO F		IONIZATION ENERGIES (kcal/mole): FLUORINE	BOND TYPE
FF	F—F	F	401.5	401.5	*Covalent*
OF_2	O—F	O	313.8	401.5	slightly ionic
NF_3	N—F	N	335	401.5	increasing ionic character ↓ / ↑ increasing covalent character
CF_4	C—F	C	259.5	401.5	
BF_3	B—F	B	191.2	401.5	
BeF_2	Be—F	Be	214.9	401.5	slightly covalent
LiF	Li—F	Li	124.3	401.5	*Ionic*

16-3.3 Ionic Character in Bonds to Hydrogen

In Chapter 6 the element hydrogen was characterized as a family by itself. Often its chemistry distinguishes it from the rest of the periodic table. We find this is the case when we attempt to predict the ionic character of bonds to hydrogen.

The ionization energy of the hydrogen atom, 313.6 kcal/mole, is quite close to that of fluorine, so a covalent bond between these two atoms in HF is expected. Actually the properties of HF show that the molecule has a significant electric dipole, indicating ionic character in the bond. The same is true in the O—H bonds of water and, to a lesser extent, in the N—H bonds of ammonia. *The ionic character of bonds to hydrogen cannot be predicted from its measured ionization energy.*

Examination of the properties of a number of compounds involving hydrogen indicates that the ionic character of bonds to hydrogen are roughly like those of an element with ionization energy near 200 kcal/mole. Thus the hydrogen fluorine bond in HF is ionic and chemists believe that the electrons are spilled toward the fluorine atom, leaving the hydrogen atom with a partial positive charge. Hydrogen acts like an element with lower ionization energy than fluorine. The same is true but in decreasing amount for hydrogen when it is bonded to oxygen and to nitrogen. The carbon–hydrogen bond has only a slight ionic character. At the other end of the periodic table, gaseous lithium hydride is known to have a significant electric dipole but now with the electric dipole turned around. In LiH the electrons are spilled toward the hydrogen atom, leaving the lithium atom with a partial positive charge. This is in accord with the low ionization energy of lithium, 124.3 kcal/mole, well below the value of 200 kcal/mole that we have assigned to hydrogen. For our purposes, *it suffices to discuss the bonding of hydrogen in terms of an apparent ionization energy near 200 kcal/mole.*

16-3.4 Bond Energies and Electric Dipoles

It is found experimentally that a bond between two atoms with very different ionization energies tends to be stronger than a bond between atoms with similar ionization energies. Since electric dipoles are caused by differences in the ionization energies of bonded atoms, we can conclude that strong bonds are expected in molecules with electric dipoles.

For example, contrast the bond energies of the gaseous molecules Na_2, Cl_2, and NaCl:

$$Na_2(g) \longrightarrow 2Na(g) \qquad \Delta H(Na_2) = 17 \text{ kcal}$$
$$Cl_2(g) \longrightarrow 2Cl(g) \qquad \Delta H(Cl_2) = 57 \text{ kcal}$$
$$NaCl(g) \longrightarrow Na(g) + Cl(g) \qquad \Delta H(NaCl) = ?$$

A rough estimate of the bond energy of NaCl could be based upon the bond energies of Na_2 and Cl_2, 17 kcal and 57 kcal. Since the 17 kcal bond energy of Na_2 is derived from the sharing of an electron pair between *two* sodium atoms, the *one* sodium atom in NaCl might be expected to contribute one-half this amount to the bond energy of NaCl, $\frac{17}{2} = 8.5$ kcal. In a similar way, the single chlorine atom in NaCl might contribute one-half the bond energy of Cl_2, $\frac{57}{2} = 28.5$ kcal. Thus we arrive at an estimate of $\Delta H(\text{NaCl})$:

$$\Delta H(\text{NaCl}) \text{ (estimated)} = \frac{\Delta H(Na_2) + \Delta H(Cl_2)}{2}$$
$$= 8.5 + 28.5 = 37.0$$

Experimentally we discover that $\Delta H(\text{NaCl})$ is much larger, 98.0 kcal—a discrepancy of $98 - 37 = 61$ kcal. This discrepancy is explained in terms of the large difference in ionization energy of sodium and chlorine atoms:

$$Cl(g) \longrightarrow Cl^+(g) + e^- \qquad E_1 = 300 \text{ kcal}$$
$$Na(g) \longrightarrow Na^+(g) + e^- \qquad E_1 = 118 \text{ kcal}$$

The large difference implies that the energy is lowered *even more than 37 kcal* because the electron pair need not remain equally shared between the two atoms in NaCl. Instead they can concentrate nearer the atom that holds electrons more tightly (the chlorine atom) if a net lowering of the energy results.

Table 16-III collects some data. The existence of a correlation between the ionization energy difference, $E_1(X) - E_1(Y)$, and the bond energy discrepancy, $\Delta H_{XY} - \frac{1}{2}(\Delta H_{X_2} + \Delta H_{Y_2})$, is obvious.

Needless to say, if ionic character affects the energy stability of a chemical bond it also affects the chemistry of that bond. The tendency toward minimum energy is one of the factors that determine what chemical changes will occur. As a bond becomes stronger, more energy is required to break that bond to form another compound. Hence we see that ionic bonds are favored over covalent bonds and that ionic character in a bond affects its chemistry.

EXERCISE 16-6

From the following bond energy data and the ionization energies given in Table 15-III, calculate the entries in the last two columns of Table 16-III for the compounds LiF and LiBr. The ionization energy, E_1, for bromine atom is 273 kcal/mole.

$$Li_2(g) \rightleftarrows 2Li(g) \qquad \Delta H = 25 \text{ kcal}$$
$$F_2(g) \rightleftarrows 2F(g) \qquad \Delta H = 36$$
$$Br_2(g) \rightleftarrows 2Br(g) \qquad \Delta H = 45.5$$
$$LiF(g) \rightleftarrows Li(g) + F(g) \qquad \Delta H = 137$$
$$LiBr(g) \rightleftarrows Li(g) + Br(g) \qquad \Delta H = 101$$

Table 16-III. **IONIZATION ENERGY DIFFERENCES AND BOND ENERGIES**

X	Y	MOLECULE XY	$E_1(X) - E_1(Y)$	$\Delta H_{XY} - \frac{1}{2}(\Delta H_{X_2} + \Delta H_{Y_2})$
Na	K	NaK(g)	18 kcal/mole	0 kcal/mole
Cl	Br	ClBr(g)	27	+1
Cl	Li	LiCl(g)	176	58
Cl	Na	NaCl(g)	182	61
Cl	K	KCl(g)	200	66.5
F	Li	LiF(g)	277	106

16-4 MOLECULAR ARCHITECTURE

The properties of a molecule are primarily determined by the bond types which hold it together and by the molecular "architecture." By architecture we mean the structure of the molecule—

the shape of the molecule. We shall investigate what is known about the molecular structures of the second-row hydrides and fluorides.

16-4.1 The Shapes of H_2O and F_2O

The orbital representations of the bonding in H_2O and in F_2O suggest that two *p* orbitals of oxygen are involved in the bonding [see representation (*24*)]. Figure 16-9 shows the spatial arrangement we assign to the *p* orbitals (assuming they are like hydrogen atom orbitals). If the

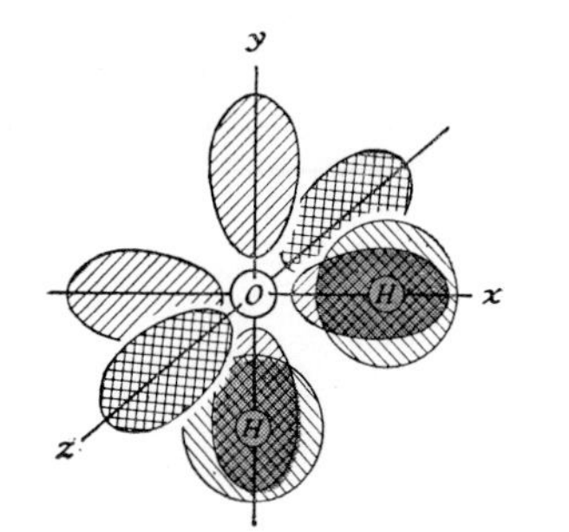

Fig. 16-9. **The expected shape** *of the H_2O molecule: p^2 bonding.*

spatial arrangement persists after the bonds form, the molecular shape would be fixed, as shown. The molecule would be bent, with an angle near 90°. The same would be true for F_2O. The measured bond angles are as follows:

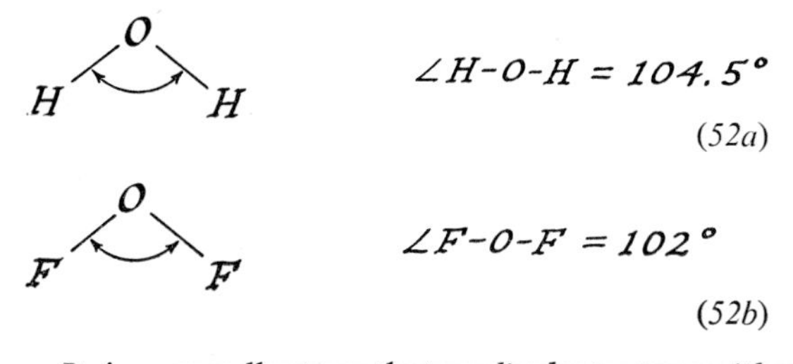

(*52b*)

It is generally true that *a divalent atom with two p orbitals as valence orbitals forms an angular molecule.* Since this prediction is reliable, the bonding is usually characterized by identifying the valence orbitals. Oxygen is said to use p^2 (read, "*p* two") bonding in water and F_2O.

Notice that the structural formulas (*52a*) and (*52b*) use another representation of the bonding. It is quite familiar, of course, since it corresponds to the ball-and-stick model of the molecule. A line is drawn between the oxygen atom and each hydrogen atom to indicate that a chemical bond holds these two atoms together. No line is drawn between the two hydrogen atoms since we feel they are not directly bonded to each other. Now we can apply our discussion of the role of electrons in bonding to add to the meaning of this line representation. A line is drawn between two atoms to indicate that a pair of electrons is shared between these two atoms, resulting in a chemical bond.

16-4.2 The Shapes of NH_3 and NF_3

In NH_3 and NF_3, three *p* orbitals are involved in the bonding [see representation (*30*)]. Figure 16-10 shows the spatial arrangement implied by assuming persistence of the hydrogen atom orbitals after bonding. We expect, then, that am-

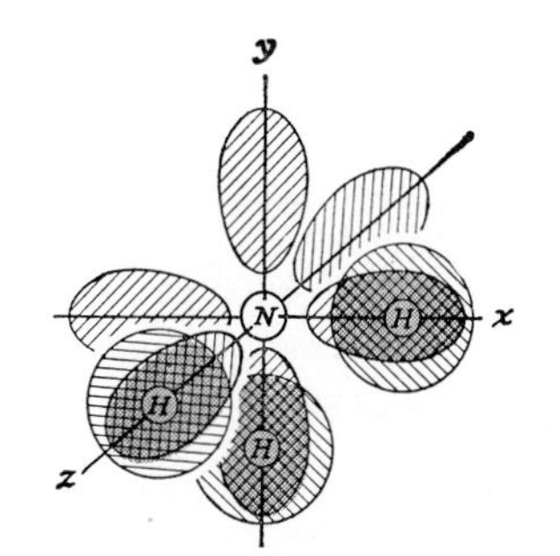

Fig. 16-10. **The expected shape** *of the NH_3 molecule: p^3 bonding.*

monia has a pyramidal shape (a pyramid with a three-sided base). The bond angles should be near 90°. Both NH_3 and NF_3 do have pyramidal shapes. The measured bond angles are as follows:

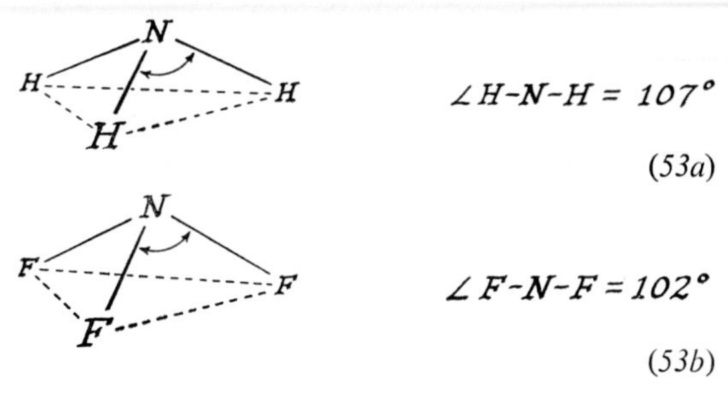

(*53b*)

Again, experiments show that it is generally true that *a trivalent atom with three p orbitals as valence orbitals forms a pyramidal molecule*. The bonding is called p^3 bonding (read, "*p* three").

16-4.3 The Shapes of CH_4 and CF_4

The bonding of methane, CH_4, and that of carbon tetrafluoride, CF_4, involve four valence orbitals, the 2*s* orbital and the three 2*p* orbitals. Four bonds are formed by carbon and, as before, we characterize the bonding by naming the valence orbitals: sp^3 (read, "*sp* three"). This time, however, the assumption of the persistence of the spatial distributions of the hydrogen atom orbitals does not indicate directly what bond angles to expect. Experiment shows, however, that *sp^3 bonding always gives bond angles which are exactly or very close to tetrahedral angles*. This means that the angle between any two carbon–hydrogen bonds is 109°28′. The structure is called tetrahedral because the four hydrogen atoms occupy the positions of the corners of a regular tetrahedron (a four-sided figure with equal edges). The structure is shown in Figure 16-11.

16-4.4 The Shape of BF_3

The boron atom in BF_3 uses the 2*s* and two 2*p* orbitals in bonding. Therefore the bonding is called sp^2. Again we must let experiment tell us the bond angles which are found with sp^2 bonding. The structure of BF_3 is that of an equilateral triangle. The structure, shown in Figure 16-12, is planar and each of the three fluorine atoms is the same distance from the boron atom as are the others.

Fig. 16-11. **The tetrahedral bonding** *of carbon: sp^3 bonding.*

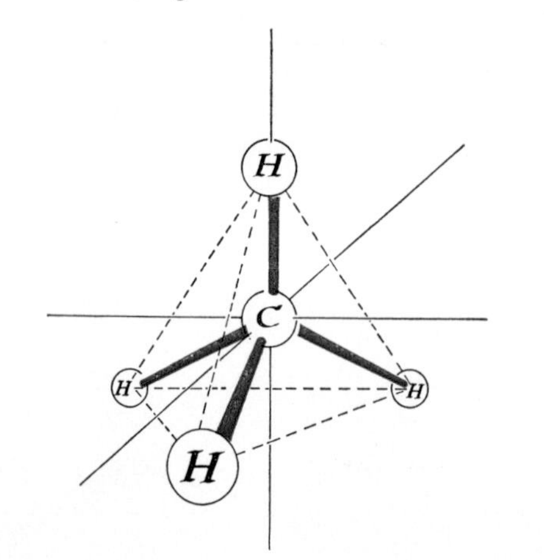

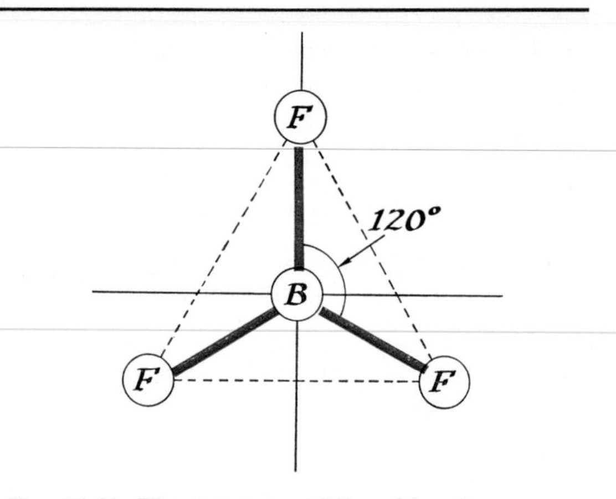

Fig. 16-12. **The structure** *of BF_3: sp^2 bonding.*

16-4.5 The Shape of BeF_2

Beryllium atom in gaseous BeF_2 uses the 2*s* and only one 2*p* orbital in bonding. The bonding is called *sp*. Experiment shows that the molecule is linear and symmetrical, as shown in Figure 16-13. The structure of gaseous BeH_2 is undoubtedly linear and symmetric as well, by analogy to BeF_2.

Fig. 16-13. **The structure** *of BeF_2: sp bonding.*

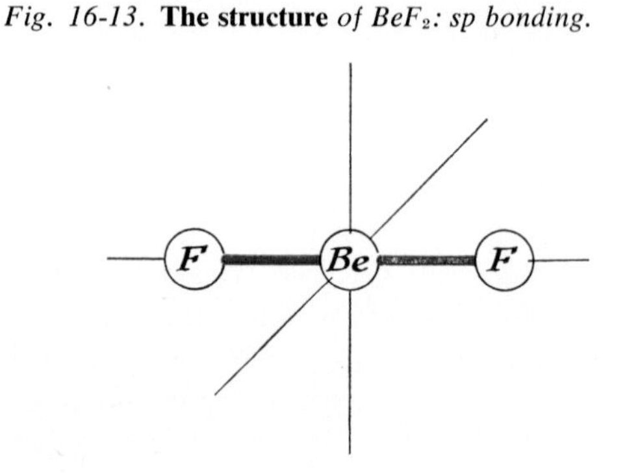

16-4.6 Summary of Bonding Orbitals and Molecular Shape

From the data presented here, the orbitals involved in bonding correlate with the molecular architecture. The relationships are summarized in Table 16-IV.

THE MOLECULAR DIPOLE OF LiF

The lithium fluoride bond is highly ionic in character because of the large difference in ionization energies of lithium and fluorine. Consequently, gaseous lithium fluoride has an unusually high electric dipole.

Table 16-IV. **BONDING ORBITALS, BONDING CAPACITY, AND MOLECULAR SHAPE**

ELEMENT	BONDING ORBITALS	BONDING CAPACITY	MOLECULAR SHAPE OF THE FLUORIDE	EXAMPLE
He	none	0	monatomic	He
Li	*s*	1	linear, diatomic molecule	LiF
Be	*sp*	2	linear	BeF_2
B	sp^2	3	planar, triangular	BF_3
C	sp^3	4	tetrahedral	CF_4
N	p^3	3	pyramidal	NF_3
O	p^2	2	bent	OF_2
F	*p*	1	linear, diatomic molecule	F_2
Ne	none	0	monatomic	Ne

16-4.7 Molecular Shape and Electric Dipoles

Consider the fluorides of the second-row elements. There is a continuous change in ionic character of the bonds fluorine forms with the elements F, O, N, C, B, Be, and Li. The ionic character increases as the difference in ionization energies increases (see Table 16-II). This ionic character results in an electric dipole in each bond. The molecular dipole will be determined by the sum of the dipoles of all of the bonds, taking into account the geometry of the molecule. Since the properties of the molecule are strongly influenced by the molecular dipole, we shall investigate how it is determined by the molecular architecture and the ionic character of the individual bonds. For this study we shall begin at the left side of the periodic table.

THE MOLECULAR DIPOLE OF BeF_2

The beryllium–fluorine bond is also highly ionic in character. However, there are two such Be–F bonds and the electrical properties of the entire molecule depend upon how these two bonds are oriented relative to each other. We must find the "geometrical sum" of these two bond dipoles.

The geometrical sum of two arrows can be understood simply with the aid of Figure 16-14. Figure 16-14A shows how two arrows pointing in the same direction combine to give a longer arrow. Figure 16-14B shows how two arrows oppositely directed combine to give a shorter

Fig. 16-14. **The geometrical sum of dipoles:** *both length and direction are important.*

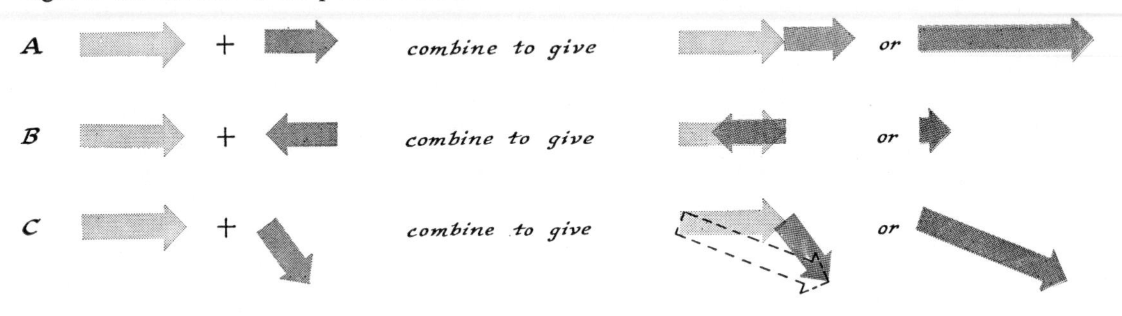

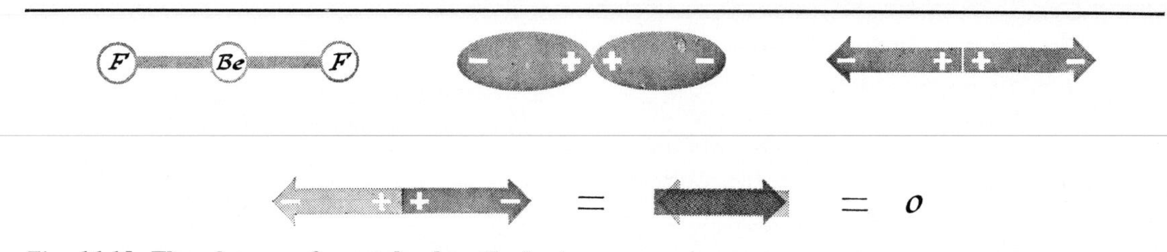

Fig. 16-15. **The absence of a molecular dipole** *in* BeF_2.

arrow. Figure 16-14C shows how two arrows that are not parallel add to give an arrow in a new direction.

Now we can apply the process of combination shown in Figure 16-14 to BeF_2. In the linear, symmetric BeF_2 molecule, the two bond dipoles point in opposite directions. Since the two bonds are equivalent, there is a complete cancellation, as shown in Figure 16-15. Hence the molecule has no net dipole; the molecular dipole is zero.

THE MOLECULAR DIPOLES OF BF_3 AND CF_4

Both of these molecules are thought to have moderate amounts of ionic character in each bond. Yet the molecular dipoles are each exactly zero. Careful consideration of the geometry shows that there is a complete cancellation by the bond dipoles in each molecule. This cancellation is shown in Figure 16-16 for BF_3. The molecular dipole is zero.

THE MOLECULAR DIPOLE IN F_2O

Since F_2O, with p^2 bonding, is a bent molecule, the two bond dipoles do not cancel each other as they do in BeF_2. On the other hand, the ionization energies of oxygen and fluorine are not very different, so the electric dipole of each bond is small in magnitude. These add together, according to the geometry, to give a polar molecule, as shown in Figure 16-17.

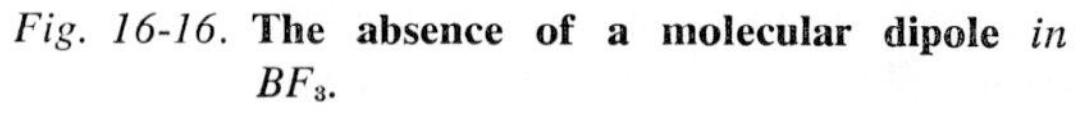

Fig. 16-16. **The absence of a molecular dipole** *in* BF_3.

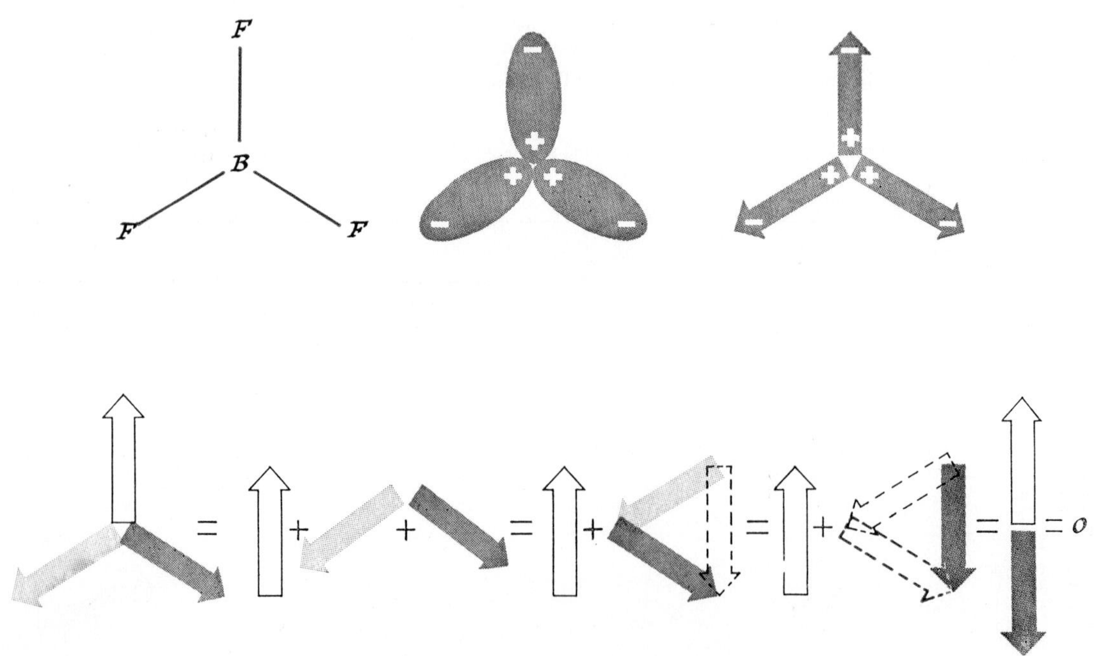

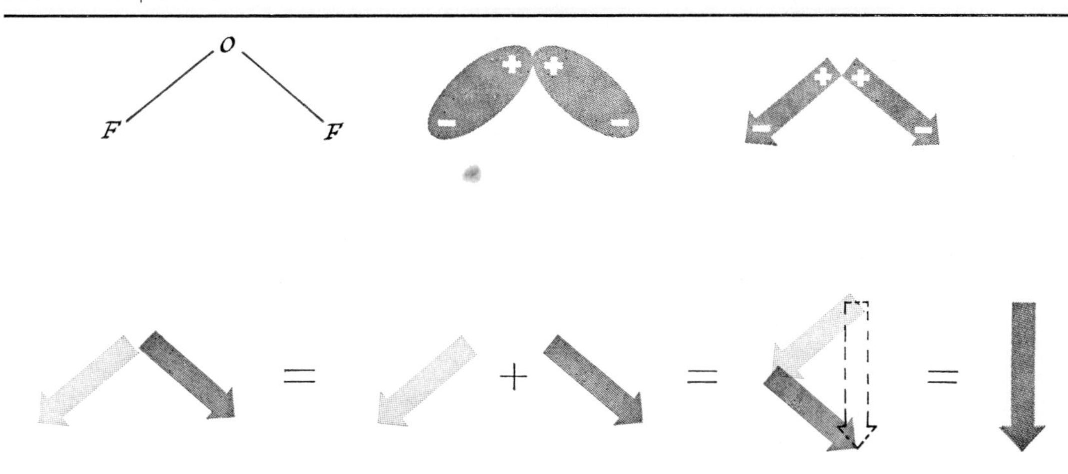

Fig. 16-17. **The molecular dipole** *of* F_2O.

16-5 DOUBLE BONDS

In deciding the bonding capacity of a given atom from the second row, we have counted the number of hydrogen atoms or fluorine atoms with which it would combine. Thus oxygen combines with two hydrogen atoms to form water, H_2O. Oxygen is said to be divalent. Oxygen shares two pairs of electrons, one pair with each hydrogen atom. Each of these shared pairs forms a *single bond.*

16-5.1 Bonding in the Oxygen Molecule

Now let us investigate the oxygen molecule, which experiment tells us has the molecular formula O_2. We might begin by considering formation of a single bond between two oxygen atoms, as represented by the orbital representation

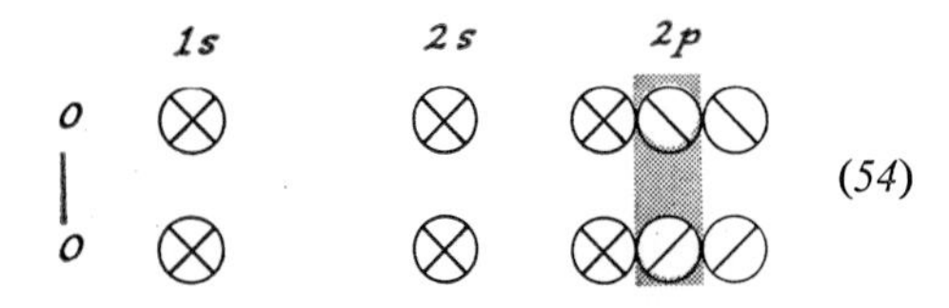

(54)

We see that each oxygen atom has residual bonding capacity. Each atom could, for example, react with a hydrogen atom to form hydrogen peroxide, as shown in electron dot representation (26). Each oxygen atom could react with a fluorine atom to form F_2O_2. In short, each oxygen atom is in need of another atom with an electron in a half-filled valence orbital so that it can act as a divalent atom.

But suppose oxygen can find no hydrogen atoms or fluorine atoms. Then, it does the lazy thing: the two atoms, already bound by one bond, form a second bond with each other. The result might be shown in the orbital representation (55).

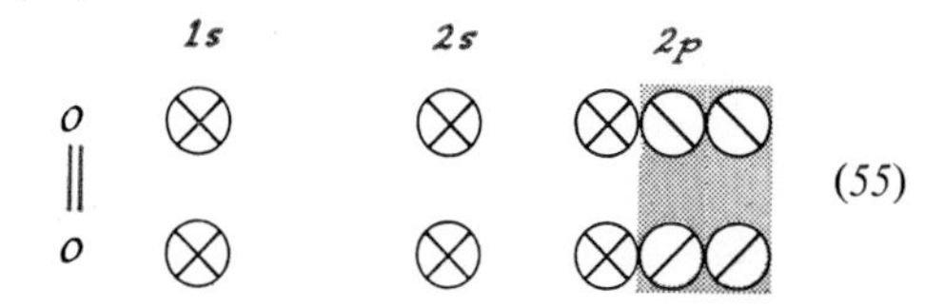

(55)

There is strong evidence supporting this proposal. The bond in the oxygen molecule is stronger than the oxygen–oxygen bond in hydrogen peroxide (more energy is required to break it). The vibrational frequency of the oxygen molecule is higher than that of a normal single bond, showing that there is additional bonding (see Section 14-3.4). The bond length in the O_2 molecule is 1.21Å. In the gaseous hydrogen peroxide molecule, the oxygen–oxygen distance is 1.48Å. The short bond length in the O_2 molecule shows that the two oxygen atoms are drawn together more effectively than in HOOH, suggesting there are extra bonding electrons in O_2.

Because all of the evidence we have examined is consistent with the orbital representation (55), the bond in O_2 is called a *double bond.* An electron dot representation can be written as follows

$$:\dot{\underset{\cdot}{\mathrm{O}}}::\dot{\underset{\cdot}{\mathrm{O}}}: \qquad (56)$$

The representation (*56*) shows two pairs of electrons shared. Each oxygen atom finds itself near eight electrons. There is, on the one hand, a stable molecule, because all of the bonding capacity of each oxygen atom is in use. On the other hand, this special aspect of the bonding of oxygen undoubtedly contributes to the reactivity of oxygen.

16-5.2 Ethylene: A Carbon–Carbon Double Bond

Ethylene is a simple compound of carbon and hydrogen with the formula C_2H_4. Thus it has two less hydrogen atoms than does ethane, C_2H_6. This means that to write a structure of ethylene we must take account of two electrons that are not used in C—H bond formation. Suppose we write an electron dot representation involving only single bonds

$$\begin{matrix} & \mathrm{H} & \mathrm{H} & \\ \mathrm{H}: & \underset{\cdot}{\overset{\cdot\cdot}{\mathrm{C}}} & :\underset{\cdot}{\overset{\cdot\cdot}{\mathrm{C}}} & :\mathrm{H} \end{matrix} \qquad (57)$$

This formula has two unpaired electrons, representing unused bonding capacity. This objectionable situation can easily be rectified by allowing the two unpaired electrons to pair, and thus form an additional two-electron bond. Now the carbon atoms are joined by a *double bond*, just as the oxygen atoms in O_2 are double bonded to each other

$$\begin{matrix} \mathrm{H} & & & \mathrm{H} \\ & \mathrm{C}:: & \mathrm{C} & \\ \mathrm{H} & & & \mathrm{H} \end{matrix} \qquad (58)$$

CHEMICAL REACTIVITY OF ETHYLENE

In ethane, C_2H_6, all of the bonds are normal single bonds. Experiment shows that ethane is a fairly unreactive substance. It reacts only when treated with quite reactive species (such as free chlorine atoms), or when it is raised to excited energy states by heat (as in combustion).

Ethylene, on the other hand, reacts readily with many chemical reagents. Having four electrons forming the carbon–carbon bond, the electrons of the double bond seem to be accessible to attack. We find that the typical reactions of ethylene are those with reagents that *seek* electrons. For example, oxidizing agents are electron-seeking, and we would expect the double bond to be readily oxidized. This is indeed the case. Ethylene will reduce (that is, be oxidized by) such oxidizing agents as potassium permanganate or potassium dichromate at ordinary temperatures. Under these same mild conditions ethane is completely unreactive to the same reagents.

GEOMETRICAL FEATURES OF ETHYLENE

The shape of the ethylene molecule has been learned by a variety of types of experiments. Ethylene is a planar molecule—the four hydrogen and the two carbon atoms all lie in one plane. The implication of this experimental fact is that there is a rigidity of the double bond which prevents a twisting movement of one of the CH_2 groups relative to the other. Rotation of one CH_2 group relative to the other—with the C—C bond as an axis—must be energetically restricted or the molecule would not retain this flat form.

CIS-TRANS ISOMERISM OF ETHYLENE DERIVATIVES

It is possible to replace hydrogen atoms of ethylene by halogen atoms. For example, one such compound has the formula $C_2H_2Cl_2$. In preparing such a compound, chemists discovered long ago that they could obtain three different pure substances with this same formula, $C_2H_2Cl_2$. *Different compounds with the same molecular formulas are called* ***isomers.*** The existence of three separate $C_2H_2Cl_2$ isomers is readily explained in terms of the molecular geometry. The three structures possible with this formula are shown in Figure 16-18. They are all called dichloroethylene.

Two types of isomerism are involved. Formula (*59*) differs from (*60*) and (*61*). In formula (*59*) both chlorine atoms are attached to the same carbon atom. In both (*60*) and (*61*) there is one chlorine atom attached to each carbon atom. The difference between (*59*) and the other pair, (*60*) and (*61*), is indicated by calling these molecules *structural isomers.*

The pair of isomers (*60*) and (*61*) differ in another way. Though each has one chlorine atom attached to each carbon atom, in (*60*) they are on the "same side" of the double bond. This relationship is called the *cis* form. In (*61*) the chlorine atoms are across from each other. This relationship is called the *trans* form. Formulas (*60*) and (*61*) identify *cis* and *trans isomers* of dichloroethylene.

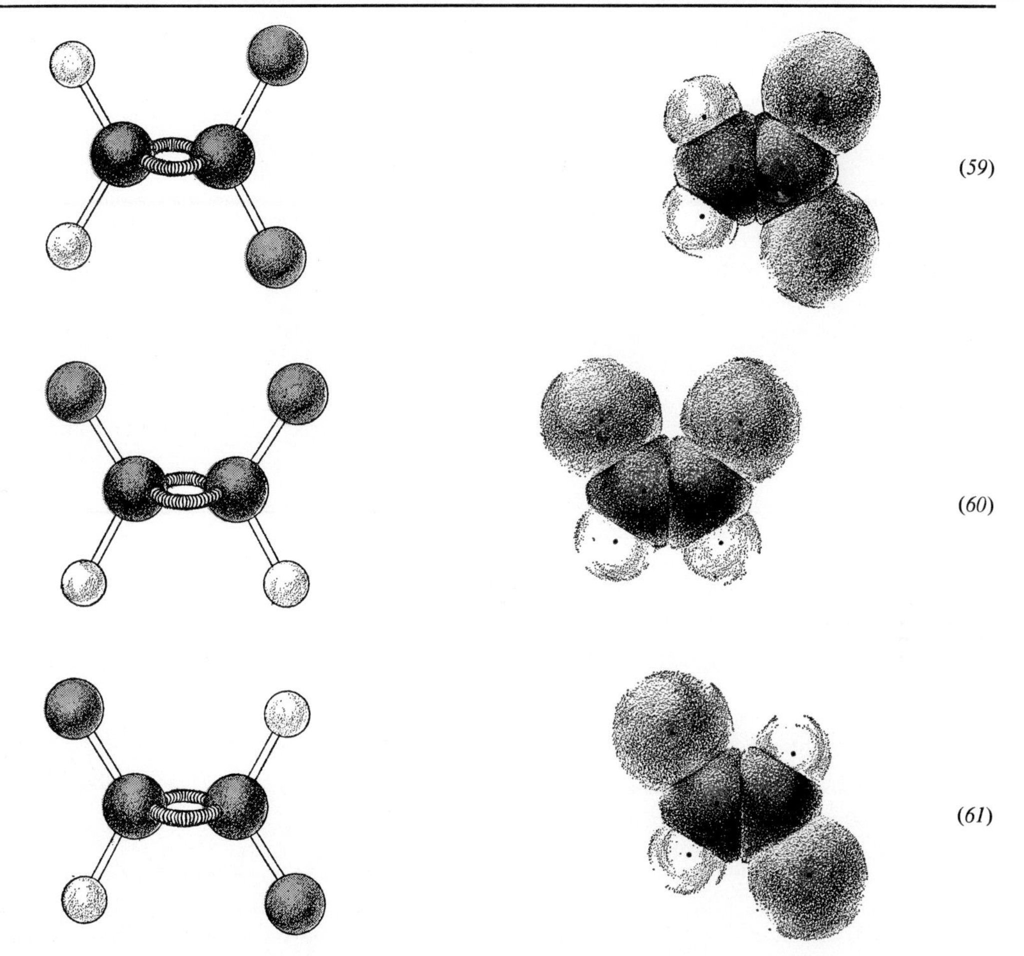

Fig. 16-18. **The isomers of dichloroethylene.**

Experiment shows that it is exceedingly difficult to convert *(59)* into *(60)* or *(61)*. To make such a conversion, bonds must be broken and reformed. Such reactions are almost always quite slow because the activation energies must be almost as large as the energies of the bonds being broken. In contrast, the conversion of *(60)* into *(61)* (or the reverse) can be accomplished merely by heating the substance. No bonds must be broken completely—only a rotation around the carbon–carbon double bond is necessary. This process has a much lower activation energy and the reaction occurs at moderate temperatures.

QUESTIONS AND PROBLEMS

1. Which one of the following statements is FALSE as applied to this equation?

   $H_2(g) \rightleftharpoons H(g) + H(g) \qquad \Delta H = 103.4 \text{ kcal}$

   (a) The positive ΔH means the reaction is endothermic.
   (b) Two grams of $H(g)$ contain more energy than 2 grams of $H_2(g)$.
   (c) Weight for weight, $H(g)$ would be a better fuel than $H_2(g)$.
   (d) The spectrum of $H_2(g)$ is the same as the spectrum of $H(g)$.

2. What are the molecular species present in gaseous neon, argon, krypton, and xenon? Explain.

3. Determine the number of attractive forces and the number of repulsive forces in LiH.

4. What energy condition must exist if a chemical bond is to form between two approaching atoms?

5. What valence orbital and valence electron conditions must exist if a chemical bond is to form between two approaching atoms?

6. Give the orbital and also the electron dot representations for the bonding in these molecules: Cl_2, HCl, Cl_2O.

7. Using the electron dot representation, show a neutral, a negatively charged, and a positively charged OH group.

8. Draw the orbital representation of the molecule N_2H_4, hydrazine.

9. Knowing the orbitals carbon uses for bonding, use the periodic table to predict the formula of the chloride of silicon. What orbitals does silicon use for bonding?

10. Draw the orbital representations of

    (a) sodium fluoride,
    (b) beryllium fluoride, BeF_2.

11. In general, what conditions cause two atoms to combine to form:

    (a) a bond that is mainly covalent;
    (b) a bond that is mainly ionic;
    (c) a polar molecule?

12. What type of bonding would you expect to find in MgO? Explain.

13. Considering comparable oxygen compounds, predict the shape of H_2S and H_2S_2 molecules. What bonding orbitals are used?

14. Predict the formula and molecular shape of a hydride of phosphorus.

15. Draw an electron dot representation for the NH_4^+ ion. What shape do you predict this ion will have?

16. Predict the type of bonding and the shape of the ion BF_4^-.

17. Consider the two compounds CH_3CH_3 (ethane) and CH_3NH_2 (methylamine). Why does CH_3NH_2 have an electric dipole while CH_3CH_3 does not?

18. Consider the following series: CH_4, CH_3Cl, CH_2Cl_2, $CHCl_3$, CCl_4. In which case(s) will the molecules have electric dipoles? Support your answer by considering the bonding orbitals of carbon, the molecular shape of the molecules, and the resulting symmetry.

19. Predict the structure of the compound N_2F_2 from the electron dot representation of the atoms and the molecule.

20. Which of the isomers of dichloroethylene shown in Figure 16-18 will be polar molecules?

21. Draw structural formulas for all the isomers of ethylene (C_2H_4) in which two of the hydrogen atoms have been replaced by deuterium atoms. Label the *cis* and the *trans* isomers.

LINUS C. PAULING, 1901–

No other living chemist has contributed more to our understanding of chemical bonding than Linus C. Pauling. His ideas pervade every aspect of chemistry. These ideas have won him some seventeen medals and high awards, including the 1954 Nobel Prize in Chemistry. His international renown is bespoken by his election to honorary membership in sixteen scientific societies in ten different countries.

Linus Pauling was born in Portland, Oregon, and his hobbies as a boy were largely scientific. At 11 he began an insect collection, which led to reading books on entomology. At 13 Linus found a chemistry book in the family library and founded a laboratory in the family basement. By the time he entered Oregon State College he was determined to become a chemical engineer. In 1922 Pauling received the B.S. degree, and this was followed by the Ph.D. at California Institute of Technology. By now his interest had turned to the fundamental aspects of chemistry and after a year of post-doctoral study in Europe he returned to the faculty at California Institute of Technology. There he established and pursued his illustrious career.

Linus Pauling's prodigious scientific productivity has broadly influenced the face of chemistry. His interest focused on the chemical bond and he was one of the earliest of chemists to recognize the importance of the quantum mechanical point of view. He gave quantitative meaning to the electronegativity concept. He discussed the mixing of the ionic and covalent character of chemical bonds and introduced the term "resonance"—a concept that many chemists criticize but that most chemists use regularly. Pauling gave detailed consideration to the effective sizes of atoms in molecules and crystals. He became an authority on hydrogen bonding and he proposed a theory of metallic bonding. He advo ated and, with his colleagues, established the existence of helical structures in proteins. And, while publishing over 300 papers, he authored several books that won wide acceptance: Introduction to Quantum Mechanics (*with E. B. Wilson, Jr.*), The Nature of the Chemical Bond, General Chemistry, *and* College Chemistry.

Pauling has a deep sensitivity to the welfare of mankind and he has worked energetically to awaken the conscience of society to its new responsibilities in a nuclear age. Attempting to inform the public of the pressing need for lasting peace, he has written a book entitled No More War. *He has engaged in many public forums and debates, even though these activities sometimes attracted ridicule in times when fear made his cause an unpopular one. In recognition of these activities he was awarded the 1962 Nobel Peace Prize, and thus became the second person in history to receive the Nobel Prize twice.*

There is no chemistry course given today that is not influenced by the ideas of Linus C. Pauling. He is a man of broad imagination, dramatic personality, and boundless inspiration. Mankind will long benefit because he chose to explore the frontiers of science.

CHAPTER 17

The Bonding in Solids and Liquids

It is · · · possible to discuss the structure of any substance · · · by describing the types · · · of its bonds and in this way to account for its characteristic properties.

LINUS PAULING, 1939

Any pure gas, when cooled sufficiently, will condense to a liquid and then, at a lower temperature, will form a solid. There is great variance in the temperature at which this condensation occurs. Apparently there is a corresponding variance of the forces in liquids and solids. For example, lithium fluoride gas at one atmosphere pressure condenses when cooled below 1949°K. When the temperature is lowered to 1143°K, the liquid forms a clear crystal. In contrast, lithium gas at this pressure must be cooled to 1609°K before it forms a liquid and this liquid does not solidify until the temperature reaches 459°K. The solid is a white, soft metal, not resembling crystalline lithium fluoride at all. Fluorine gas is equally distinctive. At one atmosphere pressure it must be cooled far below room temperature before condensation occurs, at 85°K. Then the liquid solidifies to a crystal at 50°K. Why do these three materials behave so differently? Can we understand this great variation? Let us begin by finding a common point of departure.

Two or more atoms remain near each other in a particular arrangement because the energy favors that arrangement. This is true whether the cluster of atoms is strongly or weakly bound, whether it contains a few atoms or 10^{23} atoms, whether the arrangement is regular (as in a crystal) or irregular (as in a liquid). The cluster of atoms is stable if and only if the energy is lower when the atoms are together than when they are apart.

Furthermore, there is but one reason that two or more atoms have lower energy when they are in proximity. In this way electrons can be close to two or more positive nuclei simultaneously. However, *the magnitude of the attractive forces varies greatly*, depending on how close the electrons are able to approach these positive nuclei. This approach distance is fixed by the electron occupancy of the valence orbitals.

Thus the occupancy of the valence orbitals is the clue we shall follow in our attempt to predict when to expect a substance to be a high-melting, salt-like crystal, when to expect a metal, when to expect a low-melting, molecular crystal. This is an ambitious program. Let's see how far we can go, beginning with the pure elements.

17-1 THE ELEMENTS

The examples just mentioned include two elements, fluorine and lithium. Fluorine forms a weakly bound molecular solid. Lithium forms a metallic solid. Let us see how we can account for this extreme difference, applying the principles of bonding treated in Chapter 16.

17-1.1 van der Waals Forces

The diatomic molecule of fluorine does not form higher compounds (such as F_3, F_4, $\cdots$) because each fluorine atom has only one partially filled valence orbital. Each nucleus in F_2 is close to a number of electrons sufficient to fill the valence orbitals. Under these circumstances, the diatomic molecule behaves like an inert gas atom toward other such molecules. The forces that cause molecular fluorine to condense at 85°K are, then, the same as those that cause the inert gases to condense. These forces are named *van der Waals* forces, after the Dutch scientist who studied them.

When the outer orbitals of all of the atoms in the molecules are filled—giving inert gas configurations—then the electrons of another molecule cannot approach the nuclei closely. When molecules of this sort approach each other, the energy is lowered only a few tenths of a kilocalorie per mole. This weak interaction is typical of van der Waals forces.

We have, now, a simple rule for predicting when a weakly bound molecular liquid and a low-melting crystal will be formed by a given element. *If the element forms a molecule that gives each atom the orbital occupancy of an inert gas, then only van der Waals interactions among such molecules remain.*

We shall take up later in this chapter the factors that determine the magnitude of van der Waals forces. For the moment, we will merely observe that the elements forming van der Waals liquids and solids are concentrated in the upper right-hand corner of the periodic table (see Figure 17-1). These are the elements able to form stable molecules that satisfy completely the bonding capacity of each atom.

EXERCISE 17-1

Gaseous phosphorus is made up of P_4 molecules with four phosphorus atoms arranged at the corners of a regular tetrahedron. In such a geometry, each phosphorus atom is bound to three other phosphorus atoms. Would you expect this gas to condense to a solid with a low or high melting point? After making a prediction on the basis of the valence orbital occupancy, check the melting point of phosphorus in Table 6-VIII, p. 101.

Fig. 17-1. **Elements that form molecular crystals** *bound by van der Waals forces.*

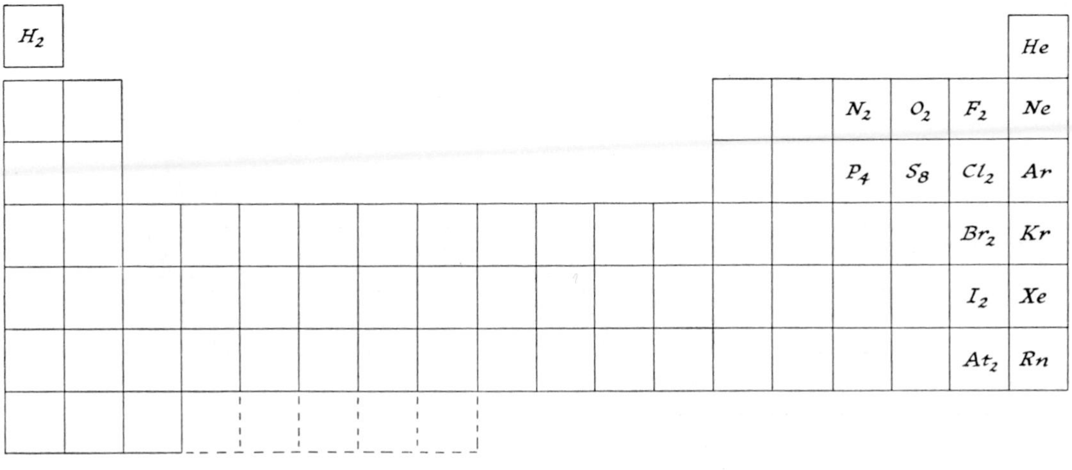

H_2 He
N_2 O_2 F_2 Ne
P_4 S_8 Cl_2 Ar
Br_2 Kr
I_2 Xe
At_2 Rn

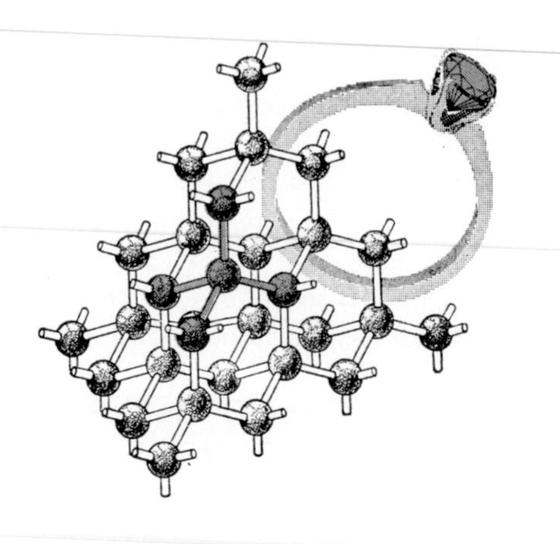

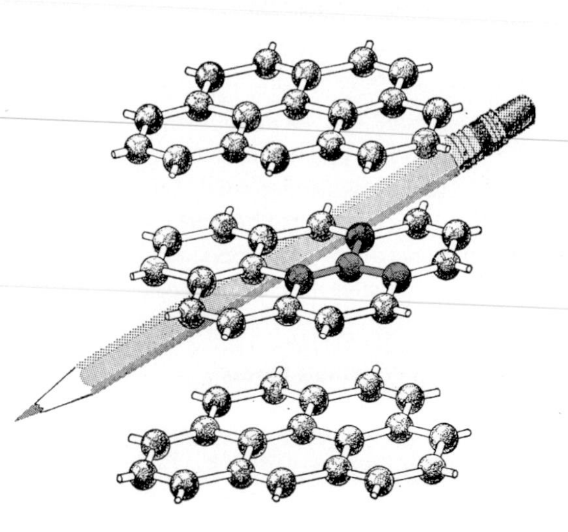

Fig. 17-2. **Carbon forms network solids:** *diamond and graphite.*

17-1.2 Covalent Bonds and Network Solids

Fluorine, F_2, oxygen, O_2, and nitrogen, N_2, all form molecular crystals but the next member of this row of the periodic table, carbon, presents another situation. There does not seem to be a small molecule of pure carbon that consumes completely the bonding capacity of each atom. As a result, it is bound in its crystal by a network of interlocking chemical bonds.

With one 2*s* and three 2*p* orbitals available for the bonding of carbon, we can expect it to form a lattice in which each atom forms four bonds. Furthermore, sp^3 bonding is connected with tetrahedral bond angles (as in Figure 16-11). These expectations are consistent with the experimentally determined structure of diamond, shown in Figure 17-2.

Diamond is a naturally occurring form of pure, crystalline carbon. Each carbon atom is surrounded by four others arranged tetrahedrally. The result is a compact structural network bound by normal chemical bonds. This description offers a ready explanation for the extreme hardness and the great stability of carbon in this form.

Fig. 17-3. **Elements that form solids involving covalent bonding.**

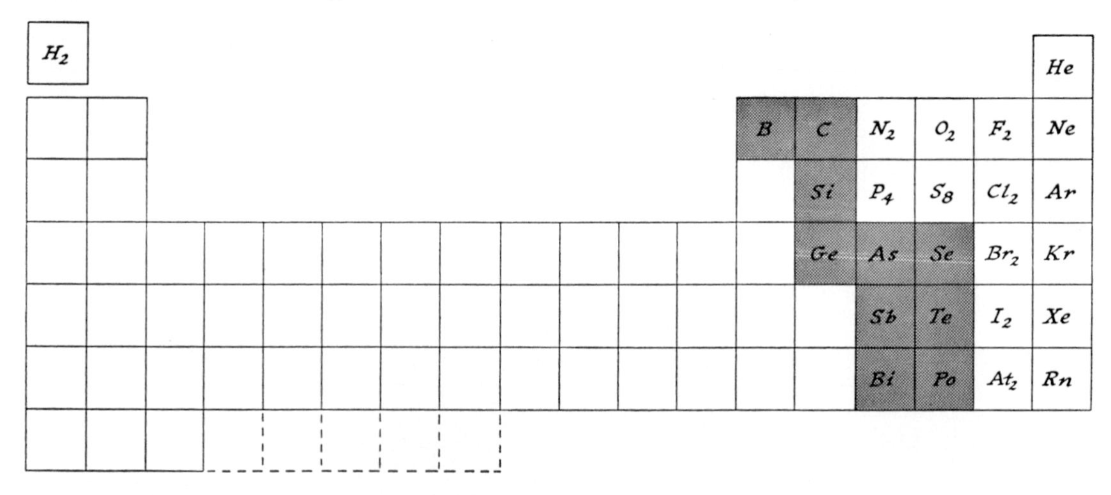

Graphite is another solid form of carbon. In contrast to the three-dimensional lattice structure of diamond, graphite has a layered structure. Each layer is strongly bound together but only weak forces exist between adjacent layers. These weak forces make the graphite crystal easy to cleave, and explain its softness and lubricating qualities.

The elements that form network solids lie on the right side of the periodic table, bordering the elements that form molecular crystals on one side and those that form metals on the other. Thus they are intermediate between the metals and the nonmetals. In this borderline region classifications are sometimes difficult. Whereas one property may suggest one classification, another property may lead to a different conclusion. Figure 17-3 shows some elements that form solids that are neither wholly metallic nor wholly molecular crystals.

17-1.3 Metallic Bonding

We have considered solid forms of the elements fluorine, oxygen, nitrogen, and carbon. In each case, a solid is formed in which the bonding capacity is completely satisfied. The remaining elements of the second row, that is, beryllium, and lithium, are metallic. These elements do not have enough electrons to permit the complete use of the valence orbitals in covalent bonding. Furthermore, the ionization energies of these elements are quite low. We find *there are two conditions necessary for* ***metallic bonding:*** *vacant valence orbitals and low ionization energies.*

CHARACTERISTIC PROPERTIES OF METALS

Perhaps the most obvious metallic property is reflectivity or luster. With few exceptions (gold, copper, bismuth, manganese) all metals have a silvery white color which results from reflecting all frequencies of light. We have said previously that the electron configuration of a substance determines the way in which it interacts with light. Apparently the characteristic reflectivity of metals indicates that all metals have a special type of electron configuration in common.

A second characteristic property of metals is high electrical conductivity. The conductivity is so much higher than that of aqueous electrolyte solutions that the charge movement cannot involve the same mechanism. Again we find a

Fig. 17-4. **The metallic elements.**

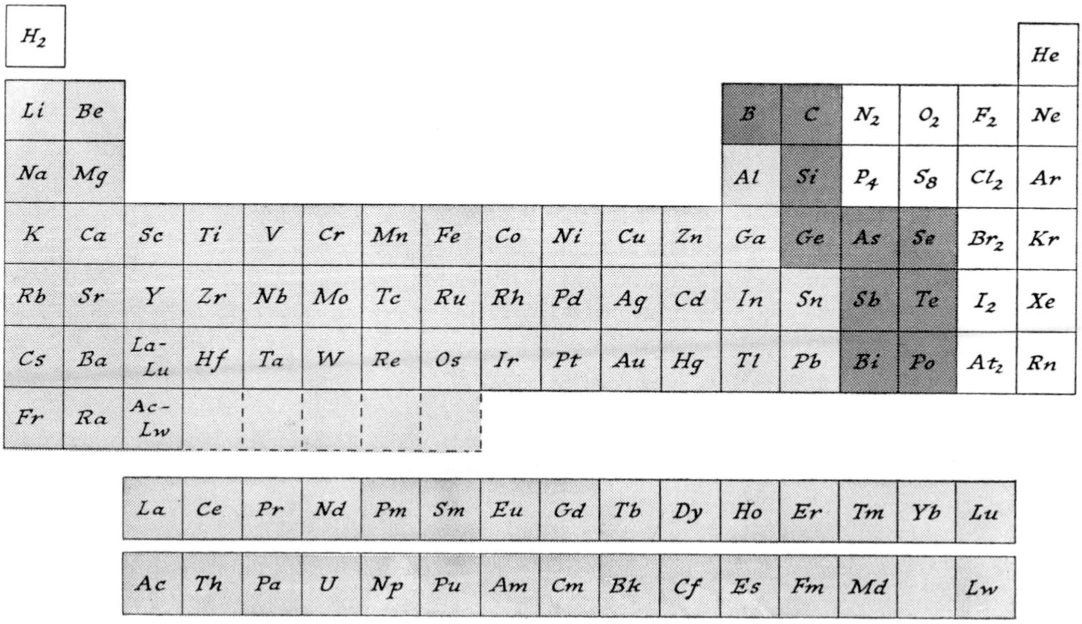

H_2																	He
Li	Be											B	C	N_2	O_2	F_2	Ne
Na	Mg											Al	Si	P_4	S_8	Cl_2	Ar
K	Ca	Sc	Ti	V	Cr	Mn	Fe	Co	Ni	Cu	Zn	Ga	Ge	As	Se	Br_2	Kr
Rb	Sr	Y	Zr	Nb	Mo	Tc	Ru	Rh	Pd	Ag	Cd	In	Sn	Sb	Te	I_2	Xe
Cs	Ba	La-Lu	Hf	Ta	W	Re	Os	Ir	Pt	Au	Hg	Tl	Pb	Bi	Po	At_2	Rn
Fr	Ra	Ac-Lw															

La	Ce	Pr	Nd	Pm	Sm	Eu	Gd	Tb	Dy	Ho	Er	Tm	Yb	Lu
Ac	Th	Pa	U	Np	Pu	Am	Cm	Bk	Cf	Es	Fm	Md		Lw

metallic behavior that suggests there is a special electron configuration.

Metals also possess unusually high thermal conductivity, as anyone who has drunk hot coffee from a tin cup can testify. It is noteworthy that among metals the best electrical conductors are also the best thermal conductors. This is a clue that these two properties are somehow related and, again, the electron configuration proves to be responsible.

Though the mechanical properties of the various metals differ, all metals can be drawn into wires and hammered into sheets without shattering. Here we find a fourth characteristic property of metals: they are malleable or workable.

LOCATION OF METALS IN THE PERIODIC TABLE

The location of the metals in the periodic table is shown in Figure 17-4. We see that the metals are located on the left side of the table, while the nonmetals are exclusively in the upper right corner. Furthermore, the elements on the left side of the table have relatively low ionization energies. We shall see that the low ionization energies of the metallic elements aid in explaining many of the features of metallic behavior.

ELECTRON BEHAVIOR IN METALS

What is the nature of the metallic bond? This bond, like all others, forms because the electrons can move in such a way that they are simultaneously near two or more positive nuclei. Our problem is to obtain some insight into the special way in which electrons in metals do this.

Consider a crystal of metallic lithium. In its crystal lattice, each lithium atom finds around itself eight nearest neighbors. Yet this atom has only one valence electron, so it isn't possible for it to form ordinary electron pair bonds to all of these nearby atoms. However, it does have four valence orbitals available so its electron and the valence electrons of its neighbors can approach quite close to its nucleus. Thus each lithium atom has an abundance of valence orbitals but a shortage of bonding electrons.

Consider the dilemma of the valence electron of a particular lithium atom. It finds eight neighbor nuclei nearby and complete freedom of movement in the empty valence orbitals around its parent nucleus. Everywhere the electron moves it finds itself between two positive nuclei. All of the space around a central atom is a region of almost uniformly low potential energy. Under these circumstances, it is not surprising that an electron can move easily from place to place. Each valence electron is virtually free to make its way throughout the crystal.

This type of argument leads us to picture a metal as an array of positive ions located at the crystal lattice sites, immersed in a "sea" of mobile electrons. The idea of a more or less uniform electron "sea" emphasizes an important difference between metallic bonding and ordinary covalent bonding. In molecular covalent bonds the electrons are localized in a way that fixes the positions of the atoms quite rigidly. We say that the bonds have directional character—the electrons tend to remain concentrated in certain regions of space. In contrast, the valence electrons in a metal are spread almost uniformly throughout the crystal, so the metallic bond does not exert the directional influence of the ordinary covalent bond.

We can obtain some idea of the effectiveness of this electron "sea" in binding the atoms together if we compare the energy necessary to vaporize one mole of a metal to the free atoms with the energy required to break one mole of ordinary covalent bonds. We find that the energy necessary to vaporize a mole of one of the alkali metals is only one-fourth to one-third of the energy needed to break a mole of ordinary covalent bonds. This is not too surprising. The ionization energy of a free alkali metal atom is small; this means that the valence electron in the free atom does not experience a strong attraction to the nucleus. Since the electron is not strongly attracted by one alkali metal atom, it is not strongly attracted by two or three such atoms in the metallic crystal. Thus, the binding energy between electrons and nuclei in the alkali metal crystals is rather small, and the resulting metallic bonds are rather weak. We might expect, however, that the metallic bond would become stronger in those elements which have a greater number of valence electrons and a greater nuclear charge. In these cases there are more electrons in the "sea," and each electron is more strongly bound, owing to the increased nuclear charge. This argument is in accord with the experimental heats of vaporization shown in Table 17-I.

Table 17-I

HEATS OF VAPORIZATION OF METALS (kcal/mole)

Second Row	Li 32.2	Be 53.5	B 129
Third row	Na 23.1	Mg 31.5	Al 67.9
Fourth Row	K 18.9	Ca 36.6	Sc 73
Fifth Row	Rb 18.1	Sr 33.6	Y 94
Sixth Row	Cs 16.3	Ba 35.7	La 96

To pick a specific case, let us compare the heats of vaporization of magnesium and aluminum. The higher value for aluminum shows that the metallic bond is indeed stronger when the number of valence electrons and the charge on the nucleus increase. Thus the strength of the metallic bond tends to increase as we go from left to right along a row in the periodic table. The transition metal elements are harder and melt and boil at higher temperatures than the alkali or alkaline earth metals.

EXPLANATION OF THE PROPERTIES OF METALS

The nonlocalized or mobile electrons account for the many unique features of metals. Since metallic bonds do not have strong directional character, it is not surprising that many metals can be easily deformed without shattering their crystal structure. Under the influence of a stress, one plane of atoms may slip by another, but as they do so, the electrons are able to maintain some degree of bonding between the two planes. Metals can be hardened by alloying them with elements which do have the property of forming directed covalent bonds. Often just a trace of carbon, phosphorus, or sulfur will turn a relatively soft and workable metal into a very brittle solid.

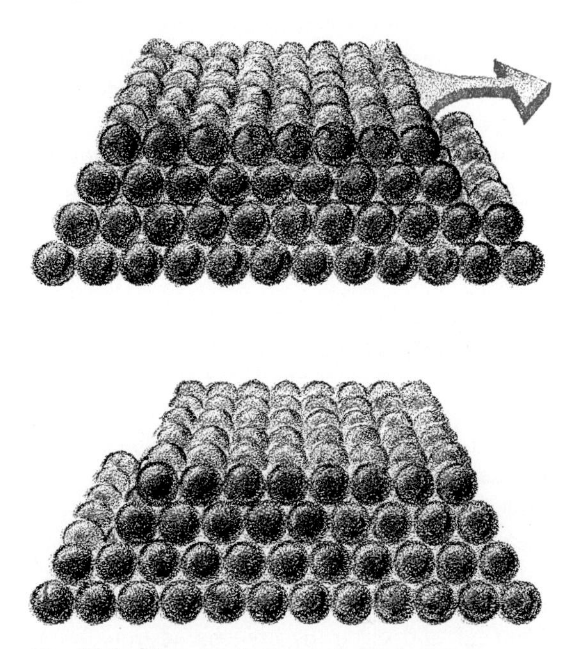

Fig. 17-5. **Slippage of planes of metal atoms.**

Metals conduct electricity because some valence electrons are free to move throughout the solid. At the same time, these mobile electrons are effective in holding the crystal together because wherever they move, they are simultaneously close to two or more nuclei. In covalently bonded solids the electrons are strongly localized in the space between a particular pair of atoms. In order for these substances to conduct electricity, a great deal of energy must be supplied to remove the electrons from this region between the atoms. This energy is not available in normal electric fields, so covalent substances do not normally conduct electricity.

The excellent heat conductivity of metals is also due to the mobile electrons. Electrons which are in regions of high temperature can acquire large amounts of kinetic energy. These electrons move through the metal very rapidly and give up their kinetic energy to heat the crystal lattice in the cooler regions. In substances where the electrons are highly localized, heat is conducted as small amounts of energy are transferred from one atom to its immediate neighbor; this is a slower process than electron energy conduction.

To complete our discussion of metallic bonding we must explain why metallic properties eventually disappear as we proceed from left to right along a row in the periodic table.

We have seen that the reasons for the mobility of electrons in metals are that they are readily removed from the atom (the ionization energy is low) and that they can be close to two or more positive nuclei just about anywhere in the crystal (there are numerous vacant valence orbitals). As the nuclear charge on atoms increases and the vacant orbitals become filled, the regions immediately between two nuclei become relatively more attractive to the electron, compared with all other regions. Electrons tend to be more and more localized in these regions, and normal covalent bonds with their directional character appear.

In summary we can say that the metallic bond is a sort of nondirectional covalent bond. It occurs when atoms have few valence electrons compared with vacant valence orbitals and when these valence electrons are not held strongly.

17-2 COMPOUNDS

We have seen that the pure elements may solidify in the form of molecular solids, network solids, or metals. Compounds also may condense to molecular solids, network solids, or metallic solids. In addition, there is a new effect that does not occur with the pure elements. In a pure element the ionization energies of all atoms are identical and electrons are shared equally. In compounds, where the most stable electron distribution need not involve equal sharing, electric dipoles may result. Since two bonded atoms may have different ionization energies, the electrons may spend more time near one of the positive nuclei than near the other. This charge separation may give rise to strong intermolecular forces of a type not found in the pure elements.

17-2.1 van der Waals Forces and Molecular Substances

Though charge separations are possible in compounds, there are many molecules that do *not* have appreciable electric dipoles. On cooling, these molecules behave much like the molecules of pure elements. If the bonding capacity of each atom is completely satisfied, then only the weak van der Waals forces remain between molecules. These weak interactions give low melting solids and low boiling liquids that retain many of the properties of the gaseous molecules.

There are three factors that seem to be particularly important in determining the magnitudes of van der Waals forces: the number of electrons, the molecular size, and the molecular shape. These factors are effective both for elements and compounds, though greater variety is found for compounds.

VAN DER WAALS FORCES AND NUMBER OF ELECTRONS

We have already observed in Chapter 6 that the melting and boiling points of the inert gases increase as the number of electrons increases (see Figure 6-3). Elements and compounds with covalent bonding behave in the same way. Figure 17-6 shows this in a graphical presentation. Figure 17-6A shows the melting and boiling point trends among the inert gases and among the halogens. The horizontal axis shows the row number, which furnishes an index of the total number of electrons of the respective elements. Figure 17-6B refers to compounds with formulas CX_4. Again the horizontal axis shows the row number but now of the outermost atoms in the

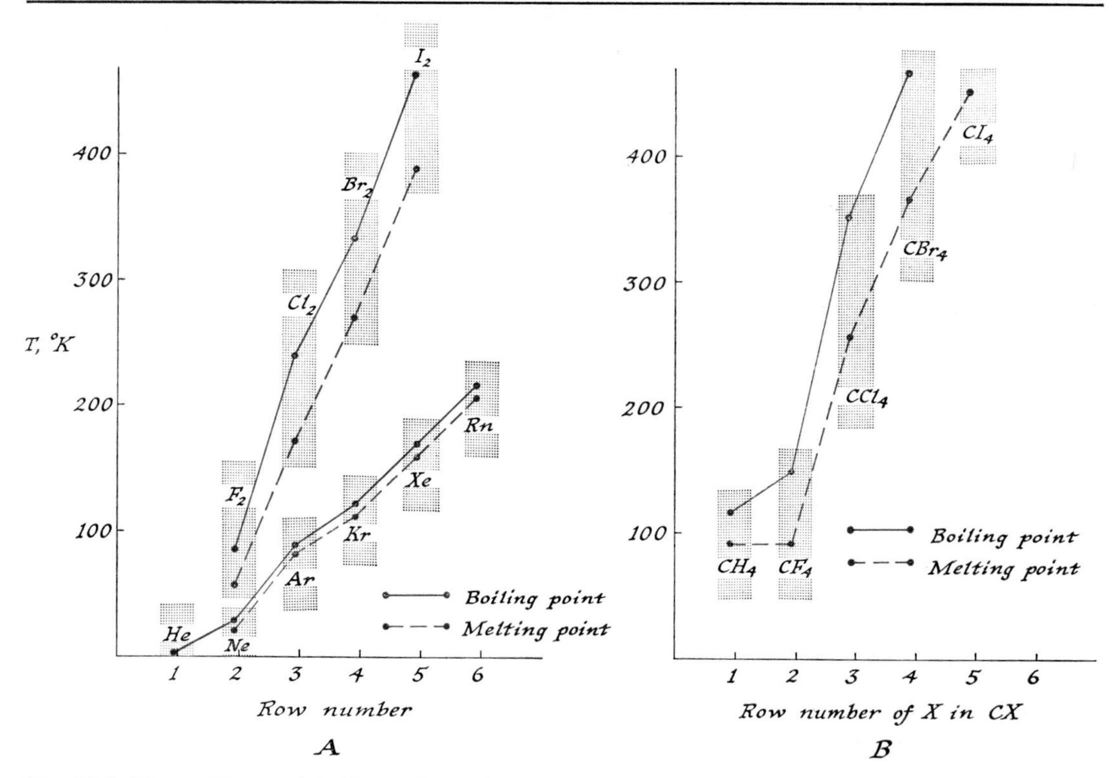

Fig. 17-6. **The melting and boiling points** *of some molecular compounds and the halogens.* **A.** *The inert gases and the halogens.* **B.** *The carbon compounds of formula CX_4.*

molecule since these are the atoms which "rub shoulders" with neighboring molecules. As far as van der Waals forces are concerned, it is quite important that CBr_4 has atoms from the fourth row of the periodic table on the "surface" of the molecule and somewhat less important that the central atom, carbon, is from the second row. The outermost atoms are most influential in fixing intermolecular forces.

VAN DER WAALS FORCES AND MOLECULAR SIZE

If comparisons are made among similar molecules, then the larger the molecule, the higher is its melting point. For example, if we compare methane, CH_4, and ethane, C_2H_6, the exterior atoms are the same—hydrogen atoms. Still, the boiling point of ethane, 185°K, is higher than that of methane, 112°K. This difference is attributed to the fact that there must be greater contact surface between two ethane molecules than between two methane molecules. The same effect is found for C_2F_6 (boiling point, 195°K) and CF_4 (boiling point, 145°K); for C_2Br_6 (this substance decomposes at 483°K before it reaches its boiling point) and CBr_4 (boiling point, 463°K).

Notice that the two factors just mentioned, number of electrons and molecular size, might lead to another generalization—that the boiling point goes up in proportion to molecular weight. The molecular weight, the molecular size, and the number of electrons all tend to increase together. This molecular weight–boiling point correlation has some usefulness among molecules of similar composition and general shape but chemists do not feel that there is a direct causative relation between molecular weight and boiling point.

VAN DER WAALS FORCES AND MOLECULAR SHAPE

Substances whose structures have a high degree of symmetry generally have higher melting points than closely related compounds that lack this symmetry. There are striking examples of this

Table 17-II. **THE EFFECT OF MOLECULAR SHAPE ON MELTING POINT: cis- AND trans- ISOMERS**

NAME	FORMULA	m.p. *cis-*	m.p. *trans-*
1,2-dichloroethylene	ClCH=CHCl	−80°C	−50°C (*trans-* 30° higher m.p.)
butenoic acid	$CH_3CH{=}CHCOOH$	15°	72° (*trans-* 57° higher m.p.)
fumaric, maleic acids	HOOCCH=CHCOOH	130°	290° (*trans-* 160° higher m.p.)

among the double-bonded compounds. The *cis-* and *trans-* isomers of many such compounds have melting point and boiling point differences that can be traced to the differences in molecular shape. For example, the higher melting point of *trans*-1,2 dichloroethylene than that of the isomeric *cis*-1,2 dichloroethylene may be partly explained by arguing that the long and symmetrical *trans-* form can pack into an orderly crystal lattice in a neater and more compact fashion than the "one-sided" *cis-* form molecules. This and two other examples of this are shown in Table 17-II. (The dichloroethylene structures are shown in Figure 16-18.)

Another example of the influence of molecular symmetry on physical properties is found in two structural isomers of the formula C_5H_{12}. These are called normal pentane and neopentane and their molecular shapes differ drastically as shown in Figure 17-7.

The extended molecule, *n*-pentane, has a zig-zag shape. We see that van der Waals forces act between the external envelope of hydrogen atoms of one molecule and those of adjacent molecules. This large surface contact gives a relatively high boiling point. On the other hand, this flexible, snake-like molecule does not pack readily in a regular lattice, so its crystal has a low melting point.

Contrast the highly compact, symmetrical neo-

Fig. 17-7. **Molecular shape,** *a factor that influences melting and boiling points.*

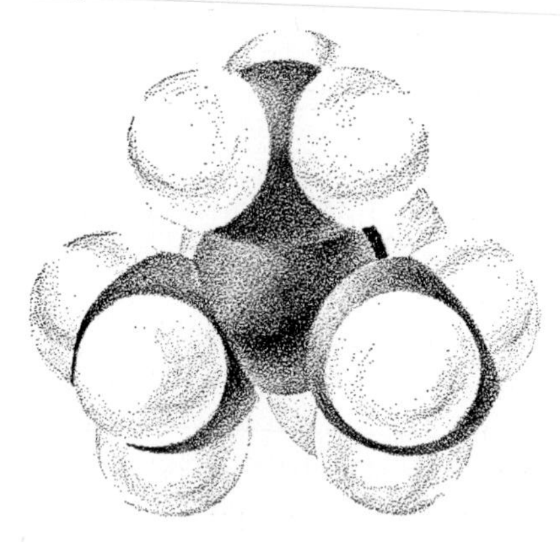

H_3C—CH_2—CH_2—CH_2—CH_3

Normal pentane

b.p. 36°C
m.p. -130°C

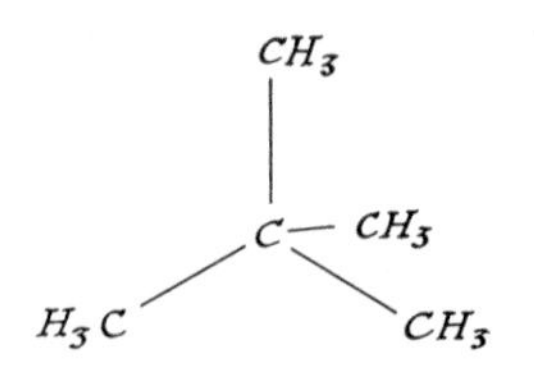

Neopentane

b.p. 9°C
m.p. -20°C

pentane. This ball-like molecule readily packs in an orderly crystal lattice which, because of its stability, has a rather high melting point. Once melted, however, neopentane forms a liquid that boils at a temperature below the boiling point of *n*-pentane. Neopentane has less surface contact with its neighbors and hence is more volatile.

It is well to add that most of the compounds of carbon condense to molecular liquids and solids. Their melting points are generally low (below about 300°C) and many carbon compounds boil below 100°C. The similar chemistry of the liquid and solid phases shows the retention of the molecular identities.

17-2.2 Covalent Bonds and Network Solid Compounds

Compounds can form network solids and, since two or more different atoms are involved, there is much greater variety among the network solid compounds than among the network solid elements. Silica, with empirical formula SiO_2, is a network solid. Silica and other silicon-oxygen compounds make up about 87% of the earth's crust. Almost all common minerals contain substantial amounts of silicates, the general term for silicon-oxygen solids. These are network solids but with interesting and important variations. Figures 17-8, 17-9, and 17-10 show in a schematic way the three types of network solids formed by silicon. The silicon is always tetravalent but in some of its compounds it forms infinite silicon-oxygen-silicon chains; in some it forms infinite interlinked sheets; and, in some, it forms an infinite three-dimensional network solid.

Many properties of silicates can be understood in terms of the type of network lattice formed. In the "one-dimensional" networks, shown in Figure 17-8, the atoms within a given chain are strongly linked by covalent bonds but the chains interact with each other through much weaker forces. This is consistent with the thread-like properties of many of these silicates. The asbestos minerals are of this type.

In a similar way, the sheets of the "two-dimensional" network silicates, shown in Figure 17-9, are held together weakly. Hence these minerals cleave readily into thin but strong sheets. The micas have this type of structure. Clays also have this structure, and their slippery "feel" when wet can be explained in terms of the hydration of the planes on the outside of the crystals. The three-dimensional network shown in Figure 17-10 is silica (quartz). Like diamond, it is hard and it has a high melting point. The various minerals that make up granite are of this type.

17-2.3 Metallic Alloys

We have already learned that metals may be deformed easily and we have explained this in terms of the absence of directional character in metallic bonding. In view of this principle, it is not surprising that two-element or three-element metallic crystals exist. In some of these, regular arrangements of two or more types of atoms are found. The composition then is expressed in simple integer ratios, so these are called metallic compounds. In other cases, a fraction of the atoms of the major constituent have been replaced by atoms of one or more other elements. Such a substance is called a solid solution. These *metals containing two or more types of atoms are called* ***alloys***.

ELECTRICAL CONDUCTIVITY

Electrical conductivity in metals apparently depends upon the smooth and uninterrupted movement of electrons through the lattice. This is suggested by the fact that small amounts of impurities reduce the conductivity very much. We shall see, in Chapter 22, that copper is purified commercially to 99.999% and the reason is directly connected to the consequent gain in electrical conductivity.

Table 17-III shows some conductivities of copper with

Table 17-III

CONDUCTIVITY OF COPPER ALLOYS (ALL AT 20°C UNLESS NOTED)

PERCENT COPPER	ALLOYING ELEMENT	PERCENT ALLOYING ELEMENT	CONDUCTIVITY $(\text{ohm-cm})^{-1}$
100.00	—	—	5.9×10^5
99	Mn (0°C)	0.98	2.1×10^5
95.8	Mn	4.2	0.56×10^5
97	Al (0°C)	3	1.2×10^5
90	Al (0°C)	10	0.79×10^5
88	Sn	12	0.56×10^5

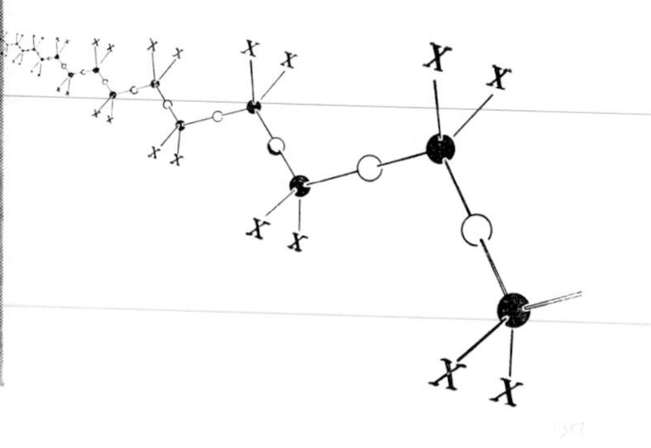

Fig. 17-8. **One-dimensional network silicates:** *the asbestos minerals.*

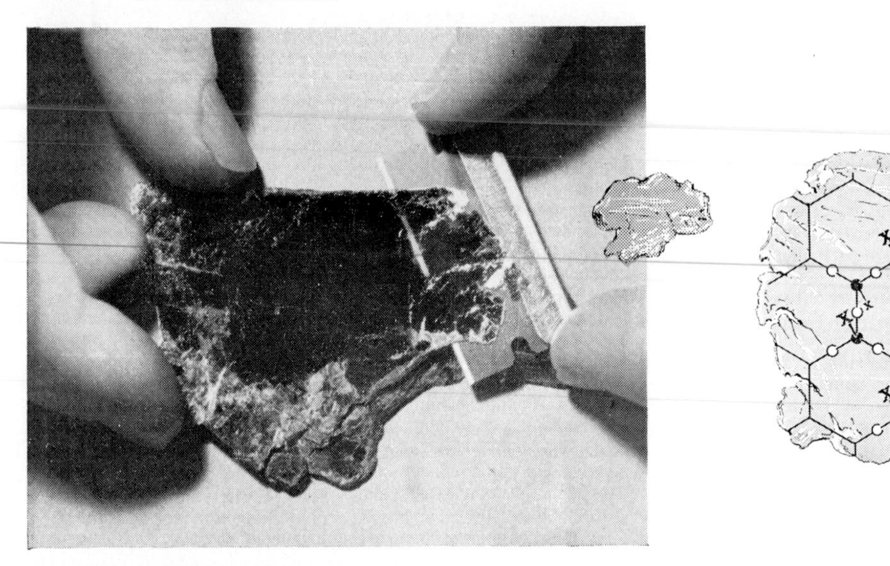

Fig. 17-9. **Two-dimensional network silicates:** *the mica and clay minerals.*

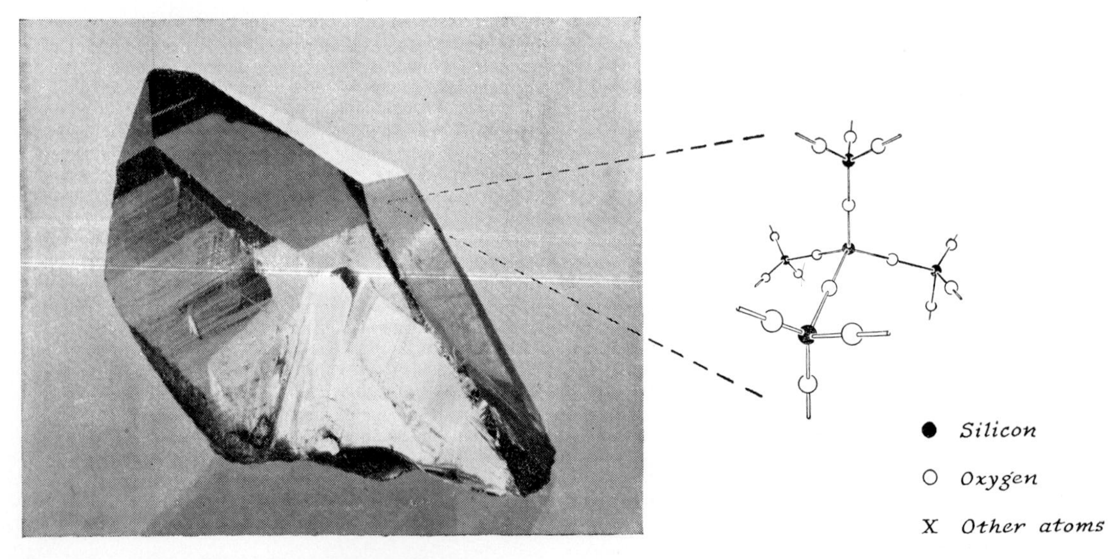

Fig. 17-10. **Three-dimensional network silicates:** *granitic minerals.*

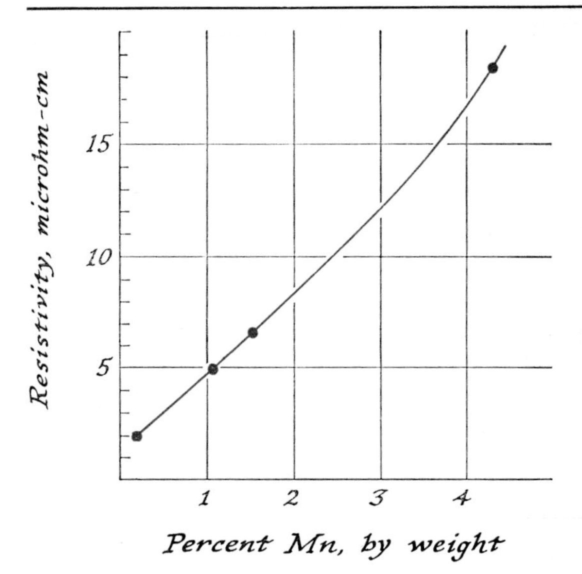

Fig. 17-11. **Electrical resistivity** *of copper containing manganese.*

various impurities.* Figure 17-11 shows the data for copper-manganese alloys graphically. The figure shows resistivity, the reciprocal of conductivity, plotted against percent manganese (by weight). The importance of purification of copper for electrical wire is evident in this figure if we remember that the power lost in a conductor is proportional to resistance (for a given current). In a conductor hundreds of miles long, a factor of two reduction in resistivity is a lucrative gain to a company selling electrical power.

EXERCISE 17-2

Use Figure 17-11 to estimate the resistivities of two metal samples, one made of pure copper and the other of a copper-manganese alloy containing one atom of manganese for every one hundred copper atoms. Calculate the ratio of the cost due to power loss from wire of the impure material to the cost due to the power loss from wire of the pure material.

HARDNESS AND STRENGTH

Alloys are harder and stronger than pure metals as usually prepared. The most familiar example is steel and pure iron. The tensile strength of pure iron can be increased ten-fold by the addition of only a percent of carbon and smaller amounts of nickel or manganese. The tensile strength of brass (65–70% Cu, 35–30% Zn) is more than twice that of copper and four times that of zinc.

The hardness and strength of alloys can be explained in terms of bonding. The impurity atoms added may form localized and rigid bonds. These tend to prevent the slippage of atoms past each other, which results in a loss of malleability and an increase in hardness.

17-2.4 Ionic Solids

Thus far we have not considered the effects that arise from charge separations. The most extreme case is represented by the formation of ionic solids. Usually, these can be looked on as arrays of positive and negative ions, neatly stacked so that each positive ion has only negative ion neighbors and each negative ion has only positive ion neighbors. Figure 5-10 (p. 81) shows such a crystal arrangement, that of sodium chloride. Why does such a solid form and what are its properties? These are the questions we shall try to answer here.

THE STABILITY OF IONIC CRYSTALS

In discussing the bonding in the gaseous LiF molecule, the electric dipole of the molecule is explained in terms of the different ionization energies of Li and F atoms. Though the molecule holds together because the bonding electrons are near both nuclei, the energy favors an electron distribution concentrated toward the fluorine. A stable and polar molecule is formed. Stable, perhaps, but in the gaseous state, reactive! The valence orbitals of the lithium atom are almost vacant. According to our experience (for example, with CH_2, BH_3, carbon atoms, metal atoms) the presence of empty valence orbitals implies that additional electron sharing can occur. Lithium fluoride molecules are, then, more stable when they condense so as to place each lithium atom simultaneously near several fluorine atoms. Just as in metals, an atom with vacant orbitals is more stable with several neighbors. Then the electrons held by the neighbor atoms can be near two or more nuclei at once. There is, however, a significant difference from metals—in solid lithium fluoride, half of the atoms have high

* The conductivities are given in units $(\text{ohm-cm})^{-1}$. The reciprocal of this number is the resistance one would find (in ohms) for a wire 1 cm long and with a cross-section of 1 cm^2.

Fig. 17-12. **Sodium chloride crystals.**

ionization energies. Fluorine atoms hold their electrons tightly. Therefore the characteristic electron mobility of metals is not present in the ionic solids. The absence of mobile electrons implies that none of the metallic properties is expected. Let us see what properties such a solid does have.

PROPERTIES OF IONIC CRYSTALS

Ionic solids, such as lithium fluoride and sodium chloride, form regularly shaped crystals with well defined crystal faces. Pure samples of these solids are usually transparent and colorless but color may be caused by quite small impurity contents or crystal defects. Most ionic crystals have high melting points.

Molten lithium fluoride and sodium chloride have easily measured electrical conductivities. Nevertheless, these conductivities are lower than metallic conductivities by several factors of ten. Molten sodium chloride at 750°C has a conductivity about 10^{-5} times that of copper metal at room temperature. It is unlikely that the electric charge moves by the same mechanism in molten NaCl as in metallic copper. Experiments show that the charge is carried in molten NaCl by Na^+ and Cl^- ions. This electrical conductivity of the liquid is one of the most characteristic properties of substances with ionic bonds. In contrast, molecular crystals generally melt to form molecular liquids that do not conduct electricity appreciably.

17-2.5 Effects Due to Charge Separation

We have considered the weak van der Waals forces that cause the condensation of covalent molecules. The formation of an ionic lattice results from the stronger interactions among molecules with highly ionic bonds. But most molecules fall between these two extremes. Most molecules are held together by bonds that are largely covalent, but with enough charge separation to affect the properties of the molecules. These are the molecules we have called polar molecules.

Chloroform, $CHCl_3$, is an example of a polar molecule. It has the same bond angles as methane, CH_4, and carbon tetrachloride, CCl_4. Carbon, with sp^3 bonding, forms four tetrahedrally oriented bonds (as in Figure 16-11). However, the cancellation of the electric dipoles of the four C—Cl bonds in CCl_4 does not occur when one of the chlorine atoms is replaced by a hydrogen atom. There is, then, a molecular dipole remaining. The effects of such electric dipoles are important to chemists because they affect chemical properties. We shall examine one of these, solvent action.

SOLVENT PROPERTIES AND MOLECULAR DIPOLES

The forces between molecules are strongly affected by the presence of molecular dipoles. Two molecules that possess molecular dipoles tend to attract each other more strongly than do molecules without dipoles. One of the most important results of this is found in solvent properties.

Table 17-IV shows some solubility data of various solutes in the two solvents, carbon tetrachloride, CCl_4 and acetone, CH_3COCH_3. These two solvents differ in their polar properties. In CCl_4 the central carbon atom is surrounded by four bonds that form a regular tetrahedron like that pictured in Figure 16-11. With this molecular shape, CCl_4 has a zero molecular dipole. In contrast, acetone has a bent structure and the oxygen atom gives it a significant electric dipole.

Table 17-IV

SOLUBILITIES IN CARBON TETRACHLORIDE, CCl_4, AND IN ACETONE, CH_3COCH_3 (25°C, moles/liter)

		SOLVENT	
SOLUTE	SOLUTE POLARITY	CCl_4 (*nonpolar*)	CH_3COCH_3 (*polar*)
CH_4, methane	nonpolar	0.029	0.025
C_2H_6, ethane	nonpolar	0.22	0.13
CH_3Cl, chloromethane	polar	1.7	2.8
CH_3OCH_3, methyl ether	polar	1.9	2.2

Contrast the solubilities in Table 17-IV. The first two substances, CH_4 and C_2H_6, have zero molecular dipoles. In each case, the solubility in CCl_4 exceeds the solubility in CH_3COCH_3. The next two substances, CH_3Cl and CH_3OCH_3, have nonzero molecular dipoles. In each of these cases, the solubility in acetone is the larger.

There is a reasonable explanation of the data in Table 17-IV. When a solute dissolves, the solute molecules must be separated from each other and then surrounded by solvent molecules. Furthermore, the solvent molecules must be pushed apart to make room for the solute molecules. Since dipoles interact strongly with each other, a polar molecule such as CH_3Cl is energetically more stable when surrounded by solvent molecules that are also polar. Hence, CH_3Cl has the higher solubility in the polar solvent acetone than in carbon tetrachloride. On the other hand, a nonpolar molecule such as CH_4 will find it difficult to wedge in between the strongly interacting molecules of a polar solvent —more difficulty than it will encounter in dissolving in a nonpolar solvent. Hence CH_4 has higher solubility in the nonpolar solvent, CCl_4.

SOLUBILITY OF ELECTROLYTES IN WATER

The dissolving of electrolytes in water is one of the most extreme and most important solvent effects that can be attributed to electric dipoles. Crystalline sodium chloride is quite stable, as shown by its high melting point, yet it dissolves readily in water. To break up the stable crystal arrangement, there must be a strong interaction between water molecules and the ions that are formed in the solution. This interaction can be explained in terms of the dipolar properties of water.

When an electric dipole is brought near an ion, the energy is lower if the dipole is oriented to place unlike charges in proximity. Hence water molecules tend to orient preferentially around ions, the positive end of the water dipole pointing inward if the ion carries negative charge and the negative end pointing inward if the ion carries positive charge. Figure 17-13 shows this process schematically: it is called **hydration.**

There are two effects of the orientation of water dipoles around the ions. First, the energy is lowered because the orientation serves to bring unlike charges near each other. This tends to encourage the ions to leave the sodium chloride crystal and enter the solution. Also, however, there is an effect on randomness whose magnitude is difficult to predict. The orientation of the water molecules around the ion, fixing them relative to the ion, constitutes an orderly arrangement. Since all systems tend toward maximum randomness, the orientation effect works against molecules leaving the crystal to enter the solu-

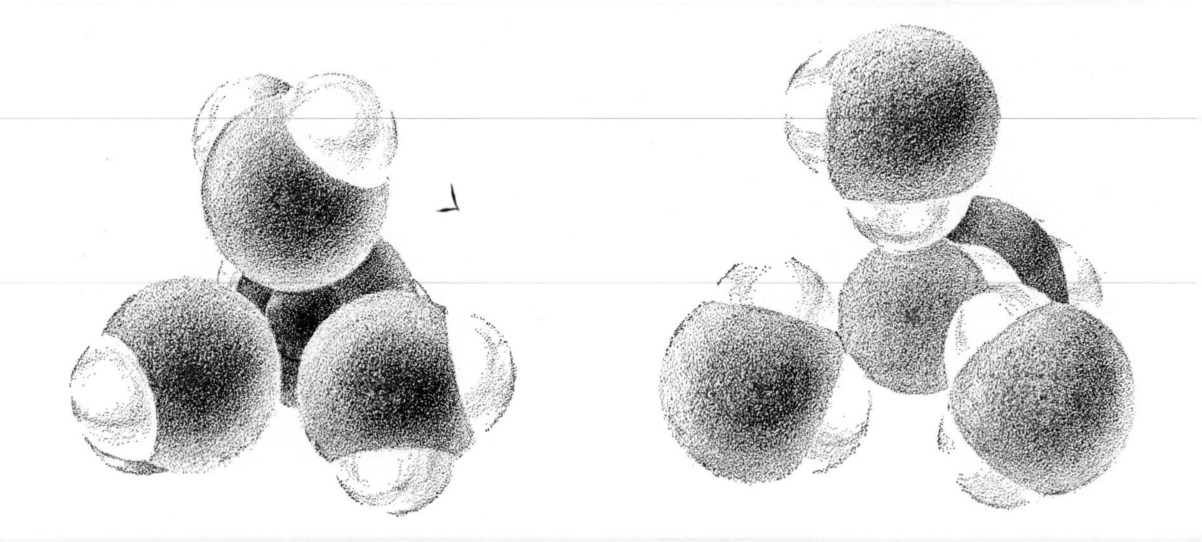

Fig. 17-13. **Hydration of ions:** *orientation of water dipoles around ions in aqueous solutions.*

tion. These two effects of ion-hydration, lowering the energy of an electrolyte solute while decreasing the change in randomness as it dissolves, give water distinctive properties as an electrolyte solvent. It helps explain, for example, why some salts absorb heat as they dissolve in water (for example, NH_4Cl) while some release heat as they dissolve (for example, NaOH). For most solvents the crystal has lower energy than the solution, and heat is absorbed as solid dissolves. In water, however, the hydration effects can cause the solution to have the lower energy, so heat can be evolved during the dissolving process.

17-2.6 Hydrogen Bonds

In Figure 17-6A we saw that the boiling points of symmetrical molecules increase regularly as we drop down in the periodic table. Figure 17-14 shows the corresponding plot for some molecules possessing electric dipoles.

Consider first the boiling points of HI, HBr, HCl, and HF. The last, hydrogen fluoride, is far out of line, boiling at 19.9°C instead of below −95°C as would be predicted by extrapolation from the other three. There is an even larger discordancy between the boiling point of H_2O and the value we would predict from the trend suggested by H_2Te, H_2Se, and H_2S.

Could the extremely high boiling points of HF and H_2O be due to the fact that these are the *smallest* molecules of their respective series? No, this does not appear to be the explanation, for corresponding discrepancies are not present in the data plotted in Figure 17-6A. There must be some other explanation for these exceptional boiling points. There must be forces of some new kind between the molecules of H_2O and of HF that tend to keep them in the liquid phase.

These same forces are recognized in solid compounds. The most familiar example is solid H_2O, or ice. Ice has a crystal structure in which the oxygen and hydrogen atoms are distributed in a regular hexagonal crystalline lattice that somewhat resembles the diamond lattice (see Figure 17-2). Each oxygen atom is surrounded by four other oxygen atoms in a tetrahedral arrangement. The hydrogen atoms are found on the lines extending between the oxygen atoms.

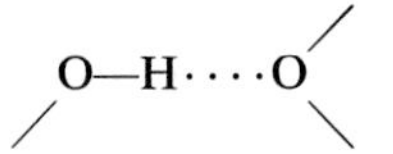

The attractive force between —OH and O must be the bond that joins the water molecules together into the crystal lattice of ice. This bond is a hydrogen bond.

ENERGY OF HYDROGEN BONDS

The hydrogen bond is usually represented by O—H····O in which the solid line represents the original O—H bond in the parent compound (as in water, HOH, or methyl alcohol, CH_3OH).

Fig. 17-14. **The boiling points** *of some hydrides.*

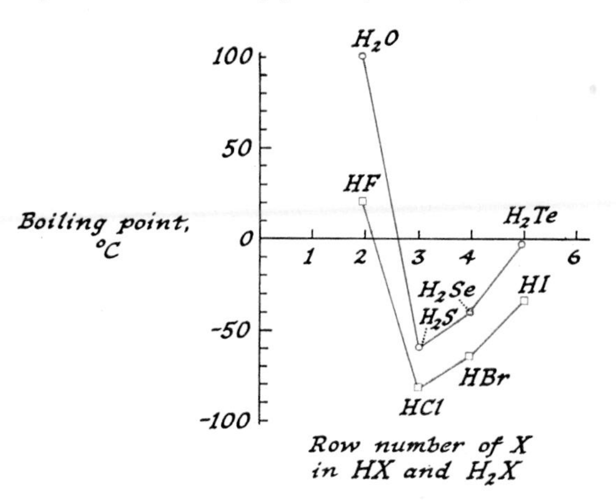

The dotted line shows the second bond formed by hydrogen, the bond called the hydrogen bond. It is usually dotted to indicate that it is much weaker than a normal covalent bond. Consideration of the boiling points in Figure 17-14, on the other hand, shows that the interaction must be much stronger than van der Waals forces. Experiments show that most hydrogen bonds release between 3 kcal/mole and 10 kcal/mole upon formation:

$$\Delta H = -3 \text{ to } -7 \text{ kcal/mole} \qquad (1)$$

The energy of this bond places it between van der Waals and covalent bonds. Roughly speaking, the energies are in the ratio

van der Waals attractions	:	hydrogen bonds	:	covalent bonds
1	:	10	:	100

WHERE HYDROGEN BONDS ARE FOUND

Hydrogen bonds are found between only a few atoms of the periodic table. The commonest are those in which H connects two atoms from the group F, O, and N, and less commonly Cl.

The hydrogen bond to fluorine is clearly evident in most of the properties of hydrogen fluoride. The high boiling point of HF, compared with those of the other hydrogen halides, is one of several pieces of data that show that HF does not exist in the liquid compound as separate HF molecules. Instead there are aggregates of molecules, which we describe in general terms as $(HF)_x$. Gaseous hydrogen fluoride contains the molecular species H_2F_2, H_3F_3, and so on up to H_6F_6 as well as some single HF molecules.

These species can be represented in a descriptive formula such as the following:

$$\text{H—F}\cdots\text{H—F}\cdots\text{H—F}\cdots\text{H—F} \qquad (2)$$

An extreme example of the fluorine-hydrogen bond is found in the hydrogen difluoride ion, HF_2^-. This ion exists in acidic solutions of fluorides,

$$H^+(aq) + F^-(aq) \rightleftarrows HF(aq) \qquad (3)$$

$$HF(aq) + F^-(aq) \rightleftarrows HF_2^-(aq) \qquad (4)$$

and in the ionic crystal lattice of salts such as KHF_2. The HF_2^- ion may be regarded as consisting of two negatively charged fluoride ions held together by a proton:

$F^- \oplus F^-$ (5)

It is not typical, however, for few hydrogen bonds form with the proton equidistant from the two atoms to which it bonds.

INTER- AND INTRAMOLECULAR HYDROGEN BONDS

One of the factors connected with the formation of strong hydrogen bonds is the acidic character of the hydrogen atom involved. Thus the hydrogen bond formed by hydrogen fluoride is one of the strongest known. Acetic acid, CH_3COOH, is a representative of an important class of acidic hydrogen bonding compounds. All of the members of this class possess the structural unit called the carboxylic acid group:

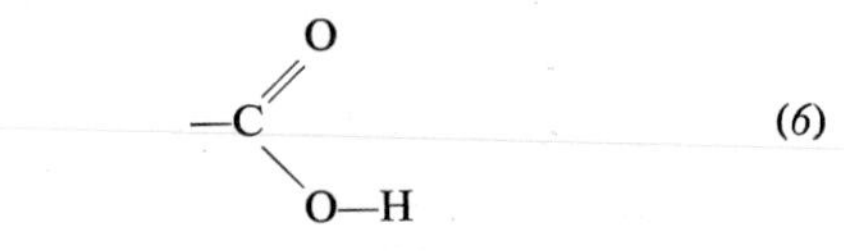

For this type of compound, the formation of hydrogen bonds can lead to the coupling of the molecules in pairs, to form a cyclic structure:

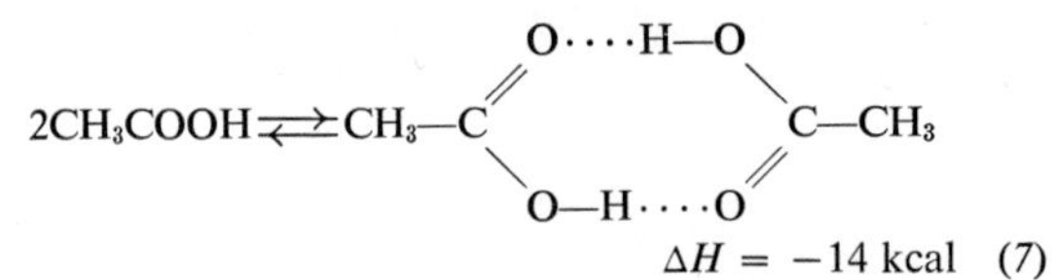

$\Delta H = -14$ kcal (7)

Here the favorable geometrical arrangement with two hydrogen bonds contributes 14 kcal to the stability of the hydrogen bonded product, (*7*). These are called *inter*-molecular hydrogen bonds (inter means between).

Hydrogen bonds can also exist when the O—H group and the other bonding atom are close together *in the same molecule* in such positions that a ring can be formed without disturbing the normal bond angles. These are called *intra*molecular hydrogen bonds (intra means within).

An example of intramolecular hydrogen bonding is provided by the *cis*- and *trans*- forms of the acid HOOC—CH═CH—COOH. The *trans*- form, fumaric acid, has a higher melting point than the *cis*- form, maleic acid. In addition to the general effect of molecular shape (mentioned earlier in this chapter), another reason for this difference is that intramolecular hydrogen bonds can exist between the two —COOH groups for maleic acid but not for fumaric acid:

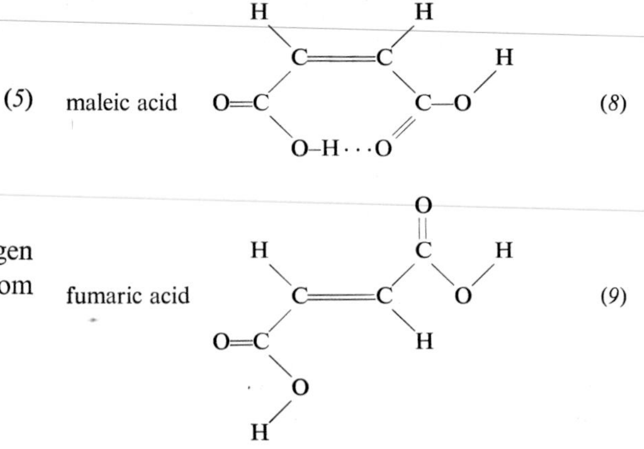

This intramolecular bonding in maleic acid, (*8*), halves its ability to form intermolecular bonds. In fumaric acid, on the other hand, all of the hydrogen bonds form *between* molecules (intermolecular bonds) to give a stronger, interlinked crystal structure.

THE NATURE OF THE HYDROGEN BOND

In the hydrogen bond we find the hydrogen atom attached to two other atoms. Yet our bonding rules tell us that the hydrogen atom, with only the 1*s* orbital for bond formation, cannot form two covalent bonds. We must seek an explanation of this second bond.

The simplest explanation for the hydrogen bond is based upon the polar nature of F—H, O—H, and N—H bonds. In a molecule such as H_2O, the electron pair in the O—H bond is displaced toward the oxygen nucleus and away from the hydrogen nucleus. This partial ionic character of the O—H bond lends to the hydrogen atom some positive character, permitting electrons from another atom to approach closely to the proton even though the proton is already bonded. A second, weaker link is formed.

THE SIGNIFICANCE OF THE HYDROGEN BOND

Hydrogen bonds play an important part in determining such properties as solubility, melting points, and boiling points, and in affecting the form and stability of crystal structures. They play a crucial role in biological systems. For example, water is so common in living matter that it must influence the chemical behavior of many biological molecules, most of which can also form hydrogen bonds. Water can attach itself by hydrogen bonding, either by providing the proton, as in

O—H····O=C (10)
H

or in accepting the proton, as in

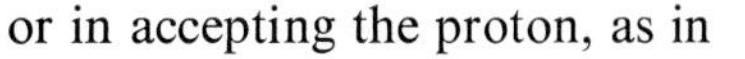

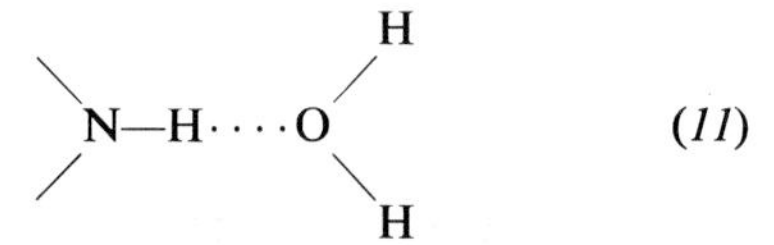

(11)

Furthermore, intramolecular hydrogen bonding is one of the chief factors in determining the structure of such important biological substances as proteins, as discussed in Chapter 24.

QUESTIONS AND PROBLEMS

1. Make a table that contrasts the melting points and boiling points of LiF, Li, and F_2, expressing the temperatures on the Centigrade scale.

2. Without looking in your textbook, do the following.

 (a) Draw an outline of the periodic table, indicating the rows but not the individual elements.
 (b) Place a number at the left of each row indicating the number of elements in that row.
 (c) Fill in the symbols for as many of the first 18 elements as you can (leave blank any that you forget).
 (d) Draw two diagonal lines across the table to separate it into three regions. Write in each region one of the words "metals," "nonmetals," "covalent solids."
 (e) Now compare your diagram to Figure 17-4.

3. Sulfur exists in a number of forms, depending upon the temperature and, sometimes, upon the past history of the sample. Three of the forms are described below. *A* is the room temperature form and it changes to *B* above the melting point of *A*, 113°C. *B* changes to *C* on heating above 160°C.

A		*B*		*C*
Crystalline solid		Liquid		Liquid
Yellow color, no metallic luster	113°C →	Clear, straw color	~200°C →	Dark color
		Viscosity (fluidity) about the same as water		Very viscous (syrupy)
m.p. = 113°C				
Dissolves in CS_2, not in water				
Electrical insulator		Electrical insulator		Electrical insulator

Which of the following structures would be most likely to account for the observed properties of each of the three forms described above?

(a) a metallic crystal of sulfur atoms;
(b) a network solid of sulfur atoms;
(c) an ionic solid of S^+ and S^- ions;
(d) a molecular crystal of S_8 molecules;
(e) a metallic liquid like mercury;
(f) a molecular liquid of S_8 molecules;
(g) a molecular liquid of S_n chains, with n = a very large number;
(h) an ionic liquid of S^+ and S^- ions.

4. Contrast the bonds between atoms in metals, in van der Waals solids, and in network solids in regard to:

   (a) bond strength;
   (b) orientation in space;
   (c) number of orbitals available for bonding.

5. Aluminum, silicon, and sulfur are close together in the same row of the periodic table, yet their electrical conductivities are widely different. Aluminum is a metal; silicon has much lower conductivity and is called a semiconductor; sulfur has such low conductivity it is called an insulator. Explain these differences in terms of valence orbital occupancy.

6. Sulfur is made up of S_8 molecules; each molecule has a cyclic (crown) structure. Phosphorus contains P_4 molecules; each molecule has a tetrahedral structure. On the basis of molecular size and shape, which would you expect to have the higher melting point?

7. Discuss the conduction of heat by copper (a metal) and by glass (a network solid) in terms of the valence orbital occupancy and electron mobility.

8. The elements carbon and silicon form oxides with similar empirical formulas: CO_2 and SiO_2. The former sublimes at $-78.5°C$ and the latter melts at about 1700°C and boils at about 2200°C. From this large difference, propose the types of solids involved. Draw an electron dot or orbital representation of the bonding in CO_2 that is consistent with your answer.

9. How do you account for the following properties in terms of the structures of the solids?

   (a) Graphite and diamond both contain carbon. Both are high melting yet the diamond is very hard while graphite is a soft, greasy solid.
   (b) When sodium chloride crystals are shattered, plane surfaces are produced on the fragments.
   (c) Silicon carbide (carborundum) is a very high melting, hard substance, used as an abrasive.

10. If you were given a sample of a white solid, describe some simple experiments that you would perform to help you decide whether or not the bonding involved primarily covalent bonds, ionic bonds, or van der Waals forces.

11. If elements *A*, *D*, *E*, and *J* have atomic numbers, respectively, of 6, 9, 10, and 11, write the formula for a substance you would expect to form between the following:

   (a) *D* and *J*;
   (b) *A* and *D*;
   (c) *D* and *D*;
   (d) *E* and *E*;
   (e) *J* and *J*.

   In each case describe the forces involved between the building blocks in the solid state.

12. Consider each of the following in the solid state: sodium, germanium, methane, neon, potassium chloride, water. Which would be an example of

   (a) a solid held together by van der Waals forces that melts far below room temperature;
   (b) a solid with a high degree of electrical conductivity that melts near 200°C;
   (c) a high melting, network solid involving covalently bonded atoms;
   (d) a nonconducting solid which becomes a good conductor upon melting;
   (e) a substance in which hydrogen bonding is pronounced?

13. Predict the order of increasing melting point of these substances containing chlorine: HCl, Cl_2, NaCl, CCl_4. Explain the basis of your prediction.

14. Identify all the types of bonds you would expect to find in each of the following crystals:

   (a) argon,
   (b) water,
   (c) methane,
   (d) carbon monoxide,
   (e) Si,
   (f) Al,
   (g) $CaCl_2$,
   (h) $KClO_3$,
   (i) NaCl,
   (j) HCN.

15. Each of three bottles on the chemical shelf contains a colorless liquid. The labels have fallen off the bottles. They read as follows.

Label No. 1	Label No. 2	Label No. 3
n-butanol	*n*-pentane	diethyl ether
$CH_3CH_2CH_2CH_2OH$	$CH_3CH_2CH_2CH_2CH_3$	$CH_3CH_2OCH_2CH_3$
mol wt = 74.12	mol wt = 72.15	mol wt = 74.12

The three bottles are marked *A*, *B*, and *C*, and a series of measurements were made on the three liquids to permit identification, as follows.

	m.p.	b.p.	density	ΔH vap'n	solubility in water
Liquid *A*	−131.5°C	36.2°C	0.63 g/cc	85 cal/g	0.036 g/100 ml
Liquid *B*	−116	34.6	0.71	89.3	7.5
Liquid **C**	−89.2	117.7	0.81	141	7.9

Which liquid should be given Label No. 1, Label No. 2, Label No. 3? Explain how each type of measurement influenced your choices.

16. Maleic and fumaric acids are *cis*- and *trans*-isomers having two carboxyl groups,

HOOC—CH═CH—COOH

Maleic acid gives up its first proton more readily than does fumaric acid. However, the opposite is the case for the second proton. Account for this in terms of structure.

PETER J. DEBYE, 1884–

During his brilliant scientific career, Peter Debye has added richly to our knowledge of the structure of physical chemistry. His research contributions have won for him awards and honorary degrees from many countries and he has earned unbounded respect wherever men seek a deeper understanding of nature.

Born in Maastrecht, the Netherlands, he graduated in electrical engineering, did his early research in theoretical physics, and received his doctorate at the University of Munich in 1908. Three years later, at the age of 27, he accepted a full professorship at the University of Zurich where his immediate predecessor was Albert Einstein. During this year he developed two of his most lasting and fundamental studies, establishing still accepted theories of the specific heat of solids and of the interactions among polar molecules. Shortly thereafter, he returned to the Netherlands as professor of theoretical physics at Utrecht. As his scientific contributions multiplied, he occupied professorships successively at the Universities of Goettingen (Germany), Zurich (Switzerland), Leipzig (Germany), and Berlin (Germany). In Berlin he was appointed director of the Max Planck Institute. During these fruitful years, his research ranged through X-ray scattering, interatomic distances, the theory of electrolytes, magnetic cooling, and dipole theory; this work won for him the Nobel Prize in Chemistry for 1936.

With the onset of World War II, politics began to interfere with his research. Debye was actually forbidden to enter the Max Planck Institute which he directed because he refused to accept German citizenship. Despite obstruction by the German government, he left Germany by way of Italy and came to the United States. In 1940 he was appointed professor of chemistry and head of the department of chemistry at Cornell University. Six years later, he became an American citizen. During the war years his research turned toward the structure and particle size of high polymers.

Attacking this new field with his usual deep insight and characteristic originality, Debye made fundamental and important contributions in the study of macromolecules. Now professor emeritus, Debye is in great demand as a consultant and lecturer. He has rare ability in presenting the most complicated subjects in a fashion that gets to the heart of the problem with penetrating clarity. Whenever he speaks at a scientific meeting, the auditorium is filled to capacity with an audience confident they will hear new and interesting ideas. Inevitably they leave inspired and stimulated by their contact with this great scientist—a man who can delve into the most profound aspects of nature and bring to them light and understanding.

CHAPTER 

The Chemistry of Carbon Compounds

The synthesis of brazilin would have no industrial value; its biological importance is problematical, but it is worthwhile to attempt it for the sufficient reason that we have no idea how to accomplish the task.

ROBERT ROBINSON, 1947

The compounds of carbon furnish one of the most intriguing aspects of all of chemistry. One reason they interest us is that they play a dominant role in the chemistry of living things, both plant and animal. Another reason is that there are innumerable carbon compounds useful to man—dyes, drugs, detergents, plastics, perfumes, fibers, fabrics, flavors, fuels—many of them tailored to suit particular needs. Manufacture of these compounds has given rise to a huge chemical industry requiring millions of tons of raw materials every year.

Where do we find the enormous quantities of carbon and carbon compounds needed to feed this giant industry? Let's begin our study of carbon chemistry by taking a look at the chief sources of carbon and carbon compounds.

18-1 SOURCES OF CARBON COMPOUNDS

18-1.1 Coal

Coal, a black mineral of vegetable origin, is believed to have come from the accumulation of decaying plant material in swamps during prehistoric eras when warm, wet climatic conditions permitted rapid growth of plants. The cycles of decay, new growth, and decay, caused successive layers of plant material to form and gradually build up into vast deposits. The accumulation of top layers of this material and of sedimentary rocks excluded air from the lower material and subjected it to enormous pressures. In time the layers were compressed into hard beds composed chiefly of the carbon that was present in the original plants, and containing appreciable amounts of oxygen, hydrogen, nitrogen, and some sulfur.

Thus, coal is not pure carbon. The "hardest" coal, anthracite, may contain from 85 to 95% carbon; the "softest," peat, is not really coal at all but one of the early stages in the geological history of coal. Peat still contains unchanged plant remains and may contain no more than 50 to 60% carbon.

When coal is heated to a high temperature *in the absence of air*, it undergoes decomposition: volatile products (coal gas and coal tar) distill away and a residue called *coke* remains. Coke is a valuable industrial material which finds its chief use in the reduction of iron ore (iron oxide) to iron for the manufacture of steel. Coke is essentially carbon that still contains the mineral substances that are present in all coals (and form the ash that results when coal or coke is burned).

About eight gallons of coal tar are obtained from a ton of coal. Coal tars are very complex mixtures; over 200 different carbon compounds have been isolated from them. While the great value of coal to mankind has been as a fuel, a source of energy, the many substances in coal gas and coal tar make coal also an important source of chemical raw materials.

18-1.2 Petroleum

Petroleum is a complex mixture which may range from a light, volatile liquid to a heavy, tarry substance. Petroleum also has its origin in living matter that has undergone chemical changes over the course of geological time. It is found in porous rock formations called oil pools, between impervious rock formations that seal off the pools. When a pool is tapped, the oil flows through the porous structure (driven by subterranean gas or water pressure) and so is brought to the surface.

18-1.3 Natural Gas

Natural gas is a mixture of low molecular weight compounds of hydrogen and carbon (hydrocarbons) found in underground "fields" of sandstone or other porous rock. This gas escapes to the surface of the earth when the field is tapped by drilling.

18-1.4 Certain Plant and Animal Products

Plants and animals are themselves highly effective chemical factories and they synthesize many carbon compounds useful to man. These include sugars, starches, plant oils and waxes, fats, gelatin, dyes, drugs, and fibers.

Because all of these sources of carbon compounds ultimately find their origin in living matter, plant or animal, the chemistry of carbon is called organic chemistry. *Compounds containing carbon are called* ***organic compounds.*** This term includes all compounds of carbon except CO_2, CO, and a handful of ionic substances (for example, sodium carbonate, Na_2CO_3, and sodium cyanide, NaCN). You may wonder how many organic substances are known. The number is actually so large it is difficult to provide a reliable estimate. A great many more compounds of carbon have been studied than of any other element except hydrogen (hydrogen is present in most carbon compounds). There are undoubtedly over one million different carbon compounds known. The number of *new* organic compounds synthesized in one year (about 100,000 per year) *exceeds the total number of compounds known* that contain no carbon!

18-2 MOLECULAR STRUCTURES OF CARBON COMPOUNDS

How can there be so many compounds containing this one element? The answer lies in the molecular structures. We shall find that carbon atoms have an exceptional tendency to form covalent bonds to other carbon atoms, forming long chains, branched chains, and rings of atoms. Each different atomic arrangement gives a molecule with distinctive properties. To understand why a particular substance has its characteristic properties, its structure must be known. Thus the determination of the molecular structure of carbon compounds is one of the central problems of organic chemistry. Let's see how it is done.

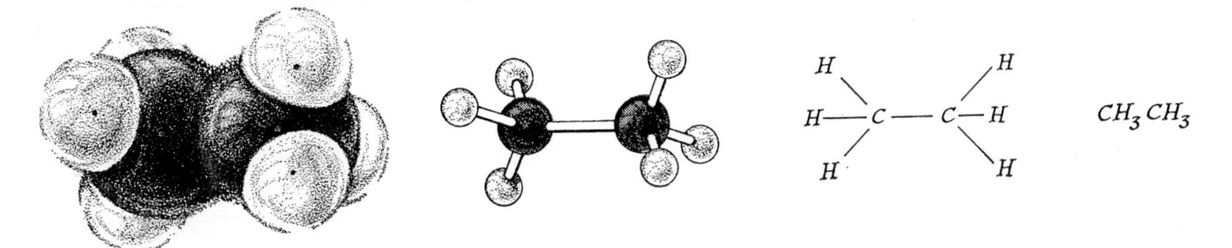

Fig. 18-1. **Structural formulas of ethane,** C_2H_6.

18-2.1 The Composition and Structure of Carbon Compounds

Ethane and ethanol* are two common carbon compounds. Ethane is a gas that usually makes up about 10% of the household gas used for heating and cooking. Its useful chemistry is almost wholly restricted to the combustion reaction. Ethanol is a liquid that takes part in a variety of useful chemical reactions. It has great value in the manufacture of chemicals and it bears little chemical resemblance to ethane. Yet, the similarity of the two names, *ethane* and *ethanol*, suggests that these compounds are related. This is so. To understand how they are related and why their chemistries are so different, we must learn their molecular structures. We must find out *what kinds of atoms* are present in each substance, *how many atoms* there are per molecule, and their *bonding arrangement*. Usually many experiments must be performed before the molecular structure of a compound is known with certainty. This fascinating problem of carbon chemistry involves three basic experimental steps: to determine the *empirical formula*, then the *molecular formula*, and finally the *structural formula*. First we shall review the information conveyed by each of these formulas, using ethane as an example. Then, in Section 18-2.2 we will consider what experiments are used in the determination of each type of formula, using ethanol as an example.

EMPIRICAL FORMULA

The empirical formula tells only the *relative* number of atoms of each element in a molecule. For example, consider ethane. Analysis shows that this is a compound of carbon and hydrogen and that there are three hydrogen atoms for every carbon atom. Its empirical formula, therefore, is CH_3.

* Ethanol is another name for the substance ethyl alcohol.

MOLECULAR FORMULA

The molecular formula tells the *total* number of atoms of each element in a molecule. Ethane is found to have a molecular weight of 30. This molecular weight together with the empirical formula tells us the molecular formula. It cannot be CH_3—this compound would have a molecular weight of 15. The molecular formula C_2H_6 also has three hydrogen atoms per carbon atom. Since it has a molecular weight of 30, it has both the correct empirical formula and the correct molecular weight. Ethane has the molecular formula C_2H_6.

EXERCISE 18-1

Write the molecular formula for the carbon-hydrogen compound containing two carbon atoms and having empirical formula CH_2. What is its molecular weight?

STRUCTURAL FORMULA

The structural formula tells which atoms are connected in the molecule. In ethane, the two carbon atoms are linked and three hydrogen atoms are attached to each carbon atom. Various ways of representing its structural formula are shown in Figure 18-1.

The formulas in Figure 18-1 all represent the same structure; the choice of which formula to use depends upon what feature of the structure is to be emphasized. The first and second drawings emphasize the three-dimensional nature of ethane; the third is a simpler way of doing the same thing; and the last formula merely shows that three hydrogens are attached to each carbon atom. It is not at all difficult to decide that CH_3CH_3 must be the structural formula for ethane. By the bonding rules developed in Chapter 16, we know that carbon is always surrounded by four electron pair bonds and that a hydrogen atom forms only one covalent bond. There is no structure other than the one shown in Figure 18-1 in which two carbon atoms and six hydrogen atoms can be bound together and satisfy all the bonding rules.

18-2.2 Experimental Determination of Molecular Structure

We have seen three steps in fixing molecular structure. What experiments are involved in each of these steps? Let's investigate the nature of these experiments using ethanol as a second example.

DETERMINING THE EMPIRICAL FORMULA

In order to determine the empirical formula of a compound, you must first find out just what elements are present in it. Sometimes this is done simply by burning some of the compound in pure oxygen. If the compound contains only carbon and hydrogen, only carbon dioxide and water will be produced. If the compound contains some nitrogen as well, nitrogen gas or one of the nitrogen oxides will be produced in the combustion and can be identified. Another way of finding out which elements are in a compound is to allow the compound to react with hot, liquid sodium metal. If the compound contains nitrogen, sodium cyanide, NaCN, will be formed; if it contains sulfur, sodium sulfide, Na_2S, will be a product. Once such reactions show which elements are in the compound, relative numbers of atoms of each element (the empirical formula) can be determined.

The usual method for finding the empirical formula is simply illustrated with ethanol, a compound containing only carbon, hydrogen, and oxygen. A weighed amount of the pure compound is completely burned in oxygen to give carbon dioxide (from the carbon) and water (from the hydrogen). The weight of the carbon dioxide reveals how much carbon was in the weighed amount of sample. The weight of water reveals how much hydrogen was in the sample. The remainder of the sample must have been oxygen. (There is no need to discuss here the procedures for compounds containing halogens, nitrogen, or sulfur, for they are quite similar.)

Suppose we burn 46 grams of ethanol. Collection of the products yields 88 grams of carbon dioxide and 54 grams of water. We wish to learn the relative numbers of carbon, hydrogen, and oxygen atoms in the compound, and we can do this by calculating the number of moles of carbon dioxide and water produced by the combustion of the 46 gram sample. Therefore, we calculate:

$$\text{Number of moles } CO_2 = \frac{88\text{ g}}{\text{mol wt } CO_2} = \frac{88\text{ g}}{44\text{ g/mole}} = 2.0\text{ moles } CO_2$$

$$\text{Number of moles } H_2O = \frac{54\text{ g}}{\text{mol wt } H_2O} = \frac{54\text{ g}}{18\text{ g/mole}} = 3.0\text{ moles } H_2O$$

Now we can make the following statements about the compound ethanol:

46 grams of ethanol yield two moles of CO_2 and three moles of H_2O

or

46 grams of ethanol contain two moles of carbon atoms and six moles of hydrogen atoms

or

46 grams of ethanol contain 24 grams of carbon atoms and 6 grams of hydrogen atoms.

Thus we have accounted for $(24 + 6) = 30$ g of the 46 g we started with. The remainder of the sample, $(46 - 30) = 16$ g, must have been oxygen. This is just

$$\frac{16\text{ g}}{\text{at. wt oxygen}} = \frac{16\text{ g}}{16\text{ g/mole}} = 1.0\text{ mole oxygen atoms}$$

Summarizing, we know that 46 grams of the compound ethanol contain

two moles of carbon atoms,
six moles of hydrogen atoms,
one mole of oxygen atoms.

Since the relative number of atoms of each element in the compound is the same as the relative number of moles of atoms of each element in the sample, we can say that the empirical formula of ethanol is

$$C_2H_6O_1 \quad \text{or} \quad C_2H_6O$$

This example has been much simplified by our selection of 46 grams of sample. In practice, less than a gram of sample is used and whole numbers of moles are not obtained. A typical set of experimentally obtained data is given in Exercise 18-2.

EXERCISE 18-2

Automobile antifreeze often contains a compound called ethylene glycol. Analysis of pure ethylene glycol shows that it contains only carbon, hydrogen, and oxygen. A sample of ethylene glycol weighing 15.5 mg is burned and the weights of CO_2 and H_2O resulting are as follows:

weight of sample burned = 15.5 mg,
weight of CO_2 formed = 22.0 mg,
weight of H_2O formed = 13.5 mg.

What is the empirical formula of ethylene glycol?

DETERMINING THE MOLECULAR FORMULA

Now we know that the relative numbers of atoms in ethanol are two carbon to six hydrogen to one oxygen. We do not know yet whether the molecular formula is C_2H_6O, $C_4H_{12}O_2$, $C_6H_{18}O_3$, or some other multiple of the empirical formula, C_2H_6O.

This returns us to the problem of the determination of molecular weight. Avogadro's Hypothesis provides one of the convenient ways of measuring molecular weight if the substance can be vaporized.

This was exactly the measurement you made in Experiment 6—it is called the vapor-density method for molecular weight determination. To apply the method to a liquid, such as ethanol, a temperature above the boiling point is needed. A weighed amount of liquid is placed in a gas collecting device held at an easily regulated temperature. For example, a steam condenser around the device provides a convenient way of holding the temperature at 100°C. When the substance has vaporized completely, its pressure and volume are measured (perhaps using equipment like that shown in Figure 9-1 of the Laboratory Manual). This provides a measurement of the weight per unit volume of gaseous ethanol at a known temperature and pressure. Again, this weight is compared with the weight of the same volume of a reference gas (usually O_2) at the same temperature and pressure.

Suppose such a vapor-density measurement shows that a given volume of ethanol at 100°C and one atmosphere weighs 1.5 times as much as the same volume of oxygen gas at 100°C and one atmosphere. Since equal volumes contain equal numbers of molecules at the same temperature and pressure (Avogadro's Hypothesis), one molecule of the unknown gas must weigh 1.5 times the weight of a molecule of O_2. Therefore,

$$\text{mol wt of the unknown gas} = (1.5) \times (\text{mol wt } O_2) = 1.5 \times 32 = 48 \text{ g/mole}$$

Even though this number is not very accurate, it will suffice for the purpose of deciding that the molecular formula is C_2H_6O (with molecular weight 46.07 g/mole), not $(C_2H_6O)_2$ (with molecular weight 92.14 g/mole), or $(C_2H_6O)_3$ (with molecular weight 138.21 g/mole), or any higher multiple of empirical formula units.

There are two other common methods for learning the molecular weight of an unknown compound. They are important in the study of compounds which are not readily vaporized (as is required in the vapor-density method). These methods are called the *boiling point elevation* and *freezing point lowering* methods. We have already mentioned in Section 5-2.1 that a solution of salt water has a higher boiling point than that of pure water. The boiling point elevation for a given solute concentration (expressed in moles) is almost independent of the choice of solute. Hence this temperature measurement can be readily interpreted in terms of a molar concentration. From the weight of sample used in preparation of the solution, the molecular weight can be calculated.

Exactly the same type of behavior is found for the freezing point of a solution except that the freezing point is lower than that of the pure solvent. Thus we have two methods for molecular weight determination which are applicable to compounds with such low vapor pressure or which decompose so readily that the vapor density method cannot be used.

EXERCISE 18-3

Ethylene glycol, the example treated in Exercise 18-2, has an empirical formula of CH_3O. (Is this what you obtained?) A sample weighing 0.49 gram is vaporized completely at 200°C and at one atmosphere pressure. The volume measured under these conditions is 291 ml. This same volume, 291 ml, of oxygen gas at 200°C and one atmosphere weighs 0.240 gram. What is the molecular formula for ethylene glycol, CH_3O, $C_2H_6O_2$, $C_3H_9O_3$, $C_4H_{12}O_4$, or some higher multiple of CH_3O?

DETERMINING THE STRUCTURAL FORMULA: ETHANOL

The determination of how the atoms of a molecule are connected is the most important problem in identifying an unknown compound. It can be as exciting as a detective story, with the chemical and physical properties furnishing the clues. With the right collection of clues, the chemist can ascertain the identity of the molecule.

What, then, is the structure of ethanol? First we must learn its empirical formula. Analysis shows that the empirical formula is C_2H_6O. The molecular formula is fixed by a vapor density measurement. The molecular weight is found to be 46, showing that the molecular formula is the same as the empirical formula, C_2H_6O. It remains to discover the arrangement and connections of the atoms. We might begin by eliminating some structures which we can be sure are incorrect. Ethanol is not simply ethane with an oxygen atom somehow attached to a carbon atom, because in ethane all of the four bonds of each carbon are satisfied, and so there is no way in which an additional bond can form. We say that ethane is a *saturated** compound. Neither can the oxygen atom just be attached somehow to a hydrogen atom. Each hydrogen atom in ethane has its bonding capacity already satisfied.

Rather than trying to find the structural formula of ethanol by tacking an oxygen atom to ethane, let us start with the oxygen atom and see how we might build a molecule around it, using the two carbon atoms and six hydrogen atoms which are at our disposal. We already know that the oxygen atom is commonly divalent, and that, as in water, it makes bonds to hydrogen atoms. Let us start our molecular construction with a bond between one hydrogen atom and the oxygen atom:

O—H

The other bond which the oxygen atom can make must be to a carbon atom, since if it were to another hydrogen atom we would simply have a water molecule. Therefore we write

C
\
O—H

The carbon atom we have added must form three additional bonds to satisfy its tetravalent bonding capacity. If all of these bonds were to hydrogen atoms, we would have the completed molecule CH_3OH, and would also have two hydrogen atoms and a carbon atom left over. Therefore, one of the bonds our first carbon atom forms must be to the other carbon atom, and the two other bonds must be to hydrogen atoms. We have then

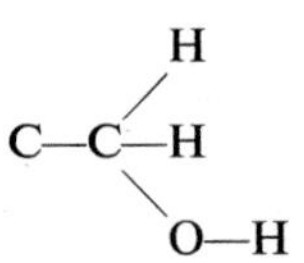

We can easily complete the structure by adding three bonds from our last carbon atom to the

* This usage of the word "saturated" shows that chemists, like other people, sometimes use the same word with two entirely different meanings. On p. 164 this word was used to describe a solution which contains the equilibrium concentration of a dissolved substance. As used here, in reference to organic compounds, it means that all bonds to carbon are single bonds and they are all formed with hydrogen or other carbon atoms.

three hydrogen atoms we have left. The result is

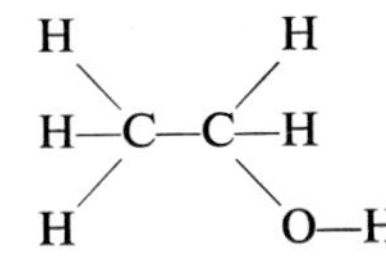

We have now used all six hydrogen atoms, the two carbon atoms and the oxygen atom which the molecular formula of ethanol requires. All the bonding rules are satisfied, so the structure we have written must be taken as a *possible* structural formula for ethanol. However, we must also decide whether this is the *only* possible structural formula for a molecule with molecular formula C_2H_6O. A little reflection shows it is not. Instead of having the oxygen atom form one bond to carbon and one to hydrogen, why not start with two oxygen-carbon bonds?

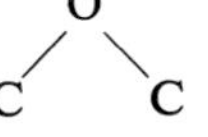

We have six hydrogen atoms at our disposal, and each of the carbon atoms must form three more bonds. Therefore we complete the structure by writing

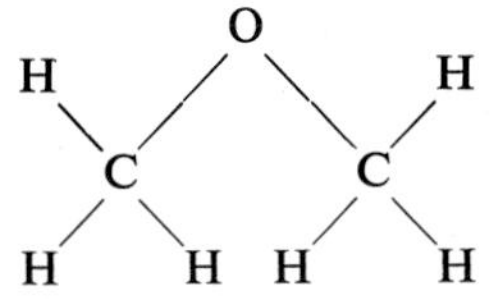

Satisfy yourself that this structure violates no bonding rules, and conforms to the empirical and molecular formula of ethanol.

We have now found all possible structural formulas for the ethanol molecule. The oxygen atom is either directly bonded to one carbon atom or to two carbon atoms. Once a choice between these two possibilities is made, the structure of the rest of the molecule can be determined from the molecular formula and the bonding rules. The two possible structures are shown in Figure 18-2. Such *compounds with the same molecular formula but different structural formulas are called* **structural isomers**. The existence of the two compounds 1 and 2 was known long before their structures were clarified. Hence the existence of these isomers perplexed chemists for decades. Now we recognize the crucial importance of learning the structural formulas (as well as molecular formulas) to identify the isomers.

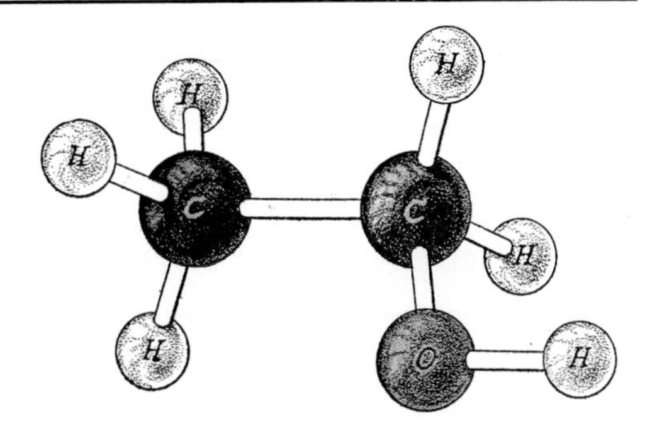

CH_3CH_2OH (1)

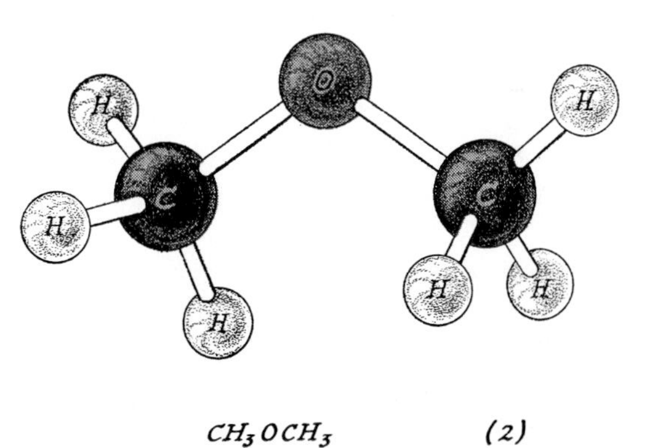

CH_3OCH_3 (2)

Fig. 18-2. **The structures** *of the C_2H_6O isomers.*

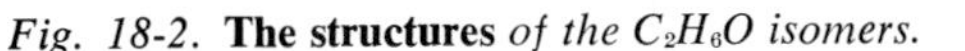

So our problem is to decide whether ethanol has structure 1 or structure 2. How can we tell which is correct? Let us see what preliminary ideas we can get from an examination of the structural formulas.

In structure 2, all of the hydrogen atoms are the same—each hydrogen atom is bonded to a carbon which is, in turn, bonded to the oxygen atom. In structure 1, one of the hydrogen atoms is quite different from any of the others: it is bonded to oxygen and not to carbon. Of the remaining five, two are similarly placed, on the carbon bonded to oxygen, and three are on the other carbon. Structures 1 and 2 should have quite different chemistries. Which one should correspond to the chemistry of ethanol?

We can offer several kinds of evidence. Some comes from the behavior of ethanol in chemical reactions and some from the determination of certain physical properties. Let's consider the reactions first.

Clean sodium metal reacts vigorously with ethanol, giving hydrogen gas and an ionic compound, sodium ethoxide, with empirical formula C_2H_5ONa. The reaction is quite similar to the behavior of sodium and water which yields hydrogen and the ionic compound sodium hydroxide, NaOH. This suggests, but certainly does not prove, that ethanol shows some structural similarity to water. In water we have two hydrogen atoms bonded to oxygen atoms, and in structure 1 we have one hydrogen atom bonded to oxygen. This bit of chemical evidence suggests ethanol has structure 1.

More quantitative evidence can be obtained by carrying out the reaction between an excess of sodium and a weighed amount of ethanol and measuring the amount of hydrogen gas evolved. When this is done it is found that 46 grams of ethanol (one mole) will produce only $\frac{1}{2}$ mole of hydrogen gas. We can therefore write a balanced chemical equation for the reaction of sodium with ethanol:

$$\mathrm{Na}(s) + \mathrm{C_2H_6O}(l) \longrightarrow \tfrac{1}{2}\mathrm{H_2}(g) + \mathrm{C_2H_5ONa}(s) \quad (1)$$

This equation expresses the fact that one mole of ethanol produces $\frac{1}{2}$ mole of hydrogen gas. Hence, one mole of ethanol must contain *one* mole of hydrogen atoms that are uniquely capable of undergoing reaction with sodium. Apparently one molecule of ethanol contains one hydrogen atom that is capable of reacting with sodium and five that are not. Let us now consider structures 1 and 2 in the light of this information. In structure 2 all six of the hydrogen atoms are structurally equivalent, whereas in structure 1 there is one hydrogen atom in the molecule which is structurally unique—that is, the one bonded to the oxygen atom. Structure 1 is therefore consistent with the experimental fact that only one hydrogen atom per molecule of ethanol will react with sodium and structure 2 is not.

We can find further evidence that structure 1, CH_3CH_2OH, is the correct structural formula for the substance known as ethanol. It is known that compounds which contain only carbon and hydrogen (such as ethane, C_2H_6) do not react at all readily with metallic sodium to produce hydrogen gas. In these compounds the hydrogen atoms are all bonded to carbon atoms (see Figure 18-1), so we can make the deduction that, in general, hydrogen atoms which are bonded to carbon atoms do not react with sodium to produce hydrogen gas. In structure 2, CH_3OCH_3, all the hydrogen atoms are bonded to carbon atoms, so we do not expect a compound with this structure to react with sodium. Ethanol reacts with sodium, so it is unlikely that ethanol has structure 2.

Let us consider one other reaction of ethanol. If ethanol is heated with aqueous HBr, we find that a volatile compound is formed. This compound is only slightly soluble in water and it contains bromine: its molecular formula is found by analysis and molecular weight determination to be C_2H_5Br (ethyl bromide, or bromoethane). With the aid of the bonding rules, we can see that there is only one possible structure for this compound. This result is verified by the fact that only one isomer of C_2H_5Br has ever been discovered.

Now we can ask how this chemical reaction furnishes a clue to the structure of ethanol. Structure 1 could give structure 3 in Figure 18-3 merely by breaking the carbon-oxygen bond.

Fig. 18-3. **The structural formula of ethyl bromide (bromoethane).**

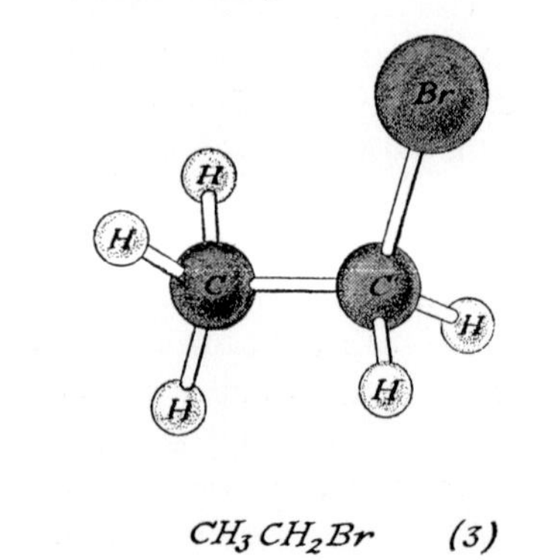

CH_3CH_2Br (3)

Convince yourself of this fact by writing an equation using the structural formulas 1 and 3. In contrast, bromoethane can be obtained from structure 2 only through a complicated rearrangement. Two carbon-oxygen and one carbon-hydrogen bond would have to be broken. Experience shows that such complicated reshufflings of atoms rarely occur. Therefore, the reaction between ethanol and hydrobromic acid, HBr, to form bromoethane provides more evidence that ethanol has structure 1.

The evidence cited so far has been associated with the chemistry of ethanol. Its boiling point provides a different sort of information also leading to structure 1. Ethanol is a liquid with a boiling point of 78°C. This can be compared with the boiling point of ethane, C_2H_6, which is −172°C, and to that of water, 100°C. Plainly, the substance ethanol is more like water than like ethane, as far as boiling point is concerned. Once again this can be understood better in terms of structure 1. Structure 1 has, in common with H_2O, oxygen linked to hydrogen. The high boiling point of water is explained in terms of an abnormally large intermolecular attraction of such an O—H group to surrounding water molecules. The interaction is called hydrogen bonding (see Section 17-2.6). If ethanol also has the O—H group (as in structure 1) then it too can exert the same abnormally large attraction to neighboring ethanol molecules. Thus structure 1 provides an explanation of the fact that the boiling point of ethanol is so high.

This possibility of forming hydrogen bonds should cause a strong attraction between water and a compound of structure 1. If there is strong attraction, then ethanol should have high solubility in water. Experiment shows that they are miscible—they dissolve in all proportions. Again the evidence tends to strengthen belief in structure 1.

Notice that our attempt to determine the structural formula of ethanol has involved the consideration of a variety of types of evidence. Others could be listed as well—for example, the infrared spectrum of the liquid and the X-ray diffraction pattern of the solid add strong support for structure 1. No one fact by itself gives absolute proof of the structure, but all the facts considered together show that 1 is unquestionably the correct structure for ethanol. A comparable set of experiments shows that *another* compound with the formula C_2H_6O has properties consistent with structure 2. This compound is called dimethyl ether.

EXERCISE 18-4

Ethylene glycol has empirical formula CH_3O and molecular formula $C_2H_6O_2$. Using the usual bonding rules (carbon is tetravalent; oxygen is divalent; hydrogen is monovalent), draw some of the structural formulas possible for this compound.

EXERCISE 18-5

Decide which of your structures in Exercise 18-4 best fits the following list of properties observed for pure ethylene glycol.

(a) It is a viscous (syrupy) liquid boiling at 197°C.
(b) It is miscible with water, that is, it dissolves, forming solutions, in all proportions.
(c) It is miscible with ethanol.
(d) It reacts with sodium metal, producing hydrogen gas.
(e) A 6.2 gram sample of ethylene glycol reacts with an excess of sodium metal to produce 2.4 liters of hydrogen gas at one atmosphere pressure and 25°C.

18-2.3 The Ethyl Group

All of the reactions and the physical properties of ethanol have been explained on the basis of the behavior of the OH group in structure 1, CH_3CH_2OH. This is true of most of the reactions of ethanol—the reaction centers at the OH group (which is called the *hydroxyl* group), and the remainder of the molecule, C_2H_5—, remains intact. The reactions lead to the suggestion that there are two parts in the molecule of ethanol, the

```
   H  H
   |  |
H—C—C—
   |  |
   H  H
```

group, *which is unchanged during*

reactions, and the —OH group, *which can change*. This concept of the structural integrity of the hydrocarbon group is an important one in organic chemistry. It focuses attention on the groups that *do* change, the so-called *functional groups*. If the chemistry of a particular functional group is understood for one compound, it is a good assumption that this same chemistry will be found for another compound containing this same functional group. Thus, *compounds with the OH group are given a family name*, ***alcohols***. The rest of the molecule, the carbon skeleton, has relatively little effect and it remains intact through the reactions of the functional group.

We have mentioned earlier that when ethanol reacts with hydrogen bromide, ethyl bromide is formed. Similar treatment of ethanol with hydrogen chloride or hydrogen iodide gives us the corresponding ethyl halides:

$$CH_3CH_2OH + HBr \longrightarrow CH_3CH_2Br + H_2O \quad (2)$$

$$CH_3CH_2OH + HCl \longrightarrow CH_3CH_2Cl + H_2O \quad (3)$$

$$CH_3CH_2OH + HI \longrightarrow CH_3CH_2I + H_2O \quad (4)$$

We say that the hydroxyl group has been *displaced*, and the halogen atom substituted for it. You can see that the group CH_3CH_2— has remained intact in all of these reactions. Indeed, this group has appeared in most of our discussion so far, sometimes attached to oxygen (as in ethanol and sodium ethoxide), sometimes attached to other atoms (as in the ethyl halides). You will recall that earlier we became acquainted with ethane, C_2H_6. Looking at the structural formula of ethane, you see that it is simply the CH_3CH_2— group attached to hydrogen:

```
   H  H
   |  |
H—C—C—H   or   CH3CH2—H
   |  |
   H  H
```

This group, CH_3CH_2— (also written C_2H_5—), is called the *ethyl group*.

Because ethyl bromide, ethyl alcohol (ethanol), etc., can be thought of as being derived from ethane by the substitution of one of its hydrogens by —Br, —OH, etc., we speak of these as derivatives of ethane, and we say that ethane is the parent hydrocarbon for a series of related compounds.

The name ethyl is derived from the name of the parent hydrocarbon, ethane. In the same way the name of the *methyl* group (CH_3—) is derived from that of methane, CH_4; the name of the *propyl* group ($CH_3CH_2CH_2$—) is derived from propane, $CH_3CH_2CH_3$; etc.

It is important to recognize that these groups are not substances that can be isolated and bottled. They are simply parts of molecules that remain intact in composition and structure during reactions. We find this way of classifying organic groups a useful and convenient one, but we must keep in mind that in the reactions we have described, the ethyl group is not actually formed as a distinct substance. Table 18-I gives more examples of group names (see p. 338).

18-3 SOME CHEMISTRY OF ORGANIC COMPOUNDS

18-3.1 Chemical Behavior of Ethyl and Methyl Bromide

We can use these bromine compounds to illustrate one kind of organic reaction. Ethyl bromide is not particularly reactive but it does react with bases such as NaOH or NH_3. If we mix ethyl bromide and aqueous sodium hydroxide solution and heat the mixture for an hour or so, we find that sodium bromide and ethanol are formed.

$$C_2H_5Br + OH^-(aq) \longrightarrow C_2H_5OH + Br^-(aq) \quad (5)$$

This reaction may seem similar to the reaction between aqueous HBr and NaOH but there are two important differences. The ethyl bromide reaction is very slow (about one hour is needed for the reaction) and it occurs between a covalent molecule (C_2H_5Br) and an ion (OH^-). In contrast, the reaction between HBr and NaOH in water occurs in a fraction of a second and it involves ions only, as shown in reaction (*6*).

$$H^+(aq) + OH^-(aq) \longrightarrow H_2O \qquad (6)$$

Let's describe the course of reaction (5) in terms of a model. We will use *methyl* bromide to make the description simpler but the reaction of ethyl bromide is of the same type.

$$CH_3Br + OH^- \longrightarrow CH_3OH + Br^- \qquad (7)$$

First of all, let us recount a few of the experimental facts.

(1) Methyl bromide is a compound in which the chemical bonds are predominantly covalent. An aqueous solution of methyl bromide does not conduct electricity, hence it does not form ions (such as CH_3^+ and Br^- ions) in aqueous solutions.
(2) The reaction takes a measurable time for completion.
(3) Experiments show that the rate of the reaction is increased by increasing the concentration of OH^- and also by raising the temperature.

These observations remind us of Chapter 8, in which we considered the factors that determine the rate of a chemical reaction. Of course, the same ideas apply here. We can draw qualitative information about the mechanism of the reaction by applying the collision theory. With quantitative study of the effects of temperature and concentration on the rate, we should be able to construct potential energy diagrams like those shown in Figure 8-6 (p. 134).

Figure 18-4 illustrates the mechanism chemists have deduced for this reaction. This picture shows: (*A*) the approach of the hydroxide ion, (*B*) the atomic arrangement thought to be the activated complex, and (*C*) the final products. In the activated complex the O—C bond is beginning to form and the C—Br bond is beginning

Fig. 18-4. **The mechanism and potential energy diagrams** *for the reaction* $CH_3Br + OH^-(aq) \longrightarrow CH_3OH + Br^-(aq)$

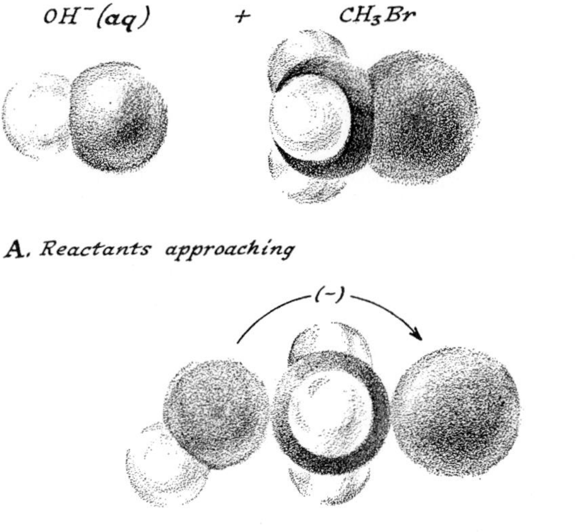

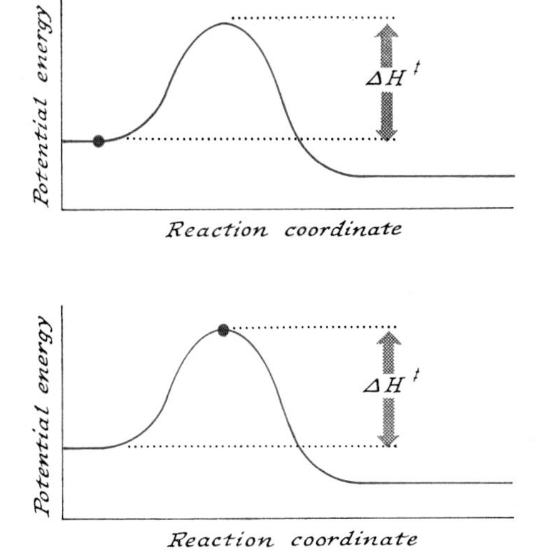

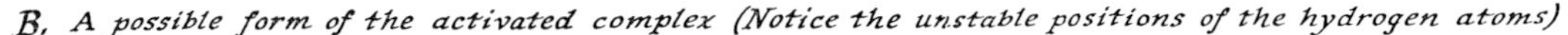
B. A possible form of the activated complex (Notice the unstable positions of the hydrogen atoms)

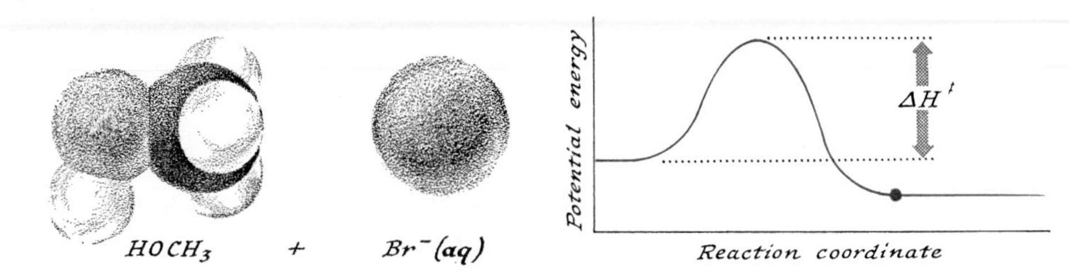

C. Products separating (Notice the new, stable positions of the hydrogen atoms)

to break. The potential energy curves for the reaction are shown alongside the molecular models. The slow rate shows that activation energy is needed. One of the reasons why activation energy must be supplied is that in the activated complex the bond angles have been distorted from their favorable (stable) configurations and forced into an unstable condition.

18-3.2 Oxidation of Organic Compounds

By far the majority of the million or so known compounds of carbon also contain hydrogen and oxygen. There are several important types of oxygen-containing organic compounds and they can be studied as an oxidation series. For instance, the compound methanol, CH_3OH, is very closely related to methane, as their structural formulas show. Methanol can be regarded as the first step in the complete oxidation of methane to carbon dioxide and water.

Fig. 18-5. **Structural formulas** *of methane and methanol.*

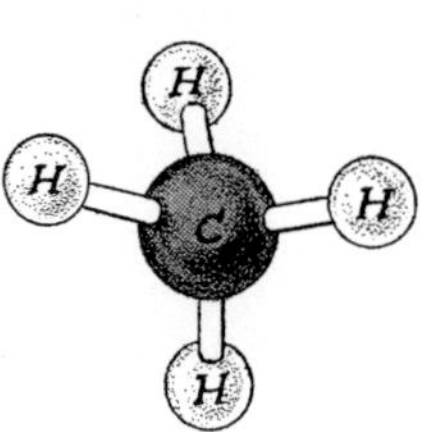

methane, CH_4

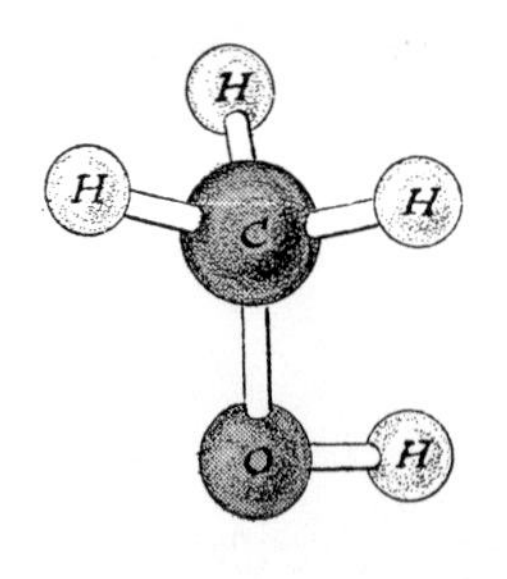

methanol, CH_3OH

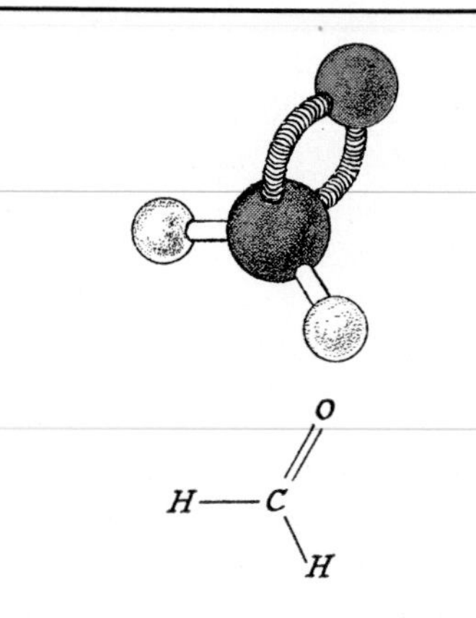

formaldehyde, HCHO

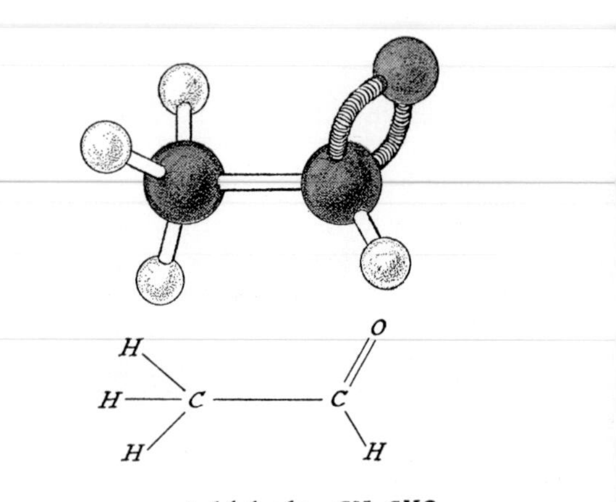

acetaldehyde, CH_3CHO

Fig. 18-6. **Structural formulas** *of formaldehyde and acetaldehyde.*

ALDEHYDES

Methanol (and other alcohols) react with common inorganic oxidizing agents such as potassium dichromate, $K_2Cr_2O_7$. When an acidic, aqueous solution of potassium dichromate reacts with methanol, the solution turns from bright orange to muddy green, owing to the production of the green chromic ion, Cr^{+3}. The solution then has a strong odor easily identified as that of formaldehyde, CH_2O. This formula represents the structure at the top in Figure 18-6. Notice that the bond between carbon and oxygen is a double bond (see Section 16.5), and that all the atoms lie in the same plane.

The balanced net reaction for the formation of formaldehyde is

$$3CH_3OH + Cr_2O_7^{-2}(aq) + 8H^+(aq) \longrightarrow 3CH_2O + 2Cr^{+3}(aq) + 7H_2O \quad (8)$$

Since the dichromate ion on the left side of the equation has been reduced to chromic ion, Cr^{+3}, on the right side, the conversion of methanol to formaldehyde must involve oxidation. To show more clearly that methanol has been oxidized, let us balance this reaction by the method of half-reactions. We have encountered the half-reaction involving dichromate and chromic ions before (Problem 20b in Chapter 12). It is

$$Cr_2O_7^{-2}(aq) + 14H^+(aq) + 6e^- \longrightarrow 2Cr^{+3}(aq) + 7H_2O \quad (9)$$

To balance the methanol-formaldehyde half-reaction we write, as a start,

$$CH_3OH \text{ gives } CH_2O \quad (10a)$$

This statement does not yet show the fact that hydrogen atoms are conserved in the reaction, since there is a deficiency of two hydrogen atoms on the right. This can be remedied by adding two hydrogen ions,

$$CH_3OH \text{ gives } CH_2O + 2H^+(aq) \quad (10b)$$

Now the equation is chemically balanced, but not electrically balanced. The addition of two electrons to the right-hand side completes the balancing procedure, and the completed half-reaction is

$$CH_3OH \longrightarrow CH_2O + 2H^+(aq) + 2e^- \quad (10c)$$

This equation shows that the methanol molecule has lost electrons and thus has been oxidized. Formaldehyde is the second member in the oxidation series of methane.

In a similar manner, ethanol can be oxidized by the dichromate ion to form a compound called acetaldehyde, CH_3CHO. The molecular structure of acetaldehyde, which is similar to that of formaldehyde, is shown at the bottom in Figure 18-6. We see that the molecule is structurally similar to formaldehyde. The methyl group, $—CH_3$, replaces one of the hydrogens of formaldehyde. The balanced equation for the formation of acetaldehyde from ethanol is

$$3CH_3CH_2OH + Cr_2O_7^{-2}(aq) + 8H^+(aq) \longrightarrow 3CH_3CHO + 2Cr^{+3}(aq) + 7H_2O \quad (11)$$

CARBOXYLIC ACIDS

Another oxidation product can be obtained from the reaction of an acidic aqueous solution of potassium permanganate with methanol. The product has the formula HCOOH, and is called formic acid. The structural formula of formic acid is shown in Figure 18-7. The structure of formic acid is also related to the structure of formaldehyde. If one of the hydrogen atoms of formaldehyde is replaced by an OH group, the

Fig. 18-7. **Structural formulas** *of formic acid and acetic acid.*

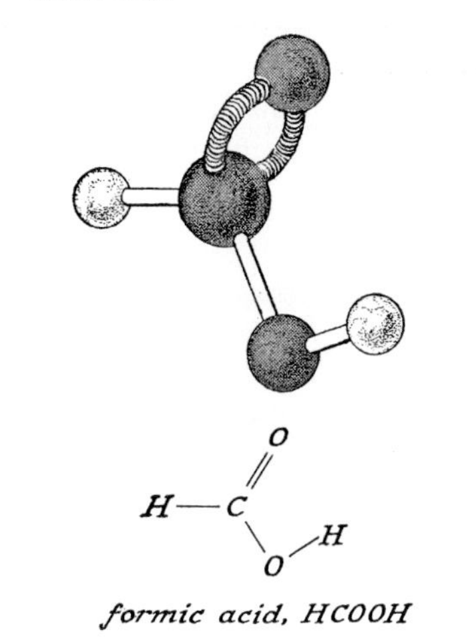

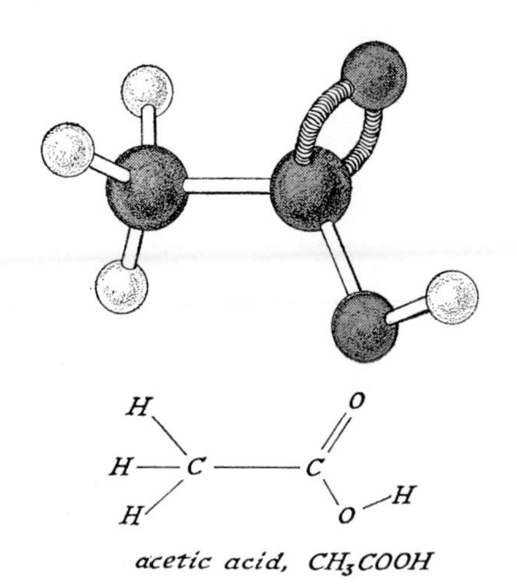

resulting molecule is formic acid. The balanced equation for its formation from methanol is

$$5CH_3OH + 4MnO_4^-(aq) + 12H^+(aq) \longrightarrow 5HCOOH + 4Mn^{+2}(aq) + 11H_2O \quad (12)$$

The half-reaction involving methanol and formic acid can be obtained by using the three steps outlined in our previous example:

$$CH_3OH \text{ gives } HCOOH \quad (13a)$$

Chemical balance:

$$CH_3OH + H_2O \text{ gives } HCOOH + 4H^+(aq) \quad (13b)$$

Charge balance: $CH_3OH + H_2O \longrightarrow HCOOH + 4H^+(aq) + 4e^- \quad (13c)$

From this completed half-reaction we see that the conversion of methanol to formic acid involves the loss of four electrons. Since the oxidation of methanol to formaldehyde was only a two-electron change, it is clear that formic acid is a more highly oxidized compound of carbon than formaldehyde or methanol.

EXERCISE 18-6

Balance the half-reaction for the conversion of formaldehyde, HCHO, to formic acid, HCOOH.

Just as methanol can be oxidized to formic acid [reaction (*12*)], ethanol can be oxidized to an acid, CH_3COOH, called acetic acid. The molecular structure of acetic acid is shown in Figure 18-7. *The atomic grouping —COOH is called the* ***carboxyl group*** *and acids containing this group are called* ***carboxylic acids.***

The balanced equation for production of acetic acid from ethanol is

$$5CH_3CH_2OH + 4MnO_4^-(aq) + 12H^+(aq) \longrightarrow 5CH_3COOH + 4Mn^{+2}(aq) + 11H_2O \quad (14)$$

Acetic acid can also be obtained by the oxidation of acetaldehyde, CH_3CHO:

$$5CH_3CHO + 2MnO_4^-(aq) + 6H^+(aq) \longrightarrow 5CH_3COOH + 2Mn^{+2}(aq) + 3H_2O \quad (15)$$

The oxidation of acetic acid is difficult to accomplish. It does not react in solutions of $K_2Cr_2O_7$ or $KMnO_4$. Vigorous treatment, such as burning, causes its complete oxidation to carbon dioxide and water. Formic acid also can be oxidized to carbon dioxide and water by combustion with oxygen.

EXERCISE 18-7

There is a compound called propanol with structural formula $CH_3CH_2CH_2OH$. If it is oxidized carefully, an aldehyde called propionaldehyde is obtained. Vigorous oxidation gives an acid called propionic acid. Draw structural formulas like those shown in Figures 18-6 and 18-7 for propionaldehyde and propionic acid.

EXERCISE 18-8

Balance the half-reaction involved in the oxidation of ethanol to acetic acid. Compare the number of electrons released per mole of ethanol with the number per mole of methanol in the equivalent reaction (*13c*). How many electrons would be released per mole of propanol in the oxidation to propionic acid?

KETONES

The bonding rules permit us to draw two acceptable structural formulas for an alcohol containing three carbon atoms, $CH_3CH_2CH_2OH$ and $CH_3CHOHCH_3$. In the first of these isomers (the one considered in Exercises 18-7 and 18-8), the OH group is attached to the end carbon atom. In the second, the OH group is attached to the second carbon atom. They are both called propanol because they are both derived from $CH_3CH_2CH_3$, propane. They are distinguished by numbering the carbon atom to which the functional group, the OH, is attached. Thus, $CH_3CH_2CH_2OH$ is called 1-propanol because the OH is on the first (the end) carbon atom in the chain. The other alcohol, $CH_3CHOHCH_3$, is called 2-propanol because the OH is on the second carbon atom. The structures of these two alcohols are shown in Figure 18-8.

We have already considered the oxidation of 1-propanol in Exercise 18-7. The second isomer,

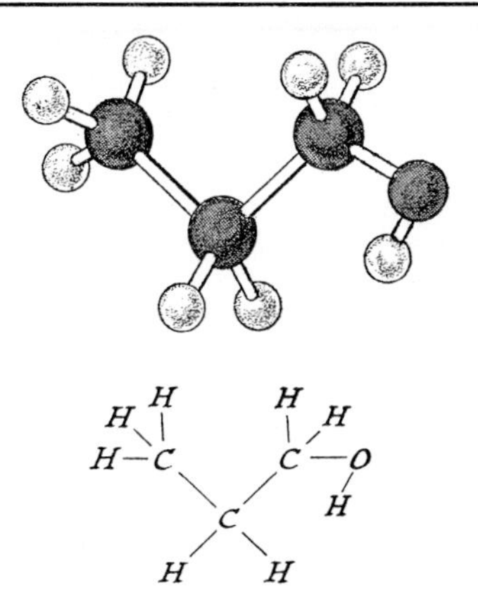

1-propanol, $CH_3CH_2CH_2OH$

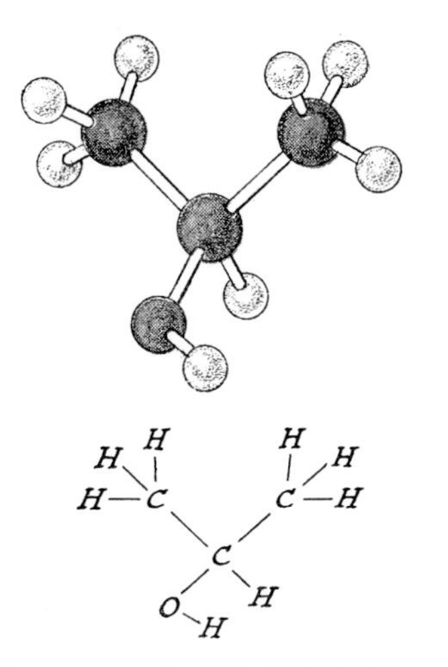

2-propanol, $CH_3CHOHCH_3$

Fig. 18-8. **The molecular structures** *of 1-propanol and 2-propanol.*

2-propanol, can also be oxidized and the product is called acetone:

$$3CH_3CHOHCH_3 + Cr_2O_7^{-2}(aq) + 8H^+(aq) \longrightarrow 3CH_3COCH_3 + 2Cr^{+3}(aq) + 7H_2O \quad (16)$$

Acetone has the structure shown in Figure 18-9.

Acetone is the simplest member of a class of compounds called *ketones*. They are quite similar in structure to the aldehydes, since each contains a carbon atom doubly bonded to an oxygen atom.* They differ in that the aldehyde has a hydrogen atom attached to this same carbon atom whereas the ketone does not. (Compare Figures 18-6 and 18-9.) Since this hydrogen atom is not present, a ketone cannot be oxidized further to an acid.

Figure 18-10 summarizes the successive oxidation products that can be obtained from alcohols. When the hydroxyl group, OH, is attached on an end carbon atom, an aldehyde and a carboxylic acid can be obtained through oxidation. When the hydroxyl group is on a carbon atom attached to two other carbon atoms, oxidation gives a ketone. Huge amounts of aldehydes and ketones are used industrially in a variety of chemical processes. Furthermore, these functional groups are important in chemical syntheses of medicines, dyes, plastics, and fabrics.

18-3.3 The Functional Group

The reactive groups we have encountered thus far, such as —Br, —OH, —CHO, —COOH, are called *functional groups*. They are the parts from which the molecules get their characteristic chemical behavior.

For example, it is a general behavior of alcohols to undergo a reaction in which the —OH group is displaced by a halogen atom, such as —Br [as in reaction (*17*)].

Fig. 18-9. **The molecular structure of acetone,** *the simplest ketone.*

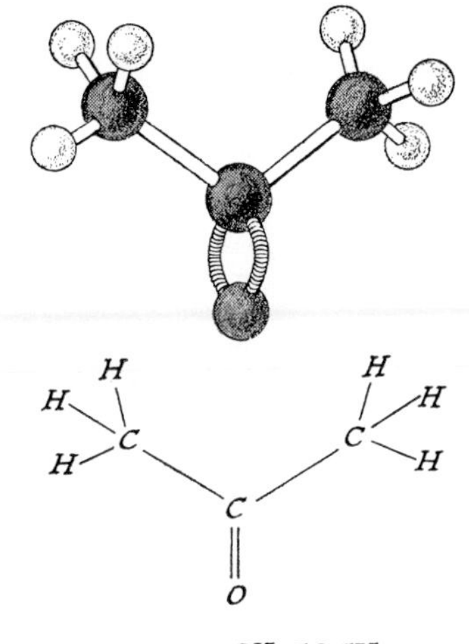

acetone, CH_3COCH_3

* The group >C=O is called the *carbonyl group*.

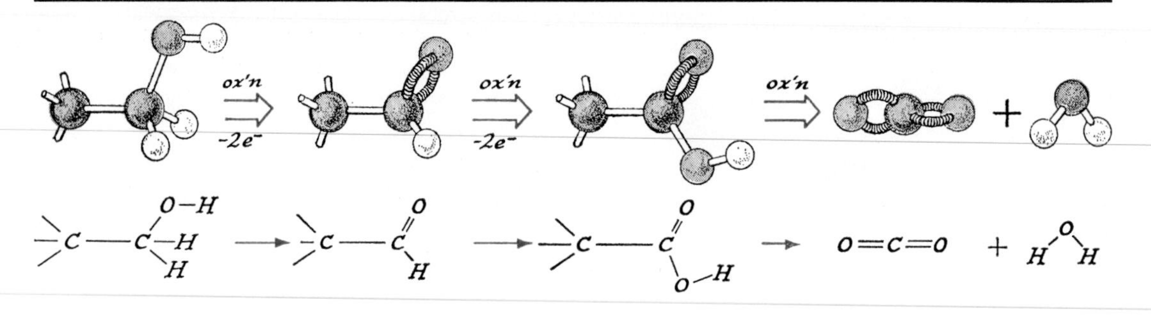

1-alcohol ⟶ *aldehyde* ⟶ *carboxylic acid* ⟶ *carbon dioxide + water*

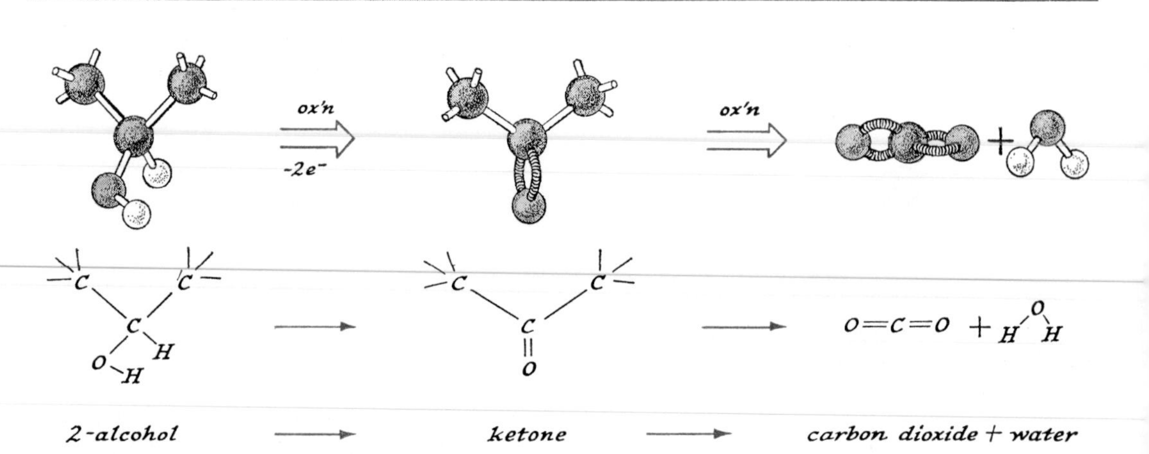

Fig. 18-10. **Summary:** *The successive steps in the oxidation of an alcohol.*

It is a general characteristic of aldehydes to be oxidizable to acids, as in reaction (*19*).

These common types of behavior are shown by using the general symbol, R—, to stand for the part of the molecule which does not change, and writing a reaction in such a way as to focus attention on the functional group. For example:

$$RCH_2OH + HBr \longrightarrow RCH_2Br + H_2O \quad (17)$$

$$RCH_2Br + OH^-(aq) \longrightarrow RCH_2OH + Br^-(aq) \quad (18)$$

$$3RCHO + Cr_2O_7^{-2}(aq) + 8H^+(aq) \longrightarrow 3RCOOH + 2Cr^{+3}(aq) + 4H_2O \quad (19)$$

The symbol R— represents any alkyl group (such as CH_3—, C_2H_5—) in these formulas.

18-3.4 Amines

Alcohols can be related to water by imagining that an alkyl group (such as $—CH_3$) has been substituted for one of the two hydrogen atoms of water. In the same way, amines are related to ammonia:

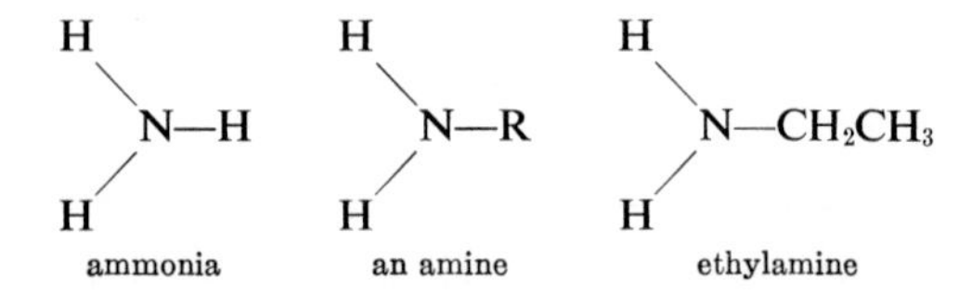

Amines can be prepared by direct reaction of ammonia with an alkyl halide, such as CH_3Br or CH_3CH_2I. Iodides react fastest and an excess of ammonia is often used to help control formation of undesired alternate products:

$$\underset{\text{ethyl iodide}}{CH_3CH_2I} + 2NH_3 \longrightarrow \underset{\text{ethylamine}}{CH_3CH_2NH_2} + NH_4I \quad (20)$$

$$R—I + 2NH_3 \longrightarrow RNH_2 + NH_4I \quad (21)$$

Equations (*20*) and (*21*) represent a net change that occurs when an excess of ammonia reacts with an alkyl iodide. The actual reaction goes by two successive steps. The first step is analogous to the attack of the hydroxide ion on an alkyl halide (see Figure 18-4):

$$NH_3 + RI \longrightarrow RNH_3^+(aq) + I^-(aq) \quad (22)$$

The second step is a proton transfer reaction (see Section 11-3.3):

$$NH_3 + RNH_3^+(aq) \longrightarrow RNH_2 + NH_4^+(aq) \quad (23)$$

18-3.5 Acid Derivatives: Esters and Amides

We see from reactions (*15*) and (*19*) that oxidation of an aldehyde gives an organic acid. All of these acids contain the functional group —COOH, the *carboxyl group*. The bonding in this group is as follows:

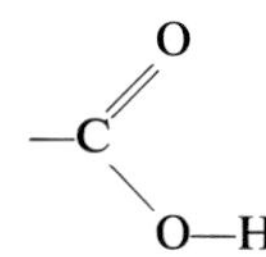

The carboxyl group readily releases a proton, so it is an acid. For example, acetic acid dissolves in water and the solution is conducting, it turns blue litmus red, it is sour, and it shows all the other properties of an acid. The reaction

$$CH_3COOH + H_2O \longrightarrow CH_3COO^-(aq) + H_3O^+(aq) \quad (24)$$

has an equilibrium constant of 1.8×10^{-5}.

In addition to this acidic behavior, an important characteristic of carboxylic acids is that the entire OH group can be replaced by other groups. The resulting compounds are called *acid derivatives*. We will consider only two types of acid derivatives, *esters* and *amides*.

ESTERS

Compounds in which the —OH of an acid is transformed into —OR (such as $—OCH_3$) are called esters. They can be prepared by the direct reaction between an alcohol and the acid. For example,

$$CH_3OH + CH_3C(=O)OH \longrightarrow CH_3C(=O)O{-}CH_3 + H_2O \quad (25)$$

methyl alcohol + **acetic** acid ⟶ **methyl** **acetate** + water

The method of naming is indicated by the bold face parts of the names of the reactants.

EXERCISE 18-9

Write equations for the reaction of (a) ethanol and formic acid; (b) propanol and propionic acid; (c) methanol and formic acid. Name the esters produced.

When equilibrium is reached in reaction (*25*), appreciable concentrations of all of the reactants may be present. If methyl acetate (the product on the right) alone is dissolved in water, it will react with water slowly to give acetic acid and methanol until equilibrium is attained:

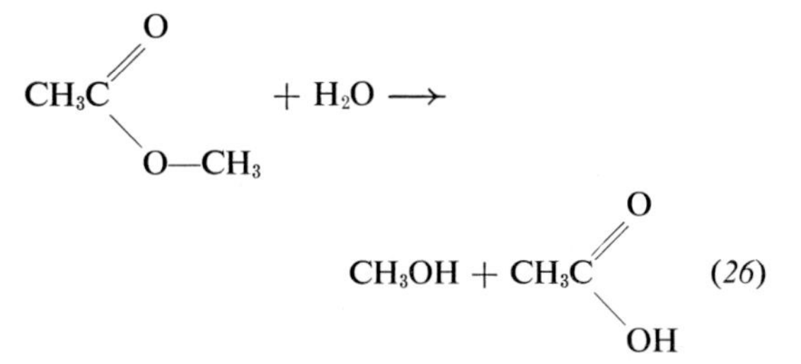

Of course, the usual equilibrium considerations apply. For example, if we add the substance methanol, equilibrium conditions will shift, consuming the added reagent (methanol) and acetic acid to produce more methyl acetate and water, in accord with Le Chatelier's Principle. Thus a large excess of methanol causes most of the acetic acid to be converted to methyl acetate.

EXERCISE 18-10

Write the equilibrium expression relating the concentrations of reactants and products in reaction (*26*). Notice that the concentration of water must be included because it is not necessarily large enough to be considered constant.

Reaction (*25*) between methanol and acetic acid is slow, but it can be speeded up greatly if a catalyst is added. Either a strong acid (such as hydrochloric or sulfuric acid) or a strong base (such as sodium hydroxide) will act as a catalyst. As mentioned in Section 9-1.4, the catalyst does not alter the equilibrium state (that is, the concentrations of the reactants at equilibrium), but only permits equilibrium to be attained more rapidly.

EXERCISE 18-11

Either a strong acid or a strong base will catalyze reaction (*25*). Explain why this implies that either a strong acid or a strong base will catalyze reaction (*26*) as well. (Consult Section 8-2.3.)

Esters are important substances. The esters of the low molecular weight acids and alcohols have fragrant, fruit-like odors and are used in perfumes and artificial flavorings. Esters are useful solvents; this is the reason they are commonly found in "model airplane dope" and fingernail polish remover.

AMIDES

A compound in which the —OH group of an acid is replaced by $-NH_2$ is called an amide, When the —OH is replaced by —NHR, the product is called a nitrogen-substituted amide or, abbreviated, an N-substituted amide. One way to make amides is to react ammonia (or an amine) with an ester:

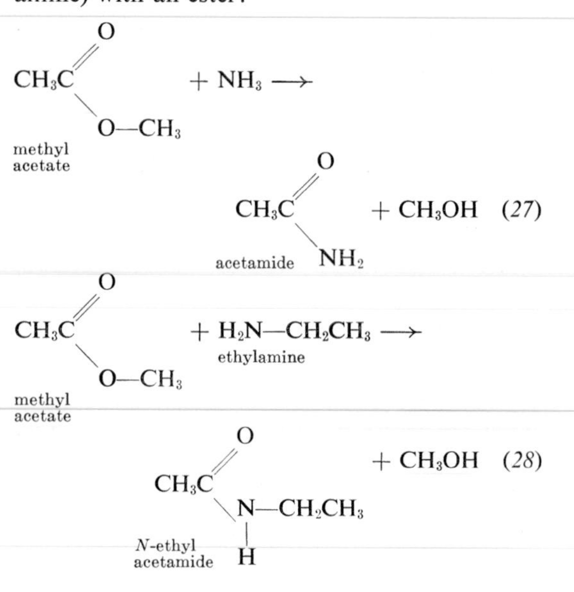

Table 18-I. **REGULARITIES IN NAMES OF ALKANES, ALCOHOLS, AND AMINES**

NUMBER OF CARBON ATOMS	ALKANES	ALCOHOLS	AMINES
1	CH_4 methane	CH_3OH methanol methyl alcohol	CH_3NH_2 methylamine
2	CH_3CH_3 ethane	CH_3CH_2OH ethanol ethyl alcohol	$CH_3CH_2NH_2$ ethylamine
3	$CH_3CH_2CH_3$ propane	$CH_3CH_2CH_2OH$ 1-propanol propyl alcohol	$CH_3CH_2CH_2NH_2$ 1-propylamine
4	$CH_3CH_2CH_2CH_3$ butane	$CH_3CH_2CH_2CH_2OH$ 1-butanol butyl alcohol	$CH_3CH_2CH_2CH_2NH_2$ 1-butylamine
8	$CH_3(CH_2)_6CH_3$ octane	$CH_3(CH_2)_6CH_2OH$ 1-octanol octyl alcohol	$CH_3(CH_2)_6CH_2NH_2$ 1-octylamine

Note the similarity of the two reactions. Amides are of special importance because the amide grouping

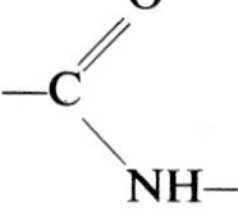

is the basic structural element in the long-chain molecules that make up proteins and enzymes in living matter. Hydrogen bonding between two amide groups helps determine the protein structure, a topic that will be dealt with later, in Chapter 24.

18-4 NOMENCLATURE

The names of organic compounds have some system. Each functional group defines a family (for example, alcohols, amines) and a specific modifier is added to identify a particular example (for example, ethyl alcohol, ethyl amine). As an alternate naming system, the family may be named by a general identifying ending (for example, alcohol names end in -ol) and a particular example is indicated by an appropriate stem (ethyl alcohol would be ethanol). These naming systems are illustrated in Tables 18-I and 18-II.

Scrutiny of these tables reveals that the key to the system is the name of the alkane which is modified in a systematic way to get the names that carry over into the acid derivatives. Starting at pentane (C_5H_{12}) and hexane (C_6H_{14}) the alkane names are themselves quite regular. They are derived from Greek words for the number of carbon atoms.

The compounds with more complicated shapes and more than one functional group are described by a straightforward numbering system that you will learn in later chemistry courses. Other functional groups will be studied then too.

Table 18-II. **REGULARITIES IN NAMES OF ACIDS, AMIDES, AND ESTERS**

NUMBER OF CARBON ATOMS	ACIDS	AMIDES	ESTERS (ACID WITH METHANOL)
1	$HCOOH$ formic acid	$HCONH_2$ formamide	$HCOOCH_3$ methyl formate
2	CH_3COOH acetic acid	CH_3CONH_2 acetamide	CH_3COOCH_3 methyl acetate
3	CH_3CH_2COOH propionic acid	$CH_3CH_2CONH_2$ propionamide	$CH_3CH_2COOCH_3$ methyl propionate
4	$CH_3CH_2CH_2COOH$ butyric acid	$CH_3CH_2CH_2CONH_2$ butyramide	$CH_3CH_2CH_2COOCH_3$ methyl butyrate
8	$CH_3(CH_2)_6COOH$ octanoic acid caprylic acid	$CH_3(CH_2)_6CONH_2$ octanamide caprylamide	$CH_3(CH_2)_6COOCH_3$ methyl octanoate methyl caprylate

18-5 HYDROCARBONS

Compounds that contain only hydrogen and carbon are called ***hydrocarbons.*** The hydrocarbons that have only single bonds all have similar chemistry and they are called, as a family, the *saturated hydrocarbons.* If there are carbon-carbon double bonds, the reactivity is much enhanced. Hence hydrocarbons containing one or more double bonds are named as a distinct family, *unsaturated hydrocarbons.* Both saturated and unsaturated hydrocarbons can occur in chain-like structures or in cyclic structures. Each of these families will be considered.

18-5.1 Saturated Hydrocarbons

We have already remarked that ethane is a member of a family of compounds called the *saturated* hydrocarbons. This term identifies compounds that contain only carbon and hydrogen in which all bonds to carbon are single bonds formed with hydrogen or other carbon atoms. They occur in chains, branched chains, and cyclic structures.

Fig. 18-11. **Structural formulas** *for some five-carbon saturated hydrocarbons.*

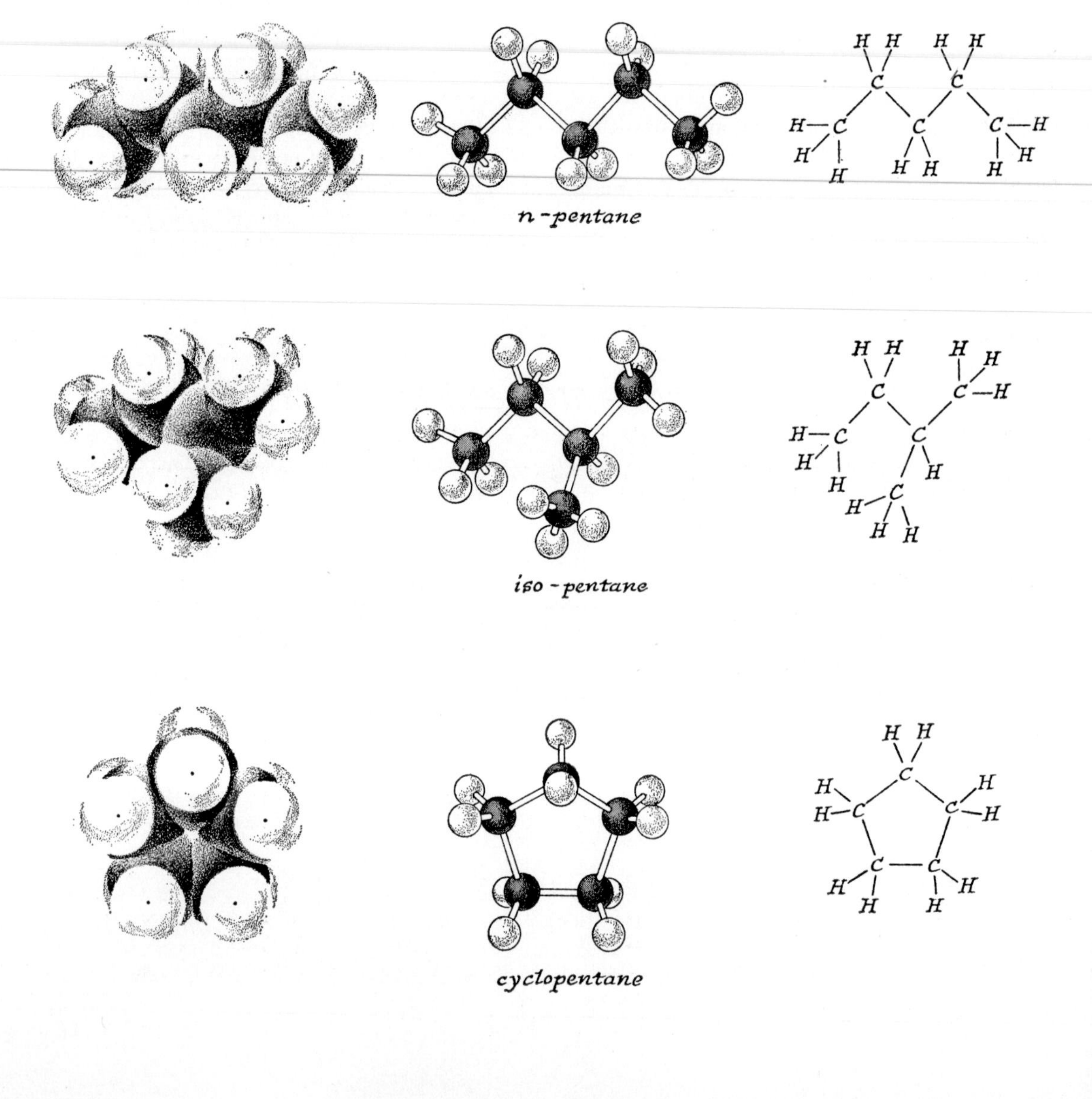

The chain and branched chain saturated hydrocarbons make up a family called the *alkanes.* Some saturated hydrocarbons with five carbon atoms are shown in Figure 18-11. The first example, containing no branches, is called normal-pentane or, briefly, *n*-pentane. The second example has a single branch at the end of the chain. Such a structural type is commonly identified by the prefix "*iso*-". Hence this isomer is called *iso*-pentane. The third example in Figure 18-11 also contains five carbon atoms but it contains the distinctive feature of a cyclic carbon structure. Such a compound is identified by the prefix "cyclo" in its name—in the case shown, cyclopentane.

EXERCISE 18-12

What are the empirical formulas of the three compounds shown in Figure 18-11? The molecular formulas? Which are structural isomers?

EXERCISE 18-13

There is one more alkane with molecular formula C_5H_{12}, called neopentane. Draw its structural formula.

The alkanes are the principal compounds present in natural gas and in petroleum. The low molecular weight compounds are gases under normal conditions—their boiling points are shown in Table 18-III. The composition of gasoline is mainly highly branched alkanes with from six to ten carbon atoms. Paraffin waxes are usually alkanes with from twenty to thirty-five carbon atoms.

The saturated hydrocarbons are relatively inert except at high temperatures. For example, sodium metal is usually stored immersed in an alkane such as kerosene (8 to 14 carbon atoms) to protect it from reaction with water or oxygen. Combustion is almost the only important chemical reaction of the alkanes. That reaction, however, makes the hydrocarbons one of the most important energy sources of our modern technology.

EXERCISE 18-14

Using the data given in the last column of Table 18-III, plot the heat released per carbon atom against the number of carbon atoms for the normal alkanes. Consider the significance of this plot in terms of the molecular structures of these compounds.

In a sense, the absence of reactivity of saturated hydrocarbons, whether cyclic or not, is a crucial aspect of their chemistry. This inertness accounts for the fact that the chemistry of

Table 18-III. **SOME PROPERTIES OF SATURATED HYDROCARBONS**

SATURATED HYDROCARBON	MOLECULAR FORMULA	MELTING POINT	BOILING POINT	HEAT OF COMBUSTION OF GAS (kcal/mole)
methane	CH_4	−182.5°C	−161.5°C	−212.8
ethane	CH_3CH_3	−183.3°C	−88.6°C	−372.8
propane	$CH_3CH_2CH_3$	−187.7°C	−42.1°C	−530.6
n-butane	$CH_3CH_2CH_2CH_3$	−138.4°C	−0.5°C	−687.7
isobutane	$CH_3—CH(CH_3)—CH_3$	−159.6°C	−11.7°C	−685.7
n-hexane	C_6H_{14}	−95.3°C	68.7°C	−1002.6
cyclohexane	C_6H_{12}	+6.6°C	80.7°C	−944.8
n-octane	C_8H_{18}	−56.8°C	125.7°C	−1317.5
n-octadecane	$C_{18}H_{38}$	+28.2°C	316.1°C	−2891.9

organic compounds is mainly concerned with the functional groups. The functional groups are usually so much more reactive than the carbon "skeleton" that it can be assumed that the skeleton will remain intact and unchanged through the reaction.

18-5.2 Unsaturated Hydrocarbons

Unsaturated compounds are those organic compounds in which less than four other atoms are attached to one or more of the carbon atoms. Ethylene, C_2H_4, is an unsaturated compound. Because ethylene involves only carbon and hydrogen, it is an unsaturated hydrocarbon. Propylene, the next most complicated unsaturated hydrocarbon, has the molecular formula C_3H_6. The structural formulas of ethylene and propylene are shown in Figure 18-12. Cyclic hydrocarbons also can involve double bonds. The structural formula of a cyclic unsaturated hydrocarbon is shown also in Figure 18-12.

Unsaturated hydrocarbons are quite reactive —in contrast to the relatively inert saturated hydrocarbons. This reactivity is associated with the double bond. In the most characteristic reaction, called "addition," one of the bonds of the double bond opens and a new atom becomes bonded to each of the carbon atoms. Some of the reagents that will add to the double bond are

Fig. 18-12. **Structural formulas** *of some unsaturated hydrocarbons.*

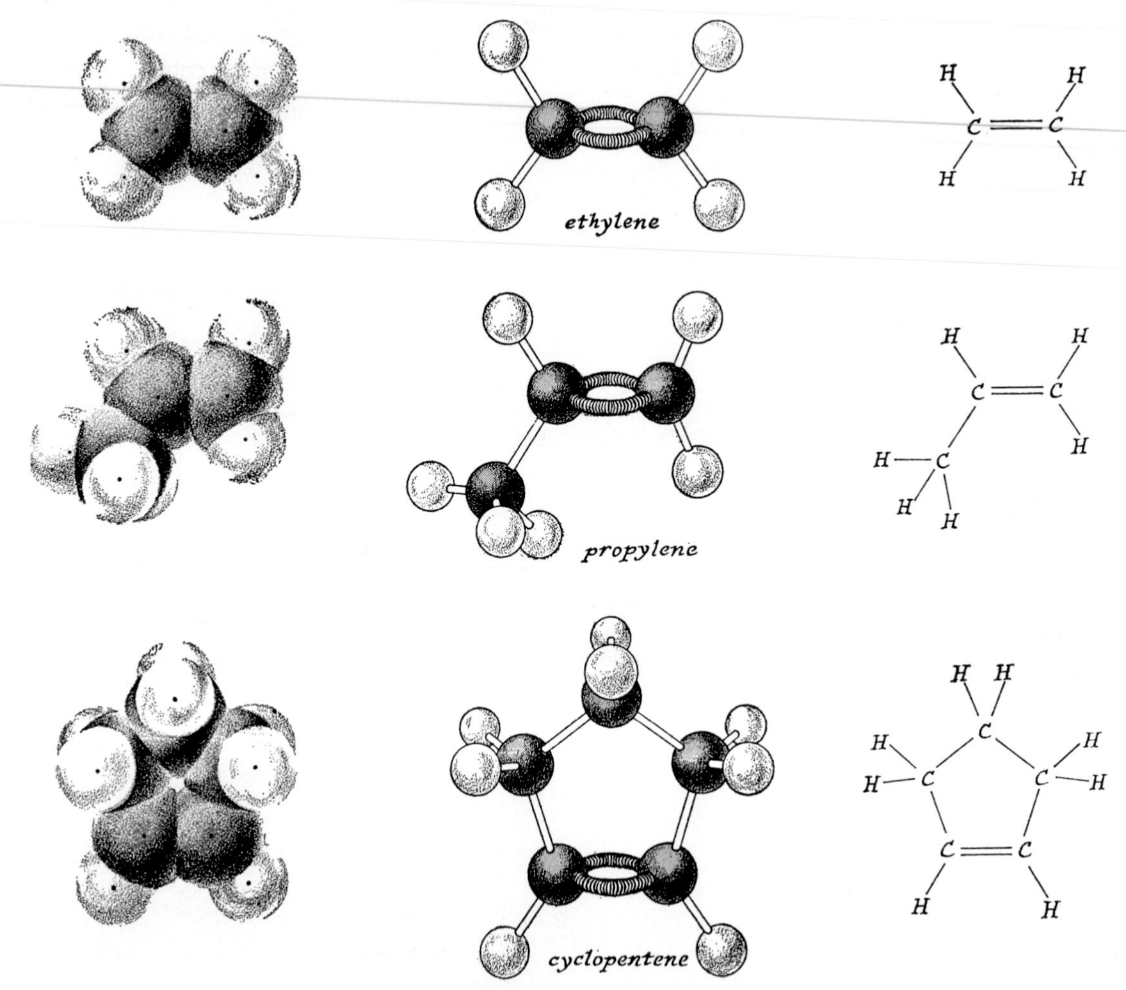

H_2, Br_2, HCl, and H_2O. Examples are shown below for ethylene.

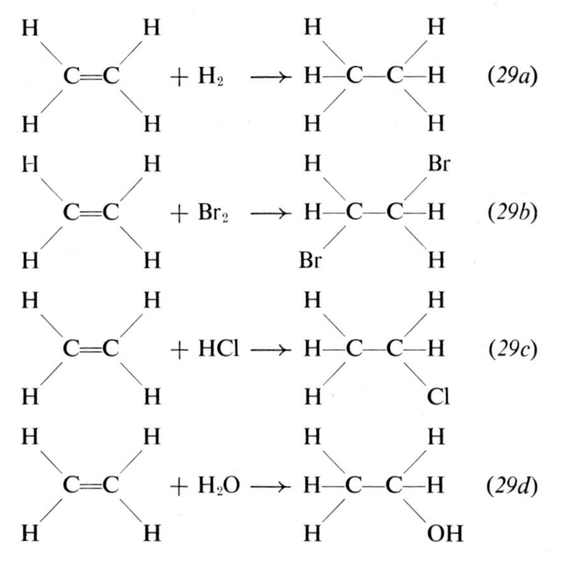

(29a) (29b) (29c) (29d)

Oxidizing agents also attack the double bond. When a reaction between an unsaturated compound and the permanganate ion occurs, the violet color of permanganate fades. This reaction, as well as reaction (*29b*) in which a color change also occurs, is used as a qualitative test for the presence of double bonds in compounds.

18-5.3 Benzene and Its Derivatives

There is another important class of cyclic compounds that is still different from the two classes just described. The simplest example is the compound benzene, a cyclic compound with six carbon atoms in the ring and formula C_6H_6. The benzene ring is found to be planar with 120° angles between each pair of bonds formed by a given carbon atom. Thus, experiment tells us the molecule is a regular hexagon with the following atomic arrangement:

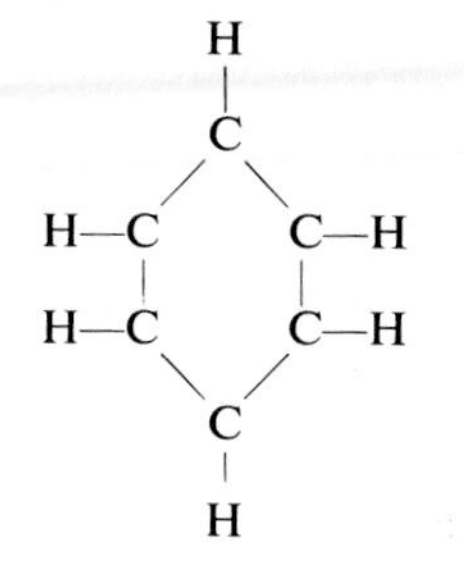

A difficulty arises when we attempt to represent the bonding in benzene and its derivatives. We might write

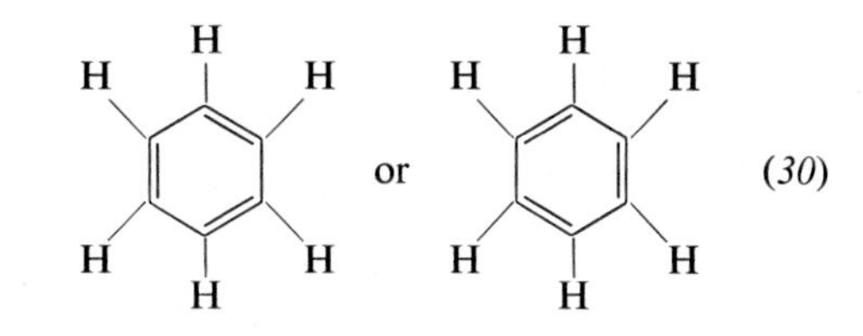

(30)

Both of these structures satisfy the formal valence rules for carbon, but each has a serious fault. Each structure shows three of the carbon-carbon bonds as double bonds, and three are shown as single bonds. There is a wealth of experimental evidence to indicate that this is not true. Any one of the six carbon-carbon bonds in benzene is the same as any other. Apparently the fourth bond of each carbon atom is shared equally with each adjacent carbon. This makes it difficult to represent the bonding in benzene by our usual line drawings. Benzene seems to be best represented as the "superposition" or "average" of the two structures. For simplicity, chemists use either one of the structures shown in (*30*) usually expressed in a shorthand form (*31*) omitting the hydrogen atoms:

or (31)

Still another shorthand symbol sometimes used is

(32)

Whichever symbol is used, (*30*), (*31*), or (*32*), the chemist always remembers that the carbon-carbon bonds are actually all the same and that they have properties unlike either simple double bonds or simple single bonds.

THE SUBSTITUTION REACTION OF BENZENE

Benzene shows neither the typical reactivity nor the usual addition reaction of ethylene. Benzene does react with bromine, Br_2, but in a different type of reaction:

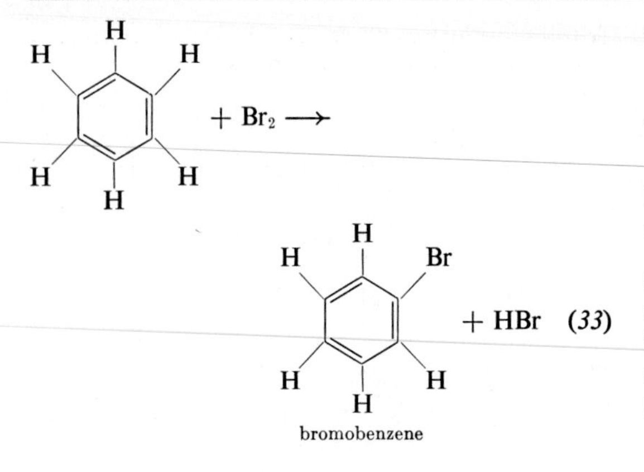

In this reaction, called bromination, one of the hydrogen atoms has been replaced by a bromine atom. Notice that the double bond structure is not affected—this is not an addition reaction. Nitric acid causes a similar reaction, called nitration:

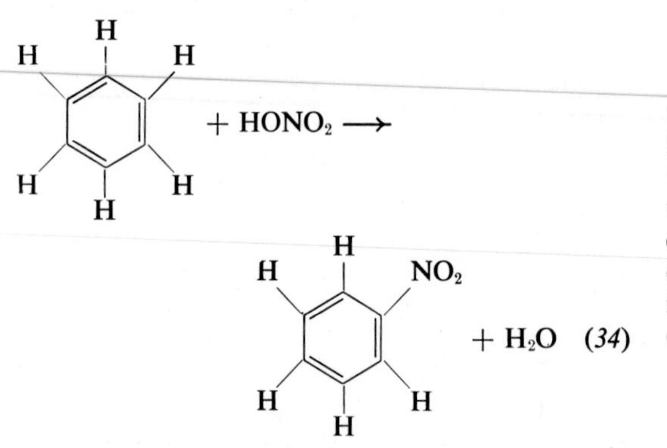

Reactions of the type shown in (*33*) and (*34*) are called *substitution* reactions. *The substitution reaction is the characteristic reaction of benzene and its derivatives* and is the way in which a multitude of compounds are prepared by the organic chemist. By this means he is able to introduce functional groups, which can then be modified in various ways. *Benzene and its derivatives are commonly called* ***aromatic compounds.***

MODIFICATION OF FUNCTIONAL GROUPS ON THE BENZENE RING

One of the most important derivatives of benzene is nitrobenzene. The nitro group is $—NO_2$. Nitrobenzene is important chiefly because it is readily converted into an aromatic amine, aniline, by reduction. One preparative procedure uses zinc as the reducing agent:

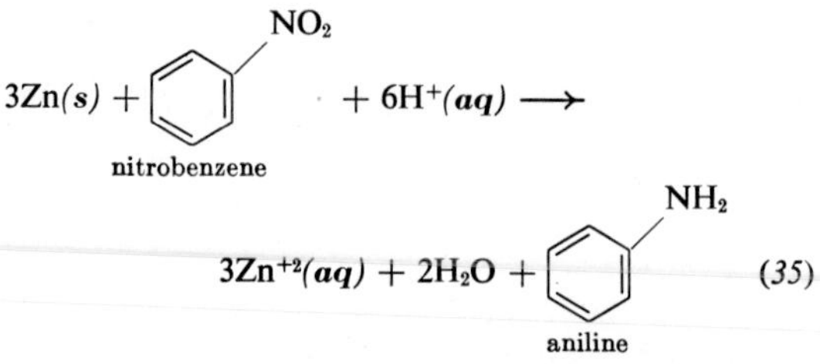

Aniline and other aromatic amines are valuable industrial raw materials. They form an important starting point from which many of our dyestuffs, medicinals, and other valuable products are prepared. For example, you have used the indicator, methyl orange, in your laboratory experiments. Methyl orange is an example of an aniline-derived dye, although it is used more as an acid-base indicator than for dyeing fabrics. The structure of methyl orange is as follows:

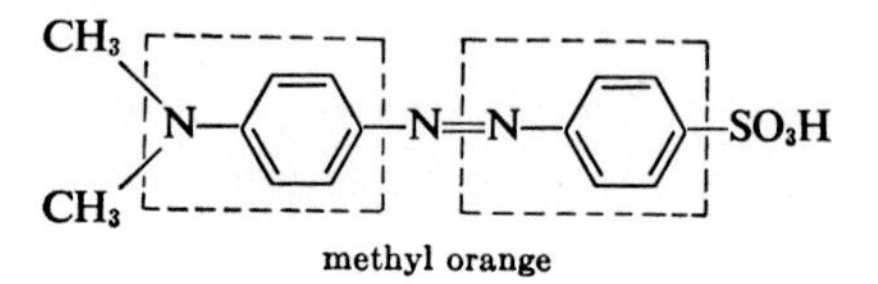

The portions of the methyl orange molecule set off by the dotted lines come from aromatic amines like aniline. Aniline is indeed the starting material from which methyl orange and related dyes ("azo dyes") are made.

Another useful aniline derivative is *acetanilide*, which is simply the amide formed from aniline and acetic acid:

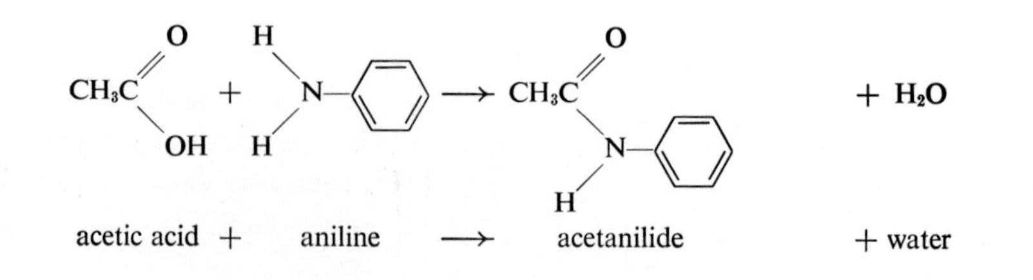

Acetanilide has been used medicinally as a pain-killing remedy.

PHENOL AND ITS USES

Another important constituent of coal tar is hydroxybenzene, or phenol:

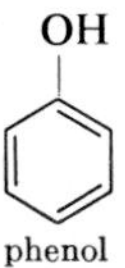

phenol

Most of our phenol is now made industrially from benzene, which is chlorinated as a first step. Reaction of chlorobenzene with base gives phenol:

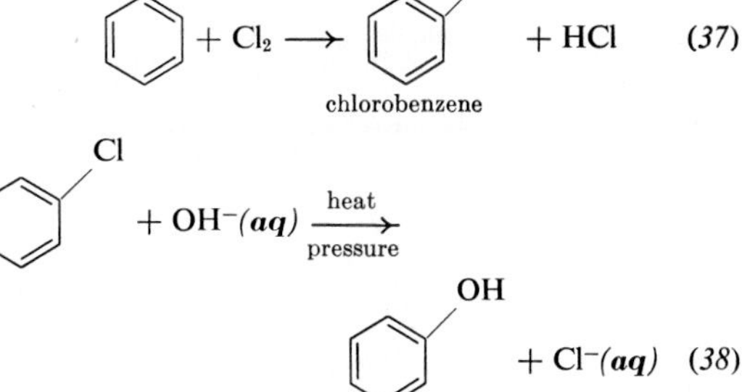

Table 18-IV. **STRUCTURES AND USES OF SOME BENZENE DERIVATIVES**

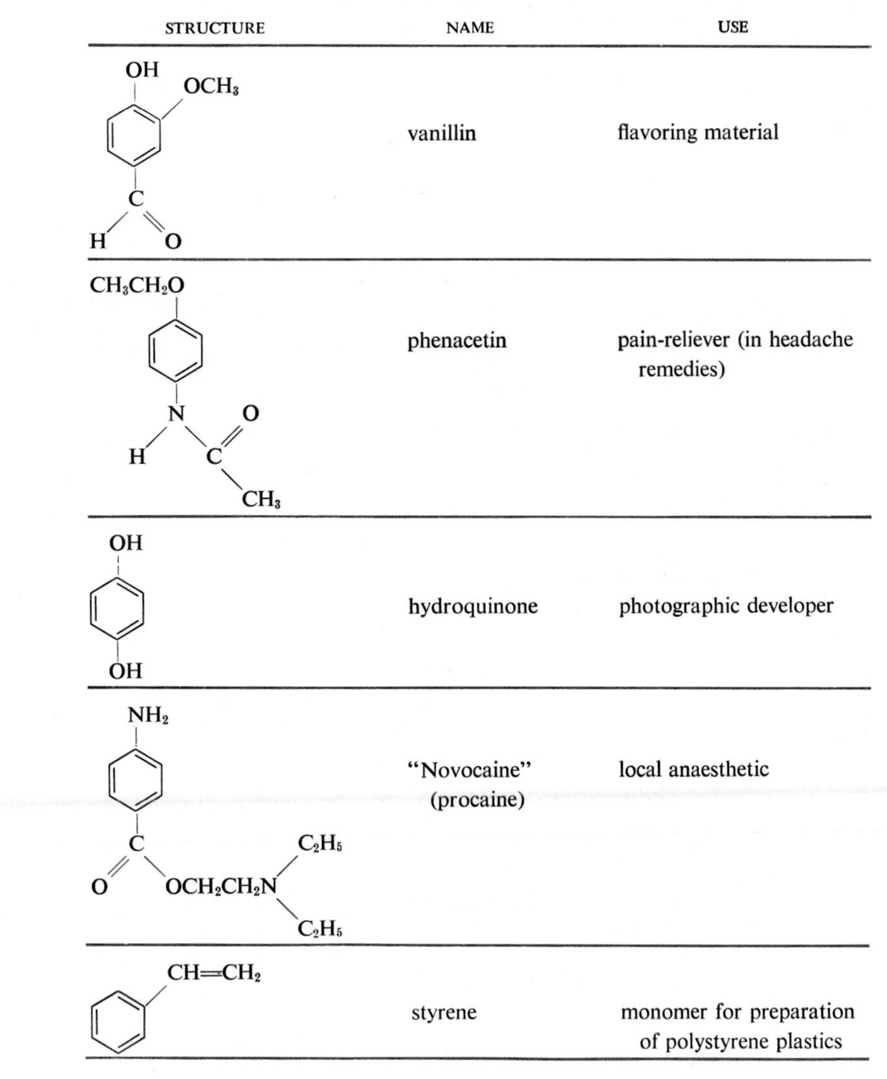

STRUCTURE	NAME	USE
	vanillin	flavoring material
	phenacetin	pain-reliever (in headache remedies)
	hydroquinone	photographic developer
	"Novocaine" (procaine)	local anaesthetic
	styrene	monomer for preparation of polystyrene plastics

Phenol is a germicide and disinfectant, and was first used by Lister in 1867 as an antiseptic in medicine. More effective and less toxic antiseptics have since been discovered.

Perhaps the most widely known compound prepared from phenol is aspirin. If phenol, sodium hydroxide, and carbon dioxide are heated together under pressure, salicylic acid is formed (as the sodium salt):

$$C_6H_5OH + CO_2 + NaOH \longrightarrow C_6H_4(OH)(COO^-Na^+) \qquad (39)$$

$$C_6H_4(OH)(COO^-Na^+) + H^+(aq) \longrightarrow C_6H_4(OH)(COOH) + Na^+(aq) \qquad (40)$$

salicylic acid

Salicylic acid is quite useful. Its methyl ester has a sharp, characteristic odor and is called "oil of wintergreen." The acid itself (or the sodium salt) is a valuable drug in the treatment of arthritis. But the most widely known derivative of salicylic acid is aspirin, which has the following structure:

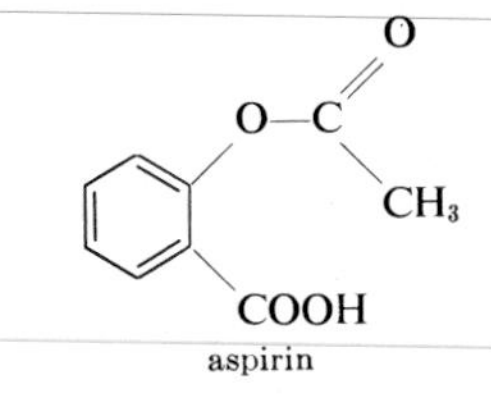

aspirin

You will see, by examining this structure, that aspirin is an ester of acetic acid. Aspirin is mankind's most widely used drug. Somewhat over 20 million pounds of aspirin are manufactured each year in the United States alone! This amounts to something like 150 five-grain tablets for every person in the country!

Table 18-IV shows the structures of a few simple benzene derivatives that are important commercial products. Study these structures so that you can see their relationship with the simple compounds from which they are derived.

18-6 POLYMERS

Table 18-III, p. 341, shows that the melting points of the normal alkanes tend to increase as the number of carbon atoms in the chain is increased. Ethane, C_2H_6, is a gas under normal conditions; octane, C_8H_{18}, is a liquid; octadecane, $C_{18}H_{38}$, is a solid. We see that desired physical properties can be obtained by controlling the length of the chain. Functional groups attached to the chain provide additional variability, including chemical reactivity. In fact, by adjusting the chain length and composition of high molecular weight compounds, chemists have produced a multitude of organic solid substances called plastics. These have been tailored to meet the needs of a wide variety of uses, giving rise to an enormous chemical industry.

The key to this chemical treasure chest is the process by which extended chains of atoms are formed. Inevitably it is necessary to begin with relatively small chemical molecules—with carbon chains involving only a few atoms. These small units, called monomers, must be bonded together, time after time, until the desired molecular weight range is reached. Often the desired properties are obtained only with giant molecules, each containing hundreds or even thousands of monomers. These giant molecules are called *polymers* and the process by which they are formed is called *polymerization.*

18-6.1 Types of Polymerization

Polymerization involves the chemical combination of a number of identical or similar molecules to form a complex molecule of high molecular weight. The small units may be combined by *addition* polymerization or *condensation* polymerization.

Addition polymers are formed by the reaction of the monomeric units without the elimination of atoms. The monomer is usually an unsaturated organic compound such as ethylene, $H_2C{=}CH_2$, which in the presence of a suitable catalyst will undergo an addition reaction to form a long chain molecule such as polyethylene. A general equation for the first stage of such a process is

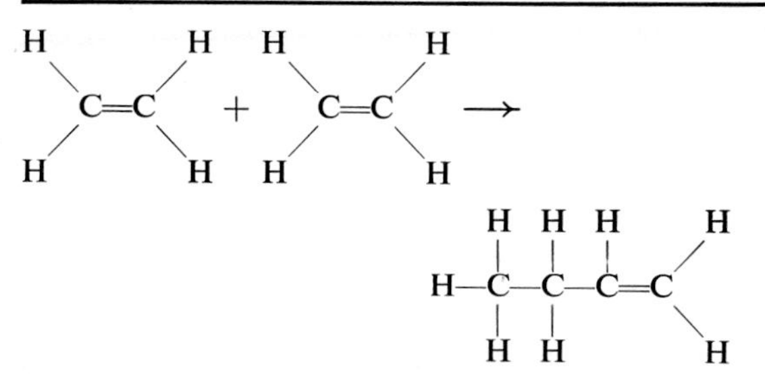

The same addition process continues and the final product is the polymer, polyethylene:

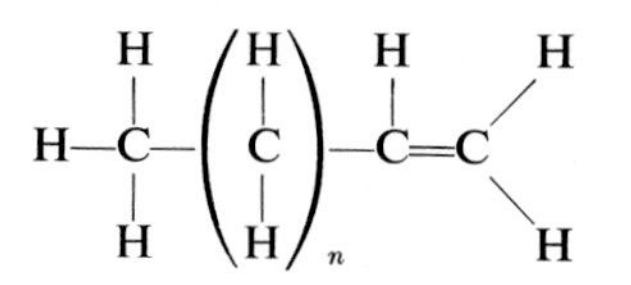

in which n is a very large number.

"Lucite," and "Plexiglas" result. It is thus possible to create molecules with custom-built properties for various uses as plastics or fibers.

Condensation polymers are produced by reactions during which some simple molecule (such as water) is eliminated between functional groups (such as alcoholic OH or acidic COOH groups). In order to form long chain molecules, two or more functional groups must be present in each of the reacting units. For example, when ethylene glycol, $HOCH_2CH_2OH$, reacts with *para*phthalic acid HOOC—⟨benzene ring⟩—COOH a polyester of high molecular weight called "Dacron" is produced. The equation below shows the first stages of this process:

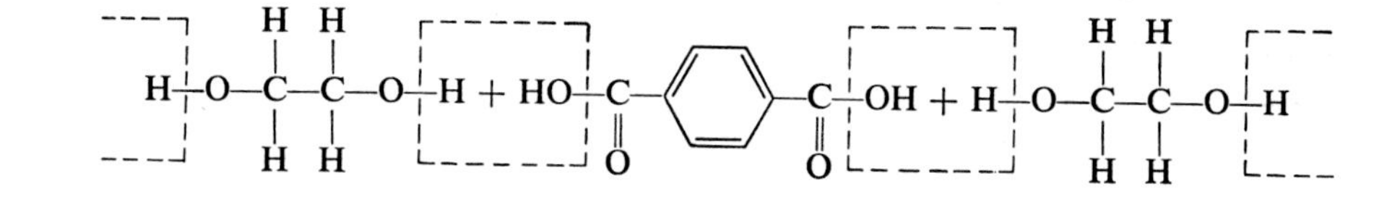

When one or more of the hydrogens are replaced by groups such as fluorine, F; chlorine, Cl; methyl, CH_3 or methyl ester, $COOCH_3$; synthetic polymers such as "Teflon," "Saran,"

Fig. 18-13. **Molecular structures** *of α-amino acids.*

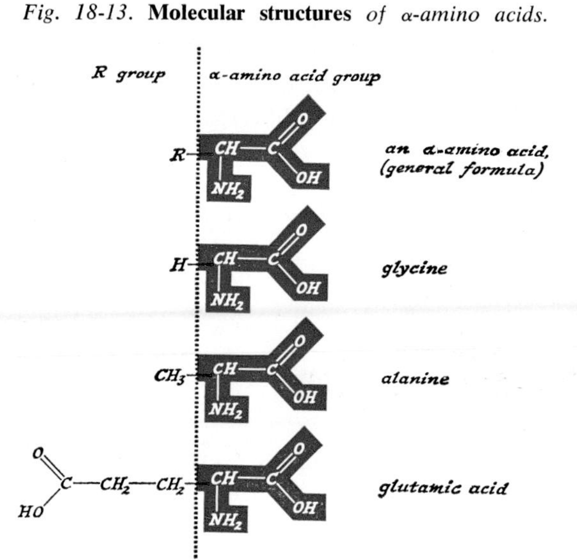

18-6.2 "Nylon," a Polymeric Amide

"Nylon," the material widely used in plastics and fabrics, is a condensation polymer. It consists of molecules of extremely high molecular weight and it is made up of small units joined one to another in a long chain of atoms. The reaction by which the units become bonded together is a conventional amide formation. There is, however, the additional requirement that the reaction must take place time after time to form an extended, repeating chain. To accomplish this, we select reactants with two functional groups. Thus polyamides can be made from one compound with two acid groups,

$$HOOC{-}CH_2{-}CH_2{-}CH_2{-}CH_2{-}COOH$$

adipic acid

and another with two amine groups,

$$H_2N{-}CH_2{-}CH_2{-}CH_2{-}CH_2{-}CH_2{-}CH_2{-}NH_2$$

1,6-diaminohexane

These molecules can react repeatedly, each time removing water, and forming amide linkages at both ends:

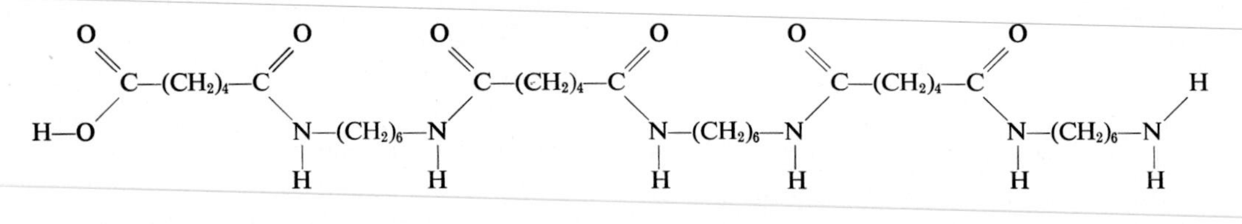

Other polyamides can be made from different acids and other amines, giving a variety of properties suited to a variety of uses.

18-6.3 Protein, Another Polymeric Amide

A most important class of polyamides is that of the *proteins*, the essential structures of all living matter. In addition, they are a necessary part in the diet of man because they are the source of the "monomeric" units, the amino acids, from which living protein materials are made.

Proteins are polyamides formed by the polymerization, through amide linkages, of α-amino acids. Three of the 25–30 important natural α-amino acids are shown in Figure 18-13. Each acid has an amine group, $—NH_2$, attached to the α-carbon, the carbon atom immediately adjacent to the carboxylic acid group.

The protein molecule may involve hundreds of such amino acid molecules connected through the amide linkages. A portion of this chain might be represented as shown in Figure 18-14.

Fig. 18-14. **The structure of protein** *showing the amide chain.*

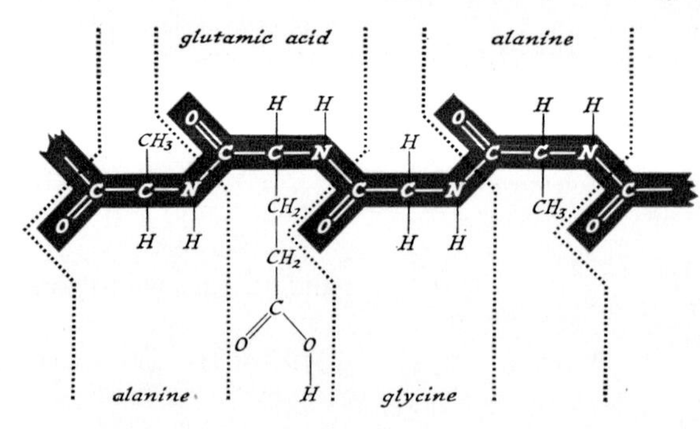

An amide is decomposed by aqueous acids to an acid and an amine (produced as the ammonium salt):

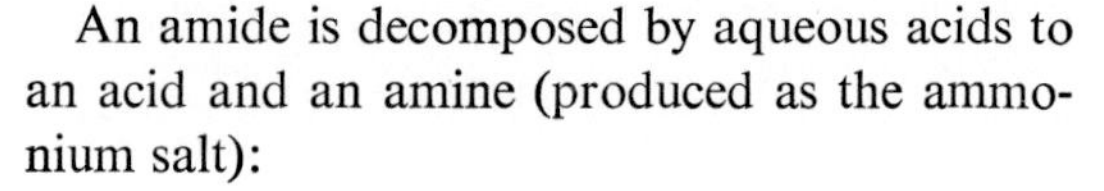
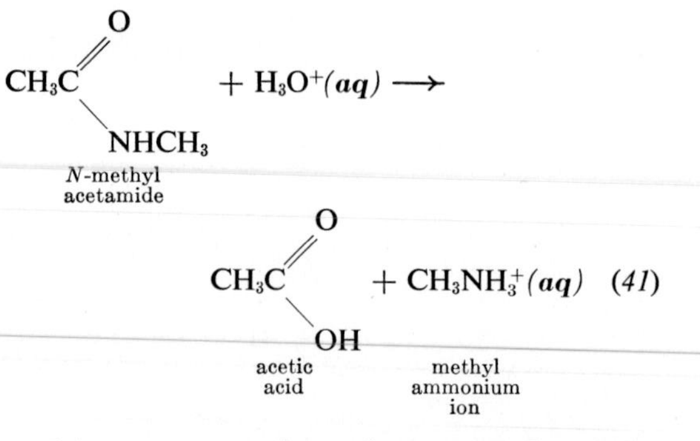

$$CH_3C(=O)NHCH_3 + H_3O^+(aq) \longrightarrow CH_3C(=O)OH + CH_3NH_3^+(aq) \quad (41)$$

In this same type of reaction, a protein can be broken down into its constituent amino acids.

It is in this way that most of what we call the "natural" amino acids have been discovered. Proteins from many sources—egg yolk, milk, animal tissues, plant seeds, gelatin, etc.—have been studied to learn what amino acids compose them. In this way about thirty of the "natural" amino acids have been identified.

When you consider how many different ways thirty (or even fewer) different amino acids can be combined in long chains of a hundred or more, you can see why there are so many proteins known, and why different living species of plants and animals can have in their tissues a great many different proteins. Enzymes, the biological catalysts, are also proteins. Each enzyme, of course, has its own particular structure, determined by the order and spatial arrangement of the amino acids from which it is formed. Perhaps the most marvelous part of the chemistry of living organisms is their ability to synthesize just the right protein structures from the myriad of structures possible.

EXERCISE 18-15

Take the letters A, B, C, and see how many different three unit combinations you can make; for example, ABC, BAC, AAC, CBC, etc. This will convince you that a chain made of hundreds of groups with up to *thirty* different kinds of units in each group can have an almost unlimited number of combinations.

QUESTIONS AND PROBLEMS

1. What information is revealed by the empirical formula? The molecular formula? The structural formula? Demonstrate, using ethane, C_2H_6.

2. Write the balanced equation for the complete burning of methane.

3. Draw the structural formulas for all the $C_2H_3Cl_3$ compounds.

4. Draw the structures of two isomeric compounds corresponding to the empirical formula C_3H_8O.

5. Draw the structural formulas of the isomers of butyl chloride.

6. What angle would you expect to be formed by the C, O, H nuclei in an alcohol molecule? Explain.

7. When 0.601 gram of a sample having an empirical formula CH_2O was vaporized at 200°C, and one atmosphere pressure, the volume occupied was 388 ml. This same volume was occupied by 0.301 gram of ethane under the same conditions. What is the molecular formula of CH_2O?

   One mole of the sample, when reacted with zinc metal, liberated (rather slowly) $\frac{1}{2}$ mole of hydrogen gas. Write the structural formula.

   Answer. The molecular formula is $C_2H_4O_2$.

8. A 100 mg sample of a compound containing only C, H, and O was found by analysis to give 149 mg CO_2 and 45.5 mg H_2O when burned completely. Calculate the empirical formula.

9. How much ethanol can be made from 50 grams of ethyl bromide? What assumptions do you make in this calculation?

10. Write the balanced equation for the production of pentanone from pentanol, using dichromate ion as the oxidizing agent.

11. One mole of an organic compound is found to react with $\frac{1}{2}$ mole of oxygen to produce an acid. To what class of compounds does this starting material belong?

12. Using the information given in Table 7-II, determine the reaction heat per mole of $C_2H_6(g)$ for the complete combustion of ethane.

13. An aqueous solution containing 0.10 mole/liter of chloroacetic acid, ClH_2CCOOH, is tested with indicators and the concentration of $H^+(aq)$ is found to be 1.2×10^{-2} M. Calculate the value of K_A (if necessary, refer back to Section 11-3.2). Compare this value with K_A for acetic acid—the change is caused by the substitution of a halogen atom near a carboxylic acid group.

14. Give simple structural formulas of

    (a) an alcohol,
    (b) an aldehyde, and
    (c) an acid,

    each derived from methane; from ethane; from butane; from octane.

15. Write the equations for the preparation of methylamine from methyl iodide.

16. Write equations to show the formation of the esters, methyl butyrate and butyl propionate.

17. Given the structural formula

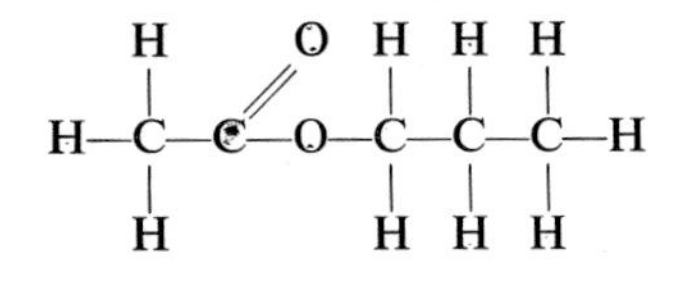

    for an ester, write the formula of the acid and the alcohol from which it might be made.

18. How much acetamide can be made from 3.1 grams of methyl acetate? See equation (*27*), p. 338. Assume the ester is completely converted.

Answer. 25 grams acetamide.

19. An ester is formed by the reaction between an acid, RCOOH, and an alcohol, R′OH, to form an ester RCOOR′ and water. The reaction is carried out in an inert solvent.

    (a) Write the equilibrium relation among the concentrations, including the concentration of the product water.
    (b) Calculate the equilibrium concentration of the ester if $K = 10$ and the concentrations *at equilibrium* of the other constituents are:

    [RCOOH] = 0.1 *M*;
    [R′OH] = 0.1 *M*;
    [H_2O] = 1.0 *M*.

    (c) Repeat the calculation of part *b* if the equilibrium concentrations are:

    [RCOOH] = 0.3 *M*;
    [R′OH] = 0.3 *M*;
    [H_2O] = 1.0 *M*.

20. Give the empirical formula, the molecular formula, and draw the structural formulas of the isomers of butene.

21. There are three isomers of dichlorobenzene (empirical formula C_3H_2Cl). Draw the structural formulas of the isomers.

22. Consider the compound phenol,

    (a) Predict the angle formed by the nuclei C, O, H. Explain your choice in terms of the orbitals used by oxygen in its bonds.
    (b) Predict qualitatively the boiling point of phenol. (The boiling point of benzene is 80°C.) Explain your answer.
    (c) Write an equation for the reaction of phenol as a proton donor in water.
    (d) In a 1.0 *M* aqueous solution of phenol, $[H^+] = 1.1 \times 10^{-5}$. Calculate *K*.

SIR ROBERT ROBINSON, 1886–

Sir Robert Robinson will always be recognized as one of the outstanding British scientists. With admiration and pride, his country knighted him in 1937. His international recognition in organic chemistry was signaled in 1947, when he received the Nobel Prize in Chemistry.

The son of a surgical dressing manufacturer, Robinson was born in Chesterfield, Derbyshire, England. He received the Ph.D. at the University of Manchester, and his early scientific promise led to a Professorial appointment at the University of Sydney when he was only 26. Three years later, he returned to England as Professor at the University of Liverpool. As his reputation grew, he moved from university to university, until, in 1930, he accepted one of the most coveted academic positions in England—Professor of Chemistry at Oxford.

During a prolific career that has produced an astounding bibliography of about 600 publications, Robinson has found leisure time to scale many of the mountain peaks in Switzerland, Norway, and New Zealand. This zeal for mountaineering couples well with his interest in photography. He is a formidable chess player and a lover of music. His wife, also a chemist of note, aided him in his career while raising his family, a son and a daughter.

Sir Robert is a master of organic synthesis. His research centered on the synthesis and the determination of structures of biochemically important substances. These include extremely complicated molecules with as many as five closed rings and a variety of functional groups whose positioning and spatial relationships are critically important in fixing biological activity. His name is linked with our knowledge of certain alkaloids—a family of nitrogen containing compounds that includes some powerful drugs, such as strychnine and morphine. His crowning achievements may have been in the total synthesis of such important compounds as cholesterol, cortisone, and other related structures, called steroids. During the second world war, Professor Robinson led the Oxford scientific team that studied the chemistry and structure of the antibiotic, penicillin.

These are but a few pinnacles of success in a brilliant scientific career. Like the Alpine peaks that Sir Robert Robinson loves to climb, these accomplishments tower high, to challenge and inspire this and future scientific generations.

CHAPTER 19

The Halogens

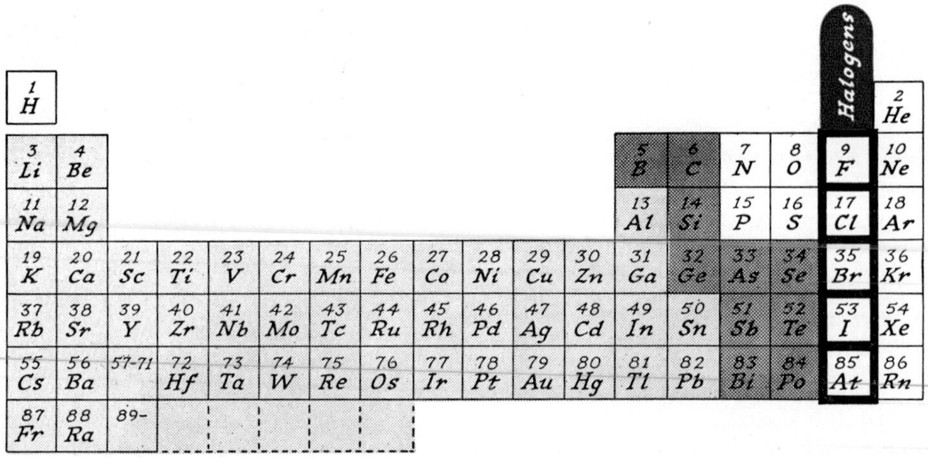

																Halogens	
1 H																	2 He
3 Li	4 Be											5 B	6 C	7 N	8 O	9 F	10 Ne
11 Na	12 Mg											13 Al	14 Si	15 P	16 S	17 Cl	18 Ar
19 K	20 Ca	21 Sc	22 Ti	23 V	24 Cr	25 Mn	26 Fe	27 Co	28 Ni	29 Cu	30 Zn	31 Ga	32 Ge	33 As	34 Se	35 Br	36 Kr
37 Rb	38 Sr	39 Y	40 Zr	41 Nb	42 Mo	43 Tc	44 Ru	45 Rh	46 Pd	47 Ag	48 Cd	49 In	50 Sn	51 Sb	52 Te	53 I	54 Xe
55 Cs	56 Ba	57-71	72 Hf	73 Ta	74 W	75 Re	76 Os	77 Ir	78 Pt	79 Au	80 Hg	81 Tl	82 Pb	83 Bi	84 Po	85 At	86 Rn
87 Fr	88 Ra	89-															

The halogens are a family of elements appearing on the right side of the periodic table, in the column just before the inert gases. The elements in this group—fluorine, chlorine, bromine, iodine, and astatine—show some remarkable similarities and some interesting trends in chemical behavior. The similarities are expected since the electron populations of the outer levels are analogous. Each element has one electron less than an inert gas arrangement. The trends, too, are understandable in terms of the increases in nuclear charge, number of electrons, and atomic size, going from top to bottom of this column of the periodic table.

19-1 PROPERTIES OF THE HALOGENS

You have already studied some properties of the halogens. They are very reactive elements that exist under normal conditions as diatomic molecules with covalent bonds. These molecules all are colored. Gaseous fluorine is pale yellow; gaseous chlorine is yellow-green; gaseous bromine is orange-red (remember the film EQUILIBRIUM); gaseous iodine is violet. The halogens are all toxic and dangerous substances. Fluorine, F_2, is the most hazardous; the danger decreases as the atomic number of the halogen becomes larger. Even iodine, I_2, should be handled with caution.*

19-1.1 Electron Configurations of the Halogens

Table 19-I shows the electron configurations of the halogens, using the symbols introduced in

* These elements produce nasty "burns" that are slow to heal. The mucous membranes are attacked especially, and chlorine "poisoning" is really a lung inflammation. Under no circumstances should inexperienced people handle these substances without close guidance.

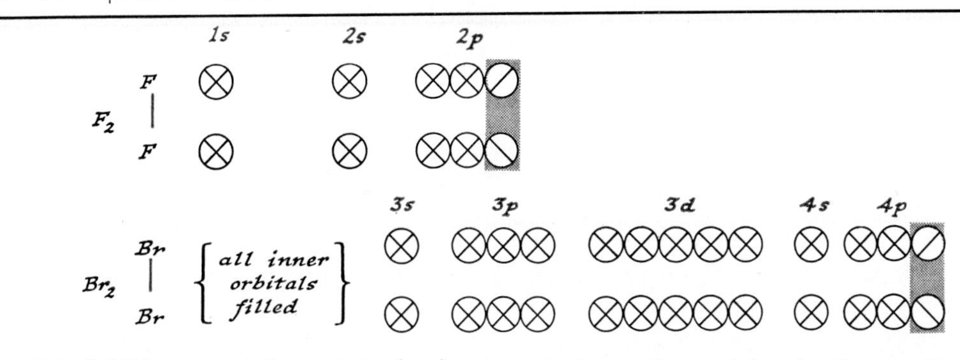

Fig. 19-1. **Orbital representations** *of the bonding in F_2 and Br_2.*

Section 15-1.6. The superscripts give the number of electrons in a particular type of orbital; the letters *s*, *p*, and *d* indicate the shape of the orbital; the number before the letter indicates the principal quantum number of the orbital. The important point to note is that each of these halogens has one less electron than the number required to fill the outermost cluster of energy levels. In each case the shortage is in the outermost *p* orbitals in which six electrons can be accommodated. Therefore, one electron could be added to each of the halogen atoms without requiring the population of additional, higher energy orbitals. Sharing an electron that spends some time in a valence orbital of another atom produces a covalent bond to the halogen atom. Organic compounds (such as ethyl bromide) and the halogen molecules (F_2, Cl_2, Br_2, I_2) contain covalently bonded halogen atoms. The similar orbital representations of the bonding in F_2 and Br_2 are contrasted in Figure 19-1.

Electron dot representations of the electron sharing in Br_2 and I_2 raise the question of how many valence electrons need be shown. Each of these two elements appears in a row of the periodic table in which the inert gas element has 18 electrons in the orbitals that fix the chemistry of the row. Nevertheless, it is a convenience to emphasize the similarity among the halogens by showing only the valence electrons in the outermost *s* and *p* orbitals. Thus the 3*d* (valence) electrons of bromine are usually omitted so that the electron dot representations of F_2 and Br_2 will appear alike, as shown in Figure 19-2.

Table 19-I also shows the ionization energies

: F : F :

: Br : Br :

Fig. 19-2. **Electron dot representations** *of the bonding in F_2 and Br_2. Note the omission, for convenience, of the 3d valence electrons of bromine.*

Table 19-I. **ELECTRON CONFIGURATIONS AND IONIZATION ENERGIES OF THE HALOGENS**

			ELECTRON CONFIGURATION		IONIZATION
ELEMENT	SYMBOL	NUCLEAR CHARGE	*inner electrons*	*valence electrons*	ENERGY, E_1 (*kcal/mole*)
fluorine	F	+9	$1s^2$	$2s^22p^5$	401.5
chlorine	Cl	+17	$1s^22s^22p^6$	$3s^23p^5$	300
bromine	Br	+35	$\cdots 3s^23p^6$	$3d^{10}4s^24p^5$	273
iodine	I	+53	$\cdots 4s^24p^6$	$4d^{10}5s^25p^5$	241

of the gaseous halogen atoms. They decrease significantly as we move downward in the periodic table. Nevertheless, all of these ionization energies are very large compared with those of the alkali metals (compare sodium, whose ionization energy is 118.4 kcal). Hence when any of the halogens reacts with an alkali metal, an ionic solid is formed. These ionic solids, or salts, contain halide ions, F^-, Cl^-, Br^-, or I^-, each with the appropriate inert gas electron population.

19-1.2 The Sizes of Halogen Atoms and Ions

The "size" assigned to an atom or ion requires a decision about where an atom "stops." From quantum mechanics we learn that an atom has no sharp boundaries or surfaces. Nevertheless, chemists find it convenient to assign sizes to atoms according to the observed distances between atoms. Thus, atomic size is defined operationally—it is determined by measuring the distance between atoms.

For example, Figure 19-3 contrasts the dimensions assigned to the halogens in the elementary state. *One-half the measured internuclear distance is called the* **covalent radius.** This distance indicates how close a halogen atom can approach another atom *to which it is bonded.* To atoms to which it is not bonded, a halogen atom seems to be larger. We can take as a measure of this size one-half the distance between neighboring molecules in the solid state. This defines an effective radius, the van der Waals radius, and it is shown by the black lines in Figure 19-3. It is an effective radius, not a real radius, because the electron distribution actually extends far out from the atom.

These distances aid us in explaining and predicting bond lengths in other covalent halogen compounds. For example, when a chlorine atom is bonded to a carbon atom (as in carbon tetrachloride, CCl_4), the bond length can be expected to be about the sum of the covalent radius of the carbon atom plus the covalent radius of the chlorine atom. The covalent radius of carbon is taken as 0.77 Å (from diamond), so the C—Cl bond length might be near $(0.77 + 0.99) =$ 1.76 Å. Experiment shows that each bond length in CCl_4 is 1.77 Å.

EXERCISE 19-1

Using the carbon atom covalent radius 0.77 Å and the covalent radii given in Figure 19-3, predict the C—*X* bond length in each of the following molecules: CF_4, CBr_4, CI_4. Compare your calculated bond lengths with the experimental values: C—F in CF_4 = 1.32 Å, C—Br in CBr_4 = 1.94 Å, C—I in CI_4 = 2.15 Å.

Fig. 19-3. **Covalent radii and van der Waals radii** *(in parentheses) of the halogens (in Ångstroms).*

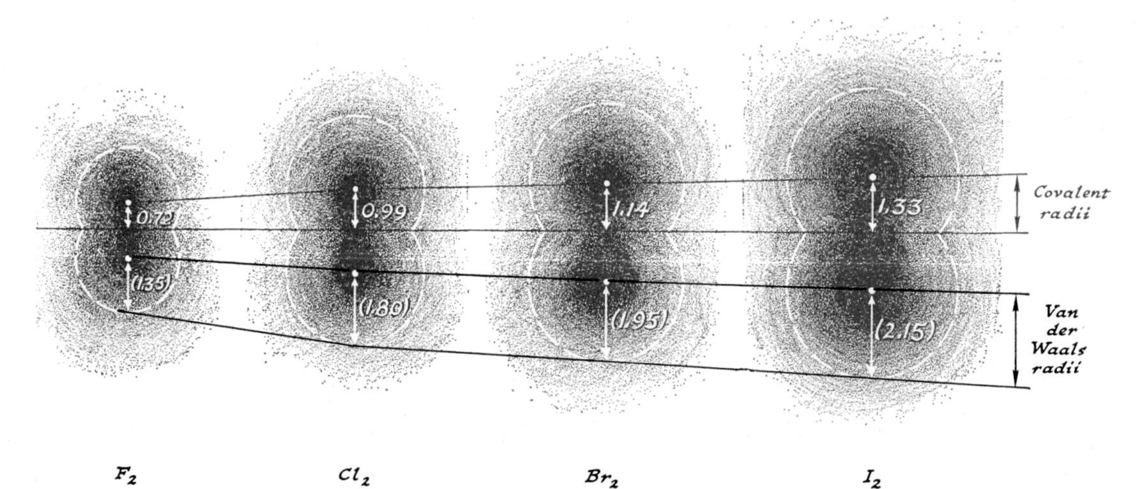

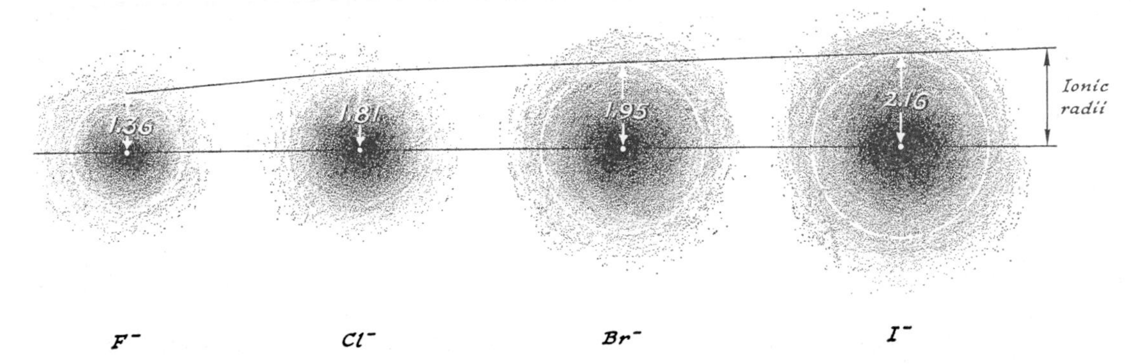

Fig. 19-4. **Ionic radii** *of the halide ions (in Ångstroms).*

Figure 19-4 contrasts the effective sizes of the halide ions. Each of these dimensions is obtained from the examination of crystal structures of many salts involving the particular halide ion. The effective size found for a given halide ion is called its *ionic radius*. These radii are larger than the covalent radii but close to the van der Waals radii of neutral atoms.

The covalent, van der Waals, and ionic radii are collected in Table 19-II together with some physical and chemical properties of the halogens. We see some interesting trends. For each type of radius we find a progressive increase from the top of the column to the bottom. This increase in size reflects the fact that as atomic number rises, higher energy levels are used to accommodate the electrons. In addition, Table 19-II shows a trend of increasing melting and boiling points as we move downward in the periodic table. This trend is appropriate for a series of molecular solids in which van der Waals forces are the principal ones holding the molecules in proximity. This type of force is higher for more complex molecules with more electrons.

The last column of Table 19-II shows the dissociation constants for the reactions of the type

$$X_2(g) \rightleftharpoons 2X(g) \qquad (1)$$

with equilibrium constant

$$K = \frac{[X]^2}{[X_2]}$$

where

$[X]$ = partial pressure of X atoms,
$[X_2]$ = partial pressure of X_2 molecules.

Again the available data show a trend; K increases in the series Cl_2, Br_2, I_2. The equilibrium conditions are fixed by two factors: tendency to minimum energy and tendency to maximum randomness. The randomness factor is about the same for the three halogens, so the trend in equilibrium constants is largely determined by the energy effects. In each dissociation reaction (of Cl_2, Br_2, and I_2) energy is absorbed. However, more energy is absorbed to break the Br_2 bond than to break the I_2 bond, and still more to break the Cl_2 bond. The energy absorbed accounts for the trend in equilibrium constants.

Table 19-II. **SIZES OF THE HALOGEN ATOMS, MELTING POINTS, BOILING POINTS, AND DISSOCIATION PROPERTIES OF THE HALOGEN MOLECULES**

HALOGEN, X	COVALENT RADIUS IN X_2	VAN DER WAALS RADIUS IN X_2	IONIC RADIUS (−1 ion)	M.P. OF X_2	B.P. OF X_2	BOND ENERGY OF X_2	DISSOCIATION CONSTANT OF X_2 AT 1000°C
fluorine, F	0.72 Å	1.35 Å	1.36 Å	55°K	85°K	36 kcal	—
chlorine, Cl	0.99	1.80	1.81	172°K	238.9°K	57.1	10^{-8}
bromine, Br	1.14	1.95	1.95	265.7°K	331.8°K	45.5	8×10^{-3}
iodine, I	1.33	2.15	2.16	387°K	457°K	35.6	10^{-1}

Why should the bond energy be greater for Cl_2 than for Br_2, and greater for Br_2 than for I_2? Presumably the size effect is a factor. Two halogen atoms remain together because a pair of electrons is simultaneously near both nuclei. However, the larger the halogen atoms, the more distant are those bonding electrons from the nuclei. Since the electrical forces decrease with distance, the bond energy lessens.

The dissociation constant of F_2 at 1000°C is not known and the bond energy of F_2 was only learned recently. Almost all chemists were surprised when experiment showed that the energy necessary to break the bond in F_2 is much lower than that in Cl_2. This is still not well explained so, rather than abandon familiar arguments concerning halogen properties based on the trend in size, chemists treat fluorine as a special case.

EXERCISE 19-2

On the basis of the trend in atomic size, what trend is expected in the ionization energy E_1 of the halogen atoms? Compare your prediction with the actual trend in E_1, given in Table 19-I.

19-2 HALOGEN REACTIONS AND COMPOUNDS

Most of the reactions of the halogens are of the oxidation-reduction type. The halogens are so reactive that they do not occur uncombined in nature and they must be made from halide compounds (salts). We shall consider briefly the preparation of the elements and then explore some of the very interesting chemistry of this family.

19-2.1 Preparation of the Halogens

Oxidation through electrolysis is used to make fluorine and chlorine. Chlorine, for example, is made by electrolysis of molten sodium chloride (dissolved in $CaCl_2$ to lower the melting point). Figure 19-5 shows the components of the electrolysis cell. At one electrode molten sodium is produced, and at the other, chlorine gas is collected.

Fig. 19-5. **Preparation of chlorine and sodium** *by electrolysis of molten NaCl.*

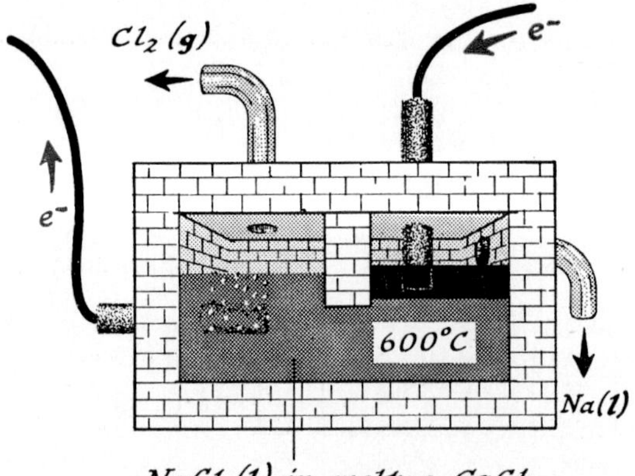

EXERCISE 19-3

At the left electrode in Figure 19-5 the half-reaction occurring is $Cl^- \longrightarrow \frac{1}{2}Cl_2(g) + e^-$, and at the right electrode the half-reaction is $Na^+ + e^- \longrightarrow Na(l)$. Which electrode is the anode and which is the cathode? With these half-reactions, balance the net reaction occurring in the electrolysis cell.

Gaseous fluorine is also prepared by electrolysis of molten fluoride salts but simpler methods are available for the preparation of bromine and iodine. Chemical oxidation, usually with chlorine as the oxidizing agent, provides Br_2 and I_2 economically because chlorine is a relatively inexpensive chemical. The reactions are

$$2I^-(aq) + Cl_2(g) \longrightarrow I_2(s) + 2Cl^-(aq) \quad (2)$$

$$2Br^-(aq) + Cl_2(g) \longrightarrow Br_2(g) + 2Cl^-(aq) \quad (3)$$

EXERCISE 19-4

From the $E°$'s for the half-reactions of the type $2X^- \longrightarrow X_2 + 2e^-$, show that Cl_2 can be used to produce Br_2 from Br^- and I_2 from I^- but not to produce F_2 from F^-.

19-2.2 Reduction of the Halogens

A substance that is readily reduced is a good oxidizing agent. The oxidizing abilities of the halogens vary in a regular manner, fluorine being the strongest and iodine the weakest. On the other hand, the iodide ion sometimes acts as a reducing agent, while fluoride ion never does. These statements are reflected in the $E°$ values for the oxidation half-reactions:

$$2F^- \longrightarrow F_2 + 2e^- \qquad E° = -2.87 \text{ volts} \qquad (4)$$

$$2Cl^- \longrightarrow Cl_2 + 2e^- \qquad E° = -1.36 \qquad (5)$$

$$2Br^- \longrightarrow Br_2 + 2e^- \qquad E° = -1.07 \qquad (6)$$

$$2I^- \longrightarrow I_2 + 2e^- \qquad E° = -0.53 \qquad (7)$$

The trend in oxidizing ability is quantitatively expressed in the trend in $E°$'s. The $E°$ for reaction (*4*), −2.87 volts, indicates that fluoride ion, F^-, has little tendency to release electrons. Conversely, F_2 has a high tendency to acquire electrons. Hence, F_2 is readily reduced—F_2 is a powerful oxidizing agent. At the bottom of the list, $E°$ for reaction (*7*) is −0.53 volt. Iodide ion, I^-, has a moderate tendency to release electrons, oxidizing I^- to I_2. Thus iodide ion has a moderate tendency to be oxidized, acting as a reducing agent. Conversely, I_2 has only a moderate tendency to acquire electrons. Iodine has a moderate tendency to be reduced—I_2 can act as an oxidizing agent.

How can this trend in half-cell potentials and oxidizing abilities be explained? Let us imagine that reaction (*7*), as an example, is carried out in a hypothetical series of simpler steps:

$$2I^-(aq) \longrightarrow 2I^-(g) \qquad \text{Dehydration} \qquad (8)$$

$$2I^-(g) \longrightarrow 2I(g) + 2e^-(g) \qquad \text{Electron removal} \qquad (9)$$

$$2I(g) \longrightarrow I_2(g) \qquad \text{Molecule formation} \qquad (10)$$

$$I_2(g) \longrightarrow I_2(s) \qquad \text{Condensation} \qquad (11)$$

$$2I^-(aq) \longrightarrow I_2(s) + 2e^-(g) \qquad \text{Net reaction} \qquad (12)$$

We can consider the energy effect that accompanies the net reaction (*12*) in terms of the energy effects of the hypothetical steps. Since the tendency toward minimum energy is one of the factors that fixes equilibrium, the energy change is one of the factors influencing $E°$. Step (*8*) is a dehydration step in which the I^- ions are pulled out of the water into the gas phase. In this step, heat is absorbed; since the system gains energy, ΔH is positive for step (*8*). In step (*9*), electrons are pulled off the gaseous I^- ions to give neutral gaseous I atoms; this step also requires energy, and again ΔH is positive. In step (*10*), two atoms of iodine come together to form a diatomic molecule. This is the opposite of breaking the bond in I_2; energy is liberated and the system loses energy, so ΔH is negative for this step. In step (*11*), the gaseous I_2 molecules form solid iodine crystals; this is the reverse of vaporizing the solid and also corresponds to liberation of energy. The question as to which halide ion is the best reducing agent depends primarily upon the *net* energy change considering all the steps involved in the half-reaction. The halide ion that requires the least energy to convert it to the halogen should be the best reducing agent. Table 19-III shows the experimentally measured requirement for each of the above steps for the four halogens (using X as a general symbol for a halogen atom).

Notice that in Table 19-III dehydration and electron removal are the steps that involve the largest energy changes. The amount of energy required for each of these processes diminishes as atomic weight increases for the three large halogens. For fluorine, dehydration accounts for more than half the total energy, but electron removal is still a major part of the positive energy change. The

Table 19-III. THE ENERGY REQUIRED FOR SOME HALOGEN REACTIONS

HALOGEN	DEHYDRATION $2X^-(aq) \longrightarrow 2X^-(g)$	ELECTRON REMOVAL $2X^-(g) \longrightarrow 2X(g) + 2e^-(g)$	MOLECULE FORMATION $2X(g) \longrightarrow X_2(g)$	CONDENSATION $X_2(g) \longrightarrow X_2$ (normal state)	OVERALL REACTION $2X^-(aq) \longrightarrow X_2 + 2e^-$
F	+246 kcal	+162 kcal	−37 kcal	−1.6 kcal	+369 kcal
Cl	+178 kcal	+171 kcal	−58 kcal	−4.4 kcal	+287 kcal
Br	+162 kcal	+161 kcal	−46 kcal	−7 kcal	+270 kcal
I	+144 kcal	+146 kcal	−36 kcal	−10 kcal	+244 kcal

values in the last column show that the energy change for the overall reaction forms a regular series, roughly comparable to the variation of $E°$. The half-reactions having the most negative $E°$ are those that require the most energy, and that show the least tendency to proceed as written from left to right. Therefore, the energies listed in the last column do help to explain why the iodide ion is a better reducing agent than the fluoride ion. This also explains why F_2 oxidizes compounds better than do the other halogens. In aqueous solution the affinity of fluorine for electrons, plus the rather strong attraction between water molecules and F^-, make F_2 a good oxidizing agent.

The halogens are reactive even without water. All the halogens react quite vigorously with most of the metals to produce simple halide salts. Copper and nickel, however, appear to be quite inert to F_2. This apparent inertness is attributed to the fact that a thin layer of the fluoride salt forms on the surface of each of those metals and protects it from further attack by fluorine.

19-2.3 Iodimetry

The I^-–I_2 half-reaction has many applications in aqueous solution chemistry. The use of I^- as a reducing agent and I_2 as an oxidizing agent, particularly for quantitative purposes, is called **iodimetry.**

This half-reaction possesses an $E°$ of -0.53 volt; neither is I^- a particularly powerful reducing agent nor is I_2 a particularly powerful oxidizing agent:

$$2I^- \longrightarrow I_2 + 2e^- \qquad E° = -0.53 \text{ volt} \qquad (13)$$

There are many half-reactions below this one in Appendix 3, so there are quite a few substances that will oxidize I^-. For example, iodide ion can be quantitatively oxidized to I_2 by Fe^{+3}, Br_2, MnO_2, $Cr_2O_7^{-2}$, Cl_2, and MnO_4^-. On the other hand, there are many half-reactions above $E° = -0.53$ volt in Appendix 3. For example, I_2 can be quantitatively reduced to I^- by Sn^{+2}, H_2SO_3, and Cr^{+2}. *The usefulness of the I^-–I_2 reaction derives from the fact that all of the substances mentioned react rapidly and without side reactions.*

To top off this versatility, iodine possesses an unusually sensitive and specific indicator. Iodine reacts with starch to give a blue-colored complex. This complex is so intensely colored that I_2 can be detected at a concentration as low as 10^{-5} M and it furnishes the basis for the qualitative test for iodine known as the "starch-iodine" test. More important, the complex serves as a sensitive indicator in oxidation-reduction titrations based upon the I^-–I_2 half-reaction.

EXERCISE 19-5

Balance the reaction that occurs when I^- is oxidized to I_2 by MnO_4^- in acid solution, producing Mn^{+2}.

19-2.4 Positive Oxidation States of the Halogens: The Oxyacids

The halogens, except fluorine, can be oxidized to positive oxidation states. Most commonly you will encounter these positive oxidation states in a set of compounds called "halogen oxyacids" and their ions.

Compounds of the type $HClO_3$, $HClO_4$, etc., are examples of the halogen oxyacids. Chlorine and iodine form a series of these acids in which the halogen oxidation number can be $+1$, $+3$, $+5$, or $+7$. For chlorine the series is made up of $HClO$ (hypochlorous acid), $HClO_2$ (chlorous acid), $HClO_3$ (chloric acid), and $HClO_4$ (perchloric acid). Although it is not easy to handle these unstable substances, aqueous solutions of these acids have been examined to find out how strong they are as proton donors. $HClO$ is a weak proton donor, $HClO_2$ is somewhat stronger, $HClO_3$ is quite strong, and $HClO_4$ is the strongest of all. (Perchloric acid, $HClO_4$, is, in fact, one of the strongest acids known.)

Now we might wonder how to account for the observed trend in terms of structure and bonding. Figure 19-6 shows the presumed positions of the atoms in these molecules. It is evident that in each case we need to break a hydrogen-oxygen bond to split off the proton. A regular decrease in the strength of the hydrogen-oxygen bond as we proceed from chlorous to perchloric acid would explain the trend in acidity. How do we account for the fact that the strength of this bond varies as we go through the sequence? Formally, we say that the oxidation number of chlorine ranges from $+1$ to $+3$ to $+5$ to $+7$

NAME	*FORMULA*	*BALL-AND-STICK MODEL*	*SPACE-FILLING MODEL*
Hypochlorous acid	HOCl		
Chlorous acid	HOClO ($HClO_2$)	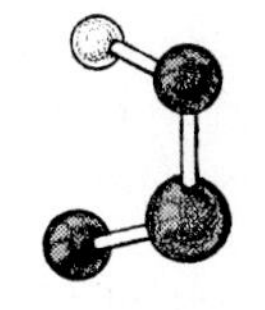	
Chloric acid	$HOClO_2$ ($HClO_3$)	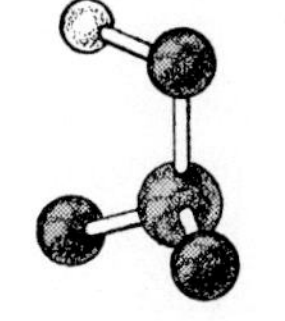	
Perchloric acid	$HOClO_3$ ($HClO_4$)	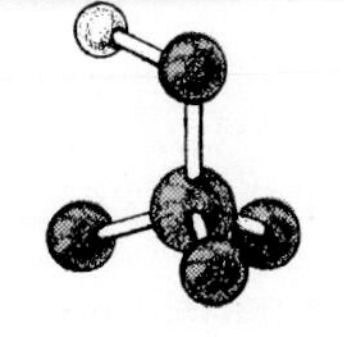	

Fig. 19-6. **Presumed structures** *of the chlorine oxyacids.*

across the set, but actually there are no Cl^{+1}, Cl^{+3}, Cl^{+5}, or Cl^{+7} ions in these acids. The oxidation number is only an artificial way of keeping count of electric charges, as we learned in Section 12-3.3. What is more to the point (and this is really why the oxidation number changes in the first place) is that there is an increasing number of oxygen atoms bonded to the central chlorine. Each time an additional oxygen bonds to the chlorine atom, some electron charge is drawn off the chlorine and hence away from the original O—Cl bond. This, in turn, draws electrons from the adjacent H—O bond and thereby weakens it.

This increase in acid strength with oxidation number is a general phenomenon. For example, nitric acid (HNO_3, in which the oxidation number of N equals +5) is stronger than nitrous (HNO_2, oxidation number +3); sulfuric acid (H_2SO_4, in which the oxidation number of S equals +6) is stronger than sulfurous (H_2SO_3, oxidation number +4). A consistent, hence useful, explanation is found. When an oxygen atom is added to the central atom, there is a reduction of the strength of O—H bonds in attached OH groups.

The halogen oxyacids and their anions are quite easily reduced—they are good *oxidizing agents*. When one of these acts as an oxidizing agent, the halogen is reduced to a lower oxidation number. Just what oxidation number is attained depends upon a variety of factors, including acidity of the solution, strength of the reducing agent, amount of reducing agent, and temperature.

For example, the common use of sodium hypochlorite solution, NaOCl, as a bleaching solution depends upon the oxidizing action of hypochlorite, OCl^-. Iodate ion, IO_3^-, also furnishes a strong oxidizing power, as shown by $E°$ for the half-reaction I_2–IO_3^-

$$I_2 + 6H_2O \longrightarrow 2IO_3^- + 12H^+ + 10e^- \qquad E° = -1.2 \text{ volts} \quad (14)$$

The chlorine and bromine counterparts, chlorate, ClO_3^-, and bromate, BrO_3^-, have $E°$'s that are even more negative. Hence these ions are even stronger oxidizing agents than iodate ion.

It is not unusual in halogen chemistry to find striking differences in the chemistry of acidic and basic solutions. For example, iodine in an acidic solution is quite stable, but in a basic solution it is spontaneously oxidized to oxidation number +5 in the IO_3^- ion. The reason for this can be seen by considering the half-reactions (*15*) and (*16*):

$$2I^- \longrightarrow I_2 + 2e^- \qquad E° = -0.53 \text{ volt} \quad (15)$$

$$I_2 + 6H_2O \longrightarrow 2IO_3^- + 12H^+ + 10e^- \qquad E° = -1.2 \text{ volts} \quad (16)$$

The $E°$ for reaction (*16*) is more negative than that of reaction (*15*), hence IO_3^- will not react with I^-. In 1 *M* acid solution, the $E°$ for the net reaction

$$6I_2 + 6H_2O \longrightarrow 2IO_3^- + 10I^- + 12H^+ \quad (17)$$

is (−1.2 volts) − (−0.53 volt) or −0.7 volt. Since $E°$ of the net reaction is so negative, we expect that the reaction will not proceed spontaneously, and it doesn't. Let us now see what happens when the H^+ concentration becomes so low that the OH^- concentration is 1 *M*. When a reaction occurs in basic solution it is conventional to show OH^- (rather than H^+) in the balanced reaction. Therefore, for the reaction of iodine in basic solution, half-reaction (*16*) becomes

$$I_2 + 12OH^- \longrightarrow 2IO_3^- + 6H_2O + 10e^- \quad (18)$$

Le Chatelier's Principle aids us in predicting how the tendency for I_2 to release electrons in (*18*) will be affected if we raise the hydroxide ion concentration. Raising the concentration of a reactant (such as OH^-) tends to favor products. Hence $E°$ for (*18*) (1 *M* OH^-) will be more positive than $E°$ for (*16*) (1 *M* H^+). In 1 *M* OH^-, $E° = 0.23$ volt. Since this $E°$ is now more positive than $E°$ for reaction (*15*) ($E° = -0.53$ volt), the net reaction (*19*) can occur:

$$6I_2 + 12OH^- \longrightarrow 2IO_3^- + 10I^- + 6H_2O \quad (19)$$

This reaction now has an $E°$ of (−0.23 volt) − (−0.53 volt) or +0.30 volt. With this positive value of $E°$, we can expect that in basic solution the reaction *will* proceed spontaneously (if the rate is rapid), and it does.

The industrial preparation of bromine takes advantage of this effect of hydrogen ion concentration on the direction of the spontaneous reaction. A dilute solution of bromine is produced by chlorination of salt well brines:

$$Cl_2 + 2Br^- \longrightarrow Br_2 + 2Cl^- \quad (20)$$

The liberated bromine is carried by a stream of air into an alkaline solution of sodium carbonate where it dissolves as a mixture of bromide and bromate [the analogy of reaction (*19*)]. This last step serves to concentrate the product, and free bromine is obtained by subsequent acidification of the solution, through the reaction

$$6H^+ + 5Br^- + BrO_3^- \longrightarrow 3Br_2 + 3H_2O \quad (21)$$

An alternative process for preparing bromine from sea water begins again with reaction (*20*). The liberated bromine is produced at a very low partial pressure and it is necessary to concentrate it. This is accomplished through the reaction between SO_2 and Br_2

$$Br_2(g) + SO_2(g) + 2H_2O(g) \longrightarrow 2HBr(g) + 2H_2SO_4(g) \quad (22)$$

The resulting acid vapors have a great affinity for water (do you remember how rapidly HCl dissolved in water in the film, GASES AND HOW THEY COMBINE?). Hence the HBr rapidly dissolves in water and concentrations as high as 0.5 *M* can be reached, a thousandfold more concentrated than original sea water. With this concentration, chlorine is again introduced to produce Br_2 by reaction (*20*).

19-2.5 Self-Oxidation-Reduction: Disproportionation

In reaction (*19*) the iodine shown on the left has an oxidation number of zero. After the reaction, some of the iodine atoms have oxidation number +5 and some −1. In other words, the iodine oxidation number has gone both up and down in the reaction. This is an example of self-oxidation-reduction, sometimes called *disproportionation*. It is a reaction quite typical of, but not at all restricted to, the halogens.

When chlorine gas is bubbled into a solution of NaOH, self-oxidation-reduction occurs to give hypochlorite ion, ClO^-, by the reaction

$$Cl_2 + 2OH^- \longrightarrow Cl^- + ClO^- + 2H_2O \quad (23)$$

EXERCISE 19-6

Show that when an aqueous solution of NaCl is being electrolyzed, vigorous stirring in the cell might permit reaction (*23*) to occur.

Going one step further, if a basic solution containing hypochlorite ion is heated, the ClO^- can again disproportionate,

$$3ClO^- \longrightarrow 2Cl^- + ClO_3^- \quad (24)$$

this time to produce the chlorate ion, ClO_3^-.

19-2.6 Special Remarks on Fluorine

Because of the small size of the atom, fluorine is rather special in the halogen group. We have already seen that it is a strong oxidizing agent in aqueous solution, and that a large part of this arises because of the large hydration energy associated with the fluoride ion. Another way in which fluorine reveals its special character is in the properties of hydrogen fluoride compared with the other hydrogen halides. These are explained in terms of a special attraction of fluorine for protons—an attraction called hydrogen bonding (reread Section 17-2.6).

The strong attraction of fluorine for protons shows up in another way. In aqueous solution, HF is a weak acid whereas HCl, HBr, and HI are strong acids. The dissociation constant of HF is 6.7×10^{-4}, so hydrofluoric acid is less than 10% dissociated in a 0.1 *M* HF solution.

Another contrast between HF and the other hydrogen halides, HCl, HBr, and HI, is found in the reactivity with glass. Hydrofluoric acid cannot be stored in glass bottles because it etches the silica, SiO_2, in glass. On the other hand, even the most concentrated hydrochloric acid solutions can be stored indefinitely in glass without any evidence of a comparable reaction. To store HF solutions, we must use either polyethylene or wax containers (rather than glass) because of this reactivity with silica. Silicon bonds more strongly to fluorine than to oxygen and hence silica dissolves in a solution of HF by the reaction

$$SiO_2 + 6HF \longrightarrow SiF_6^{-2} + 2H_3O^+ \quad (25)$$

Hydrofluoric acid is a polar material, as water is, and it behaves as an ionizing solvent when it is scrupulously free of water. Salts that dissolve readily in liquid HF include LiF, NaF, KF, AgF, $NaNO_3$, KNO_3, $AgNO_3$, Na_2SO_4, K_2SO_4, and Ag_2SO_4. Liquid HF also dissolves organic compounds and is used as a solvent for a variety of reactions.

A very stable bond involving fluorine is the carbon-fluorine bond. The strength of this C—F bond is comparable to the C—H bond, and has led to the existence of a series of compounds known as the *fluorocarbons*. These are analogous to the hydrocarbons and can be imagined as being derived from them by substituting F atoms for H atoms. For example,

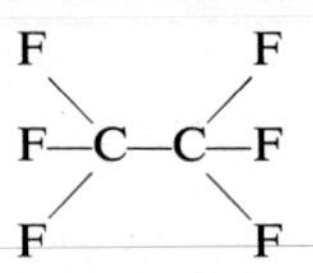

is the fluorocarbon analogue of ethane. It is called perfluoroethane. Many of the fluorocarbons are quite inert and their uses exploit this property. CCl_2F_2, Freon, is a volatile, nonpoisonous, noncorrosive material used in refrigerators and as the propellant in some aerosol cans. Increasingly important also are the polymeric fluorocarbons, such as Teflon, which are derived from perfluoroethylene, $CF_2{=}CF_2$, by polymerization. They can be used, for example, as gaskets, valves, and fittings for handling extremely corrosive chemicals.

QUESTIONS AND PROBLEMS

1. Give the electron configuration for each of the trio F^-, Ne, Na^+. How do the trios Cl^-, Ar, K^+, and Br^-, Kr, Rb^+ differ from the above?

2. Table 19-II contains values for the covalent radii and the ionic radii of the halogens. Plot both radii versus row number. What systematic changes are evident in the two curves?

3. Using the data from Table 19-II, plot on one set of axes the melting and boiling points of the halogens versus row number.

4. For astatine, use your graphs from Problems 2 and 3 as a basis for a prediction of its covalent radius, ionic radius of the −1 ion, melting point, and boiling point.

5. Predict the molecular structures and bond lengths for SiF_4, $SiCl_4$, $SiBr_4$, and SiI_4, assuming the covalent radius of silicon is 1.16 Å.

6. Explain in terms of nuclear charge why the K^+ ion is smaller than the Cl^- ion, though they are isoelectronic (they have the same number of electrons).

7. Can aqueous bromine, Br_2, be used to oxidize ferrous ion, $Fe^{+2}(aq)$, to ferric ion, $Fe^{+3}(aq)$ (use Appendix 3)? Aqueous iodine, I_2?

8. What will happen if F_2 is bubbled into 1 *M* NaBr solution? Justify your answer using $E°$ values.

9. Using $E°$ values, predict what will happen if, in turn, each halogen beginning at chlorine is added to a 1 *M* solution of ions of the next lower halogen: Cl_2 to Br^-, Br_2 to I^-. Which halogen is oxidized and which is reduced in each case?

10. Write a balanced equation for the reaction of dichromate and iodide ions in acid solution. Determine $E°$ for the reaction

$$Cr_2O_7^{-2}(aq) + I^-(aq) + H^+(aq) \text{ gives } Cr^{+3}(aq) + I_2 + H_2O$$

Answer. $E° = +0.80$ volt.

11. Balance the equation for the reaction of iodine with thiosulfate ion:

$$I_2 + \underset{\text{thiosulfate ion}}{S_2O_3^{-2}(aq)} \text{ gives } \underset{\text{tetrathionate ion}}{S_4O_6^{-2}(aq)} + I^-(aq)$$

What is the oxidation number of sulfur in the tetrathionate ion?

12. How many grams of iodine can be formed from 20.0 grams of KI by oxidizing it with ferric chloride ($FeCl_3$)? Determine $E°$.

Answer. 15.3 grams of I_2.

13. Balance the equation for the reaction between SO_2 and I_2 to produce SO_4^{-2} and I^- in acid solution. Calculate $E°$. From Le Chatelier's Principle, predict the effect on the $E°$ in this reaction if $H^+ = 10^{-7}$ *M* is used instead of $H^+ = 1$ *M*.

14. What is the oxidation number of the halogen in each of the following: HF, $HBrO_2$, HIO_3, ClO_3^-, F_2, ClO_4^-?

15. Comparable half-reactions for iodine and chlorine are shown below.

$$\tfrac{1}{2}I_2 + 3H_2O \longrightarrow IO_3^- + 6H^+ + 5e^- \qquad E^\circ = -1.195 \text{ volts}$$

$$\tfrac{1}{2}Cl_2 + 3H_2O \longrightarrow ClO_3^- + 6H^+ + 5e^- \qquad E^\circ = -1.47 \text{ volts}$$

(a) Which is the stronger oxidizing agent, iodate, IO_3^-, or chlorate, ClO_3^-?
(b) Balance the equation for the reaction between chlorate ion and I^- to produce I_2 and Cl_2.

16. Two half-reactions involving chlorine are

$$2Cl^- \longrightarrow Cl_2 + 2e^- \qquad E^\circ = -1.36 \text{ volts}$$

$$Cl_2 + 2H_2O \longrightarrow 2HOCl + 2H^+ + 2e^- \qquad E^\circ = -1.63 \text{ volts}$$

(a) Balance the reaction in which self-oxidation-reduction of Cl_2 occurs to produce chloride ion and chlorous acid, $HOCl$.
(b) What is the oxidation number of chlorine in each species containing chlorine?
(c) What is E° for the reaction?
(d) Explain, using Le Chatelier's Principle, why the self-oxidation-reduction reaction occurs in 1 M OH^- solution instead of 1 M H^+.

17. How many grams of SiO_2 would react with 5.00×10^2 ml of 1.00 M HF to produce SiF_4?

18. A water solution that contains 0.10 M HF is 8% dissociated. What is the value of its K_A?

Answer. 6.9×10^{-4}.

19. From each of the following sets, select the substance which best fits the requirement specified.

(a) Strongest acid	$HOCl$, $HOClO$, $HOClO_2$
(b) Biggest atom	F, Cl, Br, I
(c) Smallest ionization energy	F, Cl, Br, I
(d) Best reducing agent	F^-, Cl^-, Br^-, I^-
(e) Weakest acid	HF, HCl, HBr, HI
(f) Best hydrogen bonding	HF, HCl, HBr, HI

20. Describe two properties that the halogens have in common and give an explanation of why they have these properties in common.

CHAPTER 20

The Third Row of the Periodic Table

1 H																	2 He	
3 Li	4 Be											5 B	6 C	7 N	8 O	9 F	10 Ne	
11 Na	12 Mg											13 Al	14 Si	15 P	16 S	17 Cl	18 Ar	Third row
19 K	20 Ca	21 Sc	22 Ti	23 V	24 Cr	25 Mn	26 Fe	27 Co	28 Ni	29 Cu	30 Zn	31 Ga	32 Ge	33 As	34 Se	35 Br	36 Kr	
37 Rb	38 Sr	39 Y	40 Zr	41 Nb	42 Mo	43 Tc	44 Ru	45 Rh	46 Pd	47 Ag	48 Cd	49 In	50 Sn	51 Sb	52 Te	53 I	54 Xe	
55 Cs	56 Ba	57-71	72 Hf	73 Ta	74 W	75 Re	76 Os	77 Ir	78 Pt	79 Au	80 Hg	81 Tl	82 Pb	83 Bi	84 Po	85 At	86 Rn	
87 Fr	88 Ra	89-																

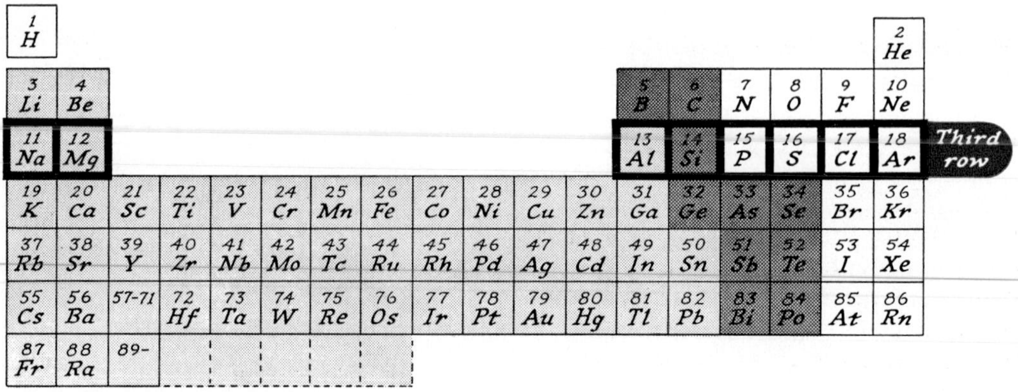

In earlier chapters we recognized that strong chemical similarities are displayed by elements which are in the same vertical column of the periodic table. The properties which chlorine holds in common with the other halogens reflect the similarity of the electronic structures of these elements. On the other hand, there is an enormous difference between the behavior of elements on the left side of the periodic table and those on the right. Furthermore, the discussions in Chapter 15 revealed systematic modification in certain atomic properties, such as ionization energy, as we proceed from left to right along a row of the periodic table. Our purpose in this chapter is to examine the chemical behavior of the elements in the third row of the periodic table to look for trends in chemical properties. Specifically, we will consider the physical properties of the elements themselves, their performance as oxidizing or reducing agents, and the acid-base behavior of their hydroxides.

20-1 PHYSICAL PROPERTIES OF THE ELEMENTS

All the elements in Row 3 are commercially available or can be easily prepared in the laboratory. Try to examine as many of these elements as possible in the laboratory as you study this chapter. If all the elements are available to you, arrange them in order of atomic number and compare them. You can hardly imagine a more varied set of appearances. At one extreme we have the metals sodium, magnesium, and aluminum. When they are freshly cut these solids show the bright luster or reflectivity typical of metals. They are soft: sodium is so soft it can be cut with a knife; magnesium and aluminum bend in your fingers and can be easily scratched by a sharp object. Silicon also shows the metallic luster, but is much harder than magnesium or aluminum. Phosphorus in the form known as white phosphorus is a yellowish, waxy solid with a distinctly nonmetallic appearance. Black phosphorus, obtained by subjecting white phosphorus to high pressure, is a dark gray solid which *does* have some of the luster which characterizes metals. By the time we get to sulfur, however, it is very clear we have a nonmetal. The two gases chlorine and argon complete the trend away from metallic appearance.

20-1.1 Sodium, Magnesium, and Aluminum: Metallic Solids

Can we explain the wide variation in appearance and physical properties of these elements? We have already said in Chapter 17 that metals are found at the left of the periodic table. The low ionization energy and vacant valence orbitals of one of these elements lead to a sea of highly mobile valence electrons. The mobile electrons hold the atoms together in the metallic crystal and, at the same time, are responsible for the ease of conduction of heat and electricity. We also remarked that the metallic bond becomes stronger as the number of valence electrons per atom and their ionization energy increase. The trend in the physical properties of the third-row metals seems to be well explained in terms of the increasing number and increasing ionization energy of the valence electrons.

EXERCISE 20-1

Write out the electron configuration of sodium, magnesium, and aluminum and find the ionization energies for all their valence electrons (Table 20-IV, p. 374). Account for the trend in the heats of vaporization and boiling points (Table 20-I) of these elements. Compare your discussion with that given in Section 17-1.3.

Table 20-I

HEATS OF VAPORIZATION AND BOILING POINTS OF METALS

	Na	Mg	Al
ΔH_{vap} (kcal/mole)	23.1	31.5	67.9
b.p. (°C)	889	1120	2327

20-1.2 Silicon: A Network Solid

Even though silicon is metallic in appearance, it is not generally classified as a metal. The electrical conductivity of silicon is so much less than that of ordinary metals it is called a semiconductor. Silicon is an example of a network solid (see Figure 20-1)—it has the same atomic arrangement that occurs in diamond. Each silicon atom is surrounded by, and covalently bonded to, four other silicon atoms. Thus, the silicon crystal can be regarded as one giant molecule.

Almost all of the valence electrons in the silicon crystal are localized in the covalent bonds and are not free to conduct heat or electricity by moving throughout the solid. On the other hand, in the solid there are always a few valence electrons which have acquired enough energy to be nonlocalized and these few electrons account for the small, but noticeable, electrical conductivity of silicon. Again, we can rationalize the behavior of silicon in terms of its atomic structure and ionization energy. The fact that the silicon atom has four electrons ($3s^2 3p^2$) in its valence orbitals accounts for its tendency to form four covalent bonds. The increase in ionization energy and

Fig. 20-1. **The crystal structures of silicon and diamond.**

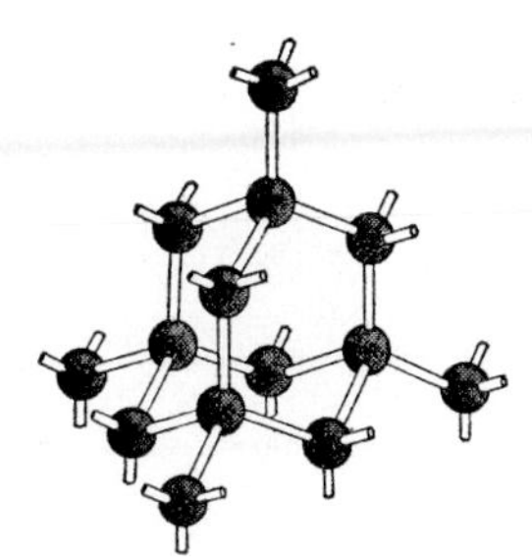

absence of vacant valence orbitals as we proceed along the row accounts for the increasing localization of the valence electrons into covalent bonds and the almost complete disappearance of electrical conductivity.

20-1.3 Phosphorus, Sulfur, and Chlorine: Molecular Solids

Since the ionization energy of the phosphorus atom is still higher than that of the silicon atom, it is not surprising that the common forms of phosphorus are nonmetallic molecular solids. White phosphorus consists of discrete P_4 molecules (see Figure 20-2) and weak van der Waals forces between the separate molecules are responsible for the stability of the solid. The electronic structure of the phosphorus atom provides an explanation of the formula and structure of the P_4 molecule. Phosphorus has the electron configuration $1s^22s^22p^63s^23p^3$, and we can suggest that since the $3p$ orbitals are half-filled, phosphorus should be able to form three covalent bonds. The geometry should be like that in ammonia, NH_3, in which the three N—H bonds form a pyramid with a three-sided base (see page 291). As shown in Figure 20-2, in the P_4 molecule each phosphorus atom does make three bonds, and each atom is at one apex of a pyramid.

Fig. 20-2. **The structure of a P_4 molecule.**

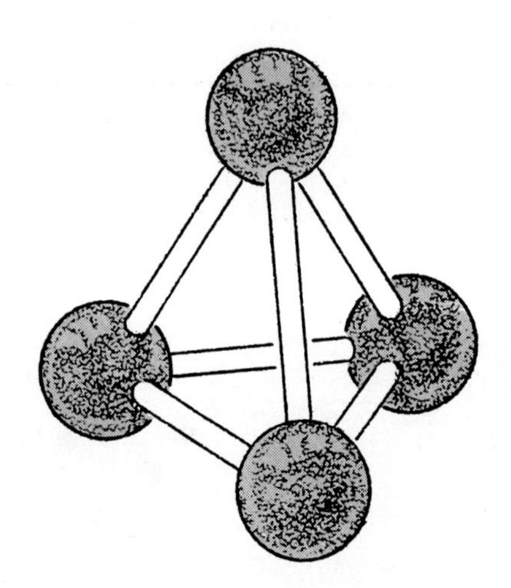

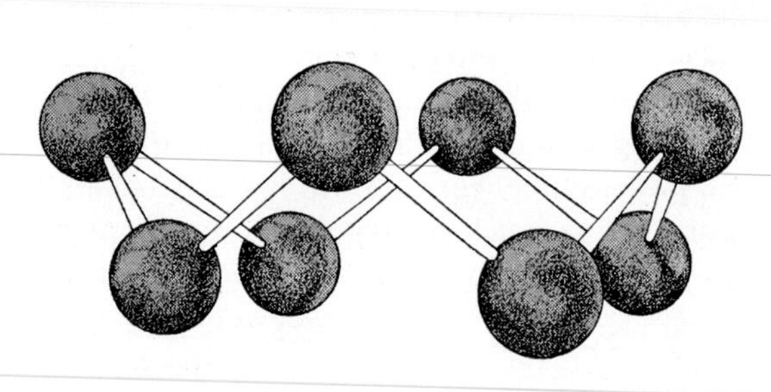

Fig. 20-3. **The structure of an S_8 molecule.**

The electron configuration in the valence orbitals of the sulfur atom ($3s^23p^4$) suggests that it will form two covalent bonds by making use of two half-filled $3p$ orbitals. This is, in fact, observed in the molecule S_8, which is present in the common forms of solid sulfur. The S_8 molecules assume the form of a puckered ring, as shown in Figure 20-3. As with the phosphorus, the stability of this crystalline form of sulfur is due to van der Waals forces between discrete molecules.

The electronic structure of the chlorine atom ($3s^23p^5$) provides a satisfactory explanation of the elemental form of this substance also. The single half-filled $3p$ orbital can be used to form one covalent bond, and therefore chlorine exists as a diatomic molecule. Finally, in the argon atom all valence orbitals of low energy are occupied by electrons, and the possibility for chemical bonding between the atoms is lost.

EXERCISE 20-2

Using the principles discussed in Chapter 17, attempt to arrange the third-row elements from silicon through argon in order of increasing boiling point, starting with the element you think has the lowest boiling point. Be prepared to defend in a class discussion your choice of the position you assign each element in this sequence.

20-2 THE ELEMENTS AS OXIDIZING AND REDUCING AGENTS

In the last section your attention was called to the extreme variation in the physical appearance and properties of the third-row elements. We might start this section with a similar statement about the behavior of the elements as oxidizing and reducing agents. The outstanding chemical characteristic common to metallic sodium, magnesium, and aluminum is strong reducing power. Their tendency to lose electrons and to react with other elements is so great that they are found in nature only in compounds, never as free elements. All three of these metals will react with water to give hydrogen. In the case of sodium this reaction is fast and all of the sodium is consumed if sufficient water is present. For magnesium and aluminum, the reaction produces a thin layer of oxide on the metallic surface. This oxide layer has low solubility in water and the oxide adheres strongly to the metal. Hence it forms a protective layer that prevents further contact between water (or air) and the metal. This protection accounts for the notable resistance of aluminum to weathering, upon which most of the structural uses of aluminum depend. If either magnesium or aluminum comes in contact with mercury, the protective layer is removed and rapid reaction takes place.

As we saw in Chapter 19, chlorine represents the other extreme in chemical reactivity. Its most obvious chemical characteristic is its ability to acquire electrons to form negative chloride ions, and, in the process, to oxidize some other substance. Since the tendency to lose or gain electrons is a result of the details of the electronic structure of the atom, let us try to explain the chemistry of the third-row elements on this basis.

20-2.1 Sodium, Magnesium, and Aluminum: Strong Reducing Agents

A glance at Appendix 3, the table of $E°$'s for half-reactions, should convince you that sodium, magnesium, and aluminum are among the strongest reducing agents available. Their $E°$'s are also listed in Table 20-II. Part of this strong tendency for metals to lose electrons and become positive ions in aqueous solutions is a result of the fact that the valence electrons of their atoms are not very strongly bound, as is shown by their low ionization energies.

Table 20-II

THE HALF-CELL POTENTIALS OF THIRD-ROW METALS

$Na(s) \longrightarrow Na^+(aq) + e^-$	$E° = 2.71$ volts
$Mg(s) \longrightarrow Mg^{+2}(aq) + 2e^-$	$E° = 2.37$ volts
$Al(s) \longrightarrow Al^{+3}(aq) + 3e^-$	$E° = 1.66$ volts

However, this is not the complete explanation for their reducing properties. Let us analyze the energy requirements of the process

$$Na(s) \longrightarrow Na^+(aq) + e^- \qquad (1)$$

in much the same way as we treated the half-reactions of the halogens in Section 19-2.2. Reaction (*1*) can be discussed in terms of a series of hypothetical steps:

$$Na(s) \longrightarrow Na(g) \qquad \text{vaporization} \qquad (2)$$

$$Na(g) \longrightarrow Na^+(g) + e^-(g) \qquad \text{ionization} \qquad (3)$$

$$Na^+(g) \longrightarrow Na^+(aq) \qquad \text{hydration} \qquad (4)$$

Step (*2*) is just the vaporization of solid sodium to a gas. This step requires energy; ΔH is positive for this reaction. In step (*3*) an electron is removed from a gaseous atom to form a gaseous ion; ΔH is positive since this step requires energy also. Step (*4*) is the hydration of a positive ion; energy is evolved in this process, so ΔH is negative. Sodium metal is a good reducing agent, primarily because the energy required to carry out reaction (*1*) is small. In other words, the energy we put in to cause steps (*2*) and (*3*) is small and is somewhat compensated by the energy we get out in step (*4*). We can use this example to set up general criteria for a good metallic reducing agent: (a) the metallic crystal must not be too stable [otherwise the energy required for step (*2*) will be large]; (b) the ionization energy of the gaseous atom should be small; (c) the hydration energy, the energy evolved in step (*4*), should be large.

We have already mentioned that the stability of the metallic crystal and the ionization energies of the atom tend to increase in the series sodium, magnesium, and aluminum. In spite of this, aluminum is still an excellent reducing agent because the hydration energy of the Al^{+3} ion is very large (Table 20-III).

Table 20-III

HYDRATION ENERGIES OF SOME THIRD-ROW IONS (kcal/mole)

Na^+	Mg^{+2}	Al^{+3}
97	460	1121

The third-row metals also show their strong reducing properties in reactions which do not take place in aqueous solution. For instance, magnesium metal ignited in air will react with carbon dioxide, reducing it to elemental carbon:

$$2Mg(s) + CO_2(g) \longrightarrow 2MgO(s) + C(s) \quad (5)$$

If aluminum metal is mixed with a metal oxide such as ferric oxide, Fe_2O_3, and ignited, the oxide is reduced and large amounts of heat are evolved:

$$2Al(s) + Fe_2O_3(s) \longrightarrow 2Fe(s) + Al_2O_3(s)$$

$$\Delta H = -203 \frac{\text{kcal}}{\text{mole } Fe_2O_3} \quad (6)$$

These reactions are possible because of the great stability, or low energy, of the oxides of magnesium and aluminum. The oxides are ionic compounds whose great stability can be attributed to strong electrostatic forces between small, positively charged Mg^{+2} ions (or Al^{+3} ions) and negatively charged oxide ions, O^{-2}. The fact that the Mg^{+2} and Al^{+3} ions are very small allows these positive ions to approach closely the negative oxide ions, resulting in strong attractive forces.

The ease of oxidation of magnesium is important in the commercial manufacture of titanium metal. Titanium, when quite pure, shows great promise as a structural metal, but the economics of production have thus far inhibited its use. One of the processes currently used, the Kroll process, involves the reduction of liquid titanium tetrachloride with molten metallic magnesium:

$$TiCl_4(l) + 2Mg(l) \longrightarrow Ti(s) + 2MgCl_2(l) \quad (7)$$

Titanium has a very high melting point (1812°C), so the magnesium chloride can be vaporized and distilled away from the solid titanium. The gaseous magnesium chloride is condensed and then electrolyzed to regenerate magnesium and chlorine:

$$MgCl_2(l) \xrightarrow{\text{electrolysis}} Mg(s) + Cl_2(g) \quad (8)$$

The magnesium metal is thus recovered for repeated use in reaction (*7*). Chlorine produced in reaction (*8*) is also put to use in the manufacture of $TiCl_4$, the other reactant in reaction (*7*).

20-2.2 Silicon, Phosphorus, and Sulfur: Oxidizing and Reducing Agents of Intermediate Strengths

Silicon also can act as a reducing agent, as we might expect from the properties of sodium, magnesium, and aluminum. It reacts with molecular oxygen to form silicon dioxide, SiO_2. This network solid is held together by very strong bonds. However, because of the rather high ionization energy of the silicon atom, and the great stability of the silicon crystal, its reducing properties are considerably less than those of the typical metals.

Phosphorus continues the trend away from strong reducing properties. Elemental phosphorus will react with strong oxidizing agents like oxygen and the halogens,

$$P_4(s) + 5O_2(g) \longrightarrow P_4O_{10}(s) \quad (9)$$

$$P_4(s) + 6Cl_2(g) \longrightarrow 4PCl_3(l) \quad (10)$$

but will also react with strong reducing agents such as magnesium to form phosphides:

$$P_4(s) + 6Mg(s) \longrightarrow 2Mg_3P_2(s) \quad (11)$$

Therefore, elemental phosphorus shows both reducing and oxidizing properties. This intermediate behavior can be explained in terms of the electron occupancy and ionization energy of the phosphorus atom. The fact that the ionization energy of phosphorus is greater than the previous elements in Row 3 suggests it will be a poorer reducing agent than, for example, aluminum. The combination of a noticeable affinity for electrons (as evidenced by the ionization energy) and three half-filled $3p$ orbitals provides an explanation for the appearance of a weak oxidizing tendency in phosphorus. Since phosphorus neither loses nor gains electrons readily, it is neither a strong reducing agent nor a strong oxidizing agent.

at 44°C. Explain this very great difference in terms of the structures of the solids.

4. Recalling the chemistry of nitrogen, write formulas for phosphorus compounds corresponding to
   (a) ammonia,
   (b) hydrazine,
   (c) ammonium iodide.

5. Write the formula for the fluoride you expect to be most stable for each of the third-row elements.

6. The heat of reaction for the formation of $MgO(s)$ from the elements is -144 kcal/mole of $MgO(s)$. How much heat is liberated when magnesium reduces the carbon in CO_2 to free carbon? See Table 7-II.

   Answer. $\Delta H = -97$ kcal/mole MgO.

7. Magnesium oxide is an ionic solid that crystallizes in the sodium chloride type lattice.
   (a) Explain why MgO is an ionic substance.
   (b) How many calories would be required to decompose 8.06 grams of MgO? (Use the data in problem 6.)
   (c) Draw a diagram of a crystal of MgO.

8. Aluminum oxide (Al_2O_3) is thought to dissociate at high temperature (1950°C) according to the equation: $2Al_2O_3(s) \longrightarrow 4AlO(g) + O_2(g)$. The total vapor pressure at 1950°C is about 1×10^{-6} atm.
   (a) Which element is oxidized and which is reduced in this reaction?
   (b) Write the equation for the equilibrium constant.
   (c) Calculate its value using partial pressure as the unit of "concentration" for the gases.

9. Explain the observation that phosphorus acts both as a weak reducing agent and as a weak oxidizing agent.

10. (a) What are the oxidation numbers of phosphorus in the two compounds phosphorous acid, H_3PO_3, and phosphoric acid, H_3PO_4?
    (b) From the $E°$ values in Appendix 3, decide which of the following substances might be reduced by phosphorous acid: Fe^{+2}; Sn^{+4}; I_2; Cr^{+3}.

    $$H_2O + H_3PO_3 \longrightarrow H_3PO_4 + 2H^+ + 2e^- \qquad E° = 0.276 \text{ volt}$$

    (c) Balance the equation for the reaction between phosphorous acid and Fe^{+3} and calculate $E°$ for the reaction.

11. Answer the following in terms of electron configuration and ionization energy:
    (a) Which elements in the second and third rows are strong
        (i) oxidizing agents?
        (ii) reducing agents?
    (b) What properties do strong oxidizing agents have?
    (c) What properties do strong reducing agents have?

12. Of the elements Na, Mg, Al, which one would you expect to be most likely to
    (a) form a molecular solid with chlorine?
    (b) form an ionic solid with chlorine?

13. One kilogram of sea water contains 0.052 mole of magnesium ion. What is the minimum number of kilograms of sea water that would have to be processed in order to obtain 1 kg of $Mg(OH)_2$?

    Answer. 3.3×10^2 kg of sea water.

14. Why is aluminum hydroxide classed as an amphoteric compound?

15. Some of the following common compounds of the third-row elements are named as hydroxides and some as acids:

NaOH	sodium hydroxide
$Mg(OH)_2$	magnesium hydroxide
$Al(OH)_3$	aluminum hydroxide
$Si(OH)_4$	silicic acid (usually written H_4SiO_4)
$P(OH)_3$	phosphorous acid (usually written H_3PO_3)
$S(OH)_2$	not known
$Cl(OH)$	hypochlorous acid (usually written HOCl)

    (a) Explain why these compounds vary systematically in their acid-base behavior.
    (b) Write equations that show the reactions of each of these substances either as acids, as bases, or both.

16. A solution containing 0.20 M H_3PO_3, phosphorous acid, is tested with indicators and the $H^+(aq)$ concentration is found to be 5.0×10^{-2} M. Calculate the dissociation constant of H_3PO_3, assuming that a second proton cannot be removed.

17. Elemental phosphorus is prepared by the reduction of calcium phosphate, $Ca_3(PO_4)_2$, with coke in the presence of sand, SiO_2. The products are phosphorus, calcium silicate, $CaSiO_3$, and carbon monoxide.
    (a) Write the equation for the reaction.
    (b) Using 75.0 kg of the ore, calcium phosphate, calculate how many grams of P_4 can be obtained and how many grams of coke (assumed to be pure carbon) will be used.

CHAPTER 

The Second Column of the Periodic Table: The Alkaline Earths

	Alkaline earths																
1 H																	2 He
3 Li	4 Be											5 B	6 C	7 N	8 O	9 F	10 Ne
11 Na	12 Mg											13 Al	14 Si	15 P	16 S	17 Cl	18 Ar
19 K	20 Ca	21 Sc	22 Ti	23 V	24 Cr	25 Mn	26 Fe	27 Co	28 Ni	29 Cu	30 Zn	31 Ga	32 Ge	33 As	34 Se	35 Br	36 Kr
37 Rb	38 Sr	39 Y	40 Zr	41 Nb	42 Mo	43 Tc	44 Ru	45 Rh	46 Pd	47 Ag	48 Cd	49 In	50 Sn	51 Sb	52 Te	53 I	54 Xe
55 Cs	56 Ba	57-71	72 Hf	73 Ta	74 W	75 Re	76 Os	77 Ir	78 Pt	79 Au	80 Hg	81 Tl	82 Pb	83 Bi	84 Po	85 At	86 Rn
87 Fr	88 Ra	89-															

In the preceding chapter we looked at the elements of the third row in the periodic table to see what systematic changes occur in properties when electrons are added to the outer orbitals of the atom. We saw that there was a decided trend from metallic behavior to nonmetallic, from base-forming to acid-forming, from simple ionic compounds to simple molecular compounds. These trends are conveniently discussed in terms of the ionization energies and orbital occupancies.

There are similar, but smaller, trends in the properties of elements in a column (a family) of the periodic table. Though the elements in a family display similar chemistry, there are important and interesting differences as well. Many of these differences are explainable in terms of atomic size.

21-1 ELECTRON CONFIGURATION OF THE ALKALINE EARTH ELEMENTS

The elements of the second column and their electron configurations are given in Table 21-I. For each element, the neutral atom has two more electrons than an inert gas. We can expect these two electrons to be easily removed, to give the stability of the inert gas electron configuration.

Table 21-I. **ELECTRON CONFIGURATIONS OF THE ALKALINE EARTH ELEMENTS**

ELEMENT	SYMBOL	NUCLEAR CHARGE	ELECTRON ARRANGEMENT: INNER LEVELS			ELECTRON ARRANGEMENT: OUTER LEVELS
beryllium	Be	4			$1s^2$	$2s^2$
magnesium	Mg	12	$1s^2$		$2s^22p^6$	$3s^2$
calcium	Ca	20	$\cdots 2s^22p^6$		$3s^23p^6$	$4s^2$
strontium	Sr	38	$\cdots 3s^23p^6$	$3d^{10}$	$4s^24p^6$	$5s^2$
barium	Ba	56	$\cdots 4s^24p^6$	$4d^{10}$	$5s^25p^6$	$6s^2$
radium	Ra	88	$\cdots 4f^{14}5s^25p^6$	$5d^{10}$	$6s^26p^6$	$7s^2$

EXERCISE 21-1

On the basis of the electron configurations and positions in the periodic table, answer the following questions.

(a) Is calcium likely to be a metal or nonmetal?
(b) Is calcium likely to resemble magnesium or potassium in its chemistry?
(c) Is calcium likely to have a higher or a lower boiling point than potassium? than scandium?

EXERCISE 21-2

Predict the chemical formula and physical state at room temperature of the most stable compound formed by each alkaline earth element with (a) chlorine; (b) oxygen; (c) sulfur.

Exercises 21-1 and 21-2 pose some of the simplest questions we can ask about the alkaline earths. The periodic table arranges in a column elements having similar electron configurations. We can expect elements on the left side of the periodic table to be metals (as magnesium is). Furthermore, we can expect that the elements in a given column will be more like each other than they will be like elements in adjacent columns. Thus, when we find that the chemistry of magnesium is almost wholly connected with the behavior of the dipositive magnesium ion, Mg^{+2}, we can expect a similar situation for calcium, and for strontium, and for each of the other alkaline earth elements. This proves to be so.

Remembering, then, that the alkaline earths are classed as a family because of general similarity, we shall investigate the detailed differences among them.

21-2 TRENDS IN PHYSICAL PROPERTIES

21-2.1 Atomic Radii in Solids

The size of an atom is defined in terms of the interatomic distances that are found in solids and in gaseous molecules containing that atom. For an atom on the left side of the periodic table, gaseous molecules are obtained only at very high temperatures. At normal temperatures, solids are found and there are two important types to consider, metallic solids and ionic solids. Table 21-II shows the nearest neighbor distances in the pure metals, in the gaseous oxide molecules, and in the solid oxides (which, except for BeO, have the sodium chloride crystal structure pictured in Figure 5-10, p. 81). This table also shows the corresponding radii assigned to each alkaline earth atom, first, in the metallic state, second, in the gaseous molecule (assuming that oxygen has the same size as in O_2) and third, in the state of a $+2$ ion (assuming that the oxide ion, O^{-2}, should be assigned a radius of 1.32 Å).

Table 21-II. **TRENDS IN INTERATOMIC DISTANCES**

ELEMENT	NEAREST NEIGHBOR DISTANCE (Ångstroms): IN THE PURE METAL	IN THE GASEOUS OXIDE	IN THE SOLID OXIDE	ALKALINE EARTH ATOMIC SIZE: METALLIC RADIUS	DOUBLE-BOND RADIUS	IONIC RADIUS
Be	2.23	1.33	1.64	1.11	0.73	0.32
Mg	3.20	1.75	2.10	1.60	1.15	0.78
Ca	3.95	1.82	2.40	1.97	1.22	1.08
Sr	4.30	1.92	2.57	2.15	1.32	1.25
Ba	4.35	1.94	2.76	2.17	1.34	1.44

We see that, no matter what type of bonding situation is considered, there is a trend in size moving downward in the periodic table. The alkaline earth atoms become larger in the sequence Be < Mg < Ca < Sr < Ba. These atomic sizes provide a basis for explaining trends in many properties of the alkaline earth elements and their compounds.

21-2.2 Ionization Energies

Table 21-III shows the first three ionization energies of the alkaline earths.

Table 21-III

IONIZATION ENERGIES OF THE ALKALINE EARTH ELEMENTS

ELEMENT	IONIZATION ENERGY (kcal/mole): E_1 (1st e^-)	E_2 (2nd e^-)	E_3 (3rd e^-)
Be	214	420	3533
Mg	175	345	1838
Ca	140	274	1173
Sr	132	253	986
Ba	120	230	811

EXERCISE 21-3

For each of the alkaline earths, calculate the ratio E_2/E_1. Account for the results in terms of the charges on the ions formed in the two ionization steps.

EXERCISE 21-4

If the ionization energy E_1 is regarded as a measure of the distance between the electron and the nuclear charge, what do the ionization energies of Be and Ba indicate about the relative sizes of the two atoms?

From Exercise 21-4 we see that the decreasing ionization energies observed for the alkaline earth atoms are readily explained in terms of their increasing size moving down in the periodic table. Notice that the ionization energy trend going *down* in the periodic table is the same as the trend going to the *left* in the periodic table.

EXERCISE 21-5

From the ionization energies, predict which solid substance involves bonds having the most ionic character: $BeCl_2$, $MgCl_2$, $CaCl_2$, $SrCl_2$, $BaCl_2$. Which substance is expected to have most covalent character in its bonds?

21-2.3 Metallic Properties

Table 21-IV shows some properties of the metals and their crystal forms. Since different crystal forms are involved in the series, trends in the properties are obscured. Figure 21-2 shows scale representations of the crystal structures of metallic beryllium, calcium, and barium.

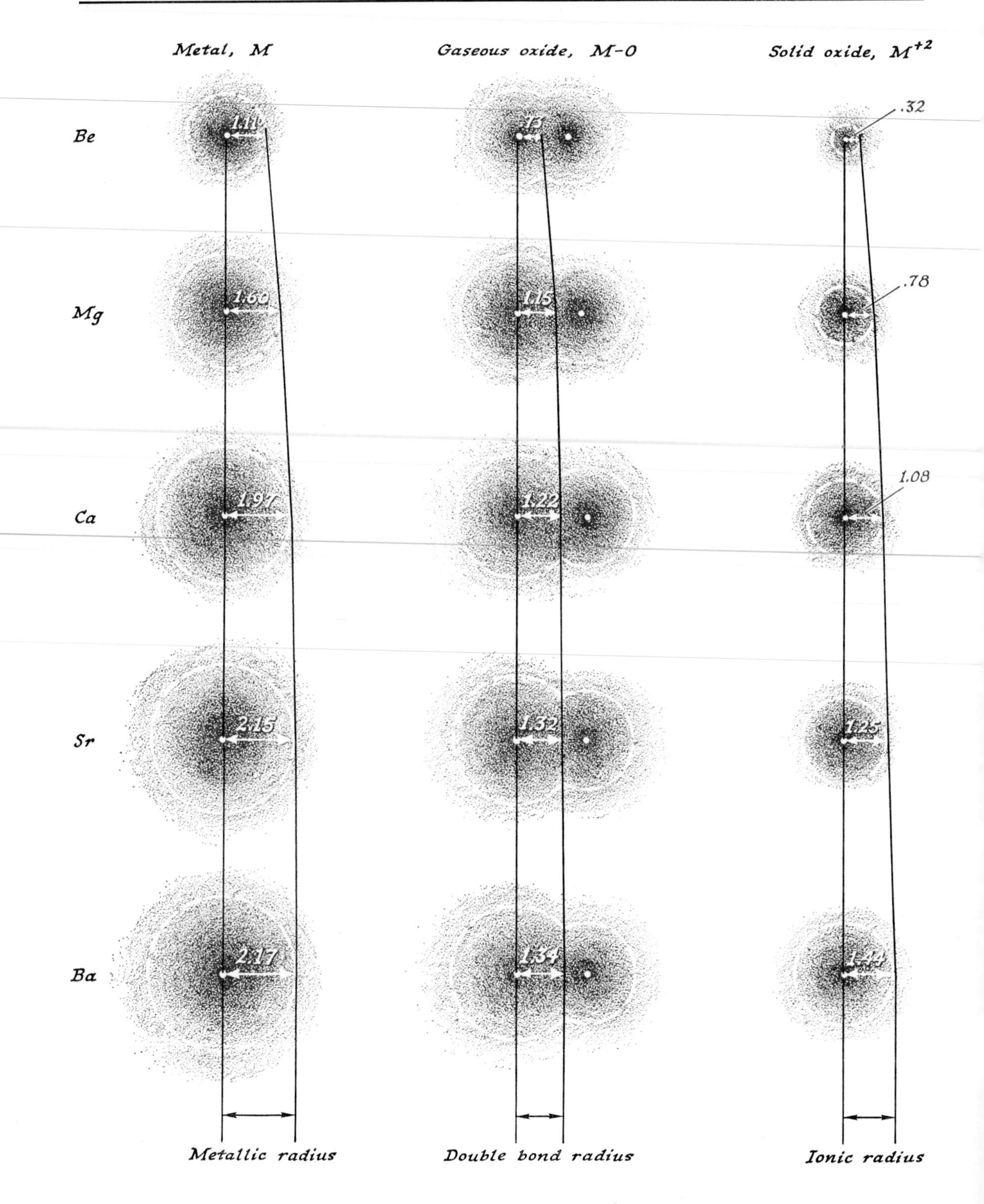

Fig. 21-1. **Sizes of the alkaline earth atoms** *with various bond types (in Ångstroms).*

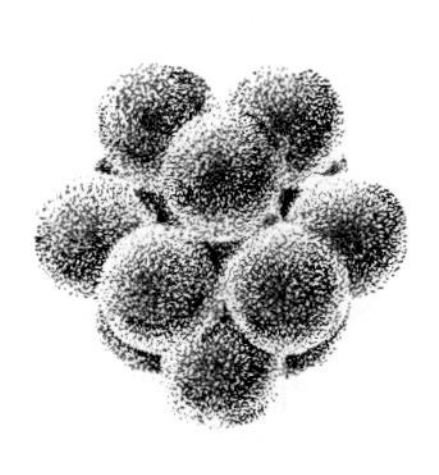

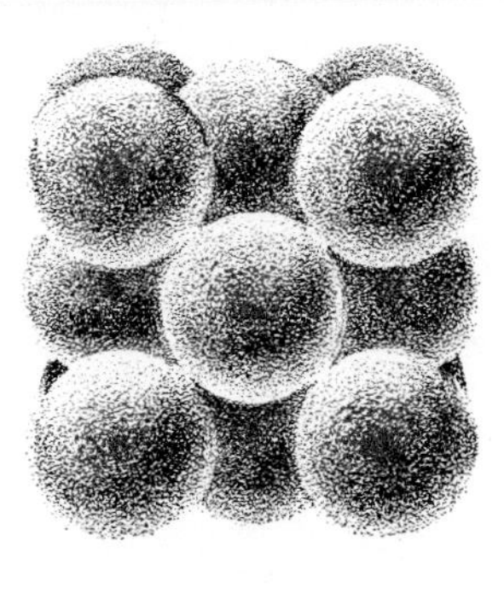

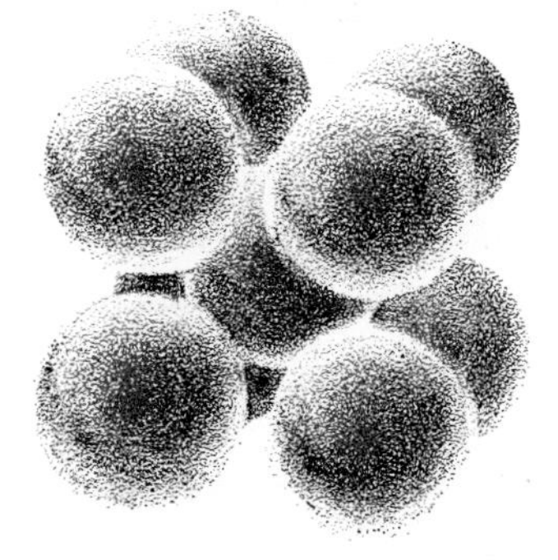

Fig. 21-2. **Scale representations of the crystal structures** *of Be, Ca, and Ba.*

Table 21-IV. **PROPERTIES OF THE ALKALINE EARTHS IN THE METALLIC STATE**

ELEMENT	CRYSTAL STRUCTURE	DENSITY (g/ml)	MELTING POINT (°C)	HEAT OF VAPORIZATION (kcal/mole)	ELECTRICAL CONDUCTIVITY $(\text{ohm-cm})^{-1}$
Be	hexagonal	1.85	1283	54	1.69×10^5
Mg	hexagonal	1.75	650	32	2.24×10^5
Ca	face-centered cubic	1.55	850	42	2.92×10^5
Sr	face-centered cubic	2.6	770	39	0.43×10^5
Ba	body-centered cubic	3.5	710	42	0.16×10^5

EXERCISE 21-6

From the density of each element, calculate the volume occupied by one mole of its atoms in the metallic state. Compare the trend in these molar volumes with the trend in the metallic radii shown in Table 21-II.

21-3 TRENDS IN CHEMICAL PROPERTIES

We have already observed (in Exercise 21-2) that the alkaline earths have similar chemistry. As shown in Table 21-I, they have similar electron configurations. Table 21-III shows that each element has two valence electrons. With these basic likenesses in mind we shall explore the chemical trends among these elements.

21-3.1 Oxidation and Reduction

All of the alkaline earths are strong reducing agents, since they readily release electrons. The values of E° are collected in Table 21-V.

Table 21-V

THE HALF-CELL POTENTIALS FOR THE ALKALINE EARTHS

$Be \longrightarrow Be^{+2} + 2e^-$	$E^\circ = +1.85$ volts
$Mg \longrightarrow Mg^{+2} + 2e^-$	$E^\circ = +2.37$
$Ca \longrightarrow Ca^{+2} + 2e^-$	$E^\circ = +2.87$
$Sr \longrightarrow Sr^{+2} + 2e^-$	$E^\circ = +2.89$
$Ba \longrightarrow Ba^{+2} + 2e^-$	$E^\circ = +2.90$

EXERCISE 21-7

The ease of removal of an electron from a gaseous atom, the ionization energy, is one of the factors that is important in fixing E°. Refer back to Table 21-III and predict the trend in E° that this factor would tend to cause.

21-3.2 Acid and Base Properties

We have explained the trends in acid-base character across the periodic table by considering the increasing ionization energy of the metal atom. As the atom M in a structure M—O—H attracts electrons more and more strongly, there is increasing tendency toward acidic properties. As the ionization energy of M goes down, there is increasing tendency toward basic properties.

Moving down in a column is equivalent in many respects to moving to the left in the periodic table. Since we find basic properties predominant at the left of the periodic table in a row, we can expect to find basic properties increasing toward the bottom of a column. Thus the base strength of the alkaline earth hydroxides is expected to be largest for barium and strontium. The greatest acid strength is expected for beryllium hydroxide.

Experimentally we find that strontium and barium hydroxides are indeed strong bases. All of the alkaline earth hydroxides dissolve readily in acidic solutions, showing that they are all bases to some extent:

$$Be(OH)_2(s) + 2H^+(aq) \rightleftharpoons Be^{+2} + 2H_2O \quad (1)$$

$$Mg(OH)_2(s) + 2H^+(aq) \rightleftharpoons Mg^{+2} + 2H_2O \quad (2)$$

$$Ca(OH)_2(s) + 2H^+(aq) \rightleftharpoons Ca^{+2} + 2H_2O \quad (3)$$

$$Sr(OH)_2(s) + 2H^+(aq) \rightleftharpoons Sr^{+2} + 2H_2O \quad (4)$$

$$Ba(OH)_2(s) + 2H^+(aq) \rightleftharpoons Ba^{+2} + 2H_2O \quad (5)$$

Only beryllium hydroxide dissolves appreciably in strong base solutions,

$$Be(OH)_2(s) + 2OH^- \rightleftharpoons BeO^{-2} + 2H_2O \quad (6)$$

These hydroxides are formed from the corresponding oxides. For example, calcium oxide, or lime, reacts with water as in reaction (*7*).

$$CaO(s) + H_2O(l) \rightleftharpoons Ca(OH)_2(s) \qquad \Delta H = -15.6 \text{ kcal} \quad (7)$$

The process is called "slaking" the lime and it is used by plasterers in preparing mortar, which requires $Ca(OH)_2$. As water is added to lime there is a considerable evolution of heat, as evidenced by wisps of steam that rise from the sample.

Because all of the alkaline earth oxides react with water to form basic hydroxides, they are called ***basic oxides.*** The reactions and their heats are as follows:

$$BeO(s) + H_2O(l) \rightleftharpoons Be(OH)_2(s) \qquad \Delta H = -2.5 \text{ kcal} \quad (8)$$

$$MgO(s) + H_2O(l) \rightleftharpoons Mg(OH)_2(s) \qquad \Delta H = -8.9 \text{ kcal} \quad (9)$$

$$CaO(s) + H_2O(l) \rightleftharpoons Ca(OH)_2(s) \qquad \Delta H = -15.6 \text{ kcal} \quad (10)$$

$$SrO(s) + H_2O(l) \rightleftharpoons Sr(OH)_2(s) \qquad \Delta H = -19.9 \text{ kcal} \quad (11)$$

$$BaO(s) + H_2O(l) \rightleftharpoons Ba(OH)_2(s) \qquad \Delta H = -24.5 \text{ kcal} \quad (12)$$

Notice the progressively increasing exothermic reaction heat, moving downward in the series.

EXERCISE 21-8

How much heat is evolved if one pound (454 grams) of lime is slaked according to reaction (*10*)? How many grams of water can be evaporated with this heat? (The heat of vaporization of water is about 10 kcal/mole.)

21-3.3 Solubilities of Alkaline Earth Compounds in Water

We encountered the solubilities of alkaline earth salts in Chapter 10 and discovered some interesting trends. Before looking back to Figures 10-5 and 10-6, see how much you can recall about these solubilities.

EXERCISE 21-9

In your notebook indicate one of the four answers

(i) none of the alkaline earth ions;
(ii) all alkaline earth ions;
(iii) Be^{+2}, Mg^{+2}, and Ca^{+2}, but not Sr^{+2}, Ba^{+2}, or Ra^{+2};
(iv) Sr^{+2}, Ba^{+2}, and Ra^{+2}, but not Be^{+2}, Mg^{+2}, or Ca^{+2}

for each of the following:

(a) ______ form compounds of low solubilities with Cl^-, Br^-, and I^-.
(b) ______ form compounds of low solubilities with sulfate, SO_4^{-2}.
(c) ______ form compounds of low solubilities with sulfide, S^{-2}.
(d) ______ form compounds of low solubilities with hydroxide, OH^-.
(e) ______ form compounds of low solubilities with carbonate, CO_3^{-2}.

Now compare your answer with Figures 10-5 and 10-6.

THE HYDROXIDES

When a hydroxide such as calcium hydroxide is added to water in sufficient amount, we get a saturated solution containing Ca^{+2} and OH^- in equilibrium with excess undissolved solid. The equilibrium

$$Ca(OH)_2(s) \rightleftharpoons Ca^{+2}(aq) + 2OH^-(aq) \quad (13)$$

can also be established by mixing Ca^{+2} ions (for example, from a solution of $CaCl_2$) with OH^- ions (for example, from a solution of NaOH) until a precipitate of $Ca(OH)_2$ forms. In either case, at equilibrium the concentrations of Ca^{+2} and OH^- ions are such that the equilibrium expression is satisfied:

$$[Ca^{+2}][OH^-]^2 = K_{sp} \quad (14)$$

In Table 21-VI the numerical values of $[M^{+2}][OH^-]^2$ are listed for some of the alkaline earth hydroxides. Small values indicate relatively few ions in solution; the larger values correspond to higher concentration in a saturated solution—that is, higher solubility.

Table 21-VI

THE SOLUBILITY PRODUCTS OF THE ALKALINE EARTH HYDROXIDES

COMPOUND	K_{sp}
$Mg(OH)_2$	8.9×10^{-12}
$Ca(OH)_2$	1.3×10^{-6}
$Sr(OH)_2$	3.2×10^{-4}
$Ba(OH)_2$	5.0×10^{-3}

EXERCISE 21-10

Suppose you have a solution in which the concentration of hydroxide ion is 1 *M*. How many moles per liter of the different alkaline earth ions listed in Table 21-VI could you have (at equilibrium) in this solution? If the concentration of hydroxide ions were 0.5 *M*, how would your answers change?

Exercise 21-10 demonstrates that there is a regular trend in the solubilities of the alkaline earth hydroxides.

THE CARBONATES AND SULFATES

Although the hydroxides of the alkaline earth elements become more soluble in water as we go down the column, the opposite trend is observed in the solubilities of the sulfates and carbonates. For example, Table 21-VII shows the solubility products of the alkaline earth sulfates.

Table 21-VII

THE SOLUBILITY PRODUCTS OF THE ALKALINE EARTH SULFATES

COMPOUND	K_{sp}
$MgSO_4$	soluble ($K_{sp} \gg 10^{-2}$)
$CaSO_4$	2.4×10^{-5}
$SrSO_4$	7.6×10^{-7}
$BaSO_4$	1.5×10^{-9}

The solubility of calcium carbonate is such that in a saturated solution the product of ion concentrations $[Ca^{+2}][CO_3^{-2}]$ is 5×10^{-9}. Though this may seem quite small, it is large enough to be important to man, especially if he lives in a region of the earth where there are extensive limestone deposits. Calcium carbonate can be dissolved in water, especially if it contains much dissolved CO_2. This is objectionable because soap added to water which contains even traces of Ca^{+2} forms a precipitate of calcium stearate. This is the ring that is so difficult to remove from the bathtub.

The dissolving of limestone by ground water is another example of chemical equilibrium. The behavior of this system depends upon the chemical equilibrium between $CaCO_3$ and its dissolved ions and the equilibrium between carbonate ion and dissolved CO_2 in the water. When $CaCO_3$ dissolves in water it establishes the equilibrium

$$CaCO_3(s) \rightleftharpoons Ca^{+2}(aq) + CO_3^{-2}(aq) \quad (15)$$

The carbonate ion, a base, can accept a proton from water, an acid,

$$CO_3^{-2}(aq) + H_2O(l) \rightleftharpoons HCO_3^-(aq) + OH^-(aq) \quad (16)$$

Thus, solutions of carbonates are found to be basic. Aqueous solutions of carbon dioxide are, on the other hand, acidic. The reactions in this equilibrium are

$$CO_2(g) + H_2O(l) \rightleftharpoons H_2CO_3(aq) \quad (17)$$

$$H_2CO_3(aq) \rightleftharpoons HCO_3^-(aq) + H^+(aq) \quad (18)$$

The combination of reaction (*18*) and (*16*) shows how carbon dioxide enhances the solubility of calcium carbonate by removing carbonate ion to form bicarbonate ion,

$$H_2CO_3(aq) + CO_3^{-2}(aq) \rightleftharpoons 2HCO_3^-(aq) \quad (19)$$

or, in the net reaction,

$$CaCO_3(s) + CO_2(g) + H_2O(l) \rightleftharpoons Ca^{+2}(aq) + 2HCO_3^-(aq) \quad (20)$$

The result is that we get an appreciable concentration of Ca^{+2} in the water, giving so-called hard water—hard on the soap and hard on the people who use it. Caves in limestone regions are formed essentially by the combination of the two equilibria above. In contrast, the weird icicle-like projections (stalactites) found hanging from the roofs of such caves are formed by the reverse of these reactions. On standing, a droplet of a saturated solution containing Ca^{+2} and HCO_3^- may lose some CO_2 and H_2O by evaporation. Loss of CO_2 and H_2O from the equilibrium (*20*) enhances the reverse-directed change, resulting in the deposit of a fleck of $CaCO_3$. The same change occurs when hard water of this kind is boiled in a pot or heated in a boiler. The white scum you may see forming on the surface of boiling water is often due to these equilibria.

Fig. 21-3. **Stalactites:** *solubility equilibria at work.*

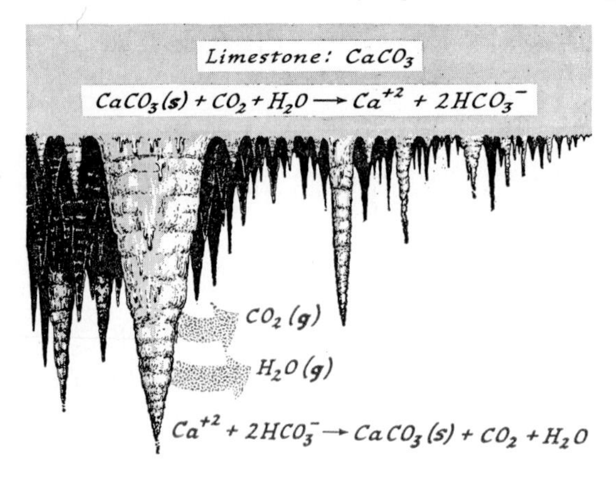

21-4 OCCURRENCE AND PREPARATION OF THE ALKALINE EARTH ELEMENTS

As we did in the preceding chapter, we conclude by summarizing some information about the occurrence in nature and the modes of preparing the alkaline earth elements.

21-4.1 Occurrence in Nature

All of the alkaline earth elements exist in nature as the M^{+2} cations.

Beryllium (forty-fourth most abundant element) is rather rare and occurs mostly as an aluminum beryllium silicate called beryl, $Be_3Al_2Si_6O_{18}$. Beryl containing traces of chromium has a beautiful green color and is called emerald.

Magnesium (eighth most abundant element) is found principally as Mg^{+2} ion in salt deposits, particularly as the slightly soluble carbonate, $MgCO_3$, and also in sea water. The natural deposits of $MgCO_3$ with $CaCO_3$ are called dolomite. Magnesium is present as a cation in the asbestos silicates.

Calcium (sixth most abundant element) is found in limestone, $CaCO_3$, and gypsum, $CaSO_4 \cdot 2H_2O$. Bones are made up of calcium phosphate, $Ca_3(PO_4)_2$.

Strontium (thirty-eighth most abundant element) is rather rare and is found principally as the mineral strontianite, $SrCO_3$.

Barium (eighteenth most abundant element) is also rather rare; it occurs as the mineral barite, $BaSO_4$.

Radium is radioactive and extremely rare. It occurs in trace amounts (one part in 10^{12}) in uranium ores such as pitchblende (mainly U_3O_8).

EXERCISE 21-11

What property held in common by the following compounds accounts for their presence in natural mineral deposits: $MgCO_3$, $CaCO_3$, $SrCO_3$, $BaSO_4$, and (in bones) $Ca_3(PO_4)_2$?

EXERCISE 21-12

What property held in common by the alkaline earth elements accounts for the fact that the free elements are not found in nature?

21-4.2 Mode of Preparation of the Element

Only magnesium is produced in any substantial quantity in the elemental form. The reaction sequence used is given in Section 20-4.2.

The general method for preparing the alkaline earth elements is to convert the mineral to a chloride or a fluoride by treatment with HCl or HF. Then the molten salt is electrolyzed or, as in the case of BeF_2, reduced with a chemical reducing agent such as Mg.

Alfred E. Stock was one of the greatest inorganic chemists of the twentieth century. His research investigations, which resulted in over 250 publications, were characterized by brilliant experimental technique and convincing thoroughness. Both of these rich qualities were needed for his hazardous studies of the hydrides of boron, an overlooked area of chemistry in which he was recognized as the single world authority for a period of at least a decade. It is fitting that his name is perpetuated in the "Stock system" of inorganic chemical nomenclature (in which Roman numerals indicate oxidation numbers).

Stock was born in Danzig, Poland, and his aptitude for science was displayed early in his boyhood collections of salamanders, butterflies, and plants. He studied at the University of Berlin where the chemistry facilities of the day were so limited that this brilliant experimentalist-to-be had to wait till his third semester to approach a laboratory bench. He received the Ph.D. at the University of Berlin in 1899, graduating magna cum laude.

Shortly after 1900, young Alfred Stock began his lifetime work: study of the chemistry of boron. He reasoned that this neighbor of the versatile carbon atom could not possibly have the dull and limited chemistry popularly assumed at the time. He entered this study stimulated by his own desire to know—despite advice by the laboratory director to select another area because the chemistry of boron was already thoroughly investigated. His persistence was rewarded by discoveries of a succession of hydrides of boron, such as diborane, B_2H_6, tetraborane, B_4H_{10}, pentaborane, B_5H_9, and decaborane, $B_{10}H_{14}$. The structures and even the very existence of these compounds baffled chemists for many years. Even to the date of Stock's death, theoreticians had no convincing explanation of the absence of the prototype molecule, BH_3, and their discussions of the nature of the bonding in diborane were based upon an assumed structure that was later shown to be incorrect. Stock's amazing exploratory study went far beyond the expectations and predictions of other inorganic chemists of his day. This work, that culminated in his book, Hydrides of Boron and Silicon, *presaged the rapidly opening field of "unusual" inorganic chemistry now so actively pursued.*

Alfred E. Stock was always zealous in recognizing the contributions and help of his coworkers and subordinates during a period in which this was an uncommon virtue. He was not only an outstanding scientist, but also a considerate and thoughtful human being as revealed by his comment: "The most important problem for the scientific mind to solve will be how to free mankind from political, social, and economic limitations and how to give it a purer, broader-minded understanding of humanity. . . ."

CHAPTER 

The Fourth-Row Transition Elements

1 H																	2 He
3 Li	4 Be											5 B	6 C	7 N	8 O	9 F	10 Ne
11 Na	12 Mg	Transition elements										13 Al	14 Si	15 P	16 S	17 Cl	18 Ar
19 K	20 Ca	21 Sc	22 Ti	23 V	24 Cr	25 Mn	26 Fe	27 Co	28 Ni	29 Cu	30 Zn	31 Ga	32 Ge	33 As	34 Se	35 Br	36 Kr
37 Rb	38 Sr	39 Y	40 Zr	41 Nb	42 Mo	43 Tc	44 Ru	45 Rh	46 Pd	47 Ag	48 Cd	49 In	50 Sn	51 Sb	52 Te	53 I	54 Xe
55 Cs	56 Ba	57-71	72 Hf	73 Ta	74 W	75 Re	76 Os	77 Ir	78 Pt	79 Au	80 Hg	81 Tl	82 Pb	83 Bi	84 Po	85 At	86 Rn
87 Fr	88 Ra	89-															

In the preceding chapters we have studied the chemistry of the elements across the top of the periodic table and down the two sides. Now we shall consider the elements in the middle. These are usually referred to as the *transition elements* because chemists once believed that some elements behaved in a way intermediate between the extremes represented by the left and right sides of the periodic table. Today, the term "transition element" remains mostly a useful way of designating elements in this particular region of the periodic table, even though we cannot pinpoint a specific set of properties and say that *all* the transition elements have *all* these properties.

22-1 DEFINITION OF TRANSITION ELEMENTS

There is some disagreement among chemists as to just which elements should be called transition elements. For our purposes, it will be convenient to include all the elements in the columns of the periodic table headed by scandium through zinc.

Across the top, as the first row of the transition region, we have the elements scandium (Sc), titanium (Ti), vanadium (V), chromium (Cr), manganese (Mn), iron (Fe), cobalt (Co), nickel (Ni), copper (Cu), and zinc (Zn). On the left, we have the scandium column which includes, besides Sc, yttrium (Y, 39), lanthanum (La, 57), and actinium (Ac, 89). For reasons that we shall

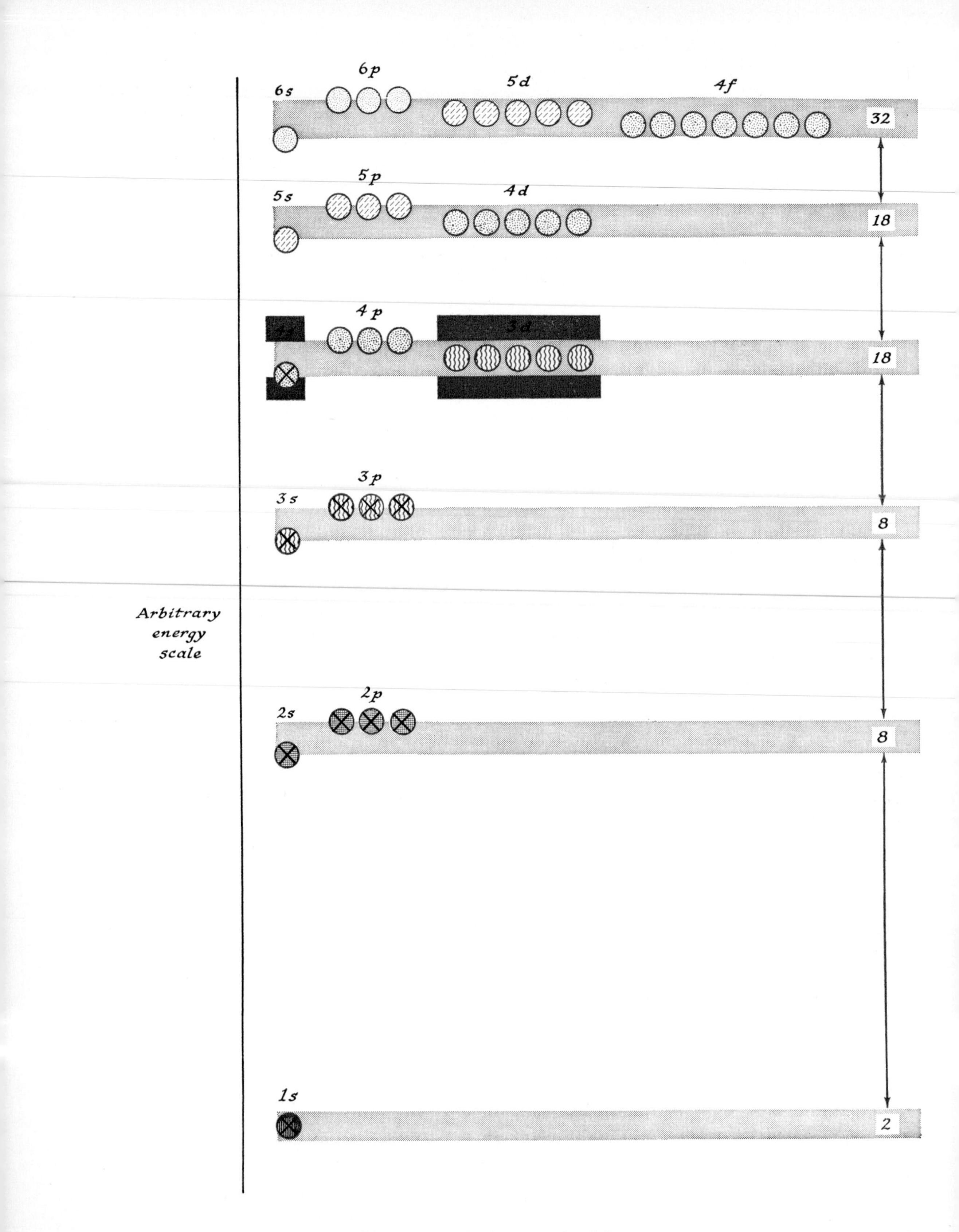

Fig. 22-1. **The fourth-row transition elements** *in the energy level diagram.*

take up in the next chapter, we group with lanthanum the fourteen elements that follow La (Z = 58 through Z = 71); these we call the *lanthanide elements*. On the right, the transition elements end with the zinc column. Besides zinc, this includes cadmium (Cd, 48) and mercury (Hg, 80). It is strongly advisable during the discussion that follows to look back at the periodic table frequently to see where each particular element is placed.

22-1.1 Electron Configuration

There are two immediate questions we ask about the transition elements once we know where they are in the periodic table: (1) Why do we consider these elements together? (2) What is special about their properties? These questions are closely related because they both depend upon the electron configurations of the atoms. What, then, is the electron configuration we might expect for these elements?

To answer this question, we need to review some basic ideas on the electronic buildup of atoms. We saw in Chapter 15 that as we progressively add electrons to build up an atom, each added electron goes into the lowest energy level that is not already fully occupied. With this principle as a guide, let us consider the electron configurations as we build up the first row of transition elements from scandium through zinc. Looking at the periodic table, we see that calcium comes just before scandium. The twenty electrons in a calcium atom are distributed as shown in the following arrangement:

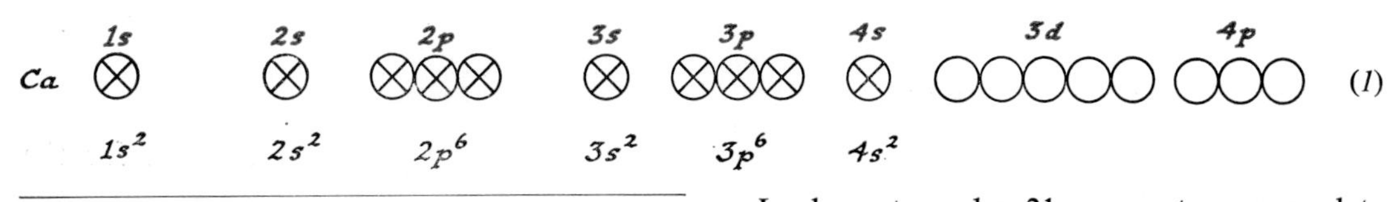

(1)

In element number 21, we must accommodate one more electron. At first sight we might predict that the 21st electron goes into the 4*p* orbital, as the next higher energy level after 4*s*. The 4*p* orbital *is* of higher energy than the 4*s* but, more important, there is a set of five 3*d* orbitals in between. The 21st electron goes into a 3*d* orbital as the level of next higher energy. This is shown in Figure 22-1 (which is just Figure 15-11 reproduced here for convenient reference).

EXERCISE 22-1

Draw on one line a set of orbitals from 1*s* through 4*d*. Under this give the orbital occupancy for Al, Sc, and Y. Account for the fact that yttrium is much more like scandium than is aluminum.

Table 22-I. **THE ELECTRON CONFIGURATIONS OF THE FOURTH-ROW TRANSITION ELEMENTS**

ELEMENT	SYMBOL	ATOMIC NUMBER, Z	ELECTRON CONFIGURATION	
scandium	Sc	21	$1s^2$ $2s^2\,2p^6$ $3s^2\,3p^6$	$3d^1\,4s^2$
titanium	Ti	22		$3d^2\,4s^2$
vanadium	V	23		$3d^3\,4s^2$
chromium	Cr	24		$3d^5\,4s^1$
manganese	Mn	25	Each fourth-row transition element has these levels filled.	$3d^5\,4s^2$
iron	Fe	26		$3d^6\,4s^2$
cobalt	Co	27		$3d^7\,4s^2$
nickel	Ni	28		$3d^8\,4s^2$
copper	Cu	29		$3d^{10}4s^1$
zinc	Zn	30		$3d^{10}4s^2$

There are five $3d$ orbitals available, all more or less of the same energy. Putting a pair of electrons in each of these five orbitals means that a total of ten electrons can be accommodated before we need to go to a higher energy level. Not only scandium but the nine following elements can be built up by adding electrons into $3d$ orbitals. Not until we get to gallium (element number 31) do we go up to another set of orbitals.

EXERCISE 22-2

Again using Figure 22-1, decide which orbital would next be used after the five $3d$ orbitals have been filled. What orbital would next be used after the $4d$ set has been filled? What element does this correspond to in the periodic table?

At this stage, with the help of Figure 22-1, or with an atomic orbital chart, you should be able to work out the electronic configuration of most of the transition elements. You will not be able to deduce them all correctly because there are some exceptions resulting from special stabilities when a set of orbitals is filled or half-filled. The fourth-row transition elements have the set of electron configurations shown in Table 22-I. Notice that chromium ($Z = 24$) and copper ($Z = 29$) provide interruptions to the continuous buildup. In the case of chromium the whole atom has lower energy if one of the $4s$ electrons moves into the $3d$ set to give a half-filled set of $3d$ orbitals and a half-filled $4s$ orbital; in the case of copper, the atom has lower energy if the $3d$ set is completely populated by ten electrons and the $4s$ orbital is half-filled, instead of having nine $3d$ electrons and two in the $4s$ orbital.

EXERCISE 22-3

Make an electron configuration table like Table 22-I for the fifth-row transition elements—yttrium ($Z = 39$) through cadmium ($Z = 48$). In elements 41 through 45, one of the $5s$ electrons moves over to a $4d$ orbital. In element 46, two electrons do this.

In the sixth-row transition elements (lanthanum through mercury) there is an additional complication. There are seven $4f$ orbitals which are very close in energy to the $5d$ orbitals. Putting electrons into these $4f$ orbitals means there will be fourteen additional elements in this row. These fourteen elements are almost identical in many chemical properties. We will discuss them in the next chapter.

22-1.2 General Properties

What properties do we actually find for the transition elements? What kinds of compounds do they form? How can the properties be interpreted in terms of the electron populations of the atoms?

Looking at a sample of each transition element in the fourth row, we see that they are all metallic. When clean, they are shiny and lustrous. They are good conductors of electricity and also of heat: some of them (copper, silver, gold) are quite outstanding in these respects. One of them (mercury) is ordinarily a liquid; all others are solids at room temperature.

So far as chemical reactivity is concerned, we find a tremendous range. Some of the transition elements are extremely unreactive. For example, gold and platinum can be exposed to air or water for ages without any change. Others, such as iron, can be polished so they are brightly metallic for a while, but on exposure to air and water they slowly corrode. Still others are vigorously reactive and, when exposed to air, produce a shower of sparks. Lanthanum and cerium, for instance, especially when finely divided, oxidize immediately when exposed to air. (Some cigarette lighters have flints containing these metals.)

It is hard to generalize about the chemical reactivities of a group of elements since reactivities depend upon two factors: (A) the relative stability of the specific compounds formed compared with the reactants used up, and, (B) the rate at which the reaction occurs. In special cases there are other complications. For example, chromium metal (familiar in the form of chrome plate) is highly reactive toward oxygen. Still, a highly polished piece of chromium holds its luster almost indefinitely when exposed to air. The explanation is that a very thin, invisible coat of oxide quickly forms on the surface and protects the underlying metal from contact with the oxygen in the air. In other words, bulk chromium is unstable with respect to oxidation by air, but the protective layer of oxide cuts the rate of conversion so much that no reaction is observed.

What about compounds of the transition elements? Suppose we go into the chemical stockroom and see what kinds of compounds are on the shelf for a particular element, say chromium. First we might find a bottle of green powder labeled Cr_2O_3, chromic oxide, or chromium(III) oxide. Next to it there would probably be a bottle of a reddish powder, CrO_3, chromium(VI) oxide. On an amply stocked chemical shelf, we might also find some black powder marked CrO, chromous oxide, or chromium(II) oxide. There would probably also be some other simple compounds such as $CrCl_3$, chromic chloride, or chromium(III) chloride, a flaky, reddish-violet solid, and maybe some green CrF_2, chromous fluoride or chromium(II) fluoride. Elsewhere in the stockroom we would run across K_2CrO_4, a bright yellow powder (potassium chromate), probably next to a bottle of orange potassium dichromate, $K_2Cr_2O_7$. Soon we would get the idea that the compounds of chromium, at least the common ones, correspond to oxidation number of +2 (CrO and CrF_2), +3 (Cr_2O_3 and $CrCl_3$), and +6 (CrO_3, K_2CrO_4, $K_2Cr_2O_7$).

EXERCISE 22-4

What is the oxidation number of chromium in each of the following compounds: $Cr_2O_7^{-2}$, CrO_4^{-2}, $Cr(OH)_3$, CrO_2Cl_2?

Along with these simple compounds, we might also find some rather more complex substances. For example, we might find next to $CrCl_3$ vials of several brightly colored solids labeled $CrCl_3 \cdot 6NH_3$, $CrCl_3 \cdot 5NH_3$, $CrCl_3 \cdot 4NH_3$, and $CrCl_3 \cdot 3NH_3$. Recalling that the dot in these formulas simply indicates that a certain number of moles of NH_3 are bound to one mole of $CrCl_3$, we would conclude that here also the oxidation number of chromium is +3. Looking further, we might find other complex compounds such as K_3CrF_6, $Na_3Cr(CN)_6$, $KCr(SO_4)_2 \cdot 12H_2O$. In all these the chromium has a +3 oxidation number. As a result of our stockroom search, we would form three conclusions: (1) chromium forms simple and complex compounds; (2) chromium forms a number of stable solids, most of them colored; (3) chromium may have different oxidation numbers, including +2, +3, and +6. *Similar conclusions would have resulted for most of the other transition elements.*

Is there any regularity to the kind of compounds the fourth-row transition elements form? Table 22-II shows what chemists have found.

Table 22-II. **TYPICAL OXIDATION NUMBERS FOUND FOR FOURTH-ROW TRANSITION ELEMENTS**

SYMBOL	REPRESENTATIVE COMPOUNDS		COMMON OXIDATION NUMBERS (most common in bold type)		NUMBER OF VALENCE 3*d*, 4*s* ELECTRONS
Sc	Sc_2O_3		**+3**		3
Ti	TiO, Ti_2O_3, TiO_2		+2, +3, **+4**		4
V	VO, V_2O_3, VO_2, V_2O_5		+2, +3, +4, **+5**		5
Cr	CrO, Cr_2O_3,	CrO_3	+2, **+3**	+6	6
Mn	MnO, Mn_2O_3, MnO_2,	K_2MnO_4, $KMnO_4$	**+2**, +3, **+4**	+6, +7	7
Fe	FeO, Fe_2O_3		+2, **+3**		8
Co	CoO, Co_2O_3		**+2**, +3		9
Ni	NiO, Ni_2O_3		**+2**, +3		10
Cu	Cu_2O, CuO		+1, **+2**		11
Zn	ZnO		**+2**		12

EXERCISE 22-5

Look through a handbook of chemistry and find one *other* compound of each oxidation state given for the elements in Table 22-II.

Several points should be noted from this table:

(1) For most of the transition elements, several oxidation numbers are possible.

(2) When several oxidation numbers are found for the same element, they often differ from each other by jumps of one unit. For example, in the case of vanadium the common oxidation numbers form a continuous series from +2 to +3 to +4 to +5. Compare this with the halogens (Chapter 19). In the case of chlorine, for example, the common states are −1, +1, +3, +5, and +7 (jumps of two units instead of one unit).

(3) The *maximum* oxidation state observed for the elements first increases and then decreases as we go across the transition row. Thus we have +3 for scandium, +4 for titanium, +5 for vanadium, +6 for chromium, and +7 for manganese. The +7 represents the highest value observed for this transition row. After manganese, the maximum value diminishes as we continue toward the end of the transition row.

What explanation can we give for these observations? Why does the combining capacity vary from one transition element to another in such a way that the above pattern of oxidation numbers develops? The combining capacity of an atom depends upon how many electrons the atom uses for bonding to other atoms. The unique feature of the transition elements is that they have several electrons in the outermost d and s orbitals, and the ionization energies of all of these electrons are relatively low. Therefore it is possible for an element like vanadium to form a series of compounds in which from two to five of its electrons are either lost to, or shared with, other elements. Consider, for example, the oxides VO and V_2O_3, which contain the V^{+2} and the V^{+3} ions, respectively. While more energy is needed to form V^{+3} than to form V^{+2}, the V^{+3} has, because of its higher charge, a greater attraction for the O^{-2} ion than does V^{+2}. This extra attraction in V_2O_3 compensates for the energy needed to form the V^{+3} ion, and both oxides (as well as VO_2 and V_2O_5) are stable compounds. Notice, moreover, that the maximum oxidation number of the transition elements never exceeds the total number of s and d valence electrons. The higher oxidation states become increasingly more difficult to form as we proceed along a row, because the ionization energy of the d and s electrons increases with the atomic number.

22-2 COMPLEX IONS

The remaining general point to be made about the transition elements is that they form a great variety of complex ions in which other molecules or ions are bonded to the central transition element ion to form more complex units. These are called **complex ions.** Take the series already mentioned: $CrCl_3 \cdot 6NH_3$, $CrCl_3 \cdot 5NH_3$, $CrCl_3 \cdot 4NH_3$, and $CrCl_3 \cdot 3NH_3$. How can we account for the existence of such a series? The answer can be seen if we consider some of the observed facts about these complex compounds. For example, if we dissolve one mole of each in water and add a solution of silver nitrate in an attempt to precipitate the chloride as AgCl,

$$Ag^+ + Cl^- \rightleftharpoons AgCl(s) \qquad (2)$$

we find that sometimes much of the chloride cannot be precipitated. The observed results are:

Compound	*Moles of Cl⁻ precipitated*	*Moles of Cl⁻ not precipitated*
$CrCl_3 \cdot 6NH_3$	3 of 3	0
$CrCl_3 \cdot 5NH_3$	2 of 3	1 of 3
$CrCl_3 \cdot 4NH_3$	1 of 3	2 of 3
$CrCl_3 \cdot 3NH_3$	0	3 of 3

Evidently, there are two ways in which chlorine is bound in these compounds, one way which allows the Cl^- to be precipitated by Ag^+ and another way which does not. In $CrCl_3 \cdot 6NH_3$, all of the chloride can be precipitated; in $CrCl_3 \cdot 3NH_3$, none of it can be. Other data also indicate different types of bonding. For instance, the freezing point lowering of an aqueous solution of $CrCl_3 \cdot 6NH_3$ indicates there are *four* moles of particles per mole of $CrCl_3 \cdot 6NH_3$; the solution is highly conducting.* On the other hand, the freezing point lowering of $CrCl_3 \cdot 3NH_3$ solution indicates there is *one* mole of particles per mole of $CrCl_3 \cdot 3NH_3$; furthermore, the solution does not conduct at all. The explanation of this behavior was provided in the early 1900's by Alfred Werner, who noted that complex compounds of chromium +3 could be accounted for by assuming each chromium is bonded to six near neighbors. In $CrCl_3 \cdot 6NH_3$, the cation consists of a central Cr^{+3} surrounded by 6 NH_3 molecules at the corners of an octahedron; the three chlorine atoms exist as anions, Cl^-. In $CrCl_3 \cdot 5NH_3$, the cation consists of the central chromium surrounded by the five NH_3 and one of the Cl atoms; the other two Cl atoms are anions. In $CrCl_3 \cdot 4NH_3$, the chromium is bound to four NH_3 and two Cl leaving one chloride anion. In $CrCl_3 \cdot 3NH_3$, all three Cl atoms and all three NH_3 molecules are tied to the central chromium. The formulas can be written

$$[Cr(NH_3)_6]Cl_3$$
$$[Cr(NH_3)_5Cl]Cl_2$$
$$[Cr(NH_3)_4Cl_2]Cl$$
$$[Cr(NH_3)_3Cl_3]$$

22-2.1 Geometry of Complex Ions

The way that atoms or molecules are arranged in space around a central atom has a great influence on whether a given complex aggregate is stable enough to be observed. What kinds of arrangements are found in complex ions? What shapes do these complex ions show? Can we find any regularity in the transition elements that will enable us to predict what complex ions will form?

First, let us introduce a concept useful in giving spatial descriptions; ***the coordination number** is the number of near neighbors that an atom has.* For example, in the complex ion AlF_6^{-3} (which is the anion present in the solid mineral cryolite), each Al atom is surrounded by six fluorine atoms at the corners of an octahedron, as shown in Figure 22-2. We say that aluminum has a coordination number of 6 with fluorine. In the complex ion $AlBr_4^-$, which seems to be an important intermediate when aluminum bromide acts as a catalyst for many organic reactions, the bromine atoms are arranged around a central Al at the corners of a regular tetrahedron. Figure 22-3 shows the arrangement. The coordination number of aluminum is 4 with bromine.

If more than simple atoms are bound to a central atom, then the coordination number still refers to the number of near neighbors. For example, in solid potassium chrome alum, $KCr(SO_4)_2 \cdot 12H_2O$, and also in its fresh aqueous solutions, the chromium-containing cation is

Fig. 22-2. **An octahedral complex:** *aluminum with coordination number 6.*

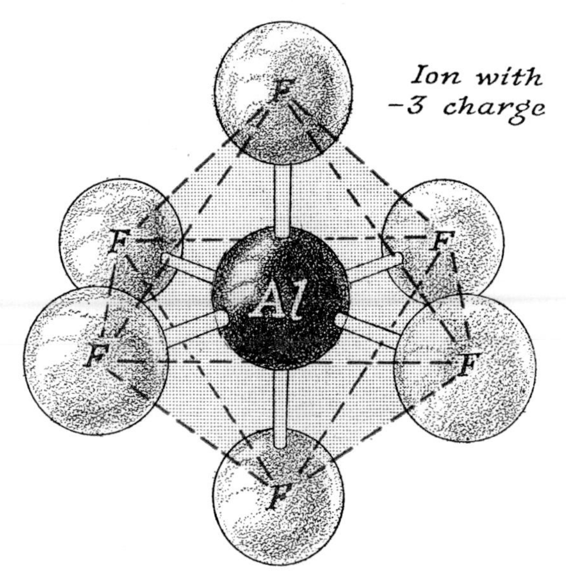

* From work on simple salts such as NaCl we expect that the "particles" are ions, and the conductivity confirms this.

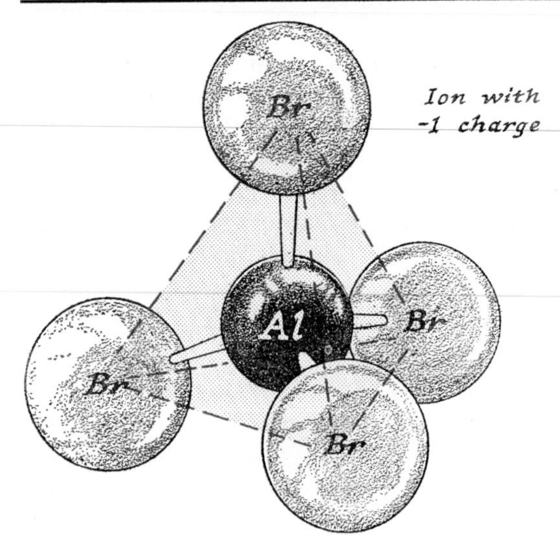

Fig. 22-3. **A tetrahedral complex:** *aluminum with coordination number 4.*

$Cr(H_2O)_6^{+3}$. It consists of a central chromium joined to six H_2O molecules, exactly as the fluorines are arranged around aluminum in Figure 22-2. The oxygen portion of each H_2O molecule is turned toward the central chromium and the H portions point away from the center. The corners of the cage that holds the central Cr atom are occupied by six oxygen atoms, each of which also holds two H atoms. The shape of this complex ion is octahedral and we say that in $Cr(H_2O)_6^{+3}$ chromium shows a coordination number of 6 to oxygen.

Notice that in an octahedral complex ion such as $[Cr(NH_3)_4Cl_2]^+$ there is a possibility of observing isomers. The two chlorine atoms may occupy octahedral positions which are next to each other on the same side of the metal atom, or positions located on opposite sides of the metal atom (see Figure 22-4). The isomer in which the two similar groups are located on the same side of the metal atom is called the *cis*-isomer, and the other is called the *trans*-isomer.

The complex ion $Fe(C_2O_4)_3^{-3}$ is formed when rust stains are bleached out with oxalic acid solution. It also has a transition element showing coordination number of 6, even though there are only three groups ($C_2O_4^{-2}$ groups) around each iron ion. Figure 22-5 shows the arrangement. Each $C_2O_4^{-2}$, the oxalate group, uses two of its oxygen atoms to bind onto the central iron atom. The number of near neighbors, *as viewed from the iron atom*, is six oxygen atoms at the corners of an octahedron. Picturesquely, *a group* such as

Fig. 22-4. **The isomers of** $[Cr(NH_3)_4Cl_2]^+$.

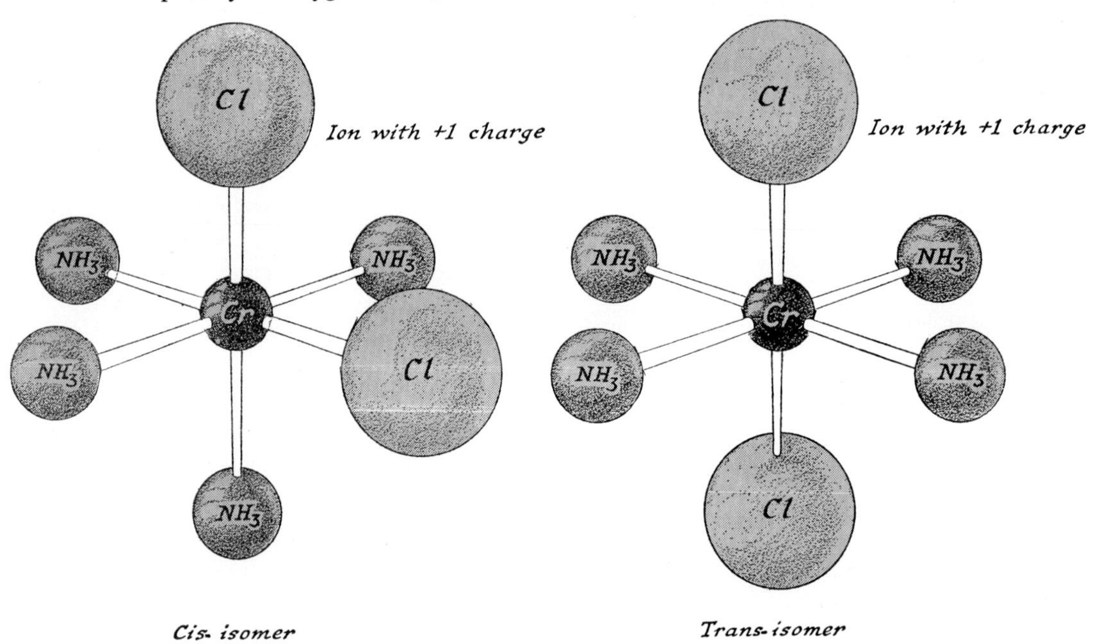

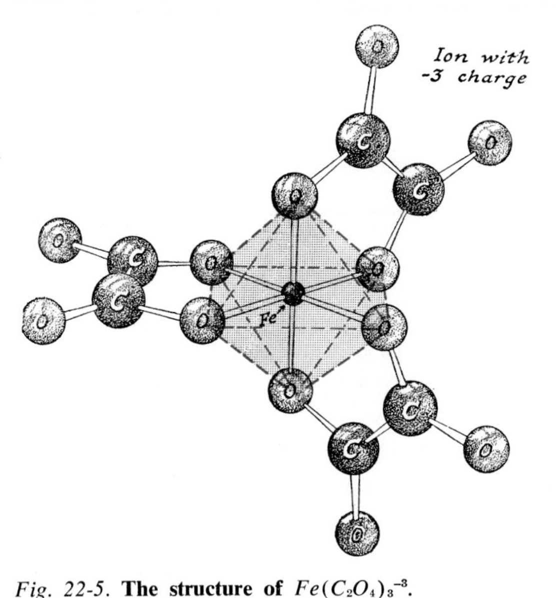

Fig. 22-5. **The structure of** $Fe(C_2O_4)_3^{-3}$.

oxalate, *which can furnish simultaneously two atoms for coordination*, is said to be **bidentate,** which literally means double-toothed.

In addition to the tetrahedral and octahedral complexes mentioned above, there are two other types commonly found—the square planar and the linear. In the square planar complexes, the central atom has four near neighbors at the corners of a square. The coordination number is 4, the same number as in the tetrahedral complexes. An example of a square planar complex is the complex nickel cyanide anion, $Ni(CN)_4^{-2}$.

In a linear complex, the coordination number is 2, corresponding to one group on each side of the central atom. The silver-ammonia complex, which generally forms when a very slightly soluble silver salt such as silver chloride dissolves in aqueous ammonia, is an example, as shown in Figure 22-6. Another example of a linear complex is $Ag(CN)_2^-$, which is formed during the leaching of silver ores with NaCN solution.

Fig. 22-6. **A linear complex,** $Ag(NH_3)_2^+$.

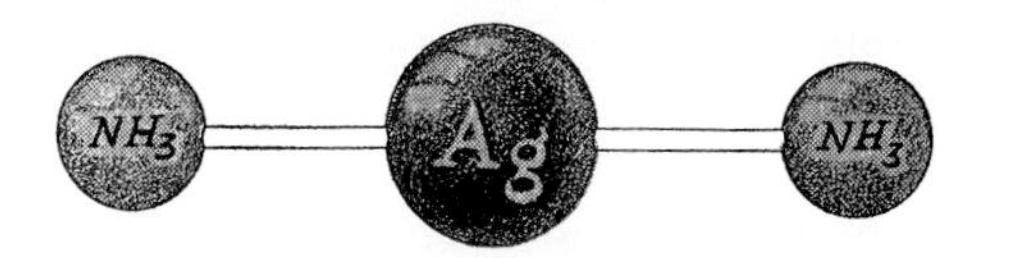

22-2.2 Bonding in Complex Ions

What holds the atoms of a complex ion together? There are two possibilities. In some complexes, as in AlF_6^{-3}, the major contribution to the bonding comes from the attraction between a positive ion (Al^{+3}) and a negative ion (F^-). The bonding is ionic. In other complexes, as in $Fe(CN)_6^{-3}$, there is thought to be substantial sharing of electrons between the central atom and the attached groups. The bonding is mainly covalent. When there is such sharing, an electron or an electron pair from the attached group spends part of its time in an orbital furnished by the central atom. In either type, as emphasized in Chapter 16, the electron is attracted to both atoms in the bond.

For transition elements there are usually empty *d* orbitals ready to accommodate electrons from attached groups. This is by no means always necessary, as witness the case of Zn^{+2}, a good complex-former even though all its 3*d* orbitals are already occupied. Any vacant orbital, low enough in energy to be populated, will serve as a means whereby complex formation can be accomplished.

The geometry of a complex ion often can be explained quite reasonably in terms of the orbitals of the central atom populated by electrons from the attached groups. If only one *s* and one *p* orbital is used, the bonding is called *sp* bonding. We have already seen in Section 16-4.5 that this bonding situation gives rise to a linear arrangement. Therefore this might explain why some complexes are linear, as is $Ag(NH_3)_2^+$. If one *s* and three *p* orbitals are used, the complex uses sp^3 bonding. Then a tetrahedral complex can be expected, as observed for $Zn(NH_3)_4^{+2}$. When *d* orbitals are involved, other geometries can be explained (for example, square planar, dsp^2; octahedral, d^2sp^3).

22-2.3 Significance of Complex Ions

Besides their occurrence in solid compounds, complex ions such as we have mentioned are important for two other reasons: (1) they may decide what species are present in aqueous solutions; and (2) some of them are exceedingly important in biological processes.

As an example of the problem of species in solution, consider the case of a solution made by dissolving some potassium chrome alum, $KCr(SO_4)_2 \cdot 12H_2O$, in water. On testing, the solution is distinctly acidic. A currently accepted explanation of the observed acidity is based upon the assumption that, in water solution, chromic ion is associated with six H_2O molecules in the complex ion, $Cr(H_2O)_6^{+3}$. This complex ion can act as a weak acid, dissociating to give a proton (or hydronium ion). Schematically, the dissociation can be represented as the transfer of a proton from one water molecule in the $Cr(H_2O)_6^{+3}$ complex to a neighboring H_2O to form a hydronium ion, H_3O^+. Note that removal of a proton from an H_2O bound to a Cr^{+3} leaves an OH^- group at that position. The reaction is reversible and comes to equilibrium:

$$Cr(H_2O)_6^{+3} + H_2O \rightleftharpoons Cr(H_2O)_5OH^{+2} + H_3O^+ \quad (3)$$

We see that $Cr(H_2O)_6^{+3}$ acts as a proton-donor, that is, an acid.

22-2.4 Amphoteric Complexes

Another reason chemists find the above complex ion picture of aqueous solutions useful is that it is easily extended to explain amphoteric behavior. Take the case of chromium hydroxide, $Cr(OH)_3$, a good example of an amphoteric hydroxide. It dissolves very little in water, but is quite soluble both in acid and in base. Presumably it can react with either. How can this behavior be explained in terms of the complex ion picture?

First, consider the equilibrium represented by equation (*3*) when NaOH is added to solution. Added OH^- combines with the H_3O^+ to form H_2O. This removes one of the species shown on the right side of the equation, so formation of $Cr(H_2O)_5OH^{+2}$ is favored. In other words, as OH^- is added to $Cr(H_2O)_6^{+3}$ the reaction is favored which corresponds to pulling a proton off $Cr(H_2O)_6^{+3}$.

What happens when enough NaOH has been added to remove *three* protons from each $Cr(H_2O)_6^{+3}$? Removal of three protons leaves the neutral species $Cr(H_2O)_3(OH)_3$, or $Cr(OH)_3 \cdot 3H_2O$. This neutral species has no charges to repel other molecules of its own kind so it precipitates. However, as more NaOH is added to this solid phase, one more proton can be removed to produce $Cr(H_2O)_2(OH)_4^-$; and the $Cr(OH)_3 \cdot 3H_2O$ dissolves. [In principle, more protons could be removed, perhaps eventually to form $Cr(OH)_6^{-3}$, but there is as yet no evidence for this.]

The following equations summarize the steps believed to occur when NaOH is slowly added to a solution of chromic ion. Step (*4c*) corresponds to formation of solid hydrated chromium hydroxide; step (*4d*) corresponds to its dissolving in excess NaOH.

$$Cr(H_2O)_6^{+3} + OH^- \rightleftharpoons Cr(H_2O)_5OH^{+2} + H_2O \quad (4a)$$

$$Cr(H_2O)_5OH^{+2} + OH^- \rightleftharpoons Cr(H_2O)_4(OH)_2^+ + H_2O \quad (4b)$$

$$Cr(H_2O)_4(OH)_2^+ + OH^- \rightleftharpoons Cr(H_2O)_3(OH)_3(s) + H_2O \quad (4c)$$

$$Cr(H_2O)_3(OH)_3(s) + OH^- \rightleftharpoons Cr(H_2O)_2(OH)_4^- + H_2O \quad (4d)$$

When acid is added to a solution such as in equation (*4d*), the above set of reactions is progressively reversed, first causing precipitation of chromium hydroxide by the reverse of reaction (*4d*) and then its subsequent dissolving by the reverse of reaction (*4c*).

22-2.5 Complexes Found in Nature

Complex ions have important roles in certain physiological processes of plant and animal growth. Two such complexes are hemin, a part of hemoglobin, the red pigment in the red corpuscles of the blood, and chlorophyll, the green coloring material in plants. The first of these, hemoglobin, contains iron and properly fits in a discussion of complex compounds of the transition elements; the second, chlorophyll, is a complex compound of magnesium. Magnesium is not a transition element, but chlorophyll is discussed here because it has some features in common with hemoglobin and because it avoids the impression that only transition elements form complexes.

Chlorophyll, as extracted from plants, is actually made up of two closely related compounds, chlorophyll A and chlorophyll B. These differ slightly in molecular structure and can be separated because they have different tendencies to be adsorbed on a finely divided solid (such as powdered sugar).

EXERCISE 22-6

If you wish to prepare some chlorophyll, grind up some fresh leaves and extract with alcohol. The alcohol dissolves the chlorophyll, as shown by the solution color.

To show the complexity of this biologically important material, the structural formula of chlorophyll A is shown on the left in Figure 22-7. You need not memorize it. The most obvious thing to note is that it is a large organic molecule with a magnesium atom in the center. Right around the magnesium atom there are four near-neighbor N atoms, each of which is part of a five-membered ring. Consider also the vast amount of knowledge and experimentation that are summed up in the statement that the structure of this complicated molecule is known.

EXERCISE 22-7

If a typical plant leaf yields 40.0 mg of chlorophyll A, how many milligrams of this will be magnesium? The molecular weight of chlorophyll A is 893.

Hemin is shown on the right in Figure 22-7. It is shown beside the model of chlorophyll A to emphasize the astonishing similarity. The portions within dotted lines identify the differences. Except for the central metal atom, the differences are all on the periphery of these cumbersome molecules. We cannot help wondering how nature managed to standardize on this molecular skeleton for molecules with such different functions. We cannot avoid a feeling of impatience as we await the clarification of the possible relationship, a clarification that will surely be provided by scientists of the next generation.

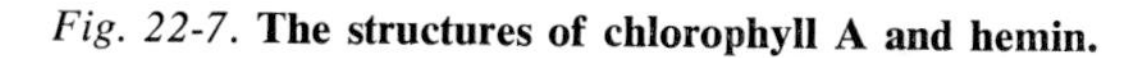

Fig. 22-7. **The structures of chlorophyll A and hemin.**

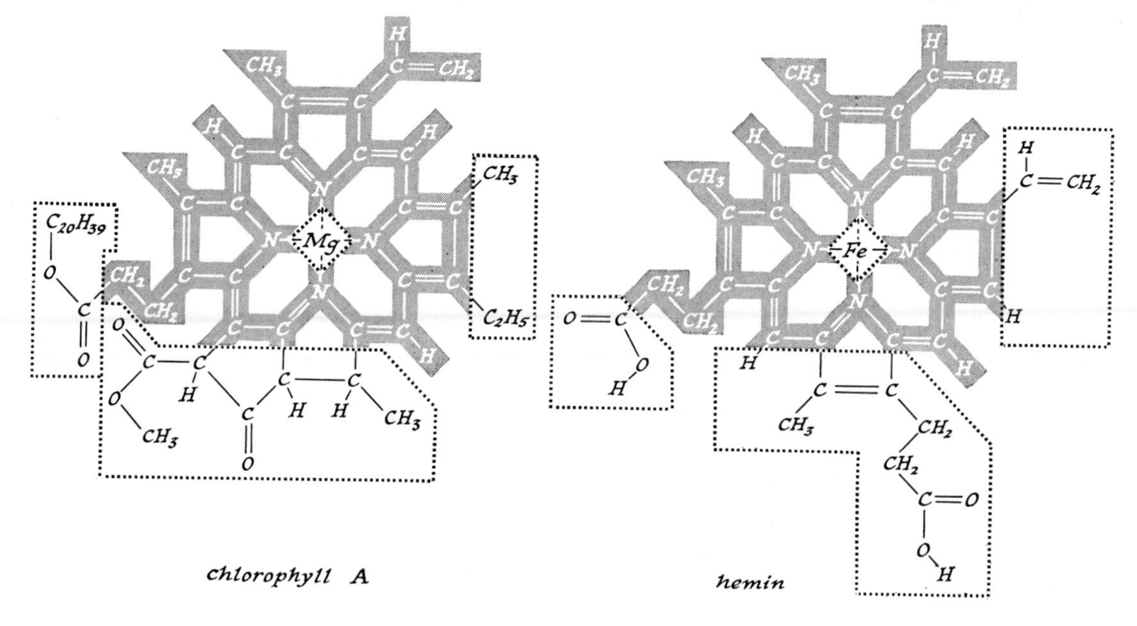

The most important function of hemoglobin in the blood is that of carrying oxygen from the lungs to the tissue cells. This is done through a complex between the iron atom of the hemin part and an oxygen molecule. Just how the O_2 is bound to the hemin is not yet clear, but it must be a rather loose combination since the O_2 is readily released to the cells. The complex is bright red, the characteristic color of arterial blood. When the O_2 is stripped off the hemin group, the color changes to a purplish red, the color of blood in the veins.

Not only O_2 molecules but also other groups can be bound to the iron atom of hemoglobin. Specifically, carbon monoxide molecules can be so attached and, in fact, CO is more firmly bound to hemoglobin than is O_2. This is one detail of the carbon monoxide poisoning mechanism. If we breathe a mixture of CO and O_2 molecules, the CO molecules are preferentially picked up by the red blood cells. Since the sites normally used to carry O_2 molecules are thus filled by the CO molecules, the tissue cells starve for lack of oxygen. If caught in time, carbon monoxide poisoning can be treated by raising the ratio of O_2 to CO in the lungs (in other words, administering fresh air or oxygen). The two reactions,

$$O_2(g) + \text{hemoglobin} \rightleftharpoons \text{complex}_1 \qquad (5a)$$

$$CO(g) + \text{hemoglobin} \rightleftharpoons \text{complex}_2 \qquad (5b)$$

have sufficiently similar tendencies to go to the right that the first reaction can be made to exceed the second if the concentration of O_2 sufficiently exceeds that of CO. Another remedial measure is to inject methylene blue directly into the blood stream. Carbon monoxide bonds more strongly to methylene blue than to hemin. Equilibrium conditions then favor the transfer of CO to the methylene blue, thus freeing hemoglobin for its normal oxygen transport function.

22-3 SPECIFIC PROPERTIES OF FOURTH-ROW TRANSITION ELEMENTS

The preceding discussion of the transition elements has been quite general, with the implication of rather wide applicability. Now we turn to a consideration of the transition elements and their compounds as specific individuals.

Table 22-III collects some of the data ordinarily found useful for the transition elements of the fourth row of the periodic table. The following are some notes on regularities observed.

The *atomic weight* increases regularly across the row except for the inversion at cobalt and nickel. We would expect the atomic weight of Ni to be higher than that of Co because there are more protons (*28*) in the Ni nucleus than in the Co nucleus (*27*). The reason for the inversion lies in the distribution of naturally occurring isotopes. Natural cobalt consists entirely of the isotope $^{59}_{27}Co$; natural nickel consists primarily of the isotopes $^{58}_{28}Ni$ and $^{60}_{28}Ni$, the 58-isotope being about three times as abundant as the 60-isotope.

Abundance in the earth's crust. With the exception of iron, which is very abundant, and titanium, which is moderately abundant, all the other elements of the first transition row are relatively scarce. However, some of them, such as copper, are quite familiar. Copper is one of the few metallic elements found free in nature. The existence of deposits of metallic copper undoubtedly accounts for the fact that man evolved through the Bronze Age before the Iron Age. Copper, the essential ingredient of bronze, did not require the difficult smelting process needed for iron.

Melting point. Except for zinc at the end of the row, the melting points are quite high. This is appropriate, since these elements have a large number of valence electrons and also a large number of vacant valence orbitals. Toward the end of the row, in zinc, the 3*d* orbitals become filled and the melting point drops.

Density. There is a steady increase in density through this row, with some leveling off toward the right. This trend is closely tied to the almost constant size of the atoms so the main effect producing density change is the increasing nuclear mass.

Ionization energy. As ionization energies go, the values found for the transition elements are neither very high nor very low. They are all rather similar in magnitude. The sequential increase in nuclear charge, which would tend to increase the ionization energy, seems to be almost offset by the extra screening of the nucleus provided by the added electrons.

Ionic radius. Ionic radii do not change much in going across a transition row. The reason for this is essentially a balance of two effects: (1) As nuclear charge increases across the row, the electrons would be pulled in, so the ions ought to shrink. (2) As more 3*d* electrons are added, these electrons repel each other and the ions ought to swell. These effects just about cancel. As expected, the size of the +3 ion is smaller than the size of the +2 ion of that same element. Keeping nuclear charge constant, removal of one additional 3*d* electron would reduce the repulsion between the 3*d* electrons remaining, thus allowing them all to be pulled closer to the nucleus.

Color. Many solid compounds of the transition metals and their aqueous solutions are colored. This color indicates light is absorbed in

Fig. 22-8. **Atomic sizes** *of the transition elements.*

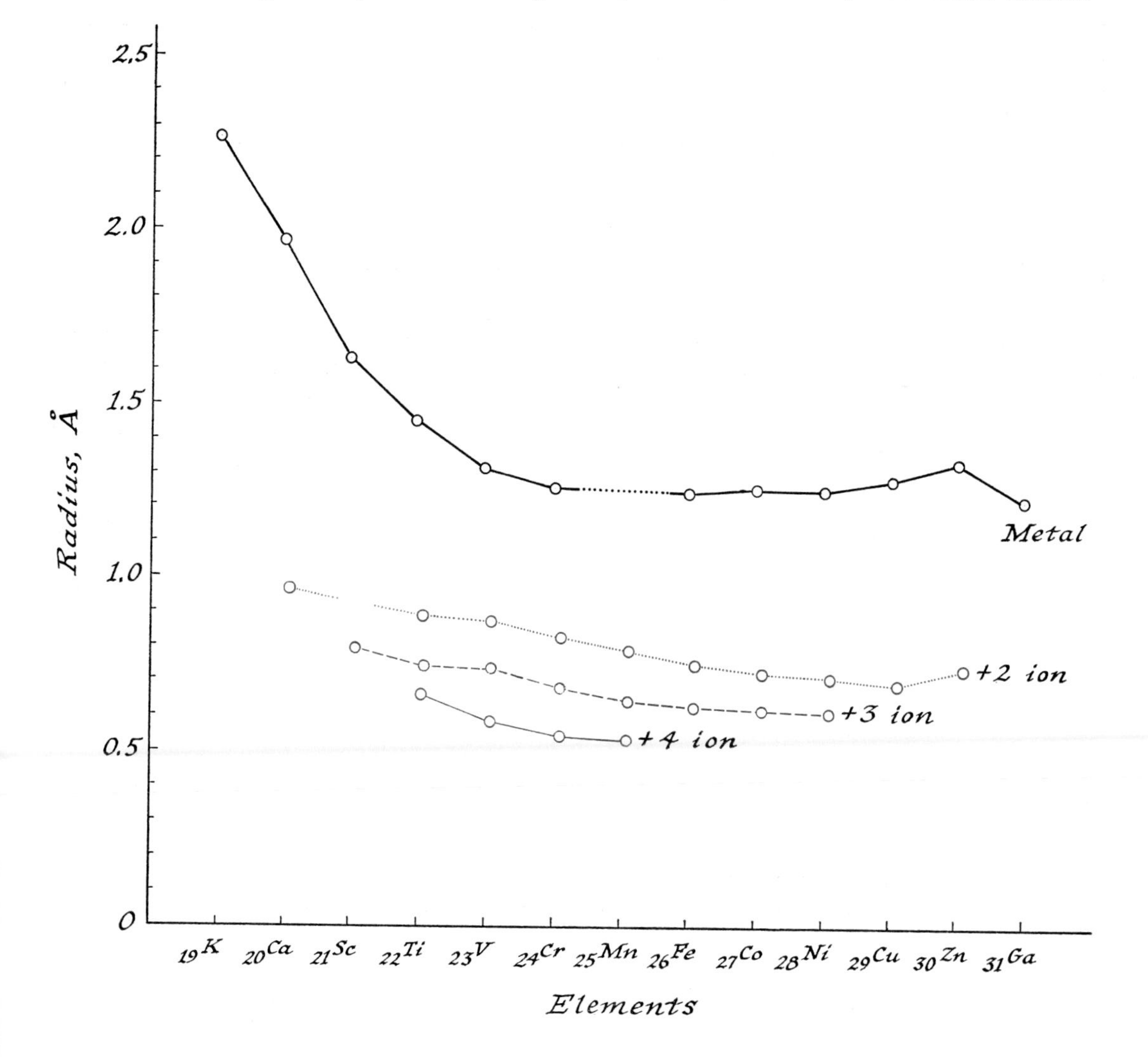

the visible part of the spectral region. The energy levels that account for this absorption are relatively close together and involve unoccupied *d* orbitals. The environment of the ion changes the spacing of these levels, thereby influencing the color. A familiar example is the $Cu^{+2}(aq)$ ion, which changes from a light blue to a deep blue when NH_3 is added. The formation of the ammonia complex alters the energy level spacing of the central Cu^{+2} ion to produce the color change.

E°. The last row of the table gives the values of the oxidation tendencies for these metals. Except for scandium (which goes to a +3 state), the values quoted correspond to the reaction

$$M(s) \longrightarrow M^{+2}(aq) + 2e^- \qquad (6)$$

As can be seen from the table, all the elements except copper have positive values, which means these metals are more easily oxidized than is hydrogen gas, for which $E°$ is zero. Thus, manganese metal should dissolve in acid to liberate hydrogen gas. The $E°$ for the overall reaction,

$$Mn(s) + 2H^+(aq) \rightleftharpoons Mn^{+2}(aq) + H_2(g) \qquad (7)$$

is +1.18 volts, so the reaction should proceed spontaneously to the right. (Note that this is an equilibrium consideration and it tells nothing about the rate. The rate may be slow.) For copper the reaction

$$Cu(s) + 2H^+(aq) \rightleftharpoons Cu^{+2}(aq) + H_2(g) \qquad (8)$$

has a negative $E°$ (−0.34 volt), so reaction to the right is not expected.

Zinc, at the end of the row, has $E°$ (+0.76), which is intermediate between the values at the beginning of the row and those toward the end. We predict, with this value of $E°$, that Zn should reduce Fe^{+2}, Co^{+2}, Ni^{+2}, and Cu^{+2} to the corresponding metals but should not be able to reduce Sc^{+3}, Ti^{+2}, V^{+2}, Cr^{+2}, or Mn^{+2} to the metals.

22-3.1 Scandium

Scandium has not yet been available in large enough amounts to have it develop interesting or important uses. Neither has it been available for much experimental work, so there remains much to be learned about this element.

22-3.2 Titanium

There is intense interest in titanium. This interest stems from an unusual combination of desirable properties in one metal. It is strong; it has low density; and it is remarkably resistant to corro-

Table 22-III. **SOME PROPERTIES OF FOURTH-ROW TRANSITION ELEMENTS**

ELEMENT	Sc	Ti	V	Cr	Mn	Fe	Co	Ni	Cu	Zn
Atomic number	21	22	23	24	25	26	27	28	29	30
Atomic weight	45.0	47.9	51.0	52.0	54.9	55.9	58.9	58.7	63.5	65.4
Abundance* (% by wt.)	0.005	0.44	0.015	0.020	0.10	5.0	0.0023	0.008	0.0007	0.01
Melting point (°C)	1400	1812	1730	1900	1244	1535	1493	1455	1083	419
Boiling point (°C)	3900	3130†	3530†	2480†	2087	2800	3520	2800	2582	907
Density (g/cm³)	2.4	4.5	6.0	7.1	7.2	7.9	8.9	8.9	8.9	7.1
First ioniz. energy (kcal/mole)	154	157	155	155	171	180	180	175	176	216
+2 ion radius (A)	—	0.90	0.88	0.84	0.80	0.76	0.74	0.72	0.72	0.74
+3 ion radius (Å)	0.81	0.76	0.74	0.69	0.66	0.64	0.63	0.62	—	—
$E°$ (volt) $M \longrightarrow M^{+2} + 2e^-$	2.1**	1.6	1.2	.90	1.18	0.44	0.28	0.25	−0.34	0.76

* In the earth's crust.
† Estimated
** $M \longrightarrow M^{+3} + 3e^-$.

sion. The difficulty has been to find an economical way of getting it out of its natural minerals: rutile, TiO_2, and ilmenite, $FeTiO_3$. This was solved in part by heating TiO_2 in chlorine gas to convert it to $TiCl_4$ and then reducing the $TiCl_4$ with magnesium metal. Two problems still stand in the way of large-scale use of this rather abundant element. One is the great sensitivity of its properties to the presence of trace impurities (especially H, O, C, and N); the other is the difficulty of forming it into useful shapes.

22-3.3 Vanadium

This element is important mainly because of its use as an additive to iron in the manufacture of steel. A few percent of vanadium stabilizes a high-temperature crystal structure of iron so that it persists at room temperature. This form is tougher, stronger, and more resistant to corrosion than ordinary iron. Automobile springs, for example, are often made of vanadium steel.

Also important is V_2O_5, divanadium pentoxide, an orange powder which is used as a catalyst for many reactions of commercial significance. For example, in the manufacture of sulfuric acid, V_2O_5 catalyzes the step in which SO_2 is oxidized to SO_3. How it works is still in dispute, but the general belief is that the catalytic action is dependent upon the ability of vanadium to show various oxidation states. One suggested mechanism is that the solid V_2O_5 absorbs an SO_2 molecule on the surface, gives it an oxygen atom to convert it to SO_3, and is itself reduced to V_2O_4, divanadium tetroxide. The V_2O_4 in turn is restored to V_2O_5 by reaction with oxygen. Catalytic reactions, especially those involving solid–gas interfaces, are not very well understood at the present time.

22-3.4 Chromium

Interest in this metal comes from its remarkable inertness to atmospheric corrosion. Also, it is very hard and thus it forms an ideal protective coating. On the basis of its $E°$ (1.18 volts higher than hydrogen) we expect chromium to be quite reactive; in fact, it is vigorously reactive with some reagents—chlorine, for instance. In air, however, it is inactive, probably because of formation of an impervious oxide coat. Other metals, such as the alkali and alkaline earth metals, also form oxide coats but they are not very effective in protecting the underlying metal from atmospheric oxidation. The main difference is that when chromium is converted to oxide there is a swelling in this oxide layer that arises from the increase in volume per chromium atom. This gives a nonporous surface coat of oxide. On the other hand, when a metal such as calcium is oxidized, the oxide layer has a smaller volume per Ca atom than the metal itself. The result is that the surface layer shrinks, tending to crack and open up fissures through which oxygen (and water vapor) can reach the underlying metal. Many of the transition elements show the kind of self-protective action found in chromium.

Most of the chromium we see is only a thin coating on iron or other metals. Such a coating called chrome plate, is put on in an electrolysis cell in which the object to be plated is the cathode of the cell. The essential ingredients of the plating bath are CrO_3, chromium(VI) oxide, and either H_2SO_4 or $Cr_2(SO_4)_3$, chromium(III) sulfate, but there are various additives, including such unlikely substances as glue or milk, which are supposed to give better coatings. Pure bulk chromium metal is fairly difficult to make. It can be done by using a reaction called the *Goldschmidt reaction* in which aluminum metal is used as a reducing agent. So much heat is released when Al_2O_3 is formed from the elements that stable oxides, as for example Cr_2O_3, can be reduced by Al. A mixture of aluminum powder and Cr_2O_3, when ignited, gives a vigorous reaction to produce Al_2O_3 and chromium.

EXERCISE 22-8

Write the equation for the reduction of Cr_2O_3 by Al. If it takes 399 kcal/mole to decompose Al_2O_3 into the elements and 270 kcal/mole to decompose Cr_2O_3, what will be the net heat liberated in the reaction you have just written?

Metallic chromium is an ingredient of several important alloys. Some forms of stainless steel, for example, contain about 12% Cr. Nichrome, which is commonly used for heating coils, has about 15% Cr in addition to 60% Ni and 25% iron. Both these alloys are quite resistant to chemical oxidation.

In compounds, the important oxidation numbers of Cr are +2, +3, and +6. In all of these states the chromium ions are colored and, in fact, the element got its name from this property (*chroma* is the Greek word for color). The +2 state is not frequently encountered but it can be made quite easily as the beautiful blue chromous ion in solution by dripping a solution containing Cr^{+3} over metallic zinc. Air has to be excluded since O_2 rapidly converts Cr^{+2} back into Cr^{+3}.

The +3 state of chromium is best represented by chromium(III) oxide, Cr_2O_3, which is a green, inert solid used as a green pigment. It can be made in rather spectacular fashion by heating ammonium dichromate. Once started, the reaction

$$(NH_4)_2Cr_2O_7(s) \rightleftharpoons N_2(g) + 4H_2O(g) + Cr_2O_3(s) \quad (9)$$

keeps itself going. The nitrogen and water are formed as hot gases which blow the light, fluffy Cr_2O_3 about. Another way of getting Cr_2O_3 is by dehydrating "chromium trihydroxide" with heat:

$$2Cr(OH)_3(s) \rightleftharpoons Cr_2O_3(s) + 3H_2O(g) \quad (10)$$

There is much argument about how to write an appropriate formula for "chromium trihydroxide." A gelatinous, green precipitate does form when base is added to a solution containing chromic ion (Cr^{+3}), but it includes a great deal of excess H_2O, so pure $Cr(OH)_3$ is never obtained. This is a common problem with the transition elements. Their hydroxides are not well-characterized, mainly because of the considerable difficulty of distinguishing between a water molecule of hydration and an OH group. For example, many chemists maintain that the formula is really $Cr_2O_3 \cdot nH_2O$, and that n usually is found to equal 3. This agrees with the empirical formula $Cr(OH)_3$.

EXERCISE 22-9

Calculate the percent chromium in $Cr(OH)_3$ and in $Cr_2O_3 \cdot 3H_2O$.

Whatever that green precipitate has for its chemical formula, it is observed to be amphoteric. It dissolves both in excess acid and in excess base, as explained earlier.

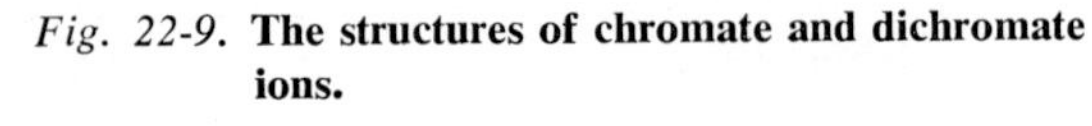

Fig. 22-9. **The structures of chromate and dichromate ions.**

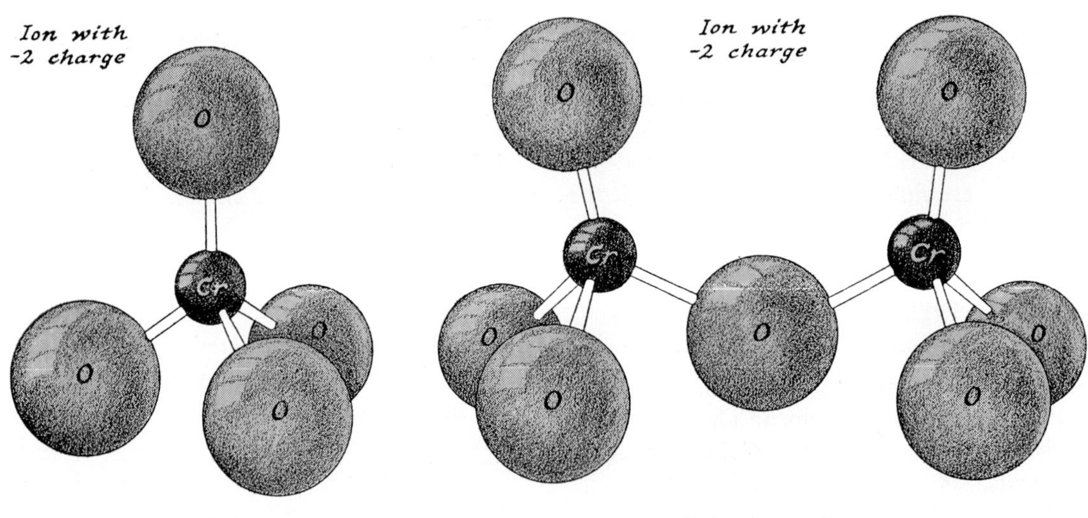

Probably the most common compound of +3 chromium is potassium chrome alum, $KCr(SO_4)_2 \cdot 12H_2O$. We know that the twelve water molecules are distributed equally, six around Cr^{+3} and six around K^+. Potassium chrome alum is just one example of the general class of solids called alums which have a +1 ion, a +3 ion, two sulfates, and twelve molecules of water. In the dyeing industry chrome alum is used for fixing dyes to fabrics.

The +6 state of chromium is represented by the chromates and dichromates. The chromate ion is a tetrahedral ion with Cr at the center; dichromate ion may be visualized as two such tetrahedra having one oxygen corner in common. Figure 22-9 shows the arrangements. Chromates can easily be converted to dichromates by addition of acid,

$$2CrO_4^{-2}(aq) + 2H^+(aq) \rightleftarrows Cr_2O_7^{-2}(aq) + H_2O \quad (11)$$

The change can be followed by noting the color change from yellow (characteristic of chromate) to orange (characteristic of dichromate). The reverse conversion from dichromate to chromate occurs on addition of base.

Both chromates and dichromates are strong oxidizing agents. One example of their use is in "cleaning solution," a mixture of $K_2Cr_2O_7$ and concentrated sulfuric acid. Laboratory glassware can be thoroughly cleaned of grease films by immersion in cleaning solution.

22-3.5 Manganese

The major use of manganese is in the production of steel, during which manganese reacts with oxygen and keeps the gas from forming bubbles when the iron solidifies. This prevents the formation of structure-weakening holes in finished steel. Manganese usually occurs in nature as a mixture of oxides along with oxides of iron. The ore, without separation, can be reduced with carbon in a high temperature furnace to give alloys of iron and manganese. This so-called "ferromanganese" is added to the crude molten iron on its way to becoming steel.

Probably the most commonly encountered compound of Mn is manganese dioxide, MnO_2, used in the ordinary flashlight cell. What goes on in a dry cell when it generates electricity is very much in dispute. The reactions are complicated and apparently change character depending upon the amount of electric current bcing drawn from the cell. When small currents are involved, which is what dry cells are designed for, the reactions are believed to be as follows:

AT THE ANODE (*the terminal indicated as the negative pole on commercial cells*) the zinc container is oxidized from the metallic state to the +2 state, probably as some complex zinc ion, but written for simplicity as Zn^{+2}

$$Zn(s) \longrightarrow Zn^{+2} + 2e^- \quad (12)$$

AT THE CATHODE the MnO_2 picks up, in a complicated reaction, an electron which reduces the manganese from the +4 to +3 state in the presence of NH_4Cl

$$2MnO_2 + 2NH_4^+ + 2e^- \longrightarrow Mn_2O_3 + 2NH_3 + H_2O \quad (13)$$

The role of the other components in the cell is not completely understood. Some of these components (such as NH_4Cl, ammonium chloride, and $ZnCl_2$, zinc chloride, in the center paste) are involved in other reactions that come into play at larger current drain.

One other important compound of manganese is potassium permanganate, $KMnO_4$. This is an intensely violet-colored material much used as an oxidizing agent in the laboratory. (It is too expensive to use on a large scale; in industry, chlorine is more likely to be used.)

22-3.6 Iron

Iron is the workhorse of the metals. It is quite abundant (ranking fourth of all elements and second of the metals, by weight) and easy to make inexpensively on a large scale; and it has useful mechanical properties, especially when alloyed with other elements. Steel, one of our most useful construction materials, is essentially iron containing a small percentage of carbon and often small amount of other elements.

NATURAL OCCURRENCE OF IRON

Most of the accessible iron is combined either with oxygen or sulfur. The oxygen compounds are the common minerals hematite, Fe_2O_3, and magnetite, Fe_3O_4, both of which are useful raw materials for producing iron. Another mineral, FeS_2, is called iron pyrites or "fool's gold," but this is not a common source since removing all the sulfur is difficult. (Sulfur impurity in steel makes it brittle. The sulfur compounds of iron are low melting and on cooling they stay liquid longer than does the mass itself and keep the iron from compacting efficiently.) Iron exists in the elemental form in some meteorites. Since meteorites are believed to come from the break-up of a planet, the existence of iron meteorites is taken as support for the theory that the core of the earth is largely iron.

MANUFACTURE OF IRON

The production of iron is an excellent example of chemical reduction on a massive scale. The process is carried out in a huge vertical reactor, called a blast furnace, which may be several stories tall. Raw materials—iron ore, limestone, and coke—are fed into the top of the furnace and oxygen is blown in at the bottom. The purpose of the iron ore (let us assume Fe_2O_3) is to provide the iron. The limestone reacts with sand, SiO_2, in the iron ore removing it as molten calcium silicate, called slag. The coke supplies the reducing agent, carbon, and as it reacts the heat released maintains the required high temperature.

Here is a simplified version of what goes on in a blast furnace. As the mixture of ore, limestone, and coke falls through the furance, it meets the updraft of oxygen. Carbon monoxide is formed,

$$2C(s) + O_2(g) \longrightarrow 2CO(g) + 52.8 \text{ kcal} \quad (14)$$

and as this carbon monoxide rises through the furnace it progressively reduces the Fe_2O_3—first to Fe_3O_4, then to FeO, and eventually to Fe. The successive reactions take place progressively as the solid descends:

$$CO(g) + 3Fe_2O_3(s) \longrightarrow 2Fe_3O_4(s) + CO_2(g) \quad (15)$$

$$CO(g) + Fe_3O_4(s) \longrightarrow 3FeO(s) + CO_2(g) \quad (16)$$

$$CO(g) + FeO(s) \longrightarrow Fe(l) + CO_2(g) \quad (17)$$

Since the reactions (*15*), (*16*), and (*17*) require successively higher temperatures, the blast furnace temperature is kept highest near the bottom of the furnace. Near the bottom, the temperature is sufficiently high that the impure iron—saturated with carbon—collects there as a molten liquid. The slag, which is mainly calcium silicate, $CaSiO_3$, removes any sand in the ore through reaction with limestone, $CaCO_3$.

$$CaCO_3(s) \longrightarrow CaO(s) + CO_2(g) \quad (18)$$

$$CaO(s) + SiO_2(s) \longrightarrow CaSiO_3(l) \quad (19)$$

Molten $CaSiO_3$ is less dense than molten iron and floats on top of it. An average furnace that produces about 750 tons of iron per day, will also yield 410 tons of slag. The slag is sometimes useful in the manufacture of cement and, when it contains sufficient phosphorus, in the manufacture of fertilizer.

When this impure iron cools, the resulting solid is called pig iron or cast iron. It is quite brittle and is not useful if high strength is needed. The impure iron is made into steel by burning out most of the carbon, sulfur, and phosphorus. Today there are three common furnace types for making steel—the open-hearth furnace (85% of U.S. production), the electric arc furnace (10%), and the Bessemer converter (5%). These furnaces differ in construction but the chemistry is basically similar.

The process of burning out the impurities is slowest in the open-hearth furnace. This implies there is plenty of time to analyze the melt and add whatever is needed to obtain the desired chemical composition. Manganese, vanadium, and chromium are frequent additives. The properties of the finished steel depend upon the amount of carbon left in and upon the identity and the quantity of other added elements. *Soft steel*, for example, contains 0.08–0.18 weight percent carbon; *structural steel*, 0.15–0.25%; *hard steel* or *tool steel*, 1–1.2%.

The electric arc furnace is used for special purpose steels. Because the environment can be

20-4 OCCURRENCE AND PREPARATION OF THE THIRD-ROW ELEMENTS

Except for argon, the third-row elements make up an important fraction (about 30%) of the earth's crust. Silicon and aluminum are the second and third most abundant elements (oxygen is the most abundant). Both the occurrence and the mode of preparation of each element can be understood in terms of trends in chemistry discussed earlier in this chapter.

20-4.1 Occurrence in Nature

Sodium (fifth most abundant element) is found principally as Na^+ ion in water soluble salt deposits, such as NaCl, and in salt waters. The element reacts rapidly with water and with atmospheric oxygen, hence is not found in an uncombined state in nature.

Magnesium (eighth most abundant element) is found principally as Mg^{+2} ion in salt deposits, particularly as the slightly soluble carbonate, $MgCO_3$, and also in sea water. The element is oxidized by atmospheric oxygen and is not found in an uncombined state in nature.

Aluminum (third most abundant element) is found as the Al^{+3} ion in oxides and as the complex ion AlF_6^{-3}. Important minerals are bauxite, which is best described as a hydrated aluminum oxide, $Al_2O_3 \cdot xH_2O$, and cryolite, Na_3AlF_6. The element is readily oxidized and is not found in an uncombined state in nature.

Silicon is the second most abundant element in the earth's crust. It occurs in sand as the dioxide SiO_2 and as complex silicate derivatives arising from combinations of the acidic oxide SiO_2 with various basic oxides such as CaO, MgO, and K_2O. The clays, micas, and granite, which make up most soils and rocks, are silicates. All have low solubility in water and they are difficult to dissolve, even in strong acids. Silicon is not found in the elemental state in nature.

Phosphorus (eleventh most abundant element) occurs mostly as the phosphate anion, PO_4^{-3}, in such minerals as "phosphate rock," which is a complex mixture of $Ca_3(PO_4)_2$ and CaF_2. Most phosphates have low solubility in water. The element is not found uncombined in nature.

Sulfur (fourteenth most abundant element) occurs in minerals either in an oxidized state as sulfate anion, SO_4^{-2}, or in a reduced state as sulfide anion, S^{-2}. Gypsum, $CaSO_4 \cdot 2H_2O$, with low solubility in water, and Epsom salt, $MgSO_4 \cdot 7H_2O$ with high solubility in water, are two common sulfate minerals. Galena, PbS, iron pyrites, FeS_2, and zinc blende, ZnS, are important sulfide minerals. Sulfur occurs as the free element in large underground deposits.

Chlorine (sixteenth most abundant element) is found as Cl^- in water soluble salt deposits, such as NaCl, and in salt waters. The element, Cl_2, is not found in the atmosphere.

Argon is found only in the elemental state. Air contains about 1% argon.

20-4.2 Mode of Preparation of the Element

Sodium is prepared by electrolysis of molten NaCl (giving chlorine as a by-product) or of molten NaOH.

Magnesium is an important structural metal. It can be prepared through the sequence of steps: precipitation of Mg^{+2} from sea water by OH^- to form $Mg(OH)_2$; conversion of $Mg(OH)_2$ to $MgCl_2$; electrolysis of molten $MgCl_2$.

Aluminum, though the third most abundant element, was quite expensive until about 1886, when a practical commercial electrolysis process was developed by a young American chemist, C. M. Hall. Bauxite, $Al_2O_3 \cdot xH_2O$, is dissolved at about 1000°C in molten cryolite, Na_3AlF_6, and electrolyzed.

Silicon in the elemental state has important electronic applications as a semiconductor that were developed only during the last decade. The discovery of these uses was possible only after methods were developed for preparing silicon of extremely high purity. Reduction of SiO_2 with carbon in an electric furnace is one process for manufacture of silicon. Very pure silicon is made by decomposing $SiCl_4$. Still further purification of the element is based upon the "zone-melting" technique in which a rod of silicon is heated to melting in a thin zone. This molten zone is gradually moved along the length of the rod. The impurities dissolve in the liquid and move along with the zone, leaving metal of ultra-high purity.

Phosphorus is prepared by heating a mixture of $Ca_3(PO_4)_2$, sand, and carbon (coke). White phosphorus, P_4, distills out and can be cooled and collected under water.

Sulfur is pumped out of natural underground deposits in the molten state after it is melted with water heated under pressure to about 170°C.

Chlorine is prepared by the electrolysis of molten NaCl or of aqueous NaCl.

Argon is obtained through fractional distillation of liquefied air.

20-4.3 Some Properties of the Second- and Third-Row Elements

The first ionization energies of elements 1 to 19 are shown in Table 15-III. The energies to remove successive electrons from gaseous Na, Mg, and Al atoms are shown in Table 20-IV.

Trends in the properties of ΔH of vaporization and boiling point for the second- and third-row elements are compared in Table 20-V.

Table 20-IV

SUCCESSIVE IONIZATION ENERGIES OF SODIUM, MAGNESIUM, AND ALUMINUM (kcal/mole)

ELEMENT	E_1 (1st e^-)	E_2 (2nd e^-)	E_3 (3rd e^-)	E_4 (4th e^-)
Na	118	1091	1653	—
Mg	175	345	1838	2526
Al	138	434	656	2767

Table 20-V. **TRENDS IN PROPERTIES OF SECOND- AND THIRD-ROW ELEMENTS**

ELEMENT	ΔH_{vap} (kcal/mole)	B.P. (°C)	ELEMENT	ΔH_{vap} (kcal/mole)	B.P. (°C)
Li	32.2	1326	Na	23.1	889
Be	53.5	2970	Mg	31.5	1120
B	128.8	~3900	Al	67.9	2327
C	170	~4000	Si	(105)	2355
N	0.67	−196	P	3.0	280
O	0.81	−183	S	2.5	445
F	0.78	−188	Cl	4.9	−34.1
Ne	0.42	−246	Ar	1.6	−186

QUESTIONS AND PROBLEMS

1. Make a graph with an energy scale extending on the ordinate from zero to 3000 kcal/mole and with the abscissa marked at equal intervals with the labels Na, Mg, and Al. Now plot and connect with a solid line the first ionization energies, E_1, of these three elements (see Table 20-IV). Plot E_2 and connect with a dashed line, E_3 with a dotted line, and E_4 with a solid line. Draw a circle around each ionization energy that identifies a valence electron.

2. Plot the ionization energy of the first electron removed from the atoms of both the second- and third-row elements against their atomic number (abscissa). What regularity do you observe?

3. Silicon melts at 1410°C and phosphorus (white)

controlled, electrically heated crucibles avoid contamination problems caused by chemical fuels.

The Bessemer converter is the oldest of the three methods and the fastest (about 15 minutes per charge). However, the speed is a mixed blessing because there is not sufficient time to make analyses and fine adjustments in the amounts of the alloying elements.

RUSTING

One well-known property of iron is the way in which it tends to go back to oxide from which it was derived. In fact, one out of every four men in the steel industry is concerned essentially with replacing iron lost by rusting! This shows how important corrosion is. What is the chemical nature of rusting, and how can it be controlled? First, rusting is a special case of corrosion in which the metal being corroded is iron and the corroding agent is oxygen. The observed facts are that H_2O and O_2 are necessary; $H^+(aq)$ speeds up the reaction; some metals such as Zn hinder the corrosion, other metals such as Cu speed it up; and strains (as are produced when a nail is bent) usually accelerate the reaction.

How can these observations be interpreted? The most promising mechanism suggested is a many-step process in which the following sequence of events occurs: (1) the iron acts as an anode to give up two electrons and form Fe^{+2} (ferrous) ion; (2) the electrons are picked up by $H^+(aq)$ ions to form transient neutral H atoms; (3) the H atoms are immediately oxidized by O_2 to form H_2O; (4) the Fe^{+2} is oxidized by O_2 in the presence of H_2O to form rust. Rust, incidentally, is not a simple compound but seems to be an indefinite hydrate of Fe_2O_3, so it is frequently given the formula $Fe_2O_3 \cdot nH_2O$.

Acids, for example those in fruit juices, catalyze rust formation because they furnish $H^+(aq)$ to accept electrons from the iron, causing it to dissolve faster. Oxygen gas is necessary to oxidize Fe^{+2} to Fe_2O_3. The presence of water facilitates the migration of Fe^{+2} from the reaction site. The resulting reduction in Fe^{+2} concentration allows more to be formed. Support for these ideas comes from the frequent observation that when the O_2 supply is restricted (as under a rivet head) the iron is eaten away at one spot (shank of the rivet) but the rust deposits where the O_2 is plentiful (where the rivet head overlaps a plate). One can surmise that the rivet shank is dissolved by the half-reaction

$$Fe(s) \longrightarrow Fe^{+2} + 2e^- \qquad (20)$$

in some acidic solution, perhaps rain water containing CO_2. Then the Fe^{+2} could have been washed out to the surface where oxidation converted Fe^{+2} to $Fe_2O_3 \cdot nH_2O$. A similar explanation would hold for the observation that iron pipes buried in the soil near cinders usually rust rapidly. Cinders generally contain acid-forming oxides, which could help speed up the dissolving of iron.

PREVENTION OF CORROSION

The observed effect of metals on the rate of rusting also supports the above theory and suggests a way to stop the corrosion. When zinc is in close contact with iron, the iron does not corrode but the zinc tends to oxidize away. The belief here is that the zinc, with a more positive $E°$ than Fe, gives electrons up to the iron, effectively preventing Fe from dissolving. This kind of protection is called cathodic protection and has a variety of applications. For example, ship hulls, particularly of tankers, are so protected in sea water. Magnesium is used rather than zinc but the principle is the same. Easily replaceable blocks of magnesium are bolted to the steel hulls and the magnesium oxidizes instead of the hull. Zinc coated iron ("galvanized iron") furnishes a second example. The zinc fortunately does not oxidize very much because when it reacts with oxygen and water in the presence of CO_2 it forms a self-protective coat of basic zinc carbonate. Thus the zinc is self-protective and at the same time gives cathodic protection to the underlying iron.

Some metals such as copper or tin, when in contact with iron, actually speed up the rate of rusting. The reason for this is that on these metals, reaction of electrons with $H^+(aq)$ is more rapid than on iron itself. Thus the effect is to draw the electrons away from the iron, speeding up the rate at which Fe goes to Fe^{+2}. Tin itself is inert to the atmosphere so a piece of iron *completely* covered with tin is safe from rusting. However, once the protective tin coat is punctured the rusting of the iron will be faster than if the tin were not there at all. This accounts for the observation that "tin cans," which are tin-covered steel, rust very quickly once they start.

One of the easiest ways to prevent rusting of iron objects is to shut out the supply of O_2 and H_2O. This can be done by painting the object or by smearing it with grease. The only caution here is to do the job thoroughly, since an exclusion of O_2 and H_2O that is only partial can do as much harm as good. Witness the rivet that would have been saved if there were a good tight coating of paint to seal the lip against entrance of the dissolving solution.

22-3.7 Cobalt

This element does not appear in the headlines very often but it is of practical importance. Probably its greatest single use is in alloys, including stainless steels. Pure cobalt is almost as magnetic as iron and, when alloyed with aluminum, nickel, copper, and iron, the resulting Alnico alloy has a permanent magnetization far exceeding that of iron.

In most of its simple compounds, cobalt has a +2 oxidation number. This includes the well known cobaltous chloride. In dilute aqueous solution this salt is an almost invisible pink; on dehydration it changes to deep blue. The color change, which is ascribed to a replacement of some of the water molecules surrounding Co^{+2} by Cl^- ions (to form a complex ion), is exploited in "invisible inks," the writing of which appears on heat application. Another use is the simple-weather forecasters in which a swatch of blotting paper turns pink when the humidity rises, suggesting that rain is supposed to be coming.

In its complex compounds, of which there are many thousands, Co almost invariably has a +3 oxidation number. Apparently, Co^{+3} ion accompanied by six coordinating groups is particularly stable. Cobalt complexes are important in biochemistry. Some enzyme reactions go through a cobalt-complexing mechanism. Although only small traces are needed, cobalt is essential to the diet.

22-3.8 Nickel

The five-cent coin, ordinarily called the "nickel," is actually 25% nickel (the other 75% is copper).

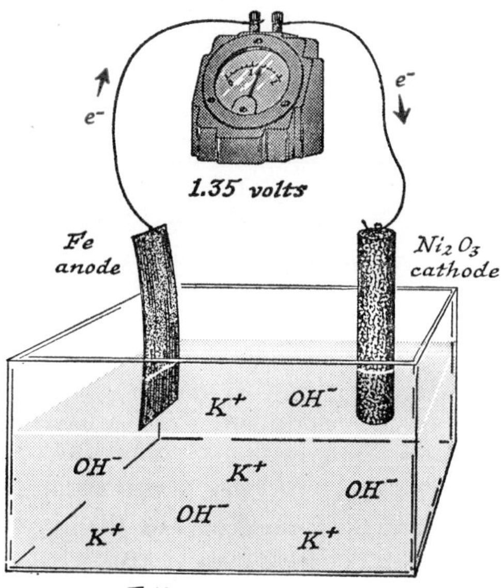

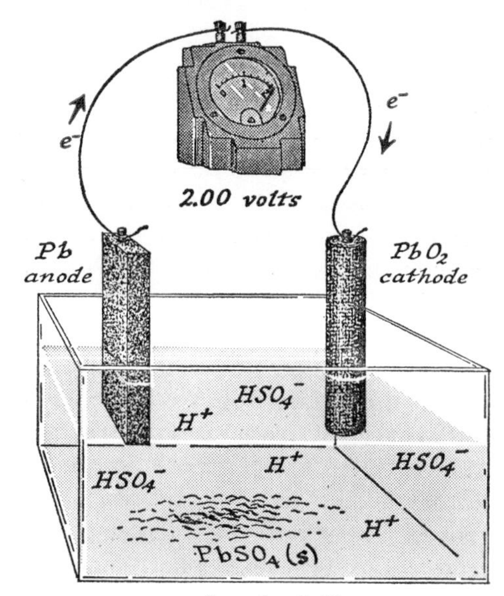

Fig. 22-10. **Cells from Edison and lead storage batteries** (*schematic*).

This familiar metallic object furnishes an important example of a nickel alloy. Other important nickel alloys include the *nickel steels*, which are tough and rust-resistant, *Monel metal* (60% Ni, 40% Cu), which is acid-proof, and *Nichrome* (60% Ni, 25% Fe, 15% Cr), mentioned in Section 22-3.4. Finely divided nickel is used as a catalyst for hydrogenation, the addition of hydrogen to double-bonded carbon atoms. This process is important in the foods industry to convert edible vegetable oils, such as cottonseed oil, into solid fats. The carbon chains of the vegetable oils have double bonds which have a high tendency to become oxidized and tend to develop an unpleasant flavor. When H_2 is added to the double bond, the carbon chain becomes saturated and the material becomes more attractive to the cook. Oleomargarine is an example of such a catalytically hydrogenated compound.

In most of its compounds nickel has a +2 oxidation number, but it is possible to get a higher state by heating $Ni(OH)_2$ with hypochlorite ion in basic solution. Hypochlorite ion, ClO^-, is one of the stronger oxidizing agents at our disposal in basic solution. There is considerable argument about the formula of the black solid that is formed, but we shall label it as Ni_2O_3 and write the equation

$$2Ni(OH)_2(s) + ClO^-(aq) \rightleftharpoons Ni_2O_3(s) + 2H_2O + Cl^-(aq) \quad (21)$$

Nickel(III) oxide is important as the oxidizing agent in the Edison storage cell, shown in Figure 22-10. Table 22-IV compares the Edison battery and the more common "lead storage battery." In both batteries the electrode products are solids and they adhere to the electrodes; the batteries can be regenerated by reversing the flow of electricity with some external device such as a direct current generator. "Charging a battery" simply means reversing the half-reaction at each electrode. It should be noted that on discharge, in the Edison cell there is no net consumption of the electrolyte, potassium hydroxide, so its concentration stays constant. In contrast, in the lead cell the sulfuric acid electrolyte is consumed during use of the cell and is regenerated during charging. This variability of H_2SO_4 concentration during use of a lead cell provides the basis for the convenient hydrometer test of the state of discharge of an automobile battery. The hydrometer measures the density of the electrolyte solution, thus indicating how much of the H_2SO_4 has been consumed. Obviously, this method cannot be used to check an Edison cell since the electrolyte concentration is constant.

Table 22-IV

COMPARISON OF EDISON AND LEAD STORAGE BATTERIES

	EDISON	LEAD
Oxidizing agent	Ni_2O_3	PbO_2
Reducing agent	Fe	Pb
Electrolyte	KOH	H_2SO_4
Voltage (one cell)	1.35 volts	2.0 volts
Features	light weight	heavy
	constant voltage	variable voltage during discharge
	expensive	inexpensive
	rugged	voltage and H_2SO_4 density indicate when recharge needed

EDISON CELL (DURING DISCHARGE)

Anode reaction

$$Fe(s) + 2OH^-(aq) \longrightarrow Fe(OH)_2(s) + 2e^- \quad (22a)$$

Cathode reaction

$$Ni_2O_3(s) + 3H_2O + 2e^- \longrightarrow 2Ni(OH)_2(s) + 2OH^-(aq) \quad (22b)$$

Net reaction

$$Fe(s) + Ni_2O_3(s) + 3H_2O \longrightarrow Fe(OH)_2(s) + 2Ni(OH)_2(s) \quad (22c)$$

LEAD CELL (DURING DISCHARGE)

Anode reaction

$$Pb(s) + HSO_4^-(aq) \longrightarrow PbSO_4(s) + H^+(aq) + 2e^- \quad (23a)$$

Cathode reaction

$$PbO_2(s) + HSO_4^-(aq) + 3H^+(aq) + 2e^- \longrightarrow PbSO_4(s) + 2H_2O \quad (23b)$$

Net reaction

$$Pb(s) + PbO_2(s) + 2H^+(aq) + 2HSO_4^- \longrightarrow 2PbSO_4(s) + 2H_2O \quad (23c)$$

Edison batteries cost more than lead storage batteries, but they have the advantage of being lighter, so the amount of electrical energy available per unit weight is greater. Also they are more rugged in standing up to mechanical shock. The difficulty of determining when recharging is needed and the expense are disadvantages.

22-3.9 Copper

This element occurs in nature in the uncombined state as *native copper* and in the combined state as various oxides, sulfides, and carbonates. The chief mineral is chalcopyrite, $CuFeS_2$, from which the element is extracted by roasting (heating in air) followed by reduction. The roasting reaction can be written

$$4CuFeS_2(s) + 9O_2(g) \longrightarrow 2Cu_2S(s) + 2Fe_2O_3(s) + 6SO_2(g) \quad (24)$$

It shows the formation of the important by-product SO_2, which may be converted to H_2SO_4. It also shows the formation of Fe_2O_3, which is subsequently removed by adding sand and heating in a furnace. The sand furnishes SiO_2 which combines with the iron oxide to form a low melting slag of iron silicate. After the slag is run off, the Cu_2S is heated in a current of air to give consecutively

$$2Cu_2S(s) + 3O_2(g) \longrightarrow 2Cu_2O(s) + 2SO_2(g) \quad (25)$$

$$Cu_2S(s) + 2Cu_2O(s) \longrightarrow 6Cu(s) + SO_2(g) \quad (26)$$

The copper obtained from this process is about 99% pure, yet this is not pure enough for most uses, especially those involving electrical conductivity. To refine the copper further, it is made the anode of an electrolytic cell containing copper sulfate solution. With careful control of the voltage to regulate the half-reactions that can occur, the copper is transferred from the anode (where it is about 99% Cu) to the cathode where it can be deposited as 99.999% Cu. At the anode there is oxidation of copper,

$$Cu(s) \longrightarrow Cu^{+2}(aq) + 2e^- \quad (27)$$

along with oxidation of any other metal (such as Fe) which is more readily oxidized than copper. The elements less readily oxidized than copper (such as silver and gold) simply crumble off into a heap under the anode. This so-called "anode sludge" can then be worked to recover and isolate these very valuable by-products. At the cathode there is reduction but, with well regulated voltage, only of copper.

$$Cu^{+2}(aq) + 2e^- \longrightarrow Cu(s) \quad (28)$$

Of course, as time goes on the copper sulfate solution in the cell has to be replaced because it collects undesirable ions such as Fe^{+2}, which have not been reduced because the voltage used is favorable to the reduction of Cu^{+2} only.

In compounds, copper usually has a +2 oxidation number, cupric, or copper(II), and occasionally +1, cuprous, or copper(I). The most common compound of cupric copper is $CuSO_4 \cdot 5H_2O$. The blue color of this solid is due to the Cu^{+2} ion hydrated by four of the five H_2O molecules in a surrounding square planar arrangement. This ion is present in aqueous solutions as well, but with two more distant H_2O molecules along the axis perpendicular to this square. When cupric solutions are treated with an excess of ammonia, they turn a deep blue. This is attributed to formation of a tetraammine copper(II) ion complex, usually written $Cu(NH_3)_4^{+2}$ but probably also containing two additional H_2O molecules coordinated to the copper. Cupric ion is toxic to lower organisms, so it is used to suppress the growth of algae in ponds and fungi and molds on vines. Bordeaux mixture used to spray grapes and potatoes is made of copper sulfate and lime.

The +1 state of copper is found only in complex compounds or slightly soluble compounds. The reason for this is that in aqueous solution cuprous ion is unstable with respect to disproportionation to copper metal and cupric ion. This comes about because cuprous going to cupric is a stronger reducing agent than copper going to cuprous. The following exercise in the use of $E°$ puts this on a more quantitative basis:

$$Cu(s) \longrightarrow Cu^{+}(aq) + e^- \qquad E° = -0.52 \text{ volt} \quad (29)$$

$$Cu^{+}(aq) \longrightarrow Cu^{+2}(aq) + e^- \qquad E° = -0.15 \text{ volt} \quad (30)$$

Since reaction (*30*) has a more positive $E°$ than reaction (*29*), it can force reaction (*29*) to reverse, thus in effect transferring an electron from one Cu^+ ion to another Cu^+ ion. The net reaction, obtained by subtracting reaction (*29*) from reaction (*30*) is

$$2Cu^+(aq) \rightleftharpoons Cu(s) + Cu^{+2}(aq) \qquad (31)$$

which has an $E°$ of -0.15 minus -0.52 volt, or $+0.37$ volt. Positive $E°$'s for net reactions mean the reaction should take place spontaneously from left to right.

22-3.10 Zinc

We have already encountered zinc as the irregular member at the end of the fourth transition row. We have also mentioned, in Section 17-2.3, its use as a constituent of the important class of alloys called brasses and its use in "galvanizing" iron to protect iron from rusting. Galvanized iron is made by dipping iron into molten zinc so as to give a thin adhering layer of Zn over the Fe. On prolonged exposure to air containing CO_2 the zinc forms a thin protective skin of basic zinc carbonate (zinc hydroxycarbonate). When a hole forms, penetrating into the iron, the iron does not rust as would be the case with tin-coated iron. On the contrary, the fresh Zn surface exposed reacts with CO_2, O_2, and H_2O of the air to form a plug of zinc hydroxycarbonate which seals the hole.

One other interesting and important compound of zinc is the sulfide, ZnS. It is the mineral zinc blende, one of the major sources of zinc and, also, it is the luminescent material on the face of many television picture tubes. Zinc sulfide is a semiconductor and, when a beam of electrons strikes the screen, electrons in the solid are energized so they can wander through the ZnS much like the electrons in a metal. When these electrons find an attractive site, usually in the vicinity of a purposely added impurity atom, they can be trapped and give off energy as visible light. This phenomenon, called fluorescence, makes possible the conversion of one frequency of light energy to another. The observed color of the fluorescence depends upon the mode of preparing the ZnS and on the nature of the impurity in the ZnS structure.

QUESTIONS AND PROBLEMS

1. Why are the elements with atomic numbers 21 to 30 placed in a group and considered together in this chapter?

2. Write the orbital representation for

   (a) chromium,
   (b) molybdenum,
   (c) tungsten.

3. What properties of the transition elements are consistent with their being classified as metals?

4. Ferrous ion, iron(II), forms a complex with six cyanide ions, CN^-; the octahedral complex is called ferrocyanide. Ferric ion, iron(III), forms a complex with six cyanide ions; the octahedral complex is called ferricyanide. Write the structural formulas for the ferrocyanide and the ferricyanide complex ions.

5. Draw the different structures for an octahedral cobalt complex containing four NH_3 and two NO_2 groups.

6. Draw the structures of the compounds

   $Cr(NH_3)_6(SCN)_3$
   $Cr(NH_3)_3(SCN)_3$

   (SCN^- is the thiocyanate ion). Consider the oxidation number of chromium to be $+3$ and the coordination number to be 6 in both compounds. Estimate

   (a) the solubility of these compounds in water;
   (b) their relative melting points;
   (c) the relative conductivity of the liquid phases.

7. Why does NH_3 readily form complexes, but NH_4^+ does not?

8. Place a piece of paper over Figure 15-13 and trace it. Extend the abscissa and add the ioniza-

tion energies of the transition elements. Complete the row with the following ionization energies: Ga, 138; Ge, 187; As, 242; Se, 225; Br, 273; Kr, 322; Rb, 96 kcal/mole.

9. The volume per mole of atoms of some fourth-row elements (in the solid state) are as follows: K, 45.3; Ca, 25.9; Sc, 18.0; Br, 23.5; and Kr, 32.2 ml/mole of atoms. Calculate the atomic volumes (volume per mole of atoms) for each of the fourth-row transition metals. Plot these atomic volumes and those of the elements given above against atomic numbers.

10. Chromic oxide, Cr_2O_3, is used as a green pigment and is often made by the reaction between $Na_2Cr_2O_7(s)$ and $NH_4Cl(s)$ to give $Cr_2O_3(s)$, $NaCl(s)$, $N_2(g)$, and $H_2O(g)$. Write a balanced equation and calculate how much pigment can be made from 1.0×10^2 kg of sodium dichromate.

11. Chromic hydroxide, $Cr(OH)_3$, is a compound with low solubility in water. It is usually hydrated and does not have the definite composition represented by the formula. It is quite soluble either in strong acid or strong base.

    (a) Write an equation showing the ions produced by the small amount of $Cr(OH)_3$ that dissolves.
    (b) Explain, using Le Chatelier's Principle, why $Cr(OH)_3$ is more soluble in strong acid than in water.
    (c) What is the significance of the fact that $Cr(OH)_3$ dissolves in base, as well as in acid?

12. What is the oxidation number of manganese in each of the following: $MnO_4^-(aq)$; $Mn^{+2}(aq)$; $Mn_3O_4(s)$; $MnO_2(s)$; $Mn(OH)_2(s)$; $MnCl_2(s)$; $MnF_3(s)$?

13. Manganese(III), $Mn^{+3}(aq)$, spontaneously disproportionates to $Mn^{+2}(aq)$ and $MnO_2(s)$. Balance the equation for the reaction.

14. Use the $E°$ values in Table 22-III to predict what might happen if a piece of iron is placed in a 1 M solution of Mn^{+2} and if a piece of manganese is placed in a 1 M solution of Fe^{+2}. Balance the equation for any reaction that you feel would occur to an appreciable extent.

15. Iron exists in one cubic crystalline form at 20°C (body centered cubic, with cube edge length 2.86 Å) and in another form at 1100°C (face centered cubic, with cube edge length 3.63 Å).

    (a) Draw a picture of each unit cell, showing the nine atoms involved in a body centered cubic cell and the fourteen atoms involved in a face centered cubic cell. (See Figure 21-2.)
    (b) Decide the number of unit cells with which each atom is involved (in each structure).
    (c) How many atoms are in each unit cell if we take into account that some atoms are shared by two or more adjoining unit cells?
    (d) Calculate the volume of the unit cell and, with your answer to part (c), the volume per atom (for each structure).
    (e) What conclusion can be drawn about the "effective size" of an iron atom?

16. One of the important cobalt ores is

$$Co_3(AsO_4)_2 \cdot 8H_2O$$

How much of this ore is needed to make 1.0 kg of Co?

17. Nickel carbonyl, $Ni(CO)_4$, boils at 43°C, and uses the sp^3 orbitals of Ni for bonding. Give reasons to justify the following:

    (a) it forms a molecular solid;
    (b) the molecule is tetrahedral;
    (c) bonding to other molecules is of the van der Waals type;
    (d) the liquid is a nonconductor of electricity;
    (e) it is not soluble in water.

18. Write balanced equations to show the dissolving of $Cu(OH)_2(s)$ on the addition of $NH_3(aq)$, and also the reprecipitation caused by the addition of an acid.

19. Cupric sulfide, copper(II) sulfide, reacts with hot nitric acid to produce nitric oxide gas, NO, and elemental sulfur. Only the oxidation numbers of S and N change. Write the balanced equation for the reaction.

20. The solubility of copper(II) iodide, CuI_2, is 0.004 g/liter. Determine the value of the solubility product.

CHAPTER

23

Some Sixth- and Seventh-Row Elements

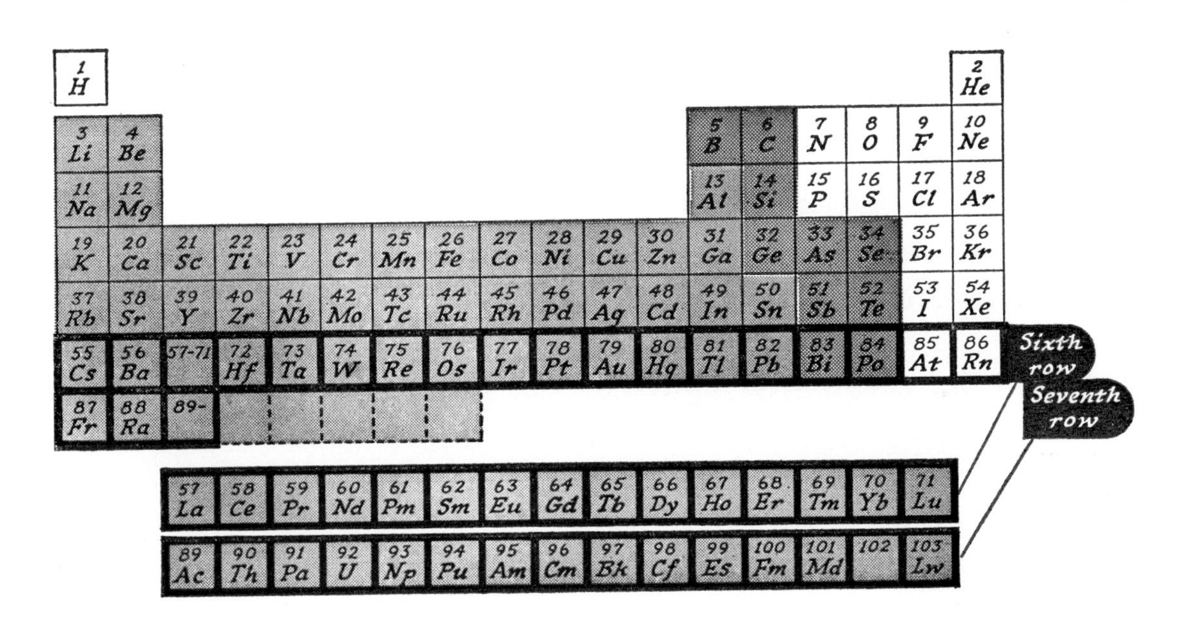

23-1 THE SIXTH ROW OF THE PERIODIC TABLE

The fifth-row transition elements have general similarity to the fourth-row transition elements. The electron structure is essentially the same except that the 4*d* orbitals are filling instead of the 3*d* orbitals. Near the beginning of the sixth-row transition elements there is a change: the *f* orbitals begin to fill to form fourteen elements before the *d* orbitals can be occupied to give the typical transition elements. This chapter will discuss these fourteen elements and some of the seventh-row elements.

23-1.1 The Lanthanides, or Rare Earths

The **lanthanides,** or **rare earths,** *are lanthanum and the fourteen chemical elements* following. These fourteen elements, all of which are very similar to La in chemical behavior, include cerium, praseodymium, neodymium, promethium, samarium, europium, gadolinium, terbium, dysprosium, holmium, erbium, thulium, ytterbium, and lutetium (atomic numbers 58 to 71). These elements are called the rare earth elements because they were extracted from oxides, for which the ancient name was "earth," and because those oxides were rather rare. During the 1940's, techniques for separating elements were developed to such a degree that the rare earth elements are no longer so rare. *The most striking property of these elements is that their chemical properties are almost identical.* For example, they are all reactive metals (about like calcium). They react with water to give a vigorous evolution of hydrogen. They all form basic trihydroxides which are only slightly soluble in water but readily soluble in acid.

For example,

$$2La(s) + 6H_2O \longrightarrow 2La(OH)_3(s) + 3H_2(g) \quad (1)$$

$$La(OH)_3(s) + 3H^+(aq) \rightleftharpoons La^{+3}(aq) + 3H_2O \quad (2)$$

And, in similar reactions,

$$2Ce(s) + 6H_2O \longrightarrow 2Ce(OH)_3(s) + 3H_2(g) \quad (3)$$

$$Ce(OH)_3(s) + 3H^+(aq) \rightleftharpoons Ce^{+3}(aq) + 3H_2O \quad (4)$$

$$2Pr(s) + 6H_2O \longrightarrow 2Pr(OH)_3(s) + 3H_2(g) \quad (5)$$

$$Pr(OH)_3(s) + 3H^+(aq) \rightleftharpoons Pr^{+3}(aq) + 3H_2O \quad (6)$$

The reason usually cited for the great similarity in the properties of the lanthanides is that they have similar electronic configurations in the outermost $6s$ and $5d$ orbitals. This occurs because, at this point in the periodic table, the added electrons begin to enter $4f$ orbitals which are fairly deep inside the atom. These orbitals are screened quite well from the outside by outer electrons, so changing the number of $4f$ electrons has almost no effect on the chemical properties of the atom. The added electrons do not become valence electrons in a chemical sense—neither are they readily shared nor are they readily removed.

The very slight differences that do exist among these elements are due to small changes in size brought about by increase of nuclear charge. The separation of the lanthanide elements from each other is based upon clever exploitation of these slight differences in properties. Table 23-I shows a comparison of some of the properties of the various lanthanide elements. As can be seen, +3 is the common oxidation number and is most characteristic of the chemistry of these elements. Another thing to note is the steady decrease in

Table 23-I. **SOME PROPERTIES OF LANTHANUM AND THE LANTHANIDE ELEMENTS**

ELEMENT	Z	OUTER ELECTRON CONFIGURATION	OXIDATION STATES	$E°$ $M \longrightarrow M^{+3} + 3e^-$	+3 ION RADIUS
La	57	$5d^1\ 6s^2$	+3	2.52 volts	1.15 Å
Ce	58	$4f^1\ 5d^1\ 6s^2$	+3, +4	2.48	1.11
Pr	59	$4f^3\ 6s^2$	+3, +4	2.47	1.09
Nd	60	$4f^4\ 6s^2$	+3	2.44	1.08
Pm	61	$4f^5\ 6s^2$	+3	2.42	1.06
Sm	62	$4f^6\ 6s^2$	+2, +3	2.41	1.04
Eu	63	$4f^7\ 6s^2$	+2, +3	2.41	1.03
Gd	64	$4f^7\ 5d^1\ 6s^2$	+3	2.40	1.02
Tb	65	$4f^9\ 6s^2$	+3, +4	2.39	1.00
Dy	66	$4f^{10}\ 6s^2$	+3	2.35	0.99
Ho	67	$4f^{11}\ 6s^2$	+3	2.32	0.97
Er	68	$4f^{12}\ 6s^2$	+3	2.30	0.96
Tm	69	$4f^{13}\ 6s^2$	+3	2.28	0.95
Yb	70	$4f^{14}\ 6s^2$	+2, +3	2.27	0.94
Lu	71	$4f^{14}5d^1\ 6s^2$	+3	2.25	0.93

the M^{+3} ionic size, shown in the last column. This decrease is called the "lanthanide contraction" and is due to the fact that the nuclear charge increases through the series (and, consequently, so is the attraction for the electrons increased), while the added electrons are not entering outer orbitals where they would tend to increase atomic size.

EXERCISE 23-1

(a) Balance the equation for the reaction between neodymium metal and chlorine gas.
(b) At 0°C and one atmosphere pressure, how many liters of chlorine gas would react with 14.4 grams of the metal?

EXERCISE 23-2

To a solution containing "tracer" amounts of radioactive gadolinium, $Gd^{+3}(aq)$ (concentration less than 10^{-12} *M*) is added 0.01 *M* lanthanum chloride, $LaCl_3$, and 0.1 *M* hydrogen fluoride, HF. A precipitate of lanthanum fluoride forms and it contains most of the radioactive gadolinium. Explain in terms of the similarity of lanthanum and gadolinium ions.

23-1.2 Occurrence and Preparation

The most important minerals of the lanthanide elements are monazite (phosphates of La, Ce, Pr, Nd and Sm, as well as thorium oxide) plus cerite and gadolinite (silicates of these elements). Separation is difficult because of the chemical similarity of the lanthanides. Fractional crystallization, complex formation, and selective adsorption and elution using an ion exchange resin (chromatography) are the most successful methods.

23-2 THE SEVENTH-ROW OF THE PERIODIC TABLE

The last row of the periodic table is unique in that all of these elements have radioactive nuclei. However, this property has no direct effect on the chemistry of these elements. The nuclear charge and nuclear mass affect the chemistry of an atom but the fact that the nucleus might explode at any moment (that is, undergo some sort of radioactive decay) isn't known by the electrons surrounding the nucleus. We might compare the situation to the packing of eggs in egg crates. How many eggs can be placed in an egg crate is fixed entirely by their shape and size. It is immaterial that eggs have the additional property that they can hatch if fertilized. The packing of fertile eggs is also fixed entirely by their shape and size, so a crate of fertile eggs is indistinguishable from a crate of infertile eggs (unless one of the fertile eggs happens to hatch while you are looking at them). In exactly the same way, the chemistry of an atom has nothing to do with the stability or instability of its nucleus.

Nevertheless, there is special interest in the chemistry of the seventh-row elements, a special interest directly tied to the nuclear instability. As we have mentioned in Chapter 7, the energy contents of nuclei are many orders of magnitude higher than chemical heat contents. If this nuclear energy content can be tapped and put to work, enormous quantities of energy become available. Thus far, the radioactive nuclei of some of these seventh-row elements have been most useful in our attempts to harness nuclear energy.

Nuclear fuels, like chemical fuels, must be purified to be most effective. The purification of the seventh-row elements has presented some fascinating and difficult problems of chemistry—so difficult, in fact, that chemists have played as big a role in the development of nuclear energy as have physicists.

23-2.1 The Occurrence of the Seventh-Row Elements

Only five of the seventh-row elements are found in nature: radium, actinium, thorium, protactinium, and uranium. It is one of the significant advances of science in the first half of this century that ten additional elements of this row have been synthesized: francium, neptunium, plutonium, americium, curium, berkelium, californium, einsteinium, fermium, and mendelevium. The methods for raising the nuclear charge of a nucleus require ingenious applications of physics. However, the synthesis of these ten elements could never have been demonstrated if it had not been for the solution of some difficult problems of chemistry, worked on by some of the most highly skilled chemists in the world. At this time, elements 102 and 103 have been prepared and the preparation of elements beyond is under study.

23-2.2 The Elements Following Actinium

The most interesting elements of the seventh row are those following actinium. For some of these elements a large amount of chemistry is known. The first four, actinium, thorium, protactinium, and uranium, used to be shown in the periodic table under lanthanum, hafnium, tantalum, and tungsten, since they resemble these elements in many chemical reactions. With the discovery that one could make elements beyond uranium (those of higher Z than 92), it was suggested that there might be a $5f$ series of elements analogous to the $4f$ series of rare earths (the lanthanides). As a result, it has become customary to place the elements following actinium under the rare earths (see inside front cover) and to call these elements the "actinides." Table 23-II collects information concerning the known oxidation states of the elements following lanthanum, those following lutetium (that is, beginning with hafnium), and those following actinium.

The most striking feature of the contrasts shown in Table 23-II is that the seventh-row elements display the multiplicity of oxidation states characteristic of transition elements rather than the drab chemistry of the +3 rare earth ions. Whereas $Ce^{+3}(aq)$ can be oxidized to $Ce^{+4}(aq)$ only with an extremely strong oxidizing agent, $Th^{+4}(aq)$ is the stable ion found in thorium salts and $Th^{+3}(aq)$ is unknown. In a similar

Table 23-II. **OXIDATION NUMBERS FOUND FOR SOME SIXTH-ROW AND SEVENTH-ROW ELEMENTS**

RARE EARTH ELEMENTS FOLLOWING LANTHANUM		SIXTH-ROW ELEMENTS FOLLOWING LUTETIUM		SEVENTH-ROW ELEMENTS FOLLOWING ACTINIUM	
Name	*Oxidation Numbers**	*Name*	*Oxidation Numbers*	*Name*	*Oxidation Numbers*
lanthanum	+3	lutetium	+3	actinium	+3
cerium	+3, +4	hafnium	+4	thorium	+4
praseodymium	+3, +4	tantalum	+5	protactinium	+4, +5
neodymium	+3	tungsten	+2, +3, +4, +5, +6	uranium	+3, +4, +5, +6
promethium	+3	rhenium	+3, +4, +5, +6, +7	neptunium	+3, +4, +5, +6
samarium	+2, +3	osmium	+2, +3, +4, +6 +8	plutonium	+3, +4, +5, +6
europium	+2, +3	iridium	+3, +4, +6	americium	+3, +4, +5, +6
gadolinium	+3	platinum	+2, +4	curium	+3
terbium	+3, +4	gold	+1, +3	berkelium	+3, +4
dysprosium	+3	mercury	+1, +2	californium	+3
holmium	+3	thallium	+1, +3	einsteinium	+3
erbium	+3	lead	+2, +4	fermium	+3
thulium	+3	bismuth	+3, +5	mendelevium	+3
ytterbium	+2, +3	polonium	+2, +4, +6	102	
lutetium	+3	astatine	(−1)	103	

* The most common oxidation numbers are in blue type.

contrast, the most important oxidation state of protactinium is +5 whereas no compound involving the +5 oxidation state of praseodymium is known.

Yet it is generally accepted today that the elements following actinium involve $5f$ energy levels, identifying them as the seventh-row equivalents of the lanthanides.

Spectroscopic investigations of the gaseous atoms and of ions in suitable crystals furnish one of the bulwarks of this view. Table 23-III contrasts the actinide electronic configurations as they are known today with the configurations of the corresponding rare earths. Although the corresponding configurations are not identical they are quite similar. They offer no hint of explanation for the varied actinide oxidation states and the monotonous similarity of the lanthanides. It may be that one of the most significant conclusions to be drawn from the investigations of the actinides is that elements of an f orbital transition series do not necessarily resemble each other strongly in chemical behavior, contrary to what had been inferred from studies of the rare earths.

Table 23-III

THE ELECTRON CONFIGURATIONS OF GASEOUS ACTINIDE AND LANTHANIDE ATOMS

LANTHANIDES				ACTINIDES			
La		$5d^1$	$6s^2$	Ac		$6d^1$	$7s^2$
Ce	$4f^1$	$5d^1$	$6s^2$	Th		$6d^2$	$7s^2$
Pr	$4f^3$		$6s^2$	Pa	$5f^2$	$6d^1$	$7s^2$
Nd	$4f^4$		$6s^2$	U	$5f^3$	$6d^1$	$7s^2$
Pm	$4f^5$		$6s^2$	Np	$5f^4$	$6d^1$	$7s^2$
Sm	$4f^6$		$6s^2$	Pu	$5f^6$		$7s^2$
Eu	$4f^7$		$6s^2$	Am	$5f^7$		$7s^2$
Gd	$4f^7$	$5d^1$	$6s^2$	Cm	$5f^7$	$6d^1$	$7s^2$

EXERCISE 23-3

Two half-reactions involving neptunium are

$$Np^{+3} \longrightarrow Np^{+4} + e^- \qquad E^\circ = -0.147 \text{ volt} \quad (7)$$

$$Np^{+4} + 2H_2O \longrightarrow NpO_2^+ + 4H^+ + e^- \qquad E^\circ = -0.75 \text{ volt} \quad (8)$$

Find an oxidizing agent in Appendix 3 that could oxidize $Np^{+3}(aq)$ to $Np^{+4}(aq)$ but not to $NpO_2^+(aq)$.

EXERCISE 23-4

Balance the equation for the reaction between permanganate, $MnO_4^-(aq)$, and plutonium(III), $Pu^{+3}(aq)$, to form manganous, $Mn^{+2}(aq)$, and plutonyl ion, $PuO_2^{+2}(aq)$, in acid solution.

EXERCISE 23-5

Plutonium(IV), $Pu^{+4}(aq)$, forms a complex ion with fluoride ion, PuF^{+3}:

$$PuF^{+3} \rightleftharpoons Pu^{+4} + F^- \qquad K = 1.6 \times 10^{-7} \quad (9)$$

If the F^- concentration is adjusted to 0.10 M in a solution containing 1.0×10^{-3} M Pu^{+4}, calculate the ratio of the concentration of Pu^{+4} to PuF^{+3}. What is the equilibrium concentration of Pu^{+4}?

Chemical investigation of the actinides is made difficult by the extreme instability of the nuclei—an instability that increases as we go to higher atomic numbers. This leads to intense radioactivity (which makes shielding precautions necessary) and small amounts of material to work with. The heavier actinides are usually made by bombarding the lighter actinides with neutrons, helium nuclei (alpha particles), or even carbon nuclei. With very unstable nuclei, it is hard to accumulate enough atoms to permit chemical studies on a macroscopic scale since the element is disintegrating at the same time it is being produced. The consequence is that much of the chemical investigation is done by tracer techniques wherein chemical reactions are performed on a bulk scale with some other, more available, element. Clues to the chemistry of the trace element are furnished by the behavior of the radioactivity—whether it follows along with the bulk carrier or is lost in the chemical operations.

There is reason to believe that not many more new elements will be created. The instability of the nuclei of the last few elements suggests that most or all of the still higher elements have nuclei so unstable that they will not survive long enough to be detected.

23-2.3 Nuclear Stability and Radioactivity

The outstanding characteristic of the actinide elements is that their nuclei decay at a measurable rate into simpler fragments. Let us examine the general problem of nuclear stability. In Chapter 6 we mentioned that nuclei are made up of protons and neutrons, and that each type of nucleus can be described by two numbers: its atomic number (the number of protons), and its mass number (the sum of the number of neutrons and protons). A certain type of nucleus is represented by the chemical symbol of the element, with the atomic number written at its lower left and the mass number written at its upper left. Thus the symbol

$$^{239}_{94}\text{Pu}$$

represents the nucleus of the plutonium isotope which contains 94 protons and $239 - 94 = 145$ neutrons. Since the forces that exist in the nucleus depend upon the number of protons and neutrons of which it is composed, we can get a rough idea of whether a nucleus is stable just by examining these two numbers.

First of all, let us define what we mean by stability. Consider an initially pure sample of $^{238}_{92}\text{U}$. Regardless of the physical or chemical state in which we find the uranium atoms, some of them will decay each instant to become thorium atoms by the spontaneous reaction:

$$^{238}_{92}\text{U} \longrightarrow {}^{4}_{2}\text{He} + {}^{234}_{90}\text{Th} \qquad (10)$$

Notice that both the electric charge and the total number of nuclear particles (nucleons) are conserved in the nuclear decomposition. Careful study of the rate of this nuclear decay shows that in a given period of time a constant fraction of the nuclei present will undergo decomposition. This observation allows us to characterize or describe the rate of nuclear decay in a very simple manner. We simply specify the length of time it takes for a fixed fraction of the nuclei initially present to decay. Normally we pick the time for half the nuclei to decay; this length of time is known as the **half-life** of the nucleus. For example, measurements show that after 4.5×10^9 years, half the atoms in any sample of $^{238}_{92}\text{U}$ will decay to $^{234}_{90}\text{Th}$. A nucleus is considered to be stable if its half-life is much longer than the age of the earth, which is about 5×10^9 years. Nuclei that are very unstable are characterized by half-lives which are quite short—in some cases only a small fraction of a second.

Now let us turn to the problem of how the composition of a nucleus affects its stability. The forces that exist between the particles in the nucleus are very large. The most familiar of the intranuclear forces is the coulomb force of repulsion which the protons must exert on one another. In order to appreciate the magnitude of this repulsive force, let us compare the force between two protons when they are separated by 10^{-8} cm, as they are in the hydrogen molecule, with the force between two protons separated by 10^{-13} cm, as they are in a helium nucleus. In the first case we have

$$\text{force} = \frac{(\text{charge on proton 1})(\text{charge on proton 2})}{(\text{distance between protons})^2}$$

$$= \frac{(+e)(+e)}{(10^{-8})^2} = \frac{e^2}{10^{-16}} = 10^{16}e^2 \qquad (11)$$

where e is the electric charge on one proton. In the second case we have

$$\text{force} = \frac{(+e)(+e)}{(10^{-13})^2} = \frac{e^2}{10^{-26}} = 10^{26}e^2 \qquad (12)$$

By comparing these two answers we can see that the repulsive force between two protons in the nucleus is about ten billion times as great as the repulsive force between two protons bound together in a hydrogen molecule. In order to overcome these enormous intranuclear coulomb repulsions and hold the nucleus together there must exist some very strong attractive forces between the nucleons. The nature of these forces is not understood and remains a very important problem in physics.

It is relatively easy to summarize how nuclear stability (and hence the attractive nuclear forces) depends upon the numbers of protons and neutrons in the nucleus. For atoms with atomic number less than 20, the most stable nuclei are those in which there are equal numbers of protons and neutrons. For atoms with atomic numbers between 20 and 83, the most stable nuclei have more neutrons than protons. For atoms of atomic number greater than 83, no nucleus can be considered stable by our definition. These

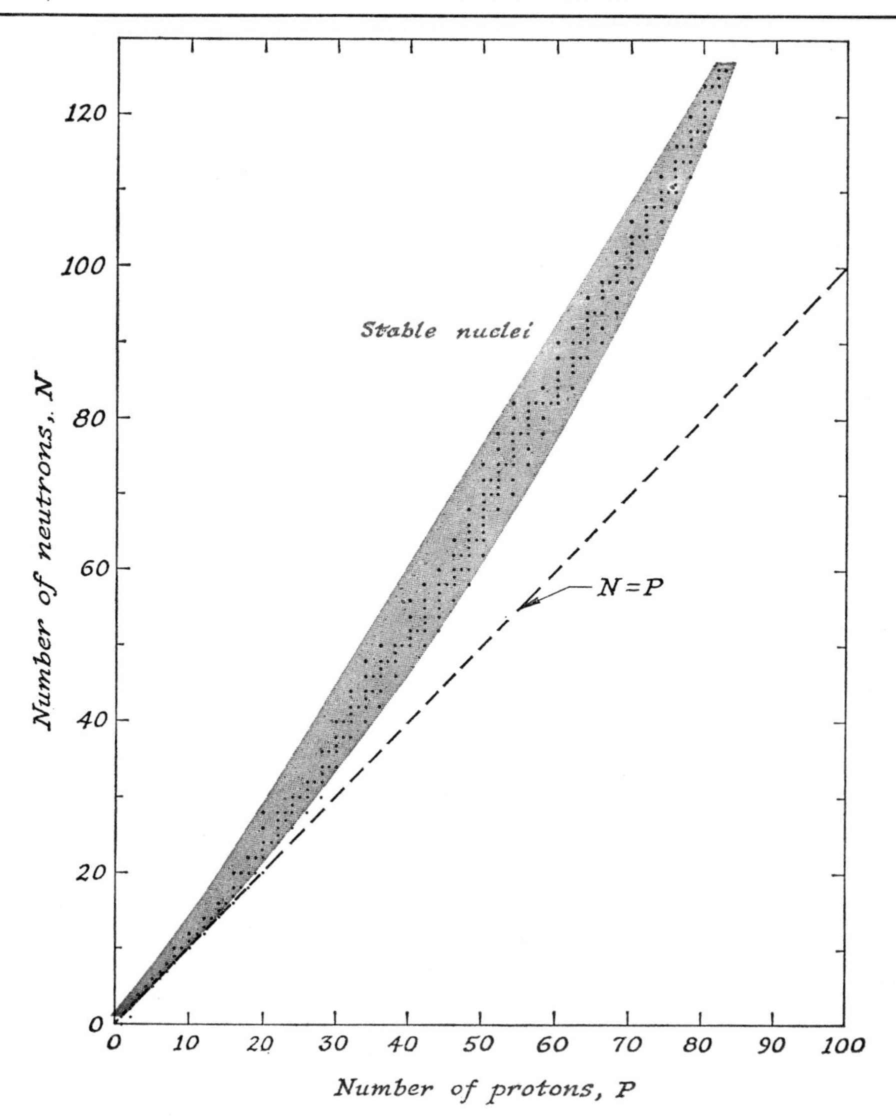

Fig. 23-1. **The relation between** *number of neutrons and protons in stable nuclei. Each dot identifies a stable nucleus.*

statements are demonstrated in Figure 23-1, where the number of neutrons is plotted against the number of protons for all stable nuclei. We see that these stable nuclei form a belt which deviates increasingly from a neutron to proton ratio of unity (the dashed line, $N = P$) as the charge on the nucleus increases. Nuclei whose neutron–proton ratio is such that they lie outside of this belt of stability are radioactive.

23-2.4 Types of Radioactivity

There are three common ways by which nuclei can approach the region of stability: (1) loss of alpha particles (α-decay); (2) loss of beta particles (β-decay); (3) capture of an orbital electron. We have already encountered the first type of radioactivity, α-decay, in equation (*10*). Emission of a helium nucleus, or alpha particle, is a common form of radioactivity among nuclei with charge greater than 82, since it provides a mechanism by which these nuclei can be converted to new nuclei of lower charge and mass which lie in the belt of stability. The actinides, in particular, are very likely to decay in this way.

For example, the most stable isotope of element 100, fermium, has a half-life of only 4.5 days:

$$^{253}_{100}\text{Fm} \longrightarrow {}^{249}_{98}\text{Cf} + {}^{4}_{2}\text{He} \qquad (13)$$

Nuclei that have a neutron–proton ratio which is so high that they lie outside the belt of stable nuclei often decay by emission of a negative electron (a beta particle) from the nucleus. This effectively changes a neutron to a proton within the nucleus. Two examples are

$$^{253}_{98}\text{Cf} \longrightarrow {}^{253}_{99}\text{Es} + {}_{-1}^{0}e \qquad (14)$$

$$^{24}_{11}\text{Na} \longrightarrow {}^{24}_{12}\text{Mg} + {}_{-1}^{0}e \qquad (15)$$

This type of radioactivity, β-decay, is found in both the light and heavy elements.

Nuclei that have too many protons relative to their number of neutrons correct this situation in either of two ways. They either capture one of their 1*s* electrons or they emit a positron (a positively charged particle with the same mass as an electron). Either process effectively changes a proton to a neutron within the nucleus.

23-2.5 Nuclear Energy

In the previous section we saw that the stability of a nucleus is affected by its neutron/proton ratio. Even among those nuclei that we consider stable, however, there is a variation in the forces which hold the nucleus together. In order to study this variation in nuclear binding energy, let us consider the process of building a nucleus from protons and neutrons. For an example, let us look at the hypothetical reaction

$$2\,{}^{1}_{1}\text{H} + 2\,{}^{1}_{0}n \longrightarrow {}^{4}_{2}\text{He} \qquad (16)$$

First we will compare the masses of the reactants with those of the product:

Mass of 2 protons	= 2 × 1.00759	= 2.01518 g/mole
Mass of 2 neutrons	= 2 × 1.00897	= 2.01794 g/mole
Total mass of reactants		= 4.03312 g/mole
Mass of $^{4}_{2}$He		= 4.00277 g/mole
Mass difference between products and reactants		0.03035 g/mole

Since there is a decrease of 0.03035 gram/mole of helium formed in this reaction, an equivalent amount of energy must be released (see Section 7-4.1). By the Einstein mass-energy relationship, $E = mc^2$, we can calculate that in the formation of 1 mole of helium nuclei there would be

$$E = mc^2 = \left(0.03\,\frac{\text{g}}{\text{mole}}\right) \times \left(3 \times 10^{10}\,\frac{\text{cm}}{\text{sec}}\right)^2$$

$$= 2.7 \times 10^{19}\,\frac{\text{g cm}^2}{\text{sec}^2\text{ mole}}$$

$$E = \left(2.7 \times 10^{19}\,\frac{\text{g cm}^2}{\text{sec}^2\text{ mole}}\right) \times \left(2.4 \times 10^{-8}\,\frac{\text{calories}}{\text{g cm}^2/\text{sec}^2}\right)$$

$$= 6.4 \times 10^{+11}\,\frac{\text{calories}}{\text{mole}} \qquad (17)$$

or 640 billion calories of energy released. This amount of energy can be considered as the binding energy of a mole of helium nuclei since this much energy must be supplied to dissociate a mole of $^{4}_{2}$He into two moles each of protons and neutrons.

Similar calculations can be made for other nuclei. A significant comparison between nuclear binding energies can be made if we divide the total binding energy of each nucleus by the number of nucleons in the nucleus. This calculation provides us with the binding energy per

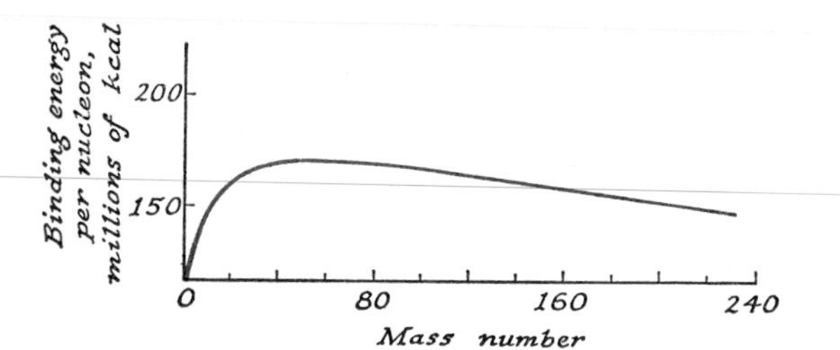

Fig. 23-2. **The binding energy** *per particle in the nucleus.*

particle in the nucleus. The binding energy per particle varies in a systematic way as the mass number of the nucleus increases. This variation is shown in Figure 23-2.

The nuclei that have a mass number of approximately 60 have the highest binding energy per nuclear particle, and are therefore the most stable nuclei. This graph helps us to understand the existence of the processes of nuclear fission and nuclear fusion. If the nuclei of the heavier elements such as uranium and plutonium are split into two smaller fragments, the binding energy per nucleon is greater in the lighter nuclei. As in every other reaction in which the products are more stable than the reactants, energy is evolved by this process of nuclear fission. Generally this fission reaction is induced by the bombardment of a particular isotope of uranium or plutonium with neutrons,

$$ {}^{1}_{0}n + {}^{235}_{92}\mathrm{U} \longrightarrow {}^{141}_{56}\mathrm{Ba} + {}^{92}_{36}\mathrm{Kr} + 3\,{}^{1}_{0}n \qquad (18) $$

The two nuclei on the right side are just two of the many possible products of the fission process. Since more than one neutron is released in each process, the fission reaction is a self-propagating, or chain reaction. Neutrons released by one fission event may induce other fissions. When fission reactions are run under controlled conditions in a nuclear reactor, the energy released by the fission process eventually appears as heat. The energy released by the fission of one pound of ${}^{235}_{92}\mathrm{U}$ is equivalent to that obtained from more than 1000 tons of coal.

Figure 23-2 shows that when very light nuclei such as ${}^{1}_{1}\mathrm{H}$ or ${}^{2}_{1}\mathrm{H}$ are brought together to form heavier elements, the binding energy per nucleon again increases and energy is released. The graph also shows that the energy released per nucleon (therefore per gram of reactant) is considerably greater in the fusion process than in the fission reaction. By use of a set of reactions in which four protons are converted into a helium nucleus and two electrons, one pound of hydrogen could produce energy equivalent to that obtained from 10,000 tons of coal. For this reason, and because of the great abundance of hydrogen, fusion reactions are potentially sources of enormous amounts of energy. Unfortunately fusion reactions take place rapidly only at temperatures that are greater than a million degrees. These temperatures have been attained briefly by use of nuclear fission explosions. At present, attempts are being made to attain the temperatures required for nuclear fusion by less destructive means, so that these reactions can be used as an energy source. The most promising processes are based upon the fusion of deuterium, ${}^{2}_{1}\mathrm{H}$, or tritium, ${}^{3}_{1}\mathrm{H}$, nuclei.

One of the major advances of science in the first half of this century was the synthesis of ten elements beyond uranium. Glenn T. Seaborg participated in the discovery of most of these, a sufficient tribute to his outstanding ability as a scientist. For the first such discoveries, those of neptunium and plutonium, he shared with Professor Edwin M. McMillan the Nobel Prize in Chemistry for 1951.

Seaborg rose from humble surroundings. He was born in a small mining town, Ishpeming, Michigan. During his boyhood, his Swedish-American parents moved to southern California. There, two important traits began to become apparent—his high intellect and his unlimited willingness to work. Attending the University of California at Los Angeles, he worked his way through, first as a stevedore, then picking apricots, then as a linotype apprentice. By his sophomore year Seaborg had distinguished himself enough to be appointed a laboratory assistant. This employment continued through his undergraduate career and it provided him his first opportunity to engage in research.

For graduate study, Seaborg moved to the Berkeley campus of the University of California. He received the Ph.D. in 1937 and, after two years of work with G. N. Lewis, began work on the attempted preparation of elements beyond uranium. This research, following earlier work of E. M. McMillan, culminated rather rapidly in the discovery of plutonium. During the war years, Seaborg had a key role in working out the chemical processes used to extract and purify plutonium. Then, after the war, he returned with full force to the discovery of new elements. There grew under his direction a large portion of the now-famous University of California Radiation Laboratory. With Seaborg's stimulation and guidance, highly skilled research teams prepared, many for the first time, americium, curium, berkelium, californium, einsteinium, fermium, and mendelevium. More elements are to come from this exciting laboratory.

In 1958, Glenn T. Seaborg was asked to become Chancellor of the Berkeley campus of the University of California, a world center of learning. Two years later his talents were demanded for national service: Seaborg was appointed Chairman of the U.S. Atomic Energy Commission. Thus he brings his deep knowledge of science and of scientists to the service of society in an age when society has desperate need for guidance by scholars of this preeminent stature.

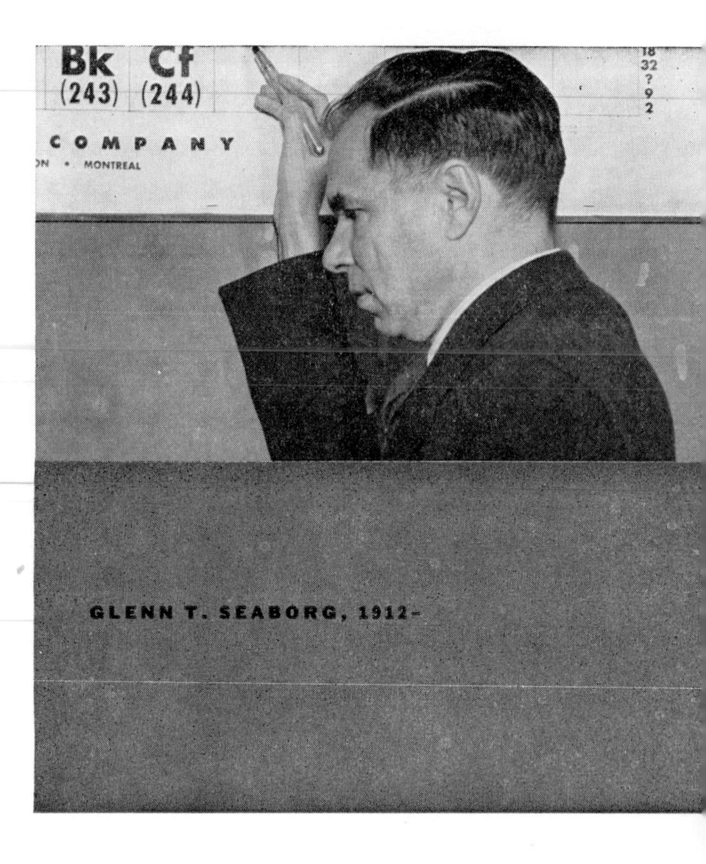

GLENN T. SEABORG, 1912–

CHAPTER 

Some Aspects of Biochemistry: An Application of Chemistry

. . . the elucidation of the structure of biochemicals . . . will necessarily lead to a deeper understanding of their function, and thus finally to the understanding of the mechanism of life itself.

ALBERTE PULLMAN AND BERNARD PULLMAN, 1962

Living organisms—bacteria, fungi, mosses, algae, plants, animals—are highly organized systems of chemical compounds. All organisms derive the energy for their activities, and produce the substances of which they are built, by means of chemical reactions.

A century and a half ago men regarded the chemistry of living organisms as something quite distinct from the chemistry of rocks, minerals, and other nonliving things. Indeed, there was in their minds at that time the inclination to believe that living things were imbued with some mysterious "vital force" that was beyond the power of men to define and understand.

As time went on, it became apparent that the mystery in the chemistry of living things was due to ignorance of the details of what went on, and with an increased understanding of chemical principles, the mystery gradually began to disappear. Compounds that were earlier known only as the products of plants and animals were produced in the laboratory from ordinary inorganic substances. By the middle of the nineteenth century the superstitious belief in a chemical "vital force" had disappeared, and now there are few chemists who believe that the chemistry of living organisms is beyond the power of men to understand.

We still, however, mark off a large area of chemical study by the term "biochemistry." This is not because biochemistry is fundamentally different from chemistry in general. It is because in order for a chemist to use his talents effectively to solve certain kinds of problems he must devote special (but not exclusive) attention to what is known about a particular field of knowledge. Biochemists are chiefly concerned with the chemical processes that go on in living organisms. These scientists must use information from all branches of chemistry to answer the questions they ask, but their questions are usually something like, "What kind of molecules make up living systems?" "How does a living system produce the energy it needs?" or "What structures do biochemicals have?"

24-1 MOLECULAR COMPOSITION OF LIVING SYSTEMS

The chemical system of even the smallest plant or animal is one of extreme complexity. It has a multitude of compounds, many of polymeric nature, existing in hundreds of interlocking equilibrium reactions whose rates are influenced by a number of specific catalysts. We will not try to study such a system. Instead we will show some parts of it, some examples that have been well studied and which illustrate the applicability of chemical principles. *All of our knowledge of biochemistry has come through use of the same basic ideas and the same experimental method you have learned in this course.*

We shall consider four classes of compounds that have great importance in biochemistry. Sugars, fats, and proteins occur in most animals and plants, while cellulose is more common in plants. These are all discussed in the following sections.

24-1.1 Sugars

SIMPLE SUGARS

The word "sugar" brings to mind the sweet, white, crystalline grains found on any dinner table. The chemist calls this substance sucrose and knows it as just one of many "sugars" which are classed together because they have related composition and similar chemical reactions. Sugars are part of the larger family of **carbohydrates**, a name given because many such compounds have the empirical formula CH_2O.

EXERCISE 24-1

Glucose, a sugar simpler than sucrose, has a molecular weight of 180 and empirical formula CH_2O. What is its molecular formula?

The structure of the glucose molecule was deduced by a series of steps somewhat like those described in Chapter 18 for ethanol. Glucose was found to contain one aldehyde group $\left(\text{—C}\begin{smallmatrix}\text{O}\\ \\ \text{H}\end{smallmatrix}\right)$ and five hydroxyl groups (—OH). These functional groups show their typical chemistry. The aldehyde part can be oxidized to an acid group. This reaction is like equation *(18–19)* (p. 336). If a mild oxidizing agent (such as the hypobromite ion in bromine water) is used, the aldehyde group can be oxidized without oxidizing the hydroxyl groups.

EXERCISE 24-2

Write the equation for the oxidation of

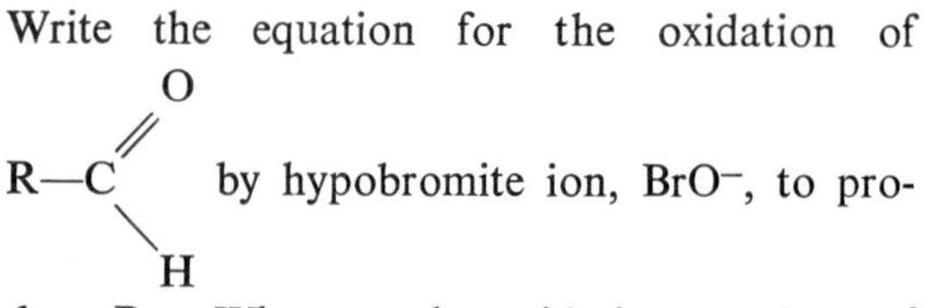

by hypobromite ion, BrO^-, to produce Br^-. What are the oxidation numbers of carbon and bromine before and after reaction? Which element is oxidized? Which is reduced?

If all the oxygen containing groups are reduced, *n*-hexane results. This test helps establish that the glucose molecule has a chain structure. One representation of the structural formula of glucose, $C_6H_{12}O_6$, is

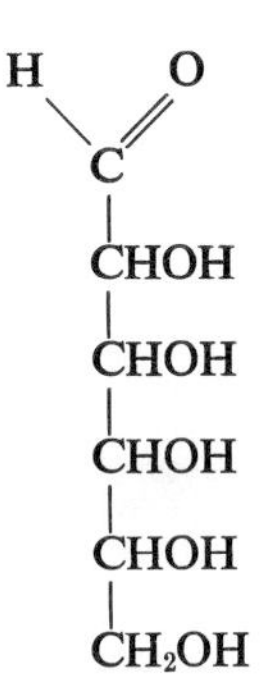

Using a delicate reduction method, the aldehyde group can be converted to a sixth hydroxyl group, giving the substance called sorbitol. This compound shows the typical behavior of an alcohol. For example, it forms esters with acids:

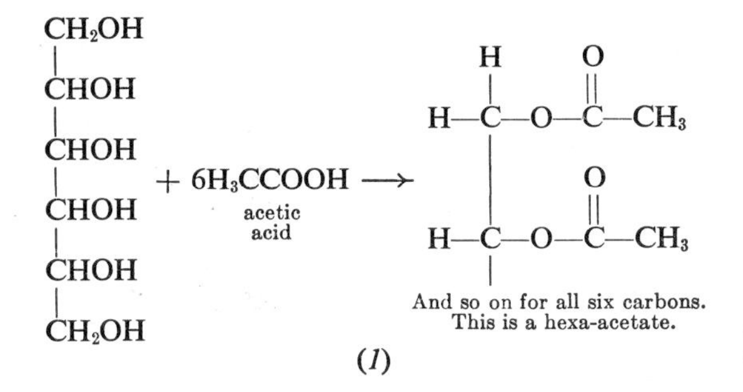

Another naturally occurring sugar is fructose, also $C_6H_{12}O_6$. It is an isomer of glucose but the carbon of the $\rangle C{=}O$ group is at the second position in the carbon chain instead of at the end. This makes fructose a ketone (see Section 18-3.2).

EXERCISE 24-3

Draw a structural formula for the fructose molecule (remember that fructose is an isomer of glucose). Explain why fructose cannot be oxidized to a six-carbon acid.

There is another aspect of the structure of glucose and fructose. They, like other simple sugars, can exist as a straight chain but this form is in equilibrium with a cyclic structure. In solutions the latter form prevails. Reaction (*2*) shows both forms of glucose.

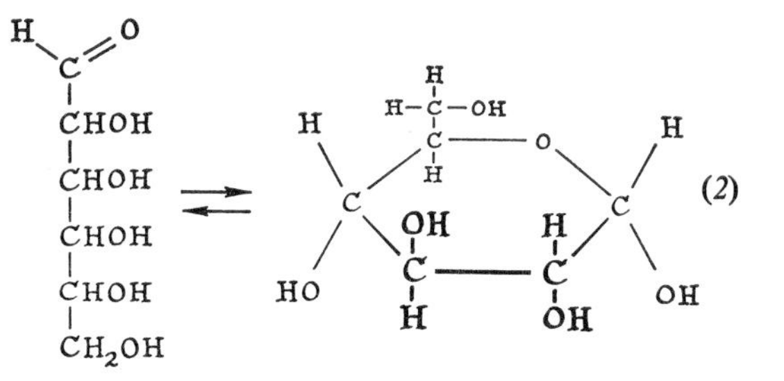

The ring form can be written in a simpler way, showing the hydrogen atoms attached to carbon atoms by lines only and omitting the symbols for the ring carbons:

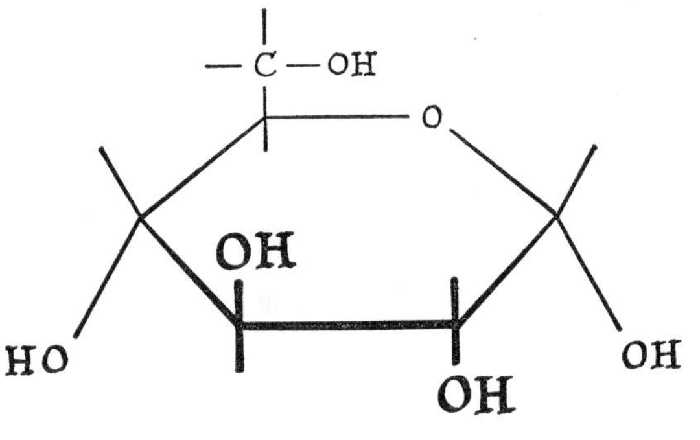

EXERCISE 24-4

At equilibrium in a 0.1 *M* solution of glucose in water, only 1% of the glucose is in the straight chain form. What is *K* for reaction (*2*)?

DISACCHARIDES

The two sugars we have discussed are monosaccharides—they have a single, simple sugar unit in each molecule. The sugar on your table is a disaccharide—it has two units. One molecule of sucrose contains one molecule of glucose and one of fructose hooked together (losing a molecule of water in the joining reaction). Fructose has a slightly different ring structure because the $\diagdown$C=O$\diagup$ group is not on the end carbon. The formation of sucrose is shown in equation (*3*).

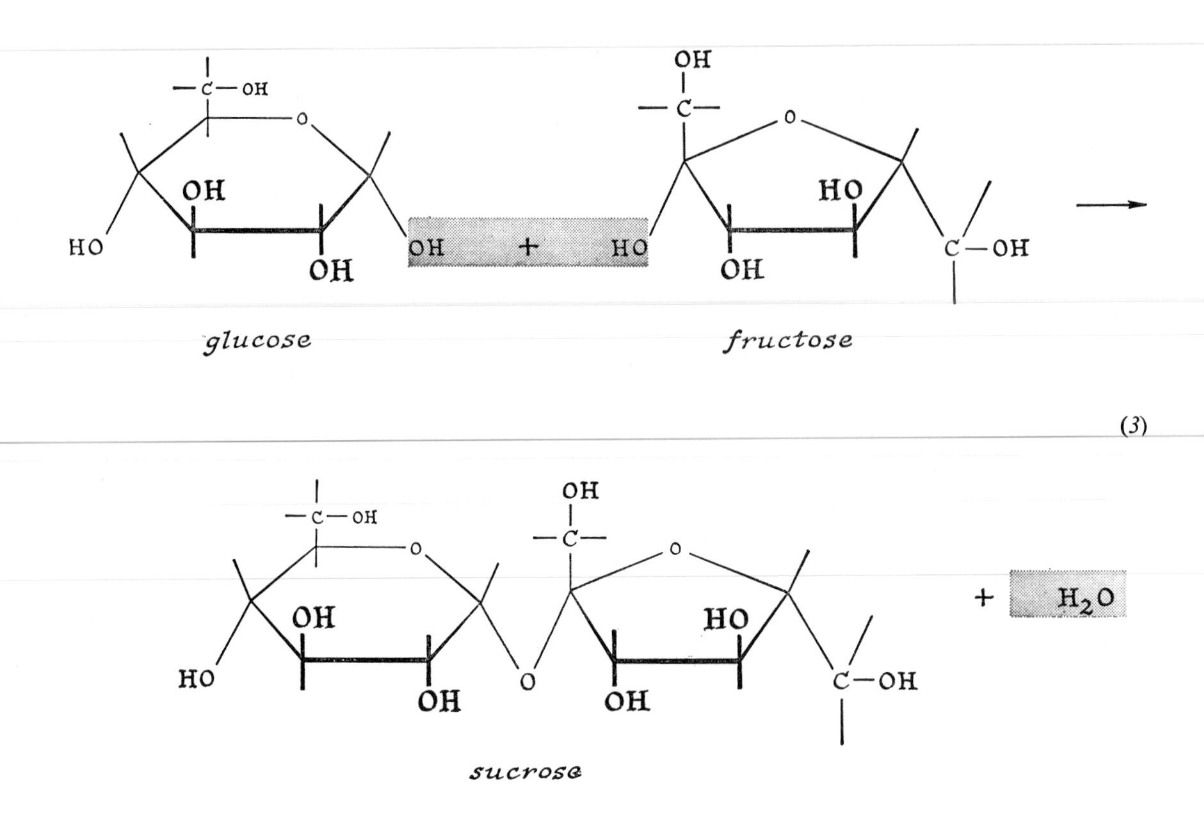

PROPERTIES

Sugars occur in many plants. Major commercial sources are sugar cane (a large, specialized grass which stores sucrose in the stem) and sugar beet (as much as 15% of the root is sucrose). In addition, fruits, some vegetables, and honey contain sugars. On the average every American eats almost 100 pounds of sugar per year. The nation requires about 2×10^{10} pounds per year of which about one-fourth is grown in the United States. Sugar, at about 10 cents a pound retail, is one of the cheapest pure chemicals produced.

Sugars are fairly soluble in water, about 5 moles dissolving per liter (solubility varies somewhat with the sugar). High solubility in water is readily explained because sugars have many functional groups that can form hydrogen bonds. From your previous study of hydrogen bonding (Section 17-2.6) you might recall that about 5 kcal are released per mole of hydrogen bonds formed. This energy can then make up part of that needed to disrupt the structure of the crystal.

Sugars are easily oxidized. One oxidation reaction that shows this involves cupric hydroxide,

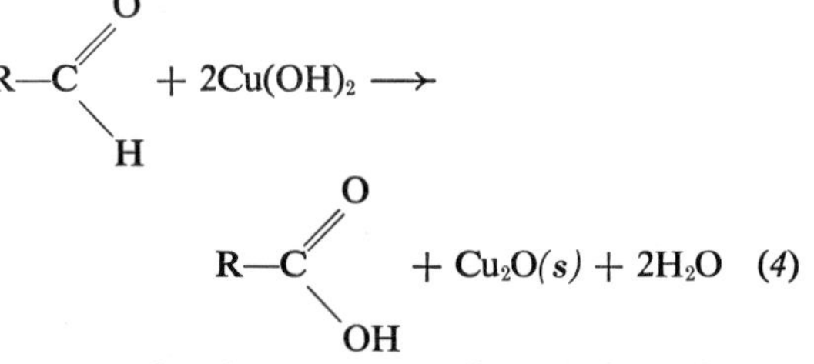

The reaction is more complicated than shown. The $Cu(OH)_2$ is not very soluble in the basic

solution used, so tartaric acid is added to form a complex ion. The $Cu_2O(s)$ is a red solid which does not form such a complex and thus precipitates from solution. The reaction is characteristic and is used as a qualitative test for simple aldehydes and for sugars.

An important metabolic reaction of disaccharides is the reverse of (*3*). Water, in the presence of $H^+(aq)$, reacts with sucrose to give glucose and fructose. *This process is called* ***hydrolysis,*** *meaning "reaction with water."*

24-1.2 Cellulose and Starch

Cellulose is an important part of woody plants, occurring in cell walls and making up part of the structural material of stems and trunks. Cotton and flax are almost pure cellulose. Chemically, cellulose is a polysaccharide—a polymer made by successive reaction of many glucose molecules giving a high molecular weight (molecular weight ~600,000). This polymer is not basically different from the polymers that were discussed in Section 18-6:

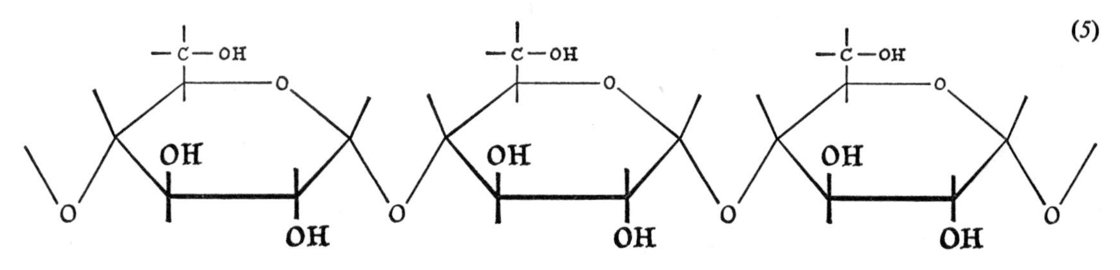

Starch is a mixture of glucose polymers, some of which are water-soluble. This soluble portion consists of comparatively short chains (molecular weight ~4000). The portion of low solubility involves much longer chains and the polymer chain is branched.

EXERCISE 24-5

The monomer unit in starch and that in cellulose each has the empirical formula $C_6H_{10}O_5$. These units are about 5.0 Å long. Approximately how many units occur and how long are the molecules of cellulose and of the soluble starch?

24-1.3 Fats

Fats, as well as animal and plant oils, are esters. Actually they are triple esters of glycerol (1,2,3-propanetriol):

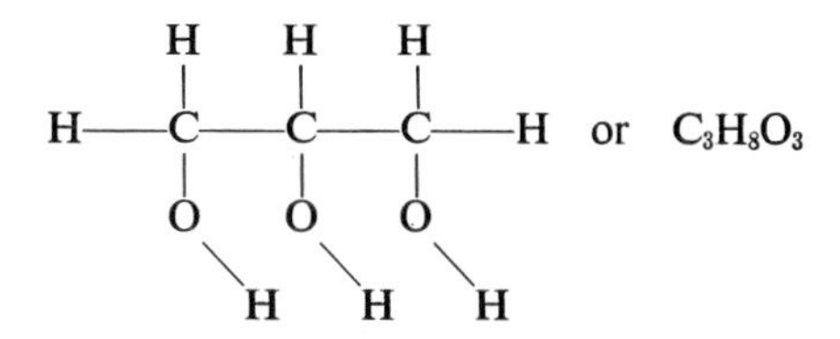

When carboxylic acids, similar to those you studied in Section 18-3.2, react with glycerol OH groups, a fat is formed. In natural fats the acids usually have twelve to twenty carbon atoms, C_{16} or C_{18} acids being most common.

EXERCISE 24-6

Write the formula for glycerol tributyrate, and then write the formula of the fat made from glycerol and one molecule each of stearic ($C_{17}H_{35}COOH$), palmitic ($C_{15}H_{31}COOH$), and myristic ($C_{13}H_{27}COOH$) acids. How many isomers are possible for the last fat? How many would be possible if all possible combinations of the three acids were used? Compare your answer with that for Exercise 18-15, p. 349.

Common fats (butter, tallow) and oils (olive, palm, and peanut) are mixed esters; each molecule has most often three, sometimes two, or, rarely, one kind of acid combined with a single glycerol. There are so many such combinations in a given sample that fats and oils do not have sharp melting or boiling points. Ranges are found instead.

An important reaction of fats is the reverse of ester formation. They hydrolyze, or react with water, just as disaccharides do. Usually hydrolysis is carried out in aqueous $Ca(OH)_2$, $NaOH$, or KOH solution. Because of long use in the preparation of soap from fats, the alkaline hydrolysis reaction (*6*) is called *saponification.* The metal salts of natural carboxylic acids, like sodium stearate, are called soaps.

Fats make up as much as half the diet of many people. Fats are a good source of energy because when they are completely "burned" in the body they supply twice as much energy per gram as do proteins or carbohydrates.

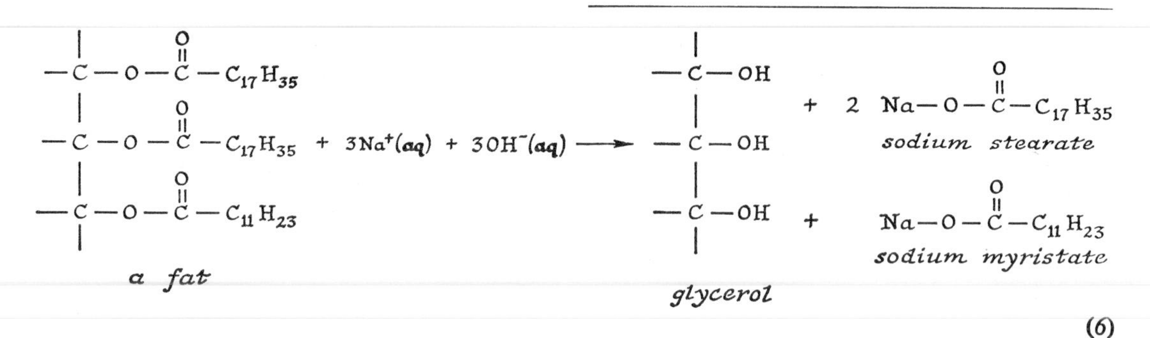

(*6*)

24-2 ENERGY SOURCES IN NATURE

24-2.1 Some Fundamental Biochemical Processes

An animal (such as man) expends energy continuously, to maintain body temperature and to perform such activities as breathing, circulating blood, and moving about. What chemical processes supply this energy?

The chief source of such energy is the combustion of carbon compounds to CO_2. You know that man exhales more carbon dioxide than he inhales in the air he breathes. This extra carbon dioxide is one of the products of the oxidation processes by which food is oxidized and energy is liberated.

One of the important foods of animal organisms is sugar. Man eats sugar in various forms: as sucrose, as glucose, and as starch, the form in which sugar is stored in many plant tissues, such as the potato. Cellulose, although a glucose polymer, is not a good food for humans because their digestive chemistry cannot hydrolyze it to sugar rapidly enough. Termites, however, can hydrolyze cellulose and they find wood products quite palatable.

We can regard sucrose and starch as sources of glucose, for these react with water to form glucose in the body:

$$\underset{\text{sucrose}}{C_{12}H_{22}O_{11}} + H_2O \longrightarrow \underset{\text{glucose}}{C_6H_{12}O_6} + \underset{\text{fructose}}{C_6H_{12}O_6} \qquad (7a)$$

$$\underset{\text{starch}}{HO(C_6H_{10}O_5)_nH} + (n-1)H_2O \longrightarrow \underset{\text{glucose}}{nC_6H_{12}O_6} \qquad (7b)$$

Glucose is one of the most important sources of energy for living creatures of all kinds. We can illustrate this with the fermentation of sugar. Yeast, a plant, uses glucose in a chemical reaction

$$C_6H_{12}O_6 \longrightarrow 2C_2H_5OH(l) + 2CO_2(g) \qquad (8)$$

In this transformation of glucose into alcohol and carbon dioxide, energy is liberated, and this energy is used by the yeast plants. Thus glucose is used as a fuel by the growing organism to furnish the energy needed for growth.

The process by which yeast breaks down glucose has been carefully studied by biochemists and the way in which this transformation occurs is now known in considerable detail. One of the reasons this process is so interesting is that a nearly identical process takes place in human muscle, in this case to furnish energy needed for muscular activity.

Both yeast and muscle break glucose down into an acid called pyruvic acid,

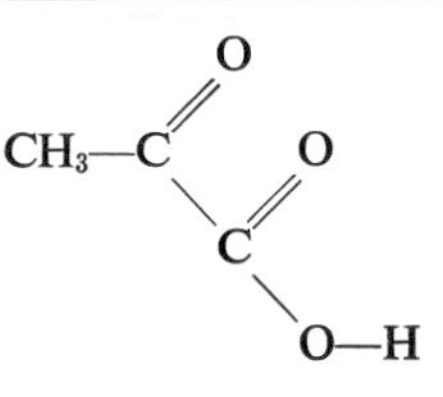

The process requires eight separate steps, *all* of which are carried out with the aid of biological catalysts called *enzymes*. We can picture the series of reactions in the following, much simplified manner:

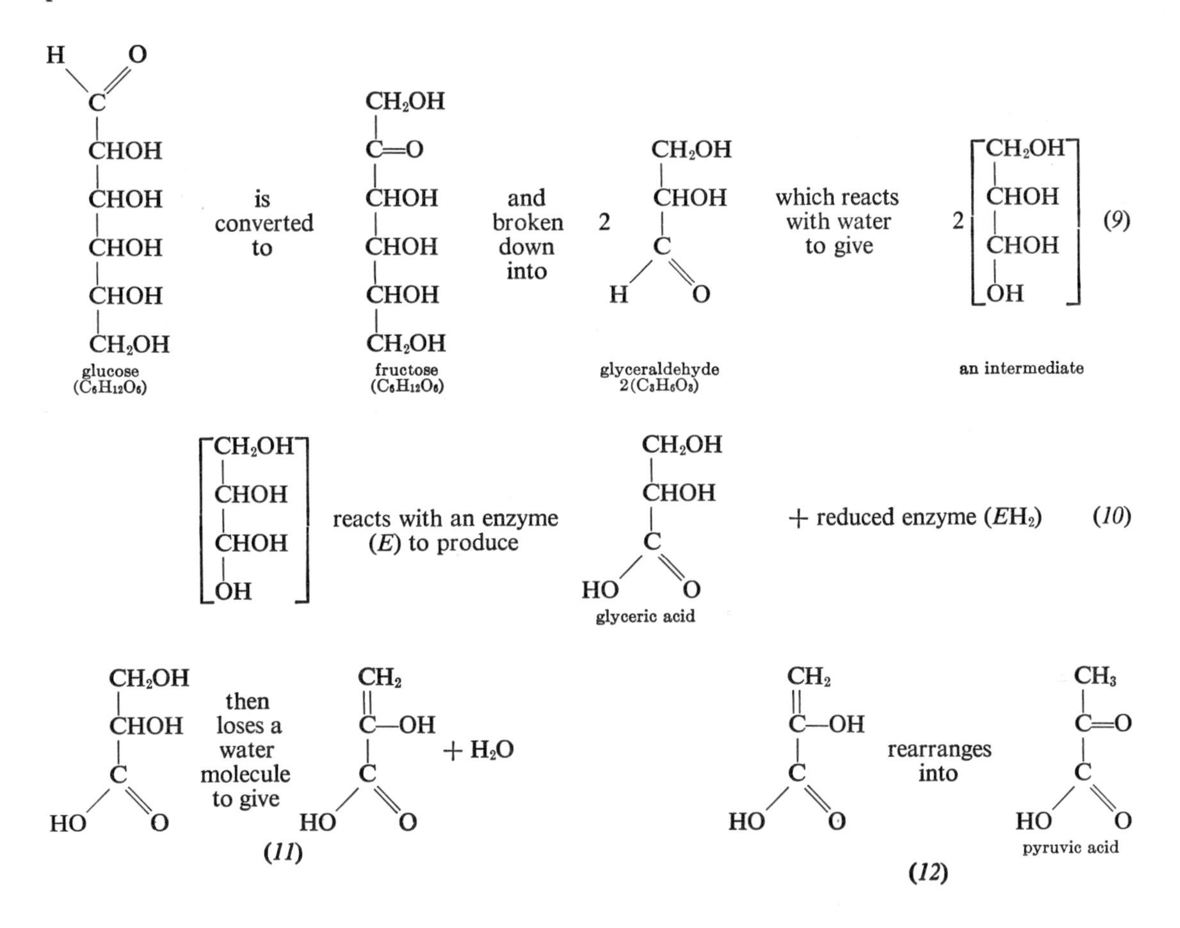

To this stage, yeast and muscle reactions are the same. Now, however, they proceed differently:

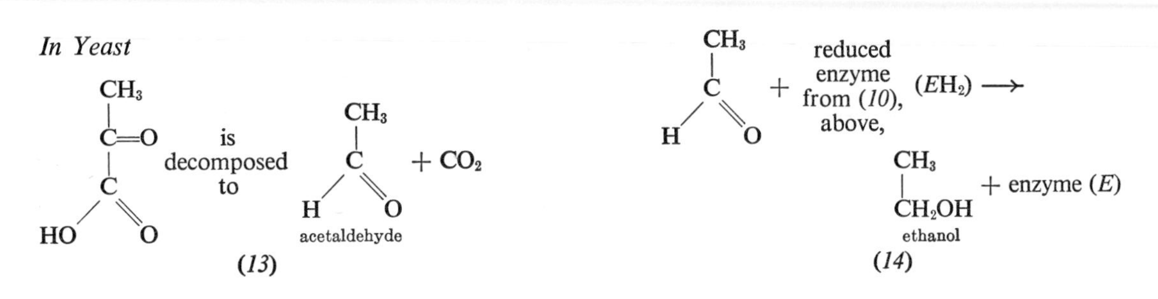

You will notice that in equation (*9*) each molecule of six-carbon sugar gave *two* molecules of the three-carbon compound, glyceraldehyde. Thus, each molecule of glucose gives *two* CO_2 and *two* ethanol molecules.

In Muscle

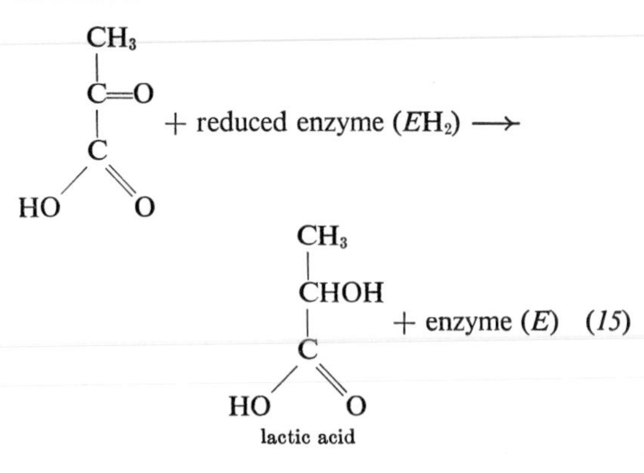

Lactic acid is commonly produced when sugar is broken down by living cells. Lactic acid is so named because it is produced when milk sours.

What happens to the energy from the oxidation of glucose? A study of the breakdown of glucose *in the absence of oxygen* shows that about 20 kcal are liberated per mole of glucose consumed:

$$C_6H_{12}O_6(s) \longrightarrow 2C_2H_5OH(l) + 2CO_2(g) \qquad \Delta H = -20 \text{ kcal} \quad (16)$$

This energy is used by the organism to synthesize other very reactive chemical compounds which are not shown in our simplified scheme. These reactive molecules then take part in other processes (such as muscle action) in which the energy is released. Part of the energy of glucose is stored as heat content (or "chemical energy") in the reactive compounds. This "storage" of energy in compounds enables living organisms to make efficient use of the energy of oxidation.

24-2.2 Oxidative Metabolism

If a mole of glucose is completely burned in a calorimeter, a great deal of energy is liberated:

$$C_6H_{12}O_6(s) + 6O_2(g) \longrightarrow 6CO_2(g) + 6H_2O(l) + 673 \text{ kcal} \quad (17)$$

Since 673 kcal/mole could be released by complete oxidation, we might wonder why the yeast cells (and muscle) extract only 20 kcal/mole and leave so much of the potentially available energy untouched. This extra energy is there in ethanol and lactic acid and could be released if these compounds were oxidized further to CO_2.

Fermentation in the absence of oxygen, equation (*16*), with the liberation of only a small fraction of the total available energy of glucose, is not the usual chemistry employed by living organisms. It is a "reserve" mechanism, useful in tiding over yeast or muscle during times of oxygen shortage. In the case of muscle, it provides a temporary source of energy when excessive demands require the performance of work at a faster rate than oxygen can be supplied by the circulation of blood to the tissue.

Ordinarily, oxygen *is* used and glucose *is* oxidized all the way to CO_2 and H_2O. Most living creatures exist in contact with a supply of oxygen, either in the air or dissolved in water. Hence, most of the metabolic activities of the living world occur in the presence of oxygen. Under these conditions, the breakdown of the glucose molecule proceeds without the use of oxygen only to the point at which pyruvic acid is formed. Then the pyruvic acid, instead of being reduced to lactic acid or ethanol and CO_2, is oxidized by oxygen to CO_2 and water. It is during this oxidation that most of the energy, originally available in glucose, is liberated.

EXERCISE 24-7

Write a balanced equation for the oxidation of pyruvic acid to CO_2 and H_2O.

Little of this energy is directly given off as heat, as it would be if we burned pyruvic acid with a match. Most of the energy is stored in new chemical compounds that can undergo reactions leading to the synthesis of fats and proteins and the other substances of which living matter is made.

Let us look at a very simple part of the overall process. The first thing that happens is the breakdown of pyruvic acid into acetic acid and CO_2 (the "oxidation" process has started!):

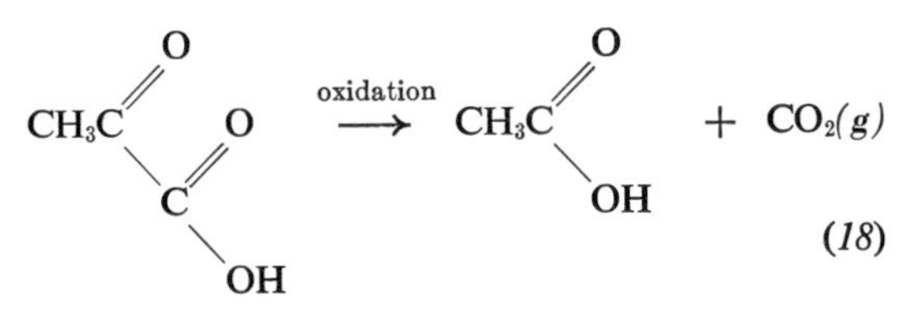

The acetic acid then enters a cycle of reactions in which it is the fuel that keeps the cycle running, and CO_2, water and energy are taken off along the course of the cycle. This cycle is shown in Figure 24-1 and its steps are represented by equations (*19a*) to (*19f*).

Each "turn" around this cycle uses up one molecule of acetic acid and produces two of CO_2 and two of water. The stages at which oxidation occurs are not shown in detail, for this is a rather complex subject and is beyond our present ability to consider in detail. However, it can be seen that the oxidation *does* occur (because acetic acid and oxygen are fed in and CO_2 and H_2O are discharged), and the energy of "burning" sugar, the source of the acetic acid, is being released.

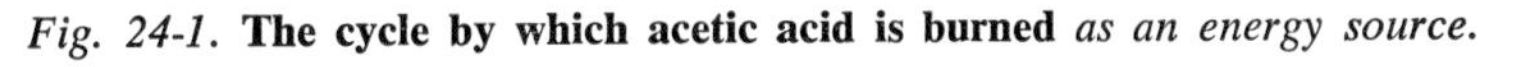

Fig. 24-1. **The cycle by which acetic acid is burned** *as an energy source.*

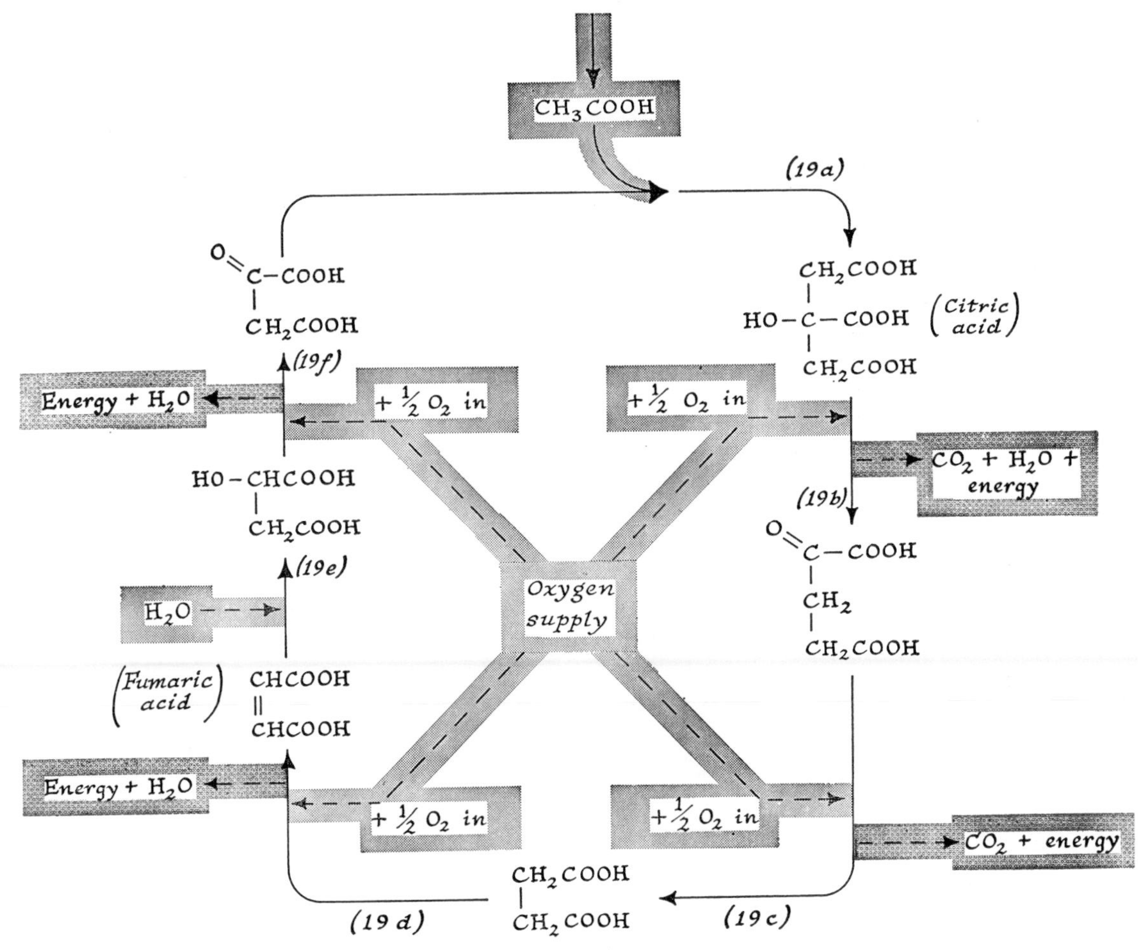

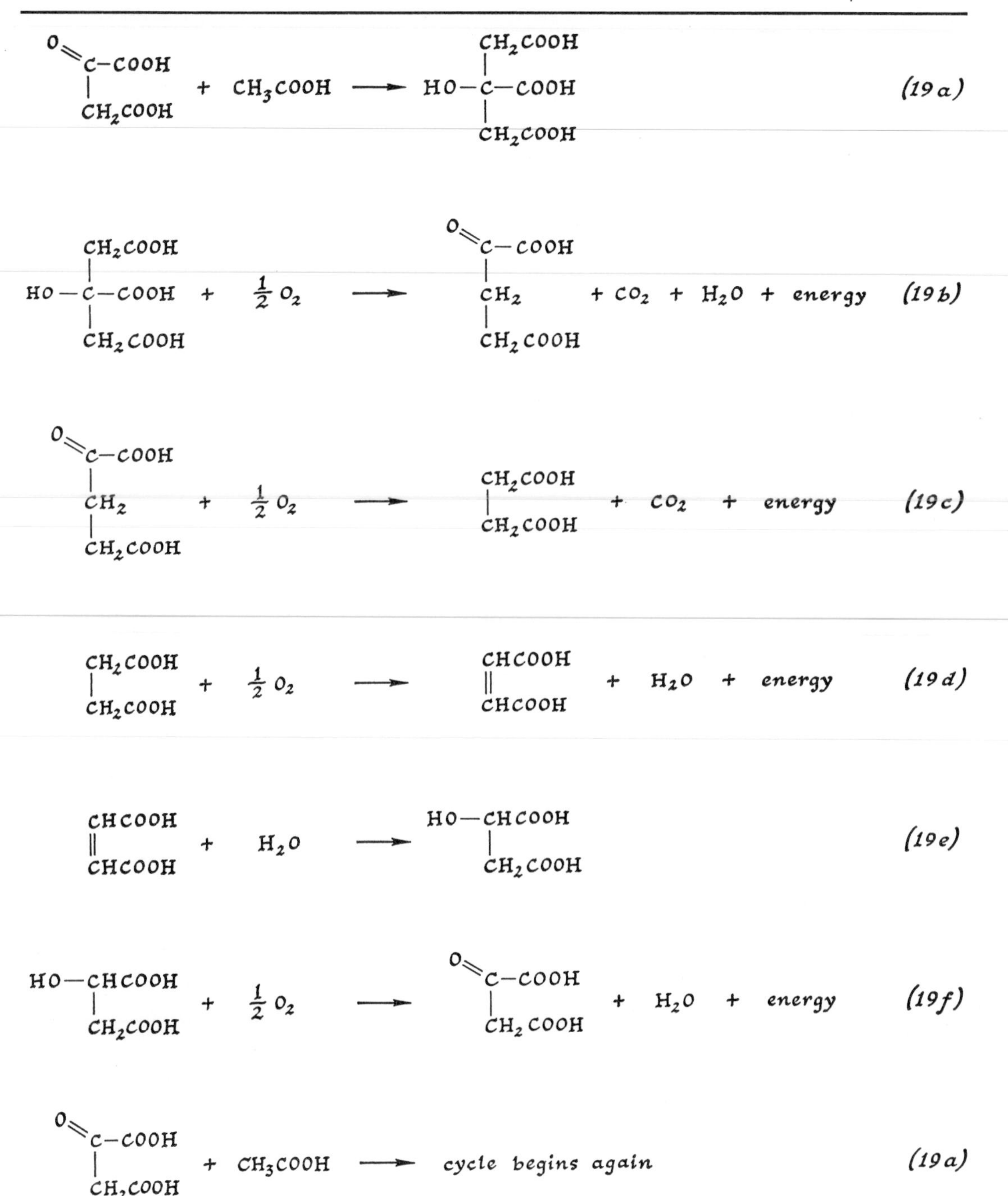

24-2.3 Photosynthesis

What is the source of the vast amount of energy consumed by our mechanized society? The largest of all of our energy sources is the sun, and energy from the sun is stored in our fuels (wood, coal, petroleum) as a result of the *photosynthesis* process.

Green plants reverse the process of sugar breakdown by synthesizing sugars from CO_2 and water:

$$6CO_2(g) + 6H_2O(l) + 673\ \text{kcal} \longrightarrow C_6H_{12}O_6 + 6O_2(g) \quad (20)$$

We cannot specify exactly how this energy is

used by the plant. We know that the reaction requires a special compound, chlorophyll, which is the green compound that gives green plants their color. Chlorophyll does not appear as a reactant in equation (*20*) because it is a catalyst.

We also know that the process of incorporating CO_2 into complex molecules (that result finally in the synthesis of sugar) is very similar to the reverse of the process of sugar breakdown. These reactions are complicated, however, and will not be discussed here.

EXERCISE 24-8

The sun provides about 0.50 calorie on each square centimeter of the earth's surface every minute. How long would it take ten leaves to make 1.8 grams of glucose if the area of each leaf is 10 cm^2 and if only 10% of the energy is used in the reaction?

EXERCISE 24-9

Normally about 0.03% of the molecules in air are carbon dioxide molecules. How many liters of air (at STP) are needed to provide enough CO_2 to form 1.8 grams of glucose in reaction (*20*)?

EXERCISE 24-10

Suppose red light of wavelength 6700 Å is absorbed by chlorophyll.

(a) Show that the frequency of this light is 4.5×10^{14} cycles per second.
(b) How much energy is absorbed per mole of photons absorbed? (See Section 15-1.1; $h = 9.5 \times 10^{-14}$ kcal sec/mole.)
(c) How many moles of photons would be needed to provide enough energy to produce one mole of glucose by reaction (*20*) if all of the energy were provided by red light?

24-3 MOLECULAR STRUCTURES IN BIOCHEMISTRY

Some of the most exciting recent advances in biochemistry have come from recognition of the importance of the structural arrangement of molecular parts. You saw in Chapter 18 that the chemistry of a C_2H_6O compound depends upon structure. Thus an ether, $CH_3—O—CH_3$, behaves quite differently from an isomeric alcohol, CH_3CH_2OH. You also learned how interactions between molecules can influence the properties of water (Sections 17-2.5, 17-2.6) and arrange the molecules in preferred positions around an ion (Figure 17-13). This kind of structure also influences the observed properties. Both the covalent bond arrangement and intermolecular interactions are involved in fixing the structure of biochemical substances. A few examples are discussed below.

24-3.1 The Structure of Starch and Cellulose

A striking example of the effect of structure is shown by cellulose and water-soluble starch. Both contain the same monomer since hydrolysis gives only glucose in each case. But the glucose ring differs slightly in the arrangement of the OH groups. This results in two different polymers. Let us represent the ring structure of equation (*2*) by this simplified symbol:

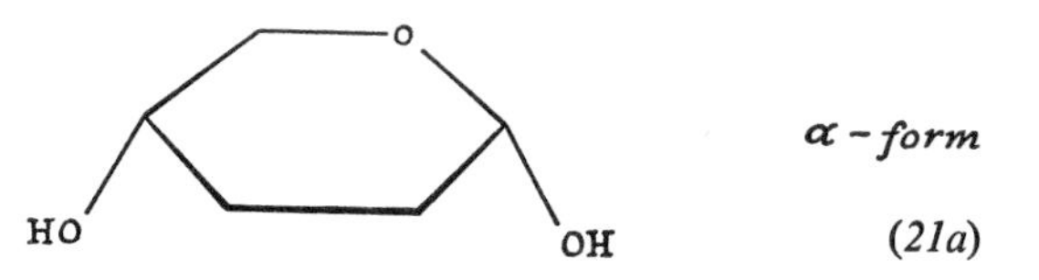

(*21a*)

There is another isomer, identical in all parts, except for the placement of the right-hand OH group. Its symbol is

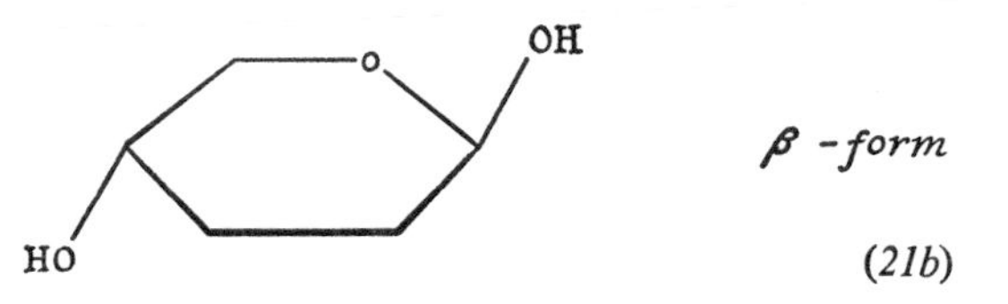

(*21b*)

If we connect a string of the α-form and allow for the normal 105° angle of bonds to oxygen we get the polymer called starch:

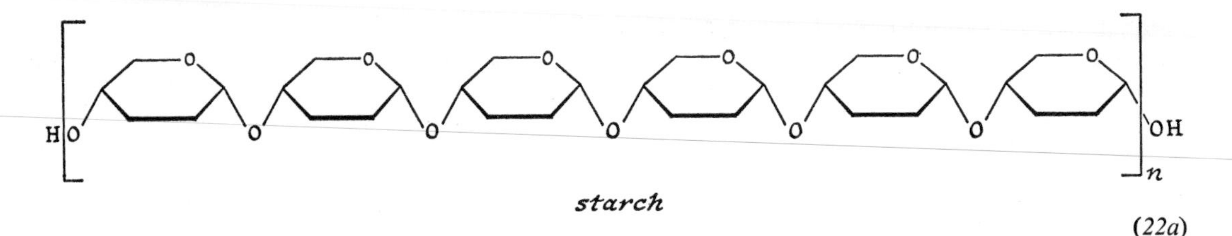

starch

(22a)

On the other hand, a chain of the β-form of glucose gives the polymer called cellulose:

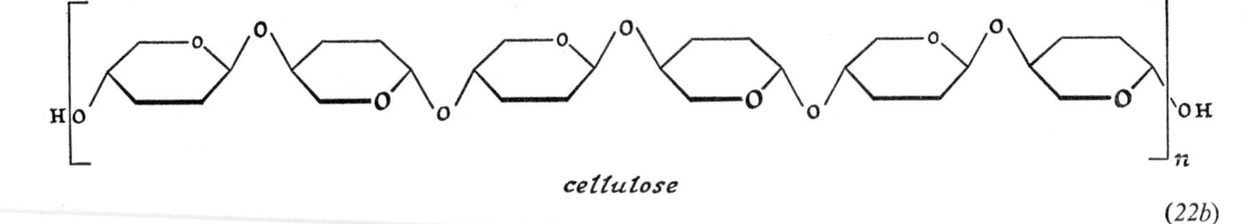

cellulose

(22b)

The very different geometry of the ether linkages in starch and cellulose causes these two polymers to have different chemical properties.

24-3.2 Proteins

In Section 18-6.3 the composition of proteins was given. They are large, amide-linked polymers of amino acids. However, the long chain formula (Figure 18-14, p. 348) does not represent all that is known about the structure of proteins. It shows the covalent structure properly but does not indicate the relative positions of the atoms in space.

The use of X-ray diffraction (Section 14-3.2) and the principles that describe hydrogen bonding have led to the recognition of a coiled form of the chain in natural proteins. This model is consistent with other tests also and has received general acceptance. It is shown in Figure 24-2. This form has a great deal of regularity—it is not at all a random shape. Order is not achieved without some energy factor to maintain it, and this order results in large part from hydrogen bonds. These are shown in the figure as dotted lines, just as they were in Section 17-2.6. When the hydrogen bonds are broken (by heating or putting the protein in alcohol), the order disappears and the coiled form loses its shape. Often this damage cannot be repaired and the coil is permanently deformed. Cooking an egg destroys the coiled form of the proteins it contains. A few moments of thought concerning the profound differences between the physical form and between the chemical potentialities of an egg before and after it is cooked will suggest the very great importance of molecular structure in biochemistry.

24-3.3 Enzymes

All of the biochemical reactions we have just been discussing proceed at ordinary temperatures and pressures. Most biochemical reactions (especially those in the human body) take place at about 37°C (98°F), and proceed at a rate adequate for the role they play, which is to make life, growth, and reproduction possible. Most of them would not proceed at a measurable rate at this temperature outside of living organisms. Glucose, pyruvic acid, acetic acid, etc., are very stable compounds and can remain in contact with oxygen without apparent change. This is so, even though their oxidation to carbon dioxide and water releases large amounts of energy. To make these reactions proceed, nature uses catalysts to provide new paths with lower activation energy hills over which the systems can pass so that measurable reaction rates are achieved.

Biological catalysts are called *enzymes*. Nearly every step of the breakdown of a complex molecule to a series of smaller ones, within living cells, is catalyzed by specific enzymes. For instance, when acetaldehyde is reduced in yeast cells to ethanol, in a process we can represent as follows,

$$CH_3{-}C(=O){-}H + 2[H] \longrightarrow CH_3{-}CH_2{-}OH \qquad (23)$$

the reaction takes place in the presence of a specific enzyme called "alcohol dehydrogenase." You can see that the *hydrogenation* of acetaldehyde is the reverse of the *dehydrogenation* of ethanol. The enzyme is named for the latter reaction, but of course it catalyzes the reaction in either direction (see Section 9-1.4). Conditions at equilibrium are not affected by the enzyme, but the rate at which the reacting substances reach the equilibrium state *is* affected by the enzyme (as with any catalyst).

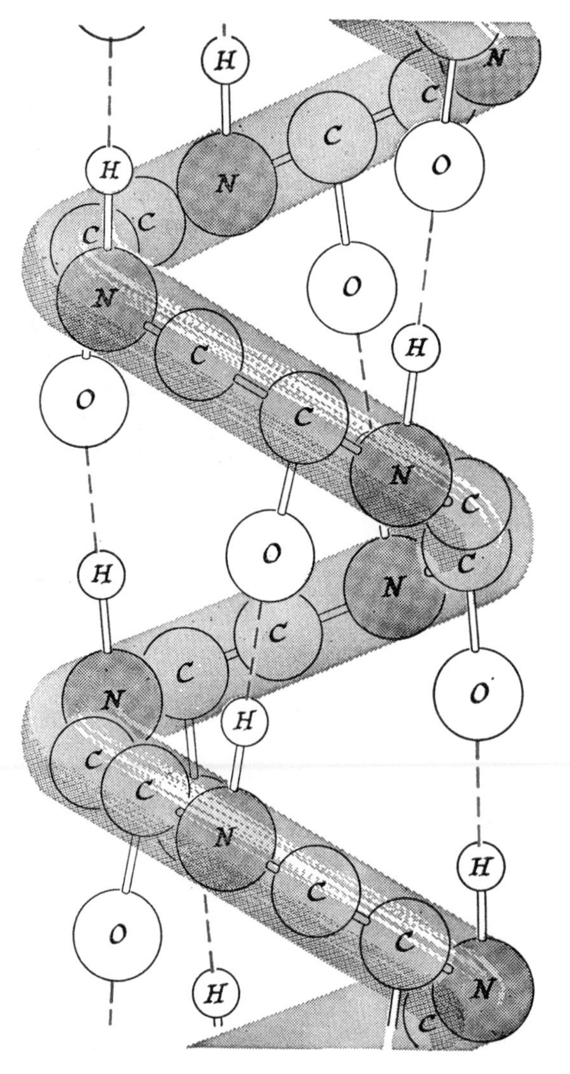

Fig. 24-2. **The coiled or helix form of a protein molecule.**

Enzymes are protein molecules. While all enzymes are proteins, we do not imply that all proteins can act as enzymes. The protein molecules of enzymes are very large, with molecular weights of the order of 100,000.* In contrast, the substance upon which the enzyme acts (called a **substrate**) is very small in comparison with the enzyme. This creates a picture of the reaction in which the small substrate molecule becomes attached to the surface of the large protein molecule, at which point the reaction occurs. The products of the reaction then dissociate from the enzyme surface and a new substrate molecule attaches to the enzyme and the reaction is repeated. We can write the following sequence:

$$\text{enzyme} + \text{substrate} \longrightarrow \text{enzyme-substrate complex} \qquad (24)$$

$$\text{enzyme-substrate complex} \longrightarrow \text{enzyme} + \text{reaction products} \qquad (25)$$

Adding these equations and cancelling gives

$$\text{substrate} \longrightarrow \text{reaction products} \qquad (26)$$

Despite the large size of an enzyme molecule, there is reason to believe that there are only one or a few spots on its surface at which reaction can occur. These are usually referred to as "active centers." The evidence for this view of enzyme reactions comes from many kinds of observations. One of these is that we can often stop or slow down enzyme reactions by adding only a small amount of a "false" substrate. A false substrate is a molecule that is so similar to the real substrate that it can attach itself to the active center, but sufficiently different that no reaction and consequently no release occurs. Thus, the active center is "blocked" by the false substrate.

* Since so many enzymes are known, this number is given only to offer a rough idea of size, because actual molecular weights may range from *considerably less* than 100,000 to considerably more.

SPECIFICITY OF ENZYMES

Most enzymes are quite specific for a given substrate. For example, the enzyme "urease" that catalyzes the reaction

$$O{=}C(NH_2)_2 + H_2O \rightleftharpoons CO_2 + 2NH_3 \quad (27)$$

urea

is specific for urea. If we try to use urease to catalyze the analogous reaction of a very similar molecule, *N*-methyl urea, no catalysis is observed:

$$O{=}C(NHCH_3)(NH_2) + H_2O \rightleftharpoons CO_2 + NH_3 + CH_3NH_2 \quad (28)$$

N-methyl urea

This suggests that on the surface of the enzyme there is a special arrangement of atoms (belonging to the amino acids of which the protein is constructed) that is just right for attachment of the urea molecule but upon which the methyl urea will not "fit."

Specificity is not always perfect. Sometimes an enzyme will work with any member of a *class* of compounds. For example, some esterases (enzymes that catalyze the reaction of esters with water) will work with numerous esters of similar, but different, structures. Usually, in cases of this kind, one of the members of the substrate class will react faster than the others, so the rates will vary from one substrate to another.

A PRACTICAL APPLICATION OF ENZYME INHIBITION BY A "FALSE" SUBSTRATE

It is now believed that many of our useful drugs exert their beneficial action by the inhibition of enzyme activity in bacteria. Some bacteria, such as *staphylococcus*, require for their growth the simple organic compound *para*aminobenzoic acid and can grow and multiply in the human body because sufficient amounts of this compound occur in blood and the tissues. The control of many diseases caused by these (and other) bacteria was one of the first triumphs of *chemotherapy*,* and the first compound found to be an effective drug of this type was sulfanilamide:

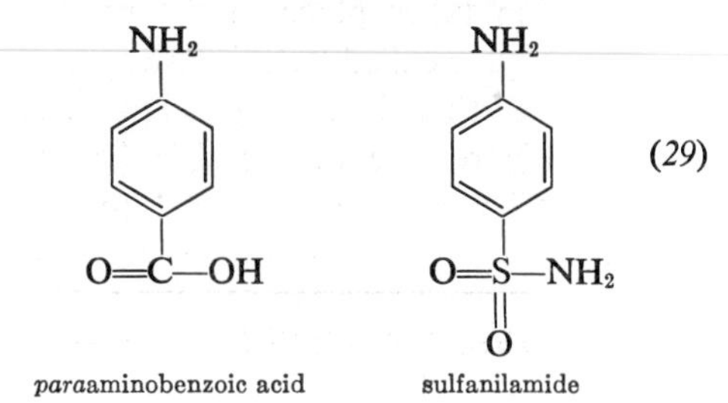

*para*aminobenzoic acid   sulfanilamide

It seems reasonable that an enzyme which used *para*aminobenzoic acid as a substrate might be deceived by sulfanilamide. The two compounds are very similar in size and shape and in many chemical properties. To explain the success of sulfanilamide, it is proposed that the amide can form an enzyme-substrate complex that uses up the active centers normally occupied by the natural substrate.

Usually fairly high concentrations of such a drug are needed for effective control of an infection because the inhibitor (the false substrate) should occupy as many active centers as possible, and also because the natural substrate will probably have a greater affinity for the enzyme. Thus the equilibrium must be influenced and, by using a high concentration of the false substrate, the false substrate-enzyme complex can be made to predominate. The bacteria, deprived of a normal metabolic process, cannot grow and multiply. Now the body's defense mechanisms can take over and destroy them.

* Chemotherapy is the control and treatment of disease by synthetic drugs. Most of these are organic compounds, often of remarkably simple structure. Sulfanilamide is one example of an organic compound synthesized by chemists for the treatment of bacterial infections.

Robert Woodward is surely one of the outstanding American synthetic organic chemists of all time. His astonishing record of successful syntheses of biologically important substances has won him ten honorary doctorate degrees and at least as many major awards here and abroad.

Woodward was born in Quincy, Massachusetts. His interest in chemistry developed at an early age and it seemed to grow without need for stimulation nor urging. By the time he entered Massachusetts Institute of Technology at the age of sixteen, he knew as much organic chemistry as the average graduating senior. M.I.T. recognized his capabilities and opened the laboratory to him. He passed course examinations at a rate and with a performance that brought him the Bachelor's Degree in three years and, in only one additional year, the Ph.D. Professor J. F. Norris, then director of the M.I.T. laboratory announced, "We saw we had a person who possessed a very unusual mind. · · · We think he will make a name for himself in the scientific world."

How richly this prophecy has come true is read in the scientific literature of organic chemistry. His first major success was the synthesis of quinine, a problem he began worrying as a high school student. This compound was to be typical of the difficult molecules he has successfully synthesized. Quinine has the formula $C_{20}H_{24}O_2N_2$, *and it includes two benzene rings, a tricyclic structure, a double bond, an OH group, and a methyl ether linkage, all combined in a very particular geometrical configuration. A host of other, comparably difficult syntheses have been achieved by Woodward and his large group of students, including some natural products whose importance has made their names familiar household words: cholesterol, cortisone, and chlorophyll. His contributions to the structure determinations of antibiotics reads like a doctor's chemical shelf, including penicillin, terramycin, and aureomycin. He has added to our knowledge of the polymerization processes by which amino acids link into proteins, and some of his synthetic protein-like polymers have physical properties quite comparable to those of silk and wool fibers.*

All of these accomplishments bespeak a dedication to chemistry and a capacity for work that are commensurate with his intellectual capability. Woodward can be found in his office or in the laboratory after midnight many days a week. However, the rewards for such intensive effort are proportionate. To Robert Woodward, as to most chemists, chemistry is an exciting adventure for which there just isn't enough time.

CHAPTER 25

The Chemistry of Earth, the Planets, and the Stars

There may, of course, be types of life with a wholly different chemical basis to our own, for example, a low temperature life on the outer planets which is based on reactions in liquid ammonia.

J. B. S. HALDANE, 1960

The advent of space exploration quickens the pulse of every scientist including, as much as any, that of the chemist. Chemists are playing many crucial roles: preparing new fuels, new metals, and new plastics to cope with a new range of performance needs; anticipating the environmental chemistry that will face the first space explorers, devising ways to permit survival under conditions so extreme that they are not even present on our planet, collecting and interpreting data that will reflect onto and illuminate age-old questions about the origin of the earth, the solar system, and life itself.

This age is heavily laden with terrestrial problems for the chemist, as well. Our planet seems to be shrinking under man's fantastic success in curbing nature's whim and in bending it to his will. The magnitude of this success is measured in an awesome and exponential population growth. This growth portends problems of food production, fuel consumption, and even of living space that dwarf those of the past. Suddenly immense urban areas are threatened by air pollution problems that were completely unknown thirty years ago. Power consumption is expanding so rapidly that some scientists anticipate the depletion of fossil fuel supplies and they urge haste in developing nuclear fuels and in harnessing more completely the vast energy of the sun.

It is an exciting age—challenging, yes, but exciting. It is an age in which we *must* understand our planet, Earth, and if we do, we can begin to venture toward our neighbor planets and beyond them toward the stars.

25-1 THE CHEMISTRY OF OUR PLANET, EARTH

The earth is the source of all the substances directly available to us. Energy comes to us from outside the earth—from our sun and to a small degree from the other stars. The earth, in turn, radiates energy into outer space. If the amount of energy radiated is more than the amount received, the earth will grow colder—if it is less, the earth will warm. Some of the solar energy is stored as chemical heat content when new substances, particularly organic compounds, are formed. For short periods of time (as measured in terms of geologic ages) we can use the energy stored in fossil fuels such as coal and oil or we can draw on nuclear fuels. In the long run, we shall have to depend upon the sun as our primary source of energy.

The substances we can use come primarily from the earth. In its movement around the sun, the earth sweeps through space and collects material from meteors and some cosmic dust, but the amount of gathered material is small compared with the amount present in the earth.* We shall consider the material of the earth and see how it is put to use by mankind.

25-1.1 The Parts of the Earth

A discussion of the chemistry of the earth is conveniently broken into three parts, each of which corresponds to one of the phases solid, liquid, or gas.

*The **lithosphere** is the solid portion of the earth.* We shall use the term to include the central core, though there remains controversy as to whether the core is solid or liquid. The lithosphere is a sphere of solid material about 4000 miles in radius. We have direct access to only a minute fraction of this immense ball. The deepest mine penetrates only two or three miles. The deepest oil wells are about five miles deep. This relatively thin shell that we can study directly is called the *earth's crust*. In view of seismic observations, we consider the thickness of this crust to be about 20 miles. The remainder we shall call the *inner lithosphere*, which includes the central part called the core.

About 80% of the earth's surface is covered with aqueous solution. *This liquid layer, the oceans, is called the **hydrosphere**.* The average depth of the hydrosphere is about three miles but at ocean "deeps" or "trenches," it changes precipitously to depths over twice that.

Surrounding the earth is the third phase, a gas. *The gas mixture surrounding the earth is called the **atmosphere**.* Over 98% of this gas (air) is less than 40 miles above the earth's surface.

25-1.2 Composition and Properties of the Atmosphere

The composition of the earth's atmosphere differs from day to day, from altitude to altitude, and from place to place. The largest variation is in the concentration of water vapor. Water evaporates continually from the hydrosphere, from the soil, from leaves, from clothes drying, etc. At intervals, parts of the atmosphere become chilled until the dew point or frost point is reached and then any vapor in excess of the saturation amount is precipitated as rain or snow.

Since the concentration of the water vapor varies so much, geochemists usually report the composition of "dry air"—that is, air from which all the water vapor has been removed. The composition of a sample of dry air is shown in Table 25-I. Notice the low concentration of hydrogen and helium in the air. The earth is a rather small object in the universe and exerts a relatively low gravitational attraction on the gases above it. Hence, most of the hydrogen and helium originally associated with the matter of the earth could be lost rather easily. Notice also that nitrogen is more abundant than oxygen in

* It has been estimated that about five tons of material are gathered per day as the earth sweeps through space. Yet the amount collected in a billion years would form a dust layer only a few millimeters thick if spread evenly over the earth's surface.

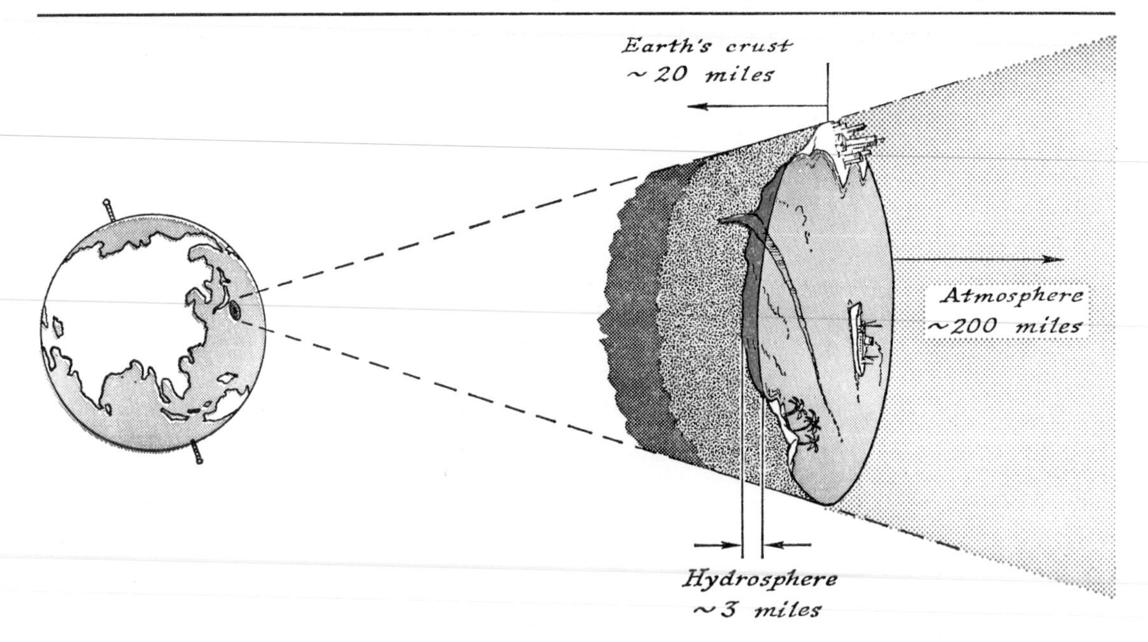

Fig. 25-1. **The parts of the earth.**

the air, even though oxygen is much more abundant in the hydrosphere and lithosphere. Notice also that, except for water and carbon dioxide, the major components of air are elements.

Table 25-I

COMPOSITION OF A CERTAIN SAMPLE OF DRY AIR*

SUBSTANCE Name	Formula	PERCENT OF MOLECULES
nitrogen	N_2	78.09
oxygen	O_2	20.95
argon	Ar	0.93
carbon dioxide	CO_2	0.03
neon	Ne	0.0018
helium	He	0.00052
krypton	Kr	0.0001
hydrogen	H_2	0.00005
xenon	Xe	0.000008

* Traces of other compounds (less than 0.0002% of the molecules) are also known to be present.

With the aid of the gas laws we can calculate the relative concentrations of the components in moist air. Suppose that on a certain day the atmospheric pressure is 750 mm and that after drying a sample of this air we find the pressure to be 738 mm. Then the partial pressure of the water vapor was 750 mm − 738 mm = 12 mm. Since the partial pressure varies directly with the number of molecules, we find that the fraction of the air molecules that are water molecules in the moist gas is 12 mm/750 mm = 0.016. In this rather moist gas, 1.6% of the molecules are water molecules.

The gravitational force on a heavy molecule exceeds that on a light molecule. Consequently, there is a tendency for "sedimentation" of the molecules of high molecular weight relative to the gas molecules of low molecular weight. This is opposed by the tendency toward maximum randomness, which tends to keep the atmosphere thoroughly mixed. The net result is a slight change of composition with altitude. Dry air at sea level contains about 78% nitrogen molecules and 21% oxygen molecules, but at 60,000 feet a sample of dry air has about 80% nitrogen molecules and only 19% oxygen molecules.

Apart from this gravitational effect, there are composition changes due to chemical reactions induced by light. These are caused by absorption of ultraviolet light in the upper atmosphere. For example, oxygen absorbs ultraviolet light and the energy taken up by the molecule exceeds the bond energy. The bond breaks, to give two oxygen atoms:

$$O_2(g) + h\nu \longrightarrow 2O(g) \qquad (1)$$

Of course the oxygen atoms so produced are very reactive. One fate of these atoms is to combine with another oxygen molecule, O_2, to form ozone, O_3:

$$O(g) + O_2(g) \longrightarrow O_3(g) \qquad (2)$$

Ozone is a highly reactive form of the element oxygen, though by no means as reactive as oxygen atoms. It is produced in the atmosphere only at high altitudes because the ultraviolet light of the frequency required in reaction (*1*) is so completely absorbed that it does not reach lower altitudes. Balloon flights have shown that the ozone concentration is negligible at sea level but that it rises to a maximum at a height of about 15 miles.

This small amount of ozone 15 miles above the earth's surface absorbs most of the ultraviolet light not absorbed by O_2. Thus, O_2 and O_3 together make the atmosphere opaque in most of the ultraviolet spectral region. Presumably the chemistry of life on this planet would have evolved quite differently if this ultraviolet light reached the earth's surface. As a single example, reflect that photosynthesis would have photons of much higher energy with which to operate if the atmosphere were transparent in the ultraviolet region.

EXERCISE 25-1

Suppose that the photosynthesis reaction (*20b*) in Chapter 24 (p. 430) could be based upon light of wavelength 2400 Å (this light is absorbed heavily by ozone). How many moles of these photons would provide the 673 kcal of energy needed to produce one mole of glucose? (Remember, $E = h\nu$, and $h = 9.5 \times 10^{-14}$ kcal sec/mole). Compare your answer with that of Exercise 24-10.

At the opposite end of the spectrum, the infrared, again the atmosphere becomes virtually opaque. This is due mainly to absorption by gaseous water and carbon dioxide. Thus we find that the air, normally regarded as transparent, actually serves to filter the sun's rays striking the earth. The very high energy photons (in the ultraviolet) and the very low energy photons (in the infrared) are removed and the spectral region between is transmitted.

EXERCISE 25-2

On this planet, of what value would be an eye that is sensitive only to light in the ultraviolet spectral region? Discuss the evolutionary significance of the facts that the human eye and the photosynthesis process are both dependent upon light in the part of the spectrum called the "visible."

25-1.3 Composition of the Hydrosphere

Waters exposed to air dissolve some of it. In water, oxygen is twice as soluble as nitrogen but since nitrogen is four times more abundant than oxygen in air, more dissolved nitrogen is present than dissolved oxygen. It is the dissolved elementary oxygen that is used by living organisms for their oxidative processes. The concentration of dissolved carbon dioxide is low because its concentration in the air is low. But dissolved carbon dioxide is necessary for photosynthesis in marine plants. Dissolved carbon dioxide is responsible for part of the pleasant taste of water. Boiled water has lost almost all of the dissolved gases. It tastes "flat."

Ocean water also contains dissolved molecules from the gases of the air. These can be removed by boiling, but other solutes remain. When a kilogram of average ocean water is distilled, 967 grams of water can be collected and 33 grams of solids (primarily salts) remain behind. Thus, we may say that 3.3% of the weight of the ocean water is due to dissolved salts. Actually, more than forty of the elements have been identified as being present in ocean water but half of these are present in very small concentrations "less than 1 gram per billion, 10^9, grams of water."

Table 25-II. **AN AVERAGE COMPOSITION OF OCEAN WATER (DISREGARDING DISSOLVED GASES)**

ELEMENT *Name*	*Symbol*	PREDOMINANT SPECIES		NUMBER OF MOLES PER KILOGRAM
hydrogen oxygen	H O	H_2O		53.7
chlorine	Cl		$Cl^-(aq)$	0.535
sodium	Na	$Na^+(aq)$		0.460
magnesium	Mg	$Mg^{+2}(aq)$		0.052
sulfur	S		$SO_4^{-2}(aq)$	0.028
calcium	Ca	$Ca^{+2}(aq)$		0.010
potassium	K	$K^+(aq)$		0.010
bromine	Br		$Br^-(aq)$	0.008

In Table 25-II are shown the concentrations (in number of moles per 1000 grams of ocean water) of water and of the most abundant ions.

Several facts become apparent. There are fewer Na^+ ions than Cl^- ions; other positively charged ions—Mg^{+2}, Ca^{+2}, and K^+—are also present. Sulfate ions, SO_4^{-2}, and bromide ions, Br^-, are other negatively charged ions present in the water. Thus, ocean water is more than a solution of sodium chloride. Another fact is that K^+ ions are much less plentiful than Na^+ ions (Na^+/K^+ is about 46) even though K^+ ions are rather plentiful in the earth (Na^+/K^+ is about 2).

25-1.4 Composition and Properties of the Lithosphere

We know very much about the outermost portion of the lithosphere because it is available for direct study. In contrast, we know almost nothing about the inner lithosphere, though it constitutes over 99.5% of the mass of the earth.

THE INNER LITHOSPHERE

Seismic observations furnish our only probe of the inner lithosphere. The shock waves initiated by an earthquake travel through the interior of the earth in paths that are bent in accordance with the elastic properties and density of the medium they penetrate. From these paths, seismologists have been able to determine the existence of zones within the lithosphere. The outermost portion, or *mantle*, is approximately 2000 miles in depth and it is generally thought to be solid. This solid has a density of about 3 grams per milliliter near the crust and it increases to about 5 grams per milliliter at the bottom of the mantle. This higher density is caused by increasing pressure in the depths of the earth. For a comparison, the pressure at this depth is thought to be over a million atmospheres, two or three times greater than the highest pressures recorded in static laboratory experiments.

The inside of the lithosphere is called the core. Still higher pressures are expected and the density may rise as high as 18 grams per milliliter at the center of the earth. Some of the core may be liquid but the evidence is not decisive.

The composition of the mantle is probably rock-like, meaning it is made up of various silicates. These minerals have densities, compressibilities, and rigidities that match those indicated by the seismic studies. The core has long been thought to be largely iron, a view suggested by the composition of meteorites. These are solid objects that plummet through the atmosphere from space and they may be pieces of exploded planets resembling earth. Hence their composition furnishes a possible clue to the composition of the inner lithosphere. Current speculation ranges from iron to a high density rock but new evidence is needed.

The temperature of the center of the earth is

thought to be a few thousand degrees. Though this would melt rock at the earth's surface, solids can remain stable at the exceedingly high pressures thought to exist in the core.

Needless to say, very much remains to be learned about the chemistry of the inner lithosphere. It is a high temperature and high pressure laboratory whose door has not yet been opened.

THE EARTH'S CRUST

Oxygen and silicon are the most abundant elements in the earth's crust. Table 25-III shows that 60% of the atoms are oxygen atoms and 20% are silicon atoms. If our sample included the oceans, hydrogen would move into the third place ahead of aluminum (remember that water contains two hydrogen atoms for every oxygen atom). If the sample included the central core of the earth, iron would probably move into second place ahead of silicon, and magnesium would be fourth. Thus, the exact order is changed by the sample (part of earth) chosen. In any of the lists of elements, *the most abundant atoms are those of elements having low atomic numbers, 26 or less.* All of the elements beyond iron (element number 26) account for less than 0.2% of the weight of the earth's crust.

Table 25-III

ABUNDANCE OF ELEMENTS IN THE EARTH'S CRUST*

RANK	ELEMENT	ATOMIC NUMBER	NO. OF ATOMS PER 10,000 ATOMS
1	oxygen	8	6050
2	silicon	14	2045
3	aluminum	13	625
4	hydrogen	1	270
5	sodium	11	258
6	calcium	20	189
7	iron	26	187
8	magnesium	12	179
9	potassium	19	138
10	titanium	22	27
11	phosphorus	15	8.6
12	carbon	6	5.5
13	manganese	25	3.8
14	sulfur	16	3.4
15	fluorine	9	3.3
16	chlorine	17	2.8
17	chromium	24	1.5
18	barium	56	0.75

* From calculations of I. Asimov (*J. Chem. Ed.*) 31, 70 (1954) on the data of B. Gutenberg (Editor) "Internal Constitution of the Earth" 2nd Ed. Dover Publications, New York, 1951, p. 87. Reprinted with permission of the publisher.

25-1.5 Availability of Elements

In our daily life most of us are more concerned with the availability of the elements than with their general abundance in earth. The air is all around us and equally accessible to all. Water is somewhat more restricted. Some regions have a surplus of water, whereas other regions are deficient. Even in regions of abundant rainfall, the use of water may be so great that the reserves become depleted gradually. Thus, as the earth's population increases and more and more water is being used, water is becoming a natural resource to be treated with respect.

Many of the metals used by ancient man—copper (cuprum, Cu), silver (argentum, Ag), gold (aurum, Au), tin (stannum, Sn), and lead (plumbum, Pb)—are in relatively short supply. Ancient man found deposits of the first three occurring as the elementary metals. These three may also be separated from their ores by relatively simple chemical processes. On the other hand, aluminum and titanium, though abundant, are much more difficult to prepare from their ores. Fluorine is more abundant in the earth than chlorine but chlorine and its compounds are much more common—they are easier to prepare and easier to handle. However, as the best sources of the elements now common to us become depleted, we will have to turn to the elements that are now little used.

During geological time, a number of separating and sorting processes—melting, crystallization, solution, precipitation—have concentrated various elements in local deposits. In these, the elements tend to be grouped together in rather stable compounds. These are called minerals. Many of the minerals have compositions similar to compounds we can make in the laboratory but most are not so pure. There are large deposits of sodium chloride, for example, apparently formed when ancient seas evaporated in locales where the solid deposits were later protected from the dissolving action of water. The ocean itself is an enormous source of sodium chloride. In contrast, potassium salts have not been concentrated in a similar way. Many of the metallic elements are concentrated as sulfide minerals (for example, lead, PbS; molybdenum, MoS_2; zinc, ZnS). Other elements occur in rather concentrated oxide deposits (for example, iron, Fe_2O_3; manganese, MnO_2). Large deposits of carbonates (for example, zinc, $ZnCO_3$; calcium, $CaCO_3$) and sulfates (for example, barium, $BaSO_4$) of some metallic elements are known. The *minerals sufficiently concentrated to act as commercial sources of desired elements are called* ***ores.***

25-1.6 The Air as a Source of Elements

We are so used to free air that we do not think of oxygen as an important chemical. For example, we buy natural gas (mostly methane) as a fuel and burn it in air to furnish heat to us. If methane were free in the air and oxygen were scarce so that we had to purchase it we would consider oxygen to be the "fuel." In either case, the amount of heat released would be that for the reaction represented by the following equation.

$$CH_4(g) + 2O_2(g) \longrightarrow CO_2(g) + 2H_2O(l) + 213 \text{ kcal} \quad (3)$$

Despite the general availability of unlimited quantities of oxygen in the air, tremendous quantities of the pure gas are prepared annually for industrial and medical use. Billions of cubic feet of oxygen gas are manufactured every year, by liquefaction of air followed by fractional distillation to separate it from nitrogen.

Ores of nitrogen are relatively rare. The best mineral is sodium nitrate, $NaNO_3$, found in large deposits in Chile. We now prepare the nitrogen compounds we desire from the nitrogen of the air. Thus, the air is our best source of oxygen and nitrogen—two very important elements.

EXERCISE 25-3

Explain in terms of energy the significance of the fact that a piece of wood is stable in air at room temperature but if the temperature is raised, it burns and releases heat.

25-1.7 The Age of the Earth

Part of looking ahead is understanding the past. One of the most interesting questions man has asked about the past is, "How old is the earth?" Of course we are not sure there is an answer. We shall see, however, that ingenious methods have been developed that date the earth's crust. Scientists generally accept the proposal that the earth's crust, as we now find it, has a finite age.

The most reliable methods for establishing the age of a long-lasting object (such as a mountain) depend upon the presence of natural radioactivity. The decay of the radioactive elements can be likened to a clock that is partially unwound. By studying the extent to which the clock has unwound, we cannot tell the age of the clock but *we can measure how long ago it was wound.*

For example, consider the chemical composition of a very old crystal of pitchblende, U_3O_8. We may presume that this crystal was formed at a time when chemical conditions for its formation were favorable. For example, it may have precipitated from molten rock during cooling. The resulting crystals tend to exclude impurities. Yet, careful analysis shows that every deposit of pitchblende contains a small amount of lead. This lead has accumulated in the crystal, beginning at the moment the pure crystal was formed, due to the radioactive decay of the uranium.

The sequences of radioactive decays that lead to lead are well-known and the rates of decay have been carefully measured. We shall consider the sequence based upon the relatively slow decomposition of the most abundant uranium isotope, mass 238 (natural abundance, 99%):

$$^{238}_{92}U \longrightarrow {}^{234}_{90}Th + {}^{4}_{2}He \quad \alpha\text{-decay, half-life, } t_{1/2} = 4.6 \times 10^9 \text{ years} \quad (4)$$

The products are an α-particle (a helium nucleus), and a thorium isotope that is unstable and that rapidly decays by emitting successively two electrons:

$$^{234}_{90}Th \longrightarrow {}^{234}_{91}Pa + {}_{-1}^{0}e \quad \beta\text{-decay } t_{1/2} = 24.1 \text{ days} \quad (5)$$

$$^{234}_{91}Pa \longrightarrow {}^{234}_{92}U + {}_{-1}^{0}e \quad \beta\text{-decay } t_{1/2} = 1.14 \text{ minutes} \quad (6)$$

Thus we have returned to an isotope of uranium, ^{234}U, but one of half-life very much shorter than that of ^{238}U. This isotope begins a succession of α-decays, each moving the product upward in the periodic table:

$$^{234}_{92}U \longrightarrow {}^{230}_{90}Th + {}^{4}_{2}He \quad \alpha\text{-decay } t_{1/2} = 2.7 \times 10^5 \text{ years} \quad (7)$$

$$^{230}_{90}Th \longrightarrow {}^{226}_{88}Ra + {}^{4}_{2}He \quad \alpha\text{-decay } t_{1/2} = 8.3 \times 10^4 \text{ years} \quad (8)$$

$$^{226}_{88}Ra \longrightarrow {}^{222}_{86}Rn + {}^{4}_{2}He \quad \alpha\text{-decay } t_{1/2} = 1.6 \times 10^3 \text{ years} \quad (9)$$

$$^{222}_{86}Rn \longrightarrow {}^{218}_{84}Po + {}^{4}_{2}He \quad \alpha\text{-decay } t_{1/2} = 3.8 \text{ days} \quad (10)$$

$$^{218}_{84}Po \longrightarrow {}^{214}_{82}Pb + {}^{4}_{2}He \quad \alpha\text{-decay } t_{1/2} = 3.1 \text{ minutes} \quad (11)$$

We have at last reached lead—but ^{214}Pb is itself radioactive! This isotope decays in a succession of β-decays:

$$^{214}_{82}Pb \longrightarrow {}^{214}_{83}Bi + {}_{-1}^{0}e \quad \beta\text{-decay } t_{1/2} = 27 \text{ minutes} \quad (12)$$

$$^{214}_{83}Bi \longrightarrow {}^{214}_{84}Po + {}_{-1}^{0}e \quad \beta\text{-decay } t_{1/2} = 20 \text{ minutes} \quad (13)$$

Again α-decay occurs, returning to the element lead, but to another isotope that decays by emitting successively two β-particles:

$$^{214}_{84}Po \longrightarrow {}^{210}_{82}Pb + {}^{4}_{2}He \quad \alpha\text{-decay } t_{1/2} = 1.5 \times 10^{-4} \text{ seconds} \quad (14)$$

$$^{210}_{82}Pb \longrightarrow {}^{210}_{83}Bi + {}_{-1}^{0}e \quad \beta\text{-decay } t_{1/2} = 22 \text{ years} \quad (15)$$

$$^{210}_{83}Bi \longrightarrow {}^{210}_{84}Po + {}_{-1}^{0}e \quad \beta\text{-decay } t_{1/2} = 5 \text{ days} \quad (16)$$

This isotope of polonium, ^{210}Po, again decays by α-decay, but *this time giving an isotope that does not decay further:*

$$^{210}_{84}Po \longrightarrow {}^{206}_{82}Pb + {}^{4}_{2}He \quad \alpha\text{-decay } t_{1/2} = 140 \text{ days} \quad (17)$$

$$^{206}_{82}Pb \longrightarrow \text{not radioactive} \quad t_{1/2} \text{ infinite} \quad (18)$$

The products in this long sequence of reactions accumulate in the stable isotope of lead, ^{206}Pb. *The amount of ^{206}Pb present depends upon how long the deposit of uranium has decayed since the crystal U_3O_8 was formed.*

There is, fortunately, a rather simple verification of the presumption that all of the lead in the U_3O_8 came from this tedious sequence of nuclear reactions, (*4*) to (*17*). Lead ores that do not contain uranium include several isotopes—^{206}Pb makes up about 26% of the total and the rest is ^{204}Pb (1.4%), ^{207}Pb (21%), and ^{208}Pb (52%). Of these, two other isotopes can be formed through radioactive decay of some other uranium or thorium isotope by a sequence like that shown for ^{238}U. Of the four stable lead isotopes, only one is *not* derived from radioactive decay, ^{204}Pb. Hence, the ratio of the amount of this isotope to that of ^{206}Pb measures the amount of ^{206}Pb present in excess of the natural abundance. This excess must have come from decay of ^{238}U. If there is *no* ^{204}Pb present, then all of the ^{206}Pb came from ^{238}U.

Thus, analysis of uranium minerals with the aid of the mass spectrograph gives information on the age of the mineral. Though many different half-lives are involved in forming the lead, only the longest half-life (the rate-determining step) is of importance. *Combining the lead content with the ^{238}U half-life provides estimates of mineral ages in the range of five billion years.*

What have we learned in this estimate? Surely we can say the age of the earth cannot be *shorter* than 5×10^9 years. That was when the uranium mineral clock was wound—but the clock could be much older. To evaluate this number further, we must look for other types of data.

Fortunately, there are other radioactive elements in nature that give similar bases for estimates. As a second example, potassium in nature includes one radioactive isotope, $^{40}_{19}K$, which decays by capturing an electron into its nucleus. The product is ^{40}Ar, a stable isotope:

$$^{40}_{19}K + {}^{0}_{-1}e \longrightarrow {}^{40}_{18}Ar \quad \text{electron capture} \qquad t_{1/2} = 1.5 \times 10^{10} \text{ years} \quad (19)$$

Once again, the ratio of the abundance in a crystal of the two isotopes, $^{40}Ar/^{40}K$, provides a clue to the age of the crystal. Mica is a mineral that has been much studied in this type of mineral age estimation. Such estimates also tend to date the minerals as a few billion years old. Similar age figures are obtained from the natural radioactivity of rubidium.

In conclusion, the agreement of all of these methods based upon radioactive decay furnishes a strong clue that the earth's crust *as we know it today* was formed about five billion years ago. What preceded is a subject of intense interest and monumental disagreement.

25-2 THE CHEMISTRY OF THE PLANETS

Our solar system consists of the Sun, the planets and their moon satellites, asteroids (small planets), comets, and meteorites. The planets are generally divided into two categories: *Earth-like* (terrestrial) planets—Mercury, Venus, Earth, and Mars; and *Giant planets*—Jupiter, Saturn, Uranus, and Neptune. Little is known about Pluto, the most remote planet from Earth.

As space exploration begins, we can look forward to a vast multiplication of our present knowledge of the planets. Conceivably we shall be analyzing samples of the moon within this decade. The distances to the other planets are such that voyages of the order of a few months suffice to reach them. Again information will accumulate rapidly.

Contrast our position ten years ago. There lay the planets—ours to see but not to touch. And not to see well, either. No, we must view the planets through the gauze curtain of the atmosphere. Small wonder we know rather little about our nearest neighbors, despite our eagerness to pry. Yet we do know enough about the other planets to say that their surface environments are totally unlike that on earth. Though much of what is today thought to be true will be known to be false tomorrow, it provides stimulating reading.

Table 25-IV begins this survey with a comparison of mass, radius, and density of the planets and the sun. These data are probably the most reliable facts known about the planets since they are deduced from the orbital movements in the solar system.

Table 25-IV

DATA ON THE SOLAR SYSTEM

	RELATIVE MASS	RADIUS (kilometers)	DENSITY (g/ml)
Sun	3.32×10^5	695×10^3	1.41
Mercury	0.05	2.5×10^3	5.1
Venus	0.81	6.2×10^3	5.0
Earth	(1.00)	6.371×10^3	5.52
Mars	0.11	3.4×10^3	3.9
Jupiter	3.18×10^2	71×10^3	1.33
Saturn	95	57×10^3	0.71
Uranus	14.6	25.8×10^3	1.27
Neptune	17.3	22.3×10^3	2.22
Pluto	0.03?	2.9×10^3	2?

25-2.1 Meteorites

We do have some direct evidence concerning the composition of solid matter outside the earth's atmosphere. Occasionally a piece of solid, a meteorite, falls through the atmosphere to give us a real sample for analysis. Such a piece of solid may become hot because of its rapid movement through the air and therefore glow brightly. Many meteors burn up or evaporate but some are large enough to survive and reach the earth's surface. These we can examine and analyze.

Meteorites are of two kinds: stony meteorites that are rock-like in character, and metallic meteorites that consist of metallic elements. The kinds of substances in the stony meteorites are very much like the substances in the crust of the earth, if we allow for the fact that the meteors could not bring gases or liquids with them. We feel that the other type, the metallic meteors, give valuable clues about the nature of the earth's central core. Experts have long believed that these meteorites are fragments from exploded planets that, perhaps, resembled the earth.

Whether this is true or not, meteorites give us one definite piece of information. Isotopic analysis shows that each element in a meteorite has the same isotopes in the same percentages that this same element has on earth. The accepted explanation for this fact is that meteorites and the earth share a common origin and that they became separated *after* the elements were created.

25-2.2 The Planetary Atmospheres

Through spectroscopic observations and sometimes tenuous deductions there has accumulated a significant picture of the makeup of the planetary atmospheres. Doubt pervades much of this picture, yet it represents our starting place in knowledge as we venture outside our own atmosphere for the first time. Table 25-V summarizes a part of this information—the maximum surface temperatures and the chemical compositions. Naturally, these compositions are incomplete:

Table 25-V

THE PLANETARY ATMOSPHERES

PLANET	MAX. SURFACE TEMPERATURE (°C)	SOME OF THE GASES PRESENT
Venus	430	CO_2
Mercury	350	none
Earth	60	O_2, N_2, H_2O, etc.
Mars	30	N_2, CO_2, H_2O
Jupiter	−138	CH_4, NH_3
Saturn	−153	CH_4, NH_3
Uranus	−184	CH_4
Neptune	−200	CH_4

these are the substances that *can* be detected and other gases are undoubtedly present as well.

There appears to be a correlation between the mass of the planets and the mass and composition of their atmospheres. Generally, only those planets of high mass were able to retain much of their atmospheres. Nitrogen, hydrogen, and helium are probably abundant, though not yet detected, on the heavier planets. Table 25-V also reveals a considerable range in the surface temperatures of the planets. The higher temperatures on the terrestrial planets also contributed to the loss of their atmospheres.

EXERCISE 25-4

Nitrogen is considered to be a likely constituent of the atmosphere of Jupiter, though it is undetected as yet. As a chemist, would you expect oxygen also to be an important constituent of Jupiter's CH_4-NH_3 atmosphere?

The composition of the planetary atmospheres is fairly constant. This is indeed surprising in view of the fact that molecules such as methane, ammonia, and carbon dioxide are easily decomposed by the ultraviolet radiation from the sun. Presumably other reactions regenerate those substances that are light sensitive.

The atmosphere of Venus is chiefly carbon dioxide in a concentration much higher than that found on Earth. Surprisingly, no evidence has been found for carbon monoxide, though ultraviolet light decomposes CO_2 to form CO. The atmosphere of Mars is thought to be largely nitrogen (around 98%) and some carbon dioxide.

Recent space-probe and earth-based spectroscopic studies of the planet Venus suggest how much remains to be learned about the other planets. Earlier estimates of the surface temperature of Venus placed it near 60°C. The more detailed studies show, however, that two characteristic temperatures can be identified, −40°C and 430°C. The lower temperature is attributed to light emitted from high altitude cloud tops. The higher temperature is likely to be the average surface temperature.

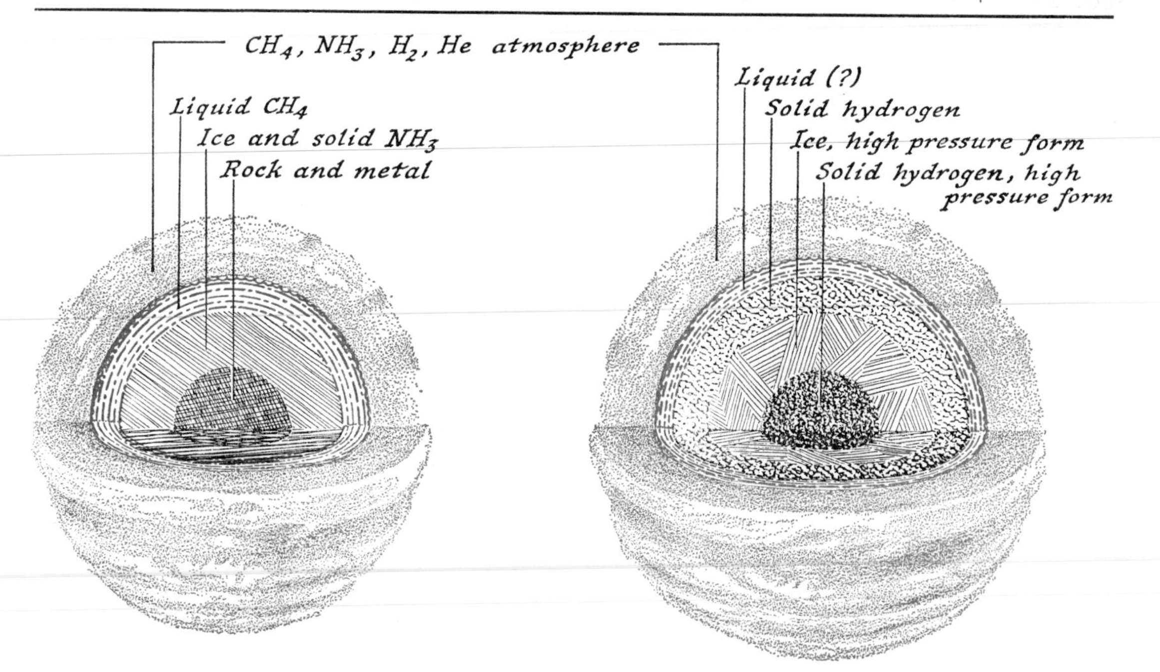

Fig. 25-2. **Two proposed structures of Jupiter.**

These latest findings provide an estimate that the atmospheric pressure at the ground surface is 10 ± 2.5 atmospheres. This is the pressure felt by a deep-sea diver at a depth of 300 feet, near the limit of human endurance. The atmospheric temperature at the surface averages around 430°C, rising in some locales possibly as high as 550°C (near the softening temperature of ordinary glass) whereas in other regions, cool breezes may blow at temperatures below 350°C (near the boiling point of mercury and the melting point of lead). Compare these extremes to the narrow range in which humans can survive comfortably. If you have ever experienced a desert temperature as high as 43°C (110°F) or a winter as cold as −35°C (−30°F), you will realize that Venus will not be plagued by too many tourists from Earth.

The giant planets possess low surface temperatures and have atmospheres that extend several thousand miles. The markings on Jupiter, the largest planet, consist of cloud formations composed of methane containing a small amount of ammonia. The atmosphere of Jupiter absorbs the extreme red and infrared portions of the spectrum. These absorptions correspond to the absorption spectra of ammonia and methane, suggesting the presence of these gases in Jupiter's atmosphere. Free hydrogen and helium are also thought to be present but this is difficult to verify. Estimates of the average molecular weight of the gases in Jupiter's atmosphere are around 3. The atmosphere of Jupiter includes belts or bands that seem to be indicative of climate variations similar to our own equatorial, temperate, and polar climates.

Jupiter has a single permanent marking called the Great Red Spot. This spot is oval in shape, about 30,000 miles long and about 7,000 miles wide. The coloring is thought to result from light reflected from the different layers in the planet's atmosphere. Theories concerning the origin of this spot and of Jupiter's brightly colored atmospheric belts are imaginative, numerous, and in general disagreement.

25-2.3 The Planetary Lithospheres

Needless to say, the extreme difficulty we experience in probing the composition of the earth beneath us suggests that little is known about the inner composition of the planets. The evidence available is indirect (average density, sur-

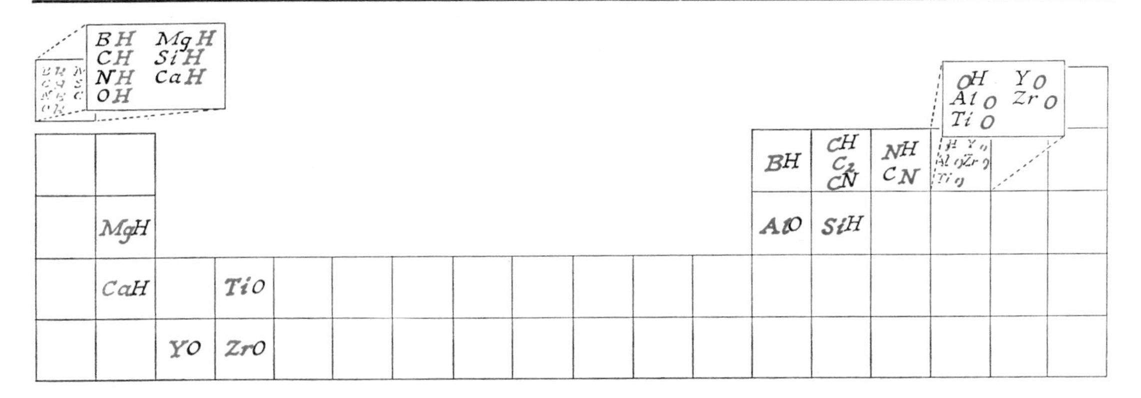

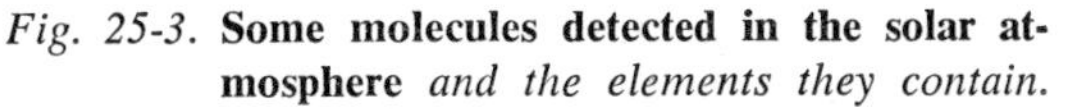
Fig. 25-3. **Some molecules detected in the solar atmosphere** *and the elements they contain.*

face composition, etc.) and the interpretations are conflicting. Nevertheless, we present in Figure 25-2 two proposals that have been offered as possible structures for the planet Jupiter. A variety of responses are evoked by these startling proposals: *dismay*—that the available data are so inconclusive; *discouragement*—that our knowledge is so incomplete; *anticipation*—for the near future when some of the uncertainties will be removed; *sympathy*—for the poor astronaut who is to step out of his space vehicle to plant his flag in an unfriendly sea of, alas, liquid methane.

25-2.4 The Sun

The surface temperature of the sun is about 5500°C. Moving inward from the surface, the temperature rises, probably above one million degrees. At these large temperatures the competition between opposing tendencies toward lowest energy (favoring molecules) and toward highest randomness (favoring atoms) is dominated by the randomness factor. As a result, only the simplest molecules are to be expected there.

The solar spectrum is, of course, as well studied as our planetary atmosphere will permit. More information will be forthcoming as spectra from man-made satellites are recorded above the atmosphere. At this time, the spectra of many diatomic molecules have been detected. These are not the familiar, chemically stable molecules we find on the stockroom shelf. These are the molecules that are stable on a solar stockroom shelf. Figure 25-3 shows some of these and the location in the periodic table of the elements represented.

Inside the sun, thermal energies are sufficient to destroy all molecules and to ionize the atoms. These ions emit their characteristic line spectra and tens of thousands of lines are observed. The lines that have been analyzed show the existence of atoms ionized as far as O^{+5}, Mn^{+12}, and Fe^{+13}. At this time, over sixty of the elements have been detected in the sun through their spectral emissions and absorptions.

25-3 THE STARS

Our knowledge of the stars and of space is entirely obtained through spectroscopy and will be for the forseeable future. That is not at all to say we know and shall know little of the other galaxies. It is to say that our information will be incomplete. But man is opportunistic and clever—small pieces of a spectroscopic jawbone may permit us to build quite a reasonable likeness of the Universe.

As evidence that this is so, consider that the element helium was detected in the sun before it was found on earth! Though oxygen contains 0.2% of the oxygen-18 isotope on earth, it, too, was first detected in a solar spectrum. Two chemical species whose spectra were first provided through photographs of comets are CO^+ and C_3. Let us look briefly then, and with respect, at the astronomers' knowledge of stellar chemistry.

25-3.1 Stellar Atmospheres

Our sun is, of course, a star. It is a relatively cool star and, as such, contains a number of diatomic molecules (see Figure 25-3). There are many stars, however, with still lower surface temperatures and these contain chemical species whose presence can be understood in terms of the temperatures and the usual chemical equilibrium principles. For example, as the star temperature drops, the spectral lines attributed to CN and CH become more prominent. At lower temperatures, TiO becomes an important species along with the hydrides MgH, SiH, and AlH, and oxides ZrO, ScO, YO, CrO, AlO, and BO.

Detailed consideration of the chemical equilibria among these species provides evidence of the presence of other molecules that cannot be observed directly. The chemical properties of the molecules mentioned provide a firm basis for predicting the presence and concentrations of such important molecules as H_2, CO, O_2, N_2, and NO. Thus the faint light by which we view the distant stars is rich with information. We need but learn how to read it.

25-3.2 Interstellar Space

In addition to the stars, with their characteristic light emissions, the space between them is part of the astronomical spectroscopy laboratory. Light from a distant star must traverse fantastic distances to reach our telescopes and the absorption of this light by the most minute concentrations of atoms and molecules in space becomes important and detectable. Absorption spectra have shown that diatomic molecules such as CH, CN, and CH^+ are present at an average concentration of about one molecule per 1000 liters. These molecules are probably concentrated in "clouds" with one molecule per 100 liters.

EXERCISE 25-5

Calculate the volume in liters of a sphere of radius 6400 kilometers (the radius of the earth). How many grams of oxygen would be needed to fill this volume to a concentration of one molecule per 1000 liters?

In addition to these molecules, atoms are present, as shown by absorptions of Ca, Na, K, Fe, and other atoms. There are some absorptions that have not been identified but these may be due to small solid particles. How these particular molecules and atoms happen to be present in these almost nonexistent "clouds" and what other molecules and atoms are there, yet to be detected, is a matter for wondering. But wondering is at once the pleasure and the driving force of science.

APPENDIX 1

A DESCRIPTION OF A BURNING CANDLE

A drawing of a burning candle is shown[1] in Figure A1-1. The candle is cylindrical in shape[2] and has a diameter[3] of about $\frac{3}{4}$ inch. The length of the candle was initially about eight inches[4] and it changed slowly[5] during observation, decreasing about half an inch in one hour[6]. The candle is made of a translucent[7], white[8] solid[9] which has a slight odor[10] and no taste[11]. It is soft enough to be scratched with the fingernail[12]. There is a wick[13] which extends from top to bottom[14] of the candle along its central axis[15] and protrudes about half an inch above the top of the candle[16]. The wick is made of three strands of string braided together[17].

A candle is lit by holding a source of flame close to the wick for a few seconds. Thereafter the source of flame can be removed and the flame sustains itself at the wick[18]. The burning candle makes no sound[19]. While burning, the body of the candle remains cool to the touch[20] except near the top. Within about half an inch from the top the candle is warm[21] (but not hot) and sufficiently soft to mold easily[22]. The flame flickers in response to air currents[23] and tends to become quite smoky while flickering[24]. In the absence of air currents, the flame is of the form shown in Figure A1-1, though it retains some movement at all times[25]. The flame begins about $\frac{1}{8}$ inch above the top of the candle[26] and at its base the flame has a blue tint[27]. Immediately around the wick in a region about $\frac{1}{4}$ inch wide and extending about $\frac{1}{2}$ inch above the top of the wick[28] the flame is dark[29]. This dark region is roughly conical in shape[30]. Around this zone and extending about half an inch above the dark zone is a region which emits yellow light[31], bright but not blinding[32]. The flame has rather

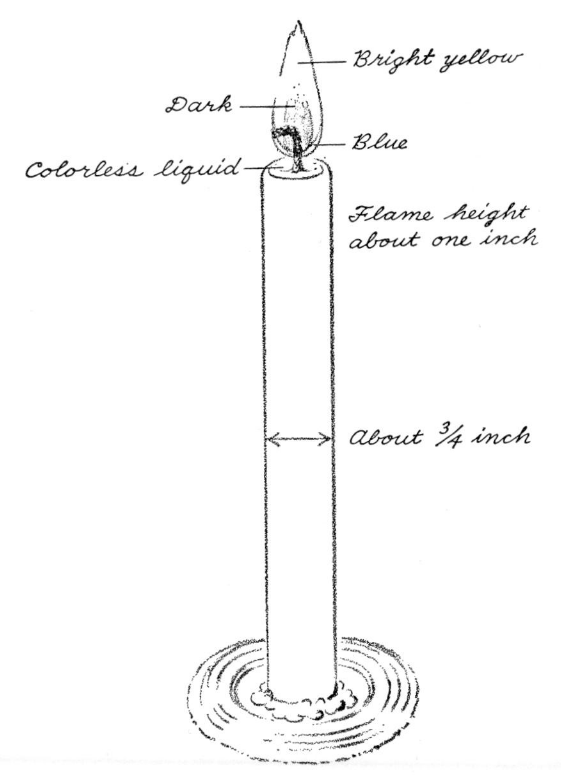

Fig. A1-1. **A burning candle.**

sharply defined sides[33], but a ragged top[34]. The wick is white where it emerges from the candle[35], but from the base of the flame to the end of the wick[36] it is black, appearing burnt, except for the last $\frac{1}{16}$ inch where it glows red[37]. The wick curls over about $\frac{1}{4}$

inch from its end[38]. As the candle becomes shorter, the wick shortens too, so as to extend roughly a constant length above the top of the candle[39]. Heat is emitted by the flame[40], enough so that it becomes uncomfortable in ten or twenty seconds if one holds his finger $\frac{1}{4}$ inch to the side of the quiet flame[41] or three or four inches above the flame[42].

The top of a quietly burning candle becomes wet with a colorless liquid[43] and becomes bowl shaped[44]. If the flame is blown, one side of this bowl-shaped top may become liquid, and the liquid trapped in the bowl may drain down the candle's side[45]. As it courses down, the colorless liquid cools[46], becomes translucent[47], and gradually solidifies from the outside[48], attaching itself to the side of the candle[49]. In the absence of a draft, the candle can burn for hours without such dripping[50]. Under these conditions, a stable pool of clear liquid remains in the bowl-shaped top of the candle[51]. The liquid rises slightly around the wick[52], wetting the base of the wick as high as the base of the flame[53].

Several aspects of this description deserve specific mention. Compare your own description in each of the following characteristics.

(1) The description is comprehensive in *qualitative* terms. Did *you* include mention of appearance? smell? taste? feel? sound? (Note: A chemist quickly becomes reluctant to taste or smell an unknown chemical. A chemical should be considered to be poisonous unless it is *known* not to be!)

(2) Wherever possible, the description is stated *quantitatively*. This means the question "How much?" is answered (the quantity is specified). The remark that the flame emits yellow light is made more meaningful by the "how much" expression, "bright but not blinding." The statement that heat is emitted might lead a cautious investigator who is lighting a candle for the first time to stand in a concrete blockhouse one hundred yards away. The few words telling him "how much" heat would save him this overprecaution.

(3) The description does not presume the importance of an observation. Thus the observation that a burning candle does not emit sound deserves to be mentioned just as much as the observation that it does emit light.

(4) The description does not confuse observations with interpretations. It is an observation that the top of the burning candle is wet with a colorless liquid. It would be an interpretation to state the presumed composition of this liquid.

APPENDIX 

RELATIVE STRENGTHS OF ACIDS

IN AQUEOUS SOLUTION AT ROOM TEMPERATURE

All ions are aquated

$$HB \rightleftharpoons H^+(aq) + B^-(aq) \qquad K_A = \frac{[H^+][B^-]}{[HB]}$$

ACID	STRENGTH	REACTION	K_A
perchloric acid	very strong	$HClO_4 \longrightarrow H^+ + ClO_4^-$	very large
hydriodic acid		$HI \longrightarrow H^+ + I^-$	very large
hydrobromic acid		$HBr \longrightarrow H^+ + Br^-$	very large
hydrochloric acid		$HCl \longrightarrow H^+ + Cl^-$	very large
nitric acid		$HNO_3 \longrightarrow H^+ + NO_3^-$	very large
sulfuric acid	very strong	$H_2SO_4 \longrightarrow H^+ + HSO_4^-$	large
oxalic acid		$HOOCCOOH \longrightarrow H^+ + HOOCCOO^-$	5.4×10^{-2}
sulfurous acid ($SO_2 + H_2O$)		$H_2SO_3 \longrightarrow H^+ + HSO_3^-$	1.7×10^{-2}
hydrogen sulfate ion	strong	$HSO_4^- \longrightarrow H^+ + SO_4^{-2}$	1.3×10^{-2}
phosphoric acid		$H_3PO_4 \longrightarrow H^+ + H_2PO_4^-$	7.1×10^{-3}
ferric ion		$Fe(H_2O)_6^{+3} \longrightarrow H^+ + Fe(H_2O)_5(OH)^{+2}$	6.0×10^{-3}
hydrogen telluride		$H_2Te \longrightarrow H^+ + HTe^-$	2.3×10^{-3}
hydrofluoric acid	weak	$HF \longrightarrow H^+ + F^-$	6.7×10^{-4}
nitrous acid		$HNO_2 \longrightarrow H^+ + NO_2^-$	5.1×10^{-4}
hydrogen selenide		$H_2Se \longrightarrow H^+ + HSe^-$	1.7×10^{-4}
chromic ion		$Cr(H_2O)_6^{+3} \longrightarrow H^+ + Cr(H_2O)_5(OH)^{+2}$	1.5×10^{-4}
benzoic acid		$C_6H_5COOH \longrightarrow H^+ + C_6H_5COO^-$	6.6×10^{-5}
hydrogen oxalate ion		$HOOCCOO^- \longrightarrow H^+ + OOCCOO^{-2}$	5.4×10^{-5}
acetic acid	weak	$CH_3COOH \longrightarrow H^+ + CH_3COO^-$	1.8×10^{-5}
aluminum ion		$Al(H_2O)_6^{+3} \longrightarrow H^+ + Al(H_2O)_5(OH)^{+2}$	1.4×10^{-5}
carbonic acid ($CO_2 + H_2O$)		$H_2CO_3 \longrightarrow H^+ + HCO_3^-$	4.4×10^{-7}
hydrogen sulfide		$H_2S \longrightarrow H^+ + HS^-$	1.0×10^{-7}
dihydrogen phosphate ion		$H_2PO_4^- \longrightarrow H^+ + HPO_4^{-2}$	6.3×10^{-8}
hydrogen sulfite ion		$HSO_3^- \longrightarrow H^+ + SO_3^{-2}$	6.2×10^{-8}
ammonium ion	weak	$NH_4^+ \longrightarrow H^+ + NH_3$	5.7×10^{-10}
hydrogen carbonate ion		$HCO_3^- \longrightarrow H^+ + CO_3^{-2}$	4.7×10^{-11}
hydrogen telluride ion		$HTe^- \longrightarrow H^+ + Te^{-2}$	1.0×10^{-5}
hydrogen peroxide	very weak	$H_2O_2 \longrightarrow H^+ + HO_2^-$	2.4×10^{-12}
monohydrogen phosphate ion		$HPO_4^{-2} \longrightarrow H^+ + PO_4^{-3}$	4.4×10^{-13}
hydrogen sulfide ion		$HS^- \longrightarrow H^+ + S^{-2}$	1.3×10^{-13}
water		$H_2O \longrightarrow H^+ + OH^-$	$[H^+][OH^-] = 1.0 \times 10^{-14}$
hydroxide ion		$OH^- \longrightarrow H^+ + O^{-2}$	$< 10^{-36}$
ammonia	very weak	$NH_3 \longrightarrow H^+ + NH_2^-$	very small

APPENDIX 3

STANDARD OXIDATION POTENTIALS FOR HALF-REACTIONS

IONIC CONCENTRATIONS, 1 *M* IN WATER AT 25°C

All ions are aquated

Very strong reducing agents (Reducing strength increases ↑)

Very weak oxidizing agents (Oxidizing strength increases ↓)

HALF-REACTION	$E°$ (volts)
$Li \longrightarrow e^- + Li^+$	3.00
$Rb \longrightarrow e^- + Rb^+$	2.92
$K \longrightarrow e^- + K^+$	2.92
$Cs \longrightarrow e^- + Cs^+$	2.92
$Ba \longrightarrow 2e^- + Ba^{+2}$	2.90
$Sr \longrightarrow 2e^- + Sr^{+2}$	2.89
$Ca \longrightarrow 2e^- + Ca^{+2}$	2.87
$Na \longrightarrow e^- + Na^+$	2.71
$Mg \longrightarrow 2e^- + Mg^{+2}$	2.37
$Al \longrightarrow 3e^- + Al^{+3}$	1.66
$Mn \longrightarrow 2e^- + Mn^{+2}$	1.18
$H_2(g) + 2OH^- \longrightarrow 2e^- + 2H_2O$	0.83
$Zn \longrightarrow 2e^- + Zn^{+2}$	0.76
$Cr \longrightarrow 3e^- + Cr^{+3}$	0.74
$H_2Te \longrightarrow 2e^- + Te + 2H^+$	0.72
$2Ag + S^{-2} \longrightarrow 2e^- + Ag_2S$	0.69
$Fe \longrightarrow 2e^- + Fe^{+2}$	0.44
$H_2(g) \longrightarrow 2e^- + 2H^+ (10^{-7}\ M)$	0.414
$Cr^{+2} \longrightarrow e^- + Cr^{+3}$	0.41
$H_2Se \longrightarrow 2e^- + Se + 2H^+$	0.40
$Co \longrightarrow 2e^- + Co^{+2}$	0.28
$Ni \longrightarrow 2e^- + Ni^{+2}$	0.25
$Sn \longrightarrow 2e^- + Sn^{+2}$	0.14
$Pb \longrightarrow 2e^- + Pb^{+2}$	0.13
$H_2(g) \longrightarrow 2e^- + 2H^+$	0.00
$H_2S(g) \longrightarrow 2e^- + S + 2H^+$	−0.14
$Sn^{+2} \longrightarrow 2e^- + Sn^{+4}$	−0.15
$Cu^+ \longrightarrow e^- + Cu^{+2}$	−0.15
$SO_2(g) + 2H_2O \longrightarrow 2e^- + SO_4^{-2} + 4H^+$	−0.17
$Cu \longrightarrow 2e^- + Cu^{+2}$	−0.34
$Cu \longrightarrow e^- + Cu^+$	−0.52
$2I^- \longrightarrow 2e^- + I_2$	−0.53
$H_2O_2 \longrightarrow 2e^- + O_2(g) + 2H^+$	−0.68

APPENDIX 3—(*Continued*)

	HALF-REACTION	$E°$ (VOLTS)	
Reducing strength increases ↑	$Fe^{+2} \longrightarrow e^- + Fe^{+3}$	−0.77	Oxidizing strength increases ↓
	$NO_2(g) + H_2O \longrightarrow e^- + NO_3^- + 2H^+$	−0.78	
	$Hg(l) \longrightarrow 2e^- + Hg^{+2}$	−0.78	
	$Hg(l) \longrightarrow e^- + \frac{1}{2}Hg_2^{+2}$	−0.79	
	$Ag \longrightarrow e^- + Ag^+$	−0.80	
	$H_2O \longrightarrow 2e^- + \frac{1}{2}O_2(g) + 2H^+ \ (10^{-7}\ M)$	−0.815	
	$NO(g) + 2H_2O \longrightarrow 3e^- + NO_3^- + 4H^+$	−0.96	
	$Au + 4Cl^- \longrightarrow 3e^- + AuCl_4^-$	−1.00	
	$2Br^- \longrightarrow 2e^- + Br_2(l)$	−1.06	
	$H_2O \longrightarrow 2e^- + \frac{1}{2}O_2(g) + 2H^+$	−1.23	
	$Mn^{+2} + 2H_2O \longrightarrow 2e^- + MnO_2 + 4H^+$	−1.28	
	$2Cr^{+3} + 7H_2O \longrightarrow 6e^- + Cr_2O_7^{-2} + 14H^+$	−1.33	
	$2Cl^- \longrightarrow 2e^- + Cl_2(g)$	−1.36	
	$Au \longrightarrow 3e^- + Au^{+3}$	−1.50	
Very weak reducing agents	$Mn^{+2} + 4H_2O \longrightarrow 5e^- + MnO_4^- + 8H^+$	−1.52	Very strong oxidizing agents
	$2H_2O \longrightarrow 2e^- + H_2O_2 + 2H^+$	−1.77	
	$2F^- \longrightarrow 2e^- + F_2(g)$	−2.87	

APPENDIX

NAMES, FORMULAS, AND CHARGES OF SOME COMMON IONS

POSITIVE IONS (CATIONS)		NEGATIVE IONS (ANIONS)	
aluminum	Al^{+3}	acetate	CH_3COO^-
ammonium	NH_4^+	bromide	Br^-
barium	Ba^{+2}	carbonate	CO_3^{-2}
calcium	Ca^{+2}	hydrogen carbonate ion, bicarbonate	HCO_3^-
chromium (II), chromous	Cr^{+2}	chlorate	ClO_3^-
chromium (III), chromic	Cr^{+3}	chloride	Cl^-
copper (I),* cuprous	Cu^+	chlorite	ClO_2^-
copper (II), cupric	Cu^{+2}	chromate	CrO_4^{-2}
hydrogen, hydronium	H^+, H_3O^+	dichromate	$Cr_2O_7^{-2}$
iron (II),* ferrous	Fe^{+2}	fluoride	F^-
iron (III), ferric	Fe^{+3}	hydroxide	OH^-
lead	Pb^{+2}	hypochlorite	ClO^-
lithium	Li^+	iodide	I^-
magnesium	Mg^{+2}	nitrate	NO_3^-
manganese (II), manganous	Mn^{+2}	nitrite	NO_2^-
mercury (I),* mercurous	Hg_2^{+2}	oxalate	$C_2O_4^{-2}$
mercury (II), mercuric	Hg^{+2}	hydrogen oxalate ion, binoxalate	$HC_2O_4^-$
potassium	K^+	perchlorate	ClO_4^-
silver	Ag^+	permanganate	MnO_4^-
sodium	Na^+	phosphate	PO_4^{-3}
tin (II),* stannous	Sn^{+2}	monohydrogen phosphate	HPO_4^{-2}
tin (IV), stannic	Sn^{+4}	dihydrogen phosphate	$H_2PO_4^-$
zinc	Zn^{+2}	sulfate	SO_4^{-2}
		hydrogen sulfate ion, bisulfate	HSO_4^-
		sulfide	S^{-2}
		hydrogen sulfide ion, bisulfide	HS^-
		sulfite	SO_3^{-2}
		hydrogen sulfite ion, bisulfite	HSO_3^-

* Aqueous solutions are readily oxidized by air.

Note: In ionic compounds the relative number of positive and negative ions is such that the sum of their electric charges is zero.

Index

Boldface numbers refer to definitions. *Italic* numbers refer to sections.

Boldface numbers refer to definitions. *Italic* numbers refer to sections.

Boldface numbers refer to definitions. *Italic* numbers refer to sections.

Boldface numbers refer to definitions. *Italic* numbers refer to sections.

Boldface numbers refer to definitions. *Italic* numbers refer to sections.

Boldface numbers refer to definitions. *Italic* numbers refer to sections.

Boldface numbers refer to definitions. *Italic* numbers refer to sections.

Boldface numbers refer to definitions. *Italic* numbers refer to sections.

Boldface numbers refer to definitions. *Italic* numbers refer to sections.

Boldface numbers refer to definitions. *Italic* numbers refer to sections.

Boldface numbers refer to definitions. *Italic* numbers refer to sections.

PERIODIC TABLE

1 H 1.008																	2 He 4.00
3 Li 6.94	4 Be 9.01											5 B 10.8	6 C 12.01	7 N 14.01	8 O 16.00	9 F 19.0	10 Ne 20.2
11 Na 23.0	12 Mg 24.3											13 Al 27.0	14 Si 28.1	15 P 31.0	16 S 32.1	17 Cl 35.5	18 Ar 39.9
19 K 39.1	20 Ca 40.1	21 Sc 45.0	22 Ti 47.9	23 V 50.9	24 Cr 52.0	25 Mn 54.9	26 Fe 55.8	27 Co 58.9	28 Ni 58.7	29 Cu 63.5	30 Zn 65.4	31 Ga 69.7	32 Ge 72.6	33 As 74.9	34 Se 79.0	35 Br 79.9	36 Kr 83.8
37 Rb 85.5	38 Sr 87.6	39 Y 88.9	40 Zr 91.2	41 Nb 92.9	42 Mo 95.9	43 Tc (99)	44 Ru 101.1	45 Rh 102.9	46 Pd 106.4	47 Ag 107.9	48 Cd 112.4	49 In 114.8	50 Sn 118.7	51 Sb 121.8	52 Te 127.6	53 I 126.9	54 Xe 131.3
55 Cs 132.9	56 Ba 137.3	57–71 See below	72 Hf 178.5	73 Ta 180.9	74 W 183.9	75 Re 186.2	76 Os 190.2	77 Ir 192.2	78 Pt 195.1	79 Au 197.0	80 Hg 200.6	81 Tl 204.4	82 Pb 207.2	83 Bi 209.0	84 Po (209)	85 At (210)	86 Rn (222)
87 Fr (223)	88 Ra (226)	89– See below															

57 La 138.9	58 Ce 140.1	59 Pr 140.9	60 Nd 144.2	61 Pm (147)	62 Sm 150.4	63 Eu 152.0	64 Gd 157.3	65 Tb 158.9	66 Dy 162.5	67 Ho 164.9	68 Er 167.3	69 Tm 168.9	70 Yb 173.0	71 Lu 175.0
89 Ac (227)	90 Th 232.0	91 Pa (231)	92 U 238.0	93 Np (237)	94 Pu (242)	95 Am (243)	96 Cm (247)	97 Bk (245)	98 Cf (251)	99 Es (254)	100 Fm (253)	101 Md (256)	102 (254)	103 Lw (257)

Parenthetical values are mass numbers of the isotopes with longest half lives

INTERNATIONAL ATOMIC WEIGHTS

NAME	SYMBOL	ATOMIC NUMBER	ATOMIC WEIGHT	NAME	SYMBOL	ATOMIC NUMBER	ATOMIC WEIGHT
Actinium	Ac	89	(227)	Mercury	Hg	80	200.6
Aluminum	Al	13	27.0	Molybdenum	Mo	42	95.9
Americium	Am	95	(243)	Neodymium	Nd	60	144.2
Antimony	Sb	51	121.8	Neon	Ne	10	20.2
Argon	Ar	18	39.9	Neptunium	Np	93	(237)
Arsenic	As	33	74.9	Nickel	Ni	28	58.7
Astatine	At	85	(210)	Niobium	Nb	41	92.9
Barium	Ba	56	137.3	Nitrogen	N	7	14.01
Berkelium	Bk	97	245	Osmium	Os	76	190.2
Beryllium	Be	4	9.01	Oxygen	O	8	16.00
Bismuth	Bi	83	209.0	Palladium	Pd	46	106.4
Boron	B	5	10.8	Phosphorus	P	15	31.0
Bromine	Br	35	79.9	Platinum	Pt	78	195.1
Cadmium	Cd	48	112.4	Plutonium	Pu	94	(242)
Calcium	Ca	20	40.1	Polonium	Po	84	210
Californium	Cf	98	(251)	Potassium	K	19	39.1
Carbon	C	6	12.01	Praseodymium	Pr	59	140.9
Cerium	Ce	58	140.1	Promethium	Pm	61	(147)
Cesium	Cs	55	132.9	Protactinium	Pa	91	(231)
Chlorine	Cl	17	35.5	Radium	Ra	88	(226)
Chromium	Cr	24	52.0	Radon	Rn	86	(222)
Cobalt	Co	27	58.9	Rhenium	Re	75	186.2
Copper	Cu	29	63.5	Rhodium	Rh	45	102.9
Curium	Cm	96	(247)	Rubidium	Rb	37	85.5
Dysprosium	Dy	66	162.5	Ruthenium	Ru	44	101.1
Einsteinium	Es	99	(254)	Samarium	Sm	62	150.4
Erbium	Er	68	167.3	Scandium	Sc	21	45.0
Europium	Eu	63	152.0	Selenium	Se	34	79.0
Fermium	Fm	100	(253)	Silicon	Si	14	28.1
Fluorine	F	9	19.0	Silver	Ag	47	107.9
Francium	Fr	87	(223)	Sodium	Na	11	23.0
Gadolinium	Gd	64	157.3	Strontium	Sr	38	87.6
Gallium	Ga	31	69.7	Sulfur	S	16	32.1
Germanium	Ge	32	72.6	Tantalum	Ta	73	180.9
Gold	Au	79	197.0	Technetium	Tc	43	(99)
Hafnium	Hf	72	178.5	Tellurium	Te	52	127.6
Helium	He	2	4.00	Terbium	Tb	65	158.9
Holmium	Ho	67	164.9	Thallium	Tl	81	204.4
Hydrogen	H	1	1.008	Thorium	Th	90	232.0
Indium	In	49	114.8	Thulium	Tm	69	168.9
Iodine	I	53	126.9	Tin	Sn	50	118.7
Iridium	Ir	77	192.2	Titanium	Ti	22	47.9
Iron	Fe	26	55.8	Tungsten	W	74	183.9
Krypton	Kr	36	83.8	Uranium	U	92	238.0
Lanthanum	La	57	138.9	Vanadium	V	23	50.9
Lead	Pb	82	207.2	Xenon	Xe	54	131.3
Lithium	Li	3	6.94	Ytterbium	Yb	70	173.0
Lutetium	Lu	71	175.0	Yttrium	Y	39	88.9
Magnesium	Mg	12	24.3	Zinc	Zn	30	65.4
Manganese	Mn	25	54.9	Zirconium	Zr	40	91.2
Mendelevium	Md	101	(256)				

Parenthetical names refer to radioactive elements; the mass number (not the atomic weight) of the isotope with largest half-life is usually given.

** Latest values recommended by the International Union of Pure and Applied Chemistry, 1961.*